T0313936

Tropical Ecology

JOHN KRICHER

PRINCETON UNIVERSITY PRESS
PRINCETON AND OXFORD

Published by Princeton University Press,
41 William Street, Princeton, New Jersey 08540

In the United Kingdom: Princeton University Press,
6 Oxford Street, Woodstock, Oxfordshire OX20 1TW

Library of Congress Cataloging-in-Publication Data

Kricher, John C.
Tropical ecology / John Kricher.
p. cm.
Includes bibliographical references and index.
ISBN 978-0-691-11513-9 (hardcover: alk. paper) 1. Rain forest ecology. 2. Ecology—Tropics. I. Title.
QH541.5.R27K75 2010
577.0913—dc22 2010037794

British Library Cataloging-in-Publication Data is available

Publication of this book has been aided by Karen Fortgang, *bookworks publishing services*

This book has been composed in Sabon LT Std by Aptara®, Inc.

Printed on acid-free paper. ∞

press.princeton.edu

Printed in the United States of America

10 9 8 7 6 5 4 3 2

Dedication

This book is lovingly dedicated
to my wife, Martha Vaughan,
and to our grandson Liam Campbell O'Toole,
who is still rather new to a world
in which we hope he will eventually enjoy and learn
from visits to the tropics.

Contents

Acknowledgments

Since this volume has in essence emerged from the two editions of *A Neotropical Companion*, all of the acknowledgments in those books apply here. In the interest of remaining concise, I shall not repeat them. Suffice it to say that I have been privileged to travel widely in my study of tropical ecology. Numerous people and organizations have made that possible and have helped with my enlightenment. I owe much to many.

This text has benefited immensely from the careful scrutiny of three outstanding reviewers: Dr. Robert A. Askins, Dr. James Dalling, and Dr. Gregory S. Gilbert. Each of them read the entire manuscript and offered numerous critiques, comments, and suggestions. Their combined effect on the final product is rather profound. They have made this a much better book, and I am deeply grateful to each of them. That said, it remains my book, and I assume full responsibility for any errors of any sort.

My good friend Dr. William E. Davis Jr. also offered helpful comments on the early chapters, as did my friend and Wheaton College colleague Dr. Scott W. Shumway. I truly appreciate their input.

The rich array of photographs that enhance the book and increase its teaching power came with the help of many good friends. I am indebted to James Castner, William E. Davis, Frederick Dodd, Bruce Hallett, Edward Harper, Carolyn Miller, Richard Payne, Scott Shumway, Pepper Trail, and James Wetterer for generously permitting me to reproduce some of their outstanding photographs. I am equally grateful to the many authors and photographers whom I do not personally know who also permitted the use of images, graphics, and maps.

I made a number of trips associated with gathering information specifically for this book. I thank Wheaton College and the American Birding Association for facilitating these travels. I want to particularly thank Raúl Arias de Para for his extraordinary generosity and hospitality at what I consider to be Panama's premier ecotourist facilities, the Canopy Tower and Canopy Lodge. I also want to thank my friends Tony White and Woody Bracie for their hospitality when I visited the Bahamas.

Researching and writing a textbook requires large blocks of time, and I thank Wheaton College for the honor of being A. Howard Meneely Professor of Biology. That five-year professorship afforded me a sabbatical leave as well as additional release time essential to the timely writing of this book.

On the home front, my wife, Martha Vaughan, helped in numerous ways. She organized materials, copied pages, and, without complaint, willingly performed other duties essential to keeping me focused and making my writing more efficient. Most importantly, she graciously assumed the daunting task of checking the accuracy of all the final page proofs. She also provided constant encouragement and listened with patience to my numerous frustrations along the tortuous path toward the final product, a book. I'm a lucky guy. Charlie the Cat also provided valued companionship, lying on the desk next to the computer as I labored through the final chapters of a long book, keeping me calm (a technique that is alleged to have also worked for Mark Twain).

The editing and production team assembled by Princeton University Press was not only awesome in its efficiency but also a pleasure to work with. Just look at the book. That's what they did! I want to thank my editor, Alison Kalett, for her patience and sound stewardship. In addition, the book could not have happened without the diligent and tenacious work of Karen Fortgang; Stefani Wexler; my most able copy editor, Jennifer Harris; and proofreader Barbara Liguori. Finally, I want to acknowledge Robert Kirk, Executive Editor and Group Publisher, Science and Reference at Princeton University Press. It was Robert who proposed the idea of a tropical text to me. It was Robert who convinced me to do it. When I asked Robert what he thought the book's title ought to be, he looked at me and said something like, "It's about tropical ecology so I think we should call it *Tropical Ecology*." Well done, Robert, and thanks.

Introduction

What Is Tropical Ecology?

Asking the question, What is tropical ecology? may seem akin to asking questions such as, Who is buried in Grant's tomb? *Tropical ecology* is the study of the ecology of tropical regions. But so what? Consider these questions: First, what is ecology? What are its paradigms, what are its principles, and what is its place in the biological sciences? What unique insights has ecology added to biology? Second and perhaps more important, is tropical ecology unique within the study of ecology? In other words, if you studied the ecology of the eastern deciduous forest, the arctic tundra, the Mojave Desert, or the seas that surround Antarctica, what paradigms and principles might you miss because none of these ecosystems are tropical? Take a few minutes to think about these questions, and then read on.

Readers of this book should be familiar with ecology and thus need little coaching as to what *ecology* is. It is, in a nutshell, the study of how organisms exist, adapt, and interact within their environments within the context of both abiotic and biotic factors. It is studied at widely different spatial and temporal scales. Ecologists reveal and attempt to explain patterns of species richness and explore how top-down and bottom-up forces influence the structure and stability of food webs, structuring whole ecological communities. Ecologists use experimental techniques and mathematical models to study competition, predation, mutualism, and the assembly, persistence, and stability of ecological communities. Ecology has matured in that disciplines such as landscape ecology, restoration ecology, and conservation ecology all use theory from basic ecology. Each of these areas of applied ecology has made an impact on issues in tropical ecology.

The various topics cited in the previous paragraph are the grist for ecological study and make up the "usual suspects" in ecology texts. Once one understands these topics, one is considered proficient in ecology. Do these topics form the paradigms of ecology? Not really. They are best considered *principles*, areas of study around which the discipline is organized. Ecology arguably has no paradigms. (I argue this point throughout my book *The Balance of Nature: Ecology's Enduring Myth*.) On close inspection, ecology is a prominent branch of evolutionary biology. Not much of what ecologists study makes sense without placing the data firmly within evolutionary context. This reality will become apparent in the study of tropical rain forest ecology, especially regarding species' origins, interactions, and interdependencies. The paradigm within which ecology is included is *organic evolution*, and it, along with *cell theory*, are the only real paradigms of biology. Cell theory and evolution are what biology is. All the rest is detail.

Where does this leave tropical ecology? What will you learn by going to Panama, or Ecuador, or Indonesia, or Queensland, Australia, that could not be learned by going to central Spain, or Norway, or Japan, or New Jersey? Yes, the species are different, but what of it? The pattern of seasonality is different; precipitation patterns are different; lots of things are different. What of it?

This "what of it?" is the subject of this book. By the conclusion of your study, you should be competent to voice an educated opinion regarding the question of just how unique tropical ecology is. Learn and enjoy the journey.

Overview

When Charles Darwin was fresh out of college, he visited a tropical rain forest in eastern Brazil as part of his famous voyage aboard the HMS *Beagle*. Like others who came before and since, his first impressions were vivid. He wrote of this experience in the tropical rain forest:

> When quietly walking along the shady pathways, and admiring each successive view, I wished to find language to express my ideas. Epithet after epithet was found too weak to convey to those who have not visited the intertropical regions the sensation of delight which the mind experiences. (Darwin 1845)

In other words, Darwin was pretty overwhelmed by what he saw, felt, heard, even what he smelled. He commented on the hazy air within the forest, evidence of the high humidity. A tropical rain forest does leave you with a vivid "first impression." It gets your attention both rationally and emotionally. But years later, a much older Charles Darwin had a different perspective when he was composing his autobiography:

> In my journal I wrote that whilst standing in the midst of the grandeur of a Brazilian forest, "it is not possible to give an adequate idea of the higher feelings of wonder, admiration, and devotion which fill and elevate the mind." I well remember my conviction that there is more in man than the mere breath of his body. But now the grandest scenes would not cause any such convictions and feelings to rise in my mind. (Darwin 1887)

What changed Darwin was, of course, evolution. When Darwin returned from the *Beagle* voyage, he quickly became convinced of the truth of organic evolution. All present species have arisen from previous species and were not specially created. This was soon followed by his discovery of the mechanism for evolutionary change, *natural selection*, and so he began to view the world differently. Of course, not everyone agreed with Darwin. Some rejected his theory outright; others agreed that evolution might be true but did not agree that a blind mechanism called natural selection caused species to evolve into different forms. From 1859, when *On the Origin of Species* was published, until Darwin's death at the age of 73 in 1882, he was strongly attacked in speeches and in print over his assertions concerning evolution.

What does all this Darwinian history have to do with tropical ecology? The great twentieth-century evolutionary biologist Theodosius Dobzhansky, who spent a fair amount of time collecting fruit flies in the tropics, said "Nothing in biology makes sense except in the light of evolution" (Dobzhansky 1973, p. 449). Darwin's big idea has become the most important paradigm in biology, and without it, the immense complexities that characterize a tropical rain forest or any other ecosystem would seem hopelessly beyond rational explanation.

Therefore, much of this book will be about evolution as it is evident in the tropics, and in rain forests in particular. There is nothing unique about this approach, as I fail to see how it could be otherwise for any kind of ecological study. If this book were about the ecology of Antarctica, evolution would still be the organizing paradigm. The noted ecologist G. Evelyn Hutchinson once titled a book about general ecology *The Ecological Theater and the Evolutionary Play*. I doubt that many who visit a rain forest or any other tropical ecosystem today need fear losing their sense of awe (as Darwin apparently did) simply because of acquiring an understanding of how the process of evolution is manifest in the multitudes of morphologies and interactions that characterize the ecological community. Darwin, when elderly, apparently lost his awe of nature, perhaps embittered by all the criticism directed toward him. You need not worry, however. Learning the science behind the landscape really adds to awe—it does not detract from it. This book aims to teach the science behind the landscape.

As author of the book you hold, I am in a unique position. It's my book, and I wrote it my way. Other books about the tropics take different approaches from what you will read within this volume. This book reflects my views from study and field experience (particularly in ornithology) over my four decades as an ecologist. Let me explain something of how the book came to be.

In 1989, Princeton University Press published my book titled *A Neotropical Companion*. The book was modest in size and had a bright green cover, which led to its nickname, the "Little Green Book." It amazed me while writing that book how much wonderful information ecologists had garnered in their collective research efforts and how little of this really cool stuff was finding its way to more popular-based audiences, including into the hands and minds of students. My goal was to be selective and make some of that amazing information available.

The success of the Little Green Book led to a major revision and expansion of *A Neotropical Companion* that appeared in 1997, again published by Princeton University Press. This volume was bigger, had new illustrations and some color photographs, and, most importantly, covered more topics in greater detail, with much additional treatment of South America.

Eager to learn from my readers, I included my e-mail address in the revised volume, and many responded. I have had a few errors of both omission and commission pointed out to me, as well as many comments as to how the book could be improved. Some want more maps; some want more illustrations; some want greater emphasis on particular topics. But the most-repeated comment (the title of the book notwithstanding) is that the book is too narrow, as it covers only the Neotropics and omits the Old World tropics of Africa, Southeast Asia, and Australia/New Guinea. How does the Neotropics compare with other tropical regions of Earth? How are these regions biogeographically distinct, one from another? Are the ecological processes that characterize Neotropical ecosystems the same as those in other regions?

Thus, it seemed both to me and to my editors at Princeton University Press that it was time for yet another revision and expansion, this time going "the full Monty" toward a more comprehensive book covering the ecology of the global tropics. And that is how *Tropical Ecology* came to be.

Ah, but as always, the devil is in the details. My previous book was specifically about the Neotropics, and the vast majority of my field experience has been in that region. True, I have visited Africa (Tanzania), Southeast Asia (Indonesia), as well as the Australian tropics in Queensland and Darwin, but most of my reading and field study is from Central and South America and the Caribbean. And, I dare say, much if not most of the published literature continues to be from studies in the Neotropics. So I cannot pretend to take an "even-handed" approach to the global tropics. This book will expose my bias, as it reflects my confidence level in my expertise—and that bias remains clearly toward the Neotropics and, in particular, birds. But such a bias has a pragmatic component in that many students will likely use this text in conjunction with an actual field experience, a college course that includes a trip to experience a tropical region. And most North American college courses in tropical ecology visit some area within the Neotropics. My continued emphasis on that region will benefit such courses.

The organization of the work is mine, my way to tell the story of tropical ecology. But the information contained herein is largely the work of others, the thousands of researchers in the field and lab whose work, often repetitive and always exacting, deals with some aspect of tropical ecology. For the most part, I rely on primary source material, papers from journals—some of which, like *Biotropica*, publish only papers about tropical ecology, and some of which, like *Ecology* and *American Naturalist*, publish many papers that deal with the tropics. I also include many citations from *Science* and *Nature*, both of which publish numerous major papers on various aspects of tropical biology. Each topic covered is illustrated by selected examples, so obviously many worthy papers are not cited. My apologies to these authors, but this book is not meant as a comprehensive review of all topics tropical. It is nice to have an embarrassment of riches when it comes to information about the tropics, and I have had to carefully pick and choose. Examples were by and large chosen for their overall interest, potential significance, clarity, and robustness. And, since researching the second edition of *A Neotropical Companion*, I am amazed by the remarkable increase in the number of tropical studies represented in professional publications of all kinds. Another point worth emphasizing is that my area of expertise is in *ornithology*, the study of birds, so whoever reads my text will encounter quite a few examples that involve birds. But consider that birds are fairly easy to observe in the field, and thus much data about species interactions in the tropics does, indeed, focus on birds.

The first question any author faces is, Who is the audience for this book? This work is aimed at students and others who take a serious interest in learning about tropical ecology. It assumes a basic knowledge of introductory biology, including ecology and the tenets of evolutionary biology. I present an overview of each topic and then note where information gaps exist and where controversies have arisen. The book is meant to be provocative and poses many questions, the answers to which are still to be revealed or are presently being hotly debated. Some of the studies I describe come to different conclusions about the same topic. Science is always a work in progress, and this book strongly takes that approach.

All science books should be subject to peer review, and this book certainly has enjoyed that advantage (see the Acknowledgments). Others, expert in various areas of tropical biology, have read chapter drafts and made numerous critical comments. I have taken many of these suggestions, but I have also chosen not to take some. No two tropical biologists would ever write the "same" book. Our views are influenced by our biases, varied interests, and experiences, to say nothing of our intellectual outlooks. We make judgments, and we often have to agree to disagree on various points. Any problems identified with this text should be directed to me (jkricher@wheatonma.edu), and I take sole responsibility for any errors that may be detected.

Some material that has remained robust has been taken and often updated from *A Neotropical Companion*, but most of the material is new to this book. That said, any book such as this is old before it is published. Right now, any number of good studies are occurring or have occurred that alter some of what is written on these pages. That is the nature of science, the strength of science, the very essence of science. Science is cumulative, and anyone entering into the study of a field of science profits from acquiring a firm foundation from which to begin critical study. I hope this book will serve that purpose for its readers.

Organization of This Book

This book consists of 15 chapters, a challenge for a one-semester course, but hopefully the writing is sufficiently engaging that the task of moving through the chapters will prove to be relatively pleasant. There are numerous figures, maps, tables, and color photographs throughout to add to the clarity of the text.

I begin (Chapter 1) with a broad overview of what and where the tropics are. This provides some information about how tropical ecology emerged from global exploration into the realm of scientific study. It also describes something of the climatic variables that are the overarching determinants of tropical ecosystem characteristics, and it sets the stage for what is to come throughout the book.

Next (Chapter 2) I discuss biogeography and evolution in the tropics. For those already well versed in these topics, this chapter may serve as a helpful review. But for many, and I expect most students, it will be essential in explaining how and why tropical regions differ, as well as in clarifying basic evolutionary principles, particularly speciation.

Then it is time to enter a rain forest (Chapter 3) and examine the physical structure of this unique ecosystem. This chapter discusses the various characteristics of rain forest vegetation, from tree shapes to buttressed roots. It should be the primer that will allow any student to see the many details and general characteristics common to all rain forests.

Chapters 4 and 5 both explore ideas as to why tropical regions and rain forests in particular are so rich in their numbers of species. *Biodiversity* is one of the most significant areas of tropical research, and thus two chapters are devoted to research focused on this complex topic. Chapter 4 discusses how diversity is partitioned within and between habitats and considers whether the high diversity of tropical regions is an example of high rates of speciation (or low rates of extinction). What factors allow the continued existence of such complex communities? Chapter 5 looks at the range of hypotheses attempting to explain the uniquely high tree species richness in tropical forests, one of the most difficult theoretical problems in tropical ecology.

The effect of disturbance is explored next (Chapter 6). Periodic natural disturbance may result in a shifting mosaic of different aged patches of forest such that high species richness is maintained over large areas. This chapter deals in part with *gap phase dynamics*. Gaps in a forest are created by forces ranging from single fallen trees to major blowdowns caused by storms or fire. Gaps are exposed to high levels of sunlight and thus have a different ecology from closed rain forest, where canopy trees strongly attenuate the light striking the forest floor. Last, this chapter considers the topic of *secondary succession*, which is of major importance in assessing how rain forest regrows following disturbance.

The richness of species in tropical forests results in numerous complex forms of interactions among species—some rather casual, some sufficiently complex that the species are evolutionarily interdependent. Chapter 7 presents examples of some of the most striking biotic interactions involving such topics as fruit consumption as it relates to seed dispersal and the propensity of many tropical species to evolve mutualistic interactions. This chapter includes a focus on *coevolution*, when two or more species become mutually interdependent.

Food webs and the adaptations associated with them form the core of Chapter 8. This chapter looks at various top-down and bottom-up interactions that contribute to structuring food webs and trophic dynamics in rain forests. Topics include the evolution of characteristics such as cryptic and warning coloration.

The high rate of photosynthesis evident in tropical rain forests results in high rates of *primary productivity*, and this is the subject of Chapter 9. How much carbon do lush tropical forests take in and how much do they emit (as carbon dioxide)? The *carbon flux* of tropical forests is an important area of research. As current climate change continues due to ongoing addition of atmospheric carbon dioxide, what is the potential of tropical forests to act as carbon sinks, absorbing and storing excess carbon? This chapter explores some of the sometimes contradictory research about how tropical forests might act to mitigate the effects of increasing atmospheric carbon dioxide.

Rich tropical forests often occur on poor soils. Chapter 10 examines nutrient cycling and tropical soils. It describes how vital nutrients such as phosphorus, nitrogen, and calcium cycle in tropical ecosystems and the essential function of decomposer organisms. This chapter includes discussion of various kinds of tropical soils and contrasts forests on *floodplains*, where annual flooding renews soil fertility, with those on *terre firme,* off floodplains.

Chapters 3 through 10 all focus on the ecology of the tropical rain forest. The remaining five chapters survey other tropical ecosystems and also examine topics having to do with conservation science in the tropics.

Areas of grassland with scattered trees are called *savanna*. Chapter 11 discusses tropical savannas and dry forest ecosystems. This discussion examines two views of how savannas and dry forests coexist in the tropics. One is that grassland, savanna, and dry forest represent a *moisture gradient effect* (where grassland is found in the most arid areas, and dry forest where there is more annual precipitation), and the other views them as *alternative stable states* dependent on local conditions. The importance of fire and grazing are also discussed.

Chapter 12 surveys other major tropical ecosystems, beginning with montane ecosystems and continuing to riverine forests and coastal mangrove forest. This chapter also considers the connectivity between montane and lowland ecosystems from the standpoint of biodiversity maintenance.

Humans evolved in tropical Africa, and Chapter 13 discusses human ecology in the tropics. Topics include hunter-gatherer communities, simple agriculture in the tropics, and some of the traditional ways that humans have affected the overall ecology of the tropics.

The last two chapters deal with conservation issues in the tropics. Chapter 14 focuses on forest fragmentation and biodiversity. Chapter 15 discusses deforestation and forest degradation, as well as the effects of fire on tropical forests. It also examines emergent pathogens and invasive species and concludes with an overview of a current debate among ecologists about the likely future of tropical forests. While not meant to be comprehensive, these two chapters taken together cover the most fundamental issues and questions facing tropical conservation science.

1 What and Where Are the Tropics?

Beginnings

The world was once a larger place, or so it seemed. Vast areas lay unexplored, and following the Renaissance, European explorers began to fan out across the Earth in a nationalistic search for conquest, treasure, and lands to claim for their respective countries. With the onset of global colonialism and imperialism came an age of exploration. By the eighteenth and nineteenth centuries, many new regions were becoming known, and many of these were in the tropics.

Science, in the philosophical sense, traces its roots back to the emergence of science in ancient Greece, the Middle East, and Asia. But in a more contemporary sense, science really began during the Renaissance, when curiosity led to inductive and deductive reasoning, observational and experimental analysis, and most importantly, empirical investigation and materialistic explanations for natural phenomena. The study of the tropics was formalized only after the ages of global exploration and discovery and when the scientific method was used to investigate phenomena.

The exploration of the world's tropics by European nations brought with it much specimen collection and cataloging. Thus, the roots of tropical ecology lie most deeply in museum cabinets housing vast numbers of carefully labeled plants and animals brought back by many devoted and indefatigable naturalists, including such figures as Charles Darwin and Alfred Russel Wallace (Plate 1-1). When Captain James Cook chose to have two naturalists, Joseph Banks and Douglas Solander,

(a)

(b)

PLATE 1-1
(a) CHARLES DARWIN
(b) ALFRED RUSSEL WALLACE

included on the voyage of the *Endeavor* (1768), he set an immensely valuable precedent. It is that action that ultimately led to Darwin's being invited on the *Beagle* voyage (1831–1836) and thus indirectly was responsible for the *Origin of Species* (1859).

One of the most noteworthy contributors to early knowledge of the tropics was Alexander von Humboldt (Plate 1-2), sometimes called the founder of the science of biogeography (Jackson 2009). *Biogeography* is the study of how organisms vary among regions, even when climate is similar. For example, there are no apes in the New World tropics, but there are no monkeys with prehensile tails in the Old World tropics. This sort of example reflects differing evolutionary histories among regions. Biogeography will be a focus of the next chapter. Humboldt led successful and bold expeditions to difficult places ranging from Russia to northern South America. He published a massive five-volume tome titled *Kosmos* in which he attempted to summarize the entire scope of scientific knowledge of the world.

Humboldt's travels in South America took him from rain forests along the Orinoco River to high-elevation Andean slopes as he traveled from Venezuela through the northern Andes. He visited central Mexico, Cuba, and the United States. Humboldt climbed some of the higher Andean peaks, reaching elevations of 5,800 meters (19,000 feet). It was Humboldt who first discovered and described the unique nocturnal oilbird (*Steatornis caripensis*) (see Chapter 7) in the caves around Caripe, Venezuela.

PLATE 1-2
ALEXANDER VON HUMBOLDT

Humboldt and his botanist colleague Aimé Bonpland described elevational changes evident on Andean slopes. They realized that as climate changed by elevation, so did plant and animal communities, and thus they were the first to elucidate what is termed the *life zone concept*. Much later, this concept was formalized by Leslie R. Holdridge (1947). A *life zone* is an ecological area defined by climatic variables such as mean annual temperature, total annual precipitation, and the ratio of mean annual evapotranspiration (a measure of heat) to mean total annual precipitation. (*Transpiration* is a term referring to the evaporative water loss from plants. Plants are adapted to pull water from soil via roots, and eventually that moisture is lost by evaporation from the surface parts of the plant.) In the Andes (as with all mountains), these variables change with elevation and result in differing ecosystems such as *páramo* (an ecosystem of wet grassland and shrubs) at high elevations and lush rain forest at low elevation (Plates 1-3 and 1-4). Holdridge's life zone concept will be further discussed later in this chapter.

Naturalists were often moved to comment on the rich and diverse nature of flora and fauna within tropical rain forests. Their observations, by today's standards, were largely (and understandably) anecdotal, though not lacking in important details. The modern student of tropical ecology will profit from reading some of the narratives of figures such as Thomas Belt, Henry Walter Bates, the aforementioned Darwin and Wallace, as well as others. For example, Henry Walter Bates discovered the form of animal mimicry (described in Chapter 8) that bears his name. These early explorers displayed much tenacity in capturing detail, transcribing vast amounts of information, and collecting and transporting specimens often under difficult circumstances.

HISTORICAL REFERENCES

These books, all still available (at least in libraries), were authored by some of the pioneers of tropical ecology and are strongly recommended as insightful, entertaining, and inspiring resources.

Bates, H. W. 1892. *The Naturalist on the River Amazons.* London: John Murray. Classic account of Amazonian natural history, and quite wonderful.

PLATE 1-3
SNOW ON PÁRAMO
Snow is common at high elevations in equatorial regions. Here, bunch grasses of the high páramo life zone are partially snow-covered. From Ecuador.

Beebe, W. 1918. *Jungle Peace.* London: Witherby. Beebe's journey from the West Indies to the rain forest ("jungle") of Guyana makes for engrossing reading, including a wonderful account of the odd Hoatzin bird (Chapter 12).

———. 1921. *Edge of the Jungle.* New York: Henry Holt. Like the previous volume, contains short, delightful essays on tropical ecology. Classic.

Belt, T. [1874] 1985 reissue. *The Naturalist in Nicaragua.* Chicago: University of Chicago Press. One of the best of the classic exploratory accounts, focused entirely on Central America.

Chapman, F. M. 1938. *Life in an Air Castle.* New York: Appleton-Century. Highly readable with much information, particularly on tropical birds.

Darwin, C. R. [1906] 1959. *The Voyage of the Beagle.* Reprint. London: J. M. Dent and Sons. One of the best classic accounts of travel throughout South America. Many reprinted editions are available. Must reading.

Wallace, A. R. 1869. *The Malay Archipelago.* London: Macmillan. This is Wallace's most famous book, an engrossing account of his eight years of travel throughout what is now Indonesia. The book abounds with information about wildlife as well as the various peoples Wallace encountered.

———. 1895. *Natural Selection and Tropical Nature.* London: Macmillan. This delightful book contains vivid descriptions, in glorious Victorian prose, of Wallace's experiences in Amazonia.

Waterton, C. [1825] 1983. *Wanderings in South America.* Reprint. London: Century Publishing. A very entertaining narrative by a rather eccentric but perceptive explorer.

PLATE 1-4
FOREST CANOPY
The dense canopy of a low-elevation, humid tropical moist forest, such as this one in the Arima Valley in Trinidad, is typically irregular, with some emergent tree species rising above others. Forest gaps created by fallen trees also add to canopy heterogeneity.

Henry Walter Bates versus Curl-Crested Aracaris

The following text from Bates's account in *The Naturalist on the River Amazons* illustrates the curl-crested aracari (*Pteroglossus beauharnaesii*), a species of toucan, and the memorable encounter that Bates had with a flock of these birds (Figure 1-1).

> *The Curl-crested Toucan (Pteroglossus Beauharnaisii).*—Of the four smaller toucans, or arassaris, found near Ega, the Pteroglossus flavirostris is perhaps the most beautiful in its colours, its breast being adorned with broad belts of rich crimson and black; but the most curious species by far is the Curl-crested, or Beauharnais Toucan. The feathers on the head of this singular bird are transformed into horny plates, of a lustrous black colour, curled up at the ends, and resembling shavings of steel or ebony wood, the curly crest being arranged on the crown in the form of a wig. Mr. Wallace and I first met with this species on ascending the Amazons, at the mouth of the Solimoens; from that point it continues as a rather common bird on the terra firma, at least on the south side of the river, as far as the Fonte Boa, but I did not hear of its being found further to the west. It appears in large flocks in the forests near Ega in May and June, when it has completed its moult. I did not find these bands congregated at fruit trees, but always wandering through the forest, hopping from branch to branch amongst the lower trees, and partly concealed amongst the foliage. None of the arassaris to my knowledge make a yelping noise like that uttered by the larger toucans (Ramphastos); the notes of the curl-crested species are very singular, resembling the croaking of frogs. I had an amusing adventure one day with these birds. I had shot one from a rather high tree in a dark glen in the forest, and entered the thicket where the bird had fallen to secure my booty. It was only wounded, and on my attempting to seize it, set up a loud scream. In an instant, as if by magic, the shady nook seemed alive with these birds, although there was certainly none visible when I entered the jungle. They descended towards me, hopping from bough to bough, some of them swinging on the loops and cables of woody lianas, and all croaking and fluttering their wings. (Bates 1892, pp. 335–336)

(a)

(b)

FIGURE 1-1
(a) Curl-crested toucan. (b) Mobbed by curl-crested toucans.

Most eminent naturalists of the nineteenth century honed their skills through participation in voyages of discovery. For example, Thomas Henry Huxley and Joseph Hooker, both close friends of Darwin, each made an extensive voyage for the express purpose of expanding his horizons. These voyages were insightful, but they normally did not permit careful, systematic study within a given region.

It was not until the twentieth century that systematic study of regions within the tropics was really possible. What was needed were permanent field stations where scientists could go and conduct their work.

In 1923, following the completion of the mammoth engineering project known as the Panama Canal, a hilltop area of about 1,500 hectares (3,706 acres) on the Panamanian isthmus remained above water, a newly formed island surrounded by the newly created Gatun Lake. This was named Barro Colorado Island. Since 1946, it has been a field station administered by the Smithsonian Institution, and it has served numerous researchers, a function that continues to the present day. As you continue through this book, you will notice numerous citations regarding Barro Colorado Island, usually referred to simply as BCI.

Another unique field station devoted to tropical biology is Simla, in the Arima Valley on the island of Trinidad. While Simla is not one of the most prominent sites today, it is noteworthy for its founder, the

William Beebe

William Beebe (1877–1962) (Plate 1-5) did far more than change the direction of research in tropical ecology. He had a long and highly distinguished career as an explorer, scientist, and writer. He spent most of his career associated with the New York Zoological Society. His productivity was amazing, having authored 24 books and about 800 scientific articles over the course of his career. Beebe began as an ornithologist, and one of his first major books bore the title *The Bird, Its Form and Function* (1906). Arguably his strongest contribution to ornithology (he had many) was his classic book *A Monograph of the Pheasants*, published in four volumes from 1918 to 1922.

PLATE 1-5
WILLIAM BEEBE

Beebe's tropical research was prolific and began in British Guiana (now Guyana) in 1916. He traveled widely, including to the Galapagos Islands, a trip that resulted in another memorable book, *Galapagos: World's End* (1924). He also authored two books about tropical ecology, one of which, *Jungle Peace* (1918), greatly impressed Theodore Roosevelt, himself an explorer of the Neotropics.

In 1928, Beebe established a field station in Bermuda, an event that led to another major chapter in his career, that of oceanic explorer. Beebe was frustrated that animals hauled up from the depths of the seas were not only dead on arrival but also mutilated by the pressure change experienced while being hauled to the surface. After consultation with Theodore Roosevelt, Beebe eventually focused on a submersible that he called a *bathysphere*, a thick metal chamber that could withstand the great pressures of the ocean depths and that could be lowered and raised from a ship. He subsequently made about 30 descents, accompanied in the tiny chamber by the bathysphere's inventor, Otis Barton. In 1934, Barton and Beebe reached a depth of 922.93 meters (3,028 feet), an amazing accomplishment that was not equaled for many years.

Beebe was a popularizer of science but, at the same time, a very productive scientific researcher. It has been said of him that he really founded the science of tropical ecology.

PLATE 1-6
The OTS field station at La Selva.

naturalist-explorer William Beebe, and for its initial purpose. Though Trinidad appears to be part of the West Indies, it is biogeographically distinct, as it is actually on the continental shelf, part of Venezuela, just as the island of Martha's Vineyard is really part of Massachusetts. Thus, the flora and fauna are less West Indian than they are South American. Recognizing this, Beebe established Simla in 1950. Beebe purchased the land at an advanced age (he was 73) and began the research station. His strong belief was that researchers must work on live animals and not merely collect specimens. At that time, specimen collection for museums essentially dominated tropical research; thus, Beebe's approach helped change the direction of ecological work from specimen collection to detailed ecological studies of organisms in the field.

Many tropical field stations exist today and more are being established. In the Western Hemisphere tropics, in addition to Barro Colorado Island, some of the most prolific research sites are La Selva (Costa Rica), Cocha Cashu (Peru), Manu Biological Preserve (Peru), Minimal Critical Size of Ecosystems (Brazil), Los Tuxtlas (Mexico), and Luquillo (Puerto Rico). The Tropical Ecology Assessment and Monitoring Network (TEAM) includes 122 sites throughout the tropical world, from Colombia to China (see http://www.teamnetwork.org/en/about). The Organization for Tropical Studies (usually referred to as OTS) maintains not only the La Selva Biological Station in Costa Rica (Plate 1-6) but also biological stations at Las Cruces and Palo Verde. Many other research stations also are found throughout the tropics, some of which will be noted for studies described within this book.

In many areas within the tropics, huge challenges ranging from marginal living conditions, lack of facilities, and political instability still face the field worker. Nonetheless, it is fair to say that many if not most field sites now have sufficient infrastructure and facilities to permit cutting edge research to be conducted. What was once the age of exploration and description has become the age of experimental design, data collection, and hypothesis testing.

Modern Tropical Ecology

Many college-level tropical ecology courses now include field opportunities at one or more research stations somewhere within the world's tropical regions. The very fact that undergraduate and graduate students now routinely visit and work at tropical research stations has vastly increased tropical ecological research.

PLATE 1-7
The harpy eagle (*Harpia Harpyja*) released on Barro Colorado Island, Panama, is monitored by radio tracking.

Professional researchers work many months at a time at tropical research stations with well-equipped laboratories, computers with Internet access, global positioning systems, and so on (Plate 1-7). Expeditions into remote areas where services are few to none still occur, however, as much of the tropics still demands field work away from major research stations. For a wonderful and insightful look into how tropical research is conducted and the dedication and drive of the researchers, read *The Tapir's Morning Bath* by Elizabeth Royte (Boston: Houghton Mifflin, 2001). It chronicles the life and dedication of researchers at Barro Colorado Island.

In the Western Hemisphere, most tropical researchers focus on the Neotropics, their work taking them anywhere from Mexico to southern Amazonia as well as to the Caribbean islands. In the Eastern Hemisphere, much research is carried on in Africa and Australasia. The intensity of tropical research varies considerably from one country to another, and thus there are some areas that are far less studied than others (Stocks et al. 2008).

Professional papers detailing tropical research are routinely published in leading science journals such as *Science* and *Nature*. In addition, journals such as *American Naturalist*, *Ecology*, *Oecologia*, *Conservation Biology*, and many others routinely include papers about research from the tropics.

The first professional society devoted to tropical research was the Bombay Natural History Society, founded in 1883. The International Society for Tropical Biology was founded in 1956 (Chazdon 2002). The Association for Tropical Biology and Conservation, founded in 1963, publishes the journal *Biotropica*. The papers published in *Biotropica* include studies about all tropical regions, but the majority of the work reported is about the Neotropics. Another important journal is the *Journal of Tropical Ecology*, first published in 1985. This journal, like *Biotropica*, publishes research papers about all tropical regions, but most papers tend to be about Africa and Asia. Taken together, a recent survey showed that the highest percentage of papers in those two journals were from studies conducted in Brazil, Costa Rica, Mexico, Panama, Malaysia, Puerto Rico, Australia, French Guiana, Venezuela, Ecuador, United States, Peru, Indonesia, Colombia, and India, in that order (Stocks et al. 2008) (Figure 1-2). The dearth of studies from African nations is obvious, as is the dominance of studies from the Neotropics.

The Tropical Rain Forest—A Pioneering Book

The study of tropical ecology became formalized in 1952 when P. W. Richards, a botany professor from Great Britain, published *The Tropical Rain Forest: An Ecological Study* (Cambridge, UK: Cambridge University Press). As the title suggests, this book focused on rain forests but also included discussions of coastal mangrove forests and savannas. Reflecting the field of ecology as it was at that time, the book is largely

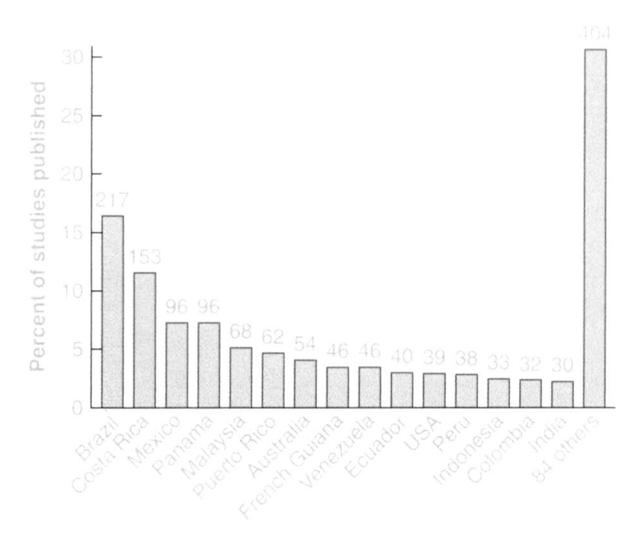

FIGURE 1-2
This graphic depicts the number of studies published in two professional journals, *Biotropica* and *Journal of Tropical Biology*, from 1995 to 2004. It shows how many studies were conducted in each of various countries. Note that most studies are focused in the Neotropics. (Numbers on bars are actual numbers of published studies.)

The View from the Canopy

FIGURE 1-3
This sketch illustrates the use of a large industrial crane to examine tropical forest canopies: *a* is the tower, *b* is the counterjib, *c* is the counterweight, *d* is the operator's cab, and *e* is the gondola, where a researcher would be located.

Tropical research has been facilitated in recent years by towers, walkways, and cranes that allow direct access to the forest canopy. It has been frustrating to tropical ecologists to realize that much of the biological diversity found in tropical forests is confined to the forest canopy, the dense and irregular layer of foliage representing numerous tree species, usually located 30 to 40 meters (about 98 to 130 feet) from the ground. Scores of arthropod and other invertebrate species as well as birds, mammals, reptiles, and amphibians spend most or all of their lives well above ground level. The same is true for *epiphytes*, the myriad species of bromeliads, orchids, cacti, and other plant species that live on the bark of tree trunks and branches. (Epiphytes live on bark and branches of other plants but are not directly parasitic.) But ecologists now do have more access to the forest canopy. The construction of high towers, some associated with tall, emergent trees, allows one to be at canopy level and make observations. Even more access is provided by canopy walkways that connect various trees or that are supported by metal towers. In some areas, tall cranes, such as are used in building construction, allow mobile access to the canopy (Parker et al. 1992) (Figure 1-3; Plates 1-8 to 1-11).

PLATE 1-8
CANOPY WALKWAY AND FOREST
This tall and lengthy canopy walkway at Sacha Lodge in Ecuador provides safe and easy access that permits study of species largely confined to the forest canopy.

Canopy research has helped not only to document patterns of biodiversity but also to demonstrate how tropical tree species vary in their response to physiological variables, such as elevated carbon dioxide levels and the production of biogenic aerosol compounds (Ozanne et al. 2003).

PLATE 1-9
HIGH CANOPY WALKWAY
Viewed from the ground, the canopy walkway is seen crossing a major forest gap.

PLATE 1-10
VIEW FROM CANOPY WALKWAY
The expanse and structural complexity of Ecuador's lowland forest is readily evident, as observed from a canopy walkway.

PLATE 1-11
WOODEN CANOPY TOWER
Canopy towers allow more restricted access than walkways but, if strategically placed, afford excellent opportunities for canopy study. Note the broad branching pattern of the tree. The tower is essentially built around the tree. From Sacha Lodge, Ecuador.

descriptive, but it does an admirable job of comparing global equatorial forests and identifying common patterns—particularly of physiognomy, climate, and soils—that they share to varying degrees (Figures 1-4 and 1-5). What is obviously missing from the discussion is treatment of the complex roles of animals in rain forest ecology. Pollination and seed dispersal—both of which, in the tropics, are usually dependent on animals—are not listed in the index, for example, and are barely touched upon. Richards's book was nonetheless the first serious attempt at summarizing knowledge about rain forest ecology as well as synthesizing concepts. It paved the way for much research, and it is still valuable and insightful. This classic book was revised and a second edition was published in 1996 by Cambridge University Press.

Geographic Definition of the Tropics

In today's world, the tropics are the region located between the Tropic of Cancer (23°27'N) and the Tropic of Capricorn (23°27'S), a latitudinal band of approximately 47°. This belt of latitude encircles the Earth at its widest circumference. At either extreme,

FIGURE 1-4
This is a sketch of a forest profile in what was then British Guiana and is now the independent nation of Guyana. This is adapted from one of several classic forest profiles included in Richards's famous book *The Tropical Rain Forest.*

FIGURE 1-5
Also adapted from Richards's classic volume, this is a profile of a mixed Dipterocarp forest in Borneo. Compare this with Figure 1-4. Both forests are structurally similar but contain entirely different tree species.

Tropical Worlds Long Ago: Rain Forest in Denver and a Very Large Snake from South America

Denver, Colorado, is located along the front range of the Rocky Mountains, an area well within the temperate zone. Forests of spruce, fir, pine, and aspen characterize the region. But had Denver existed 64.1 million years ago (a time called the Paleogene), not long after the mass extinction that ended the Cretaceous period, a very different forest would be evident—a tropical rain forest (see Appendix). At a site called Castle Rock, near Denver, fossil plants (particularly leaves) have been unearthed, plants that are clearly tropical (Johnson and Ellis 2002) (Figure 1-6). The diversity of tropical plants at this fossil site shows that tropical conditions then prevailed at a region far from today's tropics. Plant diversity had recovered quickly (within 1.4 million years) after the mass extinction that ended the Mesozoic era.

The largest snakes found in the tropics are the pythons (Pythonidae) of Africa and Australasia and the boas and anaconda (Boidae) of the Neotropics. The yellow anaconda (*Eunectes murinus*) of the Neotropics is arguably the world's largest extant snake, reaching a length of 9 meters (about 29.5 feet) (Plate 1-12). The great lengths reached by pythons and boas are facilitated by the warm tropical climate in which they live (Head et al. 2009). Snakes are poikilothermic, ectothermic vertebrates, meaning that their body temperature and thus their metabolism is dependent on ambient air

FIGURE 1-6
This drawing illustrates the morphotypes of tropical leaves found near Castle Rock, Denver, dating back 64.1 million years. Only the leaves in the upper right are nondicot leaves. By this time, modern plants, dominated by dicots, dominated terrestrial plant communities.

PLATE 1-12
ANACONDA

temperature. (*Poikilothermic* and *ectothermic* are different in meaning. Poikilothermic means specifically that body temperature changes as ambient temperature changes, whereas ectothermic means that the animal does not have physiological ability to regulate its body temperature to stay warm.) Mathematical models have been developed that correlate the difference in maximum body size of snake species occurring in different regions with mean annual temperature of each region. The warmer the mean annual temperature is, the larger the snakes are able to grow. Thus, it was of great interest when a fossil boid snake dated at 58 to 60 million years old was unearthed from a site in northeastern Colombia and the fossils (vertebrae) indicated that the snake reached an estimated length of 13 meters (about 42.6 feet). This serpent, named *Titanoboa,* was estimated to have weighed about 1.27 tons. That's a lot of snake. When included within the model that correlates today's boid snakes and air temperature, the results suggest that *Titanoboa* required a minimum annual mean temperature of between 30°C to 34°C (86°F to 93.2°F) to survive. This would make the climate at that time about 6°C to 8°C warmer than it is currently and suggests that a much higher level of carbon dioxide (a gas that contributes strongly to retention of atmospheric heat) was present in the atmosphere when *Titanoboa* lived.

These two examples show that the distribution and intensity of tropical climate have changed throughout Earth's history, as it appears to be doing today.

ecosystems become subtropical. Tropical regions represent about half of Earth's surface and thus has an immense effect on Earth's climate (Huber 2009). For much of Earth's history, even before the initial evolution of flowering plants (angiosperms) in the mid- to late Cretaceous period (144 to 65.5 million years ago), Earth has been even more tropical than it is today (see Appendix). Fossil palm trees and crocodiles have been unearthed in what is now Alaska and Siberia (Huber 2009). This is because the climate of Earth, for a variety of reasons discussed in the next chapter, is not constant. In times when climate was warmer, the distribution of the tropics was globally broader than is the case today.

Well before the evolution of flowering plants, the world was largely tropical. In the Jurassic period (206 to 144 million years ago), the giant long-necked sauropod dinosaurs lived in a largely tropical world, where there was no polar ice and where the average global temperature as well as oxygen and carbon dioxide levels exceeded those found today (Plate 1-13).

During the Cenozoic era (65 million years ago until the present), a time of proliferation of insects,

PLATE 1-13
CAMPTOSAURUS, APATOSAURUS, STEGOSAURUS, DRYOSAURUS, CAMARASAURUS AND ALLOSAURUS (LEFT TO RIGHT)
Each of these dinosaurs cohabited tropical regions in western North America during the late Jurassic period. The tropics then would bear scant resemblance to the tropics of today.

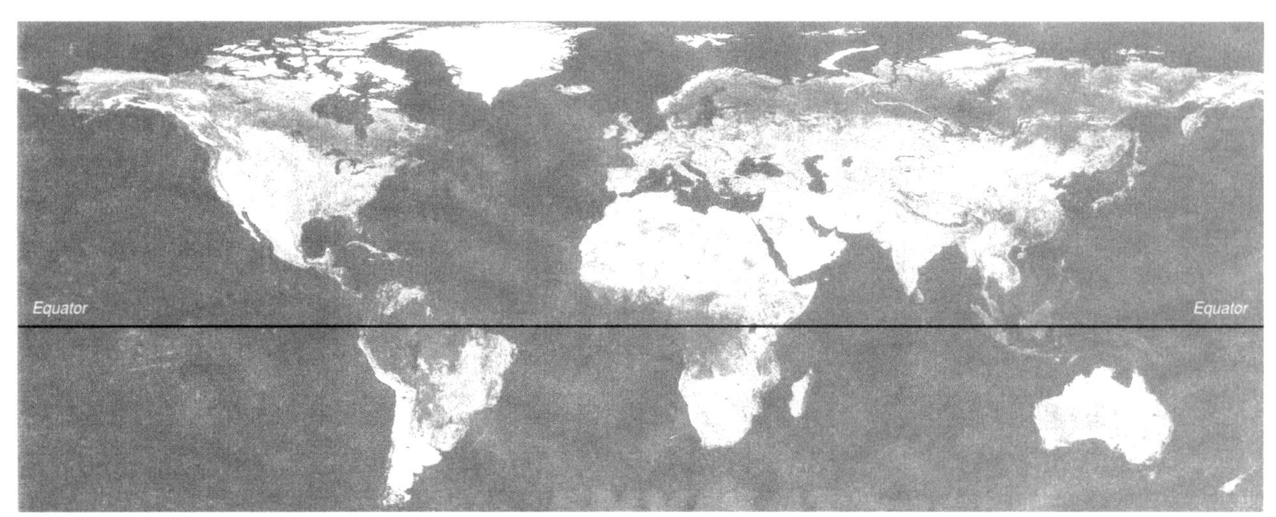

FIGURE 1-7
Satellite image of Earth, showing the band of tropics around the equator.

flowering plants, and mammalian diversity, global climate cooled, beginning strongly during the Miocene epoch (23.8 to 5.3 million years ago) and accelerating during the Pleistocene (1.8 million to 8,000 years ago), commonly referred to as *ice ages* (see Appendix). During times of cooling, tropical forests presumably shrank in area, replaced by other more seasonal tropical ecosystems such as grassland, savanna (an area where grasses predominate but with varying amounts of scattered trees), dry forest (forests that experience significant dry season), or deciduous forest (forests where most trees synchronously shed leaves for part of the year). Earth is currently considered to be within an interglacial period, and there is strong evidence for climate change due to human activities, a topic to be considered later in this book (Figures 1-7 and 1-8).

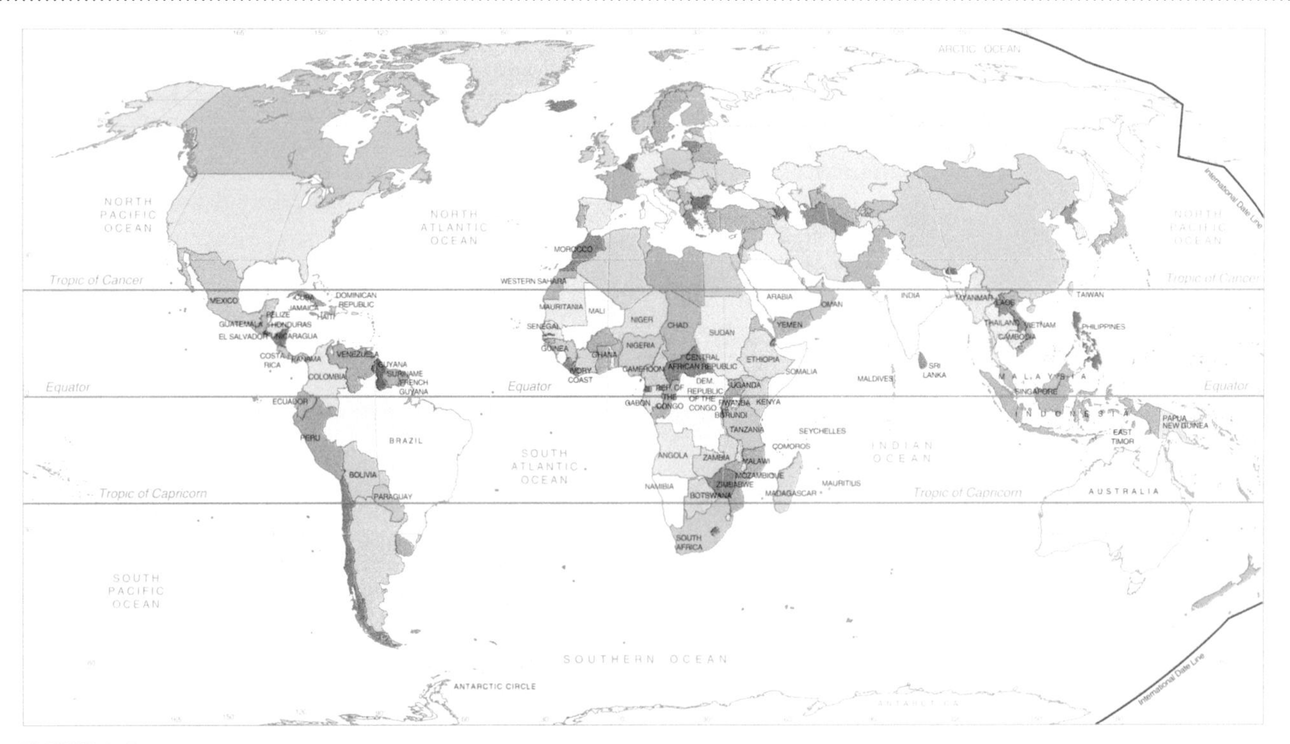

FIGURE 1-8
Political map of Earth, showing nations that occur within the tropics.

Because of the geological process of plate tectonics (described in the following chapter), the Earth's continents, united 248 million years ago at the boundary of the Paleozoic and Mesozoic eras, have gradually fragmented and separated, with large areas of ocean isolating them to varying degrees. This separation into massive continents and varying-sized islands has resulted in evolutionary consequences, particularly the evolution of endemic species. An *endemic species* is a species whose range is confined within a limited geographic area. For example, the imposing Komodo dragon lizard (*Varanus konodoensis*) is endemic to Indonesia, specifically to the Lesser Sunda Islands, most particularly to Komodo (Plate 1-14). Biogeographers recognize various biogeographic realms throughout the planet. Plate tectonics and biogeography will be discussed in greater detail in the next chapter.

Middle America and South America, along with the various islands of the Caribbean and tropical Atlantic Ocean, constitute the Neotropic realm, the so-called New World tropics. The Paleotropic realm, or Old World tropics, comprises Africa and Madagascar (Malagasy Republic), southern India, and Australasia. The latter is further divided into the Southeast Asian tropics (extreme southern China, Myanmar [Burma], Vietnam, Laos, Thailand, Malaysia, Sumatra, Borneo, Philippines) and the Australian tropics (including Sulawesi and other islands of eastern Indonesia as well as New Guinea). The point of separation between the biogeographic realms of tropical Asia and Australia is called *Wallace's Line*, after Alfred Russel Wallace, who first described it (see Chapter 2). Biogeographers differ on exactly where to place Wallace's Line because plant species are not as sharply separated by it as animal species (Whitmore 2002). The southern half of India is within the tropics, and it is unique because India separated from Gondwanaland (a massive southern continent once consisting of South America, Africa, Madagascar, India, Antarctica, Australia, and New Zealand) in the early Cretaceous period (about 140 million years ago) and drifted as an island until it united with Asia about 40 million years ago, in the early Cenozoic. Likewise, Madagascar has been isolated from the rest of Africa to the point where it has numerous endemic species such as the primate group known as lemurs (Lemuridae).

PLATE 1-14
The Komodo dragon, which sometimes reaches a length of nearly 3 meters (9.8 feet), is endemic to the Lesser Sunda Islands of Indonesia.

Each of these realms contains unique assemblages of plant and animal species (Primack and Corlett 2005).

Tropical Climate

In two words, the tropics (especially where there is lush forest) are *warm* and *wet* (Figure 1-9). Tropical regions experience high average temperature, and most (dry grassland and deserts being exceptions) experience substantial precipitation in the form of rainfall. Tropical ecosystems experience seasonality in the form of varying and often predictable amounts of rainfall throughout the year. In contrast to the seasonal pattern that typifies the temperate zone, the tropics vary primarily in precipitation amounts, not temperature fluctuation. There are no periods of protracted cold, no frost, no snow in tropical ecosystems except at high elevations. But there are distinct and pronounced dry and wet seasons. If the dry season is moderate, forests will remain evergreen throughout the year. If the dry season is protracted and severe, forests will have less biomass and more deciduous species, or the ecosystem will not be forest but will be savanna, shrubland, or grassland. The ecosystem type of a given region is also strongly influenced by soil characteristics, a point to be considered later in the text.

Leslie Holdridge and the Life Zone Concept

One of the most significant papers in the science of climatology and ecology was published in *Science* by Leslie R. Holdridge in 1947. The title of the paper, "Determination of world plant formations from simple climatic data," addressed a profound truth: It is possible by knowing three broad climatic variables to predict what major kind of terrestrial ecosystem will prevail in any given geographical region. Those variables are (1) mean annual biotemperature (defined as the average air temperature after values

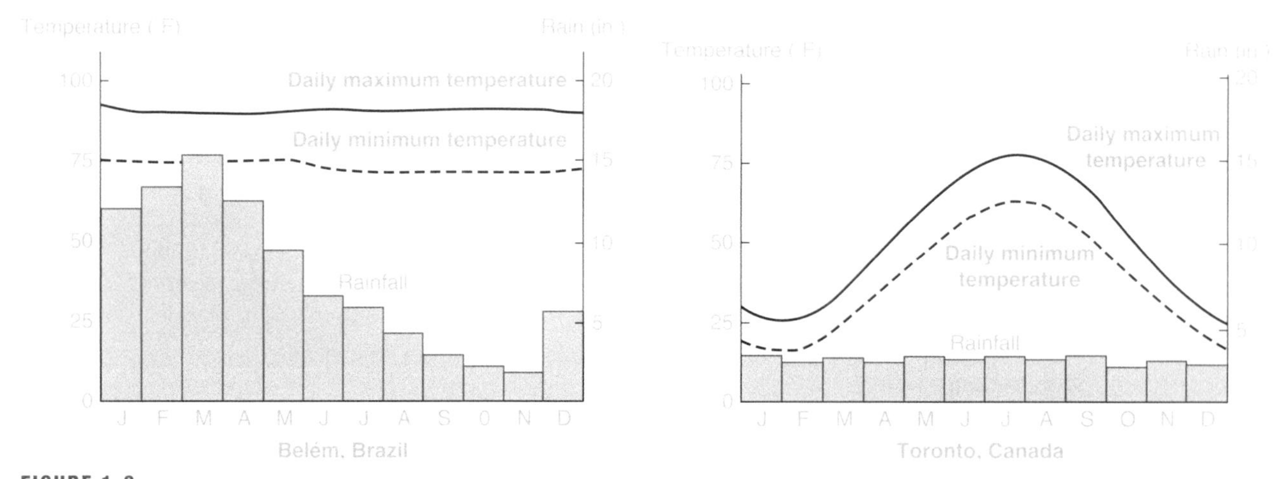

FIGURE 1-9
These graphs contrast the climate in tropical Brazil with that of temperate Canada. Note the strong seasonality of the tropics with regard to precipitation but the steady temperature that prevails in the tropics. The opposite pattern characterizes the temperate zone.

below 0°C or above 30°C are removed); (2) total annual precipitation; and (3) ratio of mean potential evapotranspiration (a function of moisture and temperature) to mean annual precipitation. Holdridge illustrated the concept with a triangle of hexagons, each of which is positioned according to where it fits among the variables, each of which represents one side of the triangle (Figure 1-10).

Holdridge's 38 life zones were broadly divided into polar, subpolar, boreal, cool temperate, warm

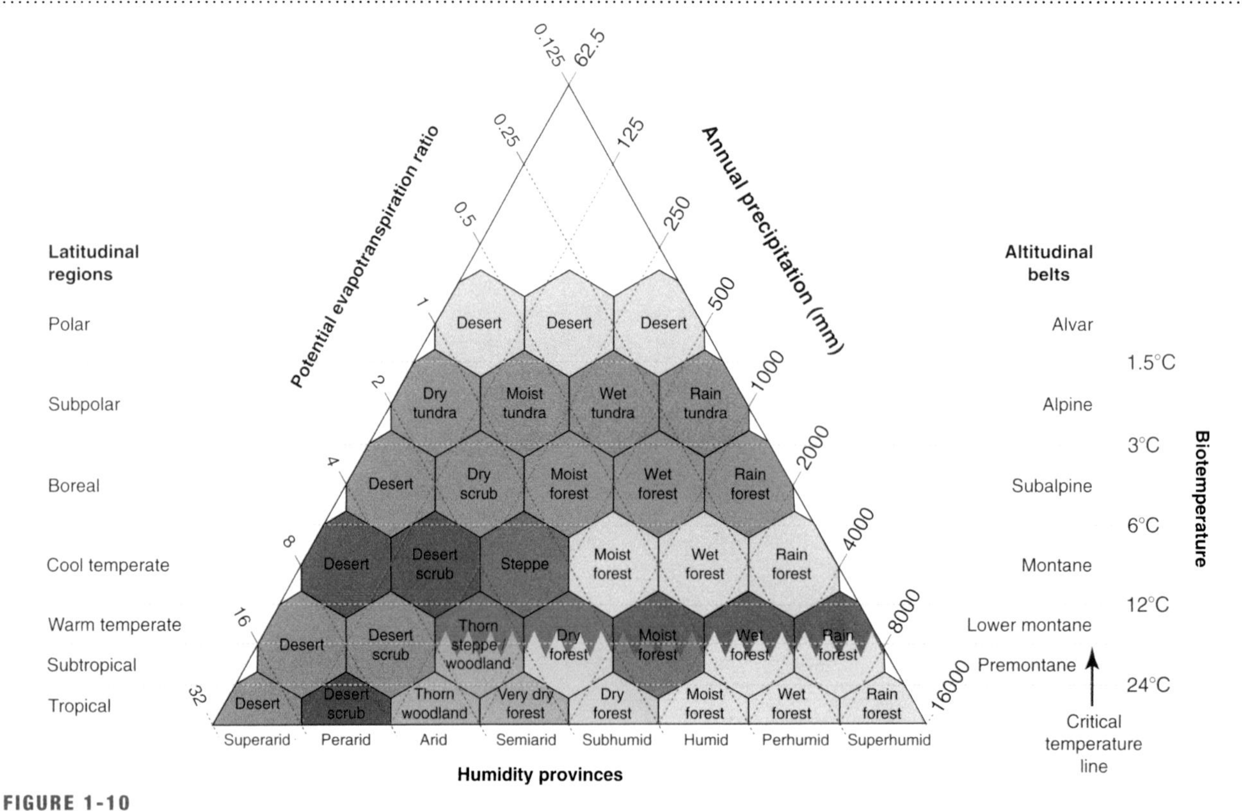

FIGURE 1-10
This is the famous Holdridge diagram that illustrates the relationship between ecosystem types as determined by latitude, elevation, and combination of precipitation and temperature.

		MEAN ANNUAL BIOTEMPERATURE (°C)	POTENTIAL EVAPORATION RATIO	HUMIDITY PROVINCE	AVERAGE TOTAL ANNUAL PRECIPITATION (CM)
Polar		<1.5	0.125–1.5		6.25–75
Subpolar (alpine)	Tundra				
	Dry	1.5–3	1–2	Subhumid	6.25–12.5
	Moist	1.5–3	0.5–1	Humid	12.5–25
	Wet	1.5–3	0.25–0.50	Perhumid	25–50
	Rain	1.5–3	0.125–0.25	Superhumid	50–100
Boreal (subalpine)	Desert	3–6	2–4	Semiarid	6.25–12.5
Boreal	Dry scrub	3–6	1–2	Subhumid	12.5–25
Boreal	Forest				
	Moist Puna	3–6	0.50–1	Humid	25–50
	Wet páramo	3–6	0.25–0.50	Perhumid	50–100
	Rain páramo	3–6	0.125–0.25	Superhumid	100–200
Cool temperate	Desert	6–12	4–8	Arid	6.25–12.5
Cool temperate	Desert scrub	6–12	2–4	Semiarid	12.5–50
Cool temperate	Tundra dry	6–12	1–2	Subhumid	25–50
Cool temperate	Forest				
	Moist	6–12	0.5–1	Humid	50–100
	Wet	6–12	0.25–0.5	Perhumid	100–200
	Rain	6–12	0.125–0.25	Superhumid	200–400
Subtropical	Desert	12–24	8–16	Perarid	6.25–12
Subtropical	Desert scrub	12–24	4–8	Arid	12–25
Subtropical	Thorn woodland	12–24	2–4	Semiarid	25–50
Subtropical	Forest				
	Dry	12–24	1–2	Subhumid	50–100
	Moist	12–24	0.5–1	Humid	100–200
	Wet	12–24	0.25–0.5	Perhumid	200–400
	Rain	12–24	0.125–0.25	Superhumid	400–800
Tropical	Desert	>24	16–32	Superarid	6.25–12
	Desert scrub	>24	8–16	Perarid	12–25
Tropical	Thorn woodland	>24	4–8	Arid	25–50
Tropical	Forest				
	Very dry	>24	2–4	Semiarid	50–100
	Dry	>24	1–2	Subhumid	100–200
	Moist	>24	0.5–1	Humid	200–400
	Wet	>24	0.25–0.5	Perhumid	400–800
	Rain	>24	0.125–0.25	Superhumid	>800

FIGURE 1-10
(continued)

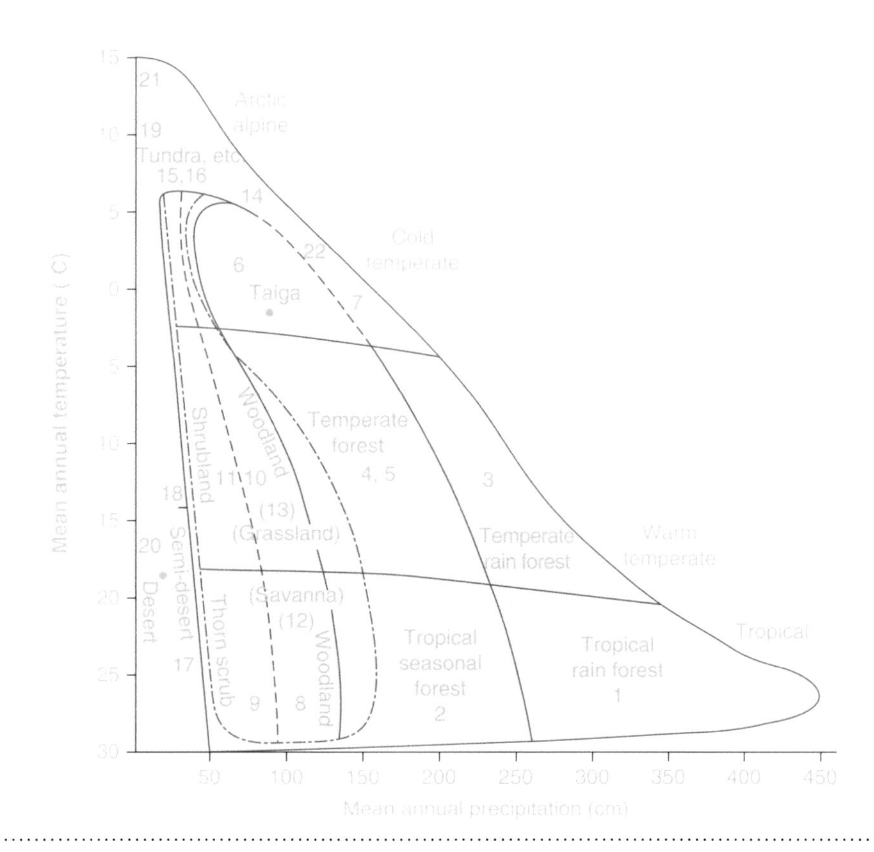

FIGURE 1-11
Diagram showing the relationship between ecosystem type, mean annual temperature (°C), and mean annual precipitation (centimeters). Note that tropical rain forest occurs where both climatic variables are highest.

temperate, subtropical, and tropical. Eight life zones are represented in subtropical and eight life zones in tropical latitudinal regions. The tropical life zones are tropical desert, tropical desert scrub, tropical thorn woodland, tropical very dry forest, tropical dry forest, tropical moist forest, tropical wet forest, and tropical rain forest (Figure 1-11).

Consider the difference between tropical rain forest and tropical dry forest, both of which may occur at the same latitude. Dry forest occurs where there is a significant dry season for part of the year. Tropical dry forest (Plates 1-15 and 1-16), under Holdridge's classification, is defined as an ecosystem that experiences a mean annual biotemperature of greater than

PLATE 1-15
Where rainfall is highly seasonal, tropical ecosystems are often dry forest, with many species of deciduous trees. Forest stature is generally small, and trees are usually widely spaced. From Venezuela.

PLATE 1-16
Northern Australia is an area of vast dry forest with tall, scattered termite mounds. This is the ecosystem typically called *outback*.

24°C (75.2°F), has a potential evaporation ratio of 1:2 (defined as the ratio of annual potential evapotranspiration [PET] to mean total annual precipitation—this means that dry ecosystems in tropical regions have higher ratios and humid ecosystems lower ratios), is subhumid, and has an average total annual precipitation of between 100 and 200 centimeters (39.4 to 78.7 inches). In contrast, rain forest has a mean annual biotemperature also greater than 24°C but has a potential evaporation ratio of only 0.125 to 0.25, is superhumid, and experiences greater than 800 centimeters (315 inches) of average total annual precipitation. Notice that Holdridge's classification makes true rain forest an extremely wet ecosystem. In fact, most rich lowland tropical forests routinely called *rain forests* tend to fall into what, in Holdridge's classification, would be termed *tropical moist* or *tropical wet forests* (Figure 1-11; Plate 1-17).

Some tropical ecosystems may be much drier than dry forest, with less annual precipitation, most of it highly seasonal. Tropical thorn woodland (under Holdridge's classification) includes ecosystems that are dominated by grasses with varying densities of trees scattered throughout. These kinds of ecosystems are called *savannas* and are found in most tropical regions. The best known savannas occur in Africa, where the last of the world's megafauna (diversity of large animals) is still found (Plates 1-18 and 1-19). Savannas and dry forests (Plate 1-20) will be discussed in detail in Chapter 11.

It is important to realize that Holdridge's life zones are not merely latitudinal but also altitudinal. A walk from low Andean rain forest up to an elevation of about 3,500 meters (about 11,483 feet) will take you through several life zones, from hot, humid, dense forest to cold, windswept páramo, a vastly different

PLATE 1-17
Rapidly growing tree species such as Cecropia trees, shown here in the foreground, typify the forest edge, where vegetation structure is dense. This would be considered tropical moist forest in the Holdridge classification. From Trinidad.

PLATE 1-18
Herds of wildebeest move with rainfall patterns in a vast seasonal migration on the East African savanna. From Tanzania.

PLATE 1-19
Savannas are open areas dominated by grasses and scattered trees, particularly in the genus *Acacia*. Here giraffes are feeding on acacia leaves on the East African savanna. From Tanzania.

PLATE 1-20
Australia has many endemic plant species, including hundreds of Eucalyptus species and grasses such as *Spinifex*. This image shows dry forest with *Spinifex* in south central Australia.

ecosystem. Likewise, life zones would markedly change if you were in Tanzania and scaled Mount Kilimanjaro (whose summit is 5,895 meters, or 19,341 feet).

Differences in elevation result in differences in average temperature and precipitation, and that, in turn, results in different ecosystems. Within the South American country of Venezuela, for example, several major elevational ecosystem types are apparent. At highest elevations (what would be described as *alpine*), a wet and windswept shrubland called *páramo* (Figure 1-12) is found. Lower in elevation is forest of low-stature gnarled trees, and lower still, is *cloud forest*, a forest shrouded in cool mist much of the time. Below cloud forest may be semievergreen forest, deciduous forest, or thorn forest, depending on soil and water availability and local seasonal patterns (Figure 1-12). Tropical rain forest occurs in lowland areas where there is ample precipitation throughout the year. Elevation, average annual temperature, and average annual precipitation coupled with seasonality all combine to determine what sort of ecosystem will be present.

What follows are some brief examples of ecosystems along an elevational gradient of the Andes Mountains, beginning at high elevation and descending to wet forest. These examples are meant only to introduce these ecosystems. They will be treated in greater detail in subsequent chapters.

1. *Páramo* is a high-elevation (between 3,800 and 5,000 meters, or 12,467 to 16,404 feet) ecosystem dominated by grasses and shrubs. Climate is generally cool and wet, with frost often occurring at night. In many regions, the dominant grasses are collectively called *tussock grasses*, their blades typically sharp (Plate 1-3). Various shrubs and ferns also are found in páramo. *Espeletias* are unique shrub-sized members of the Composite family (a huge family that includes daisies, goldenrod, and asters) (Plate 1-21). The flowers of *Espeletias*, like those of other flowering plants at high elevation, attract a diversity of hummingbird and insect species. Páramo regions are typically wet with scattered acidic bogs of heath-like vegetation. Because of the relative isolation among various mountain ranges, much endemism is evident in páramo plant species. Some high-elevation ecosystems in Africa, Indonesia, and New Guinea are structurally similar to páramo, but the species composition is quite different.

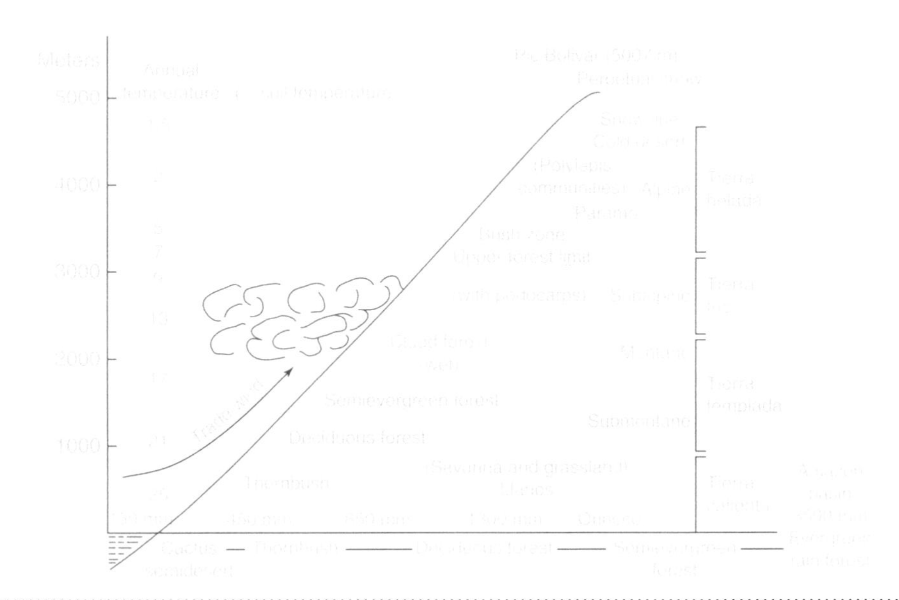

FIGURE 1-12
This diagram shows the distribution of ecosystem types in Venezuela as they relate to elevation and climatic variables.

2. *Polylepis* is a unique endemic tree genus that forms forests within relatively sheltered areas of the Andes Mountains, normally at elevations of around 4,500 meters (14,764 feet). The trees typically are short in stature, gnarled, and twisted, and they often grow on slopes, as shown in Plate 1-22. The trees are dense with epiphytes, and the forest is usually bathed in a cool mist. Many kinds of animals tend to specialize in living among *Polylepis*, and Incan peoples make much use of the trees for various medicinal purposes. There are 15 species within the genus *Polylepis,* ranging from Venezuela to southern Argentina. However, for a variety of reasons, *Polylepis* forests are much reduced from their former abundance and are considered endangered ecosystems.
3. *Puna* is a grassland ecosystem of the high Andes Mountains (mean elevation about 4,000 meters, or 13,123 feet). In Peru and Chile, it is called *altiplano.* It is much drier than páramo, with less precipitation and constant winds that stimulate continuous evaporative water loss. About 75% of all precipitation occurs from December to March. In some areas such as Chile, the mean annual

PLATE 1-21
Members of the Composite family, *Espeletias* are unique shrubs of the high páramo of the Andes Mountains. Several hummingbird species specialize on feeding on *Espeletia* nectar. From Venezuela.

PLATE 1-22
At high elevations in the Andes Mountains where moisture is normally abundant, gnarled forests of *Polylepis* trees are supported. From Ecuador.

temperature is only about 5°C (41°F). January is the warmest month, and July is the coldest (remember that south of the equator, July is a time of deep winter). Various grasses and sedges, many endemic, characterize the ecosystem. Among the most noteworthy animals of puna, the vicuña (*Vicugna vicugna*), a member of the camel family, can be found in small herds throughout much of the region.

4. *Cloud forest* is found at midelevations throughout the global tropics and varies considerably in species composition from one biogeographic region to another. As the name implies, cloud forest is typified by frequent immersion in clouds, a characteristic that keeps the ecosystem continually moist (Plate 1-23). There are many tree species, and they range in size from short stature at higher elevations to nearly the size of lowland rain forest trees at lower elevations. Masses of epiphytes of many species live on the boughs and trunks of cloud forest trees (Plate 1-24). In South America, along parts of the Andes Mountains, cloud forest is habitat for such species as the endemic spectacled bear (*Tremarctos ornatus*). In East Africa, in Uganda, cloud forest is habitat for mountain gorilla (*Gorilla beringei*), also an endemic species.
5. *Wet forest*—when most people think of what they call tropical rain forest, they are usually lumping together three categories within the Holdridge life zone classification: moist forest, wet forest, and rain forest. These forests differ in the amount of average annual precipitation they receive:

Tropical moist forest	200–400 centimeters (79–157 inches)
Tropical wet forest	400–800 centimeters (157–312 inches)
Tropical rain forest	>800 centimeters (312 inches)

PLATE 1-23
Clouds shroud much of the mountainsides along the east slope of the Andes Mountains in parts of South America, supporting dense, species-rich ecosystems called *cloud forests*. From Ecuador.

PLATE 1-24
Many tree species are found in cloud forests, along with numerous epiphytes such as orchids, cacti, bromeliads, and various vines. The ecosystem is normally cloud-covered for at least part of each day. From Ecuador.

In spite of these differences, each of these forests is lush and they share the characteristic of having uniquely high numbers of species, not only of trees, but of many other plant and animal groups (Plates 1-25 and 1-26). I will generally treat them together in this book and not attempt to strictly separate them.

The significance of Holdridge's classification schema is that it demonstrates the extreme importance of climate as an overriding evolutionary selection pressure on plant growth forms. Plants are sessile and must adapt to climate. The reason why plants look as they do, are adapted as they are, and experience differing patterns of growth changes with latitude and elevation, is ultimately due to selection pressures generated by regional climatic variables.

Seasonality in the Tropics: A Closer Look

In the Amazon Basin, the very heart of the Neotropics, climate is permanently hot and humid, with the temperature averaging 27.9°C (82.2°F) during dry season

PLATE 1-25
The complexity of structure and diversity of species characterizes tropical moist, wet, and rain forests throughout the world. This forest is in central Brazil.

and 25.8°C (78.4°F) in rainy season. It is cooler in rainy season mostly due to clouds occluding the sun. In the tropics, daily temperature fluctuation exceeds average annual seasonal fluctuation, and air humidity is high, being about 88% in rainy season and 77% in dry season (Junk and Furch 1985).

The range of seasonality within the tropics is evident when comparing places such as Singapore, in Malaysia, a region that supports tropical moist forest, and Darwin, in northern Australia, where the prevalent ecosystem is tropical savanna (Figure 1-13). In Singapore, the average temperature remains at about 27°C (80.6°F) throughout the year, every day, every month. Daily temperature fluctuations are wider ranging than annual fluctuations. Rainfall, as well, is abundant and relatively constant, averaging 2,415 millimeters (about 95 inches) annually. In contrast, Darwin is strongly seasonal, its average rainfall of 1,538 millimeters (about 60.6 inches) being largely concentrated in the wet season between the months of April and October. Rainfall drops to almost nothing in November through early March. But note that in Darwin, as in Singapore, temperature remains nearly constant, its annual average being about 28°C (82.4°F).

At La Selva Biological Station in Costa Rica, August is the month with the highest mean temperature, 27.1°C (81°F), while January has the lowest mean temperature, 24.7°C (76.5°F). Relative humidity, as noted earlier, is generally high in the tropics, especially in lowland rain forest, where humidities ranging from 90% to 95% at ground level are common. Humidity is lower in the rain forest canopy, usually no higher than 70%.

Tropical regions are warm and generally wet because the sun's radiation falls most directly and most constantly upon the equator, thus warming the Earth more in the tropics than at other latitudes. As one travels either north or south from the equator, the Earth's axial tilt of 23°27' results in part of the year being such that the sun's rays fall obliquely and for much shorter periods of time, thus the well-known cycles of day length associated with the changing seasons of temperate and polar regions. Day length varies much less in tropical regions. Equatorial regions experience close to 12 hours of daylight throughout the year. At the equator, every day lasts exactly 12 hours. North of the equator, days become a little longer in the northern summer and shorter in winter, but this means only that summer sunset is at 6:15 or 6:20 rather that 6:00 P.M.

At the equator, heat builds up, and therefore the air rises, carrying warmth (Figure 1-14). Water is evaporated, and therefore water vapor rises as well. The warm moist air is cooled as it rises, condensing the water, which then falls as precipitation, accounting for the rainy aspect of tropical climates. The normal flow of warm, moisture-laden air is from the equator to more northern and southern latitudes. As the air cools, it not only loses its moisture to precipitation, but also becomes more dense and falls, creating a backward flow toward the equator. At the equator, two major air masses, one from the north and one from the south, along with major ocean currents, form the Intertropical Convergence Zone (ITCZ), the

PLATE 1-26
The term *jungle* refers to a disturbed area where an abundance of sunlight results in a dense array of many plant species, often so thick that it is difficult to penetrate without the use of a machete. Jungles typify areas of moist forest. From Ecuador.

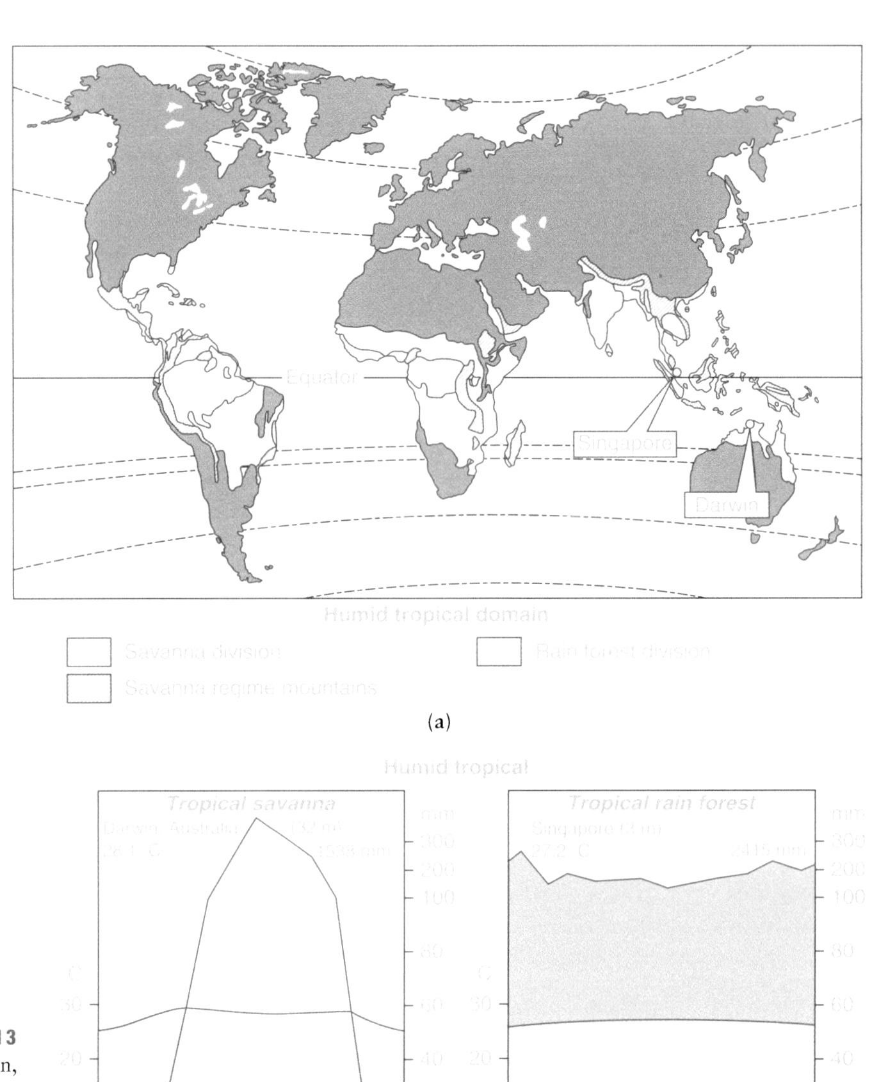

(a)

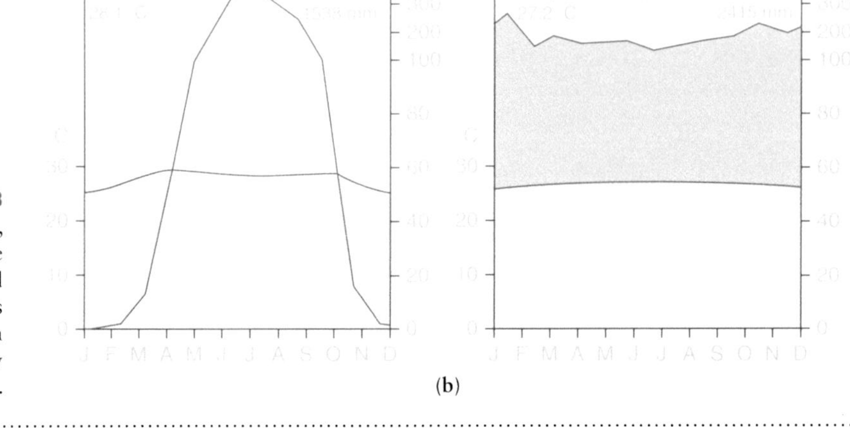

(b)

FIGURE 1-13
These climate diagrams from (a) Darwin, Australia, and (b) Singapore illustrate the extreme contrast between savanna and rain forest. But note that temperature is about the same, and relatively constant in both places. What varies is the seasonality of precipitation.

major climatic heat engine on the planet. The large air masses that drive the process are termed *Hadley cells* (Figure 1-15).

Tropical areas fall within the *trade wind belts* (so named because winds were favorable for sailing ships trading their goods) except near the equator, at the ITCZ. This region is called the *doldrums*, where winds are usually light (often becalming sailing ships). From the equator to 30°N, the eastern trade winds blow steadily from the northeast, a direction determined because of the constant rotation of the Earth from west to east. South of the equator to 30°S, the eastern trades blow from the southeast, again due to the rotational motion of the planet. As the Earth, tilted at about 23.5° on its axis, moves in its orbit around the sun, its direct angle to the sun's radiation varies with latitude, causing seasonal change, manifested in the tropics by changing heat patterns of air masses around the ITCZ that result in seasonal rainfall (Figure 1-16). In the Western Hemisphere, from July throughout October, severe wind and rain storms called *hurricanes* can occur in parts of the

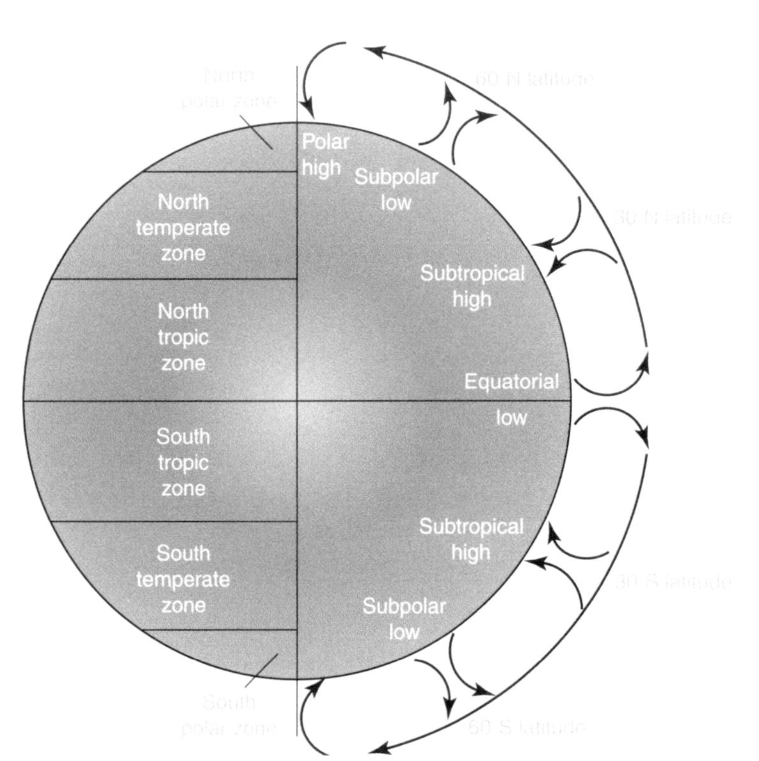

FIGURE 1-14
If Earth were not rotating, this would be the distribution of primary and secondary air movement of the planet.

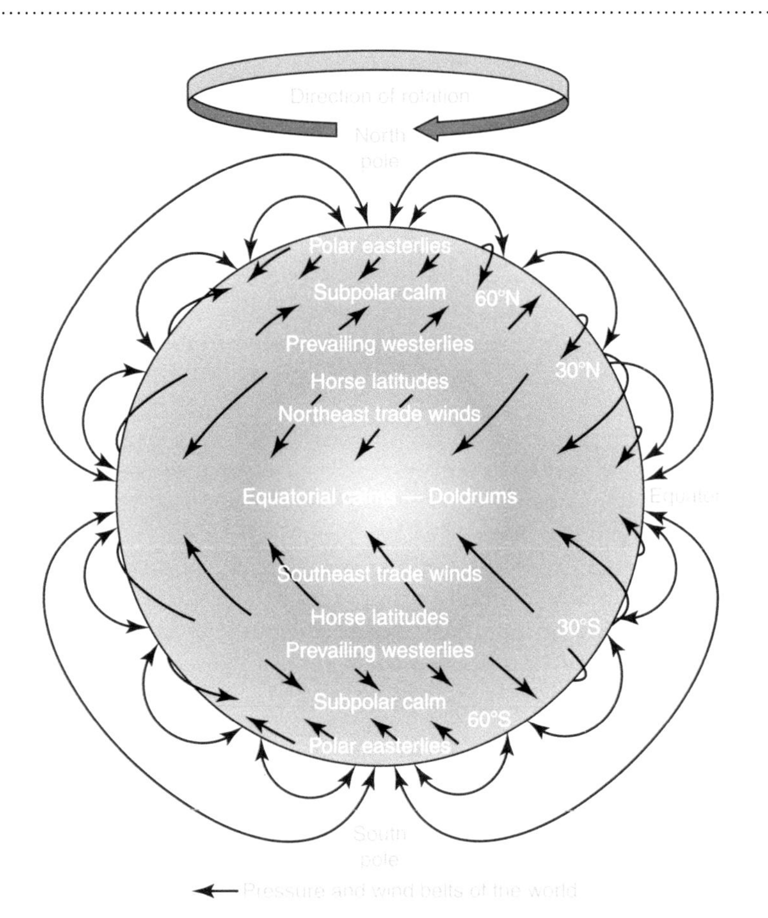

FIGURE 1-15
Earth's daily rotation deflects the Hadley cell currents and creates the trade wind belts. Note the lack of wind at the equator.

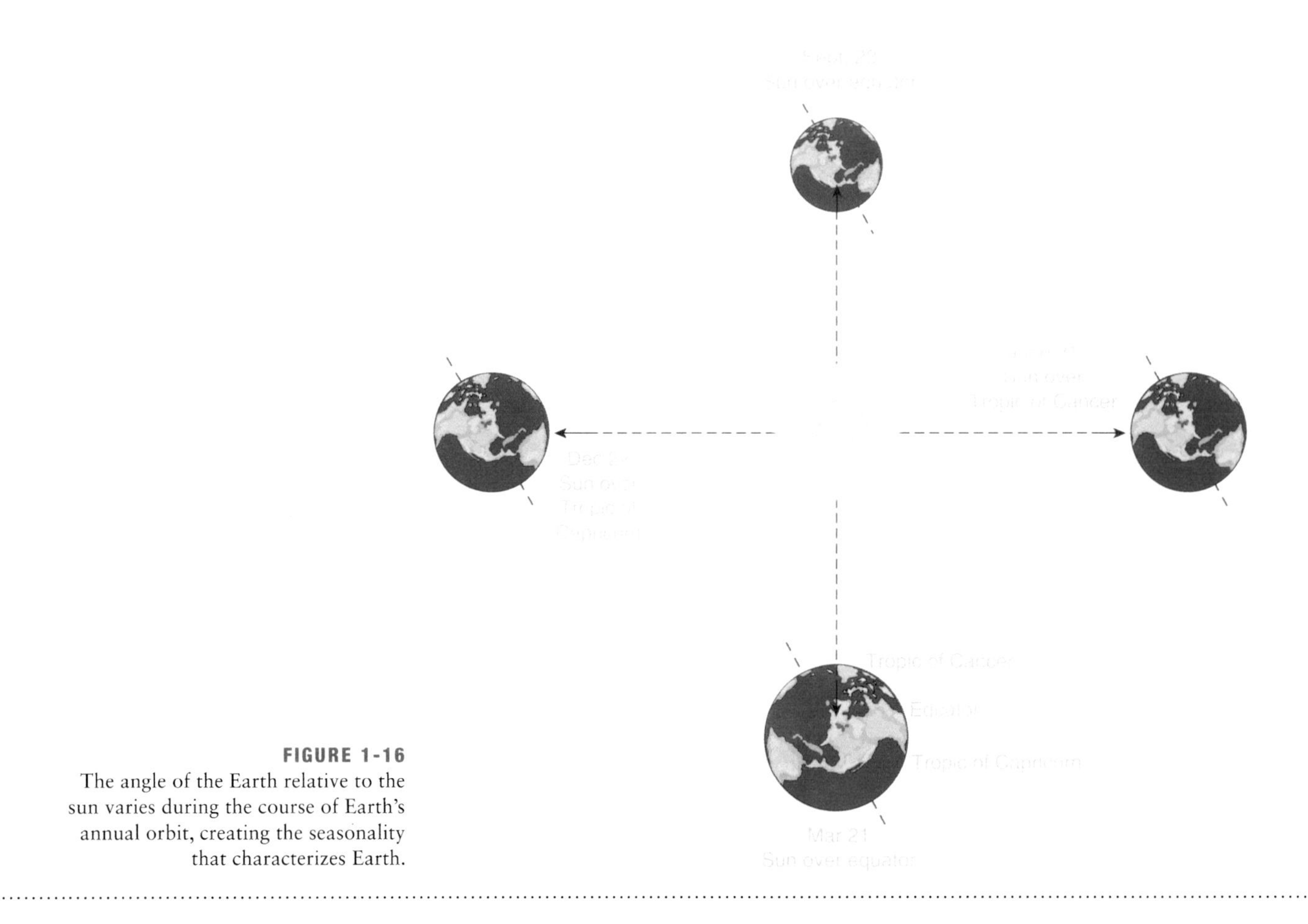

FIGURE 1-16
The angle of the Earth relative to the sun varies during the course of Earth's annual orbit, creating the seasonality that characterizes Earth.

Neotropics. Similar kinds of storms are referred to as *monsoons* in the Old World tropics.

At the point where tropics meet subtropics, at about 30° north and south latitude, deserts are common. This is because the moisture contained in the air masses (Hadley cells) moving away from the equator has been depleted by the time the air reaches 30° latitude. Thus, skies tend to be clear with low humidity and deserts form. Longitude also contributes to desert formation. If a landmass is far from the sea, or if mountain ranges block moisture circulation, moisture evaporated from the sea will not reach inland areas, and deserts will result. The combination of 30° latitude and distance from the sea produces many of the world's vast deserts, including the Sahara in Africa.

In the Amazon Basin, precipitation ranges between 1500 and 3000 millimeters (59 to 118.1 inches) annually, averaging around 2000 millimeters (78.7 inches) in central Amazonia (Salati and Vose 1984) (Figure 1-17). About half of the total precipitation is brought to the basin by the eastern trade winds blowing in from the Atlantic Ocean, while the other half is the result of evapotranspiration from the vast forest that covers the basin (Salati and Vose 1984; Junk and Furch 1985). Up to 75% of the rain falling within a central Amazonian rain forest may come directly from evapotranspiration (Junk and Furch 1985), an obviously tight recycling of water. Much of the reason for the large amount of recycled moisture lies in the nature of the forest itself. This vast assemblage of trees transpires far more moisture than other kinds of ecosystems such as savannas or agricultural land. The forest functions such that precipitation and the water-recycling system are essentially in equilibrium, though large-scale deforestation (a topic that will be discussed in more detail later in Chapters 13 to 15) could significantly upset the system (Solati and Vose 1984).

Short-Term Climate Change: ENSO

Short-term but sometimes major climatic changes also occur in the tropics, the most noteworthy attributed to the El Niño/Southern Oscillation (ENSO).

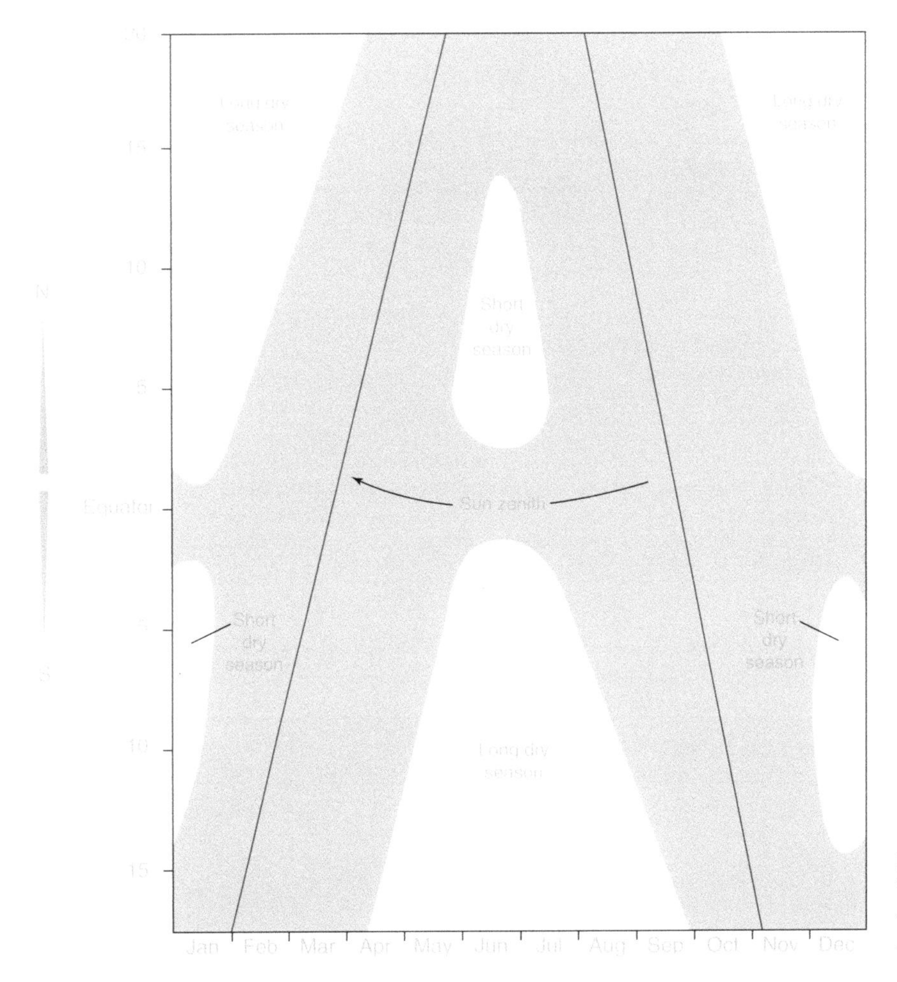

FIGURE 1-17
This diagram shows how climate varies in the tropics. Note the absence of seasonality at the equator.

This phenomenon has become increasingly evident since the mid-twentieth century, but its exact cause remains elusive. El Niño occurs periodically, approximately every two to seven years, when a high-pressure weather system that is normally stable over the eastern Pacific Ocean breaks down, destroying the pattern of the westward-blowing trade winds. Trade winds weaken, sometimes reversing from their normal westward direction. Warm water from the western Pacific flows eastward, causing an influx of abnormally warm water to the western coast of South America. The water column reaches a depth of up to about 150 meters (about 500 feet), and because it is warm, it is less dense and thus flows over and blocks the colder, more-nutrient-rich waters below. This prevents the upwelling of nutrients into the upper water column, where they would normally be available to phytoplankton. The result is that oceanic food chains are severely disrupted. When that happens, warm waters flow along the normally cold South American coast. Weather systems change, resulting in heavy downpours and flooding in some regions and droughts where there should be rainfall, effects that range ecologically from mildly stressful to highly significant.

The El Niño of 1982–1983 was considered up to then to be the most powerful of this century, estimated to have caused $8.65 billion worth of damage worldwide. There have been several other El Niños since 1982, eight major El Niño events since 1945, and at least 20 during this century. A severe El Niño occurred in 1986–1987. Another El Niño occurred in 1994–1995, comparable to those of 1982–1983 and 1986–1987. Satellite data indicated that the northern Pacific Ocean was nearly 20 centimeters (8 inches) higher than normal, due to the influx of warm surface waters. Yet another El Niño occurred in 1997–1998,

El Niño Comes to Panama

A major El Niño affected Barro Colorado Island in Panama during 1982–1983 (Leigh 1999). One of the most dramatic effects of the El Niño was the failure of trees to produce fruit during the second of two fruiting seasons. The failure of the fruit crop resulted in a cascade of severe effects on various animals. Normally wary species such as collared peccaries (*Tayassu tajacu*), coatimundis (*Nasua ilarica*), tapirs (*Tapirus baridii*), and kinkajous (*Potos flavus*) all made regular visits to the laboratory area where food had been put out for them. Some of these animals appeared emaciated, obviously under stress from lack of food (Plate 1-27).

One researcher wrote

> The spider monkeys, which normally visit the laboratory clearing at least once every day, now launched an all-out assault on food resources inside the buildings, learning for the first time to open doors and make quick forays to the dining room table, where they sought bread and bananas, ignoring the meat, potatoes, and canned fruit cocktail, and brushing aside the startled biologists at their dinner (Foster 1982).

Dead animals were also encountered far more frequently than usual:

> The most abundant carcasses were those of coatis, agoutis, peccaries, howler monkeys, opossums, armadillos, and porcupines; there were only occasional dead two-toed sloths, three-toed sloths, white-faced monkeys and pacas. At times it was difficult to avoid the stench: neither the turkey vultures nor the black vultures seemed able to keep up with the abundance of carcasses (Foster 1982).

(a)

(b)

PLATE 1-27
(a) KINKAJOU AND (b) SPIDER MONKEY

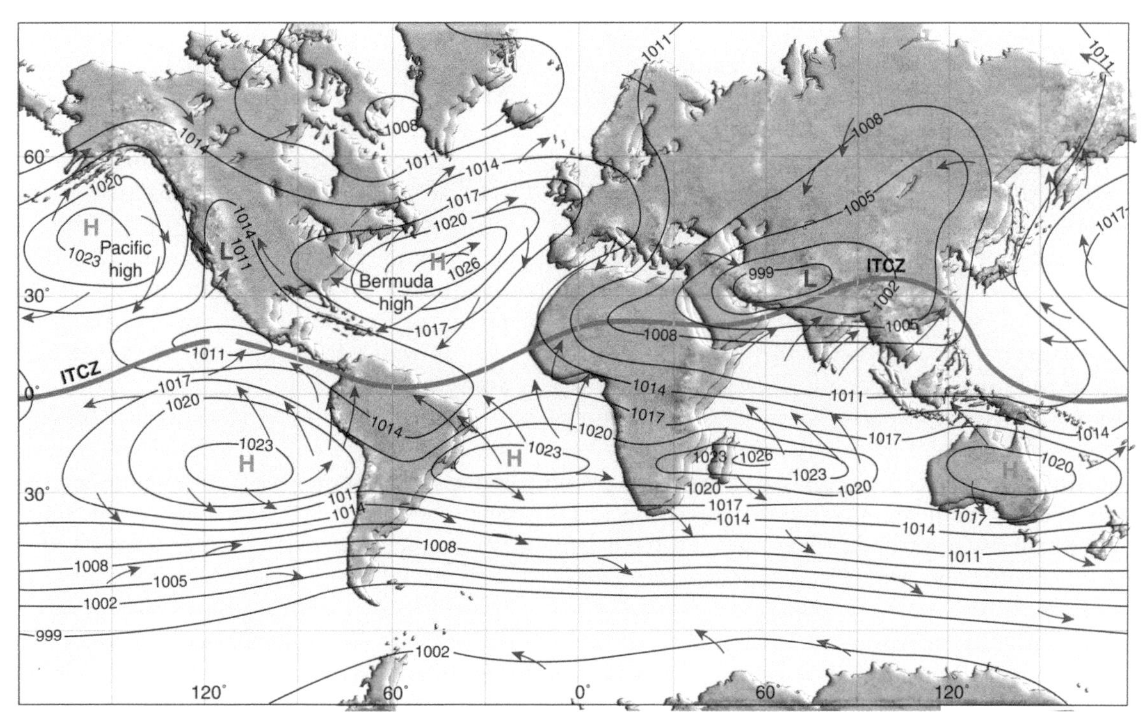

FIGURE 1-18
This map shows the general location of the ITCZ.

and its combined global effects are estimated to have resulted in 2,100 human casualties and property damage totaling a staggering $33 billion (Suplee 1999). In 2009–2010, still another El Niño altered weather patterns around the world.

The causal factors responsible for the periodicity of El Niños are thus far unknown (Canby 1984; Graham and White 1988), but it is clear that the Intertropical Convergence Zone (Figure 1-18), a complex system of oceanic and air currents, migrates to a lower latitude, raising sea surface temperatures and destroying the normal upwelling pattern along the west coast of South America. The cessation of an El Niño occurs when the ITCZ returns northward to its normal position (hence the alternative term El Niño/Southern Oscillation, or ENSO). Though El Niño has global effects, tropical ecosystems in particular can be anywhere from moderately to severely affected.

El Niños tend to alternate with another climatic phenomenon that produces largely the opposite effects and is called La Niña. Like El Niño, La Niña systems begin when normal trade winds are altered. In this case, however, the westward trade winds gain abnormal strength and move farther westward than normal, carrying warm surface water toward Asia. This creates enhanced upwelling of colder, deeper, more nutrient-rich water along the South American coast. Because the colder waters evaporate more slowly, rain is reduced and drought may result. La Niña followed the 1982–1983, 1986–1987, 1995, and 1998 El Niños.

Ecologists realize that ENSO events cause short-term but nonetheless significant perturbations in ecosystems. ENSO events represent one of several major global climatic patterns that may vary periodically, affecting tropical ecosystems. Climate variations occur at multiple time scales: short-term, as is the case with an abnormally cold or snowy winter; moderate, as is the case with ENSO events; or long-term, as is the case with major global climate change.

The Scope of Tropical Ecology

The word *tropics* is often synonymous with tropical rain forest, but such a linkage, as I hope this chapter makes clear, is too limited. Tropical rain forest is one kind of ecosystem that is found within the tropics. Because of its potentially unique nature, and because of its amazing richness of species, tropical rain forest (or more generally, tropical humid forest) will be the ecosystem most discussed in this text. But the tropics are more than tropical rain forest. Ecosystems ranging from deserts to rain forests occur in equatorial latitudes.

Because patterns of rainfall and temperature vary with elevation, because soil characteristics differ due in part to regional geological differences, and because disturbance factors such as storms (monsoons, hurricanes) and fire vary regionally in frequency, tropical ecosystems vary from one place to another. Natural disturbance may be sufficiently frequent to prevent some, perhaps most, ecosystems from attaining a point where species composition ceases to change. Add to that differences in historical biogeography from one place to another, and the reality is that ecological variation is extensive within the tropics.

Consider a comparison between savanna and rain forest in Africa. These two very different ecosystems may be found at different longitudes but at the same latitude. They have little in common except that they are both in the same biogeographic region. Now consider rain forest in Papua New Guinea as compared with that in Alta Floresta, Brazil. The structure and overall look of the two forests will be similar, but that similarity is superficial. Because these regions have been isolated for nearly 100 million years, evolutionary patterns are different and species composition is quite distinct between them. Because the species composition is different, there will be significant differences in biological interactions and evolutionary patterns (see Chapter 2).

Ecological analysis of tropical ecosystems requires full consideration of biogeographical history, climate, edaphic factors such as soil characteristics, and disturbance history, frequency, and magnitude.

Current Research Directions in Tropical Ecology

This text summarizes current research in tropical ecology, but at this juncture it is helpful to make a few generalizations. Tropical researchers study intricate interactions among interdependent species, carbon sequestration in forests and other tropical ecosystems, effects of climate change on ecosystems, mechanisms of speciation (and generation of high levels of biodiversity), food web dynamics, and maintenance of biodiversity. Tropical rain forest tree species richness is decidedly higher than that of any other kind of forest and the question of how so many species coexist is a major area of investigation, one that encompasses a lively debate. Species richness patterns in animals ranging from beetles to birds are also a major topic of investigation. Indeed, the causal patterns and function of the uniquely high biodiversity seen in many tropical ecosystems is a topic of very active research.

The role of disturbance as it relates to biodiversity continues to occupy researchers. Traditionally and naively, the tropics were once considered a region of great ecological stability. This notion followed from the assumption that climate (considered warm and wet and relatively invariable) was not a strong selective force in tropical regions, and thus species evolved to the point where they became distinct from one another. But disturbances occur constantly and at varying scales in space and time. Such disturbances force changes in ecosystems and are thus selective forces relative to both ecological trends and ultimately evolutionary direction.

It is obvious that the potential for significant and original research of tropical regions, where the abundance and diversity of both plants and animals exceeds that found anywhere else, is very high. There is still more to know than is known.

Conservation in the Tropics

The world's tropical regions include some of the densest human populations, and some of the least wealthy. The growth rate of human populations is high in most tropical regions. The human population is predicted to grow from 6.3 billion (in 2003) to 8.9 billion by 2050 (Cohen 2006) (Figure 1-19). Much of that increase will be in equatorial countries. According to the Population Reference Bureau's website (www.prb.org, 2007):

> Between 2000 and 2030, nearly 100 percent of this annual growth will occur in the less developed countries in Africa, Asia, and Latin America, whose population growth rates are much higher than those in more developed countries. Growth rates of 1.9 percent and

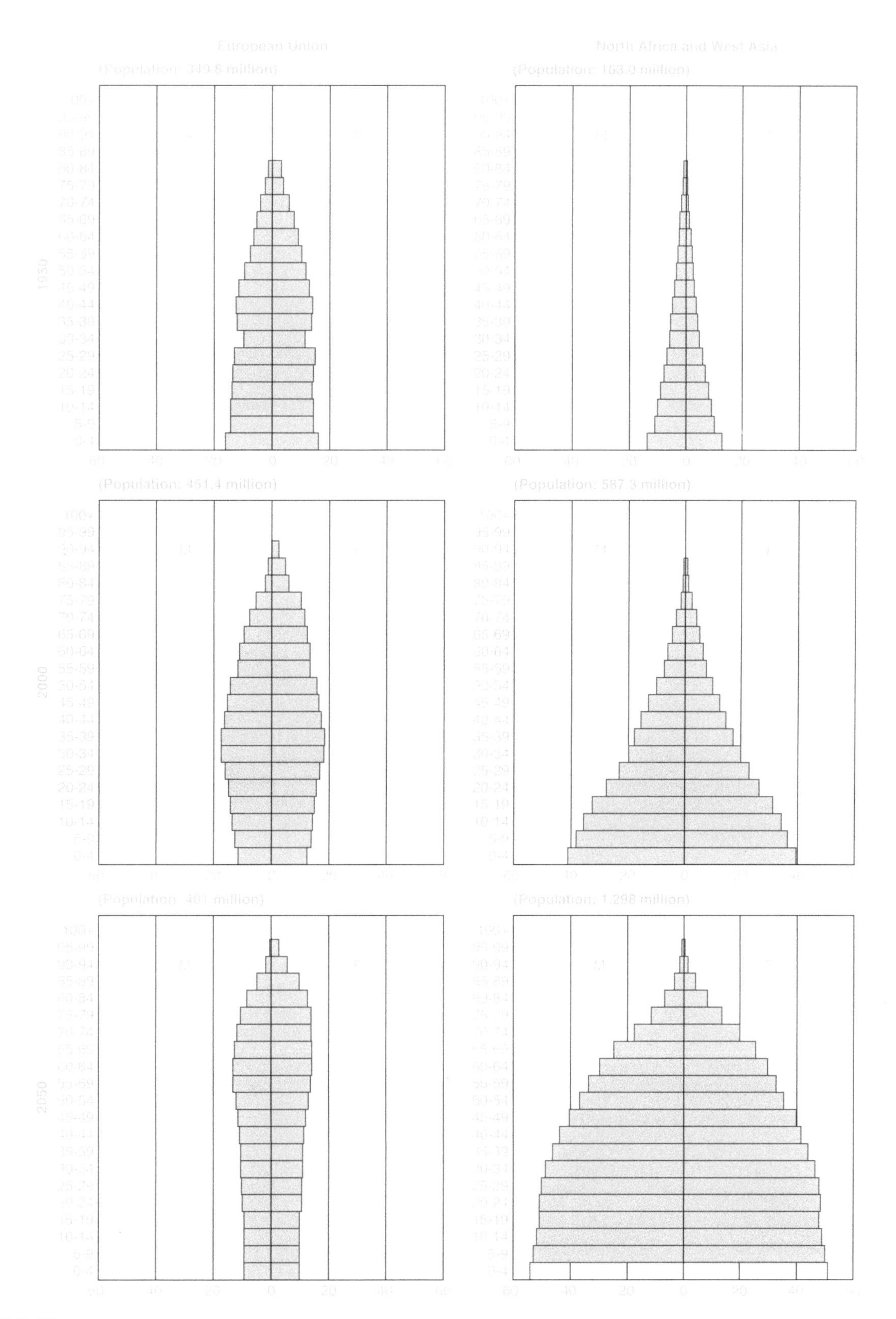

FIGURE 1-19
Human population size and age distribution contrasted between developed and underdeveloped nations, with projections to 2050.

higher mean that populations would double in about 36 years or less, if these rates continue. Demographers do not believe they will. Projections of growth rates are lower than 1.9 percent because birth rates are declining and are expected to continue to do so. The populations in the less developed regions will most likely continue to command a larger proportion of the world total. While Asia's share of world population may continue to hover around 55 percent through the next century, Europe's portion has declined sharply and could drop even more during the 21st century. Africa and Latin America each would gain part of Europe's portion. By 2100, Africa is expected to capture the greatest share.

Humans will inevitably have an increasing impact on tropical ecosystems, and as such the study of tropical ecology must consider anthropogenic factors.

Humans evolved in the tropics, emerging from tropical Africa and colonizing the remainder of the terrestrial planet. Human populations have adapted to survive and prosper in tropical regions, and much insight into ecological processes can be gained from the study of tropical anthropology.

For example, the use of the land for agriculture, ranging from slash and burn techniques to sustained terraced rice fields, demonstrates that humans can extract continuous resources from tropical ecosystems. Some communities demonstrate a highly sophisticated knowledge of the species patterns in rain forest. The use of various plants as pharmaceuticals and toxins has given rise to the science of *ethnobotany* (the study of how local cultures master and develop various uses for plant species) and the search for potent drugs from tropical plants (Chapter 13).

Tropical ecology represents an area of basic research within ecology that can be applied to conservation issues. The discipline of conservation biology is essentially derivative of basic ecology, and much research in tropical ecology deals in some way with issues related to conservation. One example is land use and its effects, particularly deforestation and forest fragmentation (Chapter 14). Deforestation rates in tropical regions are generally high. In a broad study based on global imaging from satellites, it was learned that from 1990 to 1997, 5.8 +/− 1.4 million hectares of humid tropical forest were lost annually and that an additional 2.3 +/− 0.7 hectares were visibly degraded (Achard et al. 2002).

Predicting the future of tropical forests based on current rates of deforestation is both perplexing and complex. For example, it has been argued that as the century proceeds, birth rates in tropical regions will decline and urbanization will increase, and thus rural population growth will decline, slowing the rate of deforestation (Wright and Muller-Landau 2006a and 2006b). If this were the case, tropical forests would begin to regrow on once-cleared land, much as forests in eastern North America have regrown after forest clearance in the seventeenth and eighteenth centuries. This pattern would, in essence, be a reversal of deforestation, but these forests would be secondary forests, not the same as the old-growth primary forests that preceded them. This view is highly controversial, and a lively debate has been generated. Some argue that as industrialization increases in tropical areas, even with demographic shifts from rural to urbanized, large-scale forest clearance by industrial interests will continue the loss of tropical forests (Laurance 2006; Brook et al. 2006). Because of the high biodiversity contained within tropical forests, such widespread loss would inevitably result in high rates of extinction (Brook et al. 2006). It is imperative to understand what is occurring and what can be done to mitigate species losses, but at this point, the future of tropical forests is, at best, uncertain (Gardner et al. 2006). This topic will be reviewed in Chapter 15.

The rate of forest loss in the tropics, whatever it may be, has strong implications (Table 1-1). A significant and continuing loss diminishes populations of numerous species, reduces biodiversity, and has the potential to seriously disrupt trophic dynamics and the overall functioning of remaining forest fragments. In addition, it reduces the capacity of forests to sequester excess carbon, creating a diminishing carbon sink. But the issue of climate change, carbon sequestration, and the function of tropical forests in relation to climate change is an area that is still poorly understood (Clark 2004).

Yet another conservation issue is the effect of hunting and logging on forest ecology (Chapter 15). In many tropical regions, large vertebrate animals are routinely hunted to the point where their numbers are seriously diminished. Throughout the tropical world, bushmeat, as it is termed, is widely used for food. Hunting pressure, logging, and the emergence of viral

TABLE 1-1 Humid tropical forest cover for 1990 and 1997 and mean annual change estimates during that time period. All figures are in millions of hectares.

	LATIN AMERICA	AFRICA	SOUTHEAST ASIA	GLOBAL
Total study area	1155	337	446	1937
Forest cover in 1990	669 ± 57	198 ± 13	283 ± 31	1150 ± 54
Forest cover in 1997	653 ± 56	193 ± 13	270 ± 30	1116 ± 53
Annual deforested area	2.5 ± 1.4	0.85 ± 0.30	2.5 ± 0.8	5.8 ± 1.4
Rate	0.38%	0.43%	0.91%	0.52%
Annual regrowth area	0.28 ± 0.22	0.14 ± 0.11	0.53 ± 0.25	1.0 ± 0.32
Rate	0.04%	0.07%	0.19%	0.08%
Annual net cover change	−2.2 ± 1.2	−0.71 ± 0.31	−2.0 ± 0.8	−4.9 ± 1.3
Rate	0.33%	0.36%	0.71%	0.43%
Annual degraded area	0.83 ± 0.67	0.39 ± 0.19	1.1 ± 0.44	2.3 ± 0.71
Rate	0.13%	0.21%	0.42%	0.20%

PLATE 1-28
GORILLA

diseases such as Ebola have put the African great apes, the common chimpanzee (*Pan troglodytes*), and the gorilla (*Gorilla gorilla*) (Plate 1-28), in serious peril (Walsh et al. 2003). In the African country of Gabon, by the year 2000 ape populations dropped to less than half what they had been in 1983. The authors of the study documenting the decline argue that these species should be elevated to a status of *critically endangered*. Such a statement could be construed as somewhat political, but it is not. The conclusion is based on data and follows directly from an analysis of such data, given, of course, the premise that extinction of chimpanzees and gorillas is neither in their best interests nor in the best interest of human society.

Conservation issues in tropical ecology will be treated in much greater detail in Chapter 15.

2 Biogeography and Evolution in the Tropics

Chapter Overview

For much of the past 250 million years of Earth's history, tropical ecosystems have extended north and south well beyond the latitudes they encompass today. But climate has changed over the millions of years of the planet's history, and that fact influences the distribution of tropical ecosystems. In addition, because of plate tectonics, continents have separated and moved apart as Earth has become more temperate, such that the tropical regions are currently confined within about 50° of equatorial latitudes. Climatic change, the separation of continents, and continuous tectonic activity such as mountain uplift have contributed to ongoing patterns of evolution and are responsible for complex biogeographic distributions of flora and fauna that typify today's tropical regions. This chapter will discuss *plate tectonics* (the movement and rearrangement of Earth's crust), *vicariance* (the separation and subsequent evolution of populations), and *endemism* (species restricted to a certain area) as each relates to biological evolution. It will also discuss how these events contribute to the evolution of species.

What Is Biogeography?

Biogeography is the study of the distributions of organisms as they vary from one region on Earth to another.

For example, all of the world's 71 extant lemur species occur exclusively on the island of Madagascar, off the eastern coast of Africa (Plates 2-1a and 2-1b). In addition, five familes of birds are found only on that island. Certain plant species are also confined only to Madagascar.

Such unique assemblages of species have long stimulated thought among natural historians. On a larger scale, many elements of both flora and fauna differ dramatically in diversity and distribution among tropical zones of different regions around the equator. For example, approximately 445 species of stately dipterocarp trees (Dipterocarpaceae) are distributed throughout the lowland forests of tropical Asia and New Guinea, while a mere 30 dipterocarp species are found in Africa, and but a single species is found in northeastern South America. Such a pattern begs for explanation. Charles Darwin devoted two chapters in *On the Origin of Species* (1859) to a consideration of geographical distributions of various groups of organisms, as evidence in support of his theory of evolution. Darwin and Alfred Russel Wallace (who each independently conceived of the theory of natural selection) both linked biogeography with the process of evolution.

(a)

(b)

PLATE 2-1
(a) Coquerel's Sifaka (*Propithecus coquereli*), a lemur species common to low-elevation, dry deciduous forests in Madagascar. (b) Ring-tailed lemurs (*Lemur catta*) are highly social.

Beginning with Alexander von Humboldt, biogeographers have long recognized different biogeographic realms (Lomolino, Dov, and Brown 2004) (Figure 2-1). Wallace proposed six realms encompassing the world's terrestrial environments. Four of these realms, the Neotropical, African, Oriental, and Australian, include tropical regions (Primack and Corbett 2005). Wallace is best known for denoting *Wallace's Line*, the demarcation of separation between the Oriental and Australian faunal realms.

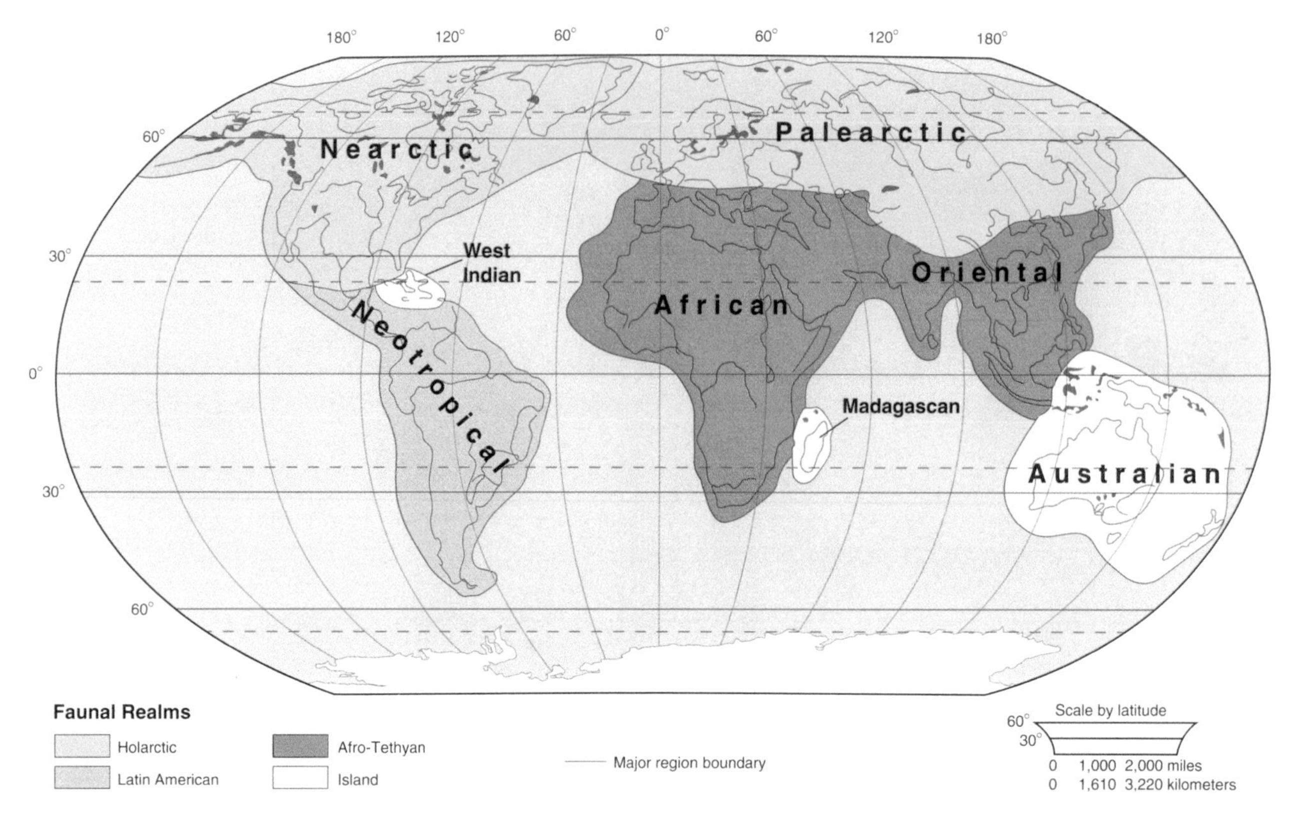

FIGURE 2-1
This map shows the major biogeographic regions of the world. Each is distinct from the others because each has various endemic groups of plants and animals.

Wallace's Line

Alfred Russel Wallace (1823–1913) devoted eight years to the study of what was then called the Malay Archipelago, an area that today is mostly within the country of Indonesia (Figure 2-2). He noted that animals found west of the large island of Sulawesi were typical of those found in southeast Asia (Figures 2-3a to 2-3c). Such

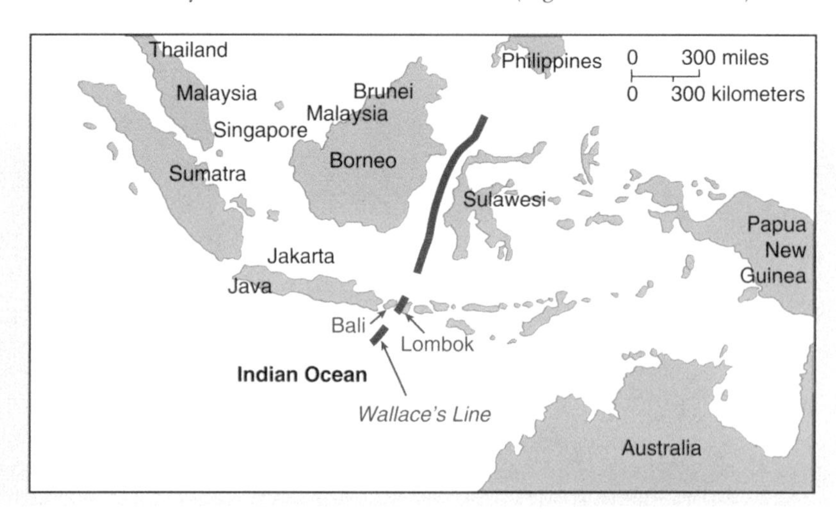

FIGURE 2-2
Wallace's Line was originally developed by Alfred Russel Wallace based on the distribution of animal groups. Those typical of tropical Asia occur on the west side of the line; those typical of Australia and New Guinea occur on the east side of the line.

FIGURE 2-3

Examples of animals found on either side of Wallace's Line. West of the line, nearer tropical Asia, one finds species such as (a) proboscis monkey (*Nasalis larvatus*), (b) flying lizard (*Draco* sp.), (c) Bornean bristlehead (*Pityriasis gymnocephala*). East of the line one finds such species as (d) yellow-crested cockatoo (*Cacatua sulphurea*), (e) various tree kangaroos (*Dendrolagus* sp.), and (f) spotted cuscus (*Spilocuscus maculates*). Some of these species are either threatened or endangered.

islands as Sumatra, Borneo, and Bali all are examples. These islands harbor various primate species, including one large ape (orangutan) as well as several gibbons and macaques. But animals found on and to the east of Sulawesi are more representative of those found in New Guinea and Australia (Figures 2-3d to 2-3f). No primates occur other than humans, but species of arboreal marsupials such as tree kangaroos and possums are found on some islands. Thus Wallace, by noting distributions of animals on various islands, surmised that the Malay Archipelago is actually two separate biogeographic realms, one representing what he termed the *Oriental province* and the other the *Australian province*. Where the two provinces come together, Wallace noted some mixing of animals from both realms, a result of natural dispersal among populations. Today biogeographers informally honor Wallace by referring to the islands from Java to New Guinea as *Wallacea*.

Because of differences in distribution patterns of plants and animals, the exact boundaries of Wallace's Line have been altered a bit since Wallace first developed the concept, but his theory remains generally accurate.

Biogeographers were initially puzzled, if not perplexed, by the complex intercontinental distributions of plants and animals. Why, for example, are marsupial mammals essentially confined to the continents of Australia and South America, which are quite widely separated? (One marsupial species, the opossum [*Didelphis*] occurs in North America, but its ancestors clearly dispersed there from South America.) The tacit assumption was made that all *taxa* (groups of organisms that are related—for example, palms, parrots) evolved initially in what were called *centers of origin*. Such areas were designated based on criteria such as where the greatest number of species were to be found (within a taxon), the greatest concentration of individuals, the presence of primitive forms, and the existence of migratory routes. From centers of origin, the belief was that taxa radiated away (to various degrees) to inhabit different regions. The dipterocarp example cited earlier would strongly suggest a tropical Asian origin for dipterocarps, with subsequent dispersal to and colonization of Africa and later South America.

Early biogeographers attemped to explain how different organisms dispersed from their centers of origin to other continents. Of course, these early biogeographers were making the assumption (quite reasonable at that time) that Earth's continents have always occupied the same locations. Plate tectonics, explained in the next section, demonstrates that such an assumption was unwarranted. Nonetheless, three dispersion models were suggested to explain the patterns of biogeography: the *broad corridor*, the *filter bridge*, and the *sweepstakes route*. Each of these models helps to explain how dispersion occurs.

A *broad corridor* would be typified by the Pleistocene (ice age) connection between northeastern Asia and extreme northwestern North America, a landmass now largely submerged and commonly called *Beringia*. Most organisms, including humans, who colonized North America from Asia at that time, would presumably be able to travel such a corridor, permitting extensive mixing between biota from different biogeographic realms—in this case, Asia and North America. Broad corridors obviously require that the oceans between the continents must be sufficiently shallow so that a global drop in sea level will expose the corridor. This is the case with the Beringia corridor.

A filter bridge would be exemplified by the Greater and Lesser Sunda Islands of Indonesia, the very area of Wallace's Line. As these islands become more distant eastward from Asia (one source of colonization) and westward from Australia and New Guinea (another source of colonization), they become increasingly species poor. This is partly because species typically must *island hop* from one landmass to another as they colonize. Thus the species richness (number of species) of various taxonomic groups on an island is related to the island's distance from its source of colonizers, a topic developed in greater detail in Chapter 4.

A *sweeptakes route* would be typical of how organisms colonize isolated islands, such as the Galápagos Archipelago (Plates 2-2a to 2-2f). Presumably, ancestral iguanid lizards from South America that would eventually evolve to be Galápagos marine and land iguanas were accidently transported over water on detached vegetation from areas in Central and South America—the ocean currents steering them, quite by chance, to landfall somewhere among the Galápagos. Currents are directionally favorable for such unique events to occur. Another sweepstakes route would be when seeds of plants contained in mud become stuck to the feet of migrating waterfowl—in the course of the birds' migration, the seeds are transported hundreds if not thousands of miles before being detached and germinating. Charles Darwin performed experiments to

(a) (b) (c) (d) (e) (f)

PLATE 2-2

These vertebrate animals are each endemic to the Galápagos Islands, but each traces its ancestry to animals living in South America. (a) and (b) Galápagos tortoise (*Geochelone nigra*). These two images show (a) a saddle-shelled tortoise and (b) a dome-shelled tortoise. Each is a distinct subspecies. Domed tortoises inhabit moist highland areas, and saddle-shelled tortoises frequent lowland, arid regions of the islands. (c) Galápagos marine iguana (*Amblyrhynchus cristatus*). The flattened face adapts the animals to grazing on marine algae attached to rocks. (d) Galápagos land iguana (*Conolophus subcristatus*). Land iguanas occupy low-elevation, arid areas of the islands, where they feed extensively on cactus. (e) Galápagos flightless cormorant (*Phalacrocorax harrisi*). This species is found only in the colder waters of the western islands. (f) Hood mockingbird (*Nesomimus macdonaldi*). This species is found only on Española (formerly Hood Island) in the Galápagos. It spends much of its time on the ground.

test the viability of seeds under varying conditions to test this concept.

The obvious difficulty with the preceding summary is that in numerous cases, it falls far short of explaining causal reasons for organismal distribution among continents. Where, for example, is the corridor between Australia and South America that the marsupials presumably used? Is it now, like the fabled island of Atlantis, lost beneath the seas? The answer is no. The sea is sufficiently deep in most places that such a corridor could never have existed. No drop in sea level would have exposed it.

The problem that faced early biogeographers, as noted earlier, was that they assumed the continents of Earth to have always been in their present positions. Some early scientists were bold enough to suggest otherwise (see the section on "Plate Tectonics"), though no one had any idea how a huge continent could move. It was not until 1957 that physical evidence was put forth validating what had come to be called *continental drift*, brought about by a dynamic process termed *plate tectonics*. The understanding of plate tectonic theory forever altered biogeography and made it into a far more powerful and predictive science.

A Sweepstakes Ticket to Madagascar

The island of Madagascar, off the east coast of Africa, is inhabited by numerous endemic plants and animals. Only four mammal groups are found on the island (not including some recently imported by humans), each having evolved into unique endemic species. These groups are the tenrecs, rodents, carnivores, and primates. While the present representatives of each of these groups (lemurs, for example, are primates) are all endemic to Madagascar,

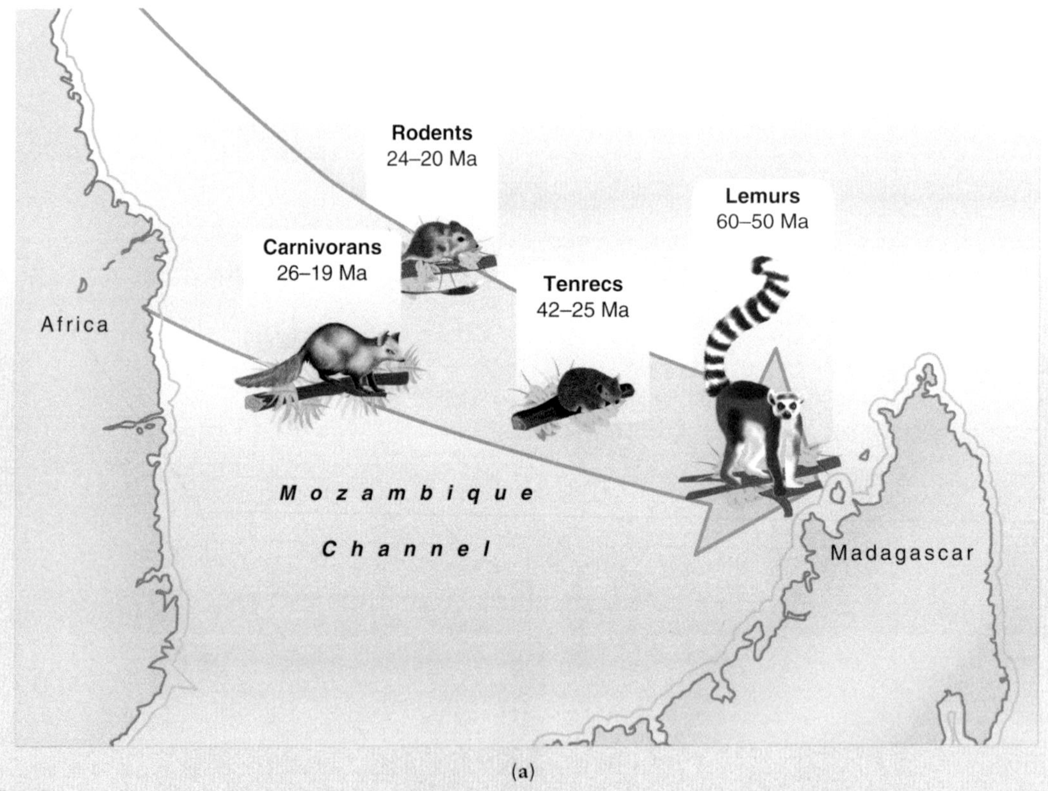

(a)

FIGURE 2-4
Figure 2-4(a) illustrates the suggested rafting route by which some mammalian groups may have colonized Madagascar. The numbers indicate estimates of when (in millions of years ago) the colonizations may have occurred. Other groups such as elephants, zebras, antelopes, and apes are considered unlikely colonists, and indeed, none are found on Madagascar. Figure 2-4(b) illustrates how today's ocean currents would minimize dispersal potential from mainland Africa to Madagascar, but such was not always the case.

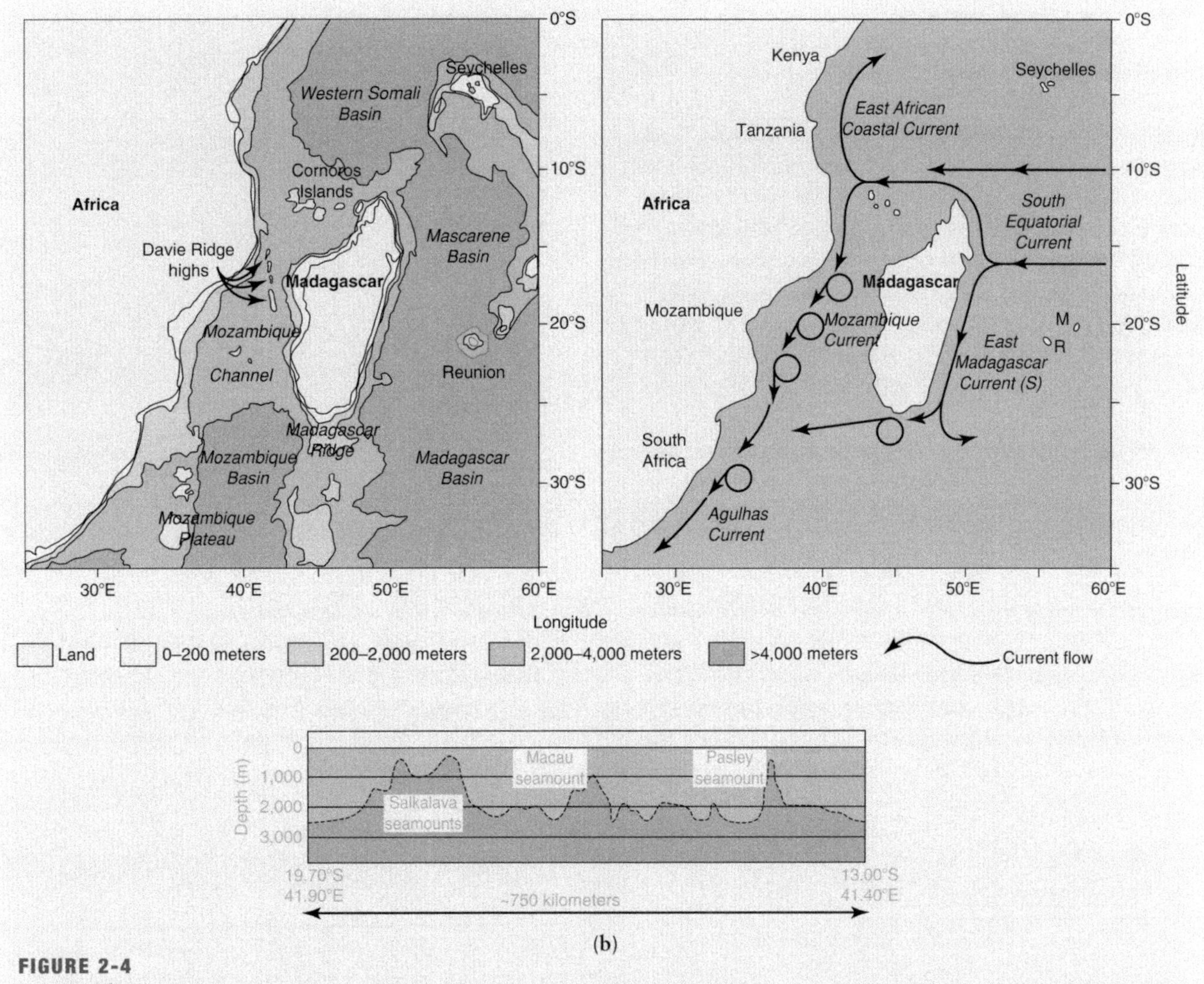

FIGURE 2-4
(continued)

the groups themselves are evolutionarily related to African mammals. The logical assumption is that Madagascar was colonized by each of these four groups at some time during its history. But how did mammals colonize Madagascar?

A close look at the mammals of Madagascar has long suggested that they evolved from Cenozoic mammals, not from the more ancient mammals of the Cretaceous period. But Madagascar had been separated from Africa by the time of the Cenozoic. So how did the mammals cross the Mozambique Channel? Perhaps they walked. Biogeographers have long suggested the possibility of a land bridge connecting Madagascar and Africa during part of the Cenozoic. Recent work on the likely topography of the seafloor during the Cenozoic argues against such a possibility. Perhaps these animals floated over on rafts of detached vegetation. This sweepstakes mode would explain why larger animals, such as elephants, zebras and other hoofed mammals, and apes never colonized the island. Only smaller creatures would have been able to fit on small rafts of vegetation. The present pattern of ocean currents would not allow an easy crossing, however.

Research modeling of Cenozoic climate and currents strongly supports the sweepstakes route hypothesis (Ali and Huber 2010). Surface water currents during the Cenozoic appear to have been different from today, moving in a direction that would facilitate sweepstakes transport. The data suggest that lemurs colonized via a sweepstakes voyage across the Mozambique Channel approximately 60 to 50 million years ago. Tenrecs arrived around 42 to 25 million years ago, carnivores 26 to 19 million years ago, and rodents 24 to 20 million years ago.

The sweepstakes model for mammalian colonization of Madagascar was first proposed by the eminent twentieth-century paleontologist George Gaylord Simpson (Krause 2010), and while still not proven, evidence now seems to support Simpson's conjecture (Figures 2-4a and 2-4b).

Plate Tectonics

On a globe, look at South America, and notice that the east coast of South America looks as though it fits rather well against the west coast of Africa. If the globe shows relief features, notice that the Himalayan Mountains form a rugged boundary all along the border of India and Asia. Australia sits by itself, alone—an immense island continent in the southern Pacific Ocean. But notice that if sea level was lower, northeastern Australia (the Cape York Peninsula) would connect to New Guinea. All along the west coast of South America extending north through Mexico to Alaska are volcanically active mountain chains where earthquakes are common. Frequent volcanic activity is also typical where Wallace's Line occurs in Indonesia. These seemingly disparate observations are each explained by the process of plate tectonics.

PLATE 2-3
Earth's moon, unlike Earth, lacks plate tectonics.

The occurrence of cataclysmic events such as earthquakes and volcanic eruptions demonstrates Earth's ceaseless active geology. Unlike Earth's moon (Plate 2-3) or planets such as Mars, Earth is geologically dynamic, continuously rearranging its surface because of processes occurring in its interior. The term for this process of continual change is *plate tectonics*. Earth is one of the few known tectonic planets. Both the planet Venus and Jupiter's large moon Europa have dynamic tectonics, but they are not organized into a system of rigid plates, as is the case with Earth.

The Discovery of Continental Drift

The intriguing distribution of continents had been noticed from the time of Francis Bacon (1561–1626), who may have been the first to muse about the possibility that South America and Africa were once joined. He is reputed to have noticed that the two continents had complementary shapes. In 1912, a German meteorologist named Alfred Wegener published *Die Entstehung der Kontinente und Ozeane* (The Movements of the Contents and the Oceans), a book in which Wegener made a bold and, to geologists at the time (and for some time thereafter), preposterous suggestion. Wegener proposed that the continents move relative to one another, drifting on the surface of the planet. If true, this would mean that the continents were not always in their present positions. By implication, they might still be drifting, each in its own particular direction (Figure 2-6).

But how could something as immense as a continent actually move? Geologists could identify no mechanism to account for such motion, and many authorities at the time ridiculed Wegener's theory, which soon came to be called *continental drift*, a name that is still applied.

Wegener's principal motive in arguing for continental drift was the compelling fit that the continents seem to have when you literally cut them out like pieces of a jigsaw puzzle and rearrange them. Prior to Wegener, in the late nineteenth century, Eduard Suess (1831–1914), an Austrian geologist, suggested that the southern continents were once fused into one gigantic continent. This huge landmass was named *Gondwana*, a romantic-sounding name that refers to an area within India with characteristic rocks that were thought to date back to when such a hypothetical supercontinent might have existed. Suess based his hypothesis on fossil ferns of the genus *Glossopteris*, found in rocks from South America, Africa, and India, now each widely separated. *Glossopteris* fossils also occur in Antarctica, which likely would have pleased Suess.

The fit between Africa and South America is fairly obvious, but Wegener was able to show how all of the continents could, if arranged carefully, roughly fit together into a single supercontinent, *Pangaea*. North of

Gondwana, and originally fused with it, was the northern supercontinent of Laurasia (containing what is today North America, Europe, and northern Asia). Wegener was at a loss for a mechanism to account for how, from one or two massive continents, today's varied continents separated and drifted apart, but he insisted that such a possibility should be considered.

In addition to Wegener's view that the continents could be assembled together like a kind of planet-sized jigsaw puzzle, there was much ecological evidence to suggest that at some time in the past, some of the continents were once literally joined. As early explorers studied the unique plants and animals discovered during the age of global exploration (see Chapter 1), their study of biogeography revealed many enigmatic examples of puzzling distributions among both plants and animals.

Numerous examples showed that closely related species occur on different continents. How could that be? For example, there are magnolias (*Magnolia* spp.) in the southeastern United States and in China. Three tapir species (*Tapirus* spp.) occur in South America, but another closely related tapir species occurs in Southeast Asia. Ostriches (*Strutio camelus*), found in Africa, are closely related to emus (*Dromaius novaehollandiae*) and rheas (*Rhea* spp.), but emus are found only in Australia and rheas in South America. The southern beech tree, *Nothofagus*, occurs in such places as southern South America, southern Australia and Tasmania, and New Zealand. Such a widespread distribution is difficult to explain if the assumption is that continents have always been in their present locations.

Let us examine one case more closely. As noted earlier, there are several extant large flightless birds that seem anatomically and behaviorally very similar: the ostrich of Africa, the two rhea species of South America, the emu of Australia, the cassowaries (*Casuarius* spp.) of Australia and New Guinea, and the kiwis (*Apteryx* spp.) of New Zealand. Collectively, this group is named the *ratites*, a term referring to their lack of a broad bony keel on the sternum, a trait typical of large flightless birds (Plates 2-4a and 2-4b). (The *keel*, or *carina*, is a prominent blade-like protrusion from the ventral surface of the breastbone. It is very well developed in all birds that fly because it is the site of attachment of the large flight muscles. In ratite birds, the carina is absent.) Each ratite species is generally large and skeletally similar, has a body covered in down-like feathers, has undersized wings, and lacks the aforementioned large bony keel on the breastbone for the support of flight muscles. These birds neither fly nor swim. How can large flightless birds disperse worldwide?

As described earlier, some biogeographers suggested the existence of now (presumably) submerged land bridges that once connected continents now separated by ocean. There was no evidence that such bridges existed, only speculation. Other biogeographers argued that examples such as the flightless birds noted earlier were examples of convergent evolution, where distantly related species evolve similar appearances due to being exposed to similar environmental selection pressures. But modern genetic studies of both mitochondrial and nuclear DNA sequences have demonstrated unequivocally that the ratites form a *monophyletic group*, meaning that they are all descended from the same common ancestor. The logical hypothesis is that ratite birds originated in and dispersed from Gondwana and rode the separating continents over time to their present positions.

(a)

(b)

PLATE 2-4
(a) Ostrich (*Struthio camelus*) and (b) emu (*Dromaius novaehollandiae*) are two examples of ratite birds. They are very similar in size and anatomy, but ostriches are endemic to Africa and emus to Australia.

Plate tectonics, now supported by multiple lines of physical evidence, recognizes that Earth's crust is divided into approximately eight large and some additional small basaltic plates. Some of these plates are entirely on the seafloor, some have continents resting atop them, and some have parts of continents. For example, the Pacific plate is huge and is covered mostly by the Pacific Ocean, except that the extreme western part of North America rests on the most eastern part of the Pacific plate. The remainder of North America sits atop the North American plate. What this means is that places such as Los Angeles and San Francisco are actually situated on the Pacific plate. The North American plate is moving ever so slightly west-southwest, and the Pacific plate is moving more directly north. The two plates meet in an area extending from the central coast of Mexico to almost the Oregon border. Part of this area is called the *San Andreas fault* (Figure 2-5). Because the plates are moving in somewhat different directions, there is friction between them, the result of which is known to most Californians: earthquakes occur.

The heat generated from Earth's interior, a region called the *asthenosphere* within Earth's mantle, produces immense subterranean convection currents that keep the basaltic plates moving in slow motion (Figure 2-6). For example, Europe and North America are moving apart by a mere 17 millimeters per year. But in a million years, they will have separated by 17 million millimeters, or about 17 kilometers, roughly 10.55 miles. Why then are there such large differences in the flora and fauna of Africa compared with South America? Geological evidence suggests that these continents began their slow but continuous separation as long ago as the early Cretaceous period, some 144 million years ago.

Tectonic motion occurs because the heat from Earth's interior forces new basalt to the surface at active ridges throughout the oceans. The lengthy mid-Atlantic ridge brings up new material from the interior, a process called *seafloor spreading* (Figure 2-7). The North American and European continents actually ride on the plates, being passively carried along as the plates move. Old seafloor plunges back into the bowels of the Earth at deep oceanic trenches,

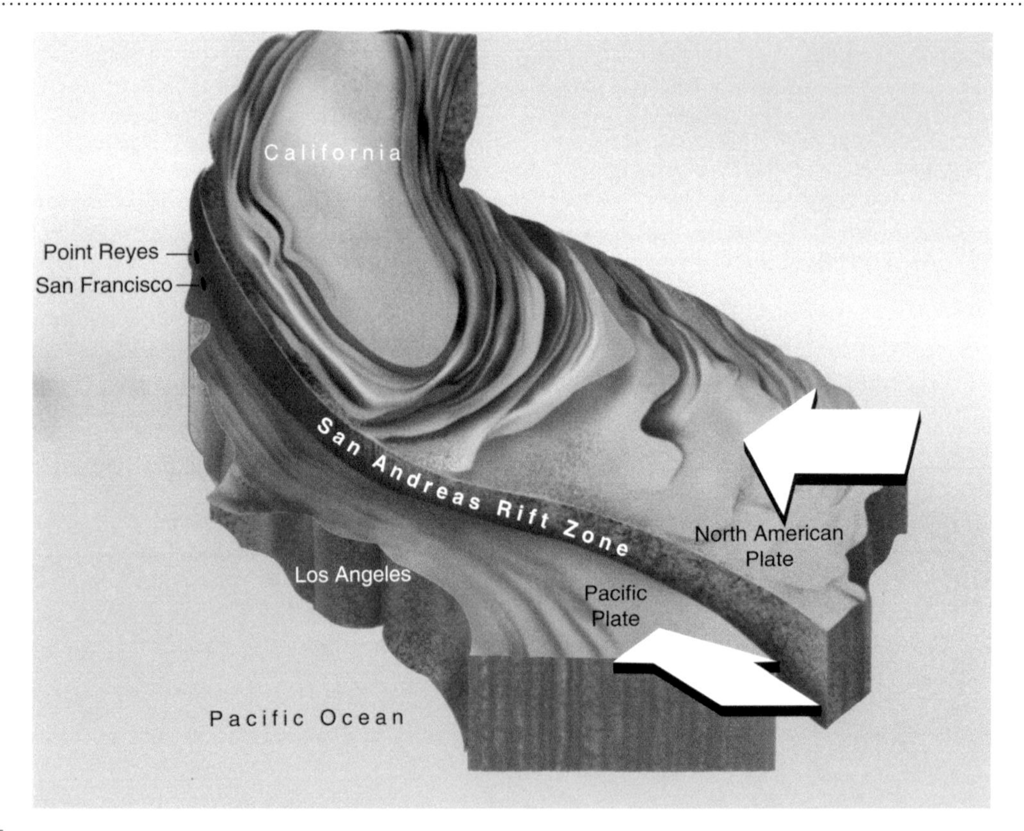

FIGURE 2-5
The San Andreas fault is well known for its relatively frequent tectonic activity, resulting in earthquakes of varying magnitudes. Note where the boundary between the North American plate and Pacific plate occurs.

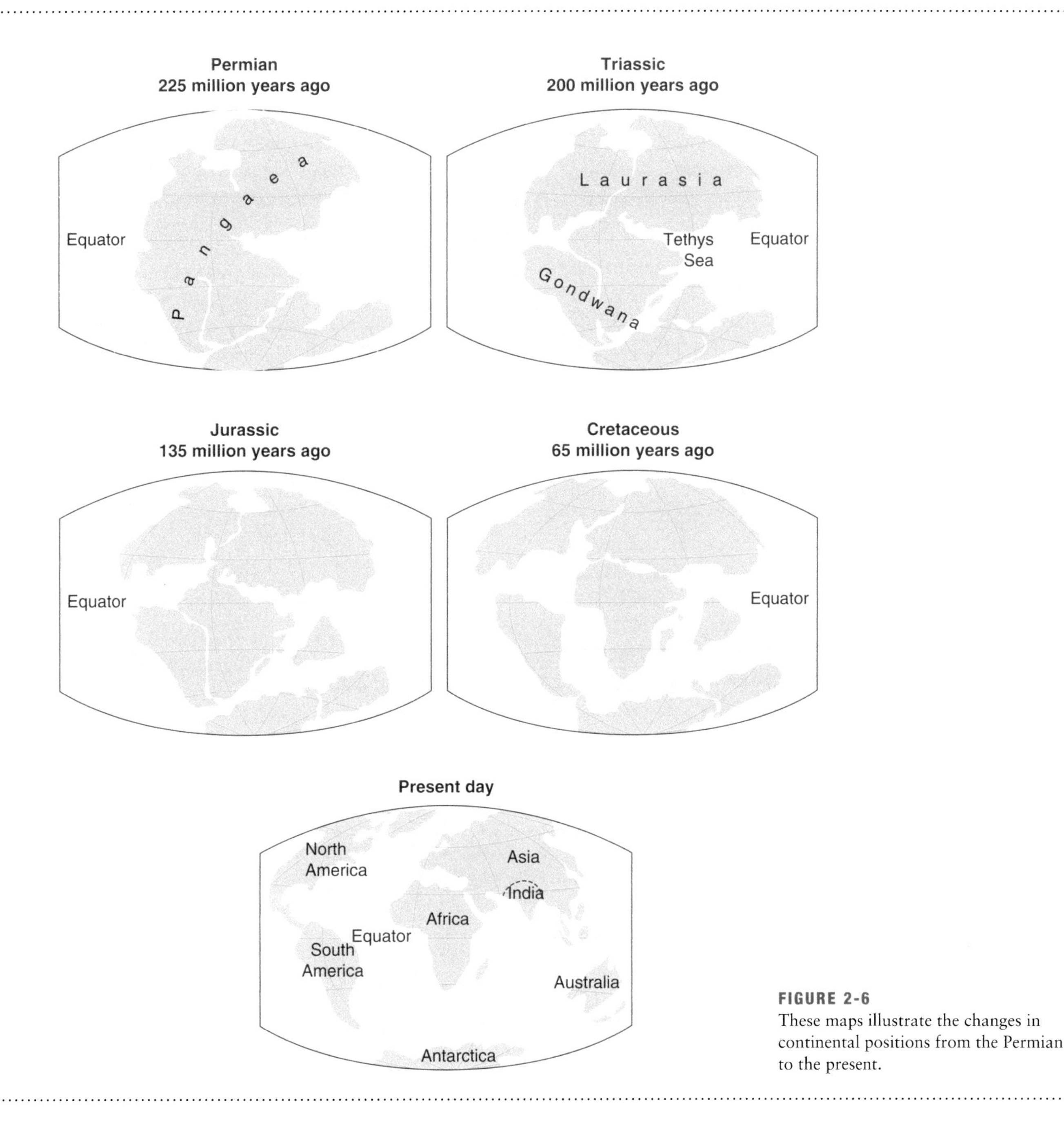

FIGURE 2-6
These maps illustrate the changes in continental positions from the Permian to the present.

some in excess of seven miles deep. The Andes Mountains have risen in large part due to the impact of tectonic plates, subducting beneath western South America and forcing the rise of the Andes and all of the subsequent volcanism and earthquakes in places such as Chile.

You can actually mimic this process in a crude way if you heat some tomato sauce and cover its surface with crackers. The convection currents from the heat rising through the sauce will cause the surface crackers to move about, somewhat like Earth's basaltic plates move in response to internal forces generating convectional heat.

Given the insights of plate tectonics, the notion that continents do move now has credence. Geologists have traced the motions of the continents back through time. About 245 million years ago, at the onset of the Mesozoic era, Pangaea began to break up, splitting into Laurasia and Gondwana. The former giant continent consisted of what today is North America, Europe, and northern Asia. Gondwana consisted of what is now South America, Africa, Madagascar, Antarctica, India, Australia, and New Zealand. Laurasia and Gondwana split throughout the Mesozoic. Antarctica, once at a temperate latitude, eventually drifted to the South Pole, where we see it today.

India drifted northeast and, about 40 million years ago, literally crashed into Asia, causing the uplift of the Himalayan Mountains. Mount Everest, Earth's tallest mountain (8,848 meters [29,028 feet]), was created in the course of this collision.

The large island of Madagascar, today a haven of endemic lemurs, birds, and plants, separated from Africa about 160 million years ago and from what would become India 90 million years ago. Thus Madagascar has been alone, unattached to any other landmass, for 90 million years, ample time for unique evolutionary trajectories to develop.

Continental movement caused by plate tectonics has profound consequences for Earth's ecology. Without such constant movement, Earth's climatic history would have been very different and less variable. As the continents move about the surface of the planet, they affect changes in ocean currents, air currents, and climate in general. As more coastline is exposed, coastal shallow-water species such as corals tend to proliferate. But when coastline is minimized, as when continents fuse, extinctions of such organisms seem common. This point cannot be overstated. Plate tectonics is a driving force in the generation of evolutionary selection pressures.

Separation of the continents acts to geographically separate organisms (known as *vicariance*; see the following section), stimulating evolutionary change and allowing evolution to proceed along varied and different pathways, evolving endemic groups varying from continent to continent and island to island. The isolation of Australia, for example, led to the further diversification of marsupial mammals (recall that marsupials also occur in South America), making Australia unique as the *land of diverse marsupials*. Likewise, eucalyptus trees of over 600 species occur in Australia and nowhere else (except when transplanted by humans). Part of the

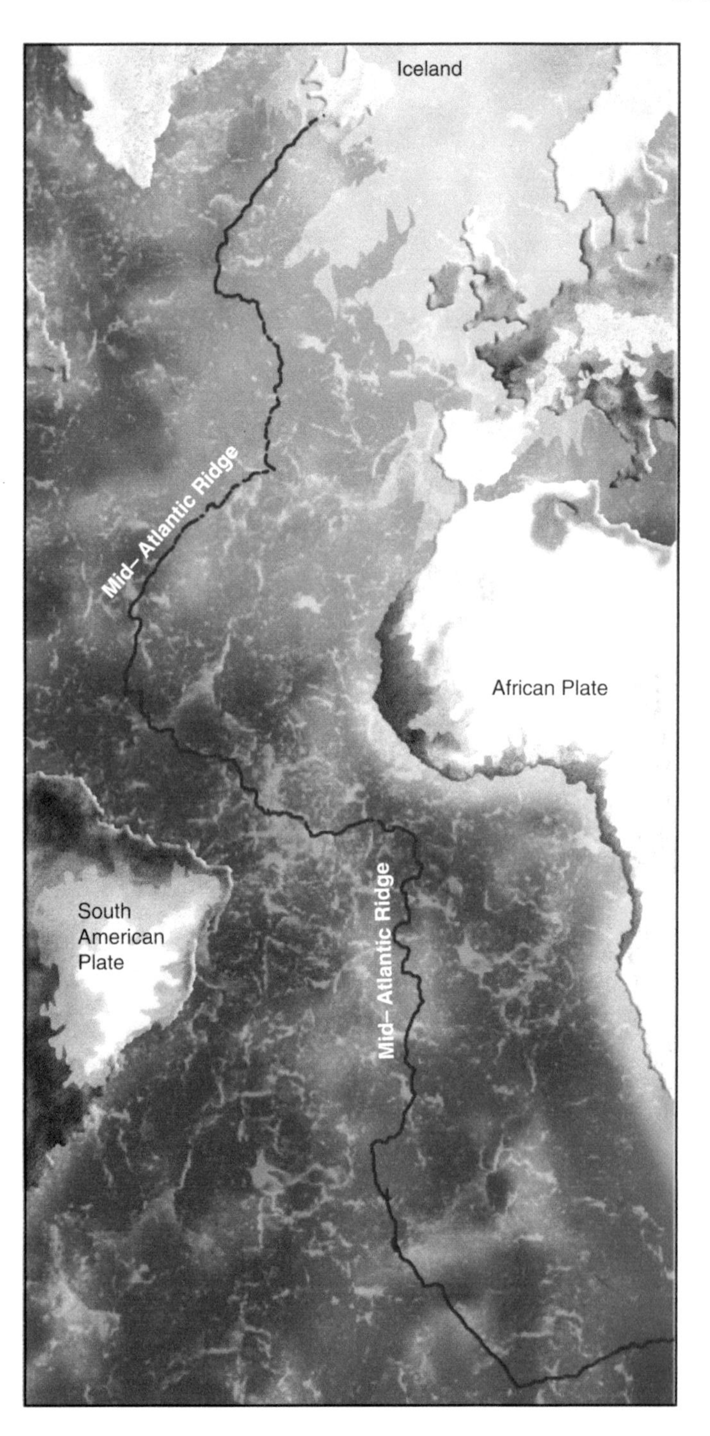

FIGURE 2-7
The mid-Atlantic ridge is part of the extensive ridge system that characterizes areas where new seafloor comes to the surface.

The Little Animal That Proved Continental Drift

A small *synapsid* (mammal-like) reptile named *Lystrosaurus* (called mammal-like for its anatomical similarities with the group that would evolve into true mammals) was abundant in Antarctica during the Triassic period of the Mesozoic era, when the continent was far from its present position (Figure 2-8). Its fossils, found in Antarctic sedimentary rock, are also found in southern Africa, India, and southern Asia. When *Lystrosaurus* was discovered in Antarctic rocks in 1969–1970, the story made the front page of the *New York Times*, so compelling was it as support for continental drift. It was obvious that *Lystrosaurus* could not have existed on Antarctica if the continent had always been at its present south polar location—the little creatures, about the size of a spaniel, would have been frozen solid.

FIGURE 2-8
Fossil remains of the synapsid (somewhat mammal-like) reptile *Lystrosaurus* have been found in such widely separated areas as Antarctica and southern Africa. When *Lystosaurus* lived, in the Permian and Triassic periods, these areas were part of the massive continent Gondwana.

great biodiversity of mammals, and in particular, placental mammals, evident in the fossil record throughout the Cenozoic is likely attributable to continental separation stimulating high levels of speciation among groups physically isolated from one another.

Vicariance and Endemism

Tropical ecosystems occur on four widely separated continents (South America, Africa, Asia, and Australia) and many islands. Among the islands, some are large, such as Madagascar, New Guinea, and Borneo, and some small, such as Komodo, New Caledonia, and Cocos.

A visit to a tropical moist forest in Queensland, Australia, will differ from a visit to a similar ecosystem south of the Orinoco River, in Venezuela, though both forests may, at first glance, appear structurally much the same. There will be impressively tall trees, buttressed roots, many vines, and epiphytes attached to the trees. There will be palms and strangler figs. Colorful birds may be seen, as well as equally striking butterflies. Ants will be common. But a comprehensive species list from these two widely separated forests will contain few (if any) species in common. Instead, not only will species differ, but also whole families of birds, mammals, insects, and others will be highly distinct between Australian rain forest and Venezuelan rain forest (Plates 2-5a to 2-5f). The forests appear structurally similar because natural selection favors certain phenotypic characters in regions of high temperature, high humidity, and abundant rainfall. Broad-leaved evergreen trees share various characteristics that confer high fitness in such areas, just as various forms of highly colorful butterflies have high fitness within broad-leaved forests. But why are the species, the genera, and indeed the families so different?

The answer is *vicariance*, a word referring to factors resulting in geographical separation (also called *allopatry*) and the evolutionary consequences of such separation. When allopatry occurs, populations are unable to interbreed and thus may diverge evolutionarily. This reality forms part of the basis for the *biological species*

(a)

(b)

(c)

(d)

PLATE 2-5
IMAGES OF VENEZUELAN ANIMALS AND AUSTRALIAN ANIMALS
From Venezuela: (a) male Guianan cock-of-the-rock (*Rupicola rupicola*); (b) royal flycatcher (*Onychorhynchus mexicanus*); and (c) Brazilian porcupine (*Coendou prehensilis*). From Australia: (d) male Victoria's riflebird; (e) noisy pitta (*Pitta versicolor*); and (f) common ring-tailed possum.

(e)

(f)

PLATE 2-5
(continued)

concept (BSC), to be described later in this chapter. Separation of continents results in allopatry on a large scale. Similarly, island flora and fauna are unique in large part due to random colonization followed by evolution in isolation. Vicariance means that taxonomic origins differ as a result of separation caused by some sort of barrier. On a continent, such a barrier could be a mountain range or a broad river, or perhaps a desert. On a broader scale, oceans form barriers between continents and islands, resulting in the evolution of *vicariants*, groups of closely related species isolated from other groups. Allopatry imposed by vicariance creates a barrier to gene exchange, and thus groups of related species evolve isolated from other groups of related species.

Evolution of species occurs wherever life exists. Thus in landforms that are isolated, one from another, speciation results in the evolution of unique assemblages of plants and animals. When understood as a product of historical vicariance biogeography, the large differences among biogeographic realms become much more clear.

When a species' evolutionary history, its very genealogy, is unique to a given region, and the species remains only in that region, it is termed *endemic*. Endemism is typically high among island species because of the relative isolation of islands. Thus it is understandable that the Galápagos Archipelago, with its large assemblage of endemic plants, birds, and reptiles, inspired Charles Darwin in 1835. *Endemism* refers to taxa that are native to and restricted to a single area. The Galápagos penguin (*Spheniscus mendiculus*) (Plate 2-6d) is endemic to the Galápagos Islands, found nowhere else. The okapi (*Okapia johnstoni*), related to giraffes, is endemic to central African forests. Whole groups such as

(a) (b) (c) (d)

PLATE 2-6
These three bird species—(a) Bahama swallow, (b) Bahama woodstar, and (c) Bahama yellowthroat—each endemic only to certain of the Bahama Islands, are each closely related to more widely ranging species. (d) The Galápagos penguin (*Spheniscus mendiculus*) is the most northern of the world's 17 penguin species and the only one to reach the equator.

the Hawaiian finches (*Drepanididae*) in Hawaii, ground antbirds (*Formicariidae*) in South America, and gibbons (*Hylobatidae*) in tropical Asia, including India, are endemic where they occur. Continents and islands all exhibit various degrees of endemism at both the species and subspecies level. For example, of the approximately 100 resident bird species in the Bahama Islands, three are endemic (Plates 2-6a to 2-6c). There are 26 endemic subspecies of birds found in the Bahamas.

Natural Selection

Modern evolutionary theory defines *natural selection* as differential reproduction among genotypes. Genes, the long, coiled molecules of DNA that contain hereditary information, were quite unknown to Darwin, although he knew that phenotypic traits were inherited. As long as a population contains genetic variability among its members, natural selection can act, since there are genetic variants that will respond differently depending on abiotic and biotic selection pressures imposed. Natural selection acts only on the present, never "planning" for the future. Those phenotypes that survive relatively better than others do so only because conditions, whatever they may be, suit them better than others in the population. Natural selection is a general statistical truth: individuals with genes that confer reproductive advantage will tend to leave the most progeny, and thus those genes will proportionally increase generation after generation relative to others in the population's gene pool. This is little different from saying that a coin weighted to heads will, when flipped, consistently come up tails more often than heads.

Genetic variability originates through the random process of *mutation*, a sudden and unpredictable change in a gene. Mutation is nondirectional: environments do not cause or produce useful mutants in response to need. The variability resulting from mutation is enhanced by recombination of alleles in sexually reproducing species. *Selection* can act only on whatever genetic variants are present. If some variants are better adapted than others, they will be passed on in higher proportion. Selection, unlike mutation, is not a random process (any more than a weighted coin flip is considered random in its outcome), because only certain members of a population are best suited for a given environment and thus have a nonrandom chance of survivorship and reproduction. In recent years, both geneticists and molecular biologists have established beyond doubt that large amounts of genetic variability exist in most populations. Thus in most cases, there is ample raw material for natural selection to act on.

Adaptation

An *adaptation* is any anatomical, physiological, or behavioral characteristic that can be shown to enhance either the survival or reproduction of an organism. Such traits, as they translate into successful reproduction, comprise what is called *evolutionary fitness*. Fitness is reflected in adaptations, all of which result from the action of natural selection.

Any organism may be viewed as a cluster of various adaptations. For instance, opossums, Neotropical porcupines, kinkajous, as well as many monkeys of the American tropics possess a prehensile tail. Such a structure functions effectively as a fifth limb, lending security and mobility to the animal as it moves through the treetops. It is easy to see intuitively that the prehensile tail is an adaptive structure. Tailless monkeys or opossums would face a smaller lifetime reproductive success because of the added risk of falling. But note that Old World monkeys also are fit within their arboreal environments, but none possess prehensile tails. Thus a prehensile tail, while adaptive to those that have evolved it, is not absolutely required to survive a life in the treetops (Plates 2-7a and 2-7b).

Not all adaptations are obvious. In tropical parts of Africa, many humans carry a gene that when present in

(a)

(b)

PLATE 2-7
(a) The woolly monkey (*Lagothrix flavicauda*) found in South America has a prehensile tail and is demonstrating how to use it. (b) The Guereza colobus monkey (*Colobus guereza*) of Africa lacks a prehensile tail; no African primates have prehensile tails.

double dose (one inherited from the mother, one from the father), produces defective hemoglobin molecules, resulting in misshapen red blood cells called *sickle cells*. These victims of sickle cell anemia usually die before reaching reproductive age. However, other individuals in the population who possess only one dose of the sickle cell gene and one dose of the normal gene for beta hemoglobin are not seriously disabled and are more resistant to malaria than others in the population, including those individuals with two normal genes for hemoglobin. Thus the sickle cell gene is adaptive (since the protozoan that causes malaria is part of this environment), but only when in a single dose. In a double dose, it is lethal.

Testing Adaptations

Because they can seem so obvious, adaptations are often inferred, and such inference may be true, but rigorous testing is the only way in which adaptation can actually be demonstrated. For example, orb-weaver spiders (family Araneidae) throughout the Neotropics and temperate zone make large webs in which there are some areas of obvious, thickened, zigzag strands called *stabilimenta* (Plate 2-8). Spiders must use considerable energy to synthesize the silk for the web, especially the dense stabilimenta. Why should spiders invest in making stabilimenta? Are these conspicuous zigzag strands adaptive? The spiders that make stabilimenta are those whose webs remain intact throughout the daylight hours. (Many spiders make new webs each evening and take them down at dawn.) Biologists have hypothesized that stabilimenta, which make webs easily visible to humans, have the same effect on birds (which are also visually oriented), thus allowing a flying bird to avoid an orb-weaver spider's web, saving the spider from having to remake the web, which would be significantly damaged should a flying bird strike it (and would yield no food for the spider—birds are too big to capture and eat). But spiders that invest energy in making stabilimenta rather than risk having to start from scratch and make a whole new web would be less prone to "bird accidents." Therefore, stabilimenta could represent an adaptation to energy saving in an environment where birds pose a risk to the security of the web. As such, the hypothesis sounds plausible, but without testing, it is just a story, an educated guess. How could it be tested?

PLATE 2-8
Orb-weaver spider female in a web with stabilamenta visible.

First, an observer could simply watch spider webs and note bird behavior. This was done, both in Panama and in Florida, and birds were observed to take short-range evasive action when approaching webs with stabilimenta (Eisner and Nowicki 1983; Craig 1989). Second, using webs that do not have stabilimenta, researchers altered some webs, adding artificial stabilimenta, keeping other webs without stabilimenta as controls. The webs without stabilimenta did not remain intact during the day, while those with stabilimenta generally did. Further direct observations implicated birds as the major threat to webs, though other large animals ranging from butterflies to deer could also be forewarned by the presence of stabilimenta. This work demonstrated the adaptiveness of stabilimenta.

Not all traits are adaptive. Organisms represent the combined effects of thousands of genes working in concert. The anatomy, physiology, and behavior of an organism represent various compromises imposed by the interactive effects of the genes that formed it. Is it adaptive that humans generally lack body hair? Quite possibly so (to allow loss of body heat in a hot, sunny, savanna climate, for example), but no generally acceptable adaptive explanation for hairlessness has been tested. Our hairlessness could be a by-product of our development, with no adaptive function now or ever in the past.

A given trait may, however, have been adaptive in the past, but not any longer. There are large fruits from about 30 species of trees and vines in Costa Rica that produce large and fleshy fruits. These fruits drop from the tree, and most essentially just rot. That is very odd considering that fruits function to attract seed dispersers (Chapter 7). So why are these fruits not consumed, and their seeds dispersed? Some years ago, Daniel Janzen and Paul Martin hypothesized that the large fruits represent what they called *ecological anachronisms*, fruits adapted to be dispersed by animals that are now extinct (Janzen and Martin 1982). During the Pleistocene, some 10,000 years ago, about 15 species of large (*megafaunal*) mammals (including horses and ground sloths) became extinct in what is now Costa Rica. These included a unique elephant-like group called *gomphotheres*. Gomphotheres may have been essential dispersers of seeds from large-fruited plants. As a result of gomphothere extinction, plants such as guanacaste (*Enterolobium cyclocarpum*) were left without any agent of dispersal. Their fruits, once extremely well adapted, are no longer well adapted.

Adaptations in Environmental Context

An adaptation must be viewed in the context of the environment. Obviously, the sickle cell gene is nonadaptive in human populations living in environments free of malaria. And a beautifully adapted creature may appear awkward, a "mistake of nature," if seen out of its proper environment. Charles Waterton (1782–1865), a rather eccentric British explorer who traversed the Amazon during the early nineteenth century, noted that the three-toed sloth (*Bradypus variegatus*) appears very poorly suited to survive when seen struggling over the ground (Plate 2-9). Tree sloths, however, do not normally perambulate on the ground, at least not in modern times. (Large ground sloths once did, and they were well adapted by size and strength to function terrestrially.) Tree sloths descend from trees only about once a week, and only then to defecate at the base of the tree. Once that is done, they climb back into the tree. Today's tree sloths are totally arboreal, skillfully moving upside down from branch to branch. In his well-known account, *Wanderings in South America* (1825), Waterton wrote of the sloth, "This singular animal is destined by nature to be produced, to live and to die in the trees; and to do justice to him, naturalists must examine him in this his upper element." Waterton then went on to describe in detail how well adapted the sloth is for its arboreal life. Waterton put adaptation in context.

PLATE 2-9
The three-toed sloth (*Bradypus variegatus*) is well adapted to an arboreal life but poorly adapted to being on the ground.

Speciation

Humanity has long recognized variability in nature and that organisms fall into natural groupings. Linnaeus generated a system organizing the many life forms based on the degree of shared similarities: kingdom, phylum, class, order, family, genus, and species. The *Linnaean system*, though much modified in modern classification studies, still prevails today, and the branch of biology that deals with it is called *systematics*, the classification of organisms.

The *species* is the basic unit of the Linnaean system. When we survey an ecosystem, we typically measure the species richness (number of species) of trees, birds, moths, ants, or whatever groups we find to be of most interest. Thus it is important for ecologists to understand what, exactly, a species happens to be. The question is complex, because it is often difficult to determine when organisms are members of the same species or are of different species.

Suppose that you were wandering the rain forest of New Guinea and you look up and observe a big red parrot with blue on its wings and belly, a strikingly handsome bird, atop a palm tree. Soon you see a second bird that looks about the same size and shape as the first but is brilliant green, with red below its wings. The two birds, both wonderfully bright, have totally different plumages. Be that as it may, you are observing a pair of eclectus parrots (*Eclectus roratus*). Males are mostly green, and females are red and blue (Plates 2-10a and 2-10b).

(a)

(b)

PLATE 2-10
MALE AND FEMALE ECLECTUS PARROTS.
(a) The male is green with an orange bill. (b) The female is mostly red and blue, with a dark bill.

PLATE 2-11
This image shows, in the bottom three rows, unpalatable butterfly model species in the family Danaidae (left) and palatable mimetic forms of female *Papilio dardanus* (right), an African swallowtail species. At top-left is the *Papilio dardanus* male; at top-right is a nonmimetic, male-like female of the same species. The polymorphic, female-limited Batesian mimicry was first described by Roland Trimen in 1869.

When males and females do not look alike, they are considered examples of *sexual dimorphism*, but they are still obviously members of the same species. Many animals, including insects, spiders, fish, reptiles, birds, and mammals (including, of course, humans), are to varying degrees externally sexually dimorphic. In some cases, sexual dimorphism is the result of *sexual selection*, which will be discussed in Chapter 7.

Now suppose that you are wandering in Africa, looking at butterflies. You see the colorful swallowtail *Papilio dardanus*, a large, conspicuous butterfly (Plate 2-11).

Nearby is a second butterfly that looks quite different. Would it surprise you to learn that it is the same species? It might very well be. In this case, the female butterflies of *P. dardanus* typically mimic other butterfly species, those that are noxious to predators such as birds. In some cases, the female looks like the male, with a "swallow-tail," but more typically, females look nothing like the males. Mimicry will be discussed in the Chapter 7.

Why are eclectus parrots and African swallowtail butterflies considered single species even though individuals within these populations are quite distinct? The answer is that they successfully interbreed. Male eclectus parrots may be green and females red and blue, but they know each other when it is time to form a pair bond. The same is true of African swallowtail butterflies. All animals are genetically adapted to recognize others within their species. The same is true, in a more passive way, for such organisms as flowering plants. The size and shapes of the flowers as well as the exact timing of when they bloom act to aid in dispersing pollinators among the various species.

This reality forms the basis of the most widely used definition of a *species*: populations of actually or potentially interbreeding organisms. In this definition, called the *biological species concept* (BSC), species designation is based on an assumption of sorts—namely, that the organisms themselves will reveal to which species they belong through their reproductive habits. Under the BSC, the species category becomes a natural unit in the Linnaean system because the organisms to which it pertains actually recognize it themselves, at least in a manner of speaking. It is unlikely that an eclectus parrot knows it is a "parrot" compared with, say, a hawk or a stork, or if it appreciates that it is an animal and not a fungus. But its genetics make it competent to recognize and interact meaningfully with another eclectus parrot. The BSC is a theoretically strong definition, heuristically satisfying, but sometimes difficult to apply in the field, especially in areas where populations are separated by allopatry. If two populations look alike and seem to be reasonably likely to be part of one species, but if these populations are separated geographically, how do we know whether they could mate and produce fertile offspring?

In spite of these obvious difficulties, the BSC has endured as the most useful definition of a species. In practical terms, it is true that characteristics other than reproductive isolation, such as anatomical differences, voice, and behavior, must be frequently employed to assess whether two populations are one or two species. Today it is common to use nucleic acid sequencing, both with mitochondrial and with nuclear

(a)

(b)

PLATE 2-12
(a) Savanna elephant compared with (b) forest elephant.

DNA, to designate species, a clear modification of the BSC. African elephants provide an example.

Elephants are widely distributed throughout East and Central Africa. As large and unmistakable as these ponderous beasts are, it may surprise you to learn that until quite recently, taxonomists were apparently mistaken about exactly how many species of elephants currently inhabit Africa. It was assumed that there is but a single species, the African elephant, *Loxodonta africana*, when there are (or at least, there may be) two. Studies by Alfred L. Roca and colleagues (2001) suggest strongly that there is a second species of African elephant, the African forest elephant, *Loxodonta cyclotis*. Compared with *L. africana*, now renamed the African savanna elephant, the forest elephant is smaller in body size, has rounded, not pointed ears, and has straighter tusks (Plates 2-12a and 2-12b). The scientists used a technique called *dart-biopsy*, allowing them to harmlessly sample the DNA of free-ranging wild elephants. They sampled 21 populations and collected data from 195 individual animals, examining DNA from mitochondria as well as sequences from four genes from the nucleus (representing 1,732 base pairs). The results of the molecular analysis suggest that forest and savanna elephants are as distinct from one another as lions are from tigers. The genetic distinction between forest and savanna elephants is 58% for the same gene sequences. This separation is approximately equal to that seen when comparing African with Asian elephants, long considered separate species, the latter in the genus *Elephas*.

Given such a large difference in DNA, there is thus little doubt that there are really two, not one, species of African elephant (Figure 2-9). The combination of genetic, morphological, and ecological differences led the researchers to conclude that there are really two species of elephants currently in Africa.

LCY4 n=1
LCY3 n=5
LCY5 n=1
LCY1 n=33
LCY6 n=1
LCY2 n=12
African forest
EMA1 n=13
Asian
LAF2 n=2
LAF1 n=101
African savanna

FIGURE 2-9
This cladogram, based on genetic similarity, illustrates that the savanna and forest elephants are about as different from one another as either is from an Asian elephant.

The biological species concept prevails throughout biology, but another species concept is gaining increasing support. Many systematists are now using a technique known as *phylogenetic systematics*, or *cladistics*. Cladistics does not depend on the establishment of reproductive isolation between two populations in order for them to be considered separate species. When applying cladistics, the biologist measures many characteristics of organisms (and the characteristics are equally weighted, not biased as more or less important) and then applies an *algorithm*, a mathematical model, which allows a computer to produce a branching diagram, called a *cladogram*, showing the degree to which individuals share various suites of characters (Figure 2-10). Characters may be anatomical or molecular: the technique is often applied to data such as was generated in the preceding elephant example. In designating species, an attempt is made to separate populations on the basis of the most recently derived genetic characteristic that two populations no longer share in common. Such a definition thus puts the focus indirectly on the most recent point, termed a *node*, when the two populations diverged genetically. This approach to species designation is called the *phylogenetic species concept* (PSC).

The key distinction between the BSC and the PSC is that reproductive isolation is not required when applying the PSC. What the PSC is based on is demonstration of a genetically distinct, diagnosable suite of characteristics (which can be anatomical or based on molecular sequence differences). *Molecular distance*, the degree of difference between molecular sequences in two populations, is measured for known species and then used as a kind of yardstick to determine whether two populations should be designated as separate species. Using the PSC, there would be many more species recognized for taxa such as birds and mammals, where many species exhibit what are termed *subspecies*, or *races*. (A subspecies [or race] is a population within a species that is genetically distinct but not reproductively isolated from other similar populations.) Most subspecies, even though not necessarily reproductively isolated, would likely attain full species status under the PSC.

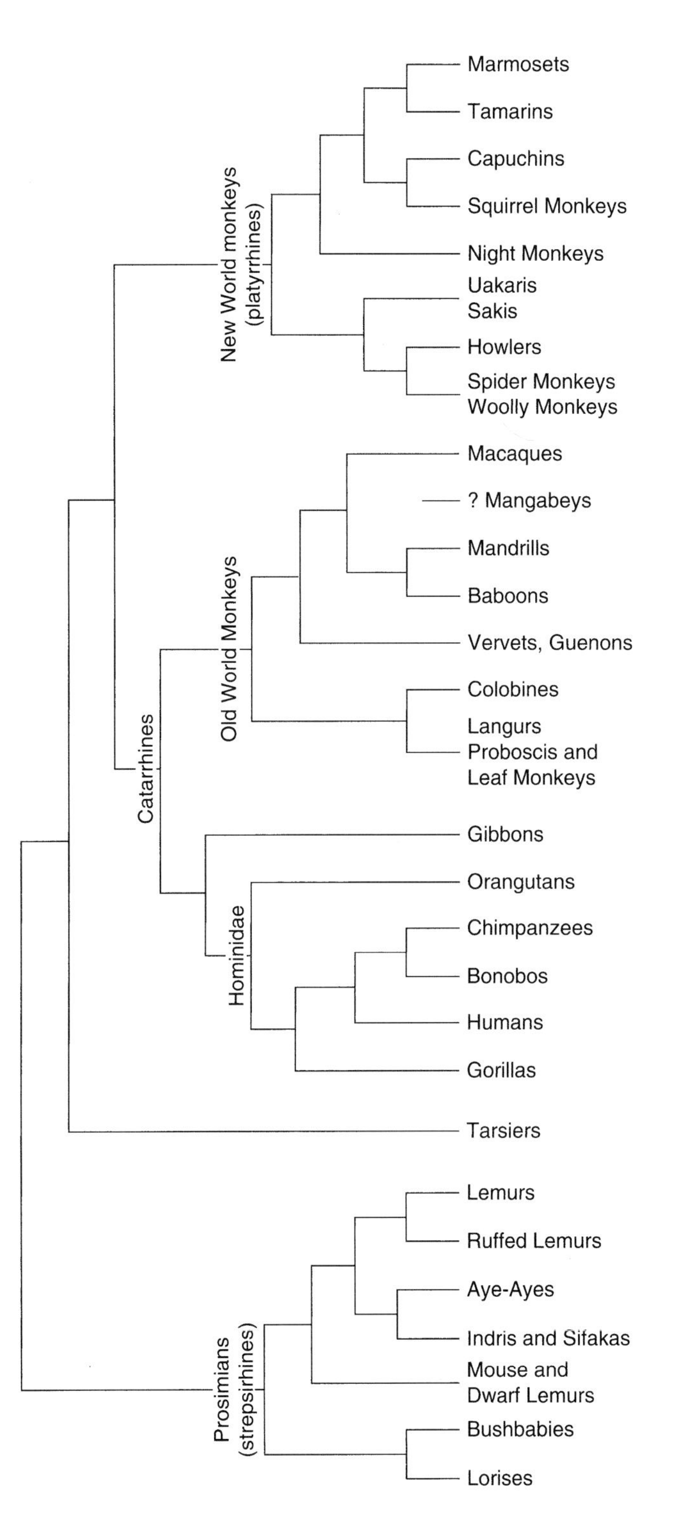

FIGURE 2-10
This is an example of a cladogram—in this case, for primates. The shorter the distance from branch to branch, the more closely related are the species. For example, note the top of the cladogram. Tamarins and marmosets are each other's closest relatives.

Speciation as a Process

Speciation is the essence of evolution, the production of new forms of life, different from ancestral populations. It can happen in two ways. A population can change with time, so much so that it deserves to be called a new species. In this model, the population follows a linear trajectory and gradually changes, under the influence of natural selection, into something else. It was once believed that human evolution was best described in this way. But another, and likely more common, method of speciation is multiplication of species. From a single species, new species form as various branches off a bush. This model seems to describe most of speciation and is now the model that seems to accurately account for human evolution. About 1.8 million years ago, the fossil record suggests that there were up to six species of hominids coexisting in Africa, all evolved from an ancestral species that dates back to just over four million years ago (Tattersal 1995; Gibbons 2009). Though *Homo sapiens* is the only descendant species, we need to understand that the history of humanity involves many genetic branches, only one of which persists to this day.

The common model for speciation, largely based on the biological species concept and on vicariance, involves several steps (Figure 2-11). First, because of an event causing vicariance, populations become geographically isolated. Second, geographic isolation prevents gene flow, allowing genetic differences between the isolates to accumulate with time. Natural selection may act to promote such genetic divergence. Last, genetic differences between an isolate population and its parental population increase to the point that, should they establish secondary contact, natural selection acts to promote reproductive isolation between them, thus establishing them as separate species. It should be emphasized that this process does not require a great deal of genetic difference to occur between the isolated populations, only enough to establish reproductive isolation. Speciation can be surprisingly quick.

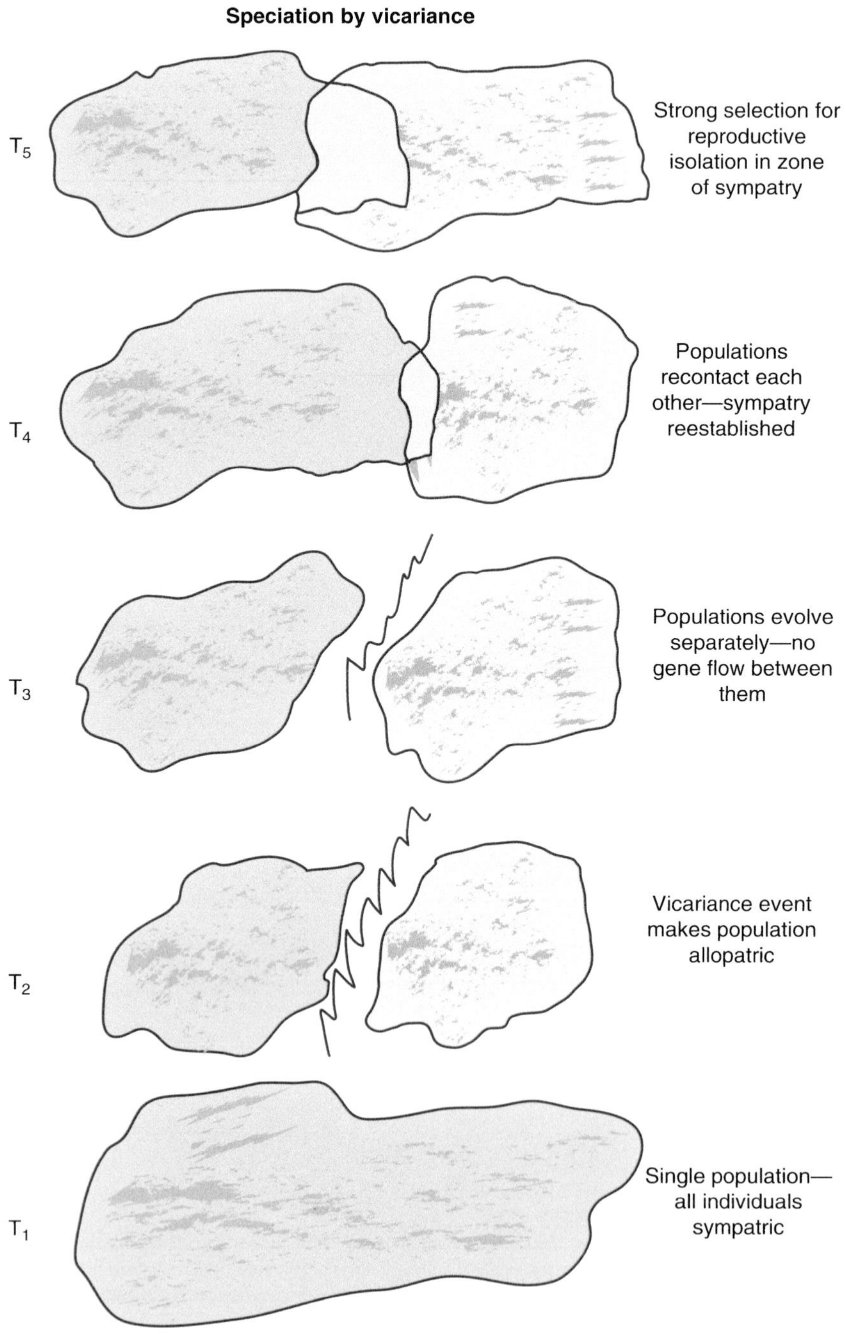

FIGURE 2-11
Beginning with T1, note how a vicariant event such as a mountain rising may separate populations geographically, thus allowing for genetic differences to develop over time between the now-isolated populations. At time T4, the populations again become partially sympatric and selection pressures act to enhance their separation. At time T5, there are reproductive-isolating mechanisms that select to prevent hybridization and to maintain the two populations as full species.

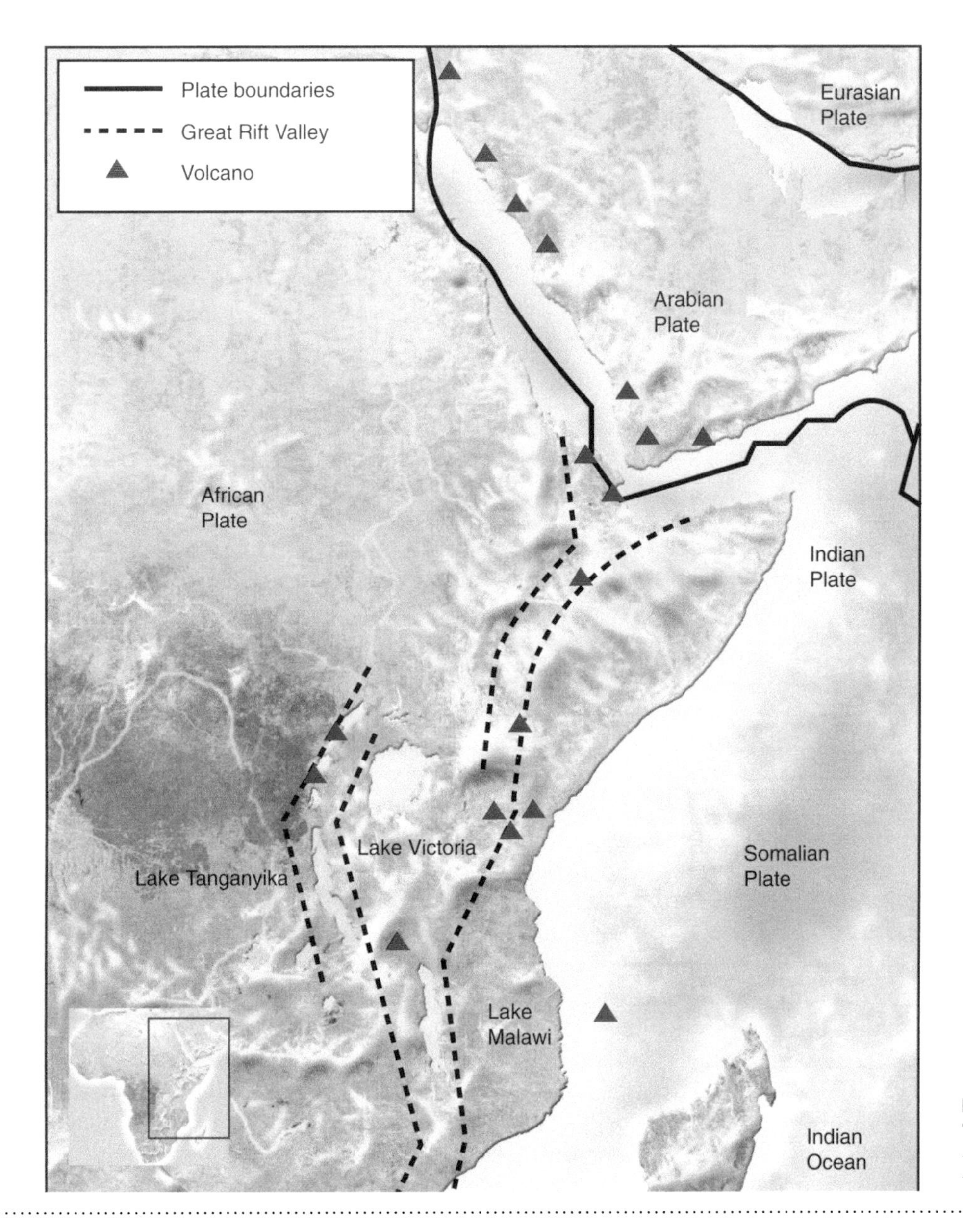

FIGURE 2-12
This map shows the location of the three major lakes of the Great Rift Valley in Africa.

In the model described here, geographic isolation promotes genetic divergence and subsequent speciation. It is believed that most speciation, especially among animals, occurs in this manner. But recent work suggests that it is possible for natural selection to drive the process of speciation and promote reproductive isolation, even in the absence of complete geographic isolation.

Cichlid Fish: An Example of Multiple Speciation

Cichlids are diverse bass-like fish found in tropical waters around the world (Barlow 2000; Stiassny and Meyer 1999). There are about 300 species in the Americas, including one that occurs as far north as Texas. But most cichlid species are found in East Africa, clustered in three lakes that formed in the Great Rift Valley: Lake Victoria (more than 400 species), Lake Tanganyika (200 species), and Lake Malawi (300 to 500 species) (Figure 2-12).

The oldest of the three African lakes is Lake Tanganyika, which formed 9 to 12 million years ago, and the youngest is Lake Victoria, whose origin is between 250,000 and 750,000 years ago. Considering only Lake Tanganyika, if one assumes it to be 12 million years old, the highest estimate of its age, and that it

PLATE 2-13
Cichlid fish of the Great Rift Valley lakes demonstrate an amazingly high species richness.

has 200 cichlid species, that is a speciation rate of about 1 species every 60,000 years. However, Lake Victoria is at most 750,000 years old, and it has 400 cichlid species. If all of them evolved in that lake, the speciation rate is about one species every 1,875 years. But even that number is too low. Researchers are in agreement that the amazing diversity of cichlids in the African rift lakes arose recently, within the past few million years, demonstrating that speciation can occur with impressive rapidity (Verheyen et al. 2003).

Studies of the mitochondrial DNA of the cichlids from Lake Victoria have demonstrated that this cluster of species is genetically very closely related, showing that they have all recently evolved from a common ancestor (Plate 2-13). What is stunning is that the total genetic variation among these 400 species is less than that found throughout *Homo sapiens*—humans! In other words, the genetic distance between you and a stranger you might meet is likely to be greater than that between two separate species of cichlids from Lake Victoria. Given the estimated rates at which mutations occur in mitochondrial DNA, researchers believe that the entire assemblage of cichlid species in Lake Victoria has arisen within 200,000 years. That is equivalent to about one species every 500 years.

One reason for such rapid speciation is that the African lakes have repeatedly experienced periods of extreme dryness, followed by replenishment. Evidence suggests that Lake Victoria, for example, was nearly dry as recently as 14,000 years ago. During dry periods, cichlid populations would become smaller and be isolated from one another, ideal conditions for genetic divergence and subsequent speciation. (Compare this with the refugia model for speciation, described later in this chapter.) With repeated waves of dryness would come repeated bouts of speciation, increasing the number of species. But that's not all.

Cichlids in the African lakes show strong evidence of natural selection acting on their sensory systems as they perceive one another, something termed *sensory drive speciation*. As males have diverged in color, female cichlids show strong preferences for male color types. This is likely due to the fact that color penetration in the various lakes varies with depth, and thus the perception of color must vary. Red becomes more predominant with depth, blue less predominant. Thus red males are easier for females to see in deeper waters. It has been learned that certain genes, termed *opsin genes*, are responsible for adjusting the color detection of the fish, and red-biased opsin genes typify fish of deeper depths. Thus females at shallow depths prefer bluer males, while females at greater depths tend to mate mostly with redder males. The study (Seehausen et al. 2008) concludes by suggesting that the selection pressure is sufficient to allow for speciation to occur as a function of depth, without requiring any form of vicariance (geographic isolation).

Natural selection has greatly influenced the species clusters of cichlids in the African lakes in ways other than speciation. Cichlids have adapted and diversified to become specialists and occupy many ecological niches. As an example, there are cichlids that feed by nipping off the scales of other cichlids. These predatory fish are actually specialized, by species, to nip scales either from the right or from the left side of their prey!

The evolutionary process whereby one kind of organism—in this case, an ancestral cichlid—evolves into numerous species, each uniquely specialized, is called *adaptive radiation*. There are numerous examples of adaptive radiation both in the fossil record and among extant animals and plants. The famous Darwin's finches of the Galápagos Islands form *the* classic example of adaptive radiation. Less famous, but no less interesting, are the *Scalesias* of the Galápagos Islands, plants in the daisy family that have evolved from a single colonizing ancestral species into some 20 species found throughout the islands. But 13 species of Darwin's finch and 20 species of *Scalesia* pale in comparison with the amazingly rapid and diverse evolution of the cichlids of the Great Rift lakes in Africa.

The African cichlids have not only undergone a stunning adaptive radiation in each of the three large lakes, but they have also evolved an amazing convergence. Species not closely related from different lakes have come to bear a remarkable resemblance to one another, occupying essentially equivalent ecological roles (niches).

The combined effects of rapid speciation and natural selection on a time scale of about 200,000 years resulted in the extraordinary diversity of cichlid fishes in the African lakes. Evolution is not sluggish.

Speciation in the Tropics

Speciation under the BSC involves establishment of reproductive isolation between two populations. For this to happen, the flow of genes between populations must be prevented or severely reduced for a sufficient number of generations to permit genetic divergence adequate to establish reproductive isolation. Once two populations have genetically diverged such that they are unable to produce fertile offspring, they become separate species. Speciation under the PSC requires that populations diverge genetically to the point where they are clearly distinct, though such a distinction may not necessarily be obvious. Cryptic species are known in which members of each species appear highly similar but are genetically divergent.

The most likely precursor to speciation in either case is *vicariance*, the occurrence of physical barriers that fragment populations and prevent gene flow. Mountains, rivers, deserts, and savannas all represent possible barriers between rain forest sites. Other factors, most notably climate changes, can fragment species' ranges, creating vicariance (see the section "Age, Endemism, and Refugia: Many Questions," later in this chapter). If a mountain is formed by uplifting of Earth's crust, as has happened extensively in the Andes Mountains in South America, what was once a contiguous area of forest will be fragmented by the mountain range. Individuals on one side of a mountain, or in an isolated valley, or even at various elevations on the mountainside are prevented from mating with other populations because they cannot easily reach them. Once populations are separated by geographic factors, there is the strong possibility of genetic divergence with time between the fragmented populations.

For instance, a population on the west side of a mountain range may be subjected to different selection pressures than a population on the east side of the range. Each population will be selected for somewhat different characteristics, and thus for different genes. Specifically, this will result in genetic differences from one population to another among alleles at specific chromosomal locations. (An *allele* is one or more alternative forms of a gene that occur at a specific site, or *locus*, on a chromosome.) Since vicariance prevents exchanging of genes, fragmented populations diverge. They may also diverge genetically due to random loss and fixation of alleles, a process called *genetic drift*. But genetic drift is not as powerful in causing divergence as natural selection, and it does not result in adaptational change.

The Andes Mountains, Amazonia, Vicariance, and Speciation

Much of South America remains geologically active. The Andes Mountain chain, which has been uplifting since at least the Mesozoic era and became particularly active during the Cenozoic era (approximately 65 million years), is responsible for the initial creation of diversified habitats as well as providing numerous climatic and physical barriers that greatly enhance geographic isolation among populations (Plate 2-14). More recently, approximately 20 million years ago,

PLATE 2-14
The complex topography of the Andes Mountain chain has resulted in numerous vicariant speciation events throughout the full range of the Andes.

additional uplift events affected dispersal patterns, isolating ecological communities and stimulating evolution (Antonelli et al. 2009).

Geologists have determined that elevational changes throughout the history of the Andes have occurred in short time spans, from 1 to 4 million years, with the mountain chain remaining stable for long intervals between such times of rapid change (Garzione et al. 2008). The topography of the central Andes is complex, consisting of eastern and western cordilleras separated by an expansive altiplano (Figure 2-13). The topographic complexity of the Andes chain makes for

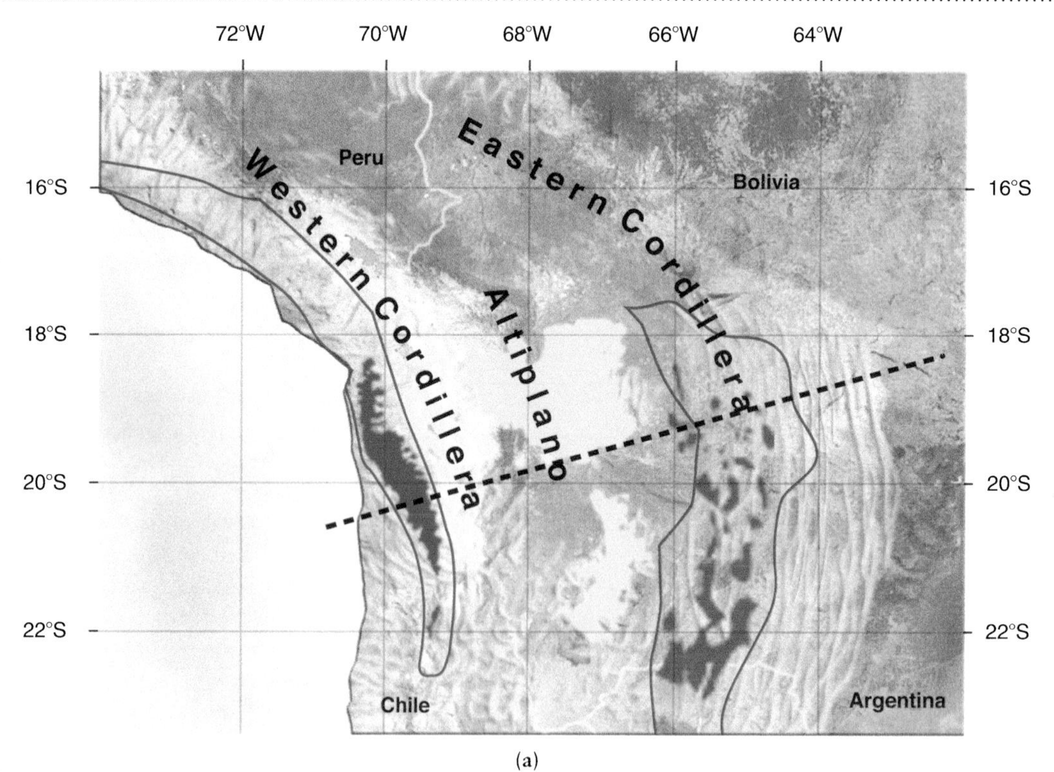

FIGURE 2-13
The altiplano region of the central Andes.

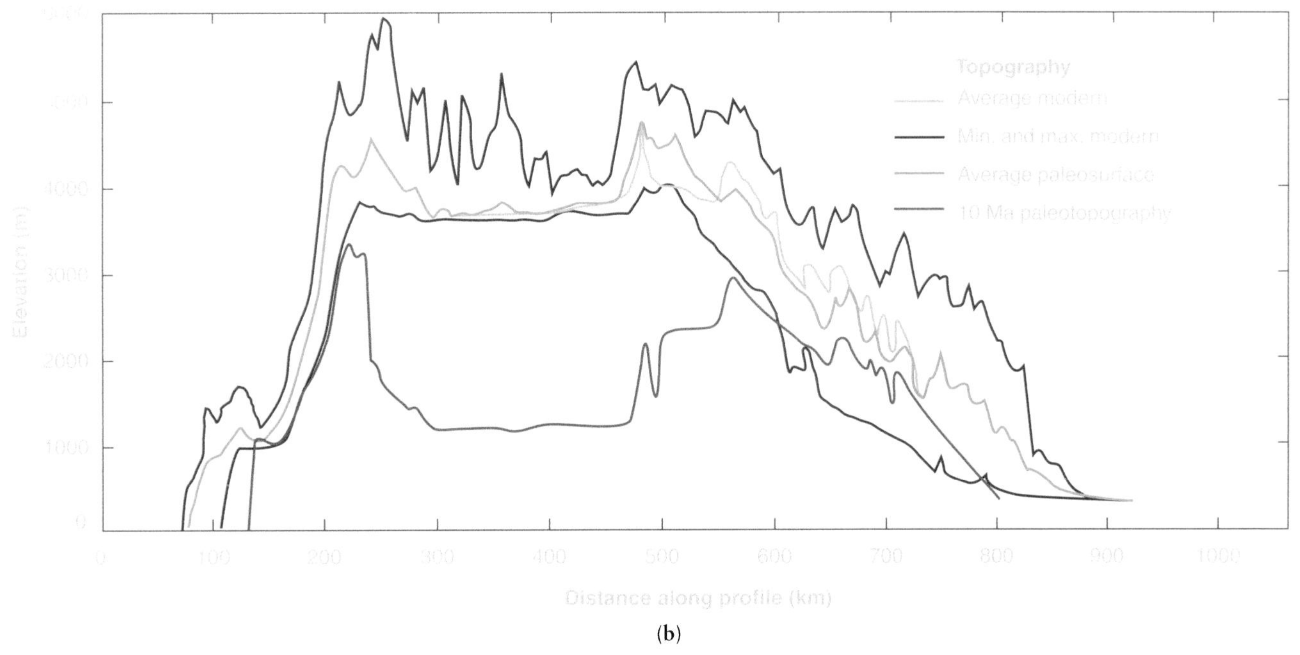

(b)

FIGURE 2-13
(continued)

a veritable engine driving speciation. Consider the following examples.

A likely example of vicariant speciation can be demonstrated for tapirs. Baird's tapir (*Tapirus bairdii*) is found only in lowland forest on the west side of the Andes, extending into Central America as far north as tropical Mexico (Figure 2-14a). The similar Brazilian tapir (*T. terrestris*) occurs only on the east side of the Andes, occupying the entire Amazon Basin (Figure 2-14b). The Andes Mountains geographically isolate these two species, and likely provided the main factor in the initial split of the ancestral species into isolate populations. Last, there is a third Neotropical tapir species, the mountain tapir (*T. pinchaque*), which as the common name suggests inhabits *montane* (meaning mountainous) forests at mid- to high elevations of the central and eastern cordilleras of the Andes, in Colombia and Ecuador (Eisenberg 1989). The mountain tapir is isolated both by range and elevation from the other two species.

Chat-tyrants are common insectivorous birds in the huge family (almost 400 species) of tyrant flycatchers (Tyrannidae). Figure 2-15 illustrates the distribution of a cluster of chat-tyrant species found in the Andes Mountains from Colombia to Bolivia. The group is composed of four closely related, phenotypically similar species (with subspecies) divided into two superspecies. A *superspecies* is a very closely related, recently evolved cluster of species, typically quite similar in morphology. Such a pattern is possible from isolation and subsequent genetic divergence of local populations at various ranges and elevations. The *frontalis* superspecies occurs on the west side of the Andes and is composed of the four birds on the left in the figure, and the *diadema* superspecies is found on the east side and consists of the three birds on the right in the figure. Note the similarity among them—such a complex pattern suggests that speciation is strongly facilitated by Andean topography.

As another example, metaltail hummingbirds (genus *Metallura*) show a rapid and complex evolutionary pattern. By comparing 345 base pairs of the mitochondrial cytochrome b gene and using cladistic analysis, researchers have determined that there are seven species at various ranges in the Andes, one of which exhibits seven subspecies (Figure 2-16). Mitochondrial DNA in birds shows an average mutation rate of 2% every million years. Using that "clock," the metaltail species and various subspecies all evolved within the past two million years. This demonstrates rapid evolution and is characteristic of other studies of Andean bird groups (Price 2008). The stimulus for speciation seems to have been mountain uplift and/or climate change, isolating some populations of hummingbirds on different mountain peaks, while at least one species occupies an extensive mid-elevational belt. The midelevational

(a)

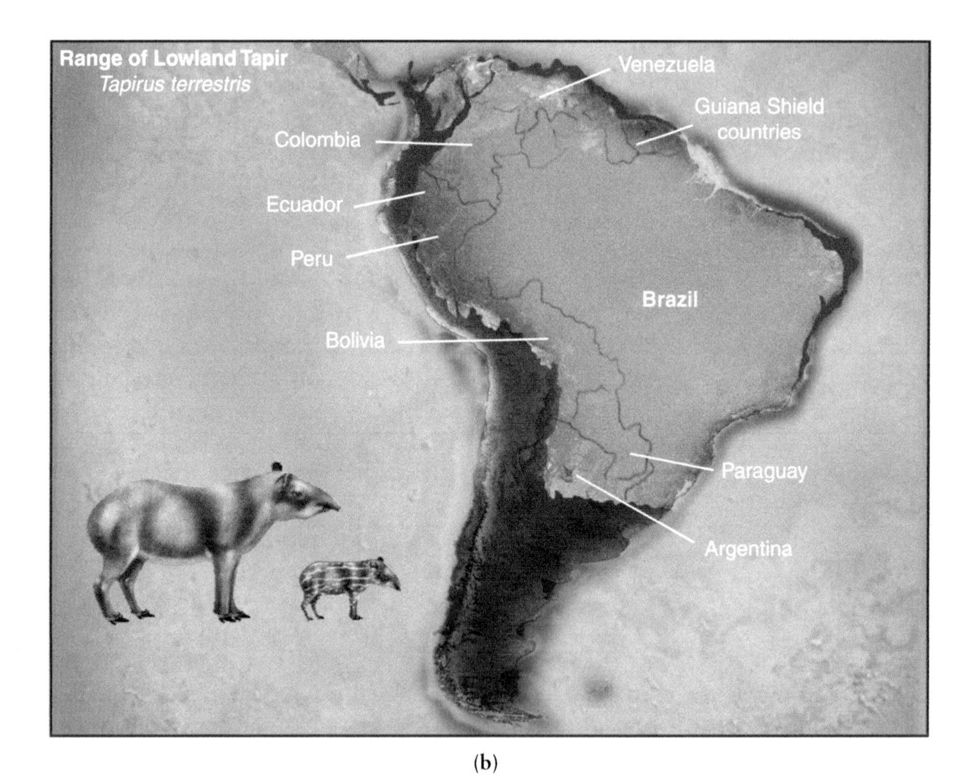

(b)

FIGURE 2-14
(a) The range of Baird's tapir. Note that it occurs throughout Central America and on the most western side of the Andes in Colombia and Ecuador. (b) The range of the lowland tapir.

FIGURE 2-15
The chat-tyrant genus *Ochthoeca* shows a complex pattern of differentiation in the Andes from southern Colombia to central Bolivia. The birds on the left are part of the *O. frontalis* group (called a *superspecies* because there are different distinct populations). The group on the right is part of the *O. diadema* superspecies group, found on the eastern slopes of the Andes. Note how similar the species are to one another, a result of recent evolutionary divergence. This is an actively evolving group of genetically closely related populations, separated to various degrees by their ranges within the Andes chain.

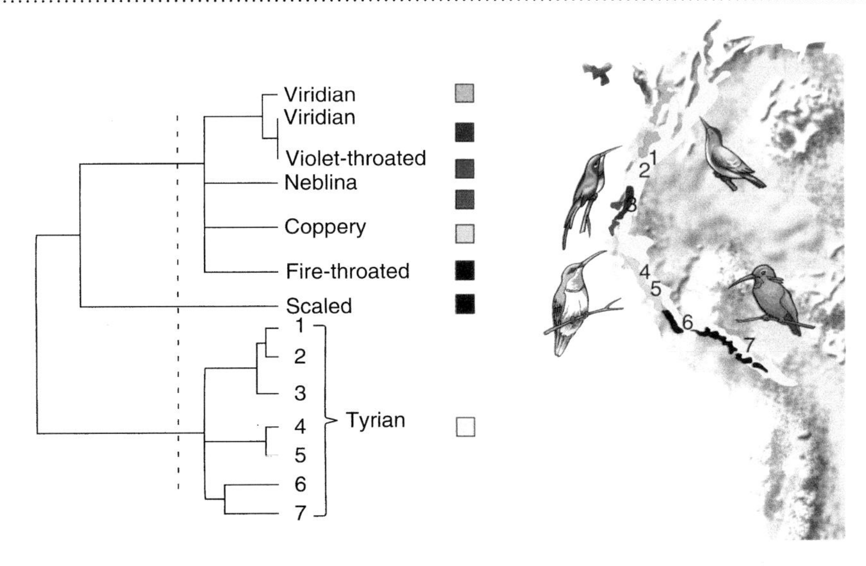

FIGURE 2-16
The cladogram on the left depicts relationships among a group of hummingbird species of the genus *Metallura* (the metaltail hummingbirds). The cladogram is based on comparisons of sections of mitochondrial DNA, a commonly used index to ascertain degrees of genetic relationship. The colored squares correspond to the ranges of the various species shown in color on the map. Species differ in elevation as well as geographic location. The dashed line indicates two million years, so the cladogram depicts a group of recently evolved species. Note the high rate of genetic differentiation occurring in the Tyrian metaltail, shown in the lower cladogram, and note the extensive range of that species shown in the map.

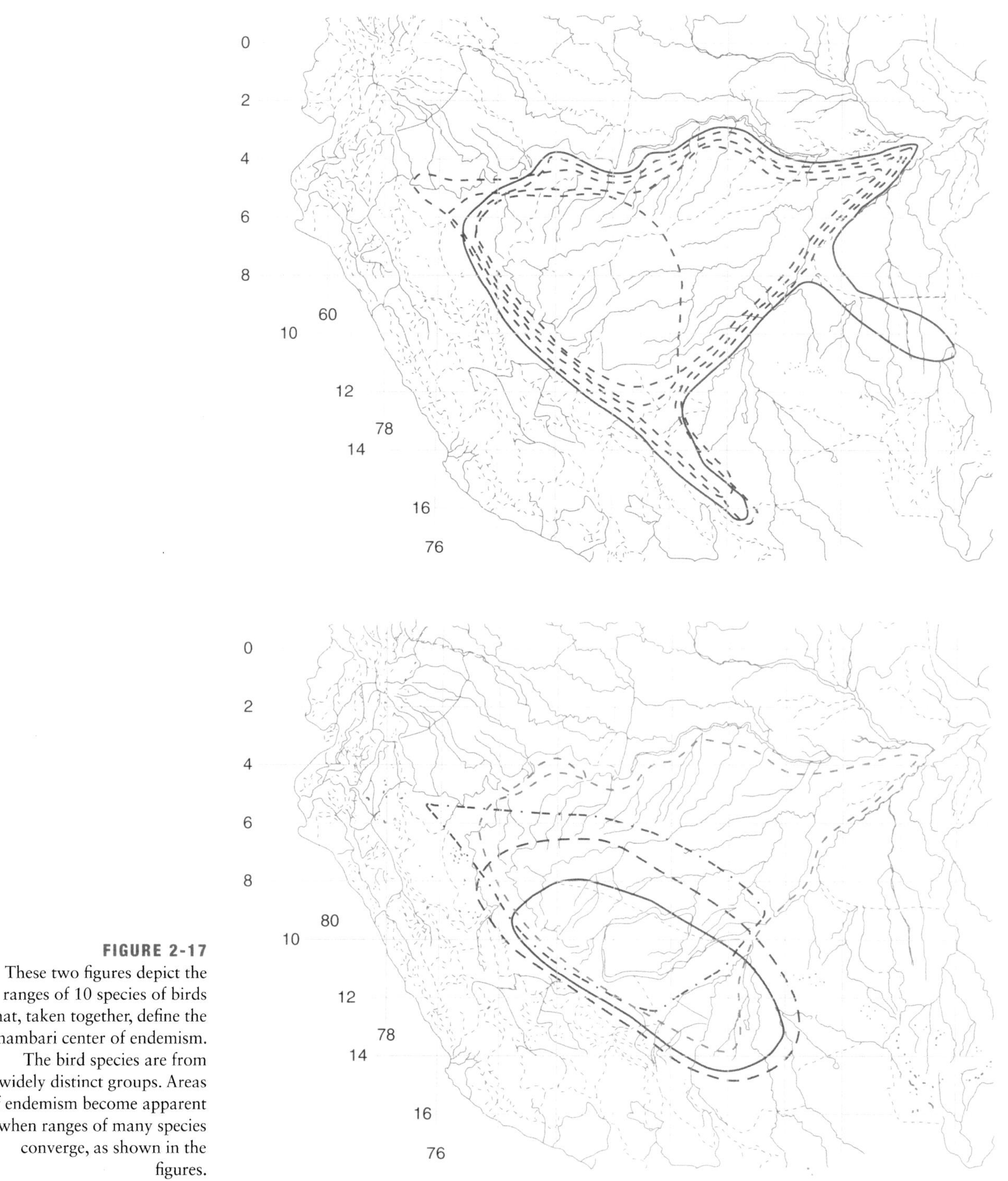

FIGURE 2-17
These two figures depict the ranges of 10 species of birds that, taken together, define the Inambari center of endemism. The bird species are from widely distinct groups. Areas of endemism become apparent when ranges of many species converge, as shown in the figures.

species could provide dispersing individuals that colonized various high montane peaks.

By examining the ranges of many species, it is possible to ascertain geographic areas where groups of species trace their origins, called *centers of endemism*. Figure 2-17 shows such an analysis for part of South America. Note the high correlation among the ranges of the 10 bird species illustrated on the maps.

Analysis of bird distribution patterns throughout South America by Joel Cracraft has demonstrated numerous areas of probable endemism throughout the continent. Each of the postulated areas of endemism contains unique assemblages of species not found in other areas. This helps to account for the uniquely high diversity of birds throughout the continent, and it suggests that evolutionary patterns have been complex (Figures 2-18a and 2-18b).

With the rise of the Andes Mountains, the Amazon River began its flow from west to east. The massive Amazon River and its numerous wide tributaries have served to isolate tracts of forest and savanna, and given the sedentary nature of many animal populations in Amazonia, rivers have probably served as important forces of geographic isolation. The width of the Amazon and its tributaries is sufficient to isolate populations of birds whose individual members are reluctant to cross such a wide expanse of water (Haffer 1985; Haffer and Fitzpatrick 1985). The Amazon River isolates two similar species of antbirds. The dusky antbird (*Cercomacra tyrannina*) occurs north of the Amazon, while the similar blackish antbird (*C. nigrescens*) occurs south of the river. Both species occur together only along a section of the northern bank of the river, but here the blackish antbird inhabits wet varzea forests, and the dusky antbird favors second-growth vegetation of the *terre firme* (Haffer 1985). (*Terre firme* is a term used for areas of forest and savanna that occur off the riverine floodplain.) A similar pattern is evident in the distribution of the razor-billed curassow superspecies (Figures 2-19a and 2-19b).

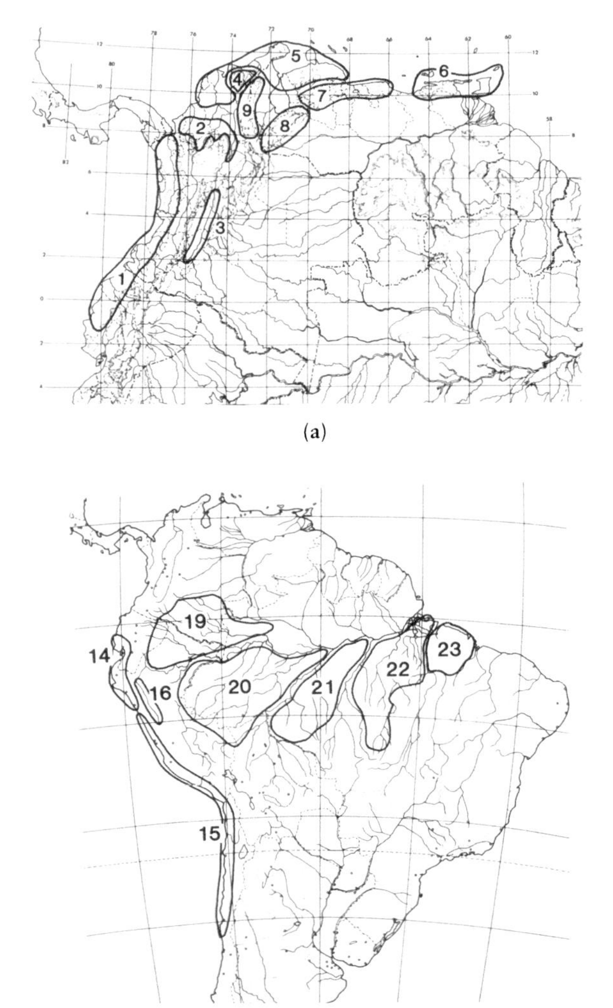

FIGURE 2-18
Postulated areas of endemism for birds in South America. (a) Areas of postulated endemism in northern South America. (b) Areas of postulated endemism for Amazonia and the Andes Mountains.

A comparison of bird species evolution in South America and Africa demonstrates the greater diversification that has occurred in South America. Over the past 12.5 million years, the rate of speciation minus extinction has been far greater in South America than in sub-Saharan Africa (which actually exceeds South America in area). South America is estimated to have 3,150 land and freshwater bird species compared with 1,623 in sub-Saharan Africa (Price 2008). Using molecular analysis, it has also been learned that bird species from Andean areas of endemism are in general more recently evolved than those of lowland Amazonia. *Young species* (in this specific context) are defined as belonging to a *clade* (a cluster of closely related species) of at least 10 species whose common ancestor dates back to about six million years or younger, and *old species* are those whose ancestry is further removed in time.

Age, Endemism, and Refugia: Many Questions

How does the high diversity within tropical regions relate to geological—long periods when speciation might exceed extinction? Contrasting South America with other major tropical regions, one finds that as with bird species richness, plant species richness is also unequivocally highest in South America. But how did that happen? What factors are responsible? South American species richness, as extreme as it is, has long posed vexing questions for evolutionary biologists and biogeographers. Since the 1970s, a debate has been ongoing between those who believe most speciation in South America (and by implication, elsewhere in the global tropics) is recent, dating primarily to events in the Pleistocene and immediately before, and those who argue that the data do not support such assertions and that much of the speciation occurred many millions of years before the Pleistocene.

It is unlikely that the equatorial tropics were climatically stable and constant throughout the Pleistocene (from 1.64 million to 10,000 years ago), undisturbed by the giant glaciers bearing down on northern temperate areas. Thomas Belt, in 1874, discussed the possible effects of northern glaciation on the tropics. More recently, studies from *geomorphology* (the historical development of present landforms), *paleobotany* (the study of past patterns of plant distribution), and biogeography suggest dramatic changes in Amazonia during the Pleistocene (Colinvaux 1989a and 1989b; Haffer 1969, 1974, 1985, and 1993; Prance 1985; Simpson and Haffer 1978). From these studies has emerged the *refugia hypothesis* for high speciation rates in the tropics, and particularly in the Neotropics.

It is hypothesized that during glacial advances in northern latitudes, the tropics may have become cooler and drier. For example, during part of the Pleistocene, temperature in the Ecuadoran foothills, east of the Andes, was between 4°C and 6°C cooler than at present (Colinvaux 1989a and 1989b). A possible ecological result of the climatic cooling was to alter and shift the distribution of ecosystems. Ecosystems such as savannas presumably enlarged, and moist forests presumably contracted. Large continuous tracts

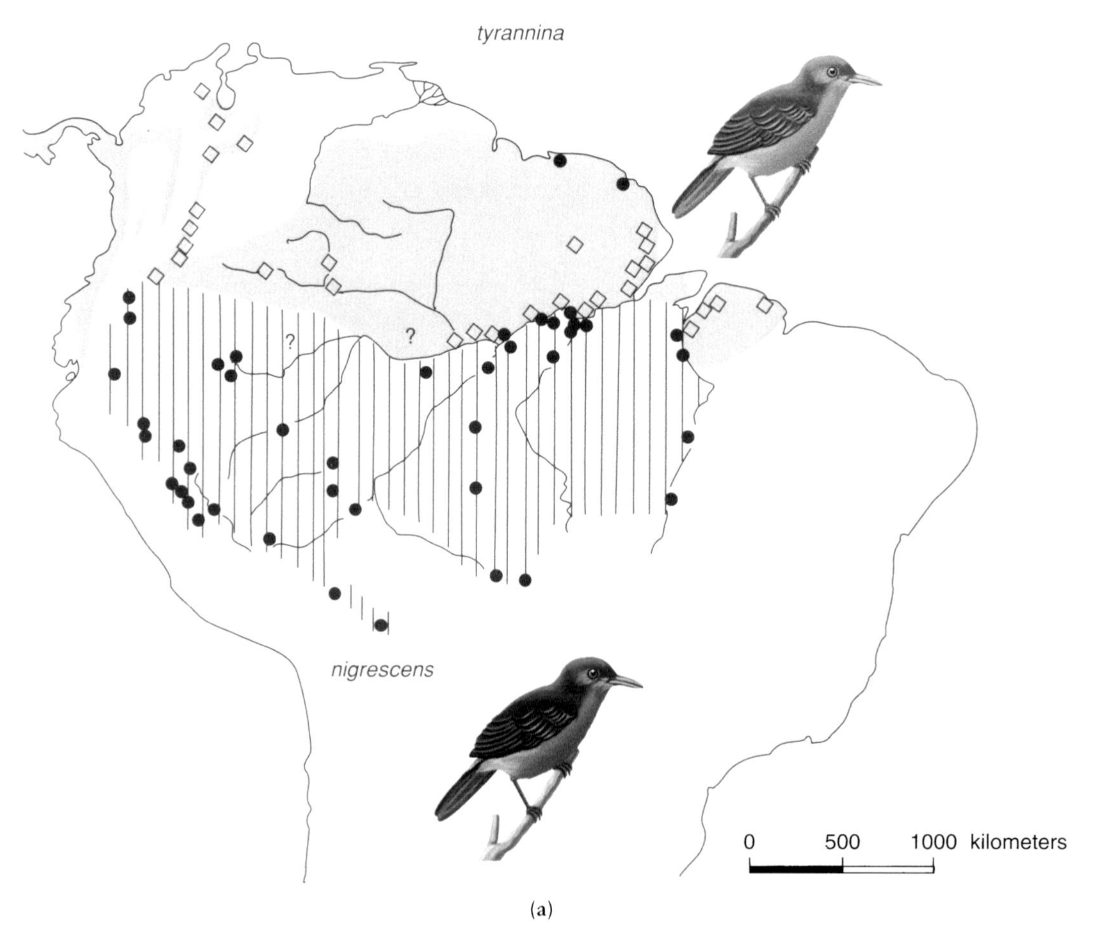

FIGURE 2-19
The distributions of two superspecies groups of Amazonian birds closely follow the geography of the river systems in Amazonia. (a) This map shows the distribution of two closely related antbird species in Amazonia. Note that they are separated based on occurrence either north or south of the Amazon River. (b) This figure depicts the distribution of the razor-billed curassow species group. The species are morphologically similar but are regionally separated. Note that the Amazon River separates the southern *Mitu tuberosa* from the more northern populations, and note the isolated *Mitu mitu*, found in the Brazilian Atlantic forest region.

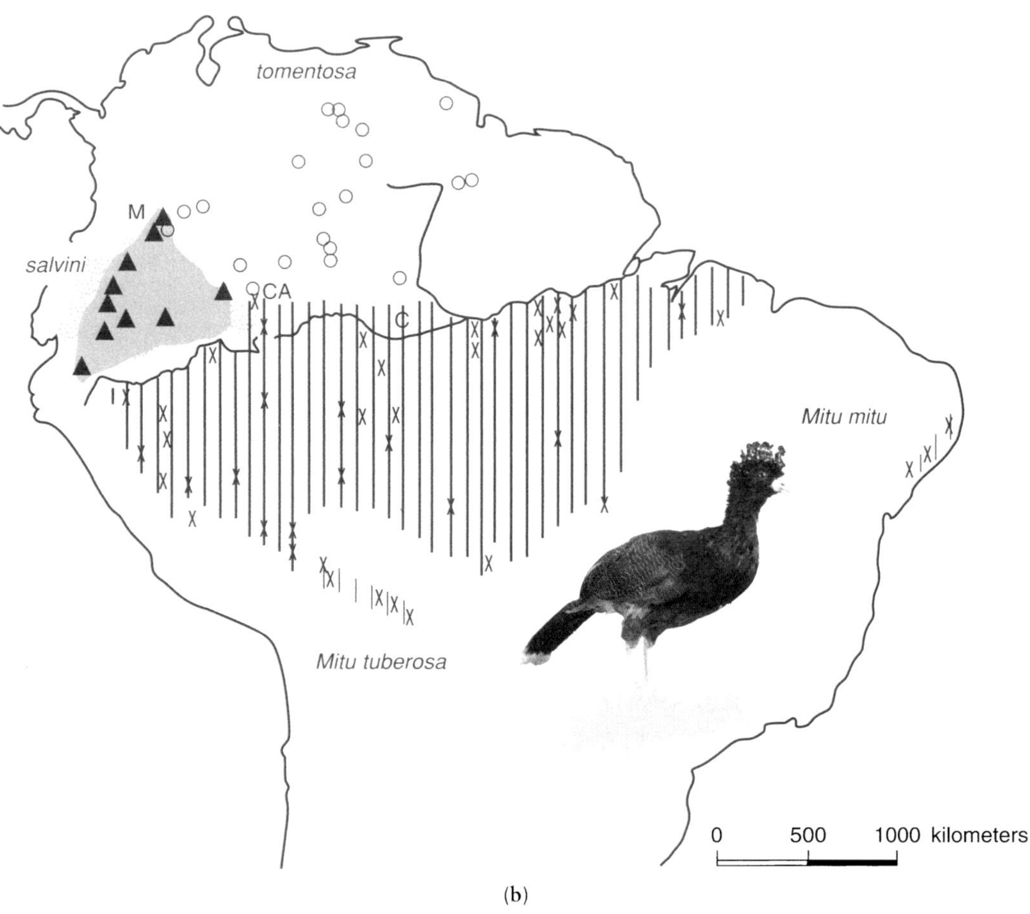

FIGURE 2-19
(continued)

of lowland rain forest were therefore fragmented into varying-sized "forest islands" surrounded by "seas" of savanna or dry woodland. Savanna expansion, in the refugia model, created vicariance among forest tracts. Because of the repeated shrinking and fragmenting of forests, forest organisms periodically became geographically isolated from populations in other forest areas, promoting speciation (Nores 1992). The Amazon Basin became a climatically dynamic "archipelago" of varying-sized rain forest islands. In other words, rain forest species were confined to fragmented refuges during the driest periods. These small, island-like "refuges" promoted speciation among plants and animals.

Studies based on the current distribution of certain kinds of birds (Haffer 1969, 1974, 1985, and 1993; Haffer and Fitzpatrick 1985; Simpson and Haffer 1978) postulate that at least nine major and numerous smaller forest island refuges were present in Amazonia during the Pleistocene (Figure 2-20). They assert that many taxonomic groups subsequently went through periods of rapid speciation because there were repeated episodes of rain forest shrinkage and expansion. During interglacial periods forests expanded, and secondary contact was established between newly speciated populations, explaining why so many extremely similar species can be found today in Amazonia.

The refugia model has been and continues to be the subject of much debate (Haffer 1993). Evidence supporting it is based mostly on present distribution and diversity patterns. For example, Prance (1982a and 1982b) has closely examined woody plant diversity and concluded that 26 probable forest refuges existed for these plants. Kinzey (1982) concluded that present primate distribution fits predictions of

FIGURE 2-20
This series of maps depicts the presumed forest refugia during the dry phases of the Pleistocene. The left-most map depicts refugia based on bird distribution, the center map is based on Amazonian lizards, and the right-most map is based on butterflies of the genus *Heliconius*. Note that the presumed refugia differ substantially among the groups.

the refugia model, and Haffer (1974), Haffer and Fitzpatrick (1985), and Pearson (1977 and 1982) have presented evidence from present bird distribution in support of the model. In a series of papers by Keith Brown, cited by Prance (1982b and 1985), heliconid butterflies have been shown to have 44 centers of endemism throughout Amazonia and surrounding areas. But it is important to note that centers of endemism, the presumed Pleistocene refuges, do not precisely overlap among taxa, which lessens support for the refugia model. Supporters of the refugia model respond that different taxa have different dispersal powers and different generation times and thus would be expected to differ somewhat with regard to the degree of regional endemism (Prance 1982a and 1985).

The refugia model has been subject to strong criticism. Colinvaux (1989a) summarizes the arguments against the Pleistocene-based refugia model, pointing out that at least one study showed that for plants, refuge locations coincide with areas in which sampling of plants for herbaria specimens has been historically most intense. This, of course, would suggest that the refuges are artifactual derivatives of uneven sampling effort.

Analysis of pollen from sediments taken from the Amazon lakes in Ecuador and Brazil do not support the refugia model (Bush and de Oliveira 2006). Pollen is resistant to decomposition, particularly in anaerobic sediments of lake basins. As pollen accumulates year after year, it forms a vertical *pollen profile*, a historical indicator of which plant species were present in a region. (Lake sediments are cored, and pollen at the bottom of the core is oldest, pollen at the top youngest.) Pollen profiles from various Amazonian lakes showed that although the region likely cooled by about 5°C, it remained fully forested, not broken up by savanna.

Other strong objections to the refugia theory have been raised (Willis and Whittaker 2000). One analysis of Amazonian sedimentary patterns, for example, suggests that there was no substantive climatic change or reduction in forest cover in Amazonia during the Pleistocene (Irion 1989). However, what is suggested by sediment patterns is not a stable Amazonia, but rather strong oscillations throughout the Pleistocene in the distribution of land, water, floodplain forests, and *terre firme*. Sea-level changes are thought to have led to the alternation of huge Amazonian lakes with strong valley cutting, geological events that would presumably have a strong impact on flora and fauna. This view sees the Amazonian rain forest as subject to vicariance, though not necessarily reduced to scattered refuges.

PLATE 2-15
These photos of fossil plants demonstrate morphological diversity in the Laguna del Hunco flora from about 52 million years ago.

Another objection to the refugia model focuses on the ages of species. The refugia model suggests that speciation has been recent. As mentioned earlier, molecular analysis (unavailable when the refugia model was proposed) indicates that many bird species in Amazonia trace their origins to more than six million years ago, well before the Pleistocene. A complex of frog species of the genus *Leptodactylus* exhibits high species richness dating back beyond the Pleistocene. Speciation appears to have occurred in the mid-Tertiary period (Maxson and Heyer 1982; Heyer and Maxson 1982). Studies of plant fossils and subsequent analysis indicate that the high plant species richness traces back fully to 52 million years ago, in the Eocene epoch of the Tertiary period (Wilf et al. 2003) (Plate 2-15).

At present, the evidence does not strongly support the refugia model as a *species pump* in tropical regions (Willis and Whittaker 2000; Mayle 2004). Rather, evidence now seems to favor the temperate zone as the region where refugia may have been important drivers of speciation.

The Great American Interchange

The Panamanian land bridge, now called the *Isthmus of Panama*, formed approximately three million years ago because of a combination of uplift of the northern Andes and a global drop in sea level, perhaps as much as 50 meters, a result of the increasing size of the polar ice caps (Marshall et al. 1982). Thus it was just prior to the onset of the ice age that the continents of North and South America were no longer separated by water. The Panamanian land bridge profoundly altered the ecology of South America, much more so than it did that of North America. Consider that the fauna of North and South America had evolved independently of one another for at least 40 million years, but their mingling, once the land bridge developed, was completed within a mere two million years (Webb 1978) (Figure 2-21).

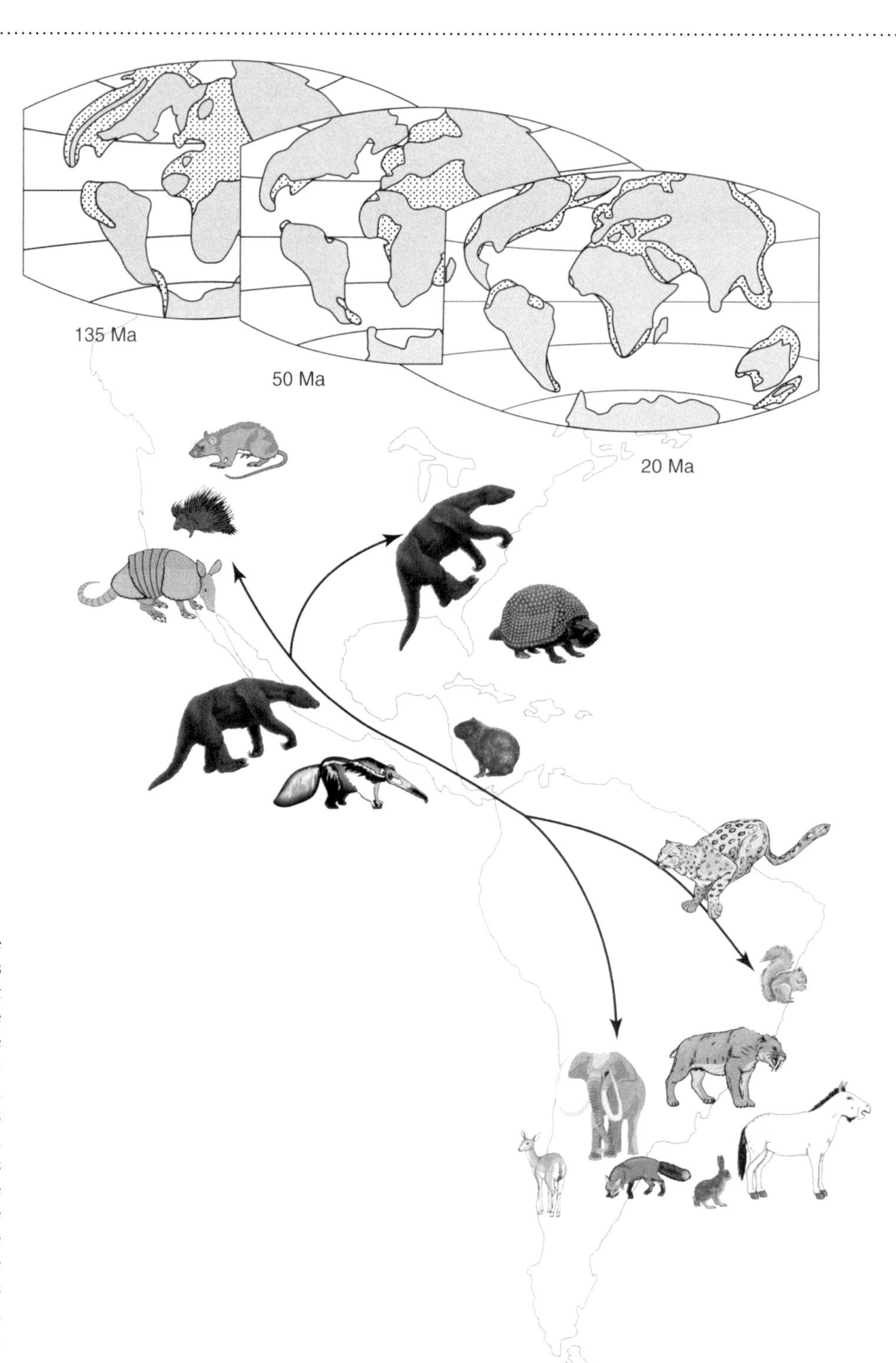

FIGURE 2-21
This figure illustrates the principal groups of mammals involved with the Great American Interchange that resulted in mixing the mammalian (and other animal) faunas of North and South America once the Isthmus of Panama became exposed approximately 3 million years ago. The upper maps illustrate how the positions of the Americas changed from the late Cretaceous through the Miocene, bringing the continents sufficiently close that when glaciation occurred, the land bridge was established.

Once the land bridge formed, some South American animals moved northward as early as 2.5 million years ago, literally walking to North America. These included two armadillo species, a glyptodont, two species of large ground sloths, a porcupine, a large capybara, and one phorusrhacoid bird. Ground sloths, now all extinct, were very large terrestrial sloths with long claws that likely fed on foliage. They were closely related to the much smaller tree sloths of the Neotropics. Glyptodonts looked like gigantic armadillos, covered with bony armor, whose tails (depending on species) sometimes terminated in a mace-like club. Capybaras are cavimorph rodents, such as chinchillas and guinea pigs. The extant capybara (*Hydrochoerus hydrochaeris*), abundant in parts of South America, is the world's largest rodent. Phorusrhacoid birds, now all extinct, have earned the nickname *terror birds*. Some stood almost as tall as a human being. They were flightless but could move quickly, running on long legs. Their huge, raptor-like beaks were easily capable of killing small to moderate-sized mammals. At the time they lived, they were among the top carnivores of South America (Figures 2-22a and 2-22b).

Other invaders included *Didelphis*, the familiar opossum and North America's only marsupial mammal. Opossums invaded approximately 1.9 million years ago (Marshall 1988). Last, from South America came at least one species of toxodon, an odd mammal that belonged to a group named *notoungulates*. Toxodons were bulky mammals whose appearance was suggestive of a cross between a cow and a hippopotamus. The collective impact of the South American invaders was modest at best. Of their host, only armadillos and opossums remain, both of which are thriving. The ground sloths were probably killed by humans as the human population spread southward. The others have simply dropped out of the fossil record.

Many North American animals walked across the land bridge to South America. Their impact on the native fauna was substantial. The list of invaders includes skunks, peccaries, horses, dogs, saber-toothed cats, other cats, tapirs, camels, deer, rabbits, tree squirrels, bears, and an odd group of elephant-like mammals, the gomphotheres. Add to this list the field mice, or cricetid rodents, whose travel route to South America is still debated, but who have since radiated into 54 living genera, and you begin to see why the effect of North American mammals on South American ecosystems was so great (Marshall 1988).

Various amphibians, reptiles, and bird groups also migrated in either direction, some from the north, some from the south. The faunal interchange altered ecological communities on both continents.

The exchange strongly altered the look of the South American fauna such that it is today much more like that of North America. As Marshall (1988) so aptly summarizes (about mammals), "the Great American Interchange resulted in a major restructuring. Nearly half of the families and genera now on the South American continent belong to groups that emigrated from North America during the last 3 million years."

It is quite unclear how the influx of North American mammals, as diverse as it was, affected the abundances of South American mammal species. For example, many hoofed mammals invaded from North America. However, their ecological counterparts in South America, the litopterns and notoungulates,

FIGURE 2-22
Phorusrhacoid bird (left) and toxodon (right). These animals represent unique components of the faunas of South America before the Great American Interchange took place.

FIGURE 2-23
Smilodon (left) and *Thylacosmilus* (right) were both morphologically similar, particularly regarding the enlarged saber-like canine teeth. But *Smilodon* was a placental mammal, whereas *Thylacosmilus* was a marsupial mammal. Their similarity is a clear case of convergent evolution.

were declining in numbers and diversity long before the camels and horses arrived. (*Litopterns* comprised a camel-like group that included that unique *Machrauchenia*, an animal that looked like a camel with an elephant-like proboscis.) It is not considered likely that the North American ungulates were the primary cause of the extinction of the various South American groups (Marshall 1988). Among predatory mammals, the large, wolverine-like South American borhyaenoids were essentially outcompeted by the terror birds, not by mammalian invaders from the north. However, the phorusrhacoid terror birds may have eventually lost out to invading placental mammals. There is really only one relatively clear-cut example where it appears that there was direct competition between a North American species and a South American species. This is the case of the marsupial saber-toothed cat, *Thylacosmilus*. This animal bore a striking anatomical similarity to the placental saber-toothed cat, *Smilodon*, an example of convergent evolution (Figures 2-23).

Smilodon crossed the land bridge into South America, and the extinction of *Thylacosmilus* coincides closely with the arrival of *Smilodon*. Much of the extinction of large mammals on both continents is probably explainable by the proliferation of humans during the Pleistocene (Marshall 1988) and not due to competition among the animal groups themselves.

The Great American Interchange demonstrates that ecological communities are highly subject to major change over time. The mixing of the two fauna created new communities. One of the most challenging questions in ecology, one related both to long-term biogeographic history as well as to present ecological interactions, is to understand how various factors structure communities.

3 Inside Tropical Rain Forests: Structure

Chapter Overview

Rain forests stand out among other tropical terrestrial ecosystems because of their complex physical structure and high biodiversity. This chapter describes the various structural attributes of rain forests and provides an introduction to rain forest structural diversity. In Chapters 4 and 5, we will take a detailed look at biodiversity.

Why Rain Forest Is Unique

Rain forest is a major terrestrial biome. A *biome* is an ecosystem type that has distinctive and characteristic plant and animal species and that is largely determined by climatic characteristics. (Compare this definition with the Holdridge life zone system described in Chapter 1.) Rain forest represents a type of biome, with characteristics distinguishing it from other biomes such as desert, savanna, tundra, or various kinds of temperate forests (e.g., boreal needle-leaved forests and broad-leaved deciduous forests). Biomes are fundamentally the result of global differences in climate (see Chapter 1). Tropical rain forest biomes occur equatorially, where the temperature remains warm and conditions are generally wet throughout the year.

There are two broad characteristics that are evident in rain forest. The first is complex physiognomy. *Physiognomy*, when used in ecology, refers to the physical structure of an ecosystem. Rain forests are voluminous, high in biomass, and structurally elaborate. The stature of the trees is usually tall (between 30 and 35 meters, and occasionally taller), and from the canopy to the ground, the physical structure is considerably more varied than in other forest types (Plate 3-1). This reality means that numerous life forms, both plant and animal, have the opportunity to evolve and specialize in different vertically structured microhabitats in rain forest. That contributes to the other major characteristic—high biodiversity of numerous taxonomic groups.

Biodiversity has various meanings, depending on precisely how the term is used. The most basic meaning is that biodiversity refers to the number of species within a habitat, termed *species richness*. But the term may also refer to how species richness is distributed relative to species' population sizes, termed *species evenness*. Concepts such as species diversity; evenness; and alpha, beta, and gamma diversity will be discussed in the next chapter. *Biodiversity* may sometimes refer to genetic diversity among populations within a species (or individuals within a population), an important parameter in much conservation research.

No matter where you are on Earth, once you enter a rain forest, biodiversity becomes higher, often much higher, than is found in adjacent or nearby non-rain-forest

PLATE 3-1
A look directly up into a tropical forest canopy shows the structural complexity and tall stature characteristic of this ecosystem.

ecosystems. No complete inventory of total biodiversity has yet been made for any rain forest, but various taxonomic groups such as trees, bromeliads, orchids, mammals, birds, reptiles, amphibians, and numerous varied arthropod groups have been well documented (Figures 3-1 and 3-2). These groups are all most diverse within rain forests.

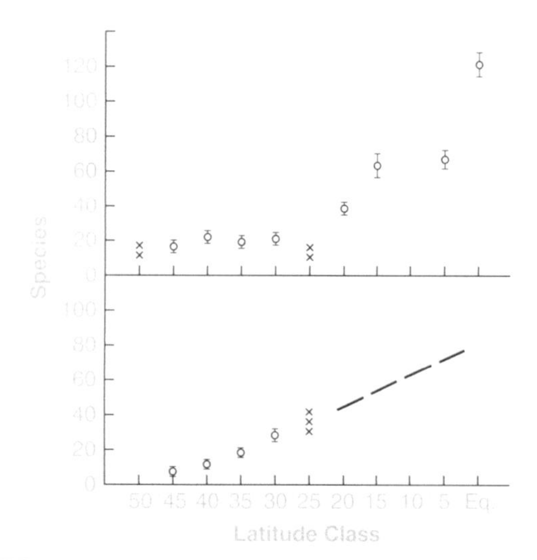

FIGURE 3-1
This figure shows the number of bird species as it changes with latitude in summer (top) and winter (bottom). Note the significant jump in number of species at 20° latitude, which is within the tropics. Also note how the number of species is by far highest at the equator, 0° latitude.

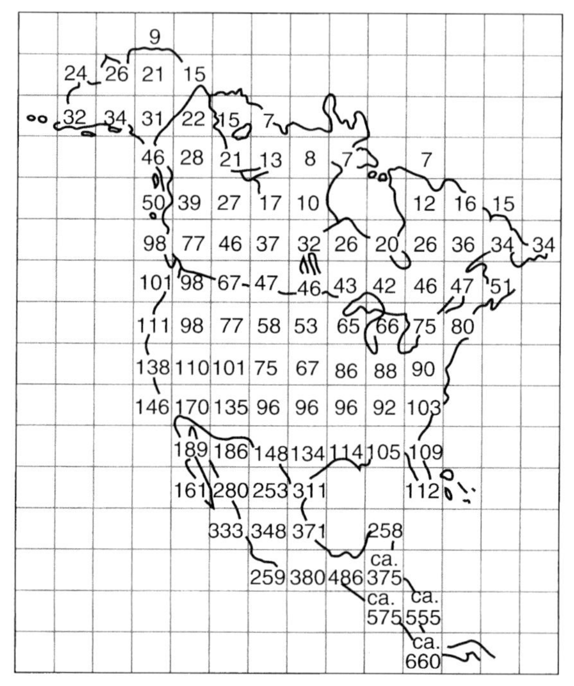

FIGURE 3-2
The squares on this map indicate the number of landbird species found within. Each square represents 500 kilometers (about 300 miles) on a side. Look at Central America and note how the number becomes much higher even though some of the land areas do not cover an entire square.

Two questions thus emerge: Why do rain forests have such complex physical structure, and why do rain forests harbor such high biodiversity?

These questions are interwoven. In fact, the two characteristics, complex physiognomy and high biodiversity, may each contribute causally to each other. It should be obvious that high structural complexity is a function of diverse growth forms evident in rain forest plant species. Rain forest physiognomy is caused by the numerous varied plant species that inhabit it, accommodating a diverse range of growth forms (trees of varying stature, vines, epiphytes). Such complexity of structure provides numerous and varied spatial and energy resources for animals (and smaller plants), acting as a stimulus to promote the evolution of animal (and plant) diversity.

Let's begin with a basic definition of rain forest and then examine rain forest physiognomy.

What Is Rain Forest?

> The observer new to the scene would perhaps be first struck by the varied yet symmetrical trunks, which rise up with perfect straightness to a great height without a branch, and which, being placed at a considerable average distance apart, give an impression similar to that produced by columns of some enormous building. Overhead, at a height, perhaps, of a hundred and fifty feet, is an almost unbroken canopy of foliage formed by the meeting together of these great trees and their interlacing branches; and this canopy is usually so dense that but an indistinct glimmer of the sky is to be seen, and even the intense tropical sunlight only penetrates to the ground subdued and broken up into scattered fragments. There is a weird gloom and solemn silence, which combine to produce a sense of the vast—the primeval—almost of the infinite. It is a world in which man seems an intruder, and where he feels overwhelmed by the contemplation of the ever-acting forces which, from some simple elements of the atmosphere, build up the great mass of vegetation which overshadows and almost seems to oppress the earth. (Wallace 1895, p. 240)

So wrote Alfred Russel Wallace (1895), who spent four years exploring along the Rio Negro and Amazon, and is credited along with Charles Darwin with formulating the theory of natural selection. Recall from the Introduction that Charles Darwin was equally moved by his first experience of the rain forest.

Though rain forest favorably impressed, indeed awoke a sense of awe in both Wallace and Darwin, it has been depicted in art and literature in ways that range widely, from hauntingly idyllic to the infamous "green hell" image that typified the writings of authors such as Joseph Conrad (Putz and Holbrook 1988). What, ecologically, is rain forest?

Neotropical rain forest was first described by Alexander von Humboldt, who called it *hylaea,* meaning "forest" in Greek. Though Humboldt meant the term for New World rain forest, it applies equally well elsewhere in the global tropics, because rain forests are generally structurally similar even when composed of different species of organisms.

A *rain forest* (Figure 3-3), in a general sense, is a nonseasonal forest dominated by broad-leaved evergreen trees, sometimes of great stature, where rainfall is both abundant and relatively constant. Recall the Holdridge life zone classifications described in Chapter 1. Under the Holdridge system, rain forest has a mean annual biotemperature greater than 24°C, a potential evaporation ratio of only 0.125 to 0.25 (superhumid), and experiences greater than 800 centimeters (315 inches) of average total annual precipitation. As noted in Chapter 1, Holdridge's classification makes true rain forest an extremely wet ecosystem. In fact, most rich lowland tropical forests routinely called rain forests tend to fall into what would be termed, in Holdridge's classification, *tropical moist* or *tropical wet forests*. But all such forests are lush, with many kinds of vines and epiphytes (air plants) growing on the tall, impressive trees. Bark is often laden with abundant lichens and other tiny forms of plant life. In general, a forest that receives at least 200 centimeters (about 80 inches) of rainfall annually, with precipitation spread relatively evenly from month to month, is sufficient to support the numerous rain forest characteristics cited in this chapter and would be called, in a general sense *rain forest* (Myers 1980). But in the strictest sense, most forest termed *rain forest* would not be so classified using the Holdridge life zone classification.

This is because most lush tropical forests experience some degree of seasonal variation in rainfall. This characteristic is what really separates (to varying degrees) true rain forest from tropical moist forest. Tropical moist forest is evergreen or partly evergreen (some trees may be deciduous), receiving not less than

FIGURE 3-3
This is a sketch of a section of Barro Colorado forest in Panama. It conveys the structural complexity and tree diversity typical of tropical lowland forest.

100 millimeters of precipitation in any month for two out of three years, frost-free, and with an annual temperature of 24°C or more (Myers 1980). Because of gradations in how much moisture is available from one region in the tropics to another, the distinction between rain forest and moist forest is often not obvious. For example, some tree species may, depending on soil moisture, exhibit partial deciduousness, with thinner leaf cover in dry season compared with rainy season. Deciduous tropical trees tend to experience growth of new leaves at the close of dry season or early rainy season.

Gaps and Tropical "Jungle"

Tropical forests, like all other ecosystems, are subject to periodic natural disturbance, a factor that influences many of the ecological patterns evident in tropical forests, including the high diversity of species. In addition to natural disturbance, human disturbance occurs with high frequency in many tropical areas. Disturbance is a major force affecting rain forest dynamics and will form the subject of Chapter 6. What follows is an introduction.

When rain forest is disturbed, such as by hurricane, lightning strike, tree-fall, or human activity, the disturbed area, called a *forest gap*, is opened, permitting high light penetration. Fast-growing plant species intolerant or less tolerant of shade are advantaged, and a dense tangle of thin-boled trees, shrubs, and vines results. Soon an irregular mass of dense greenery covers the gap created by the disturbance. Note that the term *gap* is used in ecology to refer to an area largely opened by disturbance, allowing high light penetration. Trees that grow in gaps are typically in close proximity. Palms and bamboos may abound along with various vines, creating thick tangles. Areas of extensive disturbance, large gaps, that have grown as described earlier have historically been called *jungles*, a term that means a land overgrown with vegetation. To penetrate a jungle requires the skilled use of a machete. Jungles, in an ecological sense, are successional. In ecology, the term *ecological succession* refers to a pattern of temporal change in the species composition and physiognomy of an ecosystem onward from the time it is disturbed. Gaps, once open and sunlit, become dense with jungle, and will eventually revert to shaded forest as slower-growing, more-shade-tolerant species replace colonizing heliophilic (sun-loving) species (Plates 3-2 and 3-3). This process is described in greater detail in Chapter 6.

PLATE 3-3
This extensive gap in Ecuador demonstrates the dense and irregular vegetation profile. It is an area of ecological succession.

PLATE 3-2
This forest gap shows a high density of small trees that are growing in the well-lit gap.

Rain Forest Physiognomy

While individual rain forests vary one from another, the *field marks* of rain forest can be summarized, in a general sense, as follows:

- Tall trees of many species, some reaching above 30 meters (100 feet).
- Deep shade with scattered sun flecks at ground level.
- Scattered emergent (unusually tall) trees reaching well above most of the tree canopy. This characteristic gives the forest an uneven canopy. Though emergent trees also may be seen in many temperate forests, they are usually much more apparent in tropical forests.
- No clearly delineated understory, shrub layer, or herb layer. Many tree species of various sizes typically comprise the understory and midelevation levels. Palms often dominate in the understory, but many palms may reach canopy height and persist for long time periods.
- Forest gaps ranging from single tree-falls to moderate-sized blow-downs are common. Rapidly growing sun-dependent plant species are common

to gaps, which, if large, become jungle-like. Gaps add a major component of horizontal patchiness to forests, accounting for the shifting mosaic pattern typical of rain forests. (The term *shifting mosaic* refers to patches of forest that change with time, the result of accumulated disturbances in various areas of forest.)

- Many tree species tend to rise to nearly canopy height before widely branching. This branching pattern is often umbrella-like, the branches appearing to radiate as spokes from the trunk.
- Trees most commonly have animal-pollinated flowers, though some species are wind-pollinated.
- Flowering and fruiting may be seasonal, but some flowers and fruit are evident in any month.
- Leaves are usually oval with little or no lobing and tend to be thick and waxy, with pointed *drip-tips*. Drip-tips are sharp points at the tip of a leaf that permit water to drain easily from the leaf. Compound leaves are common in many species, such as within the legume family (Fabaceae). Large leaves are common.
- Most trees are broad-leaved evergreen, but some deciduous tree species occur. Within rain forest, deciduous leaf-drop (confined to isolated trees of certain species) is often correlated with flowering and fruiting. Within moist forest, where dry season is typical, deciduous leaf-drop is associated with water stress.
- Trees often have buttressed roots, some have prop (also called *stilt*) roots, and many exhibit surface roots that radiate across the forest floor. A *buttress* is a flaring of root from above ground. *Prop roots* are root clusters that emerge from the trunk well above ground. *Surface roots*, as the name implies, wind conspicuously over the surface of the forest floor.
- Bark is variable among tree species. It may be light or dark, smooth or ridged, depending on species. There is no generalized pattern for bark color or texture.
- Many epiphytic plants grow on branches and bark, including various lichens, orchids, ferns, and others.
- Other plants, many vine-like, of many kinds (stranglers, lianas) are common and often abundant.
- Litter layer is often thin, as leaf decomposition rates are rapid.
- Soil is often reddish but may be of some other color and is typically acidic, often high in clay content. Soil characteristics are highly variable. Some tropical soils are sandy and some are rich in nutrients, but many are thin and poor in nutrient content. In general, flood plain soils are rich in nutrients, and upland soils are nutrient poor.

The preceding list characterizes the most general and obvious physiognomic characteristics of tropical rain forest. It does not matter whether you are in Borneo or Papua New Guinea, the Ivory Coast, or Brazil—at first glance, all rain forest may appear quite similar in physical structure and appearance (Plate 3-4).

This similarity results from similar selection pressures acting on diverse species groups. When climates are similar, ecosystems develop to be structurally and functionally similar. But on closer inspection, numerous differences are readily evident among rain forests both within and among various geographical areas. Much variation, as discussed in the previous chapter, is due to evolutionary differences, both ancient and recent, most caused by vicariance biogeography operating at various spatial and temporal scales. One does not find orangutans or rattan palms in Venezuela nor sloths or hummingbirds in Sumatra. But even within a given biogeographical region such as the Neotropics, rain forests in Costa Rica differ in many ways from their counterparts in Ecuador. And in Ecuador, Amazonian forests show considerable differences from site to site, some sites hosting dense rain forest, some

PLATE 3-4
A stream such as this one, meandering through rain forest and seen from this vantage point, would look similar anywhere on Earth where rain forest occurs because rain forest plant species have converged in overall structural characteristics, and numerous families have global distributions. This particular stream is in Queensland, Australia. Note the reddish soil, typical of tropical soils in many regions.

FIGURE 3-4
This drawing illustrates a mature dipterocarp forest in Belalong, Brunei. The smaller, younger dipterocarps in the understory exhibit a monopodial crown (which is taller than it is broad). The mature canopy trees exhibit a sympodial crown, broader than it is tall. Such transition in growth form is typical of many tropical tree species.

more open forest with palms, some open forest without palms, and some open forest with abundant lianas (Pires and Prance 1985). Regional differences occur because of local variability in microclimate, soil characteristics, and hydrologic factors that vary on regional geographic scales.

Rain forests on poor soils differ from those with richer soils, usually being more species-rich. Rain forests in areas not subject to periodic flooding (termed *terre firme* in Amazonia) are distinct in various ways from those on flood plains (termed *varzea* in Amazonia). However, the overall similarities, apparent as first impressions, are striking.

We will now look more closely at some of the rain forest structural characteristics cited earlier.

STRUCTURAL COMPLEXITY

Equatorial rain forest may seem immense, dark, and enclosing, as dense canopy foliage shades the forest interior, especially in the attenuated early morning light (Figure 3-4). Because of high humidity and frequent rain, the sky is often pale, making lighting conditions challenging. This reality, plus the height of the canopy, often makes it more difficult to detect various canopy-inhabiting animals such as monkeys and birds (Plates 3-5 and 3-6). At midday, if the sun is shining high overhead, only scattered flecks of sunlight dot the interior forest floor, because up to 99% of the ambient light has been absorbed within the canopy and understory. Deep shade (Plate 3-7) prevents dense undergrowth from forming, making transit through the forest fairly easy (in contrast to the difficulty of moving through recently disturbed areas such as dense jungle). Large trees tend to be widely spaced, many with prominent, flaring buttressed roots, some with draping, extended prop roots. The forest floor is typically covered with a thin litter layer and dense arrays of surface roots radiating from the tree buttresses.

For most tropical forests, virtually all trees are broad-leaved. Palms of many species often abound, especially in the understory. Gymnosperm needle-leaved

PLATE 3-5
Dark-colored monkeys such as these black howler monkeys in Panama are sometimes difficult to see well when in the canopy with pale sky behind them.

PLATE 3-6
Colorful and large birds, such as this trogon in Panama, are often difficult to detect as they sit motionless in the shaded forest understory or canopy.

PLATE 3-7
Sun flecks illuminate a seedling on an otherwise dark rain forest floor. This image illustrates how most light is absorbed well before reaching the floor of a tropical rain forest.

trees that abound throughout the temperate zone, the pines, spruces, and hemlocks, have few equivalents in the tropics. Tropical gymnosperms include plants of the cycad family (Zamiaceae), which resemble palms (Plate 3-8). Other gymnosperms, the Podocarpaceae, have flattened, needle-like leaves.

Tree boles are characteristically straight, and most rise to considerable height before spreading into crowns (Plate 3-9). The crown of a tropical canopy tree, when viewed from the ground, may be partially obscured by its epiphyte load. Epiphytes, including various orchids, many kinds of ferns, and, in the Neotropics, an abundance of pineapple-like plants with spiky leaves called *bromeliads* (Bromeliaceae), commonly adorn the widely spreading branches. To study these plants, it is useful to have binoculars, as the epiphytes are typically high in the canopy. Vines of various forms are usually present and often abundant, draping throughout and hanging from the canopy.

Tropical rain forest, in sharp contrast to many temperate zone forests, is not clearly stratified (Richards 1952), though up to five poorly defined strata are sometimes present (Klinge et al. 1975). Weak stratification contributes to the complex physiognomy of rain forest (Hartshorn 1983). Some trees, called *emergents* (a term that is descriptive of their height, not a taxonomic group), erupt above the canopy to tower over the rest of the forest. Trees of varying heights occur in both understory and canopy. Palms are often common. Disturbance events result in gaps throughout a forest where trees are

PLATE 3-8
Cycads are ancient gymnosperms. Note the palm-like leaves. The central cone resembles those of others gymnosperms such as pines and spruces. Cycads are common in many tropical regions.

PLATE 3-9
Tropical trees characteristically show major branches radiating out like spokes.

PLATE 3-10
This isolated tree along the Napo River in Ecuador demonstrates the growth form typical of many tropical tree species.

stimulated to grow quickly. Because there are so many different tree species and varying levels of light availability, trees grow at different rates, and therefore their heights are typically staggered. Thus there are two factors, *horizontal* (gap formation and aging) and *vertical* (varied growth heights at maturity of different tree species), contributing to rain forest biodiversity and physical complexity.

Most trees are monotonously green, but a few may be bursting with colorful blossoms, while a few may be essentially leafless because they are periodically deciduous and have dropped their leaves (Plates 3-10 and 3-11). Shrubs and other herbaceous plants share the heavily shaded forest floor with numerous seedling and sapling trees, ferns, and palms. It is difficult to perceive a simple pattern in the overall structure of a rain forest. Structural complexity is the overarching pattern.

STATURE

Tropical rain forests have a reputation for having huge trees. Old engravings depict trees of stunning size with up to a dozen people holding hands around the circumference of the trunk. Tropical trees may, indeed, be both wide and lofty, but bear in mind that many appear tall partly because their boles are slender (just as a thin person gives the appearance of being taller than a stocky person of equal height), and branches tend not to radiate from the trunk until canopy level, thus enhancing the tall appearance of the bole. The tallest tropical trees are found in lowland rain forests, and these trees range in height from between 25 and 45 meters (roughly 80 to 150 feet), the

PLATE 3-11
This tree has dropped its leaves but is in full flower. Lack of leaves makes the flowering tree more obvious to pollinating animals.

majority around 30 to 35 meters (100 to 115 feet). Tropical trees occasionally exceed heights of 45 meters (about 150 feet), and some emergent trees do top 60 meters (about 200 feet) and may occasionally approach 90 meters (about 300 feet), though such heights are uncommon. Some temperate zone forest equals or exceeds tropical forest in height. In North America, Sierra Nevada giant sequoia groves, coastal California redwood groves, and Pacific Northwest old-growth forests all routinely exceed the height of the majority of tropical forest. So do the temperate blue gum eucalyptus forests in southeastern Australia. Neither the tallest, the broadest, nor the oldest trees on Earth occur in rain forest: the tallest is a California redwood, at 367.8 feet; the broadest is a Montezuma cypress in subtropical Mexico, with a circumference of 160 feet; and the oldest is a bristlecone pine in the White Mountains of eastern California, about 4,600 years old.

PLATE 3-12
Buttressed roots characterize numerous species of tropical trees.

BUTTRESSES, PROP ROOTS, AND SURFACE ROOTS

A *buttress* is a root that flares out from the trunk to form a flange-like base (Plates 3-12 and 3-13). Many, if not most, rain forest trees have buttressed roots, giving tropical rain forest a distinctive look in comparison with temperate forests (though some old-growth temperate rain forest trees such as are found in the Pacific Northwest are sometimes weakly buttressed). Several buttresses radiate from a given tree, surrounding and seeming to support the bole. Buttress shape is sometimes helpful in identifying specific trees. Buttresses can be large, often radiating from the bole 2 to 3 meters (6.6 to 10 feet) from the ground.

Because buttressing is particularly common to trees near streams and on riverbanks as well as to trees lacking a deep taproot, it is likely that buttressing acts to aid in supporting the tree (Richards 1952; Longman and Jenik 1974). Some research suggests that buttress size and strength and distribution of buttresses as they occur around the tree do correlate with likely directions of wind throw or flood. In other words, the largest buttresses (and more buttresses) occur on the leeward side of the tree trunk. Buttressing clearly facilitates the pattern of surface and subsurface root spreading evident in many rain forest trees. Some trees lack buttresses but have stilt or prop roots that radiate from the tree's base, remaining above ground (Plates 3-14 and 3-15). Stilt roots are particularly common in areas such as flood plains and mangrove forests that become periodically inundated with water.

Buttresses may reduce the likelihood of a tree falling, but trees nonetheless do fall during disturbance events, even trees that are well buttressed.

PLATE 3-13
This is an example of a large buttress.

PLATE 3-14
Prop roots in an Ecuadorian forest.

PLATE 3-15
Prop roots in an Ecuadorian forest.

As mentioned earlier, surface roots are common among rain forest trees. For example, the huge Brazil nut tree (*Bertholletia excelsa*) exhibits widely spreading but shallow surface roots without buttresses. Root distribution in free-standing figs has been traced to extend over 30 meters (100 feet) (Dalling, personal communication). This pattern is a likely adaptation for rapid nutrient retrieval, as leaves decompose quickly in the humid tropics, and to provide support for the tree. Most recycling of minerals occurs on or just below the surface of the soil. Surface root spreading is especially evident in forests of thin and nutrient-poor soils (see Chapter 10).

TRUNKS, BARK, AND CROWNS

Many rain forest trees have tall, relatively slender boles. Bark, depending on species, may be smooth or rough and light-colored or dark—almost white in some cases, almost ebony in others. Bark varies among species and is sometimes a useful characteristic in identification of species. Bark is often splotchy, with pale and dark patches. There is much variability, and thus no generalizations are possible (though once it was thought that tropical trees tended to exhibit light-colored bark). Tropical tree bark may be thin, but on some trees it is thick (and the wood inside may be very hard—remember that wood-eating termites abound in the tropics). Some trees such as the chicle tree (*Manilkara zapota*) of Central America (the original source of the latex base from which chewing gum was manufactured) have particularly distinct bark. Chicle bark is nearly black and vertically ridged into narrow strips, the inner bark red, with white resin. Though bark is generally not useful in identification, the color and taste of the underlying cambium layer is sometimes useful in identifying the tree species (Richards 1952; Gentry 1993).

Many canopy trees have a spreading, flattened crown (Richards 1952). Main branches radiate in a manner resembling the spokes of an umbrella. Each of these main radiating branches contributes to the overall symmetrical crown, an architectural pattern called *sympodial construction*. Of course, the effect of crowding and thus competition for light by neighboring trees can significantly modify crown shape. Single trees left standing after adjacent trees have been felled often have oddly shaped crowns, a clear result of earlier competition for light with neighboring trees. Many trees that grow both in the canopy and in the shaded understory have foliage that is monolayered, where a single dense blanket of leaves covers the tree. Understory trees are often lollipop-shaped and monolayered. Because they have not yet reached the canopy, their crowns are composed of lateral branches from a single main trunk, a growth pattern termed *monopodial*. Lower branches will eventually drop due to self-shading as the tree grows and becomes a sympodial canopy tree. Trees growing in forest gaps where sunlight is much more abundant may be monopodial when fast-growing or are multilayered, with many layers of leaves to intercept light (Horn 1971; Hartshorn 1980 and 1983) (Figure 3-5).

A study conducted in a Liberian rain forest examined 53 tree species in an attempt to analyze tree architecture as it was affected by attaining adult stature in the canopy and by shade tolerance (Poorter et al. 2003) (Figure 3-6). It was learned that light demand and adult stature both strongly affect tree architecture and vary independently of one another. Tree architecture changes as trees grow in size. Trees that grow in suppressed light have wider crowns, an adaptation, as the authors put it, "to forage for light and intercept light over a large area." Large tree species that attain full canopy height are characterized by narrow, slender crowns because they have a higher risk of being toppled by factors such as wind. The architecture of tropical trees is discussed further in Halle et al. (1978) and King (1990 and 1996).

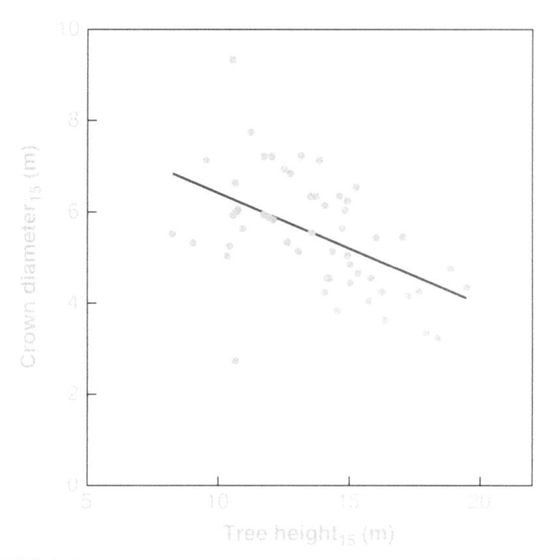

FIGURE 3-6
This graph illustrates the negative correlation between crown diameter and tree height for 53 rain forest tree species. As trees become taller, crowns widen less. Broad crowns would place trees in danger of windfall and other forms of mechanical failure, so there is fitness in having a smaller crown when a tree is tall.

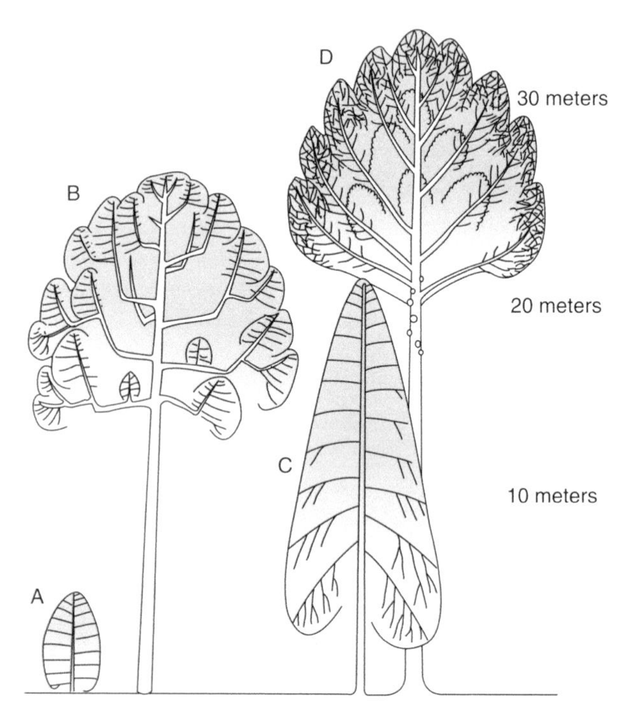

FIGURE 3-5
These two diagrams of dipterocarp species demonstrate that as young plants, they exhibit monopodial growth, but as mature trees, they are sympodial. See also Figure 3-4.

Many tropical trees exhibit a unique characteristic termed *cauliflory,* meaning that the flowers and subsequent fruits are produced directly from the wooded trunk, rather than from the canopy branches. Cauliflory is rare outside the tropics. Cacao (*Theobroma cacao*), from which chocolate is produced, is a cauliflorous understory tree (Plates 3-16 and 3-17). Some trees may be cauliflorous because the weight of their large, heavy fruits could not be supported on outer branches (though it is equally arguable that the opposite may be the case—the fruits may have grown large and heavy because they were growing from the trunk, not the outer branches). The presence of cauliflorous flowers may facilitate pollination by large animals such as bats, or cauliflorous fruiting may facilitate dispersal of seeds by large, terrestrial animals that

PLATE 3-16
Cauliflorous fruits of cacao.

PLATE 3-17
Ripened cauliflorous fruits of cacao.

could not reach canopy fruits. A similar phenomenon, *ramiflory*, is the bearing of flowers on older branches or, occasionally, underground.

LEAVES

Leaves of many tropical tree species are surprisingly similar in shape, making species identification difficult. The distinctive lobing patterns of many North American maples and oaks are missing from most tropical trees. Instead, the leaves of many species are characteristically oval, unlobed, and waxy, and leaves of many species possess sharply pointed ends, called *drip-tips*, that facilitate rapid runoff of rainwater (Plates 3-18a and 3-18b). Leaves of most species have smooth margins rather than "teeth," though serrated leaves are found in some species. Studies show that trees in temperate climates have a much higher frequency of serrated leaves than trees in the tropics

(a)

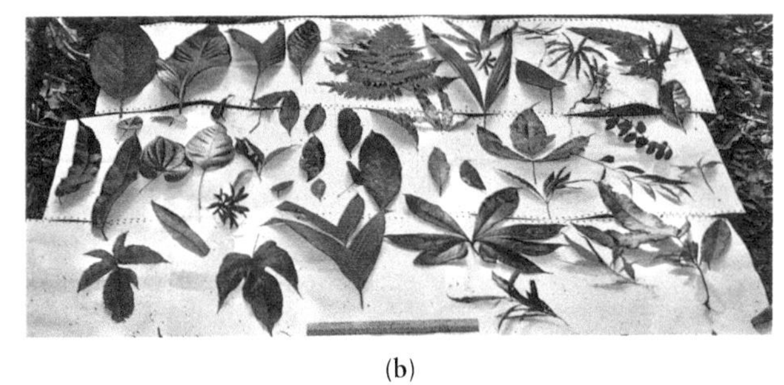

(b)

PLATE 3-18
(a) The pointed ends of these two leaves are called drip-tips because they facilitate the flow of water off the leaves. (b) This small collection of leaves from a Belizean tropical forest demonstrates the large size (note 12-inch ruler in foreground) and diversity of shapes of tropical leaves. Many of the leaves here have drip-tips.

(Wilf 1997; Weimann et al. 1998). Indeed, serration frequency on leaves has been applied in assessing past climate change (Wilf 1997; Royer and Wilf 2006). It is not known why serrated leaves typify many broadleaved tree species in temperate climates, but it has been suggested that it may enhance sap flow. It is also true that leaf length and area correlate positively with

PLATE 3-19
Leaves of the kapok tree, *Ceiba pentandra*, show the waxy appearance typical of many species.

moisture, which means that tropical leaves are generally larger and longer than those in the temperate zone (Weimann et al. 1998).

Both lowland and montane tropical forest trees produce heavy, thick, leathery, and waxy leaves that may remain on the tree for well over a year (Plates 3-19 and 3-20). Many tropical species produce *palmate leaves*, where the leaflets radiate like spokes from a center, forming a shape similar to that of a parasol. Some leaves, particularly those on plants that are found in disturbed areas such as gaps, are exceptionally large, well in excess of any temperate zone species. Though many trees have simple leaves, compound leaves are also common, particularly due to the abundance of legumes, a highly species-rich plant family. Tropical leaves often show obvious insect damage, but given the long growing season in the tropics, trees are sometimes able to repair damaged leaf tissue. Defoliation of some Neotropical trees sometimes occurs when leaf-cutter ants select a specific tree and clip most of its leaves (see Chapter 7).

Evergreen tropical tree species drop leaves and replace them throughout the year but are never without leaf cover. As mentioned earlier, some tropical tree species are deciduous, periodically shedding all leaves, and some are partially deciduous, with greater leaf-drop during the peak of dry season. Fully deciduous trees may drop leaves in correlation with flowering times or at times of water stress. In some cases, particularly in tropical dry forests, deciduousness may act as an adaptation to periodically reduce insect herbivory. This pattern is termed *leaf flushing*. Synchronous

PLATE 3-20
Leaves of *Castilla* demonstrate the simple, nonlobed, nonserrated pattern typical of many species of tropical tree leaves.

PLATE 3-21
This photo shows how large some tropical leaves become (notice the leaf lying atop the leaf) and also shows insect damage typical of tropical leaves in general.

production of new leaves is considered an adaptation to overwhelm the collective effects of herbivores in defoliating trees. For example, a study in a tropical dry forest in India showed that the peak of leaf flushing occurred just prior to the onset of the wet season. Herbivorous insects emerged with the onset of the rains. Trees that dropped and then grew new leaves later in the rainy season suffered considerably more insect damage (Murali and Sukumar 1993) (Plate 3-21).

FLOWERS

Many tropical trees have colorful, fragrant blossoms, often large in size. This pattern reflects the general characteristic of animal pollination (primarily by insects but also by birds and bats). Typical examples include such species as the coral tree (*Erythrina* spp.), pink poui (*Tabebuia pentaphylla*), cannonball tree (*Couroupita guanianensis*), frangipani (*Plumeria* spp.), and morning glory tree (*Ipomoena arborescens*). The gorgeous and widespread royal poinciana, or flamboyant tree (*Delonix regia*), the national tree of Puerto Rico, is native to Madagascar. The bottlebrush tree (*Callistemom lanceolatus*) from Australia is one of many Australian species with very colorful florescences.

Red, orange, and yellow are associated with bird-pollinated plants such as *Heliconia* (Plate 3-23), while flowers of other plants with white, purple, yellow, orange, or green are typically insect-pollinated. Many kinds of insects act as pollinators: bees, lepidopterans (butterflies and moths), beetles, various flies, and even thrips. Some trees, such as silk-cotton or kapok (*Ceiba pentandra*), flower mostly at night, producing conspicuous white flowers that, depending on species, attract bats or moths. Flowers exhibiting a strong fragrance are typically pollinated mostly by moths, bees, beetles, or other insects. Bat-pollinated flowers smell musty, somewhat like the bats themselves, an adaptation that may serve to attract the bats to the flower (see Chapter 7). In cases where pollination is carried out by large animals such as birds, bats, and large lepidopterans, flowers tend not only to be large but also to be nectar-rich and borne on long stalks or branches away from leaves, or else on the trunk (cauliflory, as described earlier). Many flowers are tubular or brush-like in shape, though some, particularly those pollinated by small insects, are shaped as flattened bowls or plates. Although most tropical plants are pollinated by animals, wind pollination occurs in some species of canopy trees.

Rafflesia, a Unique and Smelly Flower

Named for Sir Stamford Raffles, once the lieutenant-governor of Java (1811–1815), *Rafflesia* is unique among flowering plants (Plate 3-22). Found exclusively in Southeast Asia, there are some 17 species, scattered through the Malay Peninsula and islands such as Sumatra, Borneo, and the Philippines. All *Rafflesia* lack roots, leaves, and stems, but the plant has an immense flower, indeed the world's largest individual inflorescence. Each of the five petals can measure 30 centimeters (12 inches) in length, and the entire flower can measure 100 centimeters (39 inches) in diameter and weigh up to 10 kilograms (22 pounds). *Rafflesia* is a parasitic plant; its host species is a vine (in the grape family Vitaceae) named *Tetrastigma*. Fungus-like tendrils (called *haustoria*) from *Rafflesia* grow within *Tetrastigma*, allowing *Rafflesia* to obtain nutrition and water from the host plant. Only the flower of *Rafflesia* is ever visible externally, a fact that makes the plant particularly difficult to study, since without the flower, it is extremely difficult to know whether *Rafflesia* is present within a *Tetrastigma*. At infrequent intervals, the immense reddish flower appears, occasionally within the canopy, but more usually on the forest floor. The flower emits a rank, carrion-like odor, often quite strong, and lasts for only 5 to 7 days. Swarms of insects including flies, bees, beetles, and ants are attracted, but it is not clear which, if any, act as pollinators. Some must, as male and female sex organs are on different flowers, so any isolated flower must somehow "find" a flower of the opposite sex, an act that is facilitated by some sort of animal pollinator. Seed dispersal is also far from well understood. It has been suggested that large animals such as deer and elephants may trample the decaying flower, picking up seeds on their feet, and subsequently step on and break a *Tetrastigma* vine, in the process implanting the seeds of *Rafflesia* into the injured vine, where they begin growth. The very existence of this unique plant seems to defy the odds, and *Rafflesia* is becoming scarcer due to forest clearance.

PLATE 3-22
RAFFLESIA INFLORESCENCE

PLATE 3-23
Heliconias are among the most characteristic plants of Neotropical forests. The bright red structures are bracts, not flowers. The flowers are contained within the bracts and may be seen protruding from them. The red bracts attract various hummingbird species that aid in pollinating the plant.

FRUITS AND SEEDS

Animal-facilitated seed dispersal prevails throughout the global tropics. Numerous animal species consume fruits and seeds, and many are critical in dispersing seeds. Many tropical trees produce small to medium-sized fruits, but some produce large, conspicuous fruits, and the seeds contained within are large as well. The size range of tropical fruits and seeds is noteworthy, likely reflecting an evolutionary association with a wide size range of seed-dispersing and seed-consuming animals. There are many tree species that make large fruits.

The Spiky, Smelly Durians

If you have ever wondered what the great red ape, the orangutan, has in common with the great evolutionary biologist Alfred Russel Wallace, the answer is that they both enjoy (in Wallace's case, "enjoyed") the fruit of the durian (Plate 3-24). Found in lowland rain forests in Sumatra, Borneo, and Malaysia, there are about 19 species of durian, all in the genus *Durio*. The one most commonly cultivated by humans is *D. zibethinus*. Durians have distinctive large (and sharp) spiky fruits, weighing between 2 and 4 kilograms (4.5 to 9 pounds). The flowers are white and grow from the thickest branches as well as the tree trunk, an example of cauliflory. The large flowers emit the odor of sour milk, an apparent attractant to bats, particularly *Eonycteris spelaea*, a long-tongued fruit bat. Bats are thus the principal and perhaps only pollinators of wild durian. The fruit matures on the tree, eventually splitting open, but not before emitting a distinctive odor that attracts numerous animals, potential seed dispersers. Durian fruits must be eaten immediately on ripening, as the seeds within will not maintain viability otherwise. Thus the odor that attracts consumers is an adaptation for seed dispersal. It is the distinctive odor that makes durian remarkable. Alfred Russel Wallace is said to have described the taste of durian as akin to custard passed through a sewer. The rank odor emitted by the fruit as it ripens is sufficiently powerful to discourage many people from even tasting the final product. For those who have the guts, so to speak, it is said to be remarkably tasty.

PLATE 3-24
DURIAN FRUIT

PLATE 3-25
Fruits from the Lecythidaceae (Brazil nut family) are typically very large.

Many palms, the coconut (*Cocos nucifera*), for example, produce large, hard fruits in which the seeds are encased. The monkey pot tree (*Lecythis costaricensis*) produces thick, 8-inch-diameter "cannon ball" fruits, each containing up to 50 elongate 2-inch seeds. The seeds are reported to contain toxic quantities of the element selenium (Kerdal-Vargas, in Hartshorn 1983), perhaps serving to protect the tree from seed predators.

The milk tree (*Brosimum utile*) forms succulent, sweet-tasting edible fruits, each with a single large seed inside. This tree is named for its white sap (which is drinkable). Because of its usefulness, Mayan Indians planted it throughout the Yucatan Peninsula to use for food (Flannery 1982), and it remains relatively abundant there today. The famous Brazil nut comes from the forest giant, *Bertholletia excelsa* (Plate 3-25). The nuts are contained in large, woody, rounded pods that break open on dropping to the forest floor. Many tree species in the huge legume family package seeds in long, flattened pods, and the seeds tend to contain toxic amino acids. Among the legumes, the stinking toe tree (*Hymenaea courbaril*) produces 5-inch oval pods with five large seeds inside. The pods drop whole to the forest floor and often fall prey to agoutis and other forest mammals as well as various weevils.

The prevalence of large fruits with large seeds is indicative of extensive consumption by large animals. Among Neotropical mammals, for example, monkeys, bats, various rodents, peccaries, and tapirs are common consumers of fruits and seeds, sometimes dispersing the seeds, sometimes destroying them. Agoutis, which are rodents, skillfully use their sharp incisors to gnaw away the tough, protective seed-coat on the Brazil nut, thus enabling the animal to eat the seed contained within. Some extinct mammals such as the giant ground sloths and elephant-like gomphotheres (Figure 3-7) may have been important in dispersing large seeds of various tropical plants in South American tropical forests (Janzen and Martin 1982). When more than 15 genera of large herbivorous mammals became extinct about 10,000 years ago, trees dependent on them for seed dispersal no longer had effective seed dispersers. When horses and cattle were reintroduced to the Americas by the Spanish, these mammals may have replaced the extinct megafauna at least to some degree as seed dispersers for some tree species.

Many bird species throughout the world's tropics are also attracted to large fruits and the seeds within them. In Australia and New Guinea, large flightless

FIGURE 3-7
GOMPHOTHERE

Breadfruit: The Tree That Was Part of a Mutiny

The breadfruit tree (*Artocarpus altilis*) is native to the tropical Pacific, though it is very widely propagated and is a common sight throughout the American tropics. Why should breadfruits abound in the Caribbean and much of Central America? Striking in appearance, the breadfruit tree has large, deeply lobed, dark green leaves (Plate 3-26). The green, knobby-skinned fruit, which gives the tree its name, is quite large, 15 to 20 centimeters (6 to 8 inches) in diameter, and weighs about 2 to 3 kilograms (about 6 pounds). Very rich in carbohydrates, the fruit can be fried, boiled, or baked. Europeans first learned about the breadfruit tree when it was discovered in Polynesia by Captain James Cook. The botanist Joseph Banks, who accompanied Cook, judged it a suitable source of starch for British slaves in the Caribbean. Captain William Bligh was ordered to transport healthy breadfruits from Tahiti to Jamaica, but the well-documented mutiny on the HMS *Bounty* prevented the initial success of Bligh's mission. Undaunted, Bligh returned on a second mission and succeeded, bringing 1,200 trees to Port Royal aboard the *Providence* in 1793 (Oster and Oster 1985). Initially, the fruits served merely as a food source for pigs, and it was only after the emancipation of the slaves in 1838 that people found breadfruits to be respectable as human food.

PLATE 3-26
BREADFRUIT

birds named *cassowaries* disperse various large-seeded plant species. Along flooded forests, some fish species are important fruit consumers and seed dispersers. Small fruits and seeds are also consumed by animals. Many insect species, for example, are frequent predators of small seeds.

Some trees have wind-dispersed seeds, and thus the fruits are usually not consumed by animals. The huge silk-cotton or kapok tree is so named because its seeds are dispersed by parachute-like, silky fibers that give the tree one of its common names. Other plants, such as the liana *Pithecoctenium crucigerum*, have seeds with attached wings, facilitating dispersal (Plate 3-27). Mahogany trees *(Swietenia macrophylla* and *S. humilis),* famous for their superb quality wood, develop 6-inch oval, woody fruits, each containing about 40 seeds. The seeds are wind-dispersed and would be vulnerable to predation were it not for the fact that they have an extremely pungent, irritating taste.

PLATE 3-27
Seeds of *Pithecoctenium crucigerum* (Bignoniaceae), a liana, are encased within wind-dispersed, wing-like structures. From Costa Rica.

PALMS

Palms occur worldwide and are among the most distinctive tropical plants, frequenting interior rain forest, disturbed areas, and grassy savannas. They are particularly abundant as components of swamp and riverine forest (Plate 3-28). All palms are members of the family *Palmae*, and all are *monocots*, sharing characteristics of such plants as grasses, arums, lilies, and orchids. The most obvious monocot features of palms are that their growth form is a single stem and their leaves have parallel veins. (The leaves are referred to as *palm fronds*.) There are approximately 1,500 palm species in the world (Henderson et al. 1995).

PLATE 3-28
Moriche palm (*Moriche flexuosa*) is common along riverine areas throughout the Neotropics.

Alfred Russel Wallace made a detailed study of South American palms and published an important book on the subject, titled *Palm Trees of the Amazon and Their Uses*. Palms are widely used by indigenous peoples for a diversity of purposes: thatch for houses, wood to support dwellings, various ropes and strings, weavings, hunting bows, fishing line, hooks, utensils, musical instruments, and various kinds of food and drink. Indeed, many palm species have multiple uses and are thus among the most important species for wildlife and for humans.

Palms are often abundant in the forest understory, and many species are armed with sharp spines along the trunks and leaves, an adaptation that may act to discourage mammalian herbivores that would climb the stem to devour leaf tissue. Be careful not to grab a palm sapling, because the spines can introduce bacteria as they create a wound.

IDENTIFYING TROPICAL PLANT SPECIES

Palms are fairly easy to identify (see Henderson et al. 1995), at least to the genus level, but what about all those other trees and shrubs in the rain forest? For the majority of students of tropical biology, it will not be possible to accurately identify most plants (including palms) to the level of species. There are too many look-alike species, the ranges of many species are not precisely known, and thus species identification must be left to taxonomic experts. Also, there are few field guides to tropical plants, at least not at the level of species, though that is changing, especially at prominent field stations in various places throughout the tropics. Lotschert and Beese (1981) have provided a general guide to many of the most widespread and conspicuous tropical plants that is useful as an introduction. For Neotropical plants, the best guides currently available are Gentry (1993) and Smith et al. (2004). Neither of these large volumes is easily transported in the field.

It is possible to identify many, if not most, tropical plants to the level of family using combinations of characteristics such as overall leaf shape, compound versus simple leaves, opposite versus alternate leaves, presence or absence of tendrils, presence or absence of spines, smooth or serrated leaf edges, fruit and/or flower characteristics, and in many cases, latex presence and its color, odor, and taste.

Climbers, Lianas, Stranglers, and Epiphytes

While trees dominate tropical moist forests, there are numerous other plant growth forms that add a great deal to the biodiversity and structural complexity of the ecosystem. Vines abound in many tropical forests, as do air plants, or *epiphytes*.

VINES

Vines are a conspicuous and important component of most tropical rain forests, though vine density is often variable from site to site. Of course, vines of many species also occur in forests throughout much of the temperate zone, but that said, their prodigious representation in tropical forests is noteworthy. Vines occur in many plant families and exhibit a variety of growth forms. Because of their abundance, vines form a distinct and important structural feature of rain forests, in a sense literally tying the forest together (Gentry 1991). Vines exhibit high biomass in some rain forests; they compete with trees for light, water, and nutrients; and many are essential foods for various animals (Gentry 1991). Woody vines, called *lianas*, entwine elaborately as they hang from tree crowns. Others, the *bole climbers*, attach tightly to the tree trunk and ascend. Tropical vines occur abundantly in disturbed sunlit areas as well as in forest interiors and occur at varying densities on virtually all soil types (Plate 3-29). Humans make extensive use of vines for foods, medicines, hallucinogens, poisons, and construction materials (Phillips 1991). For a comprehensive account of vine biology, see Putz and Mooney (1991).

LIANAS

A *liana* is a woody vine that typically gets its start when a forest gap is created, permitting light penetration. Lianas begin as shrubs rooted in the ground but eventually become vines, with woody stems. (It is important to realize that lianas are not restricted to the tropics. There are many woody vines in the temperate zone, such as the familiar poison ivy *Toxicodendron radicans*.) Tendrils from the branches entwine neighboring trees, climbing upward, reaching the tree crown as both tree and liana grow. Lianas spread in the crown, and a single liana may eventually loop through several tree crowns. Lianas seem to

PLATE 3-29
Extensive vine growth draping throughout vegetation in an Ecuadorian rain forest.

Rattan Palms

Rattans are found exclusively in the Old World tropics of parts of Africa, Southern Asia, and Australasia (Plate 3-30). Most of the approximately 600 species are found in Indonesia, and all rattans belong to the group Calameae. Rattans are most well-known for their flexible, slender stems, used to make finely crafted furniture, canes, and other implements.

Their stems also make rattans distinctive within the palms. The growth form of most species is distinctly vine-like, instead of the more characteristic tree shape of typical palms. Rattans, some of which may grow in excess of 100 meters (330 feet), wind their way among the branches of trees and shrubs. Their ability as climbers is enhanced by a structural adaptation, the presence of sharp spines that serve as hooks. Spines also presumably afford some protection against large herbivores. Hiking through a forest abounding with rattans is daunting, as the thorny winding stems can do significant damage to arms and legs.

PLATE 3-30
Typical rattan, showing vine-like growth pattern.

The commercial value of rattans is of importance in conservation of rain forests throughout the region where rattans grow. Forest clearance, logging, and other activities reduce rattan abundance and thus income from harvesting rattans.

PLATE 3-31
This helix-shaped liana stem is typical of many types of *lianas*, woody vines that are abundant in many tropical forests.

drape limply, winding through tree crowns or hanging as loose ropes parallel to the main bole. Their stems remain rooted in the ground, often at multiple points, and are oddly shaped, often flattened, lobed, coiled like a rope, or spiraling in a helix (Plate 3-31). The thinnest have remarkable springiness and will often support a person's weight, at least for a short time. Lianas have exceptionally long vessels within their stems, and when a section of a liana is severed, water runs out.

Liana is a growth form, not a family of plants, and thus they are represented among many different plant families (for example, Leguminosae, Sapindaceae, Cucurbitaceae, Vitaceae, Smilacaceae, and Polygonaceae, to name but several). Lianas, like tropical trees, can be difficult to identify, but some lianas can be identified to the level of genera by noting the distinctive cross-sectional shapes of their stems (Gentry 1993).

In Panama, a single hectare (10,000 square meters, or about 2.5 acres) hosted 1,597 climbing lianas, distributed among 43% of the canopy trees (Putz 1984). In the understory, 22% of the upright plants were lianas, and lianas were particularly common in forest gaps. A heavy liana burden reduced the survival rate of trees, making them more likely to be toppled by winds. Fallen lianas merely grew back into other trees. When tree-falls bring lianas to the ground, they may be sufficiently dense to significantly reduce the speed with which trees reattain canopy status (Schnitzer et al. 2000). This may, for example, slow the rate at which a logged forest regenerates. For a general summary of liana ecology, see Schnitzer and Bongers (2002).

HEMIEPIPHYTES

Hemiepiphytes are among the most important and unusual of plant growth forms in the tropics. Primary hemiepiphytes begin their life cycles as epiphytes but eventually become rooted in the ground. Perhaps the best-known group of primary hemiepiphytes is the stranglers (*Ficus* spp.). There are approximately 750 species of *Ficus* (figs) throughout the global tropics. Stranglers germinate from seeds dropped by a bird or mammal in the tree crown. Tendrils grow toward the tree bole and grow downward around the bole, *anastomosing*, or fusing together, like a crude mesh. The strangler eventually reaches the ground and sends out its own root system. The host tree often dies and decomposes, leaving the strangler standing alone. The mortality of the host tree results from being girdled by the strangler, because the strangler prevents the bole from expanding. Stranglers also may shade the tree on which they grow, reducing its ability to photosynthesize. It is a common sight in tropical forests to see a mature strangler, its host tree having died and decomposed, and the strangler's trunk a dense fusion of what were once separate vines, now making a single, strong woody labyrinth that successfully supports a wide canopy, itself now laden with vines (Plates 3-32, 3-33, and 3-34; Figure 3-8).

PLATE 3-32
This *Ficus*, a strangler, is typical of how hemiepiphytes grow to eventually reach the ground.

PLATE 3-34
Eventually a strangler will completely obliterate its host tree.

PLATE 3-33
Named the *curtain fig*, this impressive *Ficus* is well known and often visited in Queensland, Australia.

FIGURE 3-8
Diagram of how a strangler fig surrounds and eventually replaces its host tree.

Secondary hemiepiphytes begin as rooted plants and eventually become climbers. The well-known ornamental arum *Monstera deliciosa* is a philodendron that begins life on the ground. Seeds germinate and send out a tendril toward shade cast by a nearby tree. The tendril soon grows up the tree trunk, attaching by aerial roots, and the vine thus moves from the forest floor to become anchored on a tree. There it continues to grow ever upward, often encircling the bole as it proliferates. In humid tropical forests, it is common to see boles partially enshrouded by the wide, thick leaves of climbers. As it grows, the plant ceases to be rooted in the ground, and its entire root system is invested on the tree bark (Plates 3-35 and 3-36).

EPIPHYTES

Epiphytes are commonly called *air plants.* As the prefix *epi* implies, these plants live *on* other plants. They are not internally parasitic, but they do claim space on a branch where they set out roots, trap soil and dust particles, and photosynthesize as canopy residents. Rain forests, both in the temperate zone (such as the Olympic rain forests of Washington and Oregon) and throughout the tropics, abound with epiphytes of many different kinds. Cloud forests also host an abundance of air plants. In a lowland tropical rain forest, nearly one-quarter of the plant species are likely to be epiphytes (Richards 1952; Klinge et al. 1975), though the representation of epiphytes varies substantially among forests. As forests become drier, epiphytes decline in both abundance and diversity.

Many different kinds of plants grow epiphytically. In Central and South America alone, there are estimated to be 15,500 epiphyte species (Perry 1984). Looking at a single tropical tree can reveal an

PLATE 3-35
Climbers often adhere tightly to the bark of the host tree.

PLATE 3-36
Large leaves of the philodendron *Monstera* grow on the trunk of this Guatemalan tree.

amazing diversity. Lichens, liverworts, and mosses, many of them tiny, grow on trunks and branches and often on leaves. Cacti (in the Neotropics), ferns, and colorful orchids line branches. In the Neotropics, bromeliads, with distinctive sharply pointed dagger-like leaves, are abundant and conspicuous on both trunk and branch alike.

Epiphytes attach firmly to a branch and survive by trapping soil particles blown to the canopy and by using the captured soil as a source of nutrients such as phosphorus, calcium, and potassium. As epiphytes develop root systems, they accumulate organic matter, and thus a soil–organic litter base, termed an *epiphyte mat*, builds up on the tree branch, adding weight to the branch. Like most terrestrial plants, some epiphytes have root systems containing fungi called *mycorrhizae*. These fungi greatly aid in the uptake of scarce minerals. Mycorrhizae are also of major importance to many trees, especially in areas with poor soil (see Chapter 10). Epiphytes efficiently take up water and thrive in areas of dense cloud cover and heavy mist.

Though epiphytes are not parasitic in the strictest sense, they may indirectly harm their host trees through competition for light, water, and minerals. Epiphytes get "first crack" at the water dripping down through the canopy. However, some temperate and tropical canopy trees develop aerial roots that grow into the soil mat accumulated by the epiphytes, tapping into that source of nutrients and water. Because of the epiphyte presence, the host tree benefits by obtaining nutrients from its own canopy (Nadkarni 1981). Perry (1978) suggests that monkeys (and, in the Old World tropics, apes) traveling regular routes through the canopy may aid in keeping branches from being overburdened by epiphytes. The accumulating weight of epiphytes may become sufficient to cause limbs to break off, damaging the tree (Plate 3-37).

Bromeliads (Bromeliaceae) are abundant in many Neotropical forests (Plate 3-38). The most well-known bromeliad is the pineapple. Approximately 3,000 bromeliad species are known among 56 genera. Leaves of many species are arranged in an overlapping rosette to form a *cistern*, or tank, that holds water and detrital material. Some species have a dense covering of hairlike trichomes on the leaves that help rapidly absorb water and minerals. Though most bromeliads are epiphytic, there are many areas where terrestrial bromeliads make up a significant portion of the ground vegetation. Epiphytic bromeliads provide a source of moisture for many canopy dwellers. Tree frogs, mosquitoes, flatworms, snails, salamanders, and even crabs complete their life cycles in the tiny aquatic habitats provided by the cuplike interiors of bromeliads (Zahl 1975; Wilson 1991). One classic study found 250 animal species occurring in bromeliads (Picado 1913, cited in Utley and Burt-Utley 1983). Some species of small colorful birds called *euphonias* use bromeliads as nest sites. Bromeliad flowers grow on a central spike and are usually bright red, attracting many kinds of hummingbirds.

Orchids are a global family (Orchidaceae) abundantly represented among Neotropical epiphytes (Dressler 1981). There are estimated to be approximately 25,000 to 35,000 species worldwide—indeed a huge plant family. In Costa Rica, approximately 88% of the orchid species are epiphytes, while the rest are terrestrial (Walterm 1983). Many orchids

PLATE 3-37
This deciduous tree in Belize demonstrates how densely epiphytic growth covers the major branches.

grow as vines, and many have bulbous stems (called *pseudobulbs*) that store water. Indeed, the origin of the name *orchid* comes from the Greek word meaning "testicle," a reference to the appearance of the bulbs (Plotkin 1993). Some have succulent leaves filled with spongy tissue and covered by a waxy cuticle to reduce evaporative water loss.

Like so many other plants, all orchids depend on mycorrhizae fungi during some phase of their life cycles. These fungi grow partly within the orchid roots and facilitate uptake of water and minerals. The fungi survive by ingesting some of the orchid's sugary products of photosynthesis. Thus the association between orchid and fungus is mutualistic—both

PLATE 3-38
The sharply spiked leaves of a tank bromeliad adorn this Ecuadorian tree.

Damselflies Fertilize Bromeliads

Tank bromeliads harbor ecosystems unto themselves. These highly localized assemblages of animals, all contained within the confines of an epiphytic bromeliad, interact and add to the diversity and function of the forest. In tank bromeliads, the element nitrogen primarily limits fertility (whereas phosphorus tends to be the most important limiting element in a rain forest in general; see Chapter 10). In a series of experiments, it was learned that damselflies (suborder Zygoptera of the order Odonata) facilitate nitrogen uptake by bromeliads (Ngai and Srivastava 2006). But how?

Researchers traced the movement of nitrogen by using the isotope ^{15}N. They manipulated bromeliads such that some contained only *detritivores*, organisms such as various insects that consume the organic matter that falls within the bromeliad. Detritivores are considered essential in the complex process of recycling basic elements back to plants. Other bromeliads contained not only detritivores but also damselflies, insects that prey on detritivores. The result was that more nitrogen entered bromeliads that had both damselflies and detritivore insects than those that contained only detritivores. This was a surprising result because it was assumed that without damselflies to reduce their numbers, detritivores would be even more effective in releasing nitrogen that could be taken up by the bromeliad.

The answer lies in the ratio of nitrogen to phosphorus (N:P) found in detritivore insects. It is higher than that found in the litter on which they feed. In other words, detritivores are preferentially taking up nitrogen (thus increasing N:P), which, in effect, makes less nitrogen available to the plants (that are already nitrogen-limited). When damselflies devour detritivores, they defecate fecal pellets that are rich in nitrogen, little pellets of fertilizer. The microbes within the bromeliad decompose the fecal pellets and release the nitrogen that is then taken up by the bromeliad. Therefore the damselflies indirectly enhance the nutrient fertilization of bromeliads by preying on detritivores, concentrating nitrogen, and releasing it in a manner in which it is promptly recycled by microbes.

The Tasty Pineapple

Pineapple (*Ananas comosus*) is a terrestrial bromeliad originally from Brazil and Paraguay. It was widely cultivated by indigenous peoples before Columbus and the Spanish arrived in the New World, and thus it spread north into Central America (Plate 3-39). After its discovery by Europeans, the pineapple was soon cultivated in various other parts of the tropical world. The spiky, sharply spined leaves protect the plant, whose single flower cluster grows in the center of the leaf rosette. Wild pineapples have flowers ranging in color from purple to red and are normally pollinated by various hummingbird species. Domestic pineapples must be artificially propagated, though some pollination by insects can occur. Most farming families throughout the tropics have a few pineapples as part of their "dessert" crops to enjoy as a sweet food. In addition, pineapples are now grown commercially in numerous tropical countries throughout the world and have become one of the leading export crops of the tropics.

PLATE 3-39
Pineapple crop in Costa Rica.

PLATE 3-40
The orchid flower of *Prostechea fragans*. The photo shows the lip petal "landing pad" used by pollinating insects.

benefit (see Chapter 7). A close look at some orchids will reveal two types of roots: those growing on the substrate, and those that form a "basket," up and away from the plant. Basket roots aid in trapping leaf litter and other organic material that, when decomposed, can be used as a mineral source by the plant (Walterm 1983). Orchid flowers are among the most beautiful in the plant world. Some, like the familiar *Cattleya,* are large, while others are delicate and tiny. Binoculars help the would-be orchid observer in the rain forest. Cross-pollination is accomplished by insects, some quite specific for certain orchid species (Plate 3-40). Bees are primary pollinators of Neotropical orchids. These include long-distance fliers, like the euglossine bees that cross-pollinate orchids separated by substantial distances (Dressler 1968). Some orchid blossoms apparently mimic insects, facilitating visitation by insects intending (mistakenly) to copulate with the blossom (Darwin 1862). Aside from their value as ornamentals, one orchid genus is of particular importance to humans: there are 90 orchid species in the genus *Vanilla*, of which two are of economic importance, their use dating back to the Aztecs (Plotkin 1993).

In many tropical moist forests, even leaves have epiphytes. Tropical leaves often are colonized by tiny lichens, mosses, and liverworts, which grow only after the leaf has been colonized by a diverse community of microbes: bacteria, fungi, algae, and various yeasts, as well as microbial animals such as slime molds, amoebas, and ciliates. This tiny community that lives on leaves is termed the *epiphyll community* (Jacobs 1988), and its existence adds yet another dimension to the vast species richness of tropical moist forests (Plate 3-41).

PLATE 3-41
Epiphylls of various kinds adorn this large leaf in the forest understory.

The Understory and Forest Gaps: An Introduction

Much of the understory of a tropical forest is so deprived of light that plant growth is limited. Low light intensity is a chronic feature of rain forest interior and is an important potential factor for plant growth and competition. This is the reason why it is fairly easy to traverse your way through a closed canopy rain forest. Many of the seedlings and shoots that surround you may or may not eventually attain full canopy status. What is interesting and far from obvious is that a small, unpretentious sapling could be several decades old.

Certain families of shrubs frequently dominate rain forest understory. In the Neotropics, for example, these include members of the family Melasomataceae (e.g., *Miconia*), the Rubiaceae (e.g., *Psychotria*), Piperaceae (e.g., *Piper*), and palms (e.g., *Geonoma*). In addition, there are often forest interior Heliconias and terrestrial bromeliads. Many ferns and fern allies, including the ancient genus *Selaginella*, can carpet much of the forest herb layer.

Note that the understory is frequently not uniform. Deep shade is interrupted by areas with conspicuously greater light intensity, allowing for more dense plant growth. The observer inevitably notices the presence of many forest gaps of varying sizes, openings created by fallen trees or parts thereof. Gaps permit greater amounts of light to reach the forest interior, providing enhanced growing conditions for many species. Though understory plants and juvenile trees are adapted to grow slowly (Bawa and McDade 1994), many are also adapted to respond with quickened growth in the presence of a newly created gap. Research at La Selva in Costa Rica has revealed a surprisingly high disturbance frequency caused by treefalls and branch-falls, so that the average square meter of forest floor lies within a gap every hundred years or so (Bawa and McDade 1994). As described by Deborah Clark (1994), "The primary forest at La Selva is a scene of constant change. Trees and large branches are falling to the ground, opening up new gaps and smashing smaller plants in the process. Smaller branches, bromeliads, and other epiphytes, 6-m-long palm fronds, smaller leaves, and fruits fall constantly as well. The lifetime risk of suffering physical damage is, therefore, high for plants at La Selva." Gap dynamics has become an important consideration in the study of plant demographics in the rain forest, and gaps will be discussed in detail in Chapter 6.

This chapter has described the fundamental characteristics of rain forest structure and introduced some of the groups, particularly among the plants, that dominate both in species richness and biomass. In the next two chapters, we will take a close look at just how much biodiversity there is in rain forests and why that may be so.

4 Inside Tropical Rain Forests: Biodiversity

Chapter Overview

Biodiversity is measured in several ways. The simplest is *species richness*, the number of species in an ecosystem. For example, how many tree species or bird species or beetle species are there per hectare? No tropical forest has as yet been surveyed for all species encompassing all taxonomic groups. That would be a truly staggering task. Biodiversity is also measured as *species diversity*, a measure (there are several in common usage—see Magurran 2003) that considers both species richness and relative abundance of each species. Last, it can be measured as *interaction diversity*, referring to the complexity of interactions among tropic levels. Regardless, tropical regions, and rain forests in particular, exhibit the highest biodiversities of any terrestrial ecosystems. Biodiversity varies with latitude. Species richness (for most taxa) declines with increasing latitudes north or south of the equator. This pervasive pattern is called the *latitudinal diversity gradient* (LDG), and it demands explanation. This chapter looks at species richness in tropical regions, examines the LDG and some of the hypotheses for why biodiversity is uniquely high in tropical rain forests, and looks at some representative patterns of diversity.

Species Richness, Sampling Effort, and Area

The term *species richness* refers to how many different species (of a particular group of organisms) inhabit a specified area—thus we speak of such things as the species richness of flowering plants in Sri Lanka, or bromeliads in Costa Rican montane forests, or birds in Tanzania, or mammals in Amazonia, or beetles in the canopy of a single kapok tree in Ecuador. High species richness among many different taxa is one of the most distinctive features of tropical moist forests worldwide. It also characterizes other tropical ecosystems as well, though usually to a lesser degree.

There are two simple but fundamental relationships that are useful in introducing how species richness is measured and how it varies from one ecosystem to another. The first is the relationship between species richness and sampling effort. Suppose that you were to survey tree species richness in a forest, or perhaps lepidopteran (butterfly and moth) species richness. In either case (trees or butterflies and moths), you would systematically expend effort in the field to tally each species. On day one, you might identify 8 species, and on day two you might identify 9 species, 6 of which you did not find the previous day. Thus in two days of effort, you identified a total of 14 species. Day three yielded 6 species, of which 2 are

TABLE 4-1 Cumulative total of new species encountered over a 10-day period. The community has 20 species, but sampling counts found an average of only 8.1 per day.

	#SPECIES	#NEW SPECIES
Day 1	8	8
Day 2	9	6
Day 3	6	2
Day 4	9	2
Day 5	7	1
Day 6	8	0
Day 7	10	1
Day 8	7	0
Day 9	6	0
Day 10	11	0
x̄ # Species per day	8.1	20 total species in the community

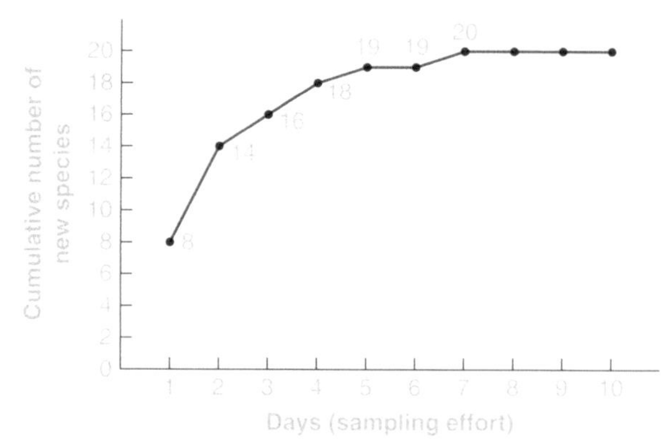

FIGURE 4-1
This graph depicts the data in Table 4-1. Note that the asymptote suggests that all species present have been found.

new to your list, now with a cumulative total species richness of 16. If you continue sampling and then plot the cumulative species richness on the y axis against the number of days sampled (sampling effort) on the x axis, you will eventually reach an asymptote, such that additional days of sampling yield no more new species (Table 4-1; Figure 4-1). At that point, you may quit, as you have found all species possible to find in that forest. Now if you were to do this sort of sampling for a temperate forest and compare that with a tropical rain forest, you would soon learn that it requires far more sampling effort to document species richness in the tropics. The curves would be quite different (Figure 4-2).

Species richness is also related to area sampled. Large areas, as logic would suggest, contain more species than comparably smaller areas. This is why the fragmentation of ecosystems is of such concern to conservation biologists. Loss of area directly translates into loss of species (see Chapter 14). The relationship between species richness and area is well established (MacArthur and Wilson 1967) and will be further discussed in the next chapter. What is important at this juncture is to realize that a given area in a tropical region will likely contain more species (often many more) than an area of equal size in a temperate region (Figure 4-3). *Area effect* is likely of major importance in considering global species richness patterns (see Chapter 5).

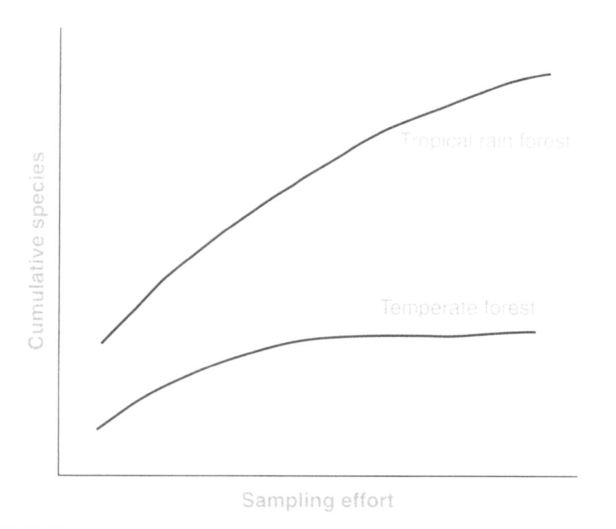

FIGURE 4-2
Species/sampling effort curves are high for tropical regions in comparison with other areas. This hypothetical graph depicts curves typically seen when comparing taxa such as birds and trees between tropical and temperate forests. Tropical forests are much more species-rich.

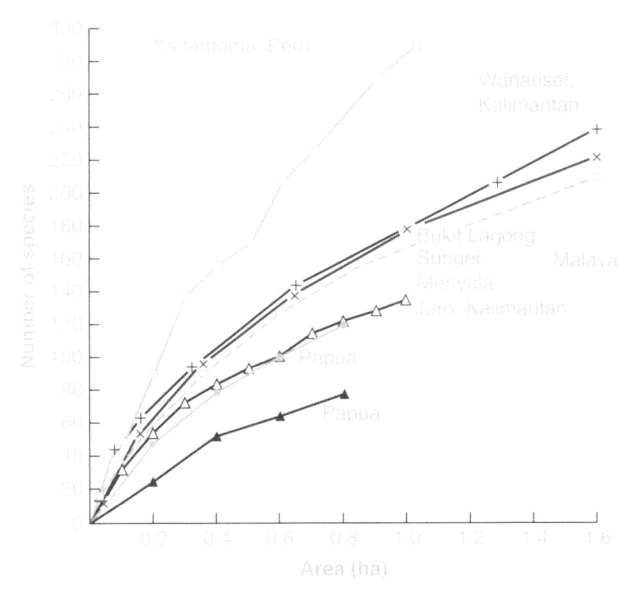

FIGURE 4-3
This graph shows the differences in species/area curves for evergreen tropical trees in several forests. The most diverse is Yanamomo, in the Peruvian Amazon.

Species Richness: Alpha, Beta, and Gamma

Ecologists distinguish among three scales of species richness. Be warned, there is some confusion possible regarding terminology. *Species richness* is the number of species present. Period. But the term *species diversity* refers to the combination of the number of species present plus the evenness in population sizes among species. For example, consider a community with three species and 100 individuals. If that community consisted of about one-third of each of the three species, it would have a much higher evenness than if one species was 90% of the total individuals and the other two were 10% combined. For much more on species diversity metrics, see Magurran 2003. But in the literature, the terms *alpha*, *beta*, and *gamma diversity* are widely used, even though in many cases they refer to species richness, not a combination of species richness and degree of relative evenness of population sizes.

Alpha richness (as noted earlier, often called *alpha diversity*) is the species richness within a specific habitat and is the most common diversity measurement in ecology. If one measures the species richness of trees in a 10-hectare moist forest or the richness of bat species inhabiting a 2-hectare moist forest, one is measuring alpha diversity. Alpha richnesses are generally highest in rain forests.

Beta richness, or *beta diversity*, measures the rate at which species change from one habitat to another within a region or along a gradient. If one looks at the differences in species from one point to another as one moves from east to west across the Brazilian Amazon, one is looking at beta diversity. This means that different species of hummingbirds, monkeys, trees, and so on are encountered as one moves from one forest site to another.

Measuring Beta Diversity

When you move from a lowland forest in Borneo to another forested area perhaps 20 kilometers away, you have changed habitats. There may be differences in elevation and/or soil characteristics. In the tropics, such differences matter. They affect species richness patterns. What is the similarity in species of trees, butterflies, leeches, beetles, birds, or mammals for the two habitats? Do the two habitats have exactly the same bird species, do they have 50% in common, or do they have none in common? The rate of turnover in species composition between two habitats is *beta diversity*. You could, of course, simply compare lists and see how many species are unique to each list and how many are shared. But ecologists have developed several mathematical approaches to beta diversity. One is called the *coefficient of community*. That is basically the percentage of species that two communities share in common. But what if you want to measure beta diversity among several communities? Another measure is $BD = Sc/S$, where BD is beta diversity, Sc is the number of species in the composite of several alpha diversity samples, and S is the mean number of species in the alpha samples. Using this measure, if you surveyed the alpha diversity (as species richness) in three communities and found a richness of 18 species in A, 12 in B, and 20 in C, you would then see how many species are found in common in each of these three communities. Suppose that 8 species are found in all three communities. Then the BD would be 8/50, or 0.16. If the three communities had 10 species in common, BD would equal 0.20.

Gamma diversity (which might be better termed *biogeographical diversity*) is the broadest diversity measure, consisting of the total species diversity of a taxon in a broadly defined biogeographic region. For example, there are 3,751 bird species in the Neotropics, the tropical regions of Central and South America, or just over one-third of all bird species known to inhabit the Earth. The importance of gamma diversity and differences in gamma diversity from one point on Earth to another is that gamma diversity is the result of long-term evolutionary processes that have resulted in biogeographic differences among the continents, islands, and so on (see Chapter 2).

ALPHA, BETA, AND GAMMA DIVERSITY: A COMPARATIVE STUDY

In a now classic study, James R. Karr (1976) examined alpha, beta, and gamma diversities of bird communities in Panama, Central America, and Liberia, Africa. He found numerous differences when comparing similar habitats from both countries.

Regarding gamma diversity, Karr found only one species in common between the bird communities of Liberia and Panama. That was the barn swallow, *Hirundo rustica*, one of the most widely dispersed bird species in the world. Otherwise, the bird communities showed no overlap at the species level and were also quite different at levels of genera and families. But given that Africa and South America separated about 100 million years ago, such a result is not surprising. Panama showed greater bird species richness in forests and in early successional shrub habitats, but Liberia had a higher richness in forest and shrub combined, suggesting that more bird species in Liberia are ecological specialists, utilizing one or the other habitat (Table 4-2).

Both alpha and beta diversities were lower in Liberia compared with Panama (Figure 4-4), and there were many differences in foraging patterns of the bird communities. For example, the number of birds feeding by probing bark, by hovering, or by *sallying* (flying a short distance to capture a prey item) were each greater in Panama than Liberia. Karr's study was important in focusing on how diversity varies both within a geographical region and between regions.

TABLE 4-2 Number of bird species found in forest, shrub, and both habitat types in Panama and Liberia.

AREA	HABITAT		
	FOREST	SHRUB	FOREST AND SHRUB
All forest species			
Panama	58	2	39
Liberia	23	3	75
Forest residents			
Panama	38	—	18
Liberia	13	—	37
Shrub residents			
Panama	—	38	12
Liberia	—	4	17

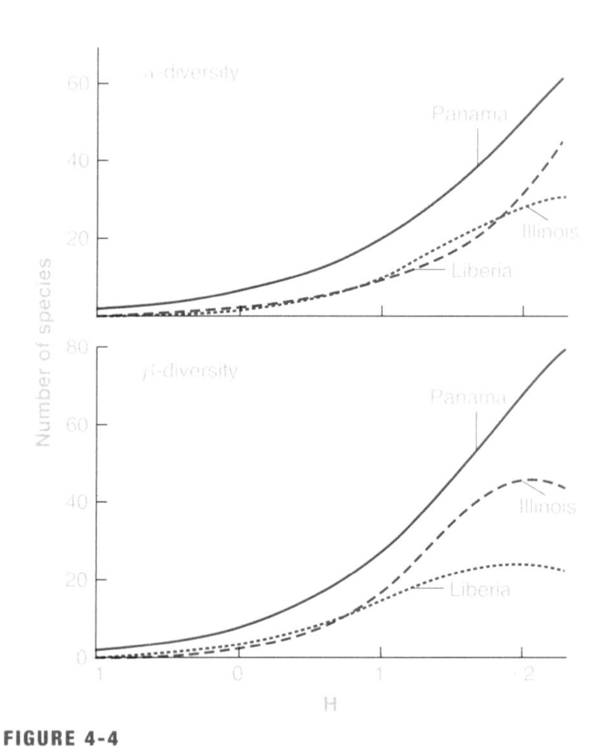

FIGURE 4-4
Alpha and beta bird diversity as they compare between Panama, Liberia, and Illinois (United States). The *x* axis is based on principal component analysis of the height and half-height of vegetation. The curves show significantly greater diversities in the two tropical countries.

AN EXAMPLE: BETA DIVERSITY OF TREES AT TWO NEOTROPICAL SITES

Beta diversity of trees is often high in tropical regions but varies from one region to another. A study

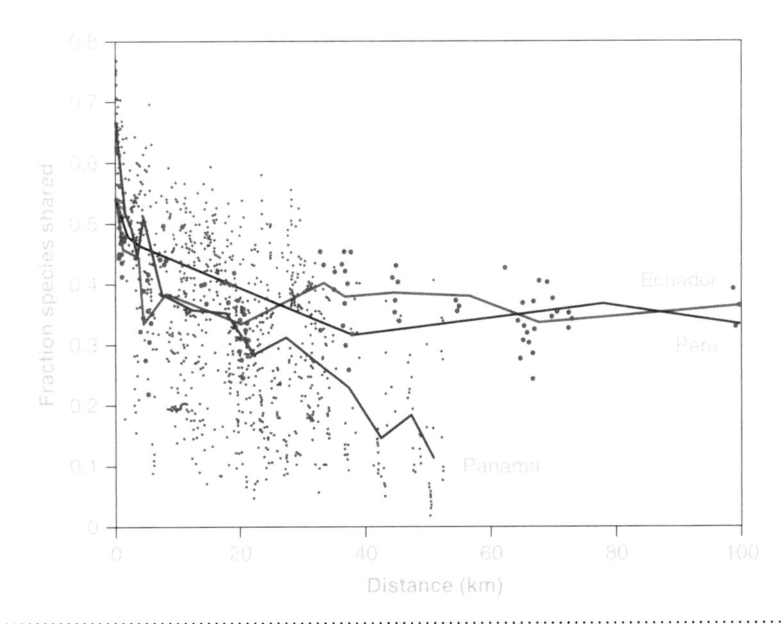

FIGURE 4-5
This graph is based on a calculation called the *Sorensen similarity index* between pairs of 1-hectare plots as a function of distance between the plots and is thus a measure of beta diversity. Note that similarity among plots declined most rapidly in Panama and more gradually in Ecuador and Peru.

comparing beta diversity of trees in Panama, Ecuador, and Peru demonstrated clear differences (Condit et al. 2002). In Panama, data were collected from 34 plots in the vicinity of the Panama Canal, an area of rich lowland moist forest. In Ecuador, 16 plots were surveyed in Yasuni National Park, and in Peru, 14 plots were surveyed in Manu Biosphere Reserve in western Amazonia. All sites were on *terre firme* (unflooded) forest. Over 50,000 trees (with stem diameter equal to or greater than 10 centimeters [3.9 inches]) were identified and measured.

At each of the regions, beta diversity was high. All three showed that similarity between two 1-hectare forest plots declined with increasing distance between them. Similarity declined rapidly with distances up to 3 to 5 kilometers in each of the three regions. In Panama, any two plots separated by 50 kilometers (30 miles) shared only between 1% to 15% of their species. South American sites showed a somewhat more gradual pattern. Similarity among plots changed less rapidly. Plots separated by as much as 100 kilometers might still have 30% to 40% of tree species in common. Last, single plots in Panama shared only 8% of their species in common with single plots in Peru and only 5% in common with plots in Ecuador (Figure 4-5).

Researchers attributed the greater beta diversity in Panama to its recent and complex geological history and complex rainfall pattern (less than 2,000 millimeters [78 inches] near the Pacific and greater than 3,000 millimeters [117 inches] near the Caribbean). In other words, the physical environment is more heterogeneous in Panama than at the other two sites, selecting for different species of trees. Soils and climate are more similar throughout both the Ecuador and Peru sites, and climate varies less.

Components of Species Richness and Species Diversity

Only in the tropics is it common to find natural communities where there are not only large numbers of species but also where population sizes among species are relatively equal—what ecologists refer to as *high evenness*. At higher latitudes, it is more typical to find a subset of species that are common to fairly common along with a number of less abundant, rare species, a distribution that produces *low evenness*. Lowland tropical forests in Amazonia contrast strongly from the pattern typical of high latitudes. For example, in a 1-hectare (2.5 acres) plot within an upper Amazonian forest, Gentry (1988) found that 63% of the tree species were represented by only single individuals and only 15% of the species were represented by more than two individuals. In a 97-hectare plot in the Peruvian Amazon, an astonishing total of 245 resident birds were found, plus 74 additional species that occasionally visited from other habitats or appeared only as migrants (Terborgh et al. 1990). Most of these species had small populations, and about 10% were considered uniformly rare throughout the study area. Up to 160 species were found at single locations in some portions of the study area.

HIGH SPECIES RICHNESS

In a temperate zone forest, there are usually fewer than 20 tree species in a hectare. Even in the most diverse North American forests, those of the lush Appalachian coves, barely over 30 tree species occur in a hectare (10,000 square meters, or about 2.5 acres). In the tropics, however, anywhere from 40 to 100 or more species of trees typically occur per hectare. Indeed, one site in the Peruvian Amazon has been found to contain approximately 300 tree species per hectare (Gentry 1988). Brazil alone has been estimated to contain somewhere around 55,000 flowering plant species (World Conservation Monitoring Centre 1992). Altogether, about 90,000 species of flowering plants (approximately 37% of the world's total) are estimated to occur in the Neotropics. This is five times the number in North America. Tropical Africa, by comparison, has about 35,000 species, and tropical Asia, about 40,000 (Plate 4-1).

British naturalist Alfred Russel Wallace commented on the difficulty of finding two of the same species of tree near each other. Wallace (1895) stated of tropical trees:

> If the traveller notices a particular species and wishes to find more like it, he may often turn his eyes in vain in every direction. Trees of varied forms, dimensions and colour are around him, but he rarely sees any one of them repeated. Time after time he goes towards a tree which looks like the one he seeks, but a closer examination proves it to be distinct.

As Wallace implies, though richness is high, the number of individuals within a single species tends to be low, which is another way of saying that rarity is common for many species in the lowland tropics. Though some plant species are abundant and widespread (for example, some *Cecropia* species in the Neotropics and certain dipterocarps in Indonesia), most are not. The concept of identifying a forest type by its numerically dominant tree species, which works well in the temperate zone (i.e., eastern white pine forest, oak-hickory forest, redwood forest), is much less useful in the tropics, though not always. On the island of Trinidad, one can visit a *Mora* forest where the canopy consists almost exclusively of but a single species, *Mora excelsa*, a tree that can reach heights

PLATE 4-1
This rain forest in Trinidad contains a high species richness of trees, birds, arthropods, and other taxonomic groups.

Barro Colorado Island, Panama

The Panama Canal was opened in 1914, and it was the construction of the Panama Canal that created Barro Colorado Island (BCI). When the canal was built, an area of lowland forest was flooded, creating Gatun Lake. The highest elevations were not flooded, so the creation of the lake also resulted in the formation of Barro Colorado "island," once merely a hill in contiguous lowland forest. BCI is a 1,500-hectare island that is part of Barro Colorado Nature Monument, which comprises not only BCI but also five adjacent peninsulas of tropical forest. BCI is the site of one of the Smithsonian Tropical Research Institute facilities and has been, since its creation, one of the most significant laboratories for tropical research. Hundreds of scientific papers and numerous books have been written on all aspects of the ecology of the region thanks to the facilities available to researchers on BCI (Plate 4-2).

PLATE 4-2
Barro Colorado Island in Panama, showing the dock and some of the facilities for researchers.

of nearly 30 meters (150 feet). The understory is also dominated by *Mora* saplings. In central Africa, the leguminous tree species *Gilbertiodendron dewevrei* occurs in almost single-species stands in forests of white-sand soils. But examples of such low tree species richness are very much an exception to the typical pattern exhibited in the tropics.

Species richness, though generally high throughout the tropics, shows regional variability. For example, consider data from some Neotropical forests, from both Central America and South America. Knight (1975), working on Barro Colorado Island (BCI) in Panama, found an average of 57 tree species per 1,000 square meters in mature forest and 58 species in young forest. Knight found that in the older forest, when he counted 500 trees randomly, he encountered an average of 151 species. In surveying 500 trees in a younger forest, he encountered an average of 115 species. Hubbell and Foster (1986b) have established a 50-hectare permanent study plot in old-growth forest at BCI. They surveyed approximately 238,000 woody plants with stem diameter of 1 centimeter diameter breast height (DBH) or more and found 303 species. They classified 58 species as shrubs, 60 as understory treelets, 71 species as midstory trees, and 114 as canopy and emergent trees. Gentry (1988), working in upper Amazonia and Choco, found between 155 and 283 species of trees greater than 10 centimeters DBH in a single hectare. When he included lianas of greater than 10 centimeters DBH, he found that the total increased to between 165 to 300 species. Prance et al. (1976) found 179 species greater than 15 centimeters

DBH in a 1-hectare plot near Manaus, on a *terre firme* forest characterized by poor soil and a very strong dry season. In each of the preceding examples, the species richness far exceeds that typical of forests outside the tropics.

If all vascular flora are taken together (trees, shrubs, herbs, epiphytes, and lianas, but excluding introduced weedy species), the inventory for BCI is 1,320 species from 118 families (Foster and Hubbell 1990; Gentry 1990). By comparison, the total number of vascular plant species documented to occur at La Selva Biological Station in nearby Costa Rica is 1,668 species from 121 families (Hammel 1990; Gentry 1990). Let's compare these totals with those from two Amazonian rain forests. A floodplain forest on rich soils at Cocha Cashu Biological Station along the Rio Manu, a whitewater tributary (*whitewater* refers to being sediment rich, meaning that the forest was located on rich soil) of the vast Rio Madeira in southeastern Peru, was found to contain 1,856 species (in 751 genera and 130 families) of higher plants (Foster and Hubbell 1990). At Reserva Ducke, a forest reserve on nutrient-poor soil near Manaus, Brazil, in central Amazonia, a total of 825 species of vascular plants from 88 families was inventoried (Prance 1990; Gentry 1990).

When the two Central American sites described earlier (BCI and La Selva) are compared with the two Amazonian sites (Cocha Cashu and Reserva Ducke), there are several important differences. Tree species richness is greater in Amazonia (Gentry 1986, 1988, and 1990), but the richness of epiphytes, herbs, and shrubs is greater in Central America. At La Selva, 23% of all vascular plant species are epiphytes, the highest percentage recorded among the sites (Hartshorn and Hammel 1994). The most species-rich of any of the four sites was Cocha Cashu, located on highly fertile *varzea* (flood plain) soils in western Amazonia. In contrast, 29 plant families that were present at BCI, La Selva, and Cocha Cashu were absent from Reserva Ducke, presumably because the poor soil conditions at that site limit species richness. However, the similarities among these four geographically separated forest sites are more compelling than the differences. The dozen most-well-represented plant families were essentially the same for each of the sites. Legumes (Leguminosae), for example, was the most-species-rich family at BCI, Cocha Cashu, and Ducke, and was the fifth most-rich family at La Selva. Of the total of 153 vascular plant families represented in at least one of the four sites, 66 (43%) were represented at all four sites (Gentry 1990) (Figures 4-6, 4-7, and 4-8).

Plants are not the only diverse groups. Insects (and arthropods in general), birds, amphibians, and most other major groups also exhibit both high species richness on a per area basis (as in number of species per hectare) and high turnover from one site to another (beta diversity). A guide to birds of Colombia lists 1,695 migrant and resident species occurring in that country (Hilty and Brown 1986). Nearly 1,800 bird species occur in Peru. At Cocha Cashu Biological Station in Amazonian Peru, in an area of approximately 50 square kilometers, the total species list of birds is approximately 550 (Robinson and Terborgh 1990). At the Explorer's Inn Reserve in the southern Peruvian Amazon, about 575 bird species have been identified within an area of approximately 5,500 hectares (Foster et al. 1994). At La Selva Biological Station in Costa Rica, an area of approximately 1,500 hect-

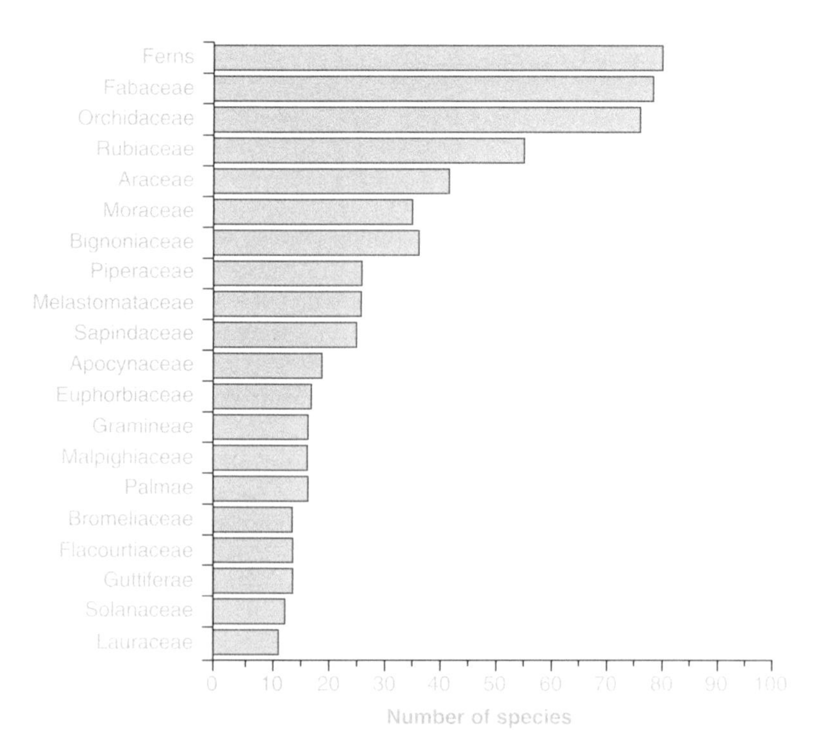

FIGURE 4-6
The relative abundance of various plant families at Barro Colorado Island.

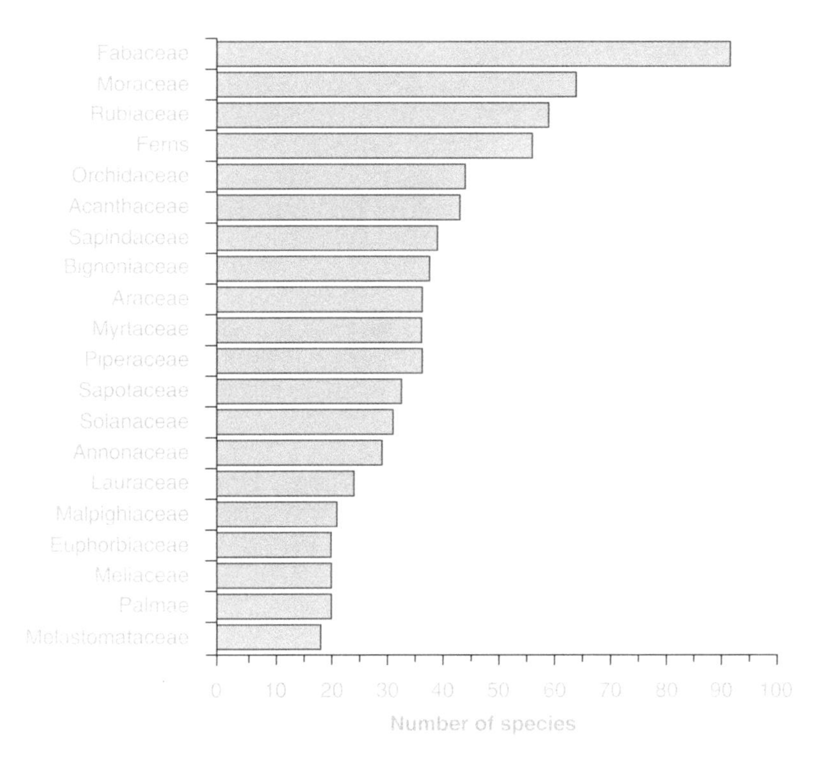

FIGURE 4-7
The relative abundance of various plant families at Manu floodplain forest in Amazonian Peru.

ares, 410 species of birds have been found (Blake et al. 1990). In comparison, barely 700 bird species occur in all of North America. The extremely high bird species richness per unit area is unique to areas of rain forest (Tramer 1974) (Plate 4-3). Bird species richness is less in other tropical ecosystems.

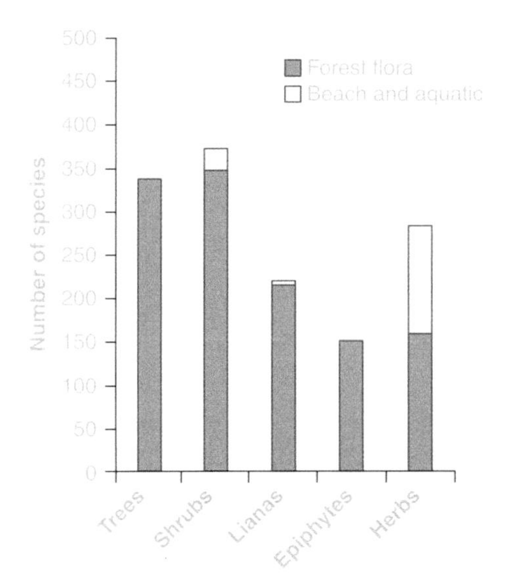

FIGURE 4-8
This graph depicts the number of species of various plant life forms at Manu floodplain forest. Note the high relative abundances of lianas (woody vines) and epiphytes.

At one site in the Ecuadorian Amazon, the species richness of frogs is 81, which is about how many species occur in all of the United States. Indeed, the researcher collected 56 different species on a single night of sampling and reports that it is routine to find 40 or more species in areas of rain forest as small as two square kilometers (Duellman 1992) (Plate 4-4).

Insect species richness can seem staggering. For the small Central American country of Costa Rica, Philip DeVries (1987 and 1997) describes nearly 800 butterfly species, representing about 5% of the world's butterfly species (Plate 4-5). At La Selva alone, 204 butterfly species have been identified, and 136 species have been documented for BCI (DeVries 1994). At Explorer's Inn Reserve in southern Peru, a total of 1,234 butterfly species have been identified from an area about 2.0 square kilometers within the reserve (Foster et al. 1994). Edward O. Wilson (1987) reported collecting 40 genera and 135 species of ants from four forest types at Tambopata Reserve in the Peruvian Amazon. Wilson noted that 43 species of ants were found in one tree, a total approximately equal to all ant species occurring in the British Isles!

In Panama, Terry Erwin, using a fogging technique to extract insects from the forest canopy, sampled 19 trees, all of which were a single species, *Luchea seemannii*. Fogging involves the spraying of insecticide into the canopy of the tree and subsequently collecting (on cloths spread on the ground) the insects that drop from the tree. Erwin identified approximately 1,200 species of beetles (including weevils) in these samples (Plates 4-6 and 4-7). Erwin considered that there are about 70 tree species per hectare. Using that estimate, Erwin judged that about 13.5% (163) of the beetles were host-specific, occurring only in *Luchea*. He calculated that perhaps as many as 11,410 host-specific beetles could be found within a hectare (70 × 163). He then multiplied this figure by the number of different tree species present in the global tropics and concluded that the potential species richness of beetles alone was over eight million! Since beetles are estimated to represent

PLATE 4-3
These golden tanagers (*Tangara arthus*) enjoying a banana represent one of 30 species of tanagers in the genus *Tangara* found in Ecuador. The tanager family Thraupidae is highly diverse, with 143 species in Ecuador alone. Tanager distribution throughout the Neotropics is an example of both high alpha and high beta diversities.

approximately 40% of all tropical terrestrial arthropod species (including spiders, crustaceans, centipedes, millipedes, as well as insects), Erwin suggested that the total arthropod species richness of the tropical canopy might be as high as 20 million, and that figure climbs to 30 million when the ground and understory arthropods are added (Erwin 1982, 1983, and 1988). Note that Erwin's estimates are indeed based on assumptions about host specificity, a characteristic that may vary considerably among species of trees and other plants. Some tropical ecologists have taken a much more conservative view of insect species richness and have been critical of Erwin's estimates (see in particular Stork 1988). Insects, as a group, remain rather poorly documented as to their actual species richness, and future research is needed. Almost without question, numerous species await discovery.

For the entire planet, only about 1.75 million species of plants, animals, and microbes have thus far been named and described. But given that about

PLATE 4-4
This anuran inhabiting the forest floor in a lowland Panamanian forest is one of 164 anuran species found in Panama.

PLATE 4-5
The butterfly *Philaethria dido* (family Heliconiinae) is found widely in lowland forest in Central and South America.

PLATE 4-6
HARLEQUIN BEETLE (*ACROCINUS LONGIMANUS*)

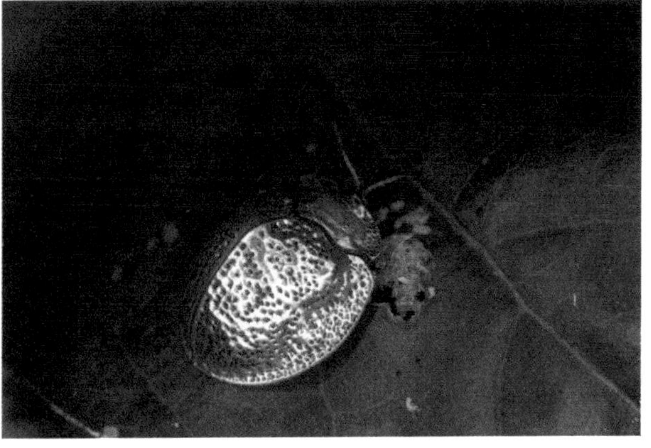

PLATE 4-7
TORTOISE BEETLE (SUBFAMILY *CASSIDINAE*)

The harlequin and tortoise beetles are just two of the more striking species of Amazonian beetles. There may be a million or more other species of beetles in the world.

300 new species (especially among arthropods) are described daily, the figure of 1.75 million may represent only about 10% of Earth's total biodiversity (Purvis and Hector 2000). For example, as a group, fungi remain poorly understood with regard to total species richness. About 100,000 species have been described, but some estimates suggest that their true diversity lies in the range of 1.5 million species (Hawksworth 2006). Basic taxonomic research remains a very high priority for many groups of organisms.

SPECIES RICHNESS AND DIVERSITY GRADIENTS

It is obviously not known with any degree of certainty just how many species inhabit the Earth (May 1988 and 1992). As noted in Chapter 2, even the definition of species is subject to some debate. But what is known, and this applies to terrestrial, freshwater, and most marine ecosystems, is that, with few exceptions, most major taxa including flowering plants, ferns, mammals, birds, reptiles, amphibians, fish, insects, spiders, millipedes, snails, and bivalves all have the greatest number of species in the tropics. Further, some tropical rain forests contain so many species in relation to any other kind of ecosystem or any other region that they exhibit what some ecologists have termed *hyperdiversity*. Why do tropical ecosystems, and rain forests in particular, contain so many species?

Charles Darwin realized that species numbers per unit area tend strongly to decline latitudinally as one travels north or south from the equator, a point he noted in Chapter 3 of *On the Origin of Species*. Alexander von Humboldt also observed this trend for plants and believed it to be related to reduced tolerance for cold in higher latitudes. The reduction in diversity with increasing latitude is termed a *latitudinal diversity gradient* or

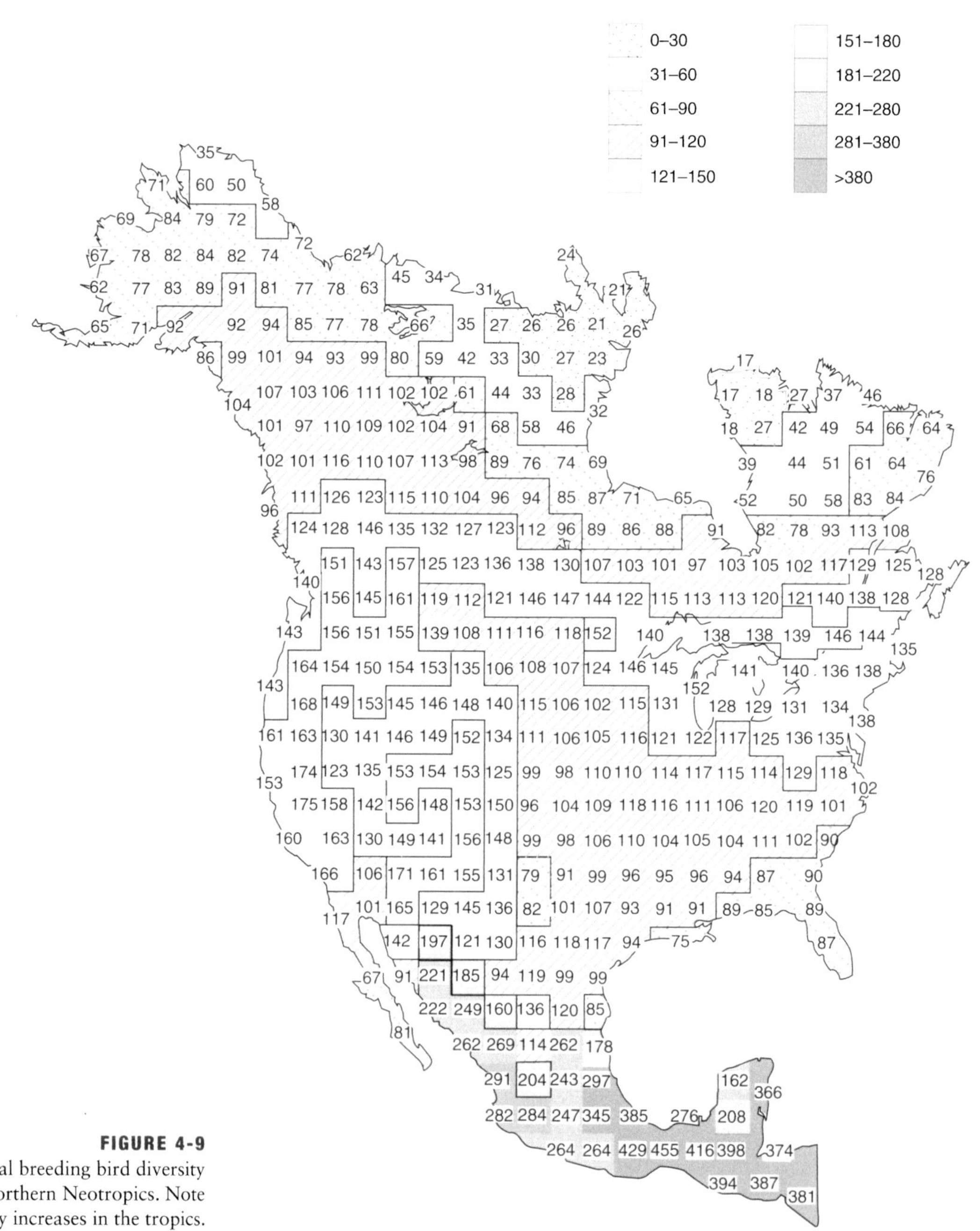

FIGURE 4-9

Patterns of terrestrial breeding bird diversity in the Nearctic and northern Neotropics. Note how quickly diversity increases in the tropics.

LDG (Connell and Orias 1964; MacArthur 1965; Pianka 1966). Dobzhansky, in a seminal paper titled "Evolution in the Tropics" (1950), noted that only 56 species of breeding birds occurred in Greenland, while New York had 195 breeding species. However, Guatemala had 469, Panama 1,100, and Colombia 1,395! Breeding bird diversity increased almost 25 times from Greenland in the arctic to Colombia on the equator. For snakes, Dobzhansky noted that 22 species occurred in all of Canada, whereas 210 were found in Brazilian forests and savannas. These figures would be revised somewhat today, as more species are known for both birds and snakes, but the extreme difference between high- and low-latitude species richness would, if anything, be greater. More new species have been discovered in the tropics (Figures 4-9 and 4-10).

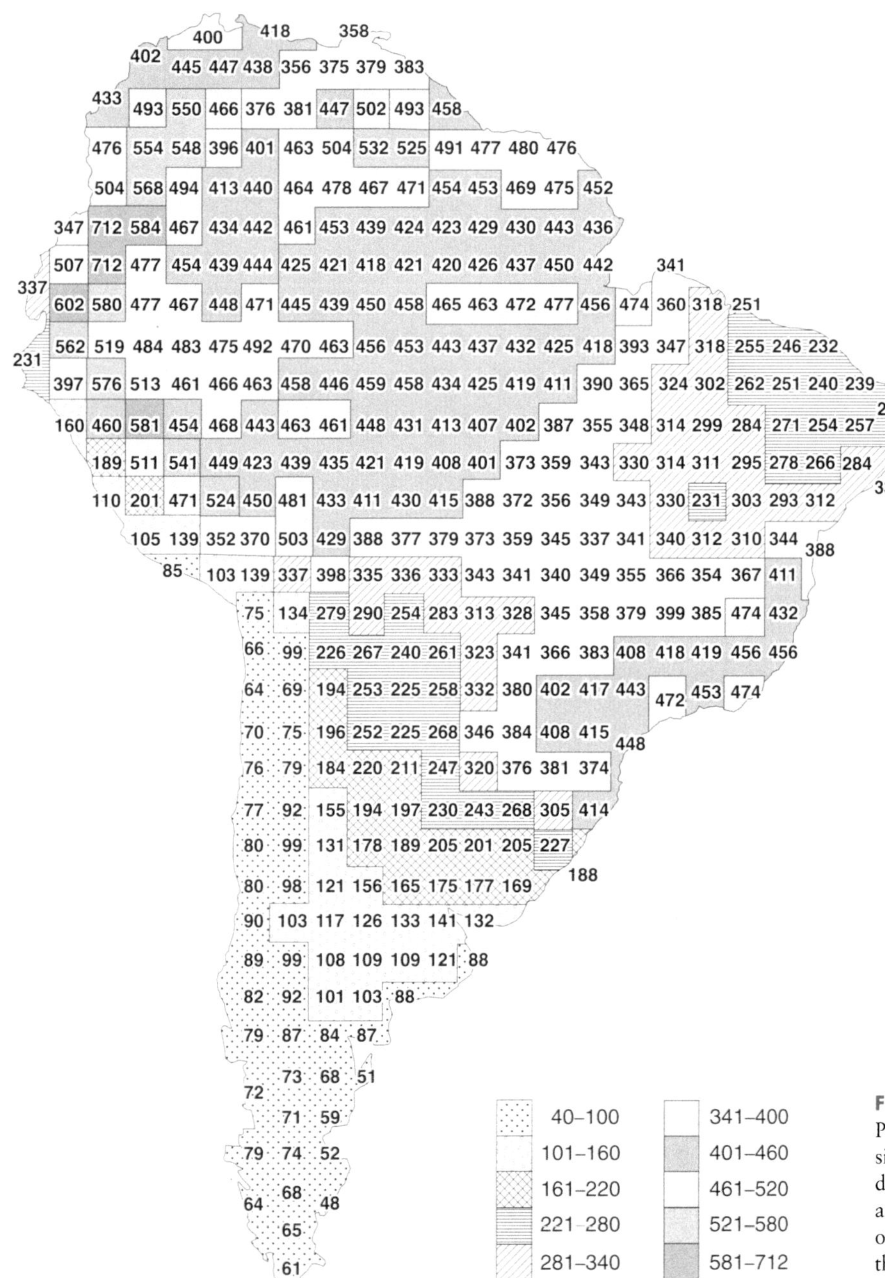

FIGURE 4-10
Patterns of terrestrial breeding bird diversity in South America. Note how diversity declines with increasing southern latitude, and note also the consistent high diversity of Amazonian locations. Last, note that the highest diversities in the central Andes Mountains.

The Lost World of the Foja Mountains

The tropics have been well explored, but the amazing species richness continues to result in the discovery of new species, ranging from numerous insects to birds and mammals. One of the most dramatic examples was unveiled following an expedition to Papua, an Indonesian province in western New Guinea led by Bruce M. Beehler, senior research scientist at Conservation International, in November and December of 2005. In a long and carefully planned expedition, researchers from several countries along with local guides combined efforts to explore the little-known Foja Mountains, an area that is difficult to access, even by helicopter. The principal research site was at an elevation of 1,650 meters (5,445 feet), a site utterly unexplored until the research team arrived for its 15 days of intensive study.

The researchers lived under very basic and challenging conditions but experienced immense success. The results of their collections and observations are still under study, but up to 20 new frog species were discovered along with between 5 and 10 new plant species, 5 new species of butterflies, and at least one new mammal species, the golden-mantled tree kangaroo (*Dendrolagus pulcherrimus*) (Plate 4-8). Bruce Beehler, the expedition leader, rediscovered both the Berepsch's bird of paradise (*Parotia berlepschi*) (Plate 4-9), also known as

PLATE 4-8
GOLDEN-MANTLED TREE KANGAROO

PLATE 4-9
BRONZE PAROTIA

bronze parotia, and the golden-fronted bowerbird (*Amblyornis flavifrons*). These species, previously known from only a few specimens, were each documented a few years earlier by Jared Diamond, another intrepid scientist and explorer. Beehler also found a bird species entirely new to science, the wattled smoky honeyeater (*Melipotes carolae*) (Plate 4-10).

The expedition's success was widely reported in many mainstream media and was even featured on the television show *60 Minutes*. Dr. Beehler's message to the public is that the area is dense with endemic species and represents a unique resource essential to conserve and protect.

PLATE 4-10
WATTLED SMOKY HONEYEATER

Dobzhansky suggested that since animals and plants are all products of evolution, any differences between tropical and temperate organisms must be the outcome of differences in evolutionary patterns. Selection pressures must vary with latitude. But, as is often the case in ecology, the devil remained in the details. Just what selection pressures and other forces might account for the greater richness and variety of the tropical faunas and floras, compared with those of temperate and polar lands? How does life in tropical environments influence the evolutionary potentialities of the various taxa? One obvious possibility is that for some reason, speciation rates exceed extinction rates in tropical regions, thus allowing a buildup of species richness with time. Another is that the tropics are ancient and stable and species have low extinction rates. But there are numerous other possibilities as well. For example, tropical rain forests, because they remain warm and wet throughout the year, fix more photosynthetic energy per unit area than other ecosystems. Does that account for greater species richness? If so, how? Is net productivity most essential to species richness or merely warm and wet climatic constancy? There is more structural complexity in rain forests. Might that not account for greater species richness in groups such as birds, mammals, and arthropods?

Dobzhansky suggested that part of the answer to high tropical richness rested in the equitable nature of the tropical climate. Echoing Humboldt, he argued that polar and even temperate climates impose such significant physical stresses that fewer organisms have been able to adapt over evolutionary time. The tropics, in contrast, offer a climate of abundant rainfall for much if not all of the year, no season of frost or cessation of plant growth, warm and relatively invariable temperature, and less overall severe meteorological fluctuation, all of which may permit speciation to exceed extinction to a degree greater than is the case in the higher latitudes. As species are added, some specialize to utilize unique resources. Dobzhansky's hypothesis was speculative and challenging to test, but it stimulated thinking about latitudinal diversity gradients (Figure 4-11).

GENESIS VERSUS MAINTENANCE OF SPECIES RICHNESS

Dobzhansky's treatment of diversity patterns focused on what could be described as the *evolutionary genesis of diversity*—how diversity develops in relation to latitude. It is extremely important to understand the speciation process (Chapter 2), which generates

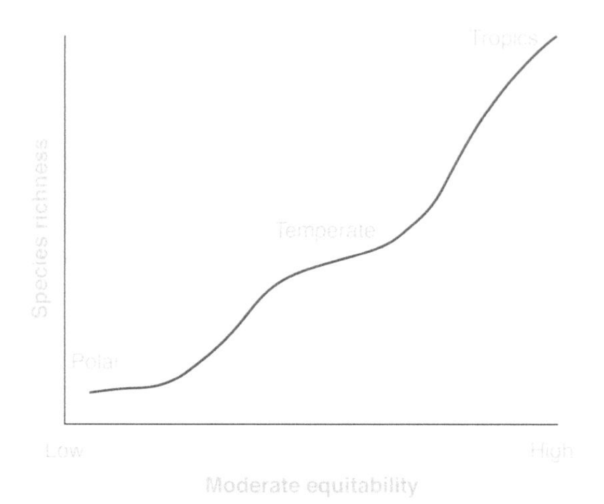

FIGURE 4-11
This graph illustrates Dobzhansky's early hypothesis for the relationship between climatic equitability and species richness as it varies with latitude.

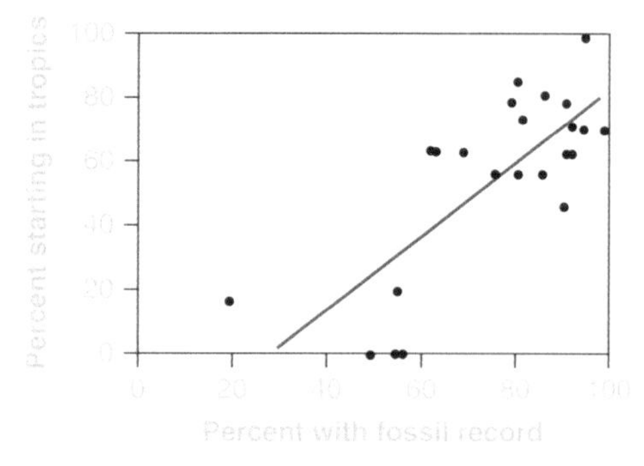

FIGURE 4-12
Sampling quality versus tropical originations in marine bivalve families since the start of the late Miocene (11 million years ago). The graph shows that families having more complete fossil records tend to show a significantly greater proportion of first occurrences of their constituent genera in the tropics.

diversity. But the question about high species richness throughout tropical ecosystems also involves another essential aspect, that of maintenance of diversity. Once new species evolve, how is it that as they are added and accumulate within tropical ecosystems, they are able to coexist with other similar species? What mechanisms allow coexistence of many more similar species in the tropics in comparison with higher latitudes? The questions of genesis and maintenance of species richness are somewhat different, and in the examples that follow, note that some focus on genesis and others on maintenance. Both are obviously important.

The Tropics as Cradle or Museum?

Ecologists have suggested two metaphoric views of the tropics to account for high species richness. One, called the *cradle*, views the tropics as uniquely suited to speciation such that species generation is high, and thus species tend to accumulate in tropical ecosystems, far more than outside of the tropics. This is a clear reflection of the Dobzhansky hypothesis, discussed earlier. The other view, called the *museum*, presumes that speciation rates are not higher in the tropics but extinction rates are quite low, so the tropics "keep" their more ancient species along with whatever new ones evolve, thus attaining a high species richness. In the museum view, extinction rates outside of the tropics are higher. Dobzhansky's reasoning also alluded to this view, but a bit less directly.

Perhaps the tropics acts as both cradle and museum. Work by Jablonski et al. (2006) suggests this possibility, at least for bivalve mollusks. The researchers examined the fossil record of 163 genera and subgenera beginning in the Miocene epoch, 11 million years ago. What they learned was that most bivalve families, including those at high temperate latitudes, originated in the tropics, support for the cradle view. Their data showed that first occurrences of bivalve mollusk genera were greatest in the tropics and that genera apparently radiated to temperate latitudes (Figures 4-12 and 4-13).

The data further showed that substantially higher extinction rates occurred at higher latitudes over the 11-million-year period covered by the study—support for the museum view of the tropics. They concluded that genera originating in tropical areas extended their ranges with time, while still occupying the tropics (Figure 4-14).

The study suggests that endemism should be most common in the tropics, decreasing with latitude. This follows if the tropics are both a cradle as well as a museum, as the data suggest. The prediction regarding endemism prevailing mostly in tropical regions was supported for bivalve mollusks. The study concludes that tropical areas must harbor both old and young taxa, while higher latitudes are inhabited by progressively older taxa that originated in the tropics. The authors warn that "a tropical diversity crisis would thus not only affect tropical biotas but also have profound

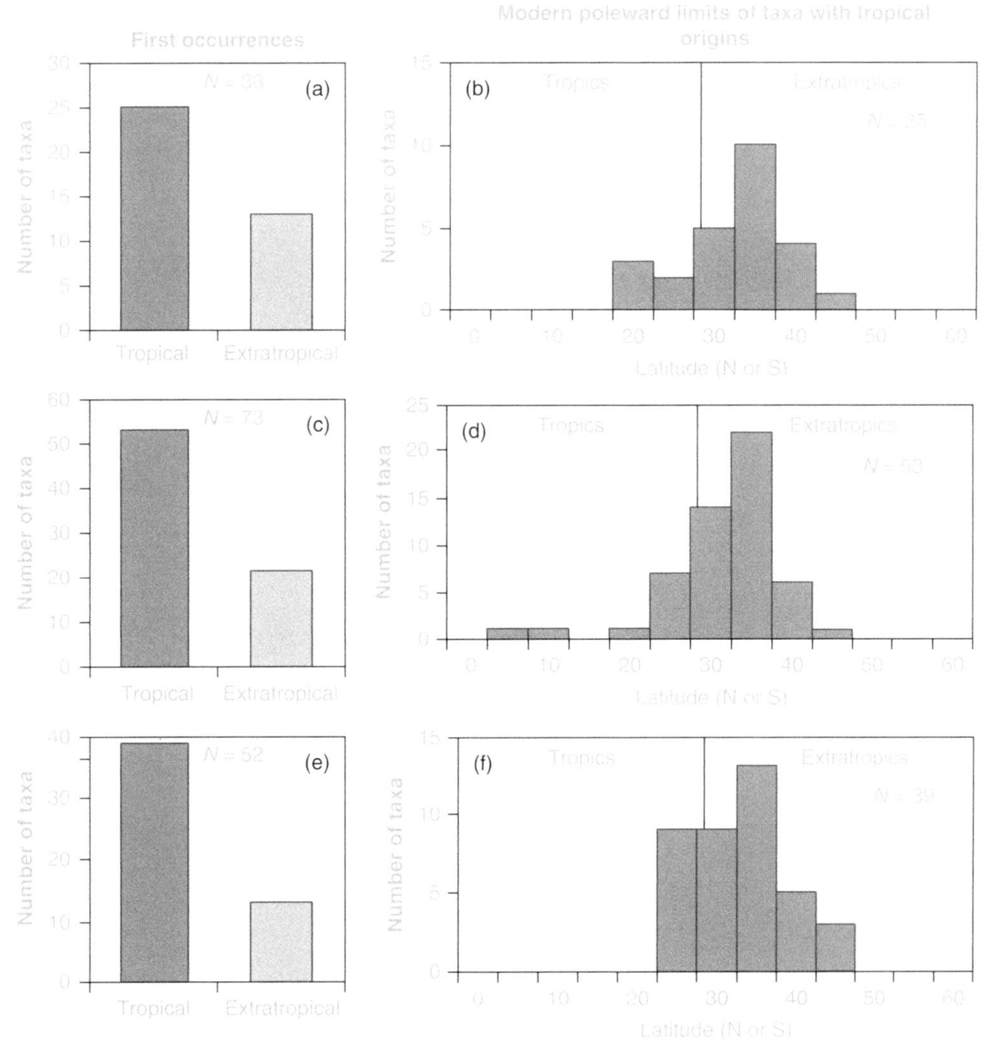

FIGURE 4-13
Latitudinal differences in originations (left) and present-day range limits (right) for marine bivalve genera first occurring in the tropics. (a) and (b) are genera first appearing in the Pleistocene; (c) and (d) are from the Pliocene; and (e) and (f) are from the Miocene. The graph shows the spread of genera away from the tropics over time, though they originated in the tropics.

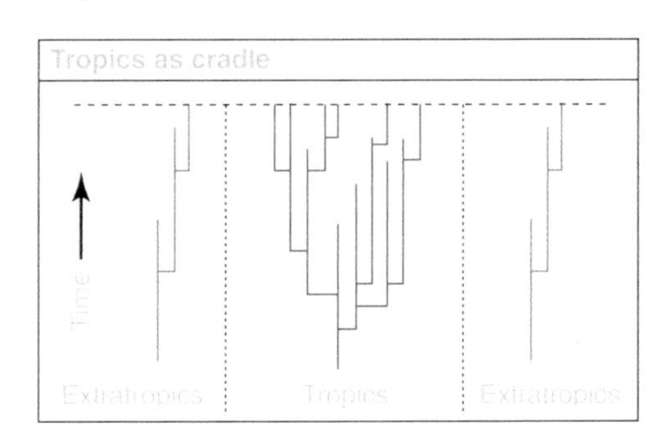

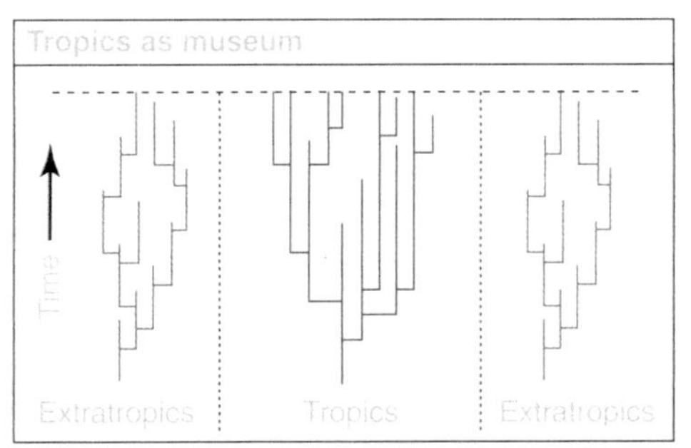

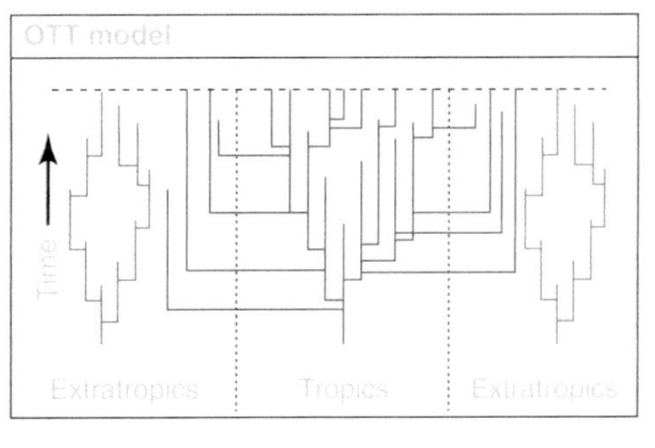

FIGURE 4-14
These figures illustrate three models of species origin: the tropics as cradle, tropics as museum, and out of tropics (OTT) models. Horizontal lines denote geographic distributions and show groups that originated in the tropics now expanding beyond the tropics (the OTT model).

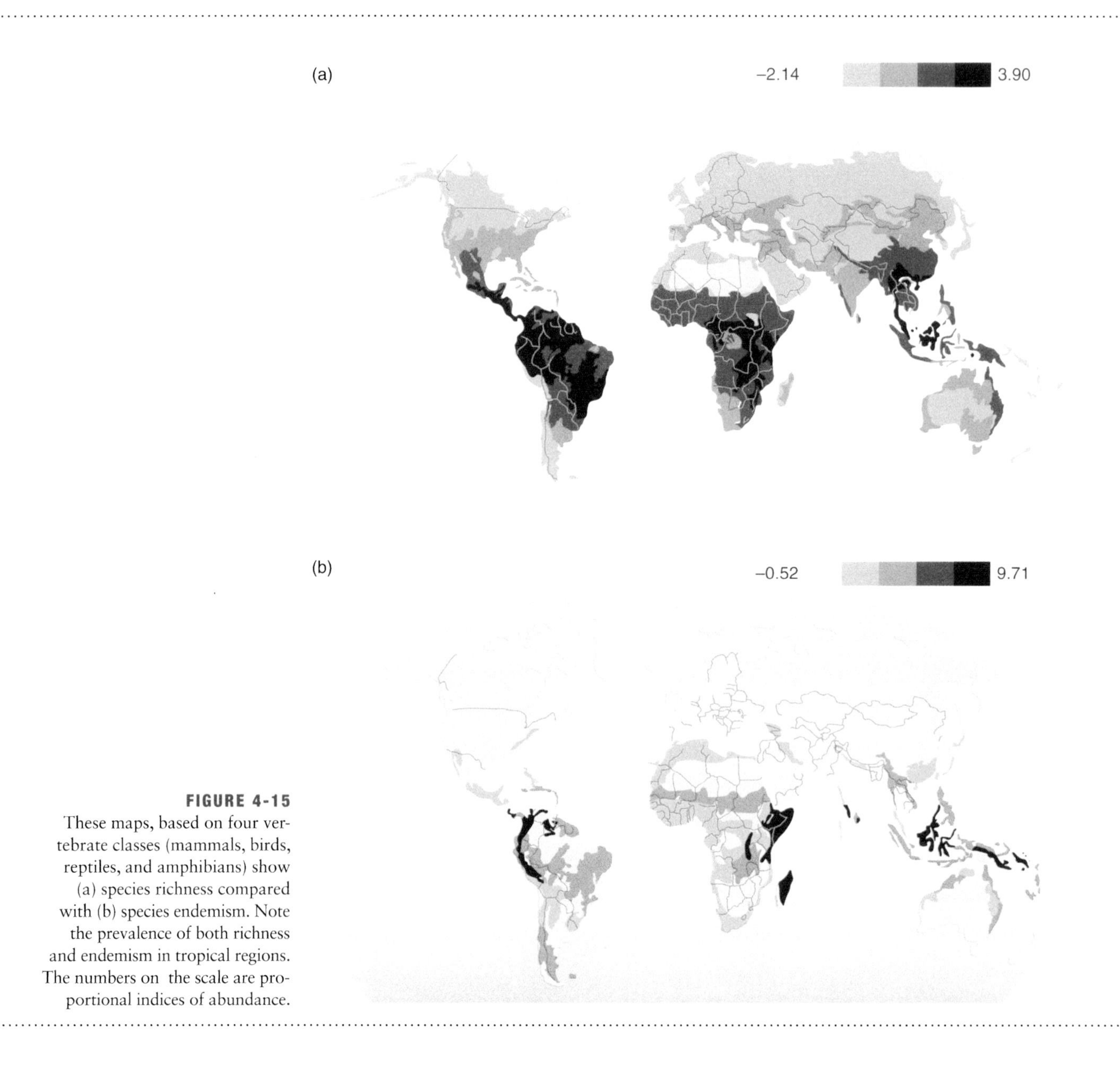

FIGURE 4-15
These maps, based on four vertebrate classes (mammals, birds, reptiles, and amphibians) show (a) species richness compared with (b) species endemism. Note the prevalence of both richness and endemism in tropical regions. The numbers on the scale are proportional indices of abundance.

long-term evolutionary consequences for biotas at higher latitudes." Note that if this is true, one could also argue that the tropics would tend to recover their diversity more rapidly than would be the case elsewhere but that such a recovery would occur over a substantially long time period (as measured in rate of speciation of various groups).

Conservation priorities (Chapter 14) should consider both high species richness and levels of endemism. This is why the tropical regions of the world have gained such a focus in conservation biology. In a sweeping study of the distribution of 26,452 vertebrates including amphibians, reptiles, birds, and mammals in 799 terrestrial ecoregions, there was no correlation between species richness and regional endemism for any of the four vertebrate classes studied (Lamoreux et al. 2006). There was, however, a strong correlation between global patterns of richness among all four classes of vertebrates, and as well, there was a strong correlation between endemism patterns among the four vertebrate classes (Figures 4-15 and 4-16; Table 4-3).

This means that patterns of high richness and high endemism (at least for the four classes sampled)

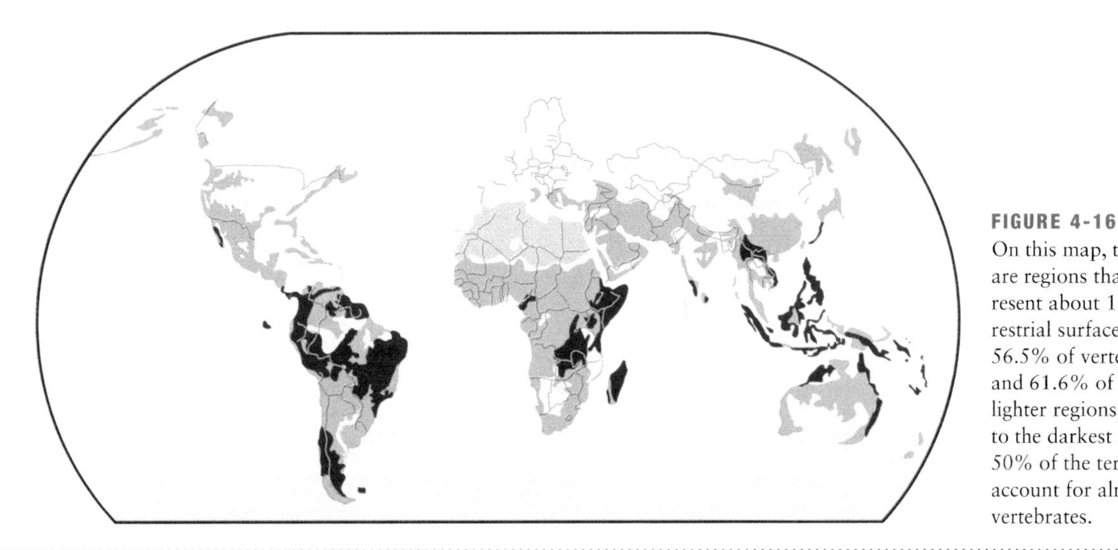

FIGURE 4-16
On this map, the darkest areas are regions that collectively represent about 10% of Earth's terrestrial surface but that include 56.5% of vertebrate endemics and 61.6% of all species. The lighter regions, in orange, added to the darkest regions represent 50% of the terrestrial Earth and account for almost all species of vertebrates.

do not totally overlap. Nonetheless, the authors produced a map showing ecoregions that, in total, represent a mere 10% of Earth's terrestrial surface but contain 56.5% of vertebrate endemics and 61.6% of all species. A look at this map shows clearly that most of these ecoregions, which are usually referred to as *hotspots* (see Chapter 14), are within the tropics.

Does speciation occur more rapidly in the tropics, thus packing the ecosystems with species? Until molecular techniques became available, this was difficult to determine. One study, using molecular techniques, suggests that speciation may actually be slower in the tropics than at higher latitudes (in contrast to the study presented earlier). In a comparison of sister species of various birds and mammals, the data indicated that within the latitudinal gradient, species divergences were most recent at higher latitudes and more distant equatorially (Weir and Schluter 2007). *Sister-species* are defined as two species that are each other's closest relatives. Mitochondrial DNA from the cytochrome b gene formed the data set. (This gene is believed to mutate at a steady rate in birds and mammals and is thus useful in measuring divergence times.) Both speciation and extinction rates were highest at higher latitudes, declining at lower latitudes (in contrast to Jablonski et al., discussed earlier for bivalve mollusks). This perhaps surprising result suggests that species turnover is considerably more rapid at higher latitudes than in the tropics. Sister-species among equatorial examples dated as far as 10 million years ago in divergence time,

TABLE 4-3 Pearson correlation coefficients between richness and endemism for four terrestrial vertebrate classes. The data show that richness correlates strongly with endemism.

	AMPHIBIANS	REPTILES	BIRDS	MAMMALS	FOUR CLASSES
Richness†	0.591*	0.380*	0.715*	0.668*	
Endemism‡	0.503*	0.587*	0.612*	0.490*	
Richness × endemism§	0.096*	0.085	−0.068*	−0.099	−0.025

*$P < 0.01$.
†Correlation between class richness and a richness index of the three remaining classes.
‡Correlation between class endemism and an endemism index of the three remaining classes.
§Correlation between richness and endemism within each class, and of the four classes combined.

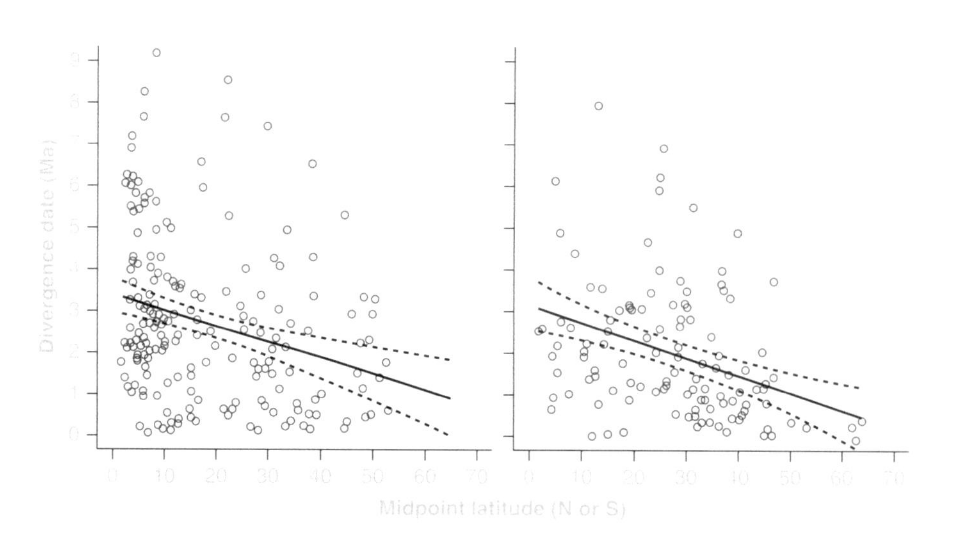

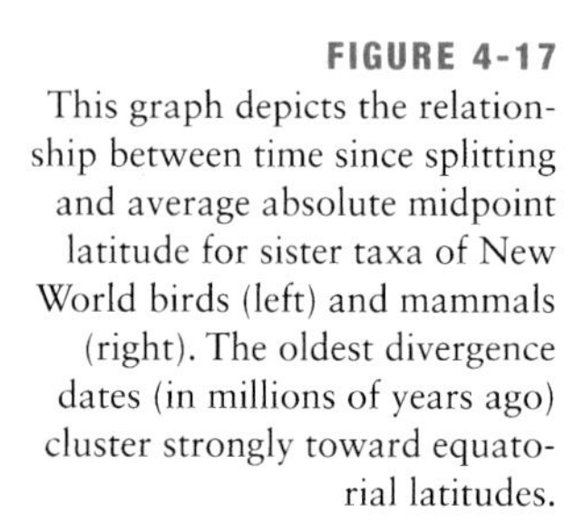

FIGURE 4-17
This graph depicts the relationship between time since splitting and average absolute midpoint latitude for sister taxa of New World birds (left) and mammals (right). The oldest divergence dates (in millions of years ago) cluster strongly toward equatorial latitudes.

with a mean of 3.4 million years ago, while at higher latitudes, sister-species divergences were less than a million years ago. If *speciation* (genesis of species) is more rapid in higher latitudes, the focus on maintenance of diversity as it accumulates in the tropics takes on added significance. Species richness could grow large with time if ecological interactions developed, permitting indefinite coexistence (Figures 4-17 and 4-18).

The authors note that rates of speciation and extinction are not constant through time in any ecosystem, particularly in the higher latitudes that have been subject to Pliocene and later Pleistocene climatic shifts. They suggest that

> bursts of diversification in tropical faunas may predate the late Pliocene and Pleistocene, and the patterns observed today may be the result of a subsequent decline in diversification either because the geological processes that promoted diversification (e.g., formation of Isthmus of Panama, marine incursions, formation of mountain ranges, and river formation) have slowed or because diversification rates declined as the number of tropical species approached a "carrying capacity." (Weir and Schluter 2007, p. 1576).

This study, like the study of bivalves discussed earlier, found higher rates of extinction in high latitudes compared with the tropics.

One study does show that the tropics may act as a cradle, at least in the case of one genus. *Inga* is one of the most widely dispersed of Neotropical tree genera, with some 300 species ranging from Mexico into Argentina. Did these 300 species accumulate over many millions of years, or are they more recent in origin? Using molecular techniques, a team of researchers led by J. E. Richardson (Richardson et al. 2001) ascertained that the diversification of the genus *Inga* is no older than six million years and may be as recent as three million years ago. They examined 32 *Inga* species from throughout the range of the genus and looked at the *internal transcribed spacer* (ITS), a noncoding part of nuclear ribosomal DNA, and the trn-LF spacer in chloroplast DNA. They compared the number of nucleotide substitutions (considered to occur randomly) that have accumulated among *Inga*

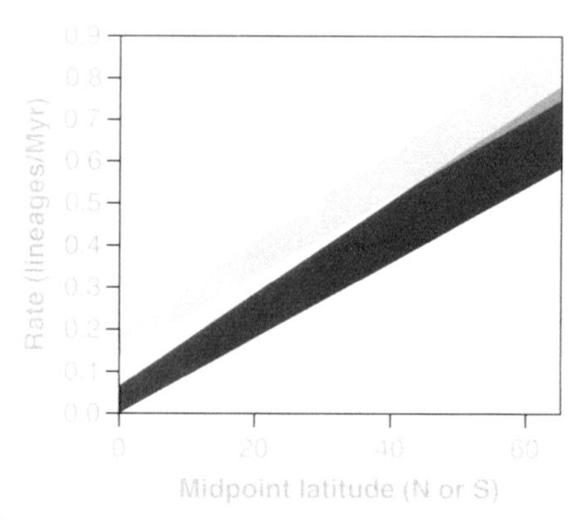

FIGURE 4-18
Estimates of speciation (light) and extinction (dark). This suggests that the highest recent speciation rates and extinction rates occur at high latitudes and decline toward the tropics, suggesting a higher *turnover rate* of speciation and extinction at high latitudes.

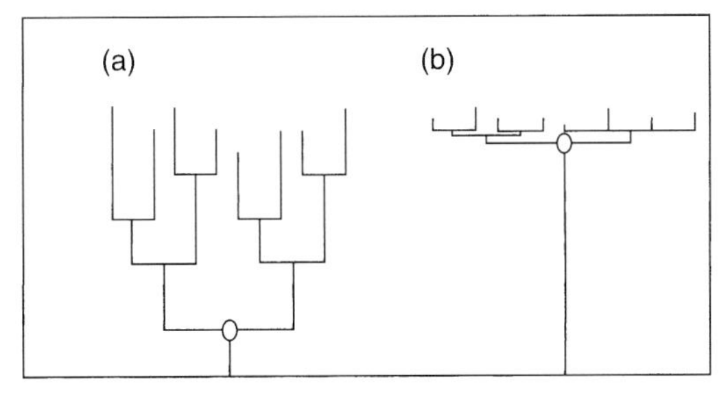

FIGURE 4-19
These figures contrast (a) the gradual accumulation of diversity from the most recent common ancestor (MRCA) with (b) a rapid and recent burst of diversification for the MRCA of an extant species. (b) results in a poorly resolved phylogeny, and (a) results in a well-resolved phylogeny.

species with those published for other plant taxa. The data showed that for both *trnL-F* and ITS, *Inga* is a *monophyletic group* (which means that each species is closely related and all are descended from a common ancestor that is not shared by other genera). In addition, there were few substitutions, indicating a very recent origin for the genus. (Substitutions accumulate with time, so the fewer there are, the more recent the divergence among the species.) Based on the ITS analysis, divergence times were estimated at between 13.4 and 2.0 million years ago, with a mean of 5.9 million years ago. The *trnL-F* data indicate that divergence ranged from 4.3 million years ago to 300,000 years ago, with a mean of 1.8 million years ago. Though the divergence time estimates differed somewhat for each of the DNA segments, they agree in that they indicate a very recent divergence. The authors suggest that *Inga* divergence is likely to be related to the recent Andean mountain building episodes as well as Pleistocene climate changes and the bridging of the Isthmus of Panama (Figures 4-19 and 4-20).

Though speciation in *Inga* has been remarkably rapid, it may not represent the "typical case" among tropical tree genera (Bermingham and Dick 2001). For example, a census of 25 hectares of rain forest in Ecuador and 25 hectares in Panama resulted in the following:

- For Ecuador, 1,104 woody plant species, 333 genera, 81 families, 43 *Inga species*.
- For Panama, 277 woody plant species, 174 genera, 56 families, 14 *Inga* species.

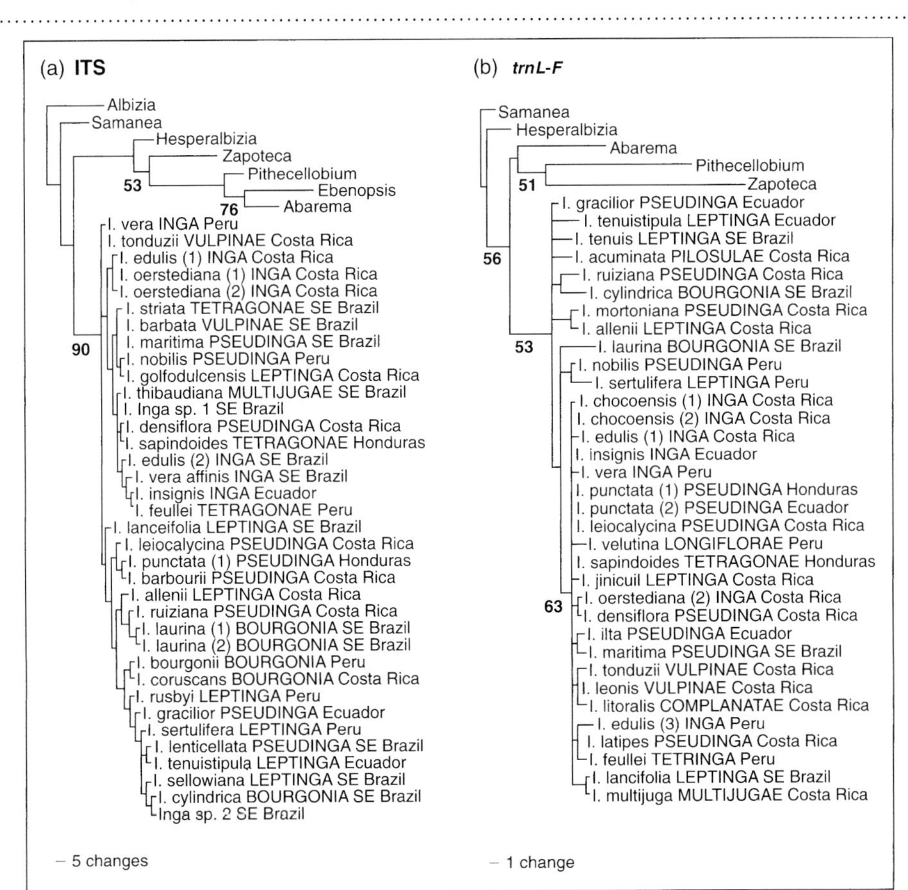

FIGURE 4-20
A comparison of the cladograms generated by ITS compared with *trnL-F* analyses. Note that many differences are evident, but divergence times are nonetheless quite rapid in each case.

What is noteworthy about those numbers is that in Ecuador, 161 of the 333 genera are represented by but a single species. In Panama, 121 genera have but a single species. Therefore, *Inga* is unusual in exhibiting a pattern of multiple closely related species throughout its range. But it is not necessarily unique. Other plant genera such as *Psychotria*, *Piper*, and *Ficus* show a similar pattern, exhibiting high species richness in relation to other genera.

Tropical regions are clearly rich not only in species but also in genera and families, and those take longer to evolve, showing that while *Inga* is clearly a cradle species cluster, many families of tropical plants may be part of a very old museum. It should be clear from the examples cited earlier that many aspects of how the tropics gain and retain species remain uncertain and that different groups of species have been subject to different patterns of diversification.

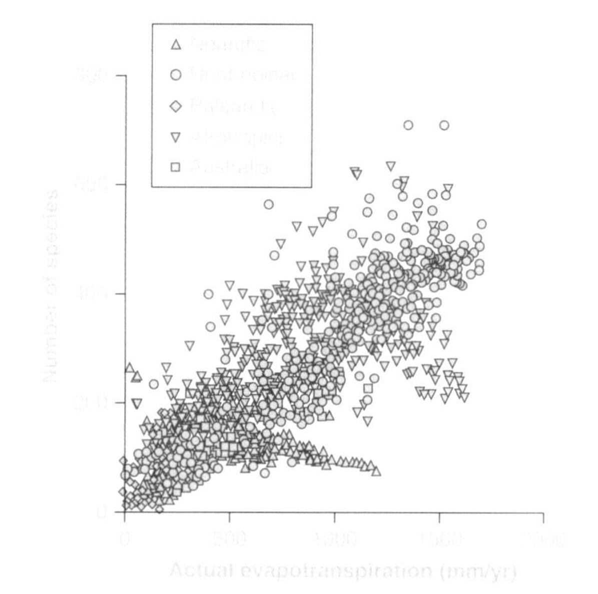

FIGURE 4-21
The relationship between evapotranspiration and number of species of terrestrial birds across five biogeographic regions. High evapotranspiration rates characterize warm regions, such as the tropics.

Climate, Energy Availability, and Species Richness

The latitudinal belt from the Tropic of Cancer to the Tropic of Capricorn, spanning approximately 47° of latitude, is by and large warm and wet. Winters are far less severe than in temperate and polar regions. Light flows more evenly and constantly, as there is less (and on the equator, no) variation in day length. Ecologists have long believed that equitable climate and an abundance of photosynthetic energy, termed *net productivity* (see Chapter 9), ultimately support greater species richness, helping account for the LDG. Let's look at climate as it pertains to the LDG of one group, birds.

Ever since Dobzhansky's 1950 paper, avian species richness has occupied the attention of researchers seeking to explain the LDG. Hawkins et al. (2003a) correlated seven variables with broad global patterns of bird species richness. The variables were (1) potential evapotranspiration (PET); (2) actual evapotranspiration (a measure of productivity); (3) mean daily temperature in the coldest month; (4) range in elevation (a measure of habitat heterogeneity); (5) annual precipitation; (6) annual mean temperature (a measure of ambient energy); and (7) biogeographic region (a measure of historical contingency). The data analysis involved various regression models that allowed partitioning of the variance among the environmental variables and biogeographic history (variable 7). The resulting analysis indicated that actual evapotranspiration (variable 2) explained 72.4% of the variance in global bird species richness. The five remaining environmental variables taken together improved the result only to 76.7%. Looking separately at a historical model based on biogeographic region, 58.2% of the global variance was explained (Figure 4-21).

The authors concluded that productivity, as reflected in actual evapotranspiration, was the best hypothesis to account for the LDG, but they acknowledged the existence of a secondary historical effect based on different evolutionary histories among regions. The data analysis also indicated that in the warm parts of the world, such as the tropical latitudes, water availability correlated most closely with avian richness (Figure 4-22).

A broader study by Hawkins et al. (2003b) examined 85 published data sets on species richness gradients (22 for plants, 44 for vertebrates, and 19 for invertebrates) that each extend in excess of 800 kilometers (480 miles) to ascertain whether species richness correlated with various climatic variables: precipitation, evapotranspiration, productivity, net primary productivity, and net above-ground productivity. Species richness did correlate strongly with climatic vari-

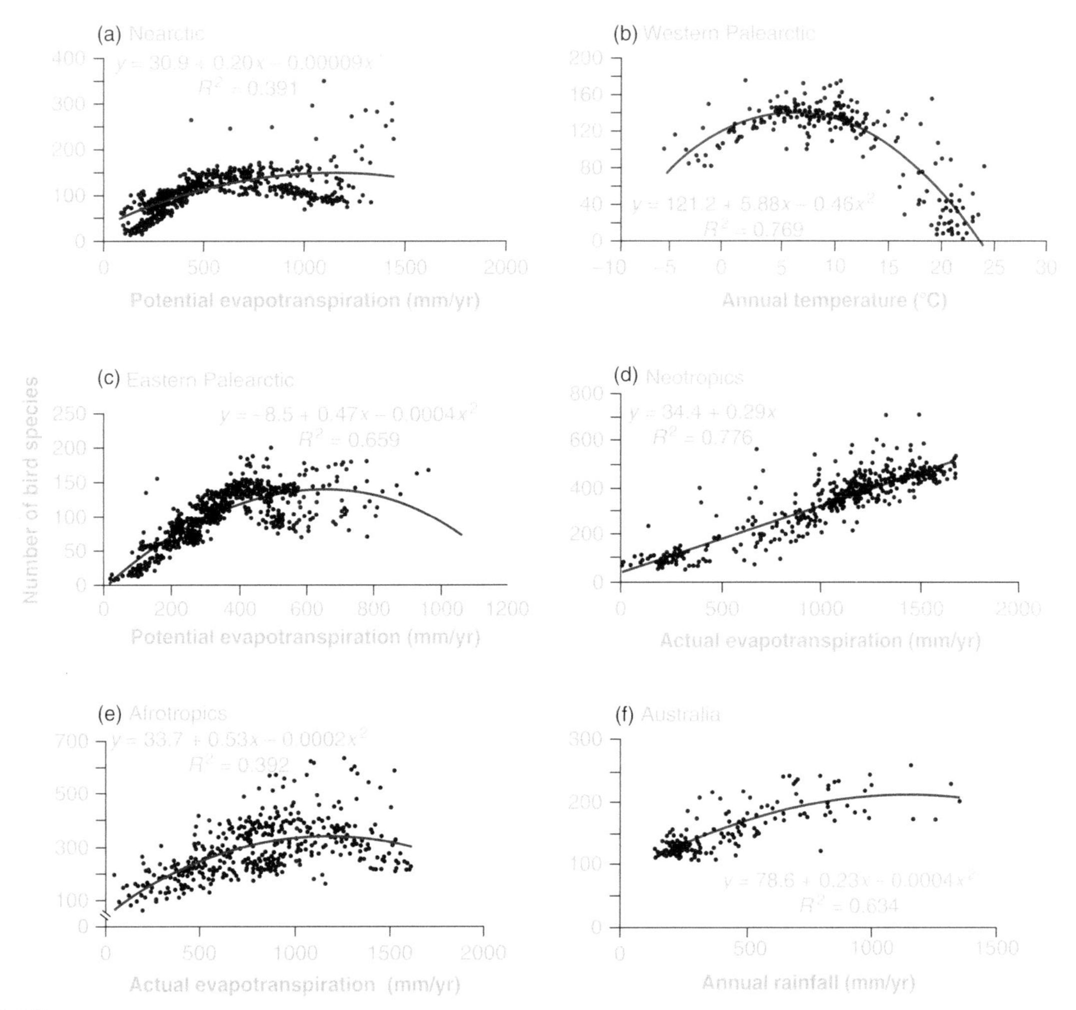

FIGURE 4-22
The relationship between avian richness and various climatic measures in six biogeographic regions.

ables in 83 of 85 data sets. What varied among taxa was whether water, energy, or their interaction best explained species richness.

For plants, water availability was the key constraint limiting species richness. This effect was most pronounced in areas of warm temperature where adequate energy would otherwise be readily available. Of course, it is well known for the tropics that dry seasons have pronounced effects on ecosystems, so this result should not be surprising.

For vertebrates, in 24 of the 44 data sets, water or *water-energy* (where energy was either directly measured or indirectly measured by plant productivity) correlated best with species richness. In 17 other data sets, direct energy measures such as ambient energy input were the best predictor of species richness. Some cases were unique. For example, the best predictor of species richness of mammals in southern Africa was not energy or water per se, but a measure of seasonality. (Dry season is very pronounced in this region.)

Invertebrate species richness also correlated closely with water or water-energy availability.

The study also showed an apparent shift with latitude in the influence exerted by climatic variables on species richness (Figures 4-23 and 4-24). In the far north (and by implication in the far south), energy was the variable that placed the strongest constraints on richness. In these regions, energy availability varies dramatically during the course of the year. But in areas of high energy input, as typify the tropics, water was the variable most responsible for constraining species richness, a result

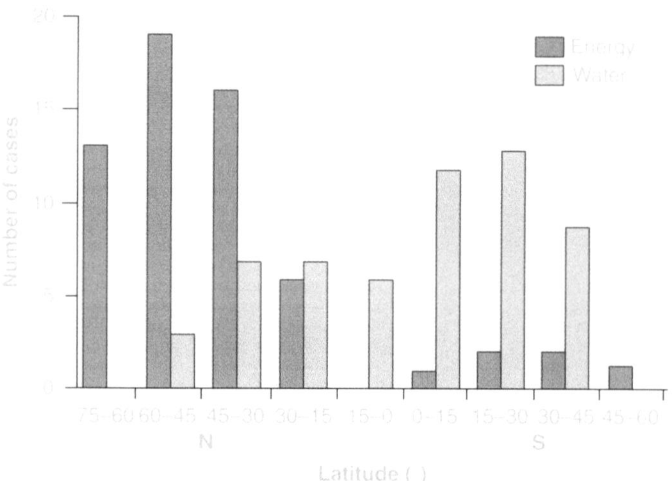

FIGURE 4-23
The bars depict the importance of energy compared with water. As to which of these primary variables best explains animal species richness as it varies with latitude, energy best explains richness for high northern latitudes, but water best explains richness in equatorial and lower latitudes.

found in the previous study with avian species richness. With regard to animal species richness, it should be noted that the study does not determine whether energy and water availability as such acts to limit species richness or whether richness is limited primarily through plant productivity, which, itself, is limited by energy and water availability. But the study does show that climatic variables do correlate broadly with species richness patterns across latitudes. Questions remain, however. The authors note that climate variables may explain up to 90% of the variance in species richness in some cases, but in other cases, they explain less than 50%. Such factors as evolutionary history (as was seen in previous studies) and biotic interactions (see the following) may also exert strong influences.

Diversity in the Tropics: The "Perfect Storm"?

Many attempts have been made to summarize the various possibilities for why tropical ecosystems are so rich with species. A now-classic paper by Eric Pianka (1966) attempted to summarize the forces that might

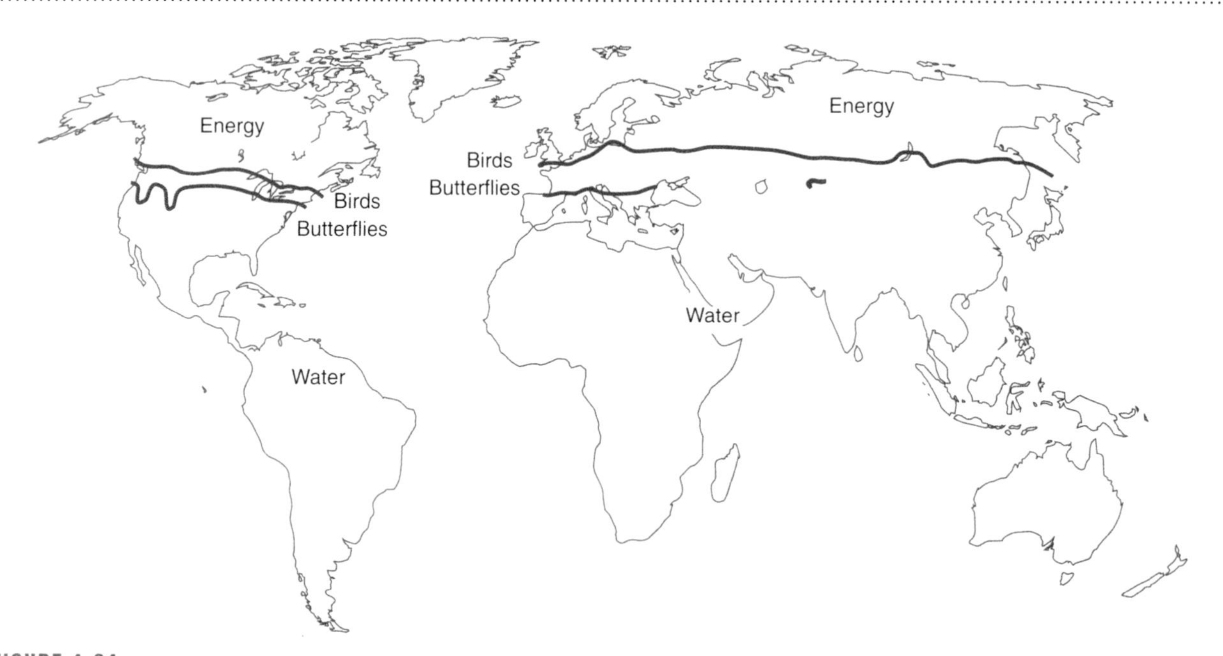

FIGURE 4-24
This map illustrates for birds and butterflies the hypothesis that energy is the limiting component of species richness in northern latitudes, while water is the limiting component in equatorial and southern latitudes. It should be noted that both variables are influential at all latitudes, but the influence of energy dominates to the north, water to the mid- and south latitudes.

account for hyperdiversity in the tropics and lowland rain forests in particular. We have already addressed some possibilities in the form of broad patterns of climate and evolutionary history, as they pertain to the LDG. But what about the effects of ecological forces within tropical ecosystems as such? What additional forces might act to increase species numbers and to maintain diversity? A review of Pianka's hypotheses for factors causal to tropical diversity suggests that the tropics may represent a "perfect storm" for speciation and diversity maintenance, in that ecological and biogeographic conditions combine in an ideal manner, like atmospheric variables that on occasion create monumental storms, to produce prodigious numbers of species and then to keep them. In this section, I discuss four contrasting hypotheses (stability–time, productivity–resources, interspecific competition, and predation intensity) that have been offered to account for why the tropics are so diverse. Together they may form the perfect storm for generation and maintenance of high species richness. Consider each and think of how each might be tested.

THE STABILITY–TIME HYPOTHESIS

This hypothesis suggests that the tropics are ancient and that such antiquity coupled with the relatively stable and equitable climate has resulted in the generation and persistence of high species richness. The idea is closely akin to what Dobzhansky believed to be the case.

A look at the fossil record of plant pollen helps focus the question of just how species-rich the tropics have been over time (Jaramillo et al. 2006). The study examined pollen and spore profiles taken from sites in central Colombia and western Venezuela. The sites spanned a time sequence from the Paleocene to the early Miocene (65 to 20 million years ago). A total of 15 stratigraphic sections, 1,530 samples, 1,411 *morphospecies* (distinct but unknown species), and 287,736 occurrences provided the data set (Figure 4-25). The data analysis showed low plant diversity during the early Paleocene. This was a time just after the end of the Cretaceous period that is considered one of the five mass extinction events in Earth's history. Thus a low diversity of plants at that time is expected.

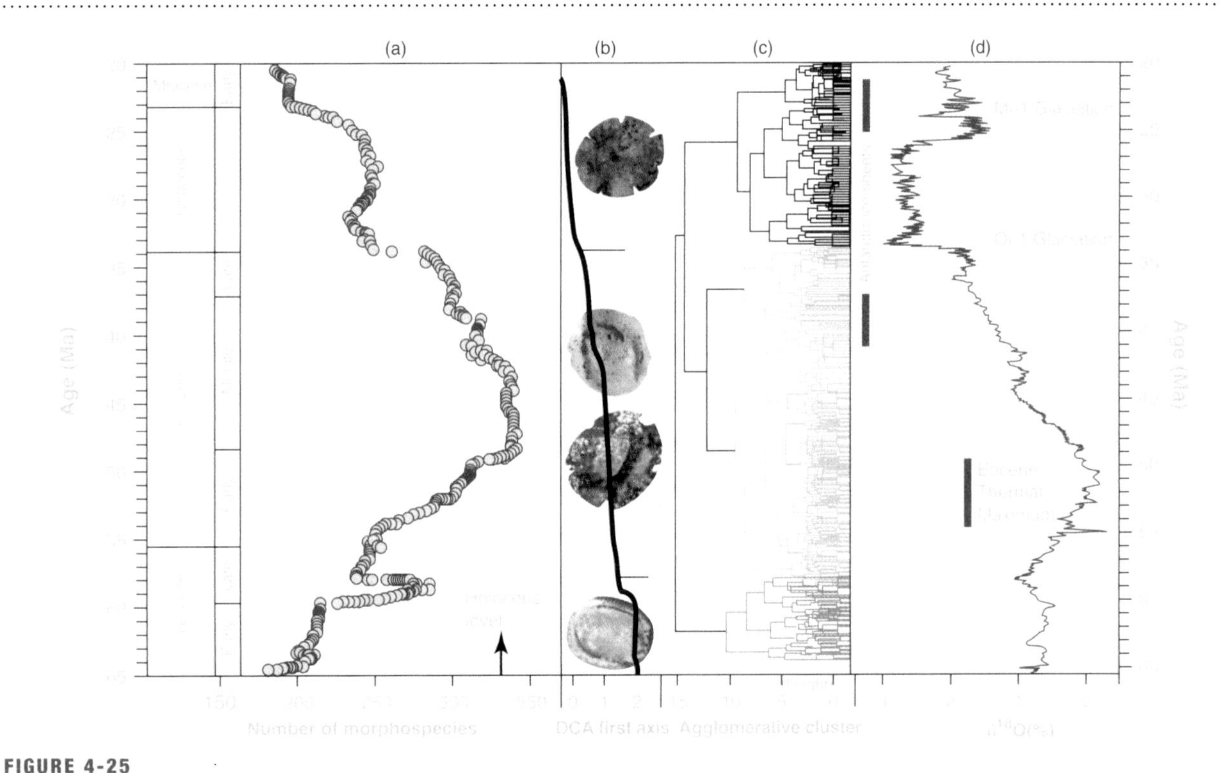

FIGURE 4-25
(a) The pattern of change in morphospecies of pollen and spores as it ranges from the Miocene through the Paleocene. (b) The change that is evident in the line shows Paleocene pollen-based floras to be clearly different from Miocene pollen-based floras. (c) A cluster analysis showing three distinct pollen-based floras through time. (d) The change in global oxygen for the Cenozoic. Note the rough correlation between the oxygen curve (d) and the morphospecies curve (a).

During the early Eocene, there was a rapid increase in plant diversity, but diversity declined beginning in the late middle Eocene and continued to drop until the early Oligocene (34 million years ago). The extinction rate was elevated during the transition from the Eocene to the Oligocene, and speciation rate increased during the early Eocene (Figure 4-26). Otherwise, both extinction and speciation rates were steady. Species richness varied closely with changes in global temperature. The authors note that global warming during the early and middle Eocene permitted the spread of tropical plant species into higher and lower latitudes and note that the increased area of tropical ecosystems might, in and of itself, be a causal factor in increasing speciation rate. (This relationship between area and speciation will be a focus of part of the next chapter.) The authors note too that the Andean uplift (described in Chapter 2) also was a major stimulus to speciation and that most of that mountain building is recent (within the past 5 million years). They conclude that "plant diversity in the Neotropics has fluctuated greatly through time, as it is sensitive to global temperature. Temperature or precipitation change in the tropics may explain the pattern" (p. 1896). The study shows that the tropics have never been really stable but have always been subject to some degree of global climatic fluctuations.

Speciation is not always dependent on vast periods of time. There are five species of kingfishers (Alcedinidae) that inhabit Neotropical rivers and streams, an assemblage that represents the total species richness of kingfishers in the Neotropics. Kingfishers are birds that dive into water and capture fish with their long beaks. The five species of Neotropical kingfishers differ fundamentally in body and bill size (see Figure 4-27). The fossil record, as well as biogeographic studies, shows that kingfishers evolved in the Old World and found their way to the Neotropics relatively recently, probably during the Pleistocene. What is interesting is that of the 91 kingfisher species in the world, only six are in the Americas (one is confined largely to North America). Four of the six (the "green-backed" kingfishers) are each other's closest relatives and likely evolved in the Neotropics from a common ancestor. Interestingly, wherever in the world kingfishers occur, there are never more than five kingfisher species together, coexisting in the same habitat (a question worth pondering in relation to how resources affect

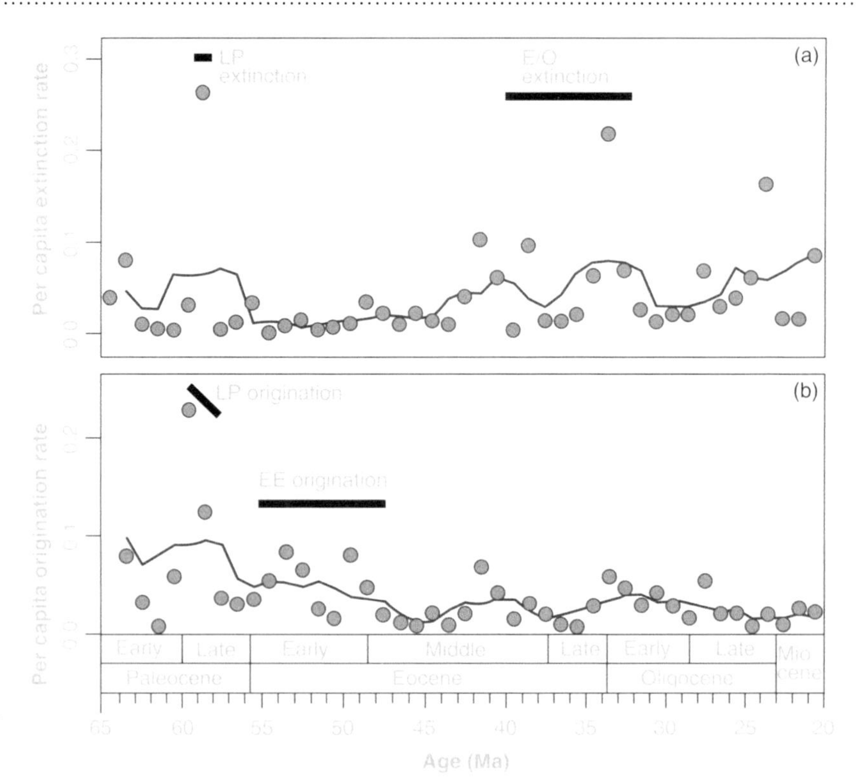

FIGURE 4-26
The upper graph depicts the per capita extinction rate (for plants) through the early and mid-Cenozoic, while the lower graph shows the per capita origination rate. There is an increase in extinction at the Eocene–Oligocene transition (upper graph), and the lower graph shows a slow decline in origination over time.

FIGURE 4-27
These are all of the kingfisher species found in all of tropical South America. All five species—pygmy, green, Amazon, green-and-rufous, and ringed—often occur in the same area.

species richness and coexistence) (Figure 4-28). The five species of Amazonian kingfishers span the same size range and within-habitat species richness as that found anywhere in the Old World (Remsen 1990). This means that in the two million or so years since kingfishers invaded the New World tropics, they have evolved a local diversity equal to that found anywhere else where kingfishers occur. This pattern, like that of *Inga* described earlier, suggests that speciation events may occur in geologically short time periods, and thus

FIGURE 4-28
These 11 kingfisher species are found in Kenya, but no more than 5 are found in any given area.

that species richness is not directly dependent on long time periods.

The major premise of the stability–time hypothesis is in all likelihood invalid, though the fact remains that the tropics have never been climatically harsh. There is almost universal agreement among climatologists, geologists, and biogeographers that tropical regions have not, in fact, been climatically stable, though they have remained warm and relatively equitable. The argument that high tropical diversity accumulates over time from tropical regions merely being old and climatically stable is not strongly supported. For one thing, it fails to address how such high diversity is ecologically maintained as it becomes ever higher.

THE PRODUCTIVITY–RESOURCES HYPOTHESIS

One frequent suggestion to explain high diversity in the tropics is that high plant productivity permits more species to be accommodated. As discussed earlier, this may indeed be the case. Variables of climate, as they affect plant productivity, strongly correlate with latitudinal species richness. Many questions remain, however. One is the question of how an abundance of plant productivity translates into resources supporting high species richness of animals. A simple measure of biomass may be insufficient to explain some patterns. Consider these examples.

Perhaps high tropical productivity and biomass translate into more space for more species. Daniel Janzen (1976), in a provocative paper titled "Why Are There So Many Species of Insects?" concluded by saying, "I think that there are so many species of insects because the world contains a very large amount of harvestable productivity that is arranged in a sufficiently heterogeneous manner that it can be partitioned among a large number of populations of small organisms." Janzen was not restricting his speculation to the tropics, but his remark fits the tropics particularly well. There is indeed a tremendous potential harvestable productivity, and there are lots of spaces for small animals in the three-dimensionally complex rain and cloud forests.

Support for this view comes from studies such as one done on birds of the Peruvian Amazon, where it was demonstrated that the total number of breeding birds was equivalent to that in many temperate forests but that their combined biomass was about five times as great in the tropical forest (Terborgh et al. 1990). What is it about tropical forests that permits such a greater avian biomass? What is the resource base?

Bird species diversity often correlates with foliage height complexity in mid- to high latitudes. The more layers of foliage there are, the more bird species. In the tropics, however, bird species diversity does not correlate as closely with foliage height diversity (Lovejoy 1974). This indicates that a tropical forest must be more spatially complex than temperate forests, offering resources not detected or measured by simple structural analysis.

Tropical forests do indeed offer unique resources for birds (Karr 1975) and, by implication, for other kinds of animals. What are these unique resources?

- Among these additional resources are large numbers of vines, high epiphyte density, and large dried leaves (which harbor many kinds of arthropods), all of which add both space and potential food resources.
- There are numerous potential prey items in addition to small prey. In a now-classic study, Schoener (1971) showed that insectivorous birds in the tropics have a much wider range of bill length (among all species) than those of the temperate zone (Figure 4-29; Plates 4-11 and 4-12). Schoener attributed

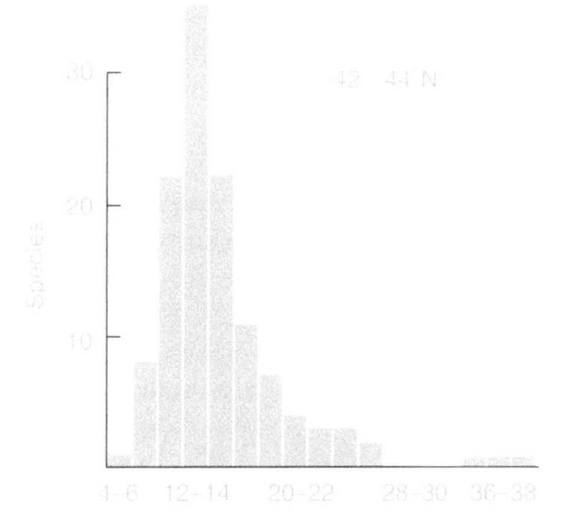

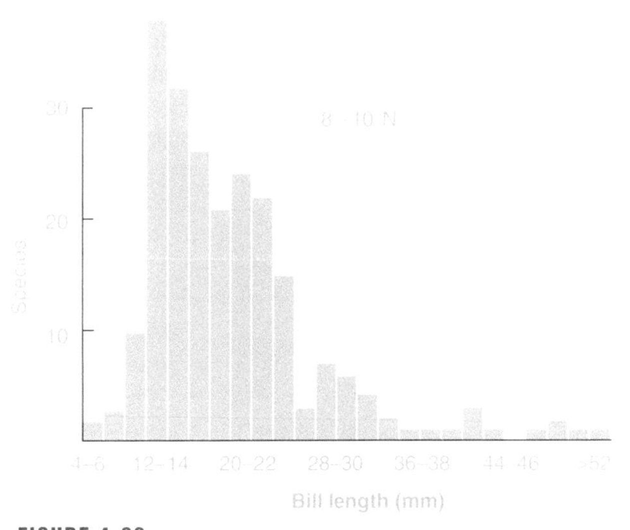

FIGURE 4-29
Comparison of bill lengths in tropical bird species and temperate bird species. Note the wider range evident in the tropics.

PLATE 4-11
Large dried tropical leaves form ideal habitats for many arthropods, which are used as food by various bird species.

this reality to the wider resource base available in the tropics, especially of large prey items.

- There is also year-round availability of nectar and fruit. Nectar specialists include hummingbirds, sunbirds, and lorikeets, and fruit-eating specialists include cotingas, figbirds, cassowaries, guans, curassows, and parrots. None of these groups can exist nearly as successfully outside the tropics, since they are so dependent on constant availability of nectar and/or fruit.
- Army ants are a unique resource in the Neotropics in that their activity provides the basis for the existence of a complex group of *professional antbirds*, a species that exclusively follows army ant swarms, feeding on the arthropods driven out by the ants. Many other bird species, not as specialized, nonetheless also become active at ant swarms, feeding on the arthropods disturbed by the swarming ants.
- Forest gaps also represent a kind of resource. A gap is created when a tree or group of trees is blown down by wind or some other disturbance event. Light availability is higher in the gap, and that permits the growth of species suppressed by forest shade. Some species of plants and animals are *gap-specialists*. Gaps are a characteristic of all forests, but tropical forest gaps may present more opportunities to specialist species than those outside the tropics. Gap dynamics will be discussed in detail in Chapter 6.

PLATE 4-12
Lanternfly, an example of one of many large tropical insects that extend the arthropod resource base more widely in the tropics.

Do additional resources translate into specialization? Specialization occurs when a species becomes uniquely adapted to a narrow resource base. For example, the abundance of Neotropical bats that inhabit rain forest has permitted the evolution of specialization in the diet of the black-and-white owl (*Ciccaba nigrolineata)*. This species, which ranges from southern Mexico to northwestern Peru, is a bat specialist, feeding almost entirely on bats (Plate 4-13). Another

PLATE 4-13
BLACK-AND-WHITE OWL

PLATE 4-14
BAT FALCON

PLATE 4-15
BLACK-CHEEKED WOODPECKER

species, the bat falcon (*Falco rufigularis*), also specializes to a degree in bat capture, feeding crepuscularly on the flying mammals (Plate 4-14).

It has long been assumed that specialization is widespread in tropical ecosystems, though that should not be taken to mean that all species have narrower foraging niches (see the following).

Specialization and narrow resource partitioning need not be the case for many, if not most, species if additional resources are uniquely available in tropical forests. In a study comparing the woodpecker communities of Maryland, Minnesota, and Guatemala, Robert A. Askins (1983) recorded seven woodpecker species in Guatemala compared with only four in each of the temperate study areas. Although there were more species of tropical woodpeckers, these were in reality no more specialized in their foraging behavior than their temperate relatives, and so their niches were not narrower in the tropics. They did utilize a wider range of resources, however. Some of the tropical species fed on termites and ants, probing into the excavations made by these insects, and the tropical species fed much more heavily on fruit than the temperate woodpeckers. Two species in particular, the black-cheeked (*Melanerpes pucherani*) (Plate 4-15) and the golden-olive (*Piculus rubiginosus*), utilized resources not available to temperate species. The black-cheeked frequently fed on fruit, and the golden-olives probed moss and bromeliads.

Novotny et al. (2006) examined why so many species of herbivorous insects are found in tropical rain forests. Using *folivorous* (leaf-eating) insects as the database, the study attempted to ascertain whether insect species in rain forests typically exhibit narrow host specificity. In other words, are species in rain forests restricted more in their dietary requirements than those in temperate forests? This hypothesis follows from the widespread belief that many tropical

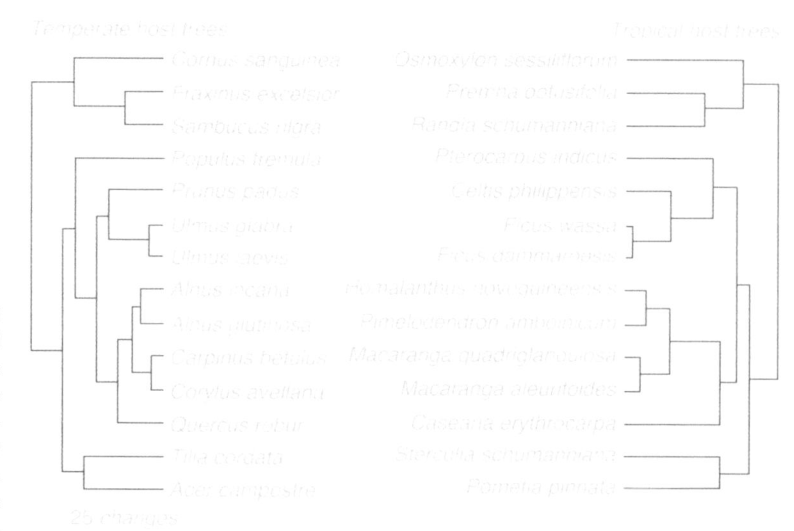

FIGURE 4-30
This cladogram shows that the temperate and tropical tree species of comparable phylogenetic distribution support similar numbers of folivorous insect species. The cladogram represents 14 tree species from Moravia (temperate zone) and Papua New Guinea (tropical), with branch lengths proportional to the number of nucleotide substitutions in rbcL sequences.

species may have narrower ecological foraging niches than their temperate counterparts. But the study, like the one by Askins discussed earlier, concluded that at least for this group of insects, specialization was no greater in the tropics than in the temperate zone (Figure 4-30). Host specificity did not differ significantly between temperate and tropical forests. Thus resource partitioning was not finer in tropical compared with temperate forests. Lepidopteran larvae (caterpillars) and adult beetles (Coleoptera), for example, did not differ in host specificity between temperate trees (a floodplain forest in Moravia in central Europe) and tropical trees (a lowland hill forest in Papua New Guinea). The study controlled for phylogenetic diversity of trees to ensure an honest comparison between the two regions. The researchers wanted to obtain comparable evolutionary histories between the trees examined at both geographic locations. Because there are far more tree species in rain forests, that alone may account for why there are so many more insect species (recall Erwin's data on beetles, discussed earlier). The average tropical tree (in New Guinea) had about 23.5 insect species per 100 square meters of leaf tissue compared with about 29 in its counterpart in temperate forests (in Moravia) (Figure 4-31). Thus they were basically the same. But there was a sevenfold increase in plant species richness between the two sites (21 species of trees per hectare in Moravia and 152 per hectare in New Guinea). And because the study selected trees based on comparable phylogenetic diversity, it included 85% of the tree species in the temperate forest but only 20% of those in the

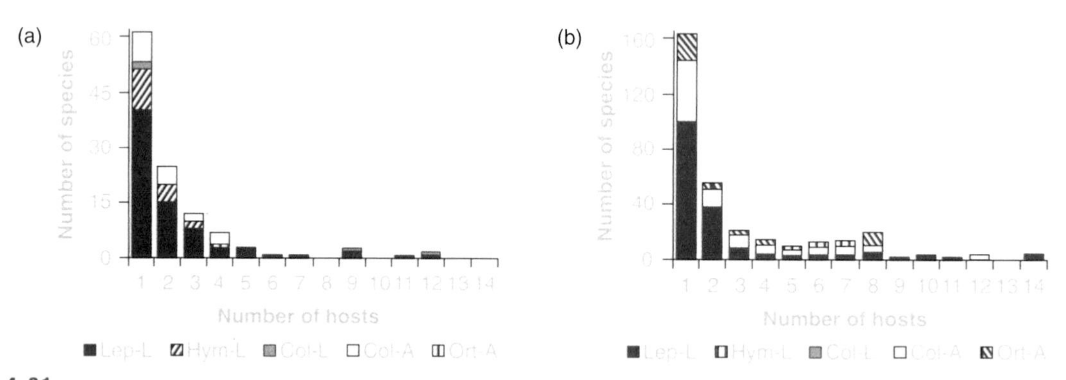

FIGURE 4-31
Host specificity in (a) temperate and (b) tropical trees. Represented are Lepidoptera larvae, Hymenoptera larvae, Coleoptera larvae, Coleoptera adults, and orthopteroid adults. Note the similarity in insect distributions between the tropical and temperate trees.

tropical forest. The authors conclude that "greater phylogenetic diversity of tropical vegetation compared to temperate forests rather than greater host specificity of tropical herbivores is the more probable explanation for the extraordinary diversity of tropical insect communities."

In contrast to woodpeckers and folivorous insects, bats (order Chiroptera) do represent a group that has specialized to various degrees in tropical ecosystems. In the Neotropics, there are many species of insectivorous bats, but there are also numerous species of fruit-eating bats as well as bats that specialize on consuming nectar, fish, frogs, and blood (the infamous vampire bats). Studies that have examined bat communities in various parts of the tropics have found species-specific differences in diet, foraging strategies, and roost sites, all of which indicate that diverse tropical bat communities are examples of specialization (Kalko and Handley 2001).

Bamboo stands provide a somewhat different example—namely, how patchy additional tropical resources enhance species richness. As such, they allow for another form of specialization (Plate 4-16). In the Neotropics, tall bamboo stands, where plants reach heights of up to 15 meters, support as many as 21 bird species specialized in some way to feed exclusively within bamboo stands. Nine species specialized in eating bamboo seeds, and 12 were insect foragers. An additional 16 species of insect foragers were found mostly in bamboo but also in other habitats (Remsen and Parker 1983).

Ephemeral Amazonian River islands of sandbar scrub (Plate 4-17) and young successional forests dominated by a single tree species, *Cecropia*, provide yet another important resource to which some bird species have specialized (Plate 4-18). For example,

PLATE 4-16
BAMBOO STAND

PLATE 4-17
Riverine sand island in the Napo River in Ecuador.

PLATE 4-18
This riverine island in Ecuador supports a young stand of rapidly growing *Cecropia* trees.

a visit to a sandbar island along the Napo River in Ecuador might, if you look carefully, turn up white-bellied spinetail (*Synallaxis propinqua*), Castelnau's antshrike (*Thamnophilus cryptoleucus*), and Parker's spinetail (*Cranioleuca vulpecula*). These three furtive and little-known bird species are each sandbar specialists, nesting only among the dense vegetation colonizing riverine sandbars.

THE INTERSPECIFIC COMPETITION HYPOTHESIS

One of the most difficult measures to make in ecological research is the degree to which competition occurs between two or more species. One must be able to

Arthropod Abundances on the Forest Floor: Space versus Nutrients

A study was performed in Panama over a 32-month period that looked at how forest floor arthropod diversity is related to the amount of litter present (a measure of the space available) compared with the nutrient quality of the litter (the quality of the habitat) (Sayer et al. 2010). Researchers manipulated the litter layer such that the litter amount was enhanced at part of the site and left unaltered as a control at part of the site. The data clearly showed that arthropod density was highest in areas with the most litter. For example, the average number of individual arthropods per square meter was 2,384 in the area with litter addition compared with 1,360 in the control. This suggests that the actual space available to arthropods is important in affecting overall arthropod biomass. However, arthropod diversity was more affected by nutrient characteristics, as they varied from site to site. Specifically, diversity correlated with phosphorus, calcium, and sodium concentrations in the area where fermentation occurs as well as by phosphorus concentration in the upper area of the litter layer (regardless of how much litter there was). The conclusion from the study was that both habitat space and habitat quality affect the arthropod community, but in different ways. The highest diversities correlated with habitat quality in the form of nutrient availability, but more arthropod biomass was found with greater depth of litter (Plate 4-19).

PLATE 4-19
This spider, part of the arthropod litter community, preys on other arthropods (mostly insects) in the litter layer within an Ecuadorian lowland moist forest.

demonstrate that two (or more) species are seeking the same limited resource and then measure the degree to which each species negatively affects the other(s) contesting for the resource. It is essential to identify the resource being contested and demonstrate the fact that it is a limited resource. If it is not limited, there will presumably be enough for both species and no competition. (So if you were to observe two different species each using the resource, you could not, automatically, conclude that they are in competition, any more than two people ordering steaks at a restaurant are in competition.) It is also essential to show how the competition affects each species. The following possibilities exist:

- One species outcompetes the other, and the loser goes extinct, a process termed *competitive exclusion*.
- The two species somehow subdivide the resource, each specializing in a part of the resource spectrum. This is called *niche partitioning*. When this

happens, each species has become more specialized. There are two ways for this to happen. One is that a *species-realized niche* (the actual resources it uses) is forced by competition to become narrower than its *fundamental niche* (the niche space it would occupy in the absence of competing species). The second way is that the actual fundamental niches may evolve to be narrower.

One hypothesis to explain high species diversity in the tropics argues that high levels of competition among species have, over time, resulted in greater *niche partitioning* (a form of increased specialization). Each species evolves into a specialist focusing on a specific resource that it and it alone is best at procuring. This trend toward specialization due to interspecific competition leads to the "packing" of greater and greater numbers of species into tropical ecosystems while at the same time reducing the intensity of competition among species as each specializes to its exclusive pool of resources. The interspecific competition hypothesis argues that many, if not most, niches are narrower in the tropics than in the temperate zone because competition has compressed them. The hypothesis is also dependent on the reality that greater specialization is possible in the tropics because particular types of resources (such as large arthropods or long tubular flowers) are consistently available throughout the year, showing seasonal stability. In the temperate zone, many specific types of resources are available during only a brief season each year, so species are forced to be generalists, particularly if they do not migrate.

There are problems with this hypothesis:

- We have already seen for woodpeckers and for herbivorous insects that the species do not exhibit narrower niches in the tropics.
- Though competition may have exerted a major influence in the past, now that specialization and niche compression have occurred, competition may be quite minimal or even nonexistent. It should be obvious that this is a very difficult hypothesis to test.

For example, both the ocelot (*Felis pardalis*) (Plate 4-20) and margay (*F. wiedii*) occur throughout most Neotropical lowland forests. These two small cats are similar in body size, the ocelot being a bit larger. Both are essentially nocturnal, and it is not unreasonable to assume that they feed on many of the same prey items: rodents, birds, snakes, lizards, and other small mammals.

PLATE 4-20
OCELOT

The ocelot and margay differ somewhat in their foraging behavior, as the ocelot is almost entirely terrestrial, while the margay routinely climbs trees in the course of its hunting behavior. Thus the ocelot and margay have foraging niches that do not precisely overlap. Did competition between these two small felines cause the divergence of foraging niche, selecting for the smaller cat to become more arboreal? There is no way to know. It cannot be known whether resources were limited, and even if they were, it cannot be known whether that limitation was, indeed, an active selection pressure. In other words, it cannot be known whether margays that climbed reproduced better than those that did not because they climbed, and even if they did, that could have been caused merely by the availability of arboreal resources presenting an opportunity for procuring more calories and protein, not because ocelots were also foraging on the forest floor.

As the preceding example was meant to illustrate, evidence for the interspecific competition hypothesis is mostly circumstantial. Direct demonstrations of interspecific competition are generally lacking in the tropics. Certain patterns suggest, however, that competition among species may be a component of tropical evolution and diversification.

Varying bill shapes and gradations in body sizes within many bird groups suggest that competition may have influenced the evolutionary history of these groups. (Review the kingfisher images earlier in this chapter [Figures 4-27 and 4-28], and see how size and bill length vary among species.) But different body sizes and bill characteristics also reflect specialization for capturing different food items. The very act of food capture per se could and probably does select for such specializations, though it is quite likely that the presence of similar species with similar ecological needs could act as an additional strong selection pressure in producing divergence among species. Insectivorous bats, for example, feed on different-sized prey items and also forage at different heights in the rain forest, a pattern possibly reflective of avoidance competition among the bat species.

One measure of interspecific competition is to demonstrate that a species exhibits a limited foraging niche, presumably because of the presence of other competing species. Each of three flatbilled flycatcher species found in Costa Rica forages at a different height in the forest, and certain flatbill species replace others in specific habitats, one species occurring in forest, another very similar species only in disturbed brushy areas (Sherry 1984). This pattern is suggestive of competitive relationships having influenced the birds' foraging choices (Figure 4-32).

Andean bird communities also show indirect evidence for interspecific competition. In a study of two

FIGURE 4-32
This amazing array of small flycatchers (Tyrannidae) are all found in Venezuela. Those most similar usually do not occur together in the same habitat. Note the differences in body size and bill characteristics.

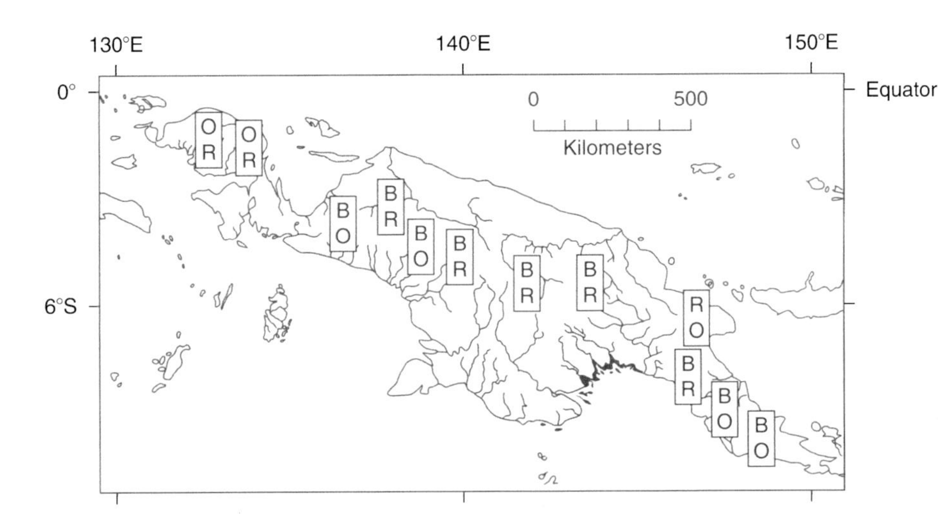

FIGURE 4-33
Compound checkerboard exclusion shown in three *Melidectes* honeyeater species in New Guinea. No region supports more than two of the three species.

Andean mountain peaks in Peru, one of which was isolated, colonization by birds was less frequent on the isolated peak, and thus this peak had a much reduced bird species richness, missing an estimated 80% to 82% of the bird species that would have occupied the isolated peak had it been part of the main range of the Andes. However, of the species that did occur, 71% had expanded their altitudinal range (compared with the other, more diverse mountain peak) presumably because of the absence of similar species that would have been competitors (Terborgh and Weske 1975). The researchers concluded that the combined effects of direct and diffuse competition account for approximately two-thirds of the distributional limits of Andean mountain-dwelling birds.

In a classic study done on bird distribution in New Guinea, Jared Diamond (1973) found evidence of competitive exclusion in several examples. In what he termed *checkerboard exclusion*, he showed that in any given area within New Guinea, there were two species of honeyeaters in the genus *Melidectes* (Figure 4-33). But the genus is a superspecies consisting of *three* closely related species. Thus any one location has two but not three species, suggesting that only two may coexist in any given area.

Diamond also provided compelling evidence for competitive exclusion along an elevational gradient. One species, *Crateroscelis murina* (a warbler) occupies the lower elevation, while another, *C. robusta,* is found higher (Figure 4-34). What is compelling is the sharpness of the separation.

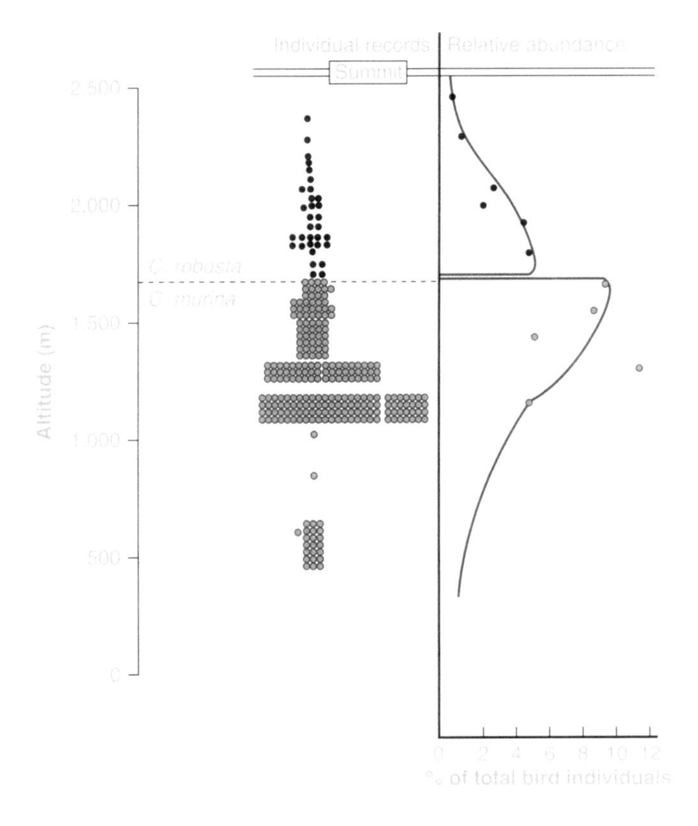

FIGURE 4-34
Two New Guinea warbler species, *Crateroscelis robusta* and *C. murina*, do not overlap altitudinally. Note that the highest relative abundance of each species occurs where they meet. The left side of the figure represents actual birds recorded, while the right side graphs relative abundance of each warbler in the entire bird community. The two species replace each other at 1,643 meters, where each species reaches its maximum abundance.

PLATE 4-21
Miconia produce fruits used by numerous bird species, many of which, like manakins, spread seeds. *Miconia* belong to the Melastomataceae family.

It is abrupt, with no overlap. Would the ranges end abruptly if there was but a single species and not two? But note that as impressive as these examples are, they demonstrate only patterns; competition was not actually measured.

Interspecific competition for pollinators and seed dispersers may have provided a selection pressure resulting in staggered flowering patterns among some plants. In Trinidad's Arima Valley, 18 species of the shrub *Miconia* (Plate 4-21) have flowering times staggered in such a way that only a few species are flowering in any given month (Snow 1966). Is this a possible evolutionary result of competition among *Miconia* for access to birds that eat the fruit and thus disperse the seeds (Plate 4-22)? Any *Miconia* species that flowered when most others did not would be able to attract more birds to disperse its seeds, and thus it would have a selective advantage compared with others of

PLATE 4-22
The white-bearded manakin (*Manacus manacus*) is a common disperser of *Miconia* seeds.

its own as well as other species. Over time, the staggered flowering pattern emerged. Staggered flowering also occurs in other species groups such as tropical Asian dipterocarps.

A similar pattern among plants that are bat-pollinated has been observed in Costa Rica (Heithaus et al. 1975). Of 25 commonly visited plant species, an average of only 35.3% flowered in any given month.

The patterns described here for inferred interspecific competition do not, in and of themselves, suggest that niches are narrower in the tropics, and none really demonstrate competition in a rigorous manner. This is a real weakness in attributing greater species richness in the tropics to greater interspecific competition. Nonetheless, patterns are compelling. High species richness of rain forests may be partly explained by competition among species, at least for some species groups. It would be very helpful if robust data replaced suggestive patterns.

THE PREDATION HYPOTHESIS

Interspecific competition involves interactions among species within the same trophic level. Interactions among species at different trophic levels also have the potential to influence species richness. One form of such interactions is termed *top-down*. This means that species at a higher trophic level (such as predators) exert a major influence on species at a lower trophic level (such as herbivores). Chapter 8 will focus on trophic dynamics and only an introduction will be provided here.

Suppose that four caterpillar species are competing for the same plant. One species is increasing much more than the others, causing the others to be driven toward extinction. What was a "four-species system" is about to become a one-species system. But suppose that birds and lizards prey on the caterpillars. Which of the four species are the predators likely to take? The most obvious and abundant species would seem the likely choice. The result of predation would be to reduce the growing population of the "winning" species, allowing the other "losing" species to regain some control of the resources and increase in population. This scenario, an example of top-down trophic dynamics, explains how predators may maintain species richness at lower tropic levels by switching among prey based solely on prey abundances.

The predation hypothesis posits that predators prevent prey species from competing within their ranks to the point where extinction occurs (Plate 4-23). Predators are thought to switch their attentions to the most abundant prey; thus the rarer the species, the safer it is from predators. This idea is a form of frequency-dependent selection, because the intensity of selection (in the form of predation) depends directly on the abundance of the prey (such that rare species have a selective advantage). The result of predator

PLATE 4-23
The Neotropical eyelash viper (*Bothriechis schlegelii*) is one of many predatory species whose collective influence may affect diversity patterns in tropical forests.

pressure is to preserve diversity by preventing extinction by competition.

Note that the predation hypothesis is basically the opposite of the interspecific competition hypothesis. The competition hypothesis suggests that competition among species promotes diversity by leading to specialization, narrower niches, and tighter species packing. The predation hypothesis suggests that predators reduce interspecific competition, thus permitting coexistence among competing species. The predation hypothesis does not predict extreme specialization. Indeed, specialization would seem less likely to occur because predators keep competition levels low. Likewise, the predator hypothesis allows for wide niche overlap among similar species.

Tropical forests contain impressive predator richness, circumstantial evidence for the possible importance of predator effects. In a study conducted in a tropical moist forest in Puerto Rico, lizards in the genus *Anolis* were shown to have strong effects on the arthropod community of the rain forest canopy (Plate 4-24). When lizards were experimentally excluded, large arthropods such as orb spiders, cockroaches, beetles, and katydids all increased significantly (Dial and Roughgarden 1995).

What if predators are nonselective, even when prey densities vary? Terborgh (1992) suggested an intriguing example of how predators, specifically cats, might be inadvertently structuring rain forest communities. Citing Emmons's (1987) work, Terborgh points out that jaguars, pumas, and ocelots all forage nonselectively, taking whatever they encounter and can catch. In other words, the populations of prey species correlate directly with the frequency upon which each prey species is taken by a cat. If agoutis represent 40% of prey species, agouti remains show up in cat scat 40% of the time. Note that in this example, predators are not automatically switching to the most abundant prey, merely taking what they find as they find it. Terborgh goes on to argue that since prey species differ in their *fecundities* (rates of reproduction), predation by nonselective cats could significantly reduce certain low-fecundity prey populations. In other words, peccaries can produce more offspring annually than pacas, so if cats do not ever discriminate between peccaries and pacas, pacas must decline more than peccaries, since they cannot replenish their losses as quickly (Plates 4-25 and 4-26). Because pacas and peccaries both eat many of the same things, an ecologist might be tempted to conclude that paca reduction

PLATE 4-24
ANOLIS LIZARD

PLATE 4-25
PACA (*CUNICULUS PACA*)

was due to losing in competition for food with peccaries, never guessing that cat predation was the real reason. Terborgh's argument demonstrates that predation can result in a loss of diversity, as well as act to maintain it, if predation is, in fact, nonselective rather than frequency dependent.

Consider also that it is possible that top-down forces actually promote specialization rather than prevent it. In a study done near Iquitos, Peru, within Amazonia, Fine et al. (2004) showed by experimental manipulation that herbivores are a key force in specialization among various tree species that inhabit either rich lateritic clay soil or nutrient-poor white sandy soil. Areas of nutrient-poor white sandy soil occur in immediate proximity to clay soils, which are far richer in nutrients. Tree species of various genera are species-specific to each soil type, a form of specialization. For example, *Oxandra xylopoides* (Annonaceae) occurs on clay soil and *O. euneua* on white sandy soil. On clay soil, trees obtain adequate

PLATE 4-26
COLLARED PECCARY (*PECARI TAJACU*)

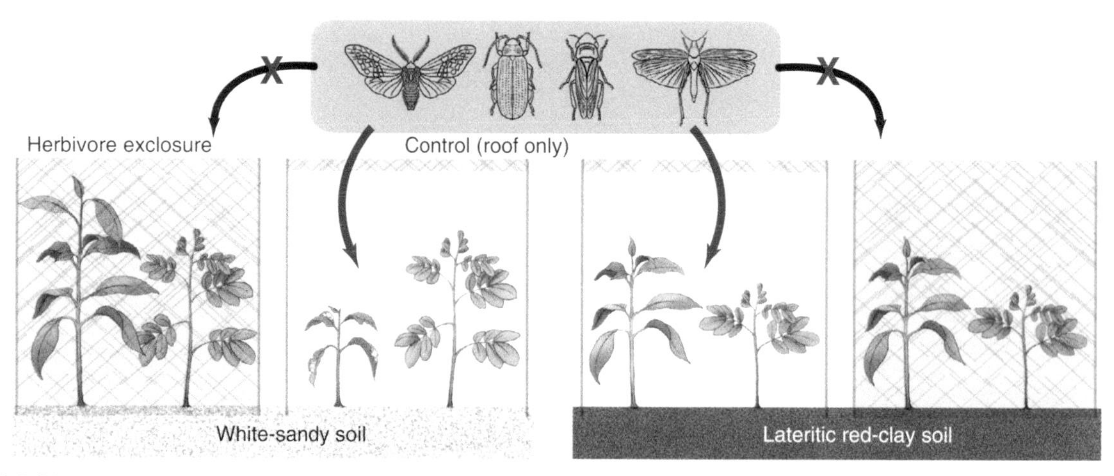

FIGURE 4-35
Experimental design for reciprocal transplant of tropical tree seedlings. Twenty species of tropical tree seedlings (two are shown) were reciprocally planted onto white-sandy soils and lateritic red-clay soils. Half the species grew naturally on one soil type or the other. Exclosures that prevent insects from access to seedlings were put in place around one-half of the seedlings of each species planted on each soil type. A control exclosure (roof only) was placed around the remainder of the transplanted seedlings, making them vulnerable to insect attack. Seedling survival and growth were monitored.

nutrients needed to grow efficiently, and an analysis of their leaves demonstrates that they contain fewer defense compounds that act to repel herbivorous insects. (The topic of defense compounds is discussed in greater detail in Chapter 7.) But on white sandy soils, plants have greater fitness if they load their leaves with defense compounds, even though doing so is metabolically costly in energy. But with such poor nutrient conditions in the white sandy soil (as well as the fact that water leaches more rapidly in such soils), leaves last far longer if they repel insects, and thus the cost of leaf manufacture is more than offset by the longer tenure of the leaves on the plant (Figure 4-35).

The researchers transplanted seedlings of 20 species of trees—9 species from white sandy soil and 11 species from clay soil—into the opposite type of soil. They protected half the seedlings from herbivores using *exclosures*, netting that prevents herbivore access. Others were covered in such a way as not to exclude herbivores. They found that where herbivores were excluded, trees from clay soil grew very well in white sandy soil, growing taller and leafier than the species normally found there. But the clay soil species fared very poorly when herbivores were not excluded. In other words, presumably due to lack of defense compounds in their leaves, the species from clay soil were subject to loss of fitness due to herbivore pressure. The conclusion from this part of the experiment is that herbivores, a top-down force, exclude clay soil tree species from white sandy soils. If that were not the case, these trees might well invade and outcompete the species that devote large amounts of energy to defense compounds. In the case of trees from white sandy soils transplanted to clay soils, they grew successfully with or without exposure to herbivores, but their growth was slower than the species normally found on clay soil. In other words, though the white sandy soil species could grow equally well in either soil type, they would be unlikely to succeed on the clay soil, because clay soil species (without the burden of synthesizing defense compounds) would grow more quickly and thus competitively exclude them (Figure 4-36).

This study demonstrates several important considerations about species richness in the tropics:

- Experimental data from studies that manipulate conditions are needed to unravel possible forces that structure communities.
- Interactions between trophic levels may be of major importance in determining patterns of species richness.
- Speciation may be promoted by trophic-level interactions. Speciation on white sandy soil may have resulted from strong selection for defense compounds, selection sufficient to overcome gene flow from individuals on clay soil, a kind of speciation called *parapatric speciation*. (*Parapatric* refers to speciation across environmental discontinuities.) Such a process may promote speciation of tropical trees if sufficient heterogeneity among

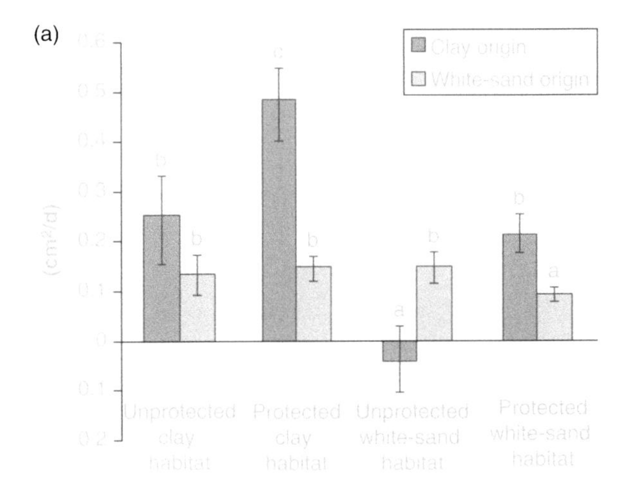

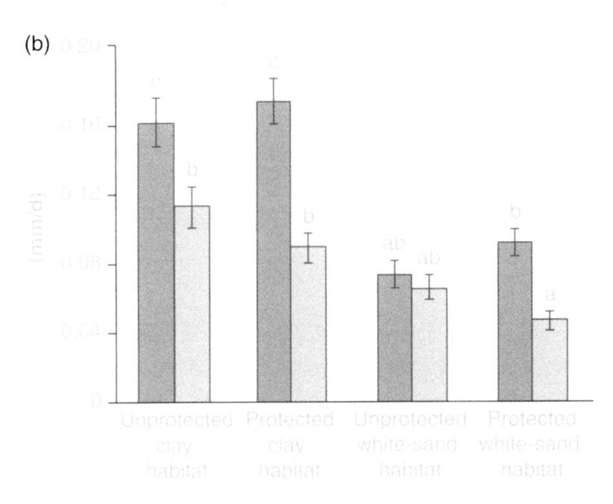

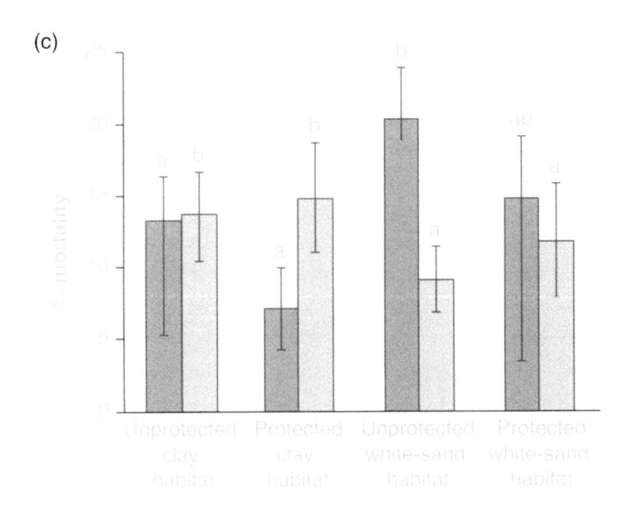

FIGURE 4-36
The effects of habitat and herbivore protection on (a) leaf area growth rate, (b) meristem height growth rate, and (c) percent mortality for white-sand and clay specialist species. Bars represent means +/− 1 standard error.

edaphic factors such as soil conditions prevails widely. This topic will be discussed further in the next chapter.

Both interspecific competition and top-down forces likely operate simultaneously in the tropics. Predators may compete among themselves and specialize on various prey groups that are prevented from competing by predation. Trophic dynamics of tropical forests, given the high species richness at all trophic levels, is extremely complex (see Chapter 7).

Some Broad Patterns of Tropical Species Richness

It should come as no surprise that animal groups such as amphibians and reptiles, unable to physiologically regulate body temperature, decline in diversity and abundance with latitude. The Arctic and Antarctic are inhospitable to crocodiles, lizards, snakes, and turtles. These animals are *ectothermic* (unable to physiologically regulate body temperature) and thus unable to obtain sufficient body heat during times of extreme cold. Such harsh conditions do not occur in subtropical or tropical latitudes, and thus reptiles thrive in such regions. Amphibians are similar in that they too are ectothermic, but they differ from reptiles because they are dependent on an aquatic or semiaquatic environment, especially for reproduction, so they are even more limited by harsh climate, and their species richness pattern is closely correlated with water availability as well as warm temperature. Thus, like reptiles, amphibians exhibit highest species richness in mid- to low latitudes. Physiological constraint undoubtedly is a major factor in the LDG for reptiles and amphibians.

Insects and other arthropods are also ectothermic, but they are small and are often able to overwinter as eggs or pupae. Thus insect numbers may be impressive during the short Arctic growing season, but arthropod species richness is far greater at lower latitudes. Like various vertebrate groups, insects (among arthropods) exhibit their highest species richness within rain forests.

Latitudinal diversity gradients show complex and nonrandom patterns. For example, most of the increase in mammalian diversity in the tropics is largely due to bats (Chiroptera) and rodents (Rodentia). In the Neotropics, monkeys, sloths, anteaters, and various marsupials all contribute to the enhanced diversity (and none of these groups are represented well outside of the

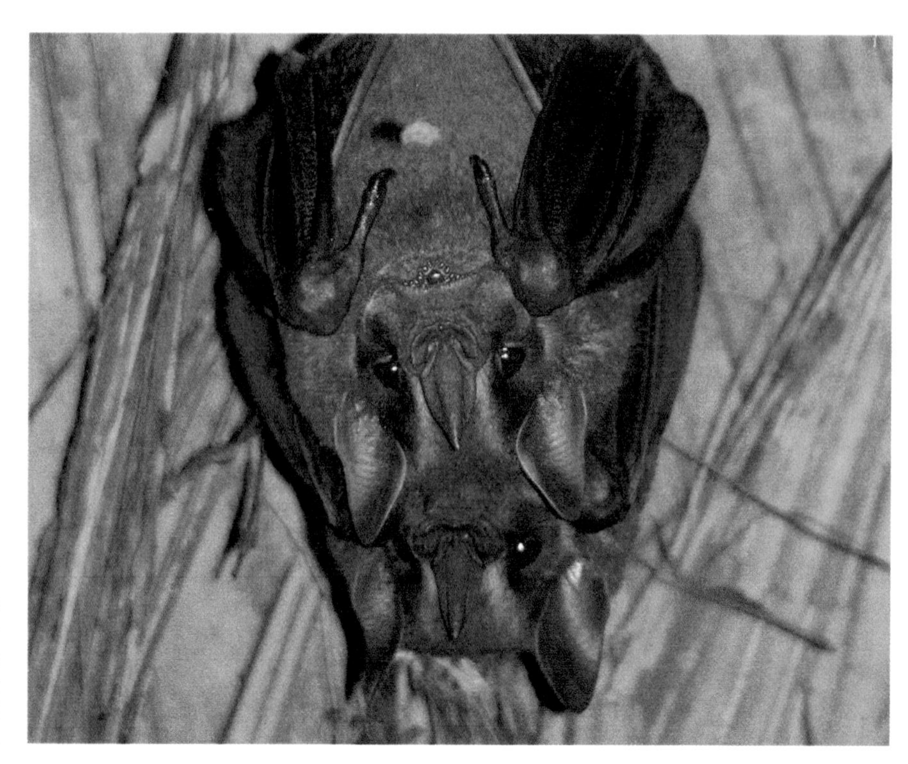

PLATE 4-27
These two roosting fruit bats (*Chiroderma* spp.) are among the many species of Chiroptera that occur in the Neotropics, where it is estimated that 39% of all mammals in Neotropical forests are bats. From Ecuador.

tropics), but bats taken alone (Plate 4-27) add by far the most species, representing about 39% of all Neotropical mammal species (Emmons 1990). Rodents come next.

Mammalian diversity in Neotropical lowland rain forests correlates with both productivity and habitat characteristics. The density and number of species of Amazonian mammals (excluding bats) correlates positively with soil fertility and undergrowth density, both a measure of plant productivity. Large mammalian species tend to range widely and maintain relatively constant densities over large areas, but small species vary dramatically in numbers and diversity from one study site to another (Emmons 1984).

Interaction Diversity: An Introduction

Ecologists look closely not only at the numbers of species present in ecological communities but also at the complexity of their collective interactions, particularly among trophic levels. Trophic dynamics will be discussed in detail in Chapter 8, but it is important here to at least introduce the topic because it provides another important avenue into the study of diversity and may relate closely to the importance of diversity at a functional level. As the preceding examples have shown, species interact within (as with interspecific competition) and among (as with the predation hypothesis) trophic levels, a complex web of interaction. Interactions link ecological communities in dynamic ways, and the study of interaction diversity has strong implications for ecologists' ultimate understanding of how ecosystems function and how biodiversity relates to such ecosystem level functions as productivity and stability (Dyer et al. 2010).

Interaction diversity can be understood by looking closely at Figure 4-37. This figure illustrates arrangements of five species, each of which results in different interaction richness. The interactions occur among three trophic levels, an enemy (predator or parasite), an herbivore, and a plant. Such interactions are typical of all terrestrial ecosystems. Experiments using the models shown in the figure have been conducted in Ecuador and Louisiana. The results show that interaction diversity is stronger in tropical Ecuador than in temperate Louisiana, even when the mean species richness per plot is about the same (Dyer et al. 2010).

Interaction diversity is becoming an important area of study in the complex questions generated from the realization that tropical forest ecosystems are so consistently rich with species.

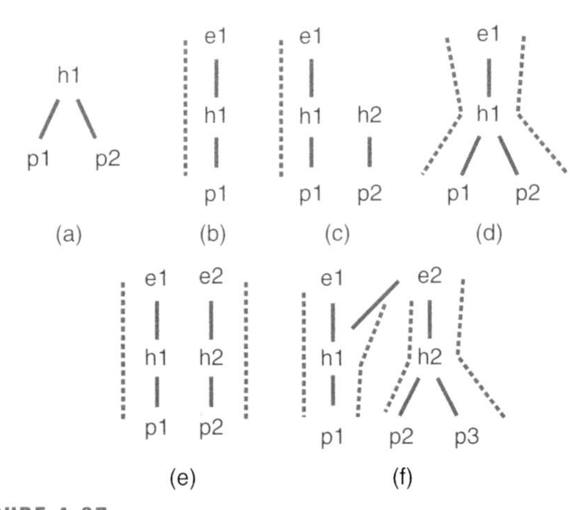

FIGURE 4-37
This diagram represents simple examples of interaction diversity in subsamples of communities: interaction richness for each subsample is (a) 2, (b) 3, (c) 4, (d) 5, (e) 6, and (f) 10.

A Big Unanswered Question

The global tropics may indeed be the perfect storm of biodiversity. Perhaps there and only there do climatic and ecological conditions combine to promote speciation to such a high degree. But we have yet to focus on a very important element of the diversity question. Given that energy availability is highest in the tropics and given that climate there is highly conducive to life, how do those realities translate into more species and not just greater population sizes of a few species? In the next chapter, using trees as the focus, this question will be considered.

THREE DIVERSE BOOKS ON BIODIVERSITY

The topic of biological diversity is immense, and the following three general references, which approach diversity in different ways, are strongly recommended:

Huston, M. A. 1994. *Biological Diversity: The Coexistence of Species on Changing Landscapes.* Cambridge, UK: Cambridge University Press. This book is a broad and thorough survey of what ecologists have learned about biodiversity as well as the various approaches taken to the topic.

Magurran, A. E. 2003. *Measuring Biological Diversity.* New York: Wiley-Blackwell. For those interested in the various metrics used to measure and calculate biodiversity, this book is ideal.

Wilson, E. O. 1992. *The Diversity of Life.* New York: W.W. Norton. Perhaps the best overview of biodiversity, this book was written for a popular audience, but it has been used in college courses. It is highly engaging and broad in scope, with a strong conservation message.

5 A Study in Biodiversity: Rain Forest Tree Species Richness

Chapter Overview

The extreme species richness of tropical trees, particularly in lowland rain forests, has proven paradoxical to ecologists. It is difficult to develop explanations for how such extreme diversities could evolve and be indefinitely maintained. Various models and hypotheses have been suggested and tested to various degrees. These range from equilibrium models based on niche partitioning to nonequilibrium models that include various levels of disturbance. Some models focus on competition among trees (at various life cycle stages), some focus on negative density dependence (when trees fare worst near others of the same species), and others focus on top-down trophic dynamics driven by pathogens, herbivores, and seed predators. One model, the *unified neutral theory* (UNT) of biodiversity and biogeography, is derived in part from the theory of island biogeography. This model assumes ecological equivalence among tree species and argues that species richness is maintained by a *random ecological walk* analogous to genetic drift. This chapter describes and reviews these various models and hypotheses.

The Paradox of the Trees

One classic paper in ecology is titled "The Paradox of the Plankton" (Hutchinson 1961). Its author was G. Evelyn Hutchinson, one of the most eminent ecologists of the twentieth century. Hutchinson was puzzled by why so many species of phytoplankton cohabit the waters. These tiny plants all appear to require pretty much the same things: light, carbon dioxide, and nutrients. Without these, they cannot photosynthesize, metabolize, and ultimately reproduce. So what is the paradox? Simplicity itself. The needs of planktonic plants seem simple, almost too simple. At the time, ecologists were focused on the principle of competitive exclusion, the concept that two or more species with identical ecological requirements cannot indefinitely coexist. Theory held that one must eventually outcompete the others. The key to coexistence was thought to be *niche partitioning*, the subdivision of niche space such that each species attains a realized niche that does not significantly overlap with other species. But how could it be that so many tiny species of plants partition what seems like a fairly homogeneous resource base such that competition is minimized among them? If all they require is light, carbon dioxide, and some minerals, shouldn't plankton compete interspecifically, as was assumed to be the case with other sympatric species with similar ecological requirements? Why shouldn't a few or even one species eventually prevail, replacing all others? That was the

Tropical Tree Species Richness: A Sampling

Surveys of 50-hectare forest plots (see the Center for Tropical Forest Science [CTFS] website, at http://www.ctfs.si.edu/) have revealed the following numbers of tree species:

- Barro Colorado Island, Panama: 305 species, 235,317 trees.
- Mudumalai, India: 68 species, 17,432 trees.
- Huai Kha Khaeng Sanctuaries, Thailand: 251 species, 96,072 trees.
- Pasoh, Malaysia: 817 species, 320,903 trees.
- Lambir, Malaysia: 1,171 species, 339,266 trees (Plate 5-1).

(Data from May and Stumpf 2000, and CTFS website.)

The forests in India and Thailand are drier and more seasonal, a fact that likely accounts for their lower species richness and tree density. The others are wet forests.

PLATE 5-1
PASOH FOREST, MALAYSIA

paradox. In other words, how was phytoplanktonic diversity maintained? What ecological factors allowed the diverse coexistence of so many species?

Now consider a tropical rain forest. How do up to 300 tree species in a hectare (in some cases, even more) indefinitely coexist (Plate 5-2)? Trees, like phytoplankton, require light, moisture, carbon dioxide, and minerals. Period. Well, not quite period. They also require pollinators and seed dispersers (at least many, likely most, of them do). They have to defend against pathogens, herbivores, and seed predators, a costly adaptation in the tropics, with no cold winter season to reduce pathogen and herbivore abundance. And most require mycorrhizal fungi in their root systems (see Chapter 10) to aid in uptake of essential soil nutrients. But the point is still well taken that the basic ecological needs of a tree are, at face value, pretty basic. Energy in the form of solar radiation is abundant

PLATE 5-2
The extreme high species richness of trees in many tropical forests such as this one in Ecuador is difficult to explain, hence the *paradox of the trees*.

Q. When Is a Tree More Than a Tree?

A. ALWAYS

When conceiving what an organism is, indeed what a species is, consider the typical tropical tree. It is not exactly "alone" in the world. It relies on mycorrhizal fungi within its root system to bring soil nutrients into it. Above ground, endophytic fungi influence relationships between tree and insects (Arnold and Lewis 2005). In most cases, a tree will have mutualistic bacteria that aid specifically in the uptake of nitrogen. It may be incapable of dispersing pollen unless it forms flowers that attract an insect or perhaps a bird or bat—sometimes a very specific insect, bird, or bat—to visit the flowers and subsequently distribute the pollen (Plate 5-3). Its seeds, the next tree generation, may not germinate if they fall too close to the parent tree, and thus seed dispersers—various animals such as birds and mammals and sometimes fish (in flooded forests)—consume fruit and either defecate or regurgitate seeds far from the mother plant. Thus a typical tropical tree is already a study in biodiversity, as it forms a series of species interlinked to varying degrees, an interdependent network ranging from microbes perhaps to orangutans and birds-of-paradise. And none of these include the various *enemy species* such as pathogens (bacteria and fungi), herbivores (ranging from the tiniest insects to elephants), and seed predators, all of which exert selection pressures on the tree.

PLATE 5-3
Part of a tree? Hummingbirds of the Neotropics, such as this white-necked Jacobin (*Florisuga mellivora*), are important cross-pollinators of many tree and other plant species.

in the tropics, as is moisture (except where dry seasons are severe). Carbon dioxide appears not to be limited (but see Chapter 9). Soil nutrients do vary considerably (see Chapter 10) from site to site, and such edaphic (soil) heterogeneity may affect tree species richness. (Recall the example of trees from poor white sandy soil and richer clay soil discussed in the previous chapter.) But wherever in the tropics one looks, it is difficult to see how so many tree species at one site somehow partition soil, water, and light so as to minimize interspecific competition and partition the niche space. That being the case, what factors are responsible not only for the high tree species richness and high species diversity, but also for the ability of such numerous species to indefinitely coexist? The previous chapter dealt in large part with the genesis of biodiversity. This chapter will emphasize ideas about how biodiversity is maintained at such high levels, particularly in tropical forests.

Tropical plant ecologists have offered several hypotheses to account for the hyperdiverse species richness of tropical forests (particularly lowland rain forest). I will introduce these hypotheses here and then present examples from field studies:

- *The niche-partitioning hypothesis* (Figure 5-1). This is a widely applied theory in community ecology and much data have been accumulated in support of it (see Finke and Snyder 2008). It argues that interspecific competition forces species into increasingly narrower realized niches until each of the competing species is sufficiently specialized in its niche dimensions such that it is no longer at risk of competitive exclusion. This hypothesis is difficult to support in tropical tree communities because it has not been possible to demonstrate how niche axes are partitioned to a sufficiently fine degree to permit the simultaneous existence of up to 300 or more species of trees in a hectare. Tropical plant ecologists have looked to other hypotheses.
- *The storage effects hypothesis* (Figure 5-2). This hypothesis argues that seeds may be stored in the ground (the *seed bed*) for prolonged time periods

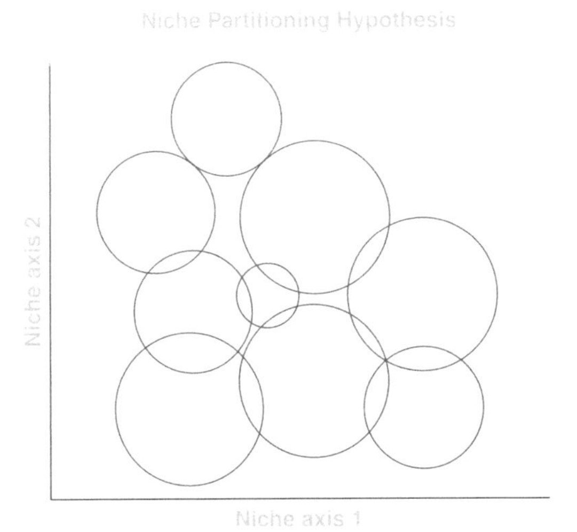

FIGURE 5-1
This diagram demonstrates how species' niches (shown in various-sized circles) attain minimal overlap when graphed against two hypothetical niche axes. This is the prediction of the niche-partitioning hypothesis.

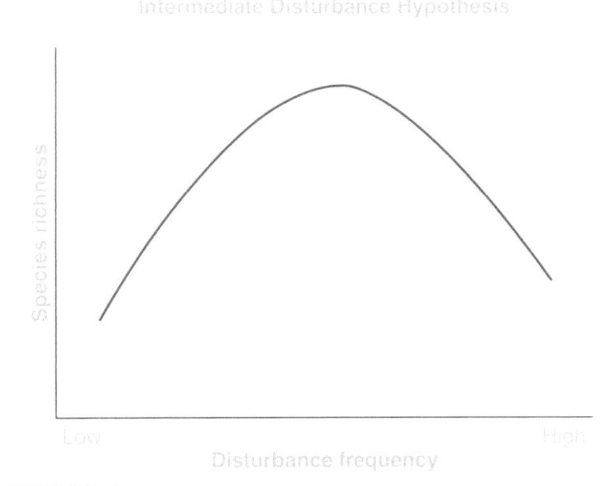

FIGURE 5-3
Species richness is maximal when disturbance frequency is intermediate. Too much disturbance reduces diversity because few species are adaptable to high disturbance. Too little disturbance results in lowering diversity due to competitive exclusion.

and that adult trees may experience high longevity, thus having ample opportunity to reproduce even after years of little or no reproductive success (Willis et al. 2006). As a result, trees are buffered from limited recruitment or disturbance. As conditions change from site to site, species richness changes by a lottery-type process. Diversity is maintained by the buffering effect that storage provides.

- *The intermediate disturbance hypothesis (IDH)* (Figure 5-3). If disturbance is severe and frequent, few species ought to be able to endure, and thus diversity is kept low. If disturbance is rare and an ecosystem experiences constant conditions, competitive exclusion among species ought to occur and diversity will decline. But if disturbance is intermediate in intensity and frequency, many species may be supported and competition among species is interrupted such that competitive exclusion is never reached (Connell 1978). The forest is maintained in a dynamic long-term state of nonequilibrium as species are added and lost by disturbance processes.

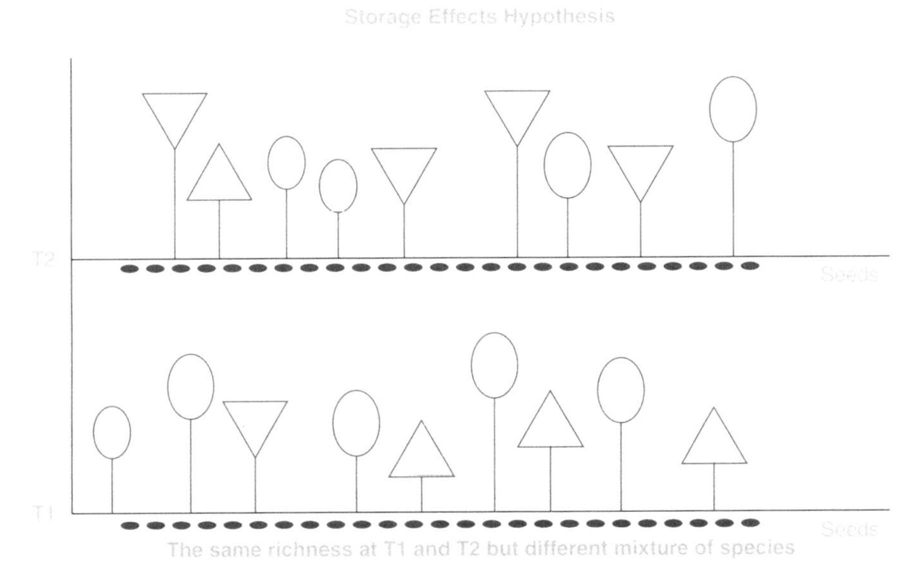

FIGURE 5-2
The storage effects hypothesis predicts that seeds are maintained in soil for long periods. Adult trees live to have many opportunities to reproduce, but a comparison of time T1 with time T2 shows that the species richness gradually changes with time.

The frequency-dependent hypotheses vary, but each is focused on the role of population size in determining likelihood of success within a species. The frequency-dependent hypotheses are:

- *The niche-complementarity hypothesis* (Figures 5-4a and 5-4b). This hypothesis is at first glance similar to the niche-partitioning hypothesis, but with a difference. In the niche-complementarity hypothesis, there is competition for resources and subsequent niche partitioning, but rare species, because they are rare, experience less competition than more abundant species and are thus at a competitive advantage. The high tree species richness that results is enhanced by environmental heterogeneity, changing competitive advantage from site to site. This model, if correct, predicts that ecological functioning of the ecosystem, as measured by such parameters as net primary productivity (see Chapter 9), would be enhanced in a diverse community compared with a community of reduced diversity (Fargione et al. 2007).

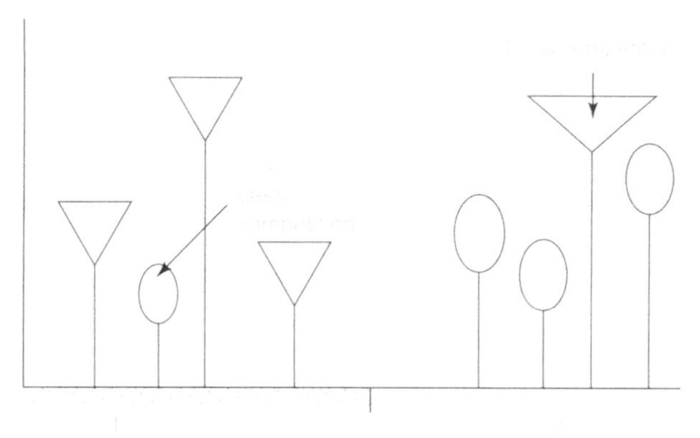

(a)

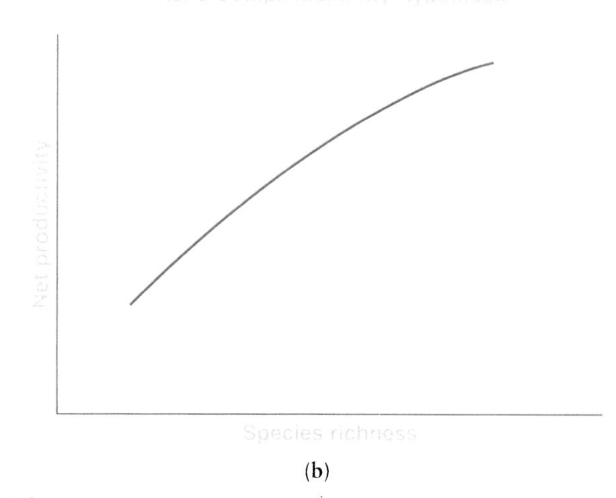

(b)

FIGURE 5-4
The niche complementarity hypothesis predicts that as species pack into a habitat that has some heterogeneity, rare species are under less competitive pressure and so enjoy an advantage and persist, adding to the species richness of the site.

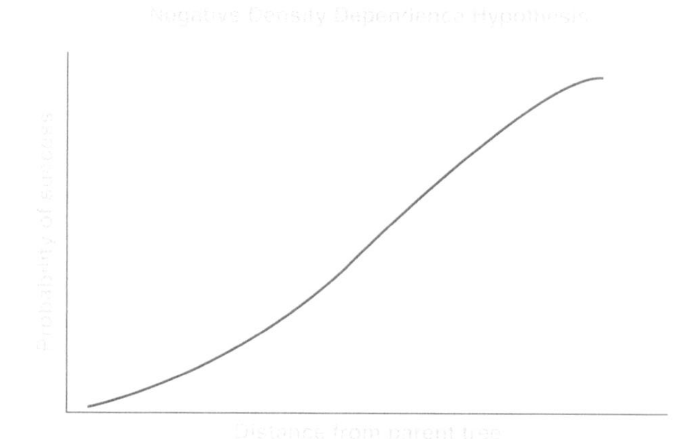

FIGURE 5-5
The negative density dependence model predicts that the probability of a seed sprouting and eventually becoming a mature reproducing tree is a function of how distant the seed is from its parent tree. Being close to the parent tree creates a *seed shadow* that reduces the chances of success due to competition with the parent tree.

- *The negative density dependence model* (Figure 5-5). It has long been thought that the worst location for a seed to fall is in the immediate vicinity of its parent tree. The parent tree would be the strongest competitor against its own seedlings and thus the seedlings would fare worse the nearer they are to the parent, an already established tree. Daniel Janzen (1969) called this the *seed shadow effect*, where the likelihood of successful germination and growth increases with distance from the parent tree. This model is similar to the next in that both hypothesize that negative selection pressures increase with population density.
- *The pathogen–herbivore–predator hypothesis* (Figure 5-6). This is sometimes called the *Janzen-Connell hypothesis* (Janzen 1970; Connell 1971) although it was anticipated in a model developed in 1962 by J. B. Gilett (Gilbert 2005). In this model, tree diversity is kept high because of density-dependent or frequency-dependent interactions between the various tree species and their pathogens, herbivores, or seed predators. The most

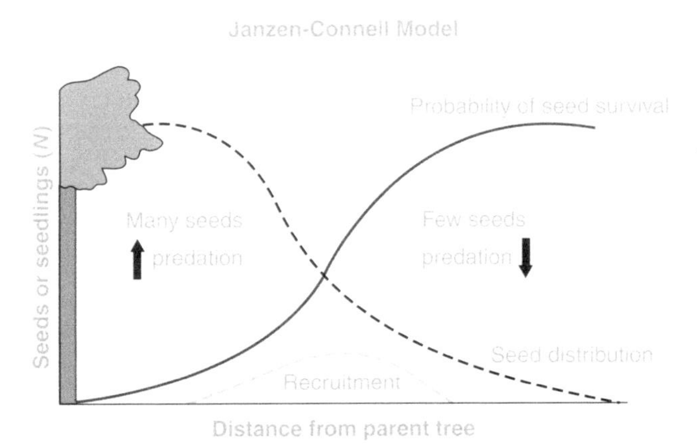

FIGURE 5-6
Predation pressure from pathogens and herbivores is most intense near mature trees, and thus seeds that are moderately distant from the tree survive best. Compare this model with the previous model. Both are similar.

common species are the targets of various attacking species (some specialized to particular species, genera, or families), and thus rarer species are at a selective advantage, as rarity itself is a defense against predation. This is a top-down model that regards trophic-level interactions as essential to maintenance of high tree species richness. There is evidence to support this model.

Last, there is one hypothesis that has created much interest and debate among ecologists:

- *The unified neutral theory (UNT) of biodiversity and biogeography* (Figure 5-7). This theory, developed by Stephen P. Hubbell (2001), is the most challenging of the various hypotheses in that its basic assumptions are clearly not true but its predictions fit well with empirical data. It argues that trees are functionally ecologically equivalent (neutral in terms of interactions among individual trees) and that diversity increases by a nondeterministic gradual influx of species (either by immigration or speciation) and that extinction is too low to balance the gain in biodiversity. Further, it argues that within an area, numbers of individuals remain constant, so the addition of species, in essence, creates increasingly low population densities. The model, though counterintuitive, is often accurate at predicting the distribution of species' abundance patterns.

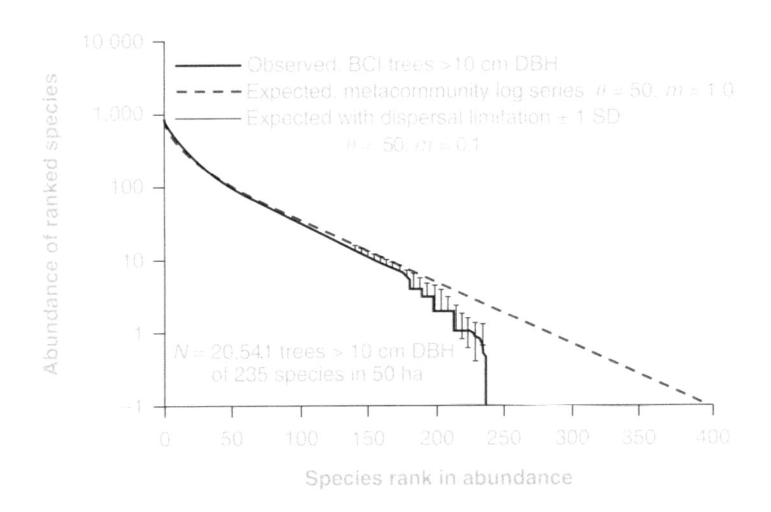

FIGURE 5-7
The neutral theory applied to known tree species abundances on BCI shows a very close fit even though the model makes none of the assumptions contained in the preceding models.

The models described here have been investigated to various degrees and are not mutually exclusive. What follows is a series of case studies that focus on various models as each attempts to investigate why species richness is so high in tropical forests.

Rain Forest Tree Richness on a Broad Scale

FLORISTIC PATTERNS ACROSS AMAZONIA

Tree species richness is high throughout the world's tropical rain forest regions and is particularly so in Amazonia. What are the major factors influencing this gamma (regional) diversity? A study was conducted to inventory tree diversity in seven of nine countries with territory in the Amazon Basin and Guiana shield (an area of ancient rock in northeastern South America) (ter Steege et al. 2006). The study analyzed seven forest inventories from across the vast region. Two dominant gradients in tree composition were evident: one reflected a gradient in soil fertility and the other a gradient in dry season length. The data set included 89 families and 513 genera. (Trees were not identified to species level.)

The most common family was the Fabaceae, also called the Leguminosae, the legumes. Fully 25% of all large trees in the

PLATE 5-4
This is one of 19,400 legume species (Fabaceae), making legumes the third largest plant family in the world. Legumes abound in the tropics, but many species range well into higher latitudes. Many legume species have feathery compound leaves, and all produce seeds in pods.

surveys were legumes (Plate 5-4). Multivariate ordination mapping explained 24% of the variance in tree community composition (of genera). One gradient extended from the Guiana shield to southwestern Amazonia (Figure 5-8). This gradient reflected soil fertility. Tree genera of the Guiana shield grew on poor soils (because nutrients are lost with soil age) and had denser wood and larger seeds. None of the 10 most abundant tree genera on the Guiana shield were among the 10 most abundant in western Amazonia. This gradient may have been in place for millions of years, because the rocks on the Guiana and Brazilian shields are Precambrian and have been slowly weathering for millions of years, whereas those that form western Amazonia are recently mineralized from the uplift of the Andes Mountains.

The second gradient reflected changes in dry season length (Figure 5-8). This showed that the greatest effect of dry season ran from northwestern (wettest) to southeastern Amazonia (driest). The researchers hypothesized that this gradient is younger than the one reflecting soil richness because of climate change during the Pleistocene.

This study shows how two major factors, climate (as discussed in the previous chapter) and soil richness (see Chapter 10) influence large-scale regional diversity patterns. Recall from the previous chapter that beta (habitat) diversity of trees varies more in Panama than in sites across Ecuador and Peru, a result attributed to greater soil and climate heterogeneity in Panama compared with Amazonia sites (Condit et al. 2002).

NONRANDOM SPECIES PATTERNS IN SIX TROPICAL FORESTS

One approach to the study of hyperdiversity of tropical tree communities is to compare communities globally (Condit et al. 2000) and ascertain how the species are spatially distributed. Are they a random spatial assemblage, or are patterns of clumping evident? The communities studied were in Panama (Barro Colorado

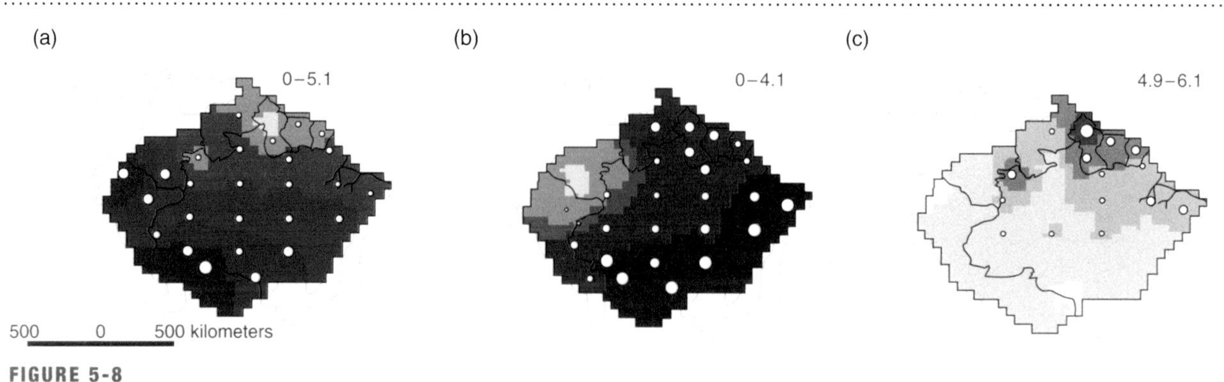

FIGURE 5-8
These maps illustrate geographic variation in community characteristics of South American tree communities. Maps (a) soil fertility gradient and (b) dry season gradient illustrate scores on two gradient axes, and (c) shows community-weighted seed mass.

Island [BCI]), Malaysia (two sites), Thailand, India, and Sri Lanka (see the sidebar "The Center for Tropical Forest Science [CTFS] and Forest Dynamics Plot [FDP] Network" later in this chapter). Plot sizes in each country were large, usually 50 hectares (the site in Sri Lanka was 25 hectares) to ensure adequate sampling of all tree species in the community. The sites varied as to length of dry season, with the site in India being quite dry and those in Malaysia wet, with no dry season. The study focused on measurement of neighborhood patterns around individual trees, and a total of 1,768 trees made up the database. When all individuals larger than 1 centimeter in stem diameter were included, virtually all species studied were more aggregated than would be expected by a random distribution. Thus the tree species tended to clump. When only trees with stem diameters greater than 10 centimeters (about 4 inches) were included, the pattern remained nonrandom, with each species aggregated to some degree. Rare species tended to aggregate more closely than common species (so it was more probable to encounter a rare species near another of its kind than it was to encounter a common species near another of its kind). Each of the six forest plots was remarkably similar in that each showed essentially the same pattern toward aggregation (Figures 5-9 and 5-10).

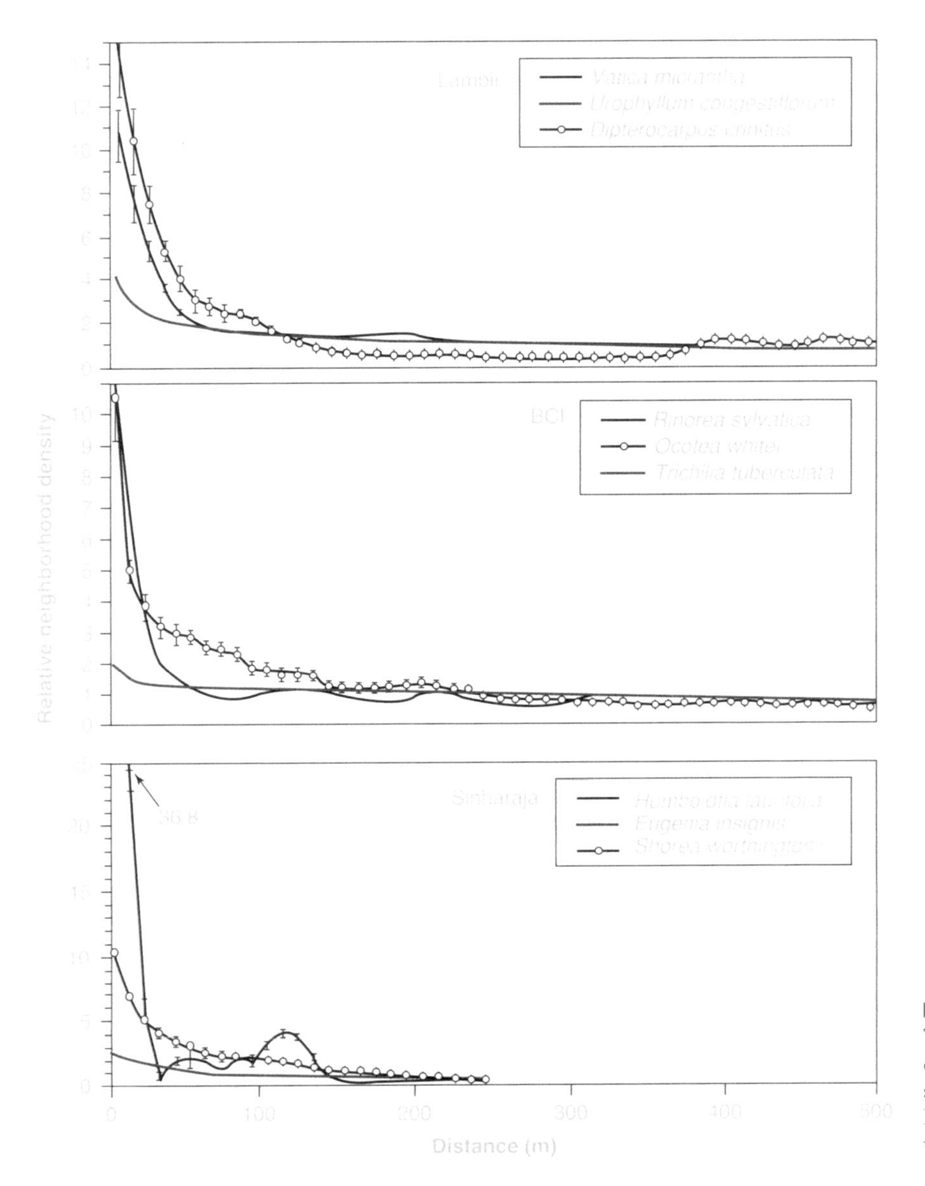

FIGURE 5-9
This figure plots relative neighborhood density against distance in meters for species in three different forest plots. Note the similarity of patterns in the three forests, each showing aggregation.

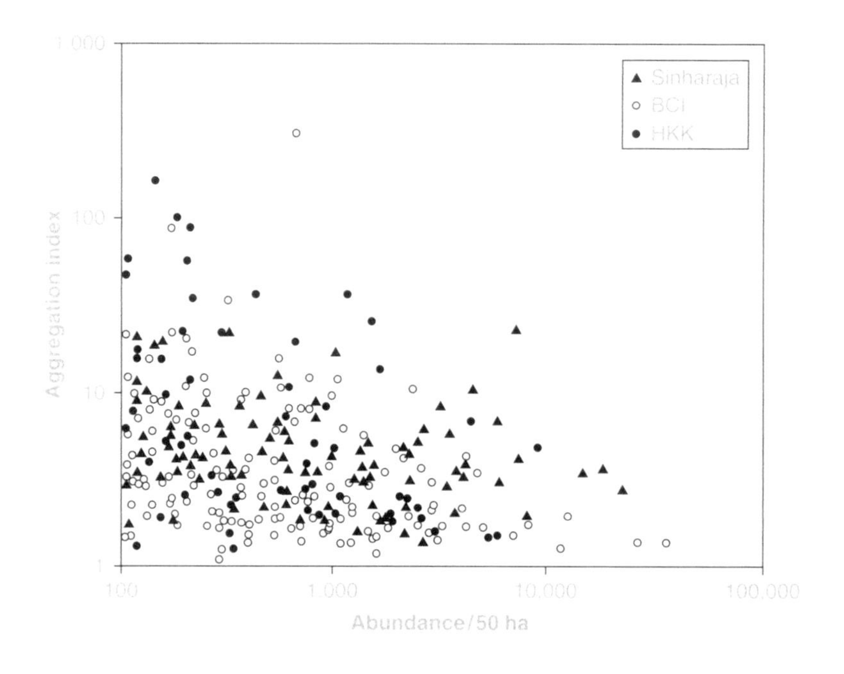

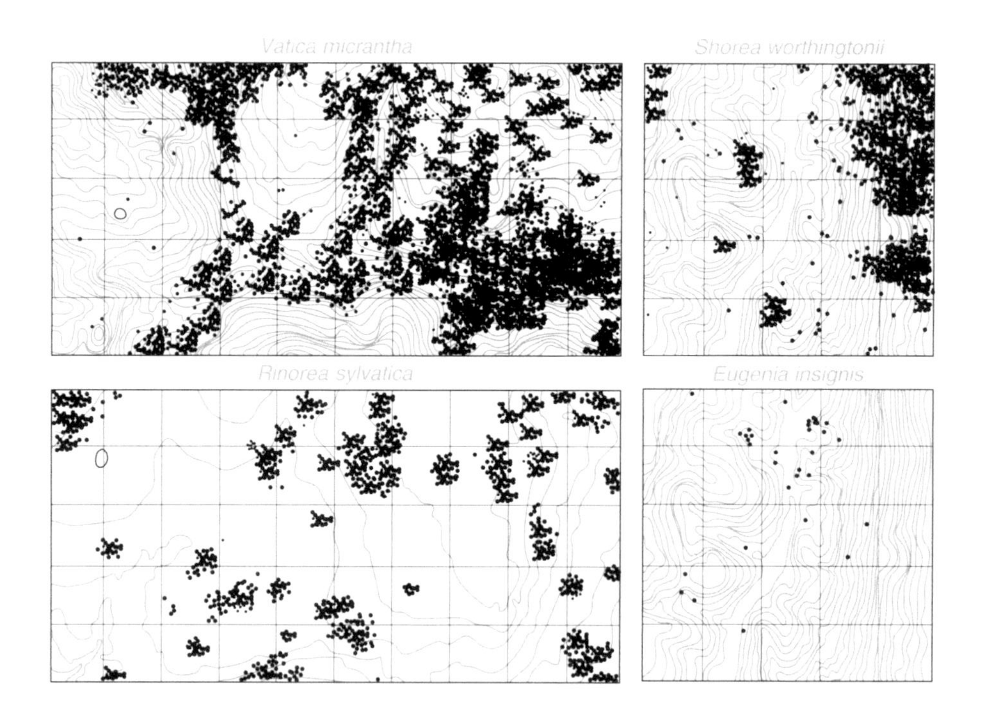

FIGURE 5-10
Graph shows the relationship between aggregation of conspecific trees and abundance. Rarer species showed more aggregation. The bottom maps illustrate four examples of species distributions. *Vatica* clumps follow ridges in Lambir. *Rinorea* clumps on BCI do not correlate with any known canopy, topographic, or soil feature and may be due to limited seed dispersal. *Shorea* follows ridge tops in Sinharaja, and *Eugenia* is a rare species in Sinharaja, with most individuals near their conspecifics.

The researchers found that:

- Rare species were substantially more aggregated than more common species.
- Those species with effective dispersal showed somewhat reduced aggregation.
- Most sites suggested that animal-dispersed seed species would show less aggregation than wind-dispersed species, and this tended to be true except at BCI.
- On the two Malaysian plots, dipterocarp trees (which tend to be numerically dominant) were more aggregated than nondipterocarps.
- Aggregation was weakest in large-diameter classes, supporting the hypothesis that herbivores and plant pathogens act to reduce aggregation. This result adds support for the seed-shadow effect hypothesis.

It is not surprising that most species were nonrandomly aggregated and clumped to varying degrees. Such a pattern is common in many plant communities throughout the world because most seeds are dispersed relatively near the parent tree.

DO RARE SPECIES SURVIVE BETTER THAN COMMON SPECIES?

In a study of six tropical forests, it was learned that when a tree species is rare within a local area, it enjoys a higher survival rate than a common tree species. The result is that the ecosystem becomes richer with rare species and increases in diversity with age and size class of trees (Wills et al. 2006). The forests in this study were located in Puerto Rico, Panama (BCI), India (two sites), Thailand, and Sarawak. (Some were the very same forests used in the previous study.) Forest plot sizes ranged from 16 to 52 hectares. In each census, locations of all trees with diameters equal to or greater than 1 centimeter at 1.3 meters above ground were determined. Each tree was identified to species. The plots were divided into quadrats of 10, 20, 30, 40, or 50 meters. Censuses were conducted on the survivorship of trees at intervals of 10 years (on two sites) and 5 years at the other sites. Diversity was quantified using a technique called *rarefaction* (which expresses richness as the average number of species to be expected in samples of a fixed number of individuals, thus correcting for different densities among plots and different plot areas). (See Sanders 1968 for an explanation of how rarefaction is used in ecology.) Researchers tabulated trees that died and those recruited during each census period and counted younger and older trees (in separate categories).

Mortality rates were lower for seedlings and saplings of tree species that are rare in the canopy than for common trees species. Locally rare species apparently enjoyed some selective advantage. Overall, the pattern of diversity increased from recruits to smaller survivors to larger survivors.

The researchers then investigated whether locally common species have higher mortality than locally rare species and whether that effect is reduced from small to larger quadrat sizes. Using a form of correlation analysis, they learned that common species had higher mortality than rare species in all quadrat sizes but that the effect was strongest in the smaller quadrats.

The result of this study was to demonstrate nonrandom, frequency-dependent survival of rarer over more common tree species. The study did not determine what forces (pathogen–herbivore pressure or environmental heterogeneity or interactions among tree species) were responsible for the pattern. For more on how pathogens and other enemies influence tropical trees, see Gilbert 2005.

The Center for Tropical Forest Science (CTFS) and Forest Dynamics Plot (FDP) Network

Barro Colorado Island (BCI) has served as an ideal laboratory in which to study tropical ecology (Leigh 1999). Under the auspices of the Smithsonian Tropical Research Institute, BCI provided the inspiration for what has become the Center for Tropical Forest Science (CTFS; http://www.ctfs.si.edu/) (Losos and Leigh 2004) and Forest Dynamics Plot (FDP) Network. The entire network consists of a series of forest plots ranging in area from 15 to 52 hectares (most are around 50 hectares). CTFS is global, dedicated to the study of tropical and temperate forest function and diversity. The multi-institutional network comprises more than 30 forest research plots across the Americas, Africa, Asia, and Europe, with a strong focus on tropical regions (Figure 5-11). CTFS monitors the growth and survival of approximately 3.5 million trees and 7,500 species. Trees are counted and measured at five-year intervals on each of the plots. All trees greater than 1 centimeter in diameter are included. The data set also includes measurements of photosynthesis (particularly as it results in carbon storage—see Chapter 9), molecular data, and economics of nontimber forest products (Burslem et al. 2001). The total databank includes in excess of 3 million trees of 6,500 different species, representing approximately 10% of all tropical tree species. Some of the studies discussed in this and other chapters involve data from CTFS surveys.

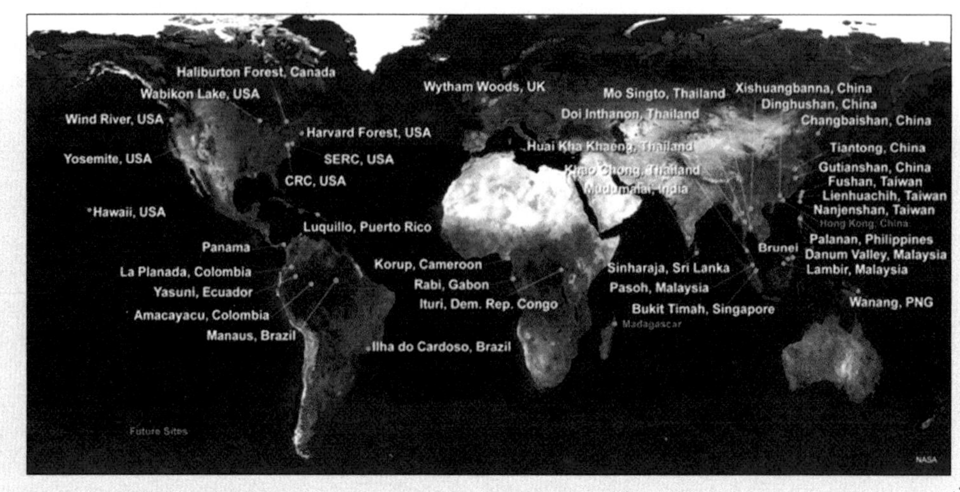

FIGURE 5-11
MAP OF CURRENT CTFS SITES

IS TROPICAL TREE SPECIES RICHNESS THE RESULT OF NEGATIVE DENSITY DEPENDENCE?

As the study described earlier showed, rarer tree species are evolutionarily more fit than common species. This pattern allows rarer species to increase more than common species until the situation reverses, maintaining a dynamic, ever-changing pattern of tree species abundances. But does such a pattern account for maintaining high species richness? Research comparing negative-density-dependent tree mortality rates in tropical compared with temperate areas did not show that negative density dependence is higher in the tropics (Hille Ris Lambers et al. 2002).

Trees were compared from forests in North Carolina (Plate 5-5a), South Carolina, and Texas with those in Australia (two forests), Indonesia (two

(a)

(b)

PLATE 5-5
These two forests, (a) one in North Carolina and (b) one in Panama, did not show differences in density-dependent mortality of trees.

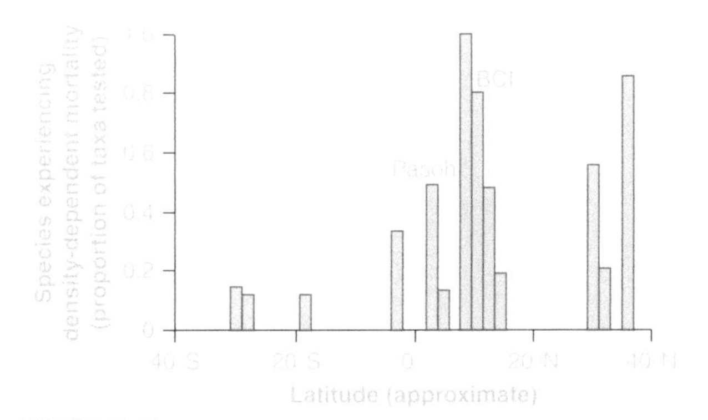

FIGURE 5-12

The proportion of tree taxa experiencing density-dependent mortality at different latitudes. Each bar represents a community of forest trees tested for density-dependent mortality. The proportion of the forest community experiencing density-dependent mortality does not increase at tropical latitudes.

forests), and Panama (BCI) (Plate 5-5b). Researchers found no evidence that the proportion of species experiencing density-dependent mortality is correlated with latitude. This means that while negative density dependence is widespread among tropical trees, it appears to be equally widespread among temperate zone tree species. The proportion of tree species exhibiting density-dependent mortality did not increase with decreasing latitude (Figure 5-12). It is important to note that this study evaluated only the frequency of studies that found density dependence and did not address the strength of density dependence, an important variable. The dynamics of tree populations at both temperate and tropical latitudes may be affected by negative density dependence, and if so, negative density dependence would not account for why there is a greater species richness of trees in the tropics. The topic remains under investigation.

Negative density dependence has been shown to be strongly asymmetric in its effects among plant species (Comita et al. 2010). In a study focused on 30,000 seedlings from 180 tree species from Barro Colorado Island (data collected from 20,000 one-square-meter seedling plots), it was shown that the intensity of negative density dependence varies strongly among individual tree species. What this means is that for some species, proximity of *conspecifics* (same species) creates a very strong negative density-dependent effect. In other cases, the effect of conspecific proximity is low. What accounts for the differences?

Tree species abundances on the study area vary by four orders of magnitude. Some species are common, and many are rare. It turns out that species abundance is directly correlated with the strength of negative density dependence. The most common tree species, such as *Dipteryx oleifera*, are those species least affected by seedling proximity. In other words, the rarer species are more negatively affected by the presence of conspecifics than are the common species. This means that negative density dependence is strongly asymmetric, affecting species very differently.

This does not mean that species such as *Dipteryx* do not experience negative density dependence—they most certainly do. The seed shadow effect applies to them as well as to rare species. But in the case of *Dipteryx*, the intensity of the negative density dependence is less than is the case with most other species, especially those that are numerically rare.

What of the proximity of *heterospecific* (different species) tree species? Suppose that seedlings of other species are growing in close proximity to an adult *Dipteryx*. The results of the study showed that heterospecific effects were weak regardless of abundance patterns. Negative density dependence is fundamentally intraspecific, not interspecific.

The possible explanation for the asymmetric pattern of negative density dependence is that natural enemies, and in particular soil pathogens, strongly affect each tree species but that the intensity of their effects varies widely. This is a hypothesis that traces back to the original Janzen-Connell model accounting for tropical tree diversity (Janzen 1970; Connell 1971). Trees more tolerant of conspecifics may be better adapted to endure exposure to pathogens. Trees less tolerant would be better off well away from conspecifics that could spread pathogens.

A complementary study, also performed at BCI, strongly supports the contention that soil pathogens, likely various bacteria and fungi, are responsible for patterns of negative density dependence (Mangan et al. 2010). Researchers employed reciprocal shade house experiments as well as field manipulation. They took seedlings of six shade-tolerant tree species and planted them in sterilized soil in the shade house. For each species, they either added soil inoculum taken from beneath an adult tree of the same species or inoculum from an adult tree of a different species. The results were striking. Those seedlings with inoculum from adult conspecifics fared much worse than those with inoculum from adult heterospecifics. Clearly, negative density dependence must be related to soil pathogens. Researchers transplanted seedlings in the field beneath conspecifics and heterospecifics. The results were the same. Further, like the study by Comita et al. discussed earlier, those species among the six tested that were numerically rarest suffered the strongest negative density dependent effects. Last, the negative effects of herbivory by arthropods or mammals was weak. It is the soil biota that is implicated most strongly in causing negative density independence.

The studies by Comita et al. and Mangan et al. both support the Janzen-Connell model for why tropical forests maintain high tree species diversity. Soil pathogens, likely various bacteria and fungi, seem to prevent tree species from becoming inordinately abundant. It remains to be seen whether this pattern will be demonstrated for tropical forests elsewhere.

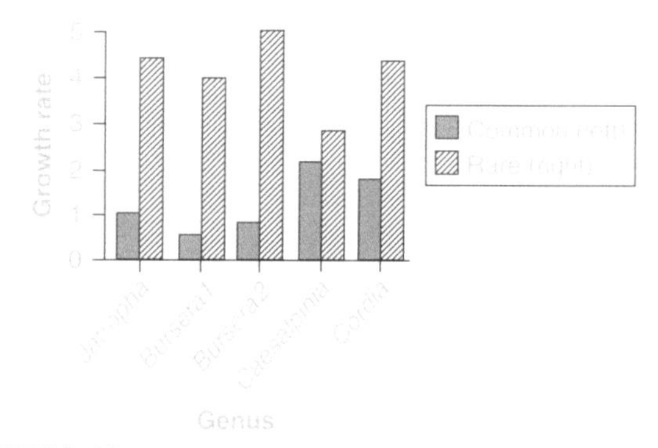

FIGURE 5-13
For five tree genera tested with the Kelly and Bowler model, growth rate was highest when the genus was rare.

STORAGE THEORY AS A MEANS TOWARD COEXISTENCE

Some seeds may be long-lived, part of a seed bank that persists in soil until conditions are favorable for growth. Seedlings also live for a long time in the forest understory, so, in effect, there are seedling banks as well in tropical forests. Trees also may be long-lived, allowing for years of depressed reproduction but persisting until conditions again become favorable for reproduction. This reality has given rise to *storage theory*, which posits that tree species may be buffered from the severe effects of competition because the combination of seed storage, adult longevity, and temporal variability results in favoring some species some of the time and others at other times, thus permitting coexistence without reaching any sort of meaningful equilibrium. (Compare this model with the intermediate disturbance hypothesis, described next.) As time changes and trees succumb to natural causes, seeds of other species sprout and are favored, producing a kind of kaleidoscopic change in species composition over time and preventing interspecific competition from reaching competitive exclusion.

Storage theory has developed into a mathematical model applied to a tropical deciduous forest in Mexico (Kelly and Bowler 2002). What the model predicted and what was supported by data was that in the case of closely related species of trees occurring in the same habitat, the less abundant species would gain a competitive advantage until its abundance increased substantially, at which time such advantage was lost. Data showed that growth rates of common species were less than those of rare species within the same genus (Figure 5-13).

The key principle governing the notion of storage theory is covariance between environment and competition. Environments fluctuate on various scales of time and space, and in doing so, those species that are most abundant are likely most negatively affected by environmental fluctuation—a form of frequency-dependent selection. The model in this study is one of a number of models that assume that environmental fluctuations result in shifting advantages among species over time, typically benefitting rare species (assuming that the fluctuation has a negative effect on more common species).

The Intermediate Disturbance Hypothesis (IDH)

Suppose that intermittent, moderate-scale disturbances such as hurricanes, lightning strikes, landslips, or fires occur with sufficient frequency to prevent interspecific competition from proceeding to the point of competitive exclusion. A disturbance, such as a hurricane, would annihilate many, perhaps most adult trees, removing the competitive edge of any one species. The "game" of competition would be "restarted" after the disturbance ceased. This hypothesis is somewhat similar to the *predator hypothesis* (see previous chapter), except that natural disturbance is the predator in this case, and the predator is nonselective (Plate 5-6). The big difference, of course, is that predators are more selective than periodic disturbance.

Disturbance intensity and frequency are critical variables in considering disturbance per se as a force for

PLATE 5-6
A severe hurricane cleared much of this mangrove forest on an offshore cay in Belize. The cay hosts a colony of magnificent frigatebirds (*Fregata magnificens*) that persist in attempting to nest.

maintaining high species richness (Connell and Slatyer 1977). Obviously, if an area is disturbed harshly and frequently, the severe physical conditions that prevail will act to preclude high species richness. Not much can persist on a site subject to an annual catastrophic hurricane! Perhaps less obviously, if an area is never disturbed, richness will theoretically decline due to interspecific competition, as suggested earlier. Disturbance frequency and intensity must be neither too severe nor too benign in order for disturbance to result in high species richness.

This model, termed the *intermediate disturbance hypothesis* (IDH), was first argued by Connell (1978) to account for high species richness in both rain forests and coral reefs and further developed by Petraitis et al. (1989). It postulates that intermediate-level disturbance is locally patchy but regionally continuous and that the overall disturbance regime is sufficient to maintain high species richness by preventing extinction of competing species (Figure 5-14). The model envisions the tropics as a mosaic of differing aged disturbance patches, some local, some regional (Condit et al. 1992; Hubbell and Foster 1986a, 1986b, and 1990). Bottom line: essentially every place is in some state of recovery from disturbance—there is no indefinite equilibrium. This means that there is no definitive assemblage of species that will form the *end point* of vegetation development in any forest. Instead, species assemblages will change as affected by disturbance events.

Some studies suggest that the frequency of local disturbance in the Neotropics is sufficiently high to maintain the tropics largely in a nonequilibrium state

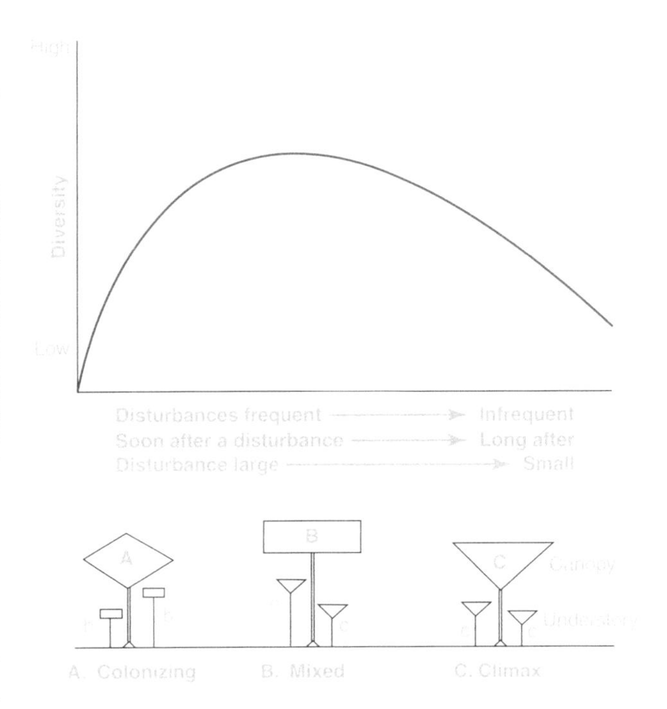

FIGURE 5-14
The Connell and Slatyer model predicts a peak in diversity at intermediate disturbance frequencies.

(Connell 1978; Boucher 1990; Clark and Clark 1992; Condit et al. 1992 and 1995).

Studies of coral reefs as well as tropical moist and dry forests provide evidence in support of IDH (Connell 1978; Boucher 1990; Tanner et al. 1994). Areas of coral reef in Belize that have not been struck by hurricanes have lower diversities of coral and other species than areas that have experienced disturbances. Disturbance frequency on the Australian Great Barrier Reef is much higher than the time required for coral communities to reach equilibrium (Tanner et al. 1994). Coral species compete for space and sunlight (because some contain photosynthesizing algae), and without disturbance, some coral species outcompete and exclude others. Disturbance, however, opens up the area and provides sites for many species to colonize. A major hurricane created highly patchy distributions on Jamaican coral reefs, supporting the notion that more mature reefs show high levels of species heterogeneity due to a past history of disturbance (Woodley et al. 1981).

Hurricane Joan, a category 4 hurricane, struck Nicaragua (on the coast of Central America) in 1988, allowing researchers to study the pattern of tree species accumulation over time and thus test the intermediate disturbance model (Vandermeer et al. 2000).

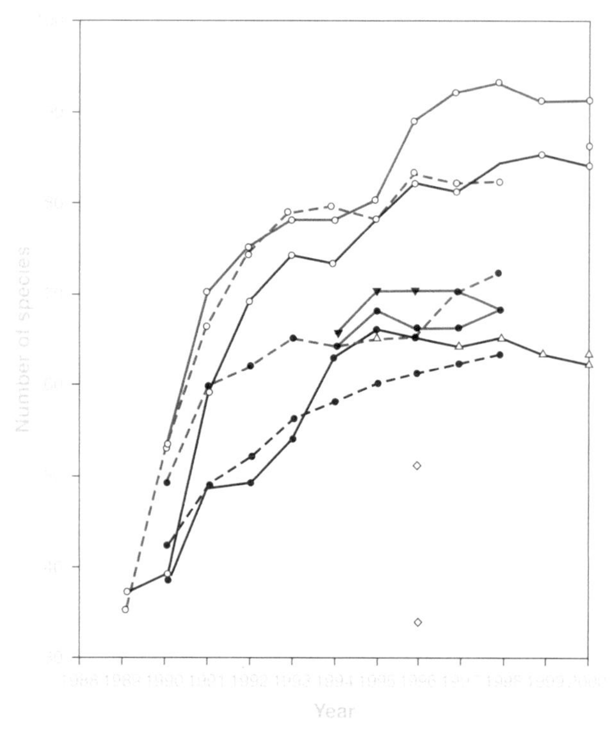

FIGURE 5-15
The eight plots in the Vandermeer et al. (2000) study each showed strong increases in species richness following the hurricane. Open circles are Bodega (three plots), solid circles are Fonseca (three plots), open triangles are La Unión (one plot), solid triangles are Loma de Mico (one plot), and open diamonds are Kurinwas (which was sampled only once).

The hurricane did major damage to about 500,000 hectares of rain forest, but other tracts of rain forest were unaffected. Researchers set up eight small plots at four sites in the damaged area and added additional plots in 1996 (Figure 5-15). Species richness increased substantially in each plot during the decade of study following the hurricane. It was estimated that between a twofold and threefold increase in species richness followed in the damaged areas over the course of the study. When comparing hurricane-damaged sites with control sites not affected by the hurricane, the rate of species accumulation was far greater in the damaged sites. All damaged sites accumulated species rapidly, some more than others, not unsurprising when light is readily available following disturbance.

The authors of the study point out that the spatial extent of damage is an important component in that with such an extensive disturbance, many species, not just pioneer species (see Chapter 6), may recolonize. They suggest a modification in the intermediate disturbance model to focus on large-scale but infrequent spatial disturbances as most influential in maintaining high tree species richness in a nonequilibrium state. In the case of small forest gap disturbances, pioneer species colonize along with seedlings that were already present but suppressed by shade (see Chapter 6). To reiterate, the study suggests that large-scale, infrequent disturbances may be key to maintaining high species richness in rain forests, rather than more frequent, intermediate-level disturbances. The authors also warn that El Niño events are becoming more frequent (see Chapter 15), and enhanced storm activity may accompany such a climate alteration. Frequent major storms may act to prevent reproduction of some species because the trees need to persist long enough to attain reproductive maturity.

The intermediate disturbance hypothesis was tested differently in French Guiana (Molino and Sabatier 2001). Transects were established in areas that had been subject to various intensities of logging and thinning. Researchers then documented the species colonizing the disturbed areas, recognizing two nested categories of species: *pioneer species* (those that typically appear after any form of small-scale disturbance such as a gap [see Chapter 6] created by tree-fall) and *heliophilic species* (those that are sun-loving, not all of which are pioneer species). Species richness peaked at intermediate disturbance level as determined by correlations between total species richness and percent of heliophilic stems, which occur in small gaps (Figure 5-16).

The intermediate disturbance model for high species richness helped focus tropical ecologists on looking closely at nonequilibrium forces to explain why tree species richness is maintained at such a high level (Reice 1994). IDH theory has become more rigorous since it was first proposed. Various models have been generated. IDH has been shown to be broad in scope in that several different mechanisms of coexistence may be accounted for within IDH theory. Models based on spatial within-patch scales, spatial between-patch scales, and purely temporal models all generate similar patterns of coexistence under IDH (Roxburgh et al. 2004).

Recall that species richness relates to the size of the area sampled. In the previous chapter, we saw that larger areas, all things considered, contain greater species richness than smaller areas. The relationship between number of species, stem density, and area, which may initially seem like a diversion

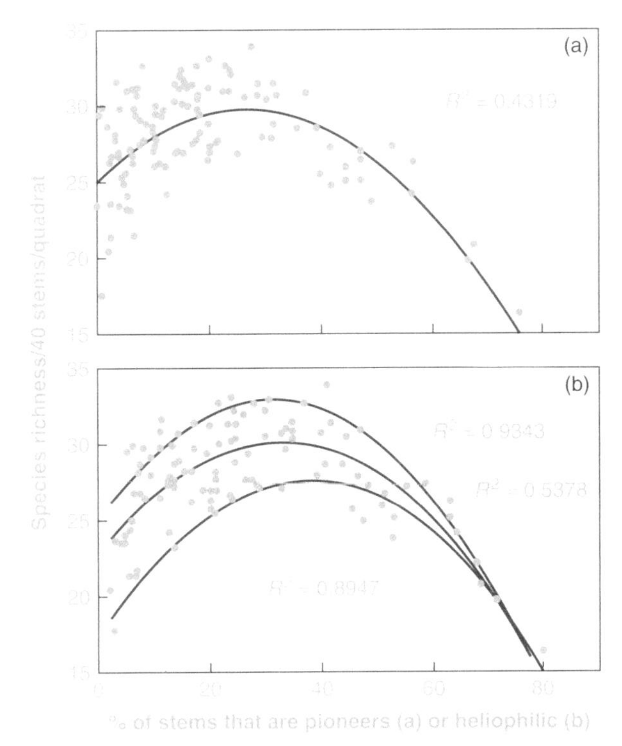

FIGURE 5-16
These graphs illustrate data obtained by Molino and Sabatier. (a) shows pioneer species, and (b) shows heliophilic species. The species richness peaks with a moderate abundance of both pioneer and heliophilic species.

from discussing the rich tree diversity within rain forests, will form a necessary prerequisite that will take us to one of the most challenging ideas in theoretical ecology, the unified neutral theory of biodiversity and biogeography.

Island Biogeography: An Important Digression

Island biogeography has been of interest to biogeographers ever since Darwin's astute observations from the Galápagos. It has long been known that organisms on islands are usually more vulnerable to extinction than mainland species. A list of species that became extinct over the past three centuries, such as the famous dodo (a large flightless pigeon) from Mauritius Island in the Indian Ocean, is strongly biased toward island-dwelling species. But beyond such realizations, the ecology of islands, with fewer species and fewer community types than are found on continents, affords ecologists an opportunity to study how ecological communities garner species and how species richness relates to habitat area.

When islands are compared using large-scale measures such as species richness, a pattern immediately becomes apparent. Larger islands contain more species than smaller ones. For example, in the Galápagos Islands (Plate 5-7), nine species of Darwin's finches inhabit the large island of Santa Cruz, but only five species occur on Plaza Sur, a much smaller island (Plate 5-8). Some of the species of finches present on Santa Cruz live in trees, but no trees occur on Plaza Sur. So it seems intuitive and not very interesting that two species of tree finches, plus the woodpecker finch and the vegetarian finch (all of which typically feed in trees), are not found on Plaza Sur.

A second general pattern is that islands closer to a mainland typically contain more species than those

PLATE 5-7
The Galápagos Islands, about 1,000 kilometers (600 miles) west of Ecuador in the Pacific Ocean, provide a good example of island biogeography.

PLATE 5-8
The large ground finch (*Geospiza magnirostris*) is one of 13 species of Darwin's finches found on the Galápagos Islands.

more isolated by distance from a mainland. Darwin noted that the Galápagos Islands were ecologically unique but nonetheless contained species most typical of South America rather than other geographic regions (Plate 5-9). The Galápagos Islands are in the Pacific Ocean about 1,000 kilometers (600 miles) from Ecuador. Again, it may seem intuitive to realize that islands close to a mainland contain more species than those at greater distances. The mainland, after all, provides colonizing species, and colonization is largely a random event. Thus random dispersal should result in more frequent colonization of closer islands.

The combined observations that large islands are more species-rich than small islands and that islands nearer a mainland have more species than those farther away were the basis of a now-classic mathematical

PLATE 5-9
This plate shows a cluster of leaves from Galápagos *Miconia* (*Miconia robinsoniana*), endemic to the Galápagos Islands and common at certain elevations on various islands. *Miconia* is a widespread genus with many species in the Neotropics, and thus *Miconia* colonized from either Central or South America. Also notice bracken fern (*Pteridium aquilinum*). Bracken fern colonizes widely, and many species are found in many regions throughout the world.

treatment of island biogeography by Robert H. MacArthur and Edward O. Wilson (MacArthur and Wilson 1967). The MacArthur and Wilson monograph inspired much research into the dynamics of island communities and has been widely applied to continental regions as well.

MacArthur and Wilson found that there was mathematical consistency to patterns of species richness in relation both to area and to distance from mainland. For example, at a given latitude on islands of, say, 1,000 square kilometers, there might be 50 species of birds. But on islands of 10,000 square kilometers, there would be 100 bird species. In other words, increasing island size by a factor of 10 doubles the number of bird species that the island supports. MacArthur and Wilson noted that the equation

$$S = CA^z$$

described the relationship between island area and species richness. In this equation, S equals number of species, A is area, and both C and z are constants that depend on the location of the source group as well as the kind of organism (so C and z differ for birds and beetles). Wilson describes z as a *parameter* dependent on proximity of the island to the mainland source of colonizing species as well as the group identity, whereas C depends on group identity alone (Wilson 1992). The area effect is easily seen in comparing the number of reptiles and amphibians on various islands within the West Indies Archipelago (Figure 5-17). The precision of the line demonstrates that the relationship between area and species richness is surprisingly predictable.

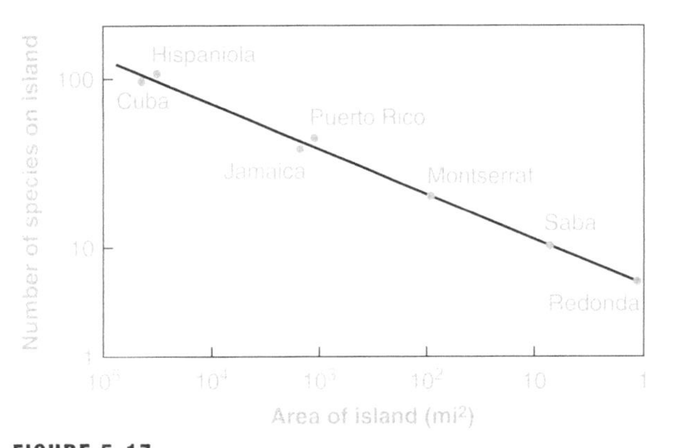

FIGURE 5-17
The relationship between number of species (in this case, of reptiles and amphibians) per island and island area.

Distance affects colonization rates in a fairly obvious way. The farther an island is from a source of immigrant species, the more slowly it accumulates species (because colonization is largely a random event and many species will not survive long while traversing open ocean). Thus the rate of colonization is predictably lower on distant islands equal in area to islands closer to mainland. As an example, a study looked at how green iguanas (*Iguana iguana*) (Plate 5-10) may have colonized the island of Anguilla, in the Caribbean Sea (Censky et al. 1998). The study showed that the occurrence of hurricanes throughout the Caribbean provided a means of transporting lizards, allowing their dispersal on rafts of dislodged vegetation and eventually the colonization of Anguilla.

Just as organisms colonize islands, populations also periodically become extinct. The risk on islands is focused on population size, which for many species tends to be small. Such small populations are subject to greater genetic drift and, because they are concentrated on islands, to greater effects of localized disturbance. Just as a severe storm such as a hurricane might bring about the colonization of an island, it also might severely reduce an island population, even to the point of extinction. Islands therefore experience both immigration and extinction of populations. This reality formed the basis of the *theory of island biogeography*.

The theory of island biogeography predicts that an island will eventually reach equilibrium in species richness for each taxonomic group (Figure 5-18a). This is the key feature of the model. It will remain at equilibrium unless disturbed in some way. In other words, initially a barren island will experience a rate of colonization far in excess of its extinction rate. The rate of extinction will gradually increase as the species richness increases because (1) there are more species that can potentially go extinct (so the overall number of extinctions per year or century will increase even if the rate per

PLATE 5-10
Green iguana, a widespread species throughout the Neotropics that has colonized the Caribbean island of Anguilla.

species is constant), and (2) negative interactions among species may increase extinction rate per species. At the same time, the colonization rate goes down steadily as more and more of the likely colonists (those that can disperse across open water) arrive on the island.

The rate of immigration and extinction will vary between near and distant islands and between large and small islands (Figure 5-18b), but each island will, nevertheless, attain an equilibrium point for any given taxonomic group. It is important to realize that the equilibrium is dynamic. In other words, there are ongoing immigrations and extinctions and thus there is species turnover, but the number of species remains the same once the equilibrium point is attained. The process of colonization and extinction is largely *stochastic*, subject to chance events.

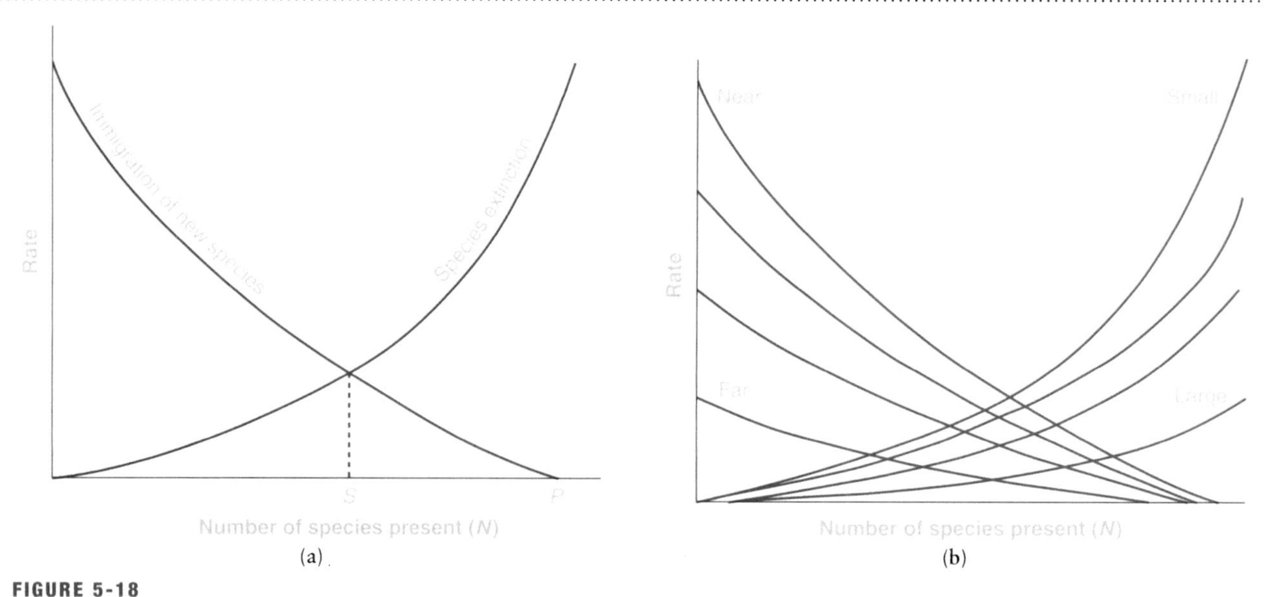

FIGURE 5-18
(a) The basic relationship between immigration rate and extinction rate on an island. The point where the curves cross is the equilibrium point for the island. (b) This graph shows how distance from colonization source and island size each affect the eventual equilibrium point. Small, distant islands have much lower equilibrium points than large islands close to the mainland.

PLATE 5-11
This is a small mangrove "island" (of *Rhizophora*) similar to the kinds used in the study by Simberloff and Wilson discussed earlier.

Rates of immigration and extinction of species (called *turnover rate*) will vary with distance. On average, islands near the source of colonization should experience more rapid turnover rates than those more distant, even though their equilibrium species richness may be identical.

The MacArthur-Wilson model was initially based on known species distributions, but a student of Wilson's, Daniel Simberloff, along with Wilson performed a creative series of experiments to test the theory (Simberloff and Wilson 1969). They selected a series of red mangrove (*Rhizophora mangle*) islets in the Florida Keys and carefully documented the arthropod diversity, confirming that both area and distance effects were present (Plate 5-11). As predicted by the theory, small islets held fewer arthropod species than larger ones, and distant islets harbored fewer species than ones closer to mangrove forests along the shoreline. Using fumigation, Simberloff and Wilson denuded various red mangrove islets of all small arthropods—insects, spiders, etc.

They documented the recolonization of the various islets such that within a year from the onset of the study, the original equilibrium levels had been reattained. However, the actual species compositions were quite dynamic, as some species colonized, became extinct, and subsequently recolonized. Islets reattained the same numbers of species as before they were denuded, but the species assemblage for any given islet was constantly changing—what Wilson described as a *kaleidoscopic effect* (Wilson 1992). In other words, the equilibrium was dynamic, with continual immigration and extinction.

Island biogeography theory is not literally confined to islands. There are "islands" of various sorts in mainland areas. For example, fragmented forests surrounded by agriculture, pasture, or some other anthropogenically created ecosystems represent islands of a sort, and island biogeography theory has been applied to such fragmented ecosystems. This topic will be further discussed in Chapter 14. In addition, high-elevation habitats such as alpine tundra and cloud forest also form islands of a sort, as they are frequently isolated elevationally.

Barro Colorado Island (BCI) in Panama was previously a high-elevation hill within a contiguous lowland rain forest. Island biogeography theory predicts loss of species from BCI because its original equilibrium species richness would have been characteristic of the vast lowland rain forest and now it is a much smaller area. Once it became an island of only about 1,500 hectares isolated in Gatun Lake, it was supersaturated with species. Island biogeography theory would predict loss of species and eventual establishment of a new equilibrium point.

Various researchers, beginning with Willis (1974), have documented the gradual extinction of bird species from BCI. About 60 bird species have been lost from BCI since it was created in 1914 (Karr 1990). Over time, from an initial diversity of 208 species,

PLATE 5-12
Great curassow (*Crax rubra*), now extinct from BCI.

48 species have been lost from the island (Plates 5-12 and 5-13). Some of this loss is due to the fact that the island itself has undergone ecological change as the forest has matured. Bird species typical of disturbed areas have disappeared as mature forest has grown up to replace brushy habitats, an ecological change that has resulted in selective loss of species. Other species have also gone extinct, however, and these species do occur in the kind of mature forest habitat now present on BCI. It is these species that illustrate the loss predicted by the island biogeography theory. Large birds such as raptors (hawks and eagles) and various ground birds are gone, likely due to their need for larger home ranges (as has been the case for large mammalian predators). Karr (1990) compared mainland survival rates of bird species that were once on BCI with those that are still on BCI and showed that species lost from BCI had, on average, lower survival rates on the mainland than species that still persist on BCI. He also suggested that nest predation may be higher on the island due to enhanced populations of various marauders such as monkeys and coatis resulting from a general decline of large predators. Biodiversity continues to gradually decline on BCI.

The equilibrium theory of island biogeography has impressive predictive powers but it is not without limits. Ted J. Case and Martin L. Cody applied the equilibrium model of island biogeography to desert-like islands in the Sea of Cortez between the Baja Peninsula and western coast of Mexico (Case and Cody

PLATE 5-13
Black-faced antthrush (*Formicarius analis*), now extinct from BCI.

1983 and 1987). They found some support for the MacArthur-Wilson model but also found patterns that varied strongly from one taxon to another. For example, they noted a difference in equilibrium patterns for highly mobile groups such as birds compared with less mobile groups such as reptiles and mammals. In approximately 10,000 years since the islands were connected to land (sea level being lower then), bird species richness has declined to the equilibrium value predicted by the MacArthur-Wilson model but lizard and mammal species richness remains *supersaturated* on the islands. Their species richness still reflects the higher value present when the islands were part of the mainland. Case and Cody concluded that land-bridge islands often remain well above equilibrium levels for certain taxonomic groups for thousands of years after the islands formed.

Case and Cody investigated what they termed an *island effect*, questioning whether isolation as such affects diversity of different taxonomic groups. They found that the richness of birds and plants on islands was, per unit area, not distinct from that on the mainland, except in extremely small islands. But when examining mammals and reptiles, that view changed. There is a markedly lower richness of these groups on islands compared with mainland richness. That is likely due to mammals and reptiles being less likely colonists, so even with low rates of extinction, the number of species steadily declines on islands.

The equilibrium theory of island biogeography began from largely intuitive observations—that small islands harbor fewer species than large islands and that isolated islands have fewer species than areas proximate to mainland. From those observations was developed a testable mathematical model for determining whether either literal or virtual islands are in equilibrium regarding species richness of various groups. Island biogeography suggested to ecologists that ecosystems are structured by assembly rules that determine the pattern of colonization and how equilibrium is reached. Are they?

Assembly Rules

It is clear that certain species, whether they colonize islands or invade forest gaps (see Chapter 6), have adaptations for dispersal and colonization. It is therefore intuitive to predict that species with good colonizing adaptations will appear first following disturbance. These will eventually be followed by more species (with lower dispersal rates) until the ecological community is more fully assembled. The patterns of colonization followed by increased species richness represent assembly rules. But are these rules—if indeed *rule* is the right word—biologically trivial in that they merely reflect obvious adaptations for dispersal and colonization? Are there actual rules by which communities are assembled?

Suppose that you look at all rain forest tree species in a tract in Borneo relative to all species present in similar habitats in that region. In other words, you examine the alpha diversity in relation to the beta diversity. To what extent is the alpha diversity a nonrandom subset of the beta diversity? In other words, are there combinations of species that are "permitted" to coexist and combinations that are "forbidden"? Or is any combination of species likely to occur?

In a study of bird communities on Southwest Pacific islands, Jared Diamond (1975) demonstrated that some combinations of species were statistically more likely to occur than others. On islands in equilibrium, some species combinations were never found, and others were more likely. Diamond demonstrated that various combinations of bird species assembled into communities—there was, in other words, more than one way to form stable combinations of species. But some species combinations were intrinsically unstable. Interspecific competition was considered the most likely factor to preclude some species from coexistence, though other factors were also considered to have possible influence. Diamond calculated what he called *incidence functions* for each bird species. Some species turned out to be *tramps*, defined as successful at colonizing but able to persist only if there were relatively few species on the island. *Super-tramps* were even better colonists but less tolerant of competition from other species. *High-S species* were poor colonizers but fared well in communities of high species richness (S). This observation suggests an evolutionary trade-off between colonizing ability and ability to persist among numerous potential competitors (Figure 5-19).

Assembly rules, to the extent that they exist, suggest that ecological communities are not random assemblages of species. It might be more appropriate to say *guidelines* rather than rules, as *rules* suggest a firm deterministic pattern that is rarely evident in nature. Most ecologists would be daunted if challenged to assemble a community from scratch in such a way that

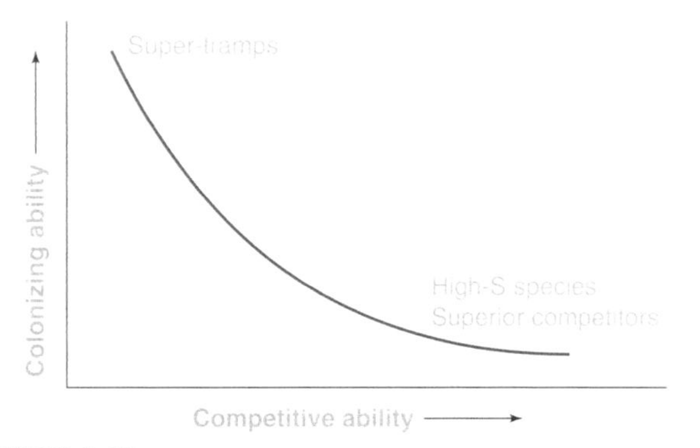

FIGURE 5-19
The possible evolutionary trade-off between colonization (dispersal) ability and ability to compete indefinitely once established. Good competitors are poor colonizers.

the ecologist could predict in advance the exact composition of the community when it reached equilibrium (Morin 1999).

Deterministic and Nondeterministic Models

Ecologists have attempted to model ecological communities using a diverse array of approaches. Each approach seeks to underpin the empirical data with an explanatory theoretical foundation. Historically, ecologists have relied on deterministic models in which, under a given set of variables, a predictable outcome will result. An example of a deterministic approach is to attempt to measure how much niche partitioning is necessary to avoid extinction by interspecific competition and thus attain a stable community.

The theory of island biogeography is deterministic in that it predicts an equilibrium point for each species group on each island, a point dependent on a balance between immigration and extinction. But it has nondeterministic, stochastic properties as well in that it does not make formal predictions about just which species will immigrate and which ones will become extinct—thus the equilibrium point is predictable but not the community assemblage itself. But consider what the word *equilibrium* really means in this context. There is no real equilibrium at all, since if you could observe the community over time, you would see species arrive and species become extinct, and you would never conclude that there was a logical end point to the process you were observing. There will always be turnover of species—always.

Some ecologists have found it enlightening to test the distribution of species in ecosystems by applying *null models*. These are nondeterministic, stochastic models to explain how species might assemble in a community. In other words, is there a random distribution of species, or is the distribution such that it is determined by some form of biological forcing? Null models were initially applied to challenge the long-held belief that ecological communities (especially those that appear to be at equilibrium) are shaped by the evolutionary outcome of interspecific competition, resulting in niche partitioning. Null models test the statistical likelihood that a particular distribution could have resulted by chance alone.

As null models have been more and more frequently applied, a vigorous debate has ensued among ecologists who favor the overall power of interspecific competition to shape ecological communities by niche partitioning and those who are skeptical of interpreting ecological patterns under the general umbrella of competition (Connor and Simberloff 1979; Salt 1984). That debate continues today, especially among plant ecologists in tropical rain forests, as you will soon learn.

Null models are frequently misunderstood in that they are assumed to suggest that only random effects are occurring in nature. It is clear that species interact in complex and nonrandom ways, and so how can it be that communities are assembled due to chance, as suggested by null models? But that is not what null models really suggest. Null models fully allow for complex deterministic interactions to structure communities. What they do is to provide a yardstick of sorts, a kind of statistical control to use in comparison with an observed pattern. Connor and Simberloff (1979) compared the actual distribution of species pairs of land birds found on various islands of the Galápagos Archipelago with those predicted using a null model. The fit was not exact, but it was close, and the null model assumed that different species were distributed independently of one another, that they were not competitively restricting each other's distribution (which was generally assumed to be the case). The data set, as interpreted by the null model, suggests that species distribution alone is not sufficient to conclude that interspecific competition (or any other force) is acting, though it very well may be. The null model does not disprove anything regarding competition. It suggests only that one needs to look further to fully demonstrate it. Now what about trees in rain forests?

The Unified Neutral Theory (UNT) of Biodiversity and Biogeography

Null models challenge assumptions that species distribution results from deterministic processes such as interspecific competition. Such an approach challenges intuitive concepts as well as long-standing ecological theory. Ecologists have thought credible the idea that competition among species inexorably results in increasingly smaller, narrower ecological niches. This increased specialization permits more species to pack together in an ecosystem. The tropical rain forest, with its remarkably high tree species richness, especially the high alpha diversities of trees, is an example of extreme species packing.

It is difficult to walk in a tropical rain forest such as those found in western Amazonia or parts of Indonesia and not wonder about the trees. Why so many species, usually in excess of 100 per hectare? As stated earlier, trees require very basic resources: light, water, carbon dioxide, and soil nutrients. Given those rather simple resources, how can it be that so many species coexist? Is it really possible that each tree species has evolved to use some kind of unique subset of resources different from all others? That seems unlikely. What seems more likely is that the trees somehow coexist even though many species require essentially the same resource base—light, water, carbon dioxide, and soil nutrients. So how can this be?

Some years ago, Stephen P. Hubbell, now at the University of California in Los Angeles, began to strongly question the prevailing opinion about how tropical tree species diversity is maintained. Working on Barro Colorado Island, Hubbell has managed an ongoing study of the forest dynamics for many years. Hubbell has also studied forest dynamics in Costa Rica, and in a paper published in 1979 (Hubbell 1979), he argued that the approximately 60 species of trees composing the dry forest were not in stable equilibrium. Further, they were not dividing the resources such that they occupied specialized, largely separate ecological niches. Hubbell asserted that periodic disturbance was sufficient to prevent any competitive exclusion and allow for the slow accumulation of species. We will return to the question of how periodic disturbance might influence rain forest tree communities later in this chapter.

In Hubbell's view, the abundance of various tree species fluctuated such as to reflect the stochastic nature of community dynamics. Hubbell saw the dynamics of the forest largely shaped by factors different from deterministic forces such as competition among species. He viewed the tree species as *ecologically egalitarian*, subject to neutral variation in abundance. No one species was consistently favored or disfavored. He further developed this model of biodiversity dynamics in his studies on Barro Colorado Island.

Hubbell's model, once formalized, became the *unified neutral theory* (UNT) of biodiversity and biogeography (Hubbell 2001). It is detailed in a Princeton University Monograph in Population Biology, the thirty-second such monograph that Princeton has published since 1967, when Princeton published *The Theory of Island Biogeography* by MacArthur and Wilson, the first of the monograph series. It is fitting that Hubbell's theory be part of the monograph series, since it builds on the MacArthur-Wilson theory, which Hubbell noted was incomplete in scope.

The theory that MacArthur and Wilson proposed does not include any component of speciation, because it does not incorporate an evolutionary time scale. It is a dynamic (as opposed to stable) equilibrium model that states that an island's species richness is entirely determined by ongoing immigration and extinction. What of speciation? Think of the Darwin's finches of the Galápagos Islands. Accounting for their very existence is not strictly part of the MacArthur-Wilson theory. The MacArthur-Wilson theory also lacks consideration of population sizes among species. Hubbell's theory includes speciation and population dynamics as well as immigration and extinction, thus combining biodiversity and biogeography in a unified construct.

A complex mathematical model accompanies Hubbell's theory. The mathematics, when applied to a data set (trees in a tropical rain forest, birds on islands, and so on) generates a single, dimensionless number that Hubbell calls the *fundamental biodiversity number*, symbolized by the Greek letter theta (θ) (Table 5-1). This number, which is a function of the speciation rate and the size of the metacommunity (all trophically similar individuals and species in a regional collection of local communities—trees, for example) is then used to predict both the relative abundance of species (RAS) and the actual species richness.

Though the derivation of θ and the model it supports is complex and beyond the scope of this text, it is possible to describe the essence of the model and to understand its potential significance. In testing his model on actual data sets, Hubbell found that it had strong predictive strength when applied to a diverse

TABLE 5-1 The fundamental biodiversity number (θ) predicted from the Hubbell model for various ecological communities.

FOREST TYPE	FOREST LOCATION	BIODIVERSITY NUMBER
Boreal forest	Flower's Cove, Newfoundland	0.15
	Clingman's Dome, Great Smoky Mountains National Park, Tennessee	0.22
	Mount Washington, New Hampshire (mid-elevation)	0.50
Northern hardwoods	Adirondacks, New York	2.0
Mixed temperate deciduous forest	Cumberland Plateau, Kentucky	5.0
	Cover forest, Sugarlands, Great Smoky Mountains National Park, Tennessee	7.1
Tropical semideciduous (dry) forest	Forest near Bagaces, Guanacaste, Costa Rica	24.0
Tropical semievergreen (moist) forest	Barro Colorado Island, Panama	50.0
Tropical evergreen forest	Pasoh Forest Reserve, Negeri Sembilan, Peninsular Malaysia	180.0

array of communities, ranging from copepods in the sea to trees in various kinds of forests. In short, the UNT fit the data (Figure 5-20).

A simple analogy will make the basic point of what Hubbell is suggesting. Evolutionists now fully accept that much genetic change through long periods of evolutionary time is due to random walk, better known as *genetic drift*. Given millennia, genes that exert no strong phenotypic effect (and thus are not subject to natural selection) tend to vary randomly in frequency. These genes are frequently called *neutral genes*, because their frequencies vary due to stochastic factors, not determinative factors such as natural selection. Now imagine this being the case with species such as trees in tropical rain forests. Hubbell is asserting that there is such a thing as *ecological drift*,

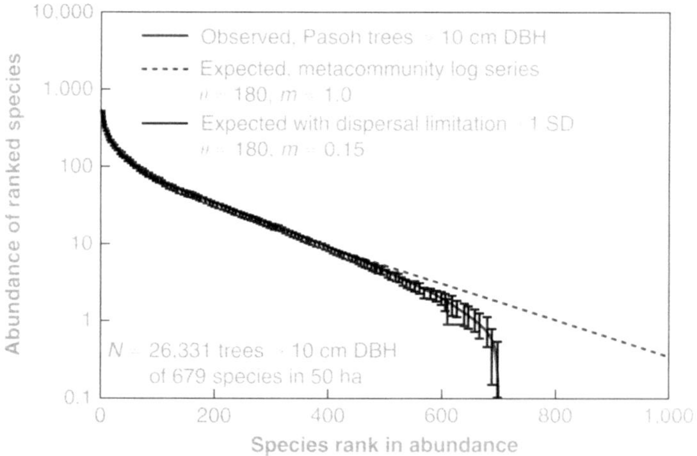

FIGURE 5-20
Fitted and observed dominance–diversity distributions for trees > 10 centimeters in diameter in a 50-hectare plot in Pasoh Forest Reserve, Malaysia. The θ value was 180. Note that very rare species do not fit the curve well, but all others do.

where species are functionally neutral relative to one another and thus vary in abundance over long time periods essentially due to stochastic factors. The rate of extinction in Hubbell's model is sufficiently slow that speciation may exceed extinction, so species are permitted to accumulate in communities. This means that evolution will add biodiversity to ecosystems, even when it results in strong ecological similarities among component species, an event that would seem to inspire intense interspecific competition.

Hubbell refers to community dynamics as described by his neutral theory as *zero-sum ecological drift*. This means that the abundance of each species follows a random walk (drift), subject only to the constraint that the total number of individuals of all species combined in that particular community remain constant. Note carefully what that statement means, as it is the very essence of Hubbell's explanation for hyperdiversity of tropical rain forest tree communities. If extinction rate is slow and speciation is permitted but also if the total number of individuals remains constant, then the trend, with evolutionary time, will be toward higher species richness and smaller population sizes—precisely what is found in tropical rain forest tree communities. Note also that it is the number of individuals that is key, not the number of species. As Hubbell states, "the number of new species arising per unit time is a function of the total number of individuals in the metacommunity, not the number of pre-existing species" (Hubbell 2001, p. 236).

A simple *neutral community model* (NCM), in its most basic sense, includes the following five variables that combine to determine the number of species in a community:

- Probability of birth b for each individual
- Probability of death d for each individual
- Probability of immigration m for each species
- Number of individuals K in each community
- Number of species N in the external species pool

It is important to note that the model is confined to species that share the same trophic level. It does not apply, for example, to top-down or bottom-up interactions comprising food webs. But, as Hubbell points out, there are far fewer trophic levels than there are species per trophic level, so a theory of biodiversity confined to a single trophic level will nonetheless include a lot of species.

Hubbell contrasts his theory with the two predominant views of community dynamics, the *niche assembly perspective* and the *dispersal assembly perspective*. The niche assembly perspective is the view that actual assembly rules govern how communities develop and that the rules themselves derive from niche differences among species. This deterministic view predicts that communities eventually attain a stable equilibrium based on species interactions such as competition and niche diversification. The dispersal assembly perspective views communities as open to invasion, not stable, but structured largely by random dispersal, chance, and disturbance effects. Any equilibrium attained is dynamic, with species changing, though the species richness may stay the same.

Hubbell recognizes that each of these views can be and has been (see the following) supported by data but that scale is critical in selecting between the two opposing views. Small-scale studies, both spatially and temporally, tend to support the niche assembly perspective, while large-scale studies tend to support the dispersal assembly perspective. The theory of island biogeography that MacArthur and Wilson proposed is an example of a dispersal assembly perspective. Immigration and extinction continue the point of dynamic equilibrium. Hubbell states clearly that "actual ecological communities are undoubtedly governed by both niche-assembly and dispersal-assembly rules, along with ecological drift, but the important question is: What is their relative quantitative importance?" (Hubbell 2001, p. 24). The neutral theory is an example of macroecology, which is concerned with very broad ecological patterns (Bell 2001).

Any ecological model begins with assumptions, and those that underlie Hubbell's theory are admittedly unrealistic. Hubbell's model envisions species as having identical birth, death, and dispersal rates, an assumption that is patently false in the real world. But Hubbell argues that though the assumptions are, indeed, unrealistic, they work for the following reason: species vary in *life-history trade-offs* such that the relative fitness of species (that is to say, the fitness of a species in relation to that of all other species of its trophic level) is equalized throughout the community, at least on the scale at which the model works. (Stop now and just think about this for a while. It is important to understand this concept before moving onward.) That statement seems counterintuitive, but in Hubbell's view, niche differences actually allow for the neutrality of the model. Why is this so? Consider a hypothetical example. If a species developed the *perfect niche* so that it could compete successfully with all other species at its trophic level, it would eventually reduce that trophic level to one of a single species, itself. But this never happens. Competitive exclusion, easy to demonstrate in lab situations, appears rare in nature. Darwin took note of the fact that populations tend to oscillate in density but only within a certain range. Year after year, decade after decade, century after century, an oak woods has more or less the same species composition, whether one looks at birds, trees, or butterflies, but the populations may vary considerably over time.

But why shouldn't a species evolve a perfect niche, such that it outcompetes all other species? Hubbell points out an obvious trade-off, between shade-intolerant and shade-tolerant tree species. He shows that shade-intolerant species, such as are found in disturbed areas and forest gaps, grow quickly and generally produce small, well-dispersed seeds. Shade-tolerant species grow more slowly, have larger seeds, but persist well. It is impossible for a tree to have both sets of traits simultaneously, and any forest contains a combination of shade-tolerant and shade-intolerant tree species (Figure 5-21).

Even if an invading species were to appear that was better adapted than all other species at its trophic level, Hubbell argues that it is much more likely that others in the species pool would adapt to the strong selection pressure asserted by the invader rather that all becoming extinct except the invader. In Hubbell's view, neutrality occurs because of niche differentiation.

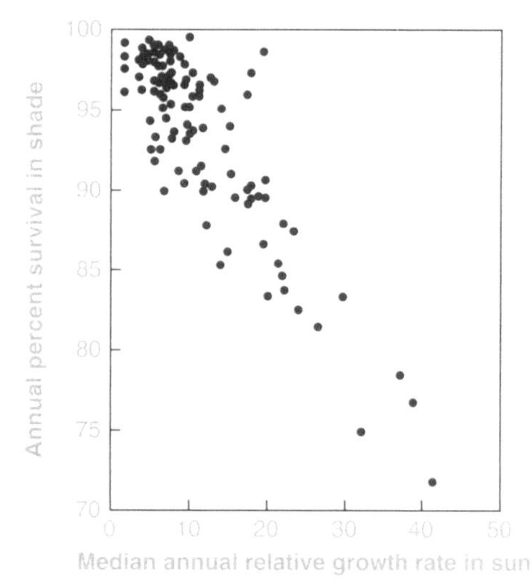

FIGURE 5-21
This graph shows niche differentiation along the shade tolerance–shade intolerance life history characteristics for BCI tree species. Dots represent species means for stems 1 to 4 centimeters in diameter at breast height. The graph shows that most species of trees on BCI are shade-adapted.

Note that there are really two rather subtle and distinct arguments here: first, that there is *demographic equivalency* among tree species based on trade-offs in life history characteristics, and second, that local tree neighborhoods are sufficiently heterogeneous and unpredictable to counterselect for specialization among species to narrow, resource-based niches. I will detail this notion more in the following.

Hubbell makes an intriguing argument. He knows that the assumptions of his model are unrealistic but that the very adaptations that differ among species seem to act to cancel out each other, so the model nonetheless works. As time proceeds, the abundances of the functionally neutral species vary, mostly because of ecological drift.

Many ecologists find Hubbell's theory controversial, and many studies have attempted to test it. Graham Bell (2001), another strong proponent of neutral macroecology, has summarized and rebutted some of the objections:

- *Species are locally adapted and resist the immigration of new species, and thus there is not a neutral ecological drift among species (with immigrant species entering communities) as posited by the model.* In an extreme form, this is obviously true. As Bell points out, coconut trees will not likely invade arctic tundra ecosystems, even if a few drift there. The neutral model is regional, not global. Within a broad ecological region such as Amazonia, differences among species are not sufficiently great, and the model works.
- *The neutral model does not take into account strong and complex interactions among organisms that are well documented to occur throughout nature.* This is clearly true in that the neutral model is confined to competition within a trophic layer and includes no top-down, bottom-up, or mutualistic component. The rebuttal is that neutral models do include strong interactions (within a trophic level) but that the individuals, regardless of species, have functionally identical properties.
- *The neutral model contains too many variables and is too complex, and with such complexity (six parameters free to vary), any result may be obtained, any pattern generated.* The rebuttal is that most ecological models have numerous variables and that neutral models are in reality no more complex that others being applied. Neutral models do predict actual distribution patterns with accuracy (Figure 5-22).

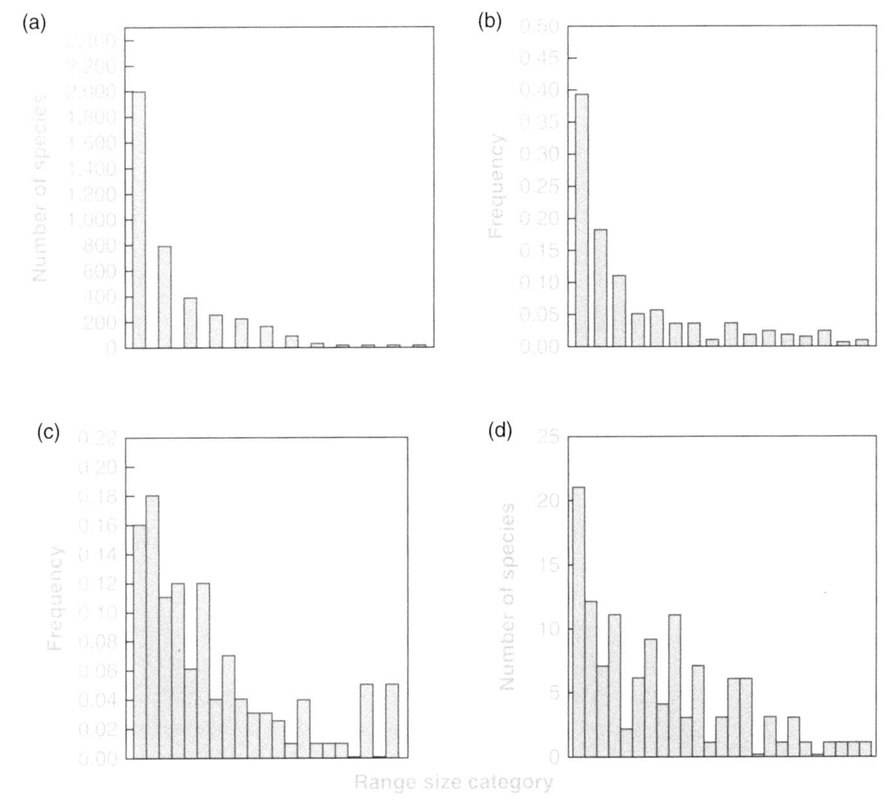

FIGURE 5-22
The actual distribution of bird species and range size for (a) New World birds, (b) passerine birds in Australia, and (c) North American birds compared with the distribution in range in (d) a neutral community. The neutral community comprised 125 species. Note that the neutral model based on local dispersal among species predicts a very similar distribution to the other data sets.

Is neutral theory a step toward formulating a unified theory of biodiversity? How useful is the model? How are rain forest tree species distributed and what factors determine species richness?

NEUTRAL THEORY EVALUATED

Stephen P. Hubbell initiated a vigorous debate among ecologists, and not merely those whose research focuses on tropical trees. Since neutral theory, now usually called universal neutral theory (UNT), was described by Hubbell in his 2001 book, there has been research supporting it (see Volkov et al. 2003 and 2005; Latimer et al. 2005) as well as research arguing against its predictive strengths (see Chave et al. 2002; McGill 2003; Wootton 2005; Dornelas et al. 2006; Ricklefs 2006). Bell (2001) described two forms of UNT, the *strong* and *weak* forms. The strong UNT reflects the true mechanisms for maintenance of high species richness—in other words, it is literally correct. The weak UNT produces mathematical patterns that correlate well with data sets but may not represent cause and effect. Regarding the weak UNT, Hubbell argues that any strong correlation of the sort produced by comparing distributions predicted from applying neutral theory with actual data sets requires some sort of explanation. In other words, critics should be able to say why such a correlation does not, in fact, represent real, underlying biological processes. Hubbell (2006) comments that "current neutral theory in community ecology is still in its infancy and very much a work in progress." But at the same time, he continues to show that neutral theory does correlate with remarkable precision with actual species distributions, and comments that such precision "cannot be easily explained away."

A key characteristic of neutral theory is to demonstrate ecological equivalence among a set of species. This is because UNT assumes ecological equivalence. It is not necessary for a species set to have identical natural history characteristics to be ecologically equivalent under the assumptions of neutral theory. They may be quite distinct in many ways, but those distinctions do not ultimately affect their abundances. Hubbell (2006) points out that the BCI tree community shows much evidence of niche differentiation, particularly between shade-intolerant and shade-tolerant species. But he also points out that the majority of tree species on BCI are shade tolerant and that it has not been possible to demonstrate resource-based niche differentiation among this set of species. Why should species that persist in low light conditions partition their niches more finely (as implied under niche theory by their greater species richness) than shade-intolerant species?

Hubbell's ideas for ecological equivalence among species have grown out of the concept of diffuse competition, offered as a potential force in permitting the indefinite maintenance of high-diversity tree communities. Diffuse competition occurs when communities are rich with species. Unlike low-richness communities, where interactions and the direction of selection between differing species may be consistent and clear, diffuse competition changes temporally and spatially within any local group of species (Figure 5-23). This directional inconsistency permits indefinite coexistence.

Hubbell argues that shady habitats in tropical forests such as BCI are more abundant than sunny areas,

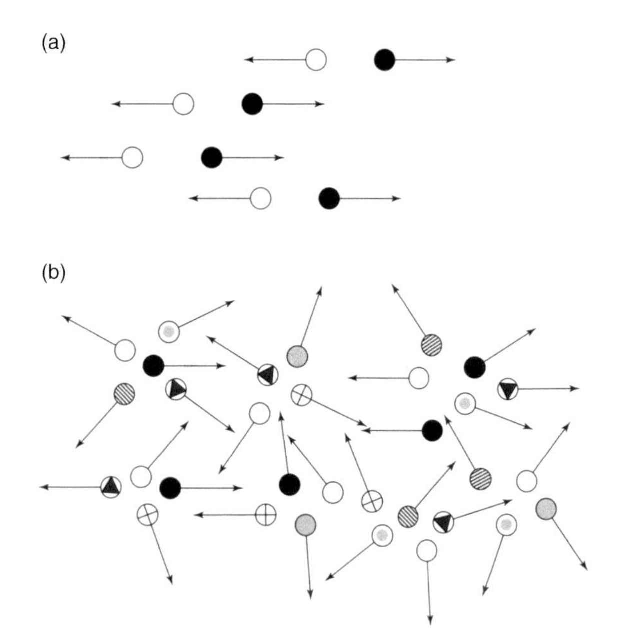

FIGURE 5-23
Hubbell's "cartoon" to illustrate diffuse coevolution in (a) a species-poor community compared with (b) a species-rich community. Each symbol represents an individual, and different symbols represent different species. Character displacement (niche separation) evolves in simple communities because pairwise species encounters are chronic and predictable. In the complex community, such encounters are much less frequent between any two species, so character displacement evolves very slowly, if at all.

and thus, with evolutionary time, more species have evolved to occupy them. But how do they coexist? Under UNT, if they were ecologically equivalent, they could persist indefinitely. He develops a model for how ecological equivalence is possible, and his starting point is that communities such as trees of BCI (and tropical trees in general) are dispersal-limited and recruitment-limited. Being *dispersal-limited* means that a tree is unable to disperse its seeds to all suitable sites where it might accomplish germination, survival, and growth. Being *recruitment-limited* means failure to recruit seedlings near adults at suitable sites.

The key concept is that if species are dispersal-limited, then many "inferior" species (with regard to competition) will nonetheless colonize sites suitable for "superior" species simply because they got there and the superior species did not. The strongest competitor failed to reach the site. Slow dispersal and recruitment (traits that are common in tropical rain forests) would greatly slow the process of competitive exclusion, making it essentially ineffective. There would be little or no selection pressure favoring niche partitioning, and instead, what Hubbell calls a *stochastic steady-state distribution* of relative species abundance will arise in the metacommunity. This type of situation could prevail especially if some factor, such as the presence of pathogens, herbivores, or seed predators, consistently reduced recruitment success and dispersal among the commonest species. This suggests that neutral theory must eventually expand to include trophic-level interactions (see Chapter 8) within the model (Plate 5-14).

A CHALLENGE TO UNT FROM THE TREES

The Yasuni Forest Dynamics Plot (25 hectares) in eastern Ecuador contains in excess of 150,000 mapped trees (equal to or greater than 1 centimeter diameter breast height [DBH]) from over 1,100 tree species. This makes it one of the most species-rich areas known. Researchers (Kraft et al. 2008) surveyed trees for the following characteristics, each of which is a reflection of each species' ecological niche:

- Specific leaf area (SLA), defined as leaf area divided by dry mass
- Leaf nitrogen concentration
- Leaf size
- Seed mass
- Maximum DBH
- Wood density (obtained from published data)

PLATE 5-14
Tropical moist forest tree species richness: deterministic or neutral?

These traits represent critical characteristics that define differences among species and that are essential to the overall fitness of the tree. Kraft et al. envisioned two potential nonneutral outcomes:

- That co-occurring species would converge in their characteristics (at any given area) because of strong abiotic selection pressures, a process termed *environmental filtering*
- That co-occurring species would diverge (at any given area) as predicted by classic niche differentiation theory

The researchers, using a null model approach, reasoned that if these traits converged, tree species would be highly similar in the preceding characteristics, but if co-occurring trees differed more than would be predicted by the null model, and thus were more ecologically distinct, then they would not represent ecological equivalents. The UNT did not explain such a distribution. Nor did the UNT explain high levels of convergence.

The data analysis supported the niche-based distribution and not the distribution predicted by UNT (Figures 5-24a and 5-24b). The range of trait variance was significantly smaller within quadrats and greater among quadrats. This is support for the habitat-filtering model. The researchers noted that the region contains two principal topographic habitats, ridgetops and valley bottoms. Species assemblages vary with soil characteristics and topographic habitat. Ridgetops have tree species

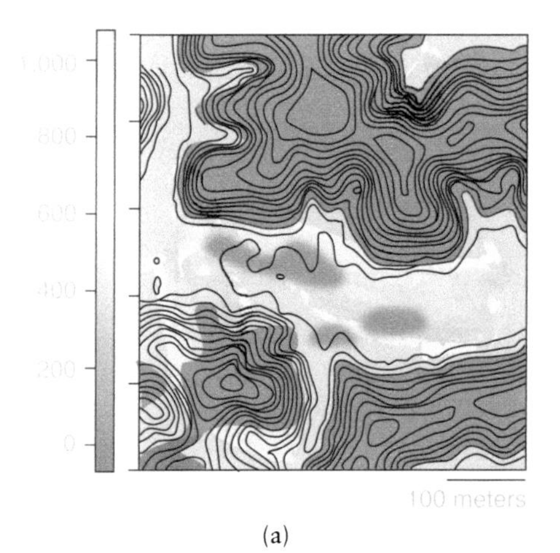

(a)

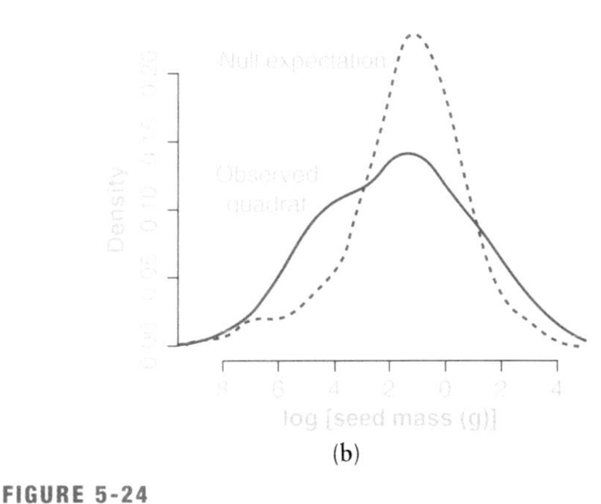

(b)

FIGURE 5-24
(a) This map of Yasuni shows contours indicating topography. The color densities indicate variation in SLA. The map shows a pattern illustrating that ridgetops have lower than expected SLAs (based on a null model) and valleys have a higher than expected SLA. (b) The graph shows the difference between the predicted distribution of seed mass in relation to density between the null expectation and the actual observed distribution.

with low SLA, smaller leaves, heavier seeds, and denser wood compared with valley communities.

The authors of the study noted that some quadrats were in concert with predictions from the UNT. But overall, they believed that patterns of dispersal, seedling establishment, or post-establishment mortality were nonrandom, producing the patterns evident in the study. They concluded:

> Although the magnitude of these processes still needs to be quantified, their existence indicates that forces included in neutral theory (such as demographic stochasticity and dispersal limitation) may not be sufficient to explain species distributions and maintenance of diversity in this forest, even though they are occurring. Taken together, our results support a niche-based view of tropical forest dynamics in which subtle but pervasive habitat specialization and strategy differentiation contribute to species coexistence. (Kraft et al. 2008, p. 582)

A CHALLENGE TO UNT FROM THE BIRDS

UNT has not found support in an analysis of avian biodiversity in South America (Ricklefs 2006). A key assumption of UNT, indeed its very core, is the concept, as described earlier, of neutral ecological drift. This means that any single species may be expected to eventually "drift" to extinction, just as genetic drift results in loss of alleles by random processes. The other side of that particular coin is that UNT predicts speciation, replacing extinction. It is through speciation that species accumulate, enhancing biodiversity, with extinction occurring at a slower cumulative rate. In the UNT, the larger the ecological community, the longer it takes for neutral ecological drift to occur (just as genetic drift is much more perceptible in small populations). The formula that predicts the average time to extinction under an assumption of neutrality when a population cannot exceed size K is equal to $2N\,[1 + \ln(K/N)]$. This translates to mean that the average time to extinction is $2N$ generations, the time a newly formed species requires to increase by drift to N individuals. For some common species with large populations, this equation suggests that it would take more time to reach extinction than the age of the Earth.

Ricklefs (2006) argues that neutral theory makes no firm prediction about the time to speciation. Rather, it is extinction that provides the "the critical test of neutrality." Ricklefs reasons that "if one assumes that extinction and speciation rates are approximately balanced, then the time to extinction will equal the time to speciation, $2J_M/S$ (where S is the species richness in the metacommunity and J is the metacommunity size)"(p. 1426). Ricklefs argues that given the large metacommunity sizes and large species richness and long generation times typical of trees, drift seems too slow (about 1×10^8 years to extinction) to allow the UNT to work. What about birds?

Using divergence times of birds based on mitochondrial DNA sequences (where the calibration is 2% sequence divergence per million years) and looking at *sister-species* (species most recently diverged from each other), Ricklefs estimated that speciation events for passerine birds would be about 2 million years. (*Passerine*, or "perching," birds belong to the order Passeriformes. They are frequently called *songbirds* for their vocal qualities. There are two major groups: *suboscines* and *oscines*. Suboscines have less complex, more primitive syrinx muscles and thus simpler vocalizations.)

Using published data on South American suboscine birds (a group of birds abundantly represented by many species in the Neotropics) (Plate 5-15), Ricklefs estimated the speciation rate (per species) to be 0.43 per million years. The average mortality he estimated to be close to 0.25 per year. He estimated generation time to be 5 years. He calculated that the average time to extinction of a newly formed species would be equal to 0.29 million generations. Looking then at the total area of semievergreen broad-leaf forest (5×10^6 square kilometers), he estimated the number of breeding birds in tropical South America to be approximately 10 billion. There are about 1,725 species in the area included. If one assumes 1,500 are in tropical forest, with an average population size of 4 million individuals, the time to extinction, using UNT, is 40 million years (for a species to go from speciation to extinction). But molecular and fossil evidence supports an extinction time of between 1.4 and 2.9 million years for any single species. Thus the predictions made based on UNT are an order of magnitude greater than what the data suggest to be the case.

Ricklefs concludes, "As shown by the present analysis, even for passerine bird populations of average size, expected times to extinction under drift greatly exceed values calculated from genetic divergence times and lineage-through-time plots" (p. 1429). Rickleffs suggests that other factors such as climate change

PLATE 5-15
The widely distributed great antshrike (*Taraba major*) is one of hundreds of species of suboscine birds found in the Neotropics.

(see Chapter 13) and coevolutionary interactions (see Chapter 7) may alter population at rates too rapid to be accounted for by neutral ecological drift. He thus rejects the contention that bird species may be regarded as ecological equivalents. But review the discussion of his study here, and note how many assumptions are included. The robustness of each assumption (or lack of it) is important in the outcome of the model. There is still much to know.

A second study on birds (Graves and Rahbek 2005) also fails to support UNT. In a continent-wide analysis of South American birds using bird distribution databases from 30 museums in 20 countries as well as satellite data to generate a global land-cover map, the researchers saw a strong pattern in evolutionary patterns of birds. Those east of the Andes typically had wider distributions (presumably due to less severe gradients in habitat) than those west of the Andes. Such a pattern argues against one assumption of UNT, that habitat requirements are to be ignored for purposes of the model. But the results of the study are not necessarily inconsistent with UNT if one recognizes that the area east of the Andes may be sufficiently different from that west of the Andes to

The Balance of Nature?

Consider the information presented on the complexities of accounting for tropical tree species richness and diversity patterns. One of the most ancient notions in human thought is that there is a *balance of nature*, a concept tracing its roots far back into the times of the ancient Greek philosophers. Ecologists for most of the twentieth century investigated equilibrium concepts that they assumed would govern ecological community dynamics. Not until late in the twentieth century did ecologists focus far more on stochastic processes, null models, and the importance of temporal and spatial scale as these affect ecosystem dynamics. Equilibrium models began to be replaced by nonequilibrium models or dynamic equilibrium models containing a strong stochastic component. It should be clear from this chapter and others that there is no such thing as a balance of nature. Instead, nature is highly dynamic in numerous manifestations. Indeed, equilibrium is likely a very rare characteristic of ecosystems. For more on the balance of nature as an ecological concept, see my book *The Balance of Nature: Ecology's Enduring Myth* (Kricher 2009).

constitute two metacommunities when looked at through the lens of UNT. In other words, no one would argue that UNT should lump species in a desert with those in a forest, and the habitat distinctions between east and west of the Andes may be a similar case.

Concluding Remarks: Why Are There So Many Tree Species in the Tropics?

The various hypotheses briefly outlined in this chapter represent the current state of thinking and research regarding one of the most difficult ecological problems, accounting for tree hyperdiversity in rain forests and in the tropics in general. It is a vexing problem. As Leigh et al. (2004) point out, "How can 0.5 km^2 of rain forest in Borneo or Amazonia contain as many tree species as the 4.2 million km^2 of temperate forest in Europe, North America, and Asia combined?" Or, to turn the point around, why are there not more species of trees in the temperate zone?

As a starting point, consider total flowering plant diversity in each of the three major tropical regions:

- 90,000 in the Neotropics
- 40,000 in tropical Asia
- 35,000 in tropical Africa

These diversities roughly correlate with area. The Neotropics is a vast area of tropical forest, and that reality may be reflected in the species richness of the region. Recall from the theory of biogeography and, as well, from empirical ecology that area matters. Larger areas consistently have more species. The relationship between species richness of trees and area in tropical regions generally follows the prediction based on the theory of biogeography (May and Stumpf 2000). There is strong evidence that extinction rate is lower and speciation rate higher (because of opportunities for various forms of isolation as well as varying selection pressures) in regions of vast area (Rosenzweig 1995; Ricklefs 2004).

It is unlikely that major vicariance events are entirely responsible for isolating gene pools of trees sufficiently to account for such speciation. It is more likely that a tree species initially forms as small populations that, with natural selection favoring rarity, eventually spreads until it becomes more common and, ironically, less fit in comparison with rarer species. Some plant ecologists believe that *sympatric* (ranges overlap) speciation occurs. More believe that speciation in tropical plants is mostly *allopatric* (divided ranges, as in some form of vicariance) and/or *parapatric* (ranges meet on a common boundary, but selection pressures differ at the boundary). Pressure from pathogens, herbivores, or seed predators, as well as adaptation to edaphic heterogeneity, could all act as selection pressures in all forms of speciation events. Various levels of disturbance could keep virtually all areas within the tropics in some state of nonequilibrium. But major vicariance matters too. Speciation patterns in Andean regions differ from those in Amazonia, for example.

Leigh et al. (2004) suggest that further work should investigate the following:

- The various ways habitat complexity can enhance tree diversity
- Whether mutual repulsion among conspecifics is usually driven by specialist pests or pathogens
- Whether the observed degree of mutual repulsion among conspecifics suffices to maintain observed tropical tree diversity
- Whether leaves of wet tropical forests have sufficiently effective antiherbivore defenses to reduce herbivory rates on young leaves and therefore to counter reduced fitness caused by excessive herbivory
- How much species turnover contributes to the latitudinal gradient in tree diversity
- Why species turnover is higher for herbs, shrubs, and treelets than for large trees
- And of great importance, whether tree speciation is usually allopatric, what factors drive speciation, and why some lineages (such as *Inga*) diversify rapidly, while other equally diverse lineages have diversified more slowly

It is clear that no one explanation will suffice in explaining hyperdiversity of tropical trees. Recall the analogy made in the previous chapter that species richness in the tropics could be a "perfect storm" of forces coming together, each of which acts in some manner that promotes speciation and maintenance of diversity. Consider one final thought: Perhaps species richness itself generates more species richness.

In a study of plants and arthropods from the Canary and Hawaiian islands, Emerson and Kolm

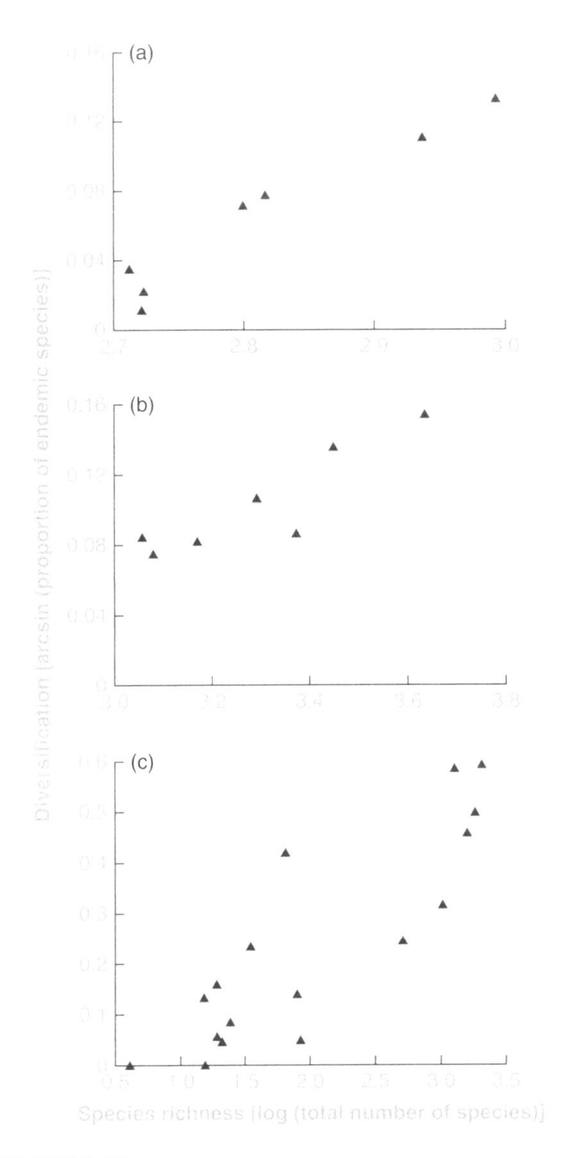

FIGURE 5-25
These three plots show the relationship between species richness and diversification (based on proportion of endemic species). (a) is plants in the Canary Islands, (b) is arthropods in the Canary Islands, and (c) is arthropods in the Hawaiian Islands.

(2005) showed a strong correlation between what they termed the *index of diversification* (proportion of species that are endemic) and species richness of the various islands (Figure 5-25).

They conclude by suggesting, "The answer to questions such as why there are so many species in the tropics might in part be because there are so many species in the tropics" (p. 1017). But the question remains, is this because of the greater (in the tropics) evolution of interdependent ecological relationships such as mutualism, parasitism, and herbivory, or is it because there are simply more species that potentially might undergo speciation? There is so much more to learn.

SOME FURTHER READING

Editors of *Nature.* 2000. Nature insight: Biodiversity. *Nature* 405: 205–253. A series of six review papers dealing with many aspects of biodiversity. It is somewhat dated now, but there is still much to be learned from reading this stimulating array of papers.

Hubbell, S. P. 2001. *The Unified Neutral Theory of Biodiversity and Biogeography.* Princeton, NJ: Princeton University Press. This is a highly technical book and will be a challenge to anyone not well-versed in mathematics. Nonetheless, it is clearly written and well argued, and anyone with a strong interest in diversity should grapple with Hubbell's theory.

MacArthur, R. H., and E. O. Wilson. 1967. *The Theory of Island Biogeography.* Princeton, NJ: Princeton University Press. This is one of the most influential monographs ever published in ecology. It is brilliantly presented and is must reading for any student of diversity.

Quammen, D. 1997. *The Song of the Dodo: Island Biogeography in an Age of Extinctions.* New York: Simon & Schuster. Quammen is known for his erudite and often amusing essays on various aspects of nature. In this book, he takes a different approach, a comprehensive overview of island biogeography that includes both historical and modern treatments. He interviews some of the most important researchers in the field and explains clearly the various aspects of debate among scientists who study this complex topic.

Rosenzweig, M. L. 1995. *Species Diversity in Space and Time.* Cambridge, UK: Cambridge University Press. This is a technical review of various aspects of species diversity. It is strongly recommended for advanced undergraduates and graduate students.

A Shifting Mosaic: Rain Forest Development and Dynamics

Chapter Overview

Natural disturbance events occur at various scales of space and time. The result is that most rain forest areas, as well as other tropical ecosystems, comprise a mosaic of various-sized disturbance patches where species composition is changing at varying rates. This is because disturbance events (whatever they may be and whenever they occur) force changes in species composition. Recruitment of plant species on disturbed sites occurs from seeds present in the soil bank, seedlings present on the site, and seed dispersal from other sites. Recruitment is strongly stochastic from site to site. Suites of plant species with differing arrays of adaptations, some requiring abundant sunlight, some intolerant of shade, and many that are shade-tolerant, are able to indefinitely persist within a temporally shifting mosaic of patches created by periodic disturbances. Small-scale forest disturbances create gaps of varying sizes, and the study of gap dynamics focuses on how species respond to gaps. Larger-scale disturbances result in ecological secondary succession, where areas are sufficiently disturbed that *heliophilic* (sun-loving) shade-intolerant plant species initially grow and thrive, to be eventually replaced by shade-tolerant, slower-growing species as biomass increases and the canopy closes. This process results in second-growth forest development, which now dominates much of the terrestrial tropics.

Disturbance, Gaps, and Ecological Succession

Tropical humid forests are composed of multiple patches that reflect the disturbance history of the site. Changes in species assemblages occur at various scales of space and time because of periodic natural or human-caused disturbance. The disturbance may be on a scale as small as a large leaf that falls on seedlings growing on the forest floor or as significant as a major hurricane flattening hundreds of hectares of forest. Small disturbances, such as tree-fall, are more frequent than large-scale disturbances. Variation in disturbance magnitude produces a range from very small-scale patches to very large-scale patches. Like the various-sized craters evident on the moon, disturbance patches sometimes overlap, adding to the heterogeneity of the site. Disturbance frequency varies and is strongly stochastic. And disturbance is normal. Large areas of tropical humid forest typically contain multiple disturbance patches of varying sizes and histories, and rates of change within each patch may vary. This results in a temporally shifting mosaic of local patch histories throughout a forest.

Forest clearance for logging (as with clear-cutting), agriculture, or pasturage represents a human-induced disturbance that varies from tiny clearings to hundreds of cleared hectares. Anthropogenic disturbance is different from natural disturbance, though both may result in secondary succession (see the following).

Local areas of natural disturbance are called *gaps*, and the pattern of plant growth that follows the creation of a gap is called *gap dynamics*. Gap dynamics is typical of temperate as well as of tropical forests (Shure et al. 2006). Disturbed areas undergo *ecological secondary succession*, a process that ecologists have studied in detail in temperate regions as well as in the tropics. Secondary succession results in *second-growth* or *secondary forest*. Because of natural forest dynamics as well as extensive clearance of tropical forest by human activities, secondary forest now covers more area within the tropics than old-growth forest (FAO 2001).

Though both result in secondary forest, natural disturbance is not at all the same as human-induced disturbance. Natural disturbance is caused by various degrees of tree-fall, which may be a single tree or large numbers of trees subjected to *windthrow* (when trees are felled by wind) from a storm such as a hurricane. Natural fire is another agent of disturbance. Fallen trees remain to decompose where they fall or where they have burned. Human disturbance is quite different in that land is normally cleared of forest, the slash typically burned on site and replaced by agriculture or pasture. Or in the case of logging, forests are penetrated by logging roads and trails, and collateral

Primary and Secondary Succession

Ecologists recognize two patterns of succession: *primary* and *secondary*. They are fundamentally different. Primary occurs in areas where there is no soil base, such as when lichens colonize rock faces, or ponds and bogs gradually fill with peat and slowly convert to terrestrial ecosystems. Because there is initially no soil base, primary succession requires hundreds of years. Primary succession has been well studied in the temperate zone on sites such as granite outcrops. Lichens colonize first, followed by mosses that take advantage of the organic base provided by the lichens. These *bare rock pioneers* are slowly but eventually replaced by various widely dispersed *forbs* (herbaceous plants) and, after that, depending on local conditions, shrubs and trees. Primary succession occurs in the tropics, such as in the case of succession on recent lava fields and regeneration following severe landslides (Plate 6-1) that result in soil removal.

PLATE 6-1
Landslides such as this along a forested area on an Ecuadorian mountainside sometimes result in soil removal, and this results in slow primary succession. If soil is not severely damaged, the result will be more rapid secondary succession.

Secondary succession occurs when soil is already present but some disturbance factor has cleared or partially cleared the dominant vegetation. Succession in the tropics, as discussed in this chapter, is generally secondary succession.

Ecological secondary succession was originally described for eastern North America. During the eighteenth and nineteenth centuries, eastern North American forests were felled so that land could be converted to agriculture and pasture. Approximately 85% of the original central New England forest was cut for homestead, agriculture, or pasture at any given time during the early to mid-1800s. Following the abandonment of agricultural land as large-scale farming moved to the Midwestern states, the open lands were recolonized in a natural way by various tree and shrub species, and second-growth forests gradually renewed their claim on the New England landscape (Foster and Aber 2006). Henry David Thoreau was one of the first authors to comment about this process of vegetation replacement dynamics, now called *ecological secondary succession.*

damage to trees ensues as select trees are removed (see Chapter 15). Secondary forest regrows when natural disturbance occurs but may or may not regrow in the case of certain kinds of human disturbance (see Chapters 13 to15).

In moderate to large-scale disturbances, a diverse community of rapidly growing, shade-intolerant, heliophilic colonizing plant species quickly appears in the disturbed area. Over time, these species are gradually replaced by slower-growing shade-tolerant species, as the once-open disturbed area regains a closed canopy. This process is characteristically viewed as one of interspecific competition among plants for light and soil resources. Ecologists sometimes use the term *recovery* when describing the pattern of how ecosystems respond to natural disturbances. This term is inappropriate as it suggests that the ecosystem was somehow *wounded* and that time heals the wound. Realize that the species colonizing and thriving following the disturbance are adapted to just such conditions, and as such, they represent part of the continuum of evolved species adaptations characteristic of that ecosystem. Their very existence as species that are clearly adapted to disturbance sites speaks to the reality of periodic natural disturbance events.

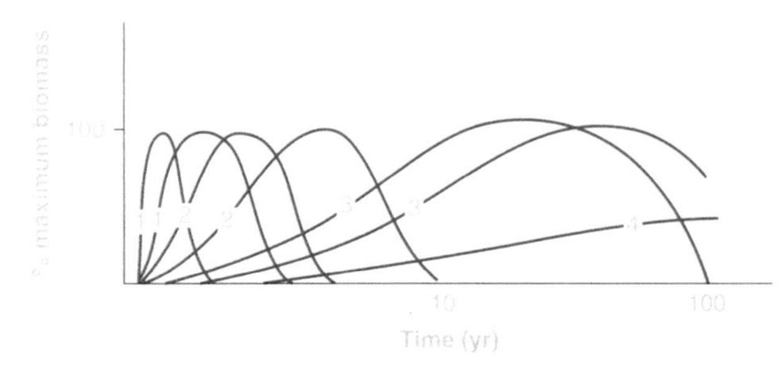

FIGURE 6-1
General description of Neotropical secondary successions. Numbers on curves represent Phases 1-4 (see text).

Throughout the tropics, there is a pattern of species replacement with time as a disturbed patch becomes increasingly higher in both biomass and species richness following the disturbance event. Patch size varies depending on the nature of the disturbance that created the patch. Secondary succession in tropical moist forests may be envisioned as a broad pattern with four phases (Figure 6-1) (Finegan 1996):

- *Phase one* is rapid colonization of cleared land by species such as herbs, shrubs, and climbers as well as seedlings from pioneer tree species (see the following text). Phase one may persist for about three years post-disturbance.
- *Phase two* is when short-lived but fast-growing shade-intolerant tree species form a canopy. Phase two may persist for 10 to 30 years from the time of disturbance.
- *Phase three* occurs when nonpioneer heliophilic tree species add to the biomass and species richness and, as well, are joined by an increasing number of shade-tolerant, slower-growing tree species. Interspecific competition is thought to drive this process. This phase, which gradually transitions into the next phase, typically lasts from 75 to 150 years.
- *Phase four* is when shade-tolerant species typical of undisturbed tracts of forest (old-growth forest) regain full canopy stature. This phase lasts indefinitely until another disturbance event reinitiates the process.

A CLOSER LOOK AT SECONDARY SUCCESSION IN THE TROPICS

Tropical secondary succession typically begins with rapidly growing heliophilic plant species, some of which germinated from the soil seed bank, others of which arrived by natural seed dispersal, and still others that were present as seedlings. Not surprisingly,

PLATE 6-2
Early secondary succession in the tropics is characterized by dense cover of sun-demanding species. From Ecuador.

most are herbaceous, many are vines, and some are shrubs. Most are nonwoody. But trees are soon apparent (Plate 6-2). Seedlings grow to sapling trees and growth rates are fast. Biomass increases. Sometimes vine growth is sufficiently dense as to envelop brushy growth of other species. Spindly trees and feathery palms push aggressively upward above the tangled mass. Biomass and species richness of plants and animals continue to increase. Sometimes clumps of huge-leaved plants are evident, typical heliophilic species. One group, named *heliconias* (Plate 6-3) for their sun-loving habit, competes for solar radiation against

PLATE 6-3
This disturbed gap in Ecuador experiences an abundance of sunlight, ideal for species such as this flowering heliconia.

PLATE 6-4
This field adjacent to a rural village in Belize, now abandoned, is undergoing secondary succession. The large-leaved plants are bananas.

scores of other fast-growing plants (Plate 6-4). The canopy is a dense, low, irregular, tangled assemblage of competing plants (Plate 6-5). Eventually, as biomass and species richness increase, more and more shade-tolerant tree species appear. Many of these tree species may have been present from the outset, but their initial growth is extremely slow. Both biomass and species richness stabilize, though species composition continues to change. Within a century and usually far sooner, a closed forest of tall-stature trees is apparent. This is second-growth forest, because its tree species composition may not yet be typical of the forest present before the initial disturbance event, or the species may be essentially the same, or very similar, but the relative abundances and size-class distributions are different.

Ecological succession, whether studied in the temperate latitudes or in the tropics, is affected by many factors, including chance dispersal and local *edaphic* (soil) conditions. Secondary succession in the tropics differs from that at higher latitudes because of a greater pool of plant species (see the sidebar "Models of Secondary Succession"). Plant species exhibit varying degrees of overlap in tolerances to light, temperature fluctuation, and soil characteristics, and thus species with effective dispersal or seeds that persist in soil banks invade first, followed by slower-growing species that are typically more shade tolerant. In the tropics, partly due to greater richness of species, partly because of variable levels of soil fertility, and partly because of differing levels of usage, successional patterns demonstrate complex and differing patterns from site to

PLATE 6-5
This abandoned house in Belize has become literally vine-covered as secondary succession has reclaimed the abandoned site. Cecropia trees are seen overtopping the house.

Models of Secondary Succession

When secondary succession was being investigated in the temperate zone, a model for the process was put forth called *initial floristic composition* (Egler 1954). This model hypothesizes that species that appear to come and go over time are all essentially present from the outset in the seed bed or as seedlings and saplings and will grow as conditions change such that different species are favored at different times. This model has been widely applied and has been supported in studies of secondary succession in the tropics (Finegan 1996; Hooper et al. 2004).

Another model has been proposed called the *nucleation model* (Yarranton and Morrison 1974). This model hypothesizes that the species present in secondary succession are derived mostly from immediately surrounding areas, those in closest proximity to the disturbed site. In a region with a large pool of species such as the tropics, and where efficacy and method of seed dispersal varies widely from one species to another, the nucleation model predicts high site-to-site variability in species composition as a characteristic of tropical secondary succession. There is evidence that this model applies to some degree to tropical succession patterns (Hooper et al. 2004). Both the initial floristic composition and nucleation models should be seen as complementary and not incompatible.

site (Ewel 1980 and 1983; Buschbacher et al. 1988; Pickett and Cademasso 2005).

Succession is often initiated by human activity such as land use for agriculture followed by abandonment. But secondary succession in the tropics is normal in the absence of human activity, as ecosystems periodically and predictably experience disturbance events. Numerous species are adapted to exploit disturbed areas; some species cannot grow unless they colonize a disturbed site. And disturbed sites are common. Heavy rainfalls, landslides, hurricanes, fires, occasional lightning strikes, and high winds typical of rainstorms destroy individual canopy trees, create forest gaps, and sometimes level whole forest tracts. Isolated branches, often densely laden with epiphytes, break off and crash down through the canopy. Natural disturbances within a forest open patches to sunlight, and a whole series of plant species are provided a fortuitous opportunity to grow much more rapidly.

EARLY SUCCESSION IN THE TROPICS

The *Oxford American Dictionary* definition of *jungle* is "land overgrown with tangled vegetation, especially in the tropics." To many people, the word *jungle* conjures up a romantic vision of dense rain forest with an abundance of thick vines (for Tarzan to swing on). But ecologically, jungles represent large patches where rain forest has been opened because of some disturbance event that has initiated ecological succession. In fact, the word *jungle* is rarely used in ecology (Plate 6-6). In early secondary succession, bare land is quickly colonized by herbaceous vegetation. Seeds dormant in the soil are now able to germinate in the high-light conditions of the disturbance patch.

Within Amazonia, a typical square meter of soil is estimated to contain from 500 to 1,000 seeds (Uhl 1988), the *seed bank*. Seeds of some species of pioneer plants may remain dormant in soil for nearly 40 years and still germinate after a disturbance occurs (Dalling and Brown 2009). Other seeds have far shorter residence time in the seed bank. In general, species with large seeds are more persistent in the seed bank than small-seeded species. The dynamics of early succession from seed germination onward are complex (Dalling and John 2008). In addition to the soil seed bank, seeds that have been transported either by wind or animals reach disturbed areas and germinate.

There is also a *seedling bank* of plants whose growth was suppressed in the dense shade that preceded the opening caused by the disturbance. Soon vines, shrubs, and quick-growing palms and trees are all competing for a place in the sun. The effect of this intensive ongoing competition for light and soil nutrients is the *tangled vegetation* of the jungle definition earlier.

Species richness and species composition vary widely from site to site (Bazzaz and Pickett 1980). Successional trajectories are more diverse in the tropics than at higher latitudes, a function of a greater overall pool of species. Though successions on rich soils usually result in the eventual redevelopment of forest, on poor soil, repeated elimination of humid forest and depletion of soil fertility can sometimes result in conversion of the ecosystem to savanna, rather than forest (see the following, and see Chapters 10 and 15).

Plant biomass increases rapidly. In onc Panamanian study, biomass increased from 15.3 to 57.6 dry tons per hectare from year 2 to year 6 (Bazzaz and Pickett 1980). This rapid growth reduces soil erosion

PLATE 6-6
This example of secondary succession reflects what the term *jungle* means in the ecological sense. Note how dense the vegetation is. The plant on the right is a tree fern.

as vegetation blankets and secures the soil. Studies in Veracruz, Mexico, have shown that young (10 months and 7 years old) secondary successional areas take up nutrients as efficiently as mature rain forest (Williams-Linera 1983).

In Amazonia, succession on abandoned pastures does not result in significant depletion of soil nutrient stocks. Successional sites have higher nutrient concentrations in their biomass than is the case in old-growth forests, and there are more extractable soil nutrients on successional sites; thus successional sites have a lower proportion of total site nutrients stored in biomass than does mature forest (Buschbacher et al. 1988). Because of the density of competing plants, the leaf area index (see the sidebar "Leaf Area Index") may reach that of a closed canopy forest within 6 to 10 years, although the vegetation is still relatively low and the species composition at that time is not what it will be as the site returns to closed forest. By the time the site is about 15 years from the onset of succession, the abiotic ground conditions can be similar to closed canopy forest, even though the species composition is quite different. In only 11 months from burning, study plots underwent a succession such that vegetation attained a height of 5 meters (about 16 feet) and consisted of dense mixture of vines, shrubs, large herbs, and small trees (Ewel et al. 1982). Large epiphytes are almost entirely absent from early successional areas.

Both major and subtle physiological changes occur in plants that live in early successional, high-light environments (Fetcher et al. 1994). Photosynthesis rates in early successional species are higher in full sun than in partial or full shade; these plants are obviously adapted to grow quickly. Some early successional plants that can grow in both shade and sun develop significantly thicker leaves in full sun, an adaptation that aids in protection from dessication. Some studies indicate that stomatal densities increase when a species is grown in full sun versus partial or full shade. More stomata permit increased rates of gas exchange, necessary when photosynthesis rate is elevated. In addition, successional species allocate considerable energy to root production, facilitating rapid uptake of soil nutrients (Uhl et al. 1990).

During early succession, many plant species termed *colonizers* tend to be small-stature and fast-growing and produce many-seeded fruits. In later succession, most plants are larger, more slow growing, and with fewer seeds per fruit. These shade-tolerant plants are adapted to persist in the closed canopy (Opler et al. 1980). While this overview is generally descriptive, these two broad categories are insufficient to describe the true complexity of succession. Because of physiological plasticity in various light regimens, distinctions between successional categories blur (Fetcher et al. 1994).

Shade-tolerant species should be thought of as exactly that. They tolerate shade well, though their rate of

Leaf Area Index

Ecologists express leaf density as *leaf area index* (LAI), the leaf area above a square meter of forest floor. In a mature temperate forest such as Hubbard Brook in New Hampshire, LAI is nearly 6, meaning that the equivalent of 6 square meters of leaves cover 1 square meter of forest floor. For tropical rain forest at Barro Colorado Island (BCI) in Panama, the figure is about 8 (Leigh 1975). Typically, LAI in the humid tropics ranges from about 5.1 on a dry shrubland on poor soil (called *Amazon Caatinga*) at San Carlos, Venezuela, to a range of 10.6 to 22.4 on a lush forest on rich soil at Darien, Panama (Jordan 1985). In forests with high LAI, it is probable that the intensity of shading is so great that many, if not most, understory leaves do not approach optimum net primary productivity (see Chapter 9) because they are severely light-limited (Plate 6-7).

PLATE 6-7
Leaf area index measures the density of leaves from ground to canopy. The leaf area index of this well-shaded Amazonian forest would be high.

growth in shade is usually very slow. Virtually all shade-tolerant species grow fastest in well-lit conditions. It is their physiological ability to persist in the dark understory until conditions favor their growth that is their key adaptation (Wright et al. 2003).

Dennis Knight (1975), in a classic study on Barro Colorado Island in Panama, found that plant species diversity of successional areas increased rapidly during the first 15 years of succession. Diversity continued to increase, though less rapidly, until 65 years. Following that, diversity still increased, but even more slowly. Knight concluded that, after 130 years of succession, the forest was still changing. He was correct, though he underestimated the actual age of the site. Hubbell and Foster (1986a, 1986b, and 1986c) note that forest at BCI is actually between 500 to 600 years old and agree with Knight that it is not yet in equilibrium. (The term *equilibrium* is sometimes used by ecologists to mean that there is no net change occurring in an ecological community.) They conclude that though initial succession is rapid, stochastic factors as well as climatic changes, periodic drought, and biological uncertainty from interactions among competing tree species act to prevent establishment of a stable equilibrium (Hubbell and Foster 1986a and 1990; Condit et. al 1992). This means that BCI remains in a dynamic state, continuing to change. This is probably the norm for tropical rain forests.

The Rio Manu floodplain forest in Amazonian Peru shows a very long-term successional pattern. Pioneer tree species such as *Cecropia* (see the following) dominate early succession and are followed by large, statuesque *Ficus* and *Cedrela*. Slow-growing emergent trees such as *Brosimum* and *Ceiba*, most of which have been present essentially since the succession began, eventually replace these. The overall pattern of succession at Manu may require as long as 600 years (Foster 1990).

A STUDY IN EARLY SECONDARY SUCCESSION

A study performed at the Montes Azules Biosphere Reserve in Chiapas, Mexico (Van Breugel et al. 2007), illustrates some important characteristics of early secondary succession in the tropics. The area is described as a mosaic of small-scale agriculture, pastures, and young secondary forest with remnant stands of old-growth forest. Researchers set up eight study plots on abandoned cornfields (locally called *milpas*) that had been established by clear-cutting old-growth forest. Each plot was 10 × 50 meters. All woody perennial plants (trees, shrubs, saplings) with a diameter equal to or greater than 1 centimeter diameter breast height (DBH) were recorded and identified.

Researchers made an initial survey and a follow-up survey 18 months later in each of the plots, looking at species density, fraction of shade-tolerant species, and the *evenness* of species (i.e., how even population sizes are among species). Each of these variables increased over the period of the study (between census 1 and census 2, depending on plot).

Tree density varied across plots from 426 to 1,786 per 500 square meters, and species richness among plots ranged from 32 to 62 species per plot (mean of 47.5). This range of variability is typical of early succession. Shade-tolerant species were initially represented by 7% of total individuals but made up 42% of all species. This shows that shade-tolerant species are present from the outset of early secondary succession (a prediction of the initial floristic composition model), even though they are not numerous. Shade-tolerant species nearly doubled to represent an average of 13% of all individual plants over the course of the brief study. Evenness values were considered low throughout the study (indicating that a few species were more numerous than most others) but increased over the course of the study. Mortality rates among the plants, measured by counting dead stems, ranged from 17% to 63%. Mortality among pioneer species (averaged across plots) was more than three times higher than among shade-tolerant species. This indicates a rapid change in the vegetation assemblage. In all plots, new species (from the time of the first to the second sample) surpassed species lost, and so species richness increased (Figure 6-2).

In comparing the early successional plots with old-growth forests, differences were apparent. The five

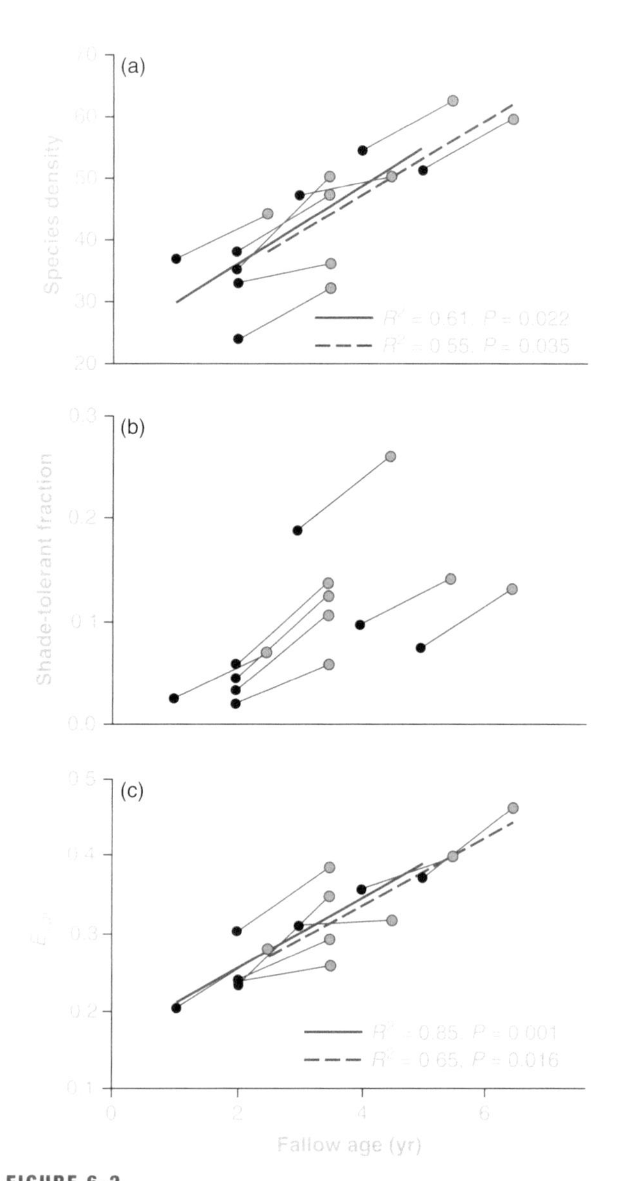

FIGURE 6-2
Early secondary succession trends in community attributes of trees greater than 1.5 meters in height at Marquéz de Comillas, Southern Mexico. (a) Species density (number of species per 500 square meters. (b) Fraction of total number of trees belonging to shade-tolerant species. (c) Smith and Wilson's measure of evenness (E_{var}). Black circles: census 1; brown circles: census 2. The bold continuous and dashed lines give the linear regressions of the variables of census 1 and 2, respectively, on fallow age.

most numerically dominant species in the secondary forest plots represented 66% of all trees. But they represented only 1.3% of the trees in old-growth forest. Five of the six most abundant species in old-growth forest were present in one to six of the eight study plots. This demonstrates that shade-tolerant species typical of old-growth forest are present early in succession even if many years are required for them to attain canopy status. Pioneer and shade-tolerant species were present from the outset, even though only pioneer species dominated. In the first decade of succession, shade-tolerant species are confined to the understory, where they grow slowly but persist through the life span of fast-growing pioneer species. The authors conclude, "Successional species turnover, resulting in replacement of pioneer species by shade-tolerant species, already started in the beginning of the succession" (p. 617).

A GENETIC STUDY OF A COLONIZING SPECIES

Iriartea deltoidea is an abundant species of shade-tolerant palm that occupies the understory of secondary forest in Central America. It is pollinated principally by stingless bees (*Trigonia* spp. and *Melapona* spp.) and more recently by introduced honeybees (*Apis* spp.). Animals ranging from toucans to peccaries, tapirs, various rodents, and white-faced monkeys all act as seed dispersers. *I. deltoidea* requires approximately 20 years to attain maturity. The genetics of colonizing plant populations has been little studied, but a study of *I. deltoidea* in Costa Rica demonstrated that most of the individual trees of *I. deltoidea* occurring in a 20-hectare second-growth forest were from the seeds of just a few parent trees in adjacent old-growth forest (Sezen et al. 2005 and 2007).

Researchers studied a founder population of *I. deltoidea* in a 24-year-old second-growth forest at La Selva Biological Field Station. Their study area measured 300 meters by 1,000 meters and included 20 hectares of second-growth and 10 hectares of adjacent old-growth forest. *I. deltoidea* is dependent on seed dispersal to colonize new sites (their seeds are not part of the soil seed bank), and all of the potential sources of seeds were outside the second-growth area in the old-growth forest.

Researchers examined 141 amplified fragment length polymorphism (AFLP) *loci* (a specific section of a gene) to make genetic comparisons among the *I. deltoidea* from the second-growth forest and those of the old-growth forest to ascertain exactly which trees in the old-growth forest parented the *I. deltoidea* in the second-growth forest. The result was perhaps surprising. Of the 66 *I. deltoidea* in old-growth forest, only two contributed 56% of the genes to the second-growth population and 23 contributed the remaining 44%. A total of 41 trees in the old-growth forest failed to reproduce.

These results clearly demonstrate that the genetic diversity of the *I. deltoidea* in the second-growth forest is far less than that apparent in old-growth forest. The mean genetic dissimilarity among individuals was 0.58 in second-growth compared with 0.73 in old-growth (using a measure called *Tanimoto's D*).

Researchers suggested that such a pattern may be common to many species that occupy second-growth forest, where chance colonization by seed dispersal is the means by which trees become recruited (a prediction that accords with the nucleation model of ecological succession). They further suggested that old-growth forest serves as a source of genetic diversity as well as a haven for mature seed-dispersing trees.

Further study of *I. deltoidea* (Sezen et al. 2007) compared two second-growth forests (including the one discussed earlier) with old-growth forest. This study involved a genetic analysis of 311 reproductively mature trees, 99 large saplings, 207 small saplings, and 601 seedlings. The results demonstrated high dispersal distances (Figure 6-3) for both pollen (greater than 3.8 kilometers) and seeds (greater than 2.3 kilometers). In a second-growth area distant (400 to 800 meters) from the border with old-growth forest, 40% of the seedlings were from seeds that originated outside of the nearby old-growth forest, and only 10% were from trees contained in the old-growth forest. These results demonstrate strong dispersal for *I. deltoidea* and a large genetic neighborhood (Figure 6-4) (meaning that the potential for high levels of gene flow is present).

Such strong dispersal ensures high levels of gene flow among local populations in general, even allowing for the reduced gene flow of early colonizers in secondary succession (as discussed in the 2005 study, earlier). Researchers pointed out that conditions are ideal for optimal pollen and seed dispersal because their second-growth forest sites are contiguous with areas of old-growth forest and the old-growth forest supports a full suite of dispersers. In this case, the chestnut-mandibled toucan (*Ramphastos swainsonii*) was considered to be a principal seed dispersal agent

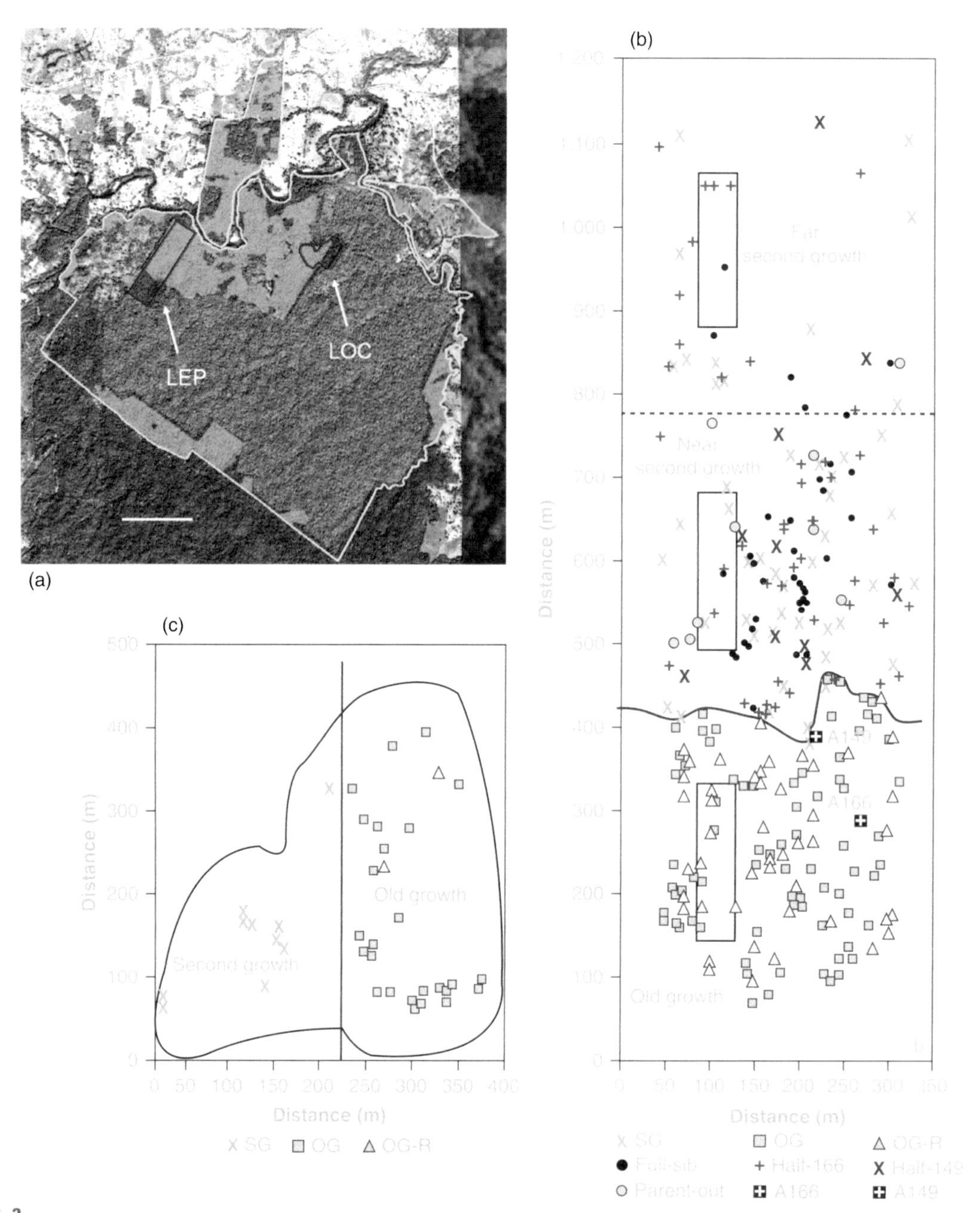

FIGURE 6-3

(a) Quickbird satellite imagery taken in 2004 showing the coverage of second-growth forest (green) and old-growth forest (dark gray) within La Selva Biological Field Station, Costa Rica. The two study sites, Lindero El Peje (LEP) and Lindero Occidental (LOC), are indicated by arrows. The horizontal bar represents 1 kilometer. (b) Map of genotyped trees in the 35-hectare LEP site. The curved line shows the boundary between the old-growth (OG) and the second-growth (SG) forest. The dashed line separates the far and near second-growth forest zones. The two reproductively dominant trees are A166 and A149. Reproductively active trees are labeled OG-R. Map positions of full-sib and half-sib (half-166 or half-149) offspring of the reproductively dominant trees are indicated. (c) Map of genotyped trees in the 8-hectare LOC site.

that might account for why genetic diversity was particularly high among the palms most distant from old-growth forest. Toucans (Plate 6-8) are large birds that typically fly long distances. They feed on many kinds of fruits. Nesting birds fly substantial distances from their tree-cavity nests to find food. Non-nesting birds move widely within old-growth and second-growth forests as well. These birds alone may have facilitated seed dispersal of *I. deltoidea* sufficiently to ensure gene flow.

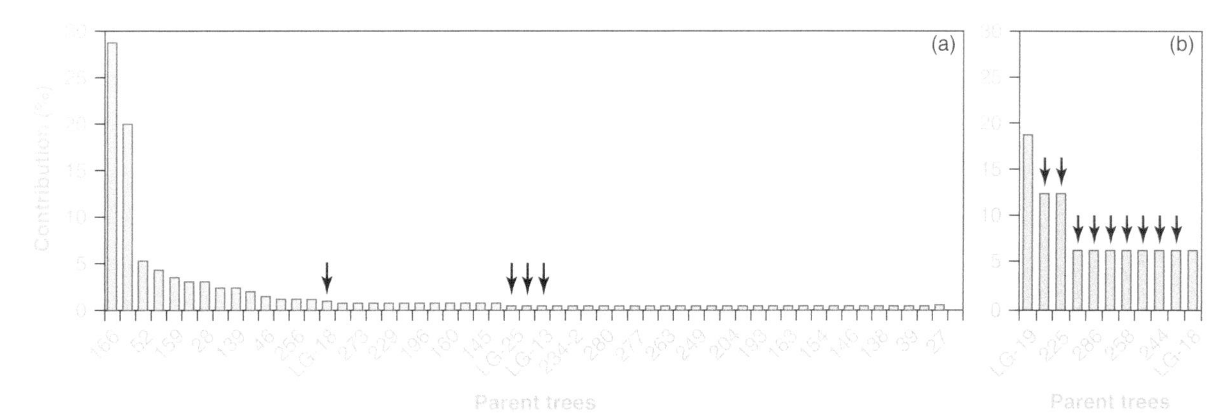

FIGURE 6-4
Reproductive contribution of parental trees to founding adult tree generation of second-growth forest in (a) LEP and (b) LOC sites. Black arrows indicate reciprocal long-distance contributions from (a) four parents found in the LOC site and (b) nine parents in the LEP site. LG represents parents in the old-growth portion of the LOC site.

ENSO EXERTS STRONG EFFECTS ON SECOND-GROWTH RAIN FORESTS

The El Niño/Southern Oscillation (ENSO) periodically affects rainfall patterns throughout the world, including the global tropics (see Chapter 1). ENSO-induced droughts have been shown to reduce canopy tree recruitment and elevate tree mortality in mature rain forests in the Neotropics as well as the Paleotropics (Curran et al. 1999; Wright et al. 1999; Gilbert et al. 2001; Aiba and Kitayama 2002). For example, in a study conducted on BCI, it was learned that drought resulting from El Niño 1982–1983 was one of several factors that was associated with nonrandom higher mortality of *Ocotea whitei* seedlings (Gilbert et al. 2001).

A study conducted at La Selva Biological Station in Costa Rica demonstrated that ENSO exerts major influence on recruitment and survivorship of trees in second-growth forest (Chazdon et al. 2005). The year 1997–1998 marked a global ENSO event, creating

PLATE 6-8
CHESTNUT-MANDIBLED TOUCAN (*RAMPHAASTOS SWAINSONII*)

a severe dry season in Costa Rica. Researchers were able to assess the impact of such a dry season on tree recruitment and mortality.

Permanent sample plots of 1 hectare were established at four sites in January to March 1997. The sites were of differing ages: 12, 15, 20, and 25 years post-abandonment. Tree censuses were conducted from March through September 1997, and all trees equal to or greater than 5 centimeters DBH were tagged, marked, and identified to species. From that point onward, annual censuses were made in July and August, and at each subsequent census, recruits into the 5-centimeter-diameter class were marked. Three size classes were established: 5 to 9.9 centimeters, 10 to 24.9 centimeters, and equal to or greater than 25 centimeters. In addition, dead trees were counted (based on stems that had no living tissue visible). Mortality and recruitment rates were calculated as percentages based on the previous year's study.

The four sites varied considerably in stem density but showed similar trends throughout the seven years of the study (Figures 6-5 and 6-6). During the ENSO year 1997–1998, stem density markedly decreased. Another decrease in stem density occurred from 2002 to 2003, a year of severe drought. Annual mortality rates varied among the study sites and among years but were much higher in drought years on all plots. In addition, mortality was highest in the smallest size class of trees (5 to 9.9 centimeters) and lowest in the size class equal to or greater than 25 centimeters in diameter. The highest size class had a mortality rate approximately equal to the 10 to 24.9 centimeters size class (Table 6-1). This pattern suggests strong competition among plants as second-growth forests age. The competition may have been largely influenced by the ability to tolerate drought among the various trees in the plot. Tree recruitment into the 5-centimeter DBH class was much lower during the ENSO year. The highest mortality in two of the plots during the ENSO year was for the tree *Cecropia insignis*, a common colonizer during early secondary succession (see page 204).

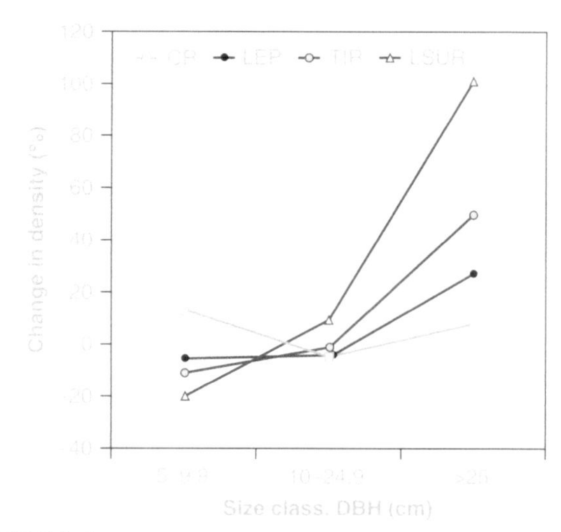

FIGURE 6-6
Percentage change in density of three diameter size classes from 1997 to 2003 in four second-growth permanent sample plots.

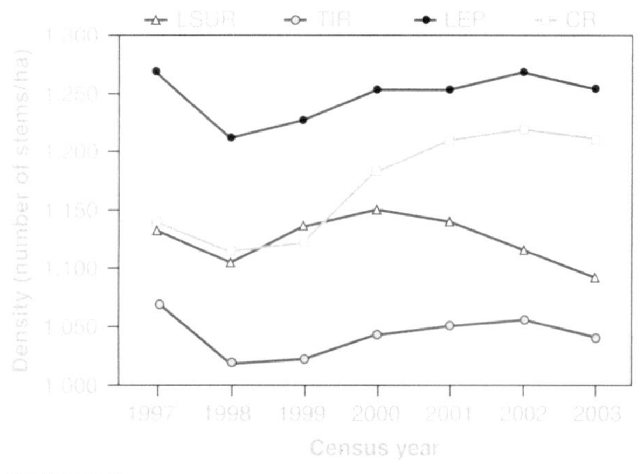

FIGURE 6-5
Absolute changes in total tree density (trees greater than 5 centimeters DBH) in four second-growth permanent sample plots from 1997 to 2003.

In the six years of the study, major changes were evident in tree size distributions and population sizes of individual species in all four plots. This is likely typical of second-growth forest development. The younger stands showed more species changes than the older stands. Tree mortality rates in all trees equal to or greater than 10 centimeters DBH were about equal to those documented for old-growth forests (see the following). The most dramatic result of the study was to demonstrate how rainfall amount during the dry season (from January to April) affects recruitment and mortality of trees (Figure 6-7). This study and others suggest that individual species of trees vary considerably in their susceptibility to mortality caused by drought. The authors cite studies comparing tree mortality in pre-ENSO years with the 1997–1998 ENSO event. In central Amazonia, pre-ENSO tree mortality was 1.1% per year compared with 1.9% in 1997–1998. In Sarawak, Malaysia, pre-ENSO tree mortality was 0.89% per year compared with 6.37% in 1997–1998. More frequent ENSO events coupled

TABLE 6-1 Annual mortality rates for four size classes (DBH) for six years and mean annual mortality rates during the five non-ENSO years (1999–2003).

CENSUS YEAR	ANNUAL MORTALITY RATE, BY SIZE CLASS (%)			
	5–9.9 CENTIMETERS	10–24.9 CENTIMETERS	≥25 CENTIMETERS	≥10 CENTIMETERS
1998	5.6 ± 3.5	3.5 ± 0.6	1.4 ± 1.1	3.1 ± 0.5
1999	2.9 ± 0.7	1.3 ± 0.6	0.5 ± 0.6	3.1 ± 0.5
2000	3.3 ± 1.9	1.5 ± 0.5	0.5 ± 0.6	1.1 ± 0.2
2001	3.9 ± 2.4	2.1 ± 0.6	0.9 ± 0.7	1.2 ± 0.4
2002	5.1 ± 2.9	2.9 ± 1.2	1.1 ± 0.8	1.9 ± 0.8
2003	3.7 ± 1.7	1.9 ± 0.6	0.9 ± 1.1	2.4 ± 0.6
Mean non-ENSO (1999–2003)	3.7 ± 1.8	1.9 ± 0.8	0.8 ± 0.7	1.6 ± 0.7
ENSO: non-ENSO ratio	1.51	1.81	1.78	1.90

Note: Data are reported as means ± SD of four second-growth stands.

with global climate change (see Chapter 15) could alter species composition considerably and accelerate species turnover during secondary succession.

Other studies have shown that seedling mortality is strongly affected by drought tolerance (Edwards and Krockenberger 2006; Poorter and Markesteijn 2008).

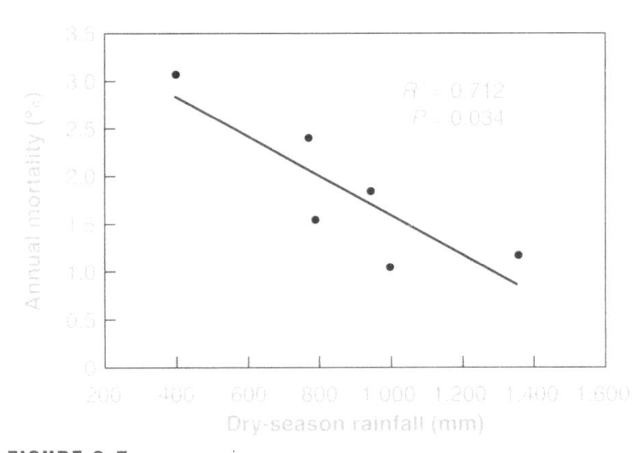

FIGURE 6-7
Regression of mean annual mortality rate of trees greater than 10 centimeters DBH on dry-season rainfall (total rainfall between January and April) across six years. Data are means for four second-growth plots.

In Bolivia, Poorter and Markesteijn performed an experiment on 38 tropical tree species from dry and moist forest. They conducted a *dry-down* experiment in which, under controlled conditions, seedlings were subjected to drought conditions. Dry forest species in general had compound leaves, high stem dry matter content, and low leaf area ratio, all of which are adaptive to reduction of water loss through transpiration. Three categories of plants were identified: *drought avoiders*, *drought resisters*, and *light-demanding forest species*. Drought survival explained 28% of the variation in species position along a rainfall gradient among evergreen tree species. The study showed that drought has the capacity to change plant community structure.

In tropical Queensland, Australia, Edwards and Krockenberger studied the effects of a severe El Niño (December 2001 to December 2002) on seedling mortality. Their study looked at two areas, one that had experienced a low-intensity fire (more common during drought—see Chapter 15) and one that hadn't. In the area subjected to fire, only 17% of individual plants and 52% of the species survived the 12-month period of the study. Survivorship was higher in the area that had not burned but was nonetheless negatively affected by the drought. Species rank abundances changed markedly, meaning that some species experienced higher mortality, while a few had lower mortality, so drought was shown to shift relative abundances of species.

PIONEER SUCCESSIONAL SPECIES: EXAMPLES FROM THE NEOTROPICS

Heliconias

Heliconias *(Heliconia* spp., family Heliconiaceae) are recognized by their huge, elongate paddle-shaped leaves (similar to banana leaves) and their distinctive colorful red, orange, or yellow bracts surrounding the inconspicuous flowers. (In some species, bracts are reminiscent of lobster claws, hence the common name, "lobster-clawed" heliconia [Plate 6-9].) Most heliconias grow best where light is abundant, in early successional patches, along roadsides, forest edges, stream banks, and forest gaps. They grow quickly, clumps spreading by underground rhizomes. Named for Mount Helicon in ancient Greek mythology (the home of the Muses), these plants are all Neotropical in origin, with approximately 150 species distributed throughout Central and South America (Lotschert and Beese 1983).

PLATE 6-9
Flowering heliconia in the understory of an Ecuadorian rain forest. Note the colorful bracts. The flower is barely visible within the second bract from the bottom.

Colorful, conspicuous bracts surrounding the smaller flowers attract hummingbird pollinators, especially a group called the *hermits*, most of which have long, down-curving bills permitting them to dip deeply into the 20 yellow-greenish flowers within the bracts (Stiles 1975). When several species of *Heliconia* occur together, they appear to flower at different times, a possible evolutionary response to competition among *Heliconias* for pollinators such as the hermits (Stiles 1975 and 1977). However, other research showed that temporal separation in flowering time was no greater than would be expected if flowering was random (Poole and Rathcke 1979). Sweet, somewhat sticky nectar oozes from the tiny flowers into the cuplike bracts where it is sometimes diluted with rainwater.

Heliconias produce green fruits that ripen and become blue-black in approximately three months. Each fruit contains three large, hard seeds. Birds attracted to heliconia fruits are important in the plant's seed dispersal. At La Selva Biological Research Station in Costa Rica, Stiles (1983) reports that 28 species of birds have been observed taking the fruits of one heliconia species. The birds digest the pulp but regurgitate the seed whole. Heliconia seeds have a six- to seven-month dormancy period prior to germination, which ensures that the seeds will germinate at the onset of rainy season.

Piper

Piper (Piper spp., family Piperaceae) is a pantropical hyperdiverse genus of plants with well over 1,000 and possibly as many as 2,000 species (Plate 6-10). The name *Piper* refers to "pepper," as spices such as black pepper, a spice derived from the South Asian species *Piper nigrum*, are derived from the cultivation of various species of *Piper*.

Piper is common in successional areas, though most *Piper* species are shade-tolerant and are part of the forest understory. Many of these reproduce both sexually and vegetatively (Dalling, personal communication). It is estimated that about 700 species occur in the American tropics, 300 species in the Asian tropics, and 15 species in Africa. Up to 40 species are distributed on various islands in the tropical Pacific. Most grow as shrubs, but some species grow as herbs, some grow as vines, and some are small trees.

PLATE 6-10
This cluster of *Piper* in Belize is ready for a nocturnal visit by piperphile bats.

In the Neotropics, the Spanish name *Candela* or *Candellillos* means "candle" and refers to the plant's distinctive flowers, which are densely packed on an erect stalk. When immature, the flower stalk droops, but it becomes stiffened and stands fully upright when the flowers are ripe for pollination. *Piper* flowers are apparently pollinated by many species of bees, beetles, and fruit flies; pollination seems unspecific. Seed dispersal, however, is dependent on bats. Small fruits form on the spike and are eaten, and the seeds are subsequently dispersed by one group of bats in the genus *Carollia,* called *piperphiles*. Several species of *Piper* may occur on a given site (at BCI, there are more than a dozen present), but they do not all flower at the same time, and like *Heliconias*, competition among them for pollinators is reduced as well as the probability of accidental hybridization (Fleming 1983, 1985a, and 1985b).

Some *Piper* species are well-defended by aggressive ants (see Chapter 7), and others lace their leaves with toxic chemicals—in particular, various *amides* (toxic compounds of nitrogen).

Piper is considered an unusually diverse (hyperdiverse) genus, and many questions about its natural history remain to be studied (Dyer and Palmer 2004).

Legumes

The legume family (family Fabaceae) is the third largest plant family in the world, with some 18,000 species. (Many older texts still use the former family name Leguminosae.) Only the orchid family is more species-rich, with an estimated 20,000 species. Growth forms include trees, herbs, shrubs, and vines. Legumes make a characteristic dry seedpod containing seeds that are often toxic to would-be seed predators. Legumes have compound leaves, and many have spines. Some, like the sensitive plant *Mimosa pudica,* have leaves that quickly lose their turgor pressure and close when touched. (*Turgor* occurs when leaves have much water, making them stiffen—with less water, they become limp.) Legumes grow in all tropical forest types, but many are found along roadsides and in wet pastures and fields.

In the Neotropics, legumes include beans, peas, and many species of trees. The colorful flamboyant tree *Delonix regia,* a native of Madagascar, has been widely introduced in the Neotropics and in tropics elsewhere in the world. Amazonian rain forests typically contain more legume species than any other plant family (Klinge et al. 1975).

Mimosa pigra (Plate 6-11), an abundant Neotropical species common in successional areas, has round pink flowers and is unusual because it flowers early in the rainy season. The flowers, which are pollinated by bees, become 8- to 15-centimeter-long flattened pods that are covered by stiff hairs. Stems and leaf stalks (*petioles*) are spiny and are not browsed by horses or cattle. Experiments with captive native mammals such as peccaries, deer, and tapir show that these creatures refuse to eat *Mimosa* stems on the basis of odor alone (Janzen 1983). Given its apparent unpalatability, it is easy to

PLATE 6-11
MIMOSA PIGRA

see why *Mimosa pigra* thrives in open areas. Seeds are spread by road construction equipment, accounting for the abundance of this species along roadsides.

Cecropias

Cecropias *(Cecropia* spp., family Urticaceae, formerly family Moraceae) are one of the most conspicuous and easily identified genera of trees in the Neotropics (Plate 6-12). There are 61 species distributed throughout the Neotropics (Berg and Rosselli 2005). Cecropias typically occur in areas of large light gaps or secondary growth, though some persist in second-growth forests. Pioneer colonizing species, cecropias are well adapted to grow quickly when light becomes abundant.

Studies on BCI have revealed that seeds remain viable in the soil for about a year and germinate only when

PLATE 6-12
Clusters of cecropia trees are a common sight in sunny areas throughout the Neotropics.

PLATE 6-13
Cecropias have large palmate leaves and flowers and fruits that dangle like fingers below the leaves, attracting numerous bird species.

a gap is created. Seeds are highly susceptible to pathogenic fungi. Experiments employing fungicide significantly increased seed survival, not only for *Cecropia* but also for *Miconia* (Dalling et al. 1998). Cecropia seeds are sometimes abundant in the seed bank. An average of 73 per square meter was present on one study site in Surinam (Holthuijzen and Boerboom 1982). Because there are so many viable seeds present, cecropias sometimes completely cover a newly abandoned field or open area. They line roadsides and are abundant along forest edges and stream banks.

Cecropias are easy to recognize. They are thin-boled, spindly trees with bamboo-like rings surrounding a gray trunk. Their leaves are large, deeply lobed, and palmate, somewhat resembling a parasol (Plate 6-13). Leaves are whitish underneath and frequently insect-damaged. Dried, shriveled cecropia leaves that have dropped from the trees are a common roadside feature in the tropics. Some cecropias have stilted roots, but the trees do not form buttresses.

Cecropias are effective colonizers. In addition to having many seeds resident in the soil seed bank, once germinated, cecropias grow quickly, up to 2.5 meters in a year. Nickolas Brokaw (1987, personal communication) recorded 4.9 meters of height growth in one year for a single cecropia. They are generally short-lived, with old ones surviving about 30 years, although Hubbell reports that once established in the canopy, *Cecropia insignis* can persist nearly as long as most shade-tolerant species, at least 50 years (Condit et al. 1993). One limit to colonization by cecropia is recruitment from distant seed sources. Without nearby adult trees, seed dispersal is very limited (Dalling et al. 1998).

Cecropias are moderate in size, rarely exceeding 25 meters in height, though Hubbell has measured emergent cecropias 40 meters tall. They are intolerant of shade, their success hinging on their ability to quickly grow above the myriads of vines and herbs competing with them for space. To this end, cecropias, like many pioneer tree species, have a very simple branching and leaves that hang loosely downward. Vines attempting to grow over a developing cecropia can be easily blown off by wind and possibly deterred by *Azteca* ants (see page 206), though I have seen many small cecropias that were vine-covered. Cecropias have hollow stems, a possible adaptation for rapid growth in response to competition for light, as it permits the tree to devote energy to growing tall rather than to the production of wood.

Cecropias have separate male and female trees and are well adapted for mass reproductive efforts. A single female tree can produce over 900,000 seeds every time it fruits, and it can fruit often! Flowers hang in finger-like catkins, with each flower base holding four long, whitish catkins. Research in Mexico (Estrada et al. 1984) showed that 48 animal species, including leaf-cutter ants, iguanas, birds, and mammals, made direct use of *Cecropia obtusifolia*. A total of 33 bird species from 10 families, including some North American migrants, feed on flowers or fruit. Mammals from bats to monkeys eat the fruit, and sloths gorge (in slow motion) on the leaves. One North American migrant bird, the worm-eating warbler (*Helmitheros vermivorus*), specializes in searching for arthropod prey in dried leaf clusters, often those of cecropias (Greenberg 1987a and 1987b).

The natural history of cecropia, like that of all species, can be thought of as a series of trade-offs. *Cecropia* makes a heavy investment in seeds (high fecundity) that are generally well-dispersed. But the seeds do not persist long in the soil and germinate only in a high-light gap. Should a gap occur, the seeds germinate, and the tree grows at a rapid rate. It may attain canopy stature, though its persistence in the canopy is less than that of many other species. Cecropias have obviously profited from human activities, as cutting the forest provides exactly the conditions it requires.

Ants (*Azteca* spp.) reside inside stems of some cecropias (Plate 6-14). These ants feed on glycogen-rich structures called *Müllerian bodies* (a form of extrafloral nectary) produced at the leaf axils. The ants sometimes protect the cecropia. Many cecropias are free of vines or epiphytes once they have reached fair size. Janzen (1969) observed that *Azteca* ants clip vines attempting to entwine cecropias. The plant "rewards" the ants by providing both room and board, a probable case of evolutionary mutualism (see Chapter 7). However, some cecropias hosting abundant ants are, indeed, vine-covered, and the ants seem to "patrol" only the stem and leaf nodes, not the main leaf surfaces (Andrade and Carauta 1982).

A similar relationship exists between the Southeast Asian tree genus *Macaranga* and the ant genus *Crematogaster* (Feldhaar et al. 2003). The symbiosis between *Macaranga* and *Crematogaster* is highly species-rich. There are about 26 species of *myrmecophytic* (ant-hosting) *Macaranga* and at least nine ant species, all in the genus *Crematogaster* (which contains many more species). Relationships between the plants and their ants begin early and are maintained from saplings to adult plants.

The Kapok, Silk-Cotton, or Ceiba Tree

One of the commonest, most widespread, and most majestic Neotropical trees is the ceiba, or kapok,

PLATE 6-14
Azteca ants on the bole of a cecropia. The brown bodies beneath the leaf axils supply the food for the ants.

tree *(Ceiba pentandra*, family Malvaceae, formerly in Bombacaceae). Ceibas are sometimes left standing when surrounding forest is felled. Throughout much of Central America, the look of today's tropics is a cattle pasture watched over by a lone kapok.

The ceiba is striking in appearance (Plate 6-15). From its buttressed roots rises a smooth gray trunk often ascending 50 meters (about 165 feet) before spreading into a wide, flattened crown. Trees may grow in excess of 60 meters (about 200 feet), though such giants are rare. Leaves are compound, with five to eight leaflets dangling like fingers from a long stalk. The major branches radiate horizontally from the trunk and are usually covered with epiphytes. Many lianas typically adorn the tree.

Kapoks originated in the American tropics but dispersed naturally to West Africa (Baker 1983). They are grown commercially (for the fiber accompanying

PLATE 6-15
Ceiba pentandra, the kapok or silk-cotton tree, is one of the most striking tropical trees.

the seeds) in Southeast Asia as well, so today they are distributed throughout the world's tropics.

Ceibas require high light intensity to grow and are most common along forest edges, river banks, and disturbed areas. Like most successional trees, they exhibit rapid growth, up to 3 meters (10 feet) annually. During the dry season, they are deciduous, dropping their leaves. When leafless, masses of epiphytes and vines stand out dramatically, silhouetted against the sky.

Leaf drop precedes flowering, and thus the flowers are well exposed to bats, their major pollinators. The five-petaled flowers are white or pink, opening during early evening. Their high visibility and sour odor probably help attract the flying mammals. Cross-pollination is facilitated, since only a few flowers open each night, thus requiring two to three weeks for the entire tree to complete its flowering. Flowers close in the morning, but many insects, hummingbirds, and mammals seeking nectar visit the remnant flowers (Toledo 1977). A single ceiba may flower only every 5 to 10 years, but each tree is capable of producing 500 to 4,000 fruits, each with approximately 200 or more seeds. A single tree can therefore produce about 800,000 seeds during one year of flowering (Baker 1983). Seeds are contained in oval fruits, which open on the tree. Each seed is surrounded by silky, cotton-like fibers called *kapok* (hence the names *kapok tree* and also *silk-cotton tree*). These fibers aid in wind-dispersing the seeds. Kapok fibers are commercially valuable as stuffing for mattresses, upholstery, and life preservers. Since the tree lacks leaves when it flowers, wind can more efficiently blow the seeds away from the parent. Seeds can remain dormant for a substantial period, germinating when exposed to high light. Large gaps are ideal for ceiba, and the tree is considered successional, though it may persist for many years once established in the canopy. (James Dalling has informed the author that there is an aerial photo in the dining hall on BCI of the island from 1920. In the photo, the southwestern part of the island is clearly composed of short-stature, regenerating secondary forest. Clearly visible in the photo is the same towering ceiba tree known today as Big Tree on the Van Tyne trail.)

Kapok leaves are extensively parasitized and grazed by insects. Leaf drop may serve not only to advertise the flowers and aid in wind-dispersing the seeds but may also help periodically rid the tree of its insect burden.

RESILIENT PASTURES—SECONDARY SUCCESSION IN AMAZONIA

The global tropics has a long history of human occupation and anthropogenic deforestation. Deforestation has dramatically increased in scope and is of major conservation concern today (see Chapter 15). Large forested areas of Amazonia have been cut and converted to pasture (Plate 6-16, and see Chapters 13 and 14). What happens when cattle pastures are abandoned? Does the natural vegetation reestablish a second-growth forest?

PLATE 6-16
Pastures such as this one in Panama have the capacity to undergo secondary succession and return to forest. The bird perched atop the cow is a smooth-billed ani (*Crotophaga ani*).

The answer depends on a number of things. Sometimes second-growth forest develops, and occasionally it does not. It depends on several variables, including intensity of usage.

Studies from Para, Brazil, in eastern Amazonia, indicate that successional patterns normally result in the growth of secondary forest (Buschbacher et al. 1988; Uhl et al. 1988a). Each of the sites in the Amazonian study had been cut, burned, and used for cattle pasture. Sites ranged in age (from abandonment) from 2 to 8 years and, depending on the site, had experienced light, medium, or heavy use for up to 13 years. Vegetation composition, structure, and biomass accumulation were documented. In areas subject to light use, succession was rapid, with a biomass accumulation of about 10 tons per hectare annually, or 80 tons after 8 years. Tree species richness was high, with many shade-tolerant forest species present. Moderately grazed pastures also underwent rapid succession when abandoned, but biomass accumulation was only about half what it was on lightly grazed sites, and tree species richness was lower as well. Heavily grazed sites remained essentially in grasses and herbaceous species, with few trees invading and a biomass accumulation of only about 0.6 ton annually per hectare.

The conclusion drawn from this study was that most Amazonian lands subjected to light or moderate grazing, once abandoned, regenerate secondary forest. Succession to secondary forest was impeded in areas subject to intensive grazing for long periods, areas that were estimated to represent less than 10% of all pastureland in northern Para. Not surprisingly, the second-growth forest subsequent to pasture abandonment contained neither the physiognomy nor the species composition found on the original undisturbed old-growth sites. Moreover, heavy, continued disturbance clearly alters the successional pattern (Uhl et al. 1988a).

DISTURBANCE IMPACT AND REGENERATION PATHWAYS

Disturbance impact and subsequent regeneration patterns are strongly influenced by duration of the disturbance as well as disturbance frequency. Extensive studies in Amazonia have demonstrated differences between regeneration patterns in small-scale compared with large-scale disturbances (Uhl and Jordan 1984; Uhl et al. 1988a and 1990). *Small-scale* is defined as a disturbance area of between 0.01 and 10 hectares. *Large-scale* is from 1 to 100,000 square kilometers, with causal factors being principally floods and fires. These scales are roughly analogous to human disturbances caused by slash-and-burn agriculture and conversion of forest to pasture (see Chapter 14).

Following disturbance, recovery and regeneration can occur from the following possible regeneration pathways:

- From seedlings and saplings already present in the forest understory (termed the *advance regeneration pathway*)

A Resilient Rain Forest: Lessons from Tikal

Tikal, now an archeological site in eastern Guatemala, was a thriving city of the classic period of Mayan civilization (Plate 6-17). It provides an example of how secondary forest reclaims a site after human abandonment. Located in the Petén region of eastern Guatemala, Tikal was founded around 600 B.C. and flourished from about 200 A.D. until it was abandoned around the year A.D. 900. Anthropologists are uncertain as to what climatic and/or social factors forced the total abandonment of the city and the subsequent deterioration of Mayan society well in advance of the Spanish conquest of the Americas (but see Diamond 2004). What is known is that when Tikal existed, the land surrounding it was largely deforested and used for agriculture and urbanization. At its peak, Tikal served as a major trade center. Maize (corn), beans, squash, chili peppers, tomatoes, pumpkins, gourds, papaya, and avocado were brought from small, widely scattered farms to be sold in the busy markets of the city.

An estimated population of 50,000 lived in Tikal, which spread over an area of 123 square kilometers, protected by earthworks and moats. The Classic Mayan society practiced sophisticated intensive agriculture (LaFay 1975; Flannery 1982; Hammond 1982; Diamond 2004). The majestic pyramid-like temples, excavated during the twentieth century, now serve as silent memorials, where tourists come to see what remains and to reflect on the past. This long-deceased civilization had developed a calendar as accurate as a modern calendar, a complex writing system that still has not been entirely deciphered, and a mathematical sophistication that included the concept of zero. The sight of the Great Plaza and Temple I, the Temple of the Giant Jaguar, enshrouded in the cool early morning tropical mist, romantically transports the mind's eye back to the time when Tikal was a focal point of civilization in Mesoamerica.

Today Tikal is enclosed by lush moist forest. Like Mayan ruins throughout the Yucatan Peninsula, the city itself had to be rediscovered and excavated, as second-growth forest had grown over and atop the ruins. Tikal was literally under forest, and much of it still is, its once-crowded plazas, thoroughfares, and temples overgrown by epiphyte-laden milk trees *(Brosimum alicastrum)*, figs, palms, mahogany trees, and chicle trees. From atop the sacred temples, one can watch spider and howler monkeys cavort in the treetops. *Agoutis* (cavimorph rodents common in Neotropical forest) and coatis shuffle through the picnic grounds, amusing tourists, while searching for food scraps. Toucans and parrots pull fruits from trees growing along what were once the central avenues leading to and from the city. This once-great metropolis covering many square miles was abandoned and subsequently reclaimed by second-growth lush tropical moist forest.

Recent studies suggest that Tikal is not an isolated case of rain forest regeneration. The Darien of Panama, a remote region is southern Panama that is today rich, diverse rain forest, was subject to extensive human disturbance. A study of the pollen and sediment profiles from the region reveals that much of the landscape was historically planted with corn and subject to frequent fires, probably set by humans. Only after the Spanish conquest was the region abandoned, allowing succession to occur. Thus the lush and seemingly pristine rain forest that defines the Darien today is only about 350 years old (Bush and Colinvaux 1994). It is second-growth forest, still regrowing.

PLATE 6-17
Tikal as it appears today.

- From vegetative sprouting from stem bases and/or roots (which remain after trees are disturbed)
- From recolonization by germination of seeds already present in soil (called the *soil seed bank*)
- From the arrival of new seeds brought by wind or animal dispersal

In cases of small-scale disturbance, the advance regeneration pathway dominates throughout the Neotropics. In Amazonia, between 10 and 20 seedlings and small saplings (less than 2 meters tall), known collectively as the *seedling bank*, are typical of every square meter of forest floor. Most of these plants persist for very long periods in the darkened understory. These account for over 95% of all trees greater than 1 meter tall four years after gap formation (Uhl et al. 1988b and 1990). The second pathway, *sprouting*, is also common in many tree species in small gaps (see the section "Forest Gaps and Tree Demographics" later in this chapter). Large-scale gaps can result in the death of understory trees, destroyed by immersion in flood or by fire. Regeneration in large gaps is from a combination of vegetative sprouting plus germination of seeds in the soil, plus import of seeds by dispersal mechanisms.

Critical to regeneration is the presence of viable seeds in the soil seedbank plus the added importation of seeds into disturbed sites (carried by either wind or animal dispersers). Further, once the seeds are so located, they must germinate and the seedlings must survive.

FIRE AS A DISTURBANCE FACTOR IN THE TROPICS

While you are standing in a rain forest experiencing 100% relative humidity, watching in wonder at the intensity of the pouring rain deluging around you, the thought of the rain forest catching fire and burning seems at best a fanciful notion. Perhaps not. Evidence has accumulated suggesting that for the past few thousand years, fire has been an important natural, large-scale disturbance factor throughout Amazonia (Uhl et al. 1988c and 1990). There is an abundance of charcoal residue in central and eastern Amazonia soils, and studies from the Venezuelan Amazon along the upper Rio Negro employing radiocarbon dating of the sediments indicate that during the past 6,000 years, there have been several major fires, occurring perhaps during periods of extended dryness (Sanford et al. 1985; Uhl et al. 1988c and 1990). Large-scale Amazonian fires, even if infrequent, add yet another

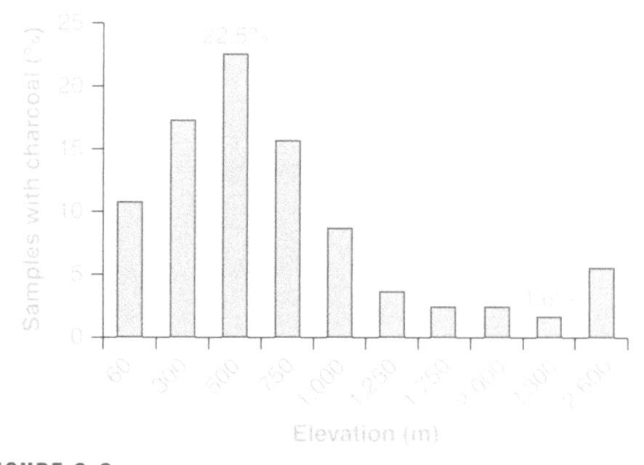

FIGURE 6-8
Charcoal mass (in grams per square meter) of soil charcoal summed for all depths (0 to 150 centimeters) and all subplots at 11 plots along the elevational gradient (240 per plot, means ± 2SE). Site 60-meter data are average of two separate sites at each elevation.

disturbance dimension to the dynamics of rain forests, a dimension that may help us to explain how the high tree biodiversity of the region originated and is maintained.

Fire may be either natural or anthropogenic in origin, and evidence from a study in Costa Rica suggests that both human and naturally occurring fires have a continuous history in the region beginning about 23,000 years before the present (Titiz and Sanford 2007). The study examined soil charcoal patterns over an elevational transect from sea level to the continental divide. Soil charcoal was most abundant in the lowland sites but was present at all sites (Figure 6-8). The authors conclude that forests in the region have, throughout the Holocene, experienced fire multiple times by both natural and human causes. Some sites showed clear evidence of multiple fires, some of which was likely related to ancient deforestation.

Today, human-set fires are a major factor in global tropical forests (Cochrane et al. 1999) and are of concern to conservation biologists. There is evidence that frequent fires may convert areas of forest to savanna (see Chapter 15). Severe fires in the tropics are more frequent in years of severe El Niño. In the El Niño of the summer of 2009, large parts of central Borneo (Plate 6-18) were subject to widespread peat fires set by people attempting to clear land in a traditional manner. *Peat* is the accumulated remains of organic matter in acidic soil. When peat becomes desiccated, as is typical of a drought year, it becomes highly combustible. The drought associated with the ENSO event had severely dried the peat, and when farmers set fire to the forest to clear land, literal conflagrations were

PLATE 6-18
Peat fire in Borneo in summer 2009.

the result (Moses 2009). The extensive fires posed a major threat to orangutans and were severe enough that the accumulated smoke closed airports.

Forest fires are also severe in Africa, particularly in places like Democratic Republic of the Congo, southern Sudan, northern Angola, and the Central African Republic. According to the United Nations Food and Agriculture Organization, Africa has the largest number of forest fires in the world (FAO 2006).

WHEN SUCCESSION DOESN'T SUCCEED

It is possible for secondary succession to become arrested, to cease vegetation development from an open, disturbed site to secondary forest. Several factors may cause arrested succession. One is the presence of invasive plant species such as exotic grasses that grow sufficiently dense to prevent other species from growing. Another factor is frequent fire or heavy grazing by livestock, both of which have occurred through human activity in areas throughout the global tropics (see Chapter 13). The combination of frequent fire and exotic grass invasion results in a grass/fire cycle that stabilizes in occupancy of the exotic grass, arresting normal succession (D'Antonio and Vitousek 1992).

A study performed on deforested, abandoned land in Panama examined the effects of exotic grass, fire, and seed dispersal on successional patterns (Hooper et al. 2004). Abandoned agricultural plots in Panama, as in other tropical regions, are often invaded by species of exotic grasses. In Panama, the grass studied was *Saccharum spontaneum,* a tall (2.5 meters) grass species native to sub-Himalayan valleys (Plate 6-19). This species has become a serious invasive throughout Indonesia, India, Thailand, the Philippines, and Puerto Rico, as well as Panama. The researchers hypothesized that the combination of low seed dispersal, fire, and presence of *S. spontaneum* would combine to affect early successional patterns. They employed a multifactoral experimental design to ascertain the influence of each of these variables.

Five sites were used, each in areas dominated by *S. spontaneum*. Each site was adjacent to a different "tree island" or forest edge that would serve as the seed source. Plots were mapped at each site, and treatments were implemented. Treatments consisted of hand-cutting *S. spontaneum* with machetes and selective burning of some plots. In order to evaluate the effect of remnant vegetation, researchers measured the distance of each regenerating tree and shrub seedling to its nearest neighbor, either a tree, sapling, shrub, or large *monocot* (such as a palm).

A total of 4,984 individuals representing 80 woody species regenerated in the *S. spontaneum* grasslands. Significantly more species were found 10 meters from forest than 35 meters at the first census. This indicates

PLATE 6-19
The invasive grass *Saccharum spontaneum*.

that proximity to a seed source strongly influenced regeneration. Community composition and recruitment differed with distance from forest (Table 6-2).

Fire affected species regeneration. Many species showed reductions in areas subjected to fire, while a few showed increases. The time since fire also significantly affected community composition.

The cutting of *S. spontaneum* had a measurable influence on community composition, but the effect depended on distance from the forest. The only major difference among areas that were cut and those that were not was that seedlings whose seeds were dispersed by agoutis were more frequent within the 85-meter distance (from forest).

Researchers concluded the following:

- Fire reduces species richness and thus is a major barrier to natural forest regeneration. It reduces the seed bank viability and reduces resprouting. The time elapsing between fires is also a critical variable affecting species composition.
- Fire resulted in wind-dispersed species being present more often in early post-fire communities, while large arboreal animal–dispersed species were present significantly more often as time since fire increased.
- Overall, fire sets back succession in grasslands dominated by *S. spontaneum*. If fire is persistent, succession will not result in second-growth forests, as woody vegetation will be prevented from becoming established. Grass presence also promotes fire, and thus the combination of grass and fire may indefinitely arrest succession.
- Seed dispersal is the critical mechanism for forest regeneration. Species richness and density are highly influenced by spatial factors such as distance from likely seed sources.
- Most plant species were significantly affected by proximity to forest. This included both animal-dispersed and wind-dispersed species.

SUMMARIZING SUCCESSION

Succession is best viewed as a continuum of life history characteristics that adapt various species of plants to grow and reproduce on a patch of habitat over time. It is very much situation dependent. Because life history characteristics vary among species, species composition on a site changes in response to changing abiotic and biotic conditions, which, of course, are in part caused by the plant species themselves. Natural selection selects for suites of traits: seed dispersal rates, establishment requirements, shade intolerance or tolerance, growth rates, longevity, size at maturity, and so on. Seeds have varying resident times in soil seed banks. Edaphic conditions vary among sites. Exotic species and fire may interact to alter conditions and prevent normal patterns of succession.

The general pattern is that light-demanding species typically appear first. These are followed by noncolonizing light-dependent species. These species are eventually replaced by shade-tolerant species. All of these groups may be present early in the succession, but they grow at different rates and to a degree they replace each other. There are many more shade-tolerant species than light-dependent tree species, and many shade-tolerant species are present from the outset of succession, though they are not evident at the time.

TABLE 6-2 The effect of fire on the number of recruits (from seeds), the number of resprouts, and the total number of individuals found naturally regenerating at burned sites at Las Pavas, Panama.

SPECIES	NUMBER OF SPECIES	RECRUITS		RESPROUTS		TOTAL	
		BEFORE FIRE	AFTER FIRE	BEFORE FIRE	AFTER FIRE	BEFORE FIRE	AFTER FIRE
Resprouts and recruits increased							
Thevetia ahouai	1	8	20	1	25	9	45
Recruits increased, no resprouts							
Trema micrantha		3	150	0	0	3	150
Byrsonima crassifolia		36	62	0	0	36	62
Total	10	52	276	0	0	52	276
Resprouts increased, no recruits							
Total	3	0	0	4	9	4	9
Recruits decreased, resprouts increased							
Gustavia superba		426	117	157	300	583	417
Cochlospermum vitifolium		105	82	22	125	127	207
Cordia alliodora		24	9	56	122	80	131
Total	11	622	225	317	751	939	976
Recruits decreased, no resprouts							
Piper marginatum		300	32	0	0	300	32
Cecropia insignis		68	3	0	0	68	3
Total	18	465	56	0	0	465	56
Recruits and resprouts decreased							
Banara guianensis		8	1	90	50	98	51
Hybanthus prunifolius		53	4	41	33	94	37
Inga vera		21	0	69	45	90	45
Swartzia simplex var. *ochnacea*		23	0	40	6	63	6
Total	14	160	6	269	140	429	146
Overall total		1307	583	591	925	1898	1508

Notes: Species are grouped into categories depending on their regeneration response to fire. Only those species that contributed significantly to the overall community abundance are listed.

Second-growth forests may be as species-rich as old-growth forests, though the two forest types may maintain species assemblages that are largely different. This is often due to long-lived pioneer species persisting in the canopy even after shade-tolerant species have attained full canopy stature. This means that species composition should be recognized as distinct from species richness when comparing secondary and old-growth forests (Finegan 1996).

Last, succession in the tropics tends to follow aspects of both the initial floristic model and the nucleation model (Hooper et al. 2004). The patterns evident in secondary succession in the tropics suggest a strongly local and strongly stochastic influences.

Central American Successional Crop System

Slash and burn (also known as *swidden*) agriculture (see Chapter 13) has much in common with ecological succession in that it mimics the successional process in restoring soil after use for farming. *Polycultures* (planting numerous species in the same plot) can effectively maximize productivity of a given plot. Instead of only one crop, several surface crops, such as corn and beans, share the same plot with root crops (such as manioc or sweet potatoes), while the border of the plot may be planted in peppers and tomatoes. Polycultures may be more resistant to insect attack because crop biodiversity provides habitat for herbivore predators and reduces the competitive effects of undesirable forbs (weeds).

Comparisons of various monoculture crops with mixed species plots have shown that the more diverse plots had significantly more root surface area, enhancing the ability to capture nutrients (Ewel et al.1983). Ewel et al. conclude that it is essential to maintain complex, well-developed root systems throughout the soil depth in designing tropical agroecosystems. This is because leaching rates are typically high, so a dense array of roots from various species is the best way to counteract loss of soil fertility due to leaching.

Robert Hart (1980), working in Costa Rica, has suggested that farming can be directly analogous to succession (Figure 6-9). He presents a scheme whereby crops are rotated into and out of plots on the basis of their successional characteristics. Using such a system, Hart claims that it would be possible to utilize a plot of land continuously and productively for at least 50 years or more. To quote Hart:

> Early successional dominance of grasses and legumes can be assumed to be analogous to maize (*Zea mays*) and common bean (*Phaseolus vulgaris*) mixtures. Euphorbiaceae, an important family in pioneer stages of early succession, can be represented by cassava (*Manihot esculenta*), a root crop in the same family. In a similar replacement, banana (*Musa sapientum*) can be substituted for *Heliconia* spp. The Palmae family can be represented by coconut (*Cocos nucifera*). Cacao (*Theobroma cacao*) is a shade-demanding crop that can be combined with rubber (*Hevea brasiliensis*) and valuable lumber crops such as *Cordia* spp., *Swietenia* spp., or other economically valuable members of the Meliaceae family to form a mixed perennial climax. (p. 77)

One example of the kind of system described here is found in Selva Lacandona in Chiapas, Mexico, the home of 450 remaining Lacandon Maya Indians (Nations 1988). These people combine agriculture with forestry, growing up to 79 varieties of food and fiber crops in 1-hectare plots in a manner that closely mimics succession. Trees are felled and burned, and a combination of

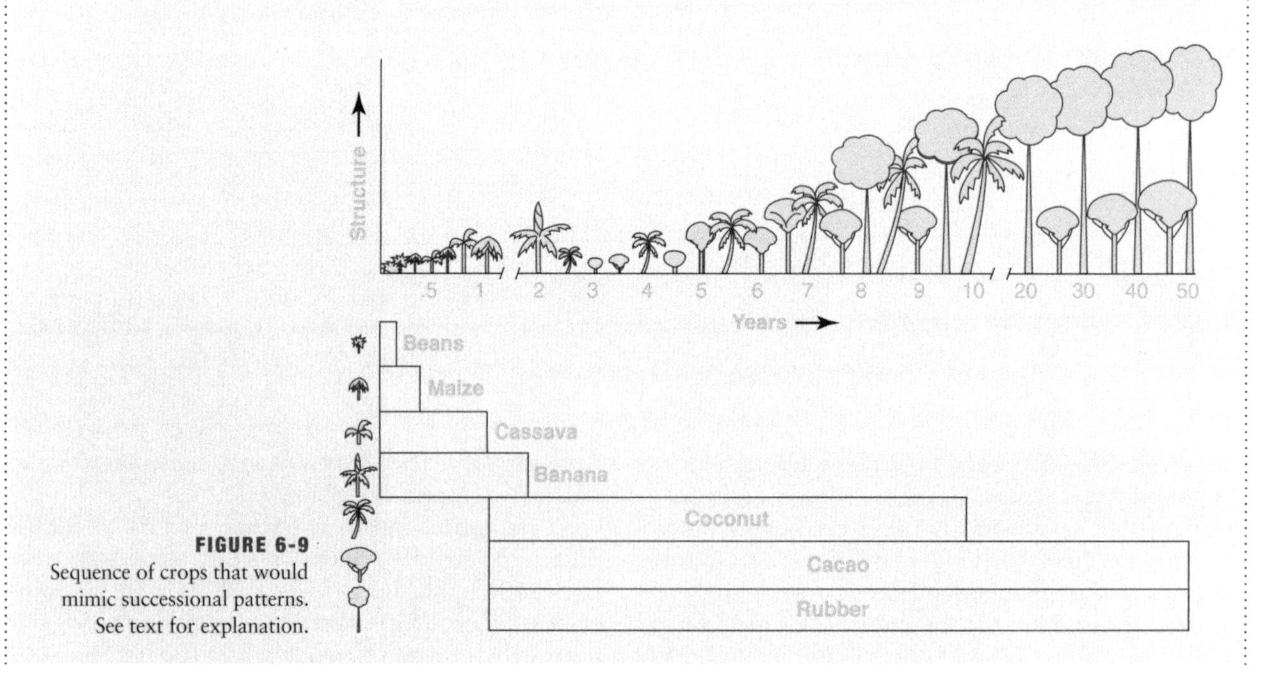

FIGURE 6-9 Sequence of crops that would mimic successional patterns. See text for explanation.

fast-growing tree and root crops such as plantains, sweet potatoes, bananas, papaya, and manioc (Plate 6-20a) are immediately planted. Once these crops become established, others are added throughout the year, depending on appropriateness of the season: corn, chiles, limes, coriander, squash, tomatoes, mint, rice, beans, sugar cane, cacao, and others. Certain natural species such as sapodilla, wild pineapple, and wild dogbane, which invade as part of normal succession, are also permitted to grow among the planted crops. These are also eventually used for various purposes. As the plot diversifies, animal species such as peccaries, brocket deer, and pacas are attracted, all of which are valuable game animals. The Lacandon system permits a single hectare to remain productive for 5 to 7 consecutive years before rotation is necessary. Eventually, trees such as rubber, cacao, banana (Plate 6-20b), citrus, and avocado are planted, and the yield from these species adds yet another 5 to 15 years of productivity to the plot.

(a)

(b)

PLATE 6-20
(a) Manioc and (b) bananas are common components of the plots described earlier.

Forest Gaps and Tree Demographics

FOREST GAPS

Rain forest trees are not immortal, and each will eventually perish. Some remain in place, becoming dead, decaying snags, and others fall immediately to the ground, some bringing their root systems to the surface as they fall, some snapping off somewhere along the trunk and thus leaving their roots in the ground. A tree may be blow down by windthrow (Plate 6-21) or may topple when weakened by termites, epiphyte load, or old age. Landslides may bring down groups of trees. Large branches commonly break off and drop. Indeed, one of the often-heard sounds when walking through rain forest is the sudden crack of a falling tree or large branch. When a rain forest tree or significant part of a tree falls, it creates a canopy opening, a *forest gap*. In gaps, light is increased, causing microclimatic conditions to differ from those inside the shaded, cooler, closed canopy. Air and soil temperatures as well as humidity fluctuate more widely in gaps than in forest understory.

Gaps vary in size. The general pattern is that most wet forests are characterized by many small gaps and few large gaps, where a *large gap* is defined as having an area in excess of 300 or 400 square meters (Clark and Clark 1992; Denslow and Hartshorn 1994). Large

PLATE 6-21
Fallen trees such as this former member of the canopy are the main cause of gaps.

gaps, though few in number, have greater total area, and thus comprise a large percentage of total gap space within a forest. An emergent tree, should it fall, may bring down several other trees, creating a large gap. *Lianas*, vines that may connect several trees together, increase the probability of multiple tree-falls. When one tree goes, its liana connections to others can result in additional trees falling, or large branches being pulled down (Putz 1984). A study in Amazonian Peru demonstrated that large lianas (equal to or greater than 10 centimeters diameter) represented less than 5% of liana stems but 80% of biomass of well-lit canopy lianas. Long-term estimates of turnover rate showed that lianas have between 5% and 8% annual turnover, three times that of canopy trees. Most importantly, the study implicated lianas in tree mortality, as liana-infested trees were three times more likely to perish than liana-free trees. Large lianas were implicated in the death of about 30% of tree basal area in the forest (Phillips et al. 2005). Lianas are clearly a strong influence in the dynamics of rain forests.

Treefalls are often correlated with seasonality. On Barro Colorado Island, tree-falls peak around the middle of rainy season, when soils as well as the trees themselves are saturated with moisture and strong gusty winds blow (Brokaw 1982). At La Selva, most gaps occur in June to July and November to January, the wettest months (Denslow and Hartshorn 1994). In parts of the Caribbean and Central America, hurricanes have periodically leveled hundreds of acres of forest, a giant gap indeed. For example, when Hurricane Hugo struck Puerto Rico in 1989, it resulted in dramatic changes that were felt throughout the food web as it opened large areas of the El Verde rain forest to high light intensity (Figure 6-10) (Reagan and Waide 1996).

Gaps occur normally in all rain forests and are reflective of the reality of continual natural disturbance. In Amazonia, for instance, from 4% to 6% of any forest will be made up of recently formed gaps (Plate 6-22) (Uhl 1988).

Hubbell and Foster (1986b and 1986c) have censused over 600 gaps in the BCI forest in Panama. Their research has shown that large gaps are less common than small gaps. They assert that a typical gap is shaped roughly like an inverted cone standing on its point, a pattern resulting in expansion of gap area as one moves higher in the canopy, and adding an additional component of structural complexity to an already complex forest physiognomy. Since both horizontal and vertical heterogeneity of a forest is significantly increased by gaps, gaps become a potentially obvious consideration in explaining high biodiversity.

Because they admit light, gaps create suitable conditions for rapid growth and reproduction. Plant species, including many shade-tolerant species, respond to gaps with a spurt of growth, and at least a few species

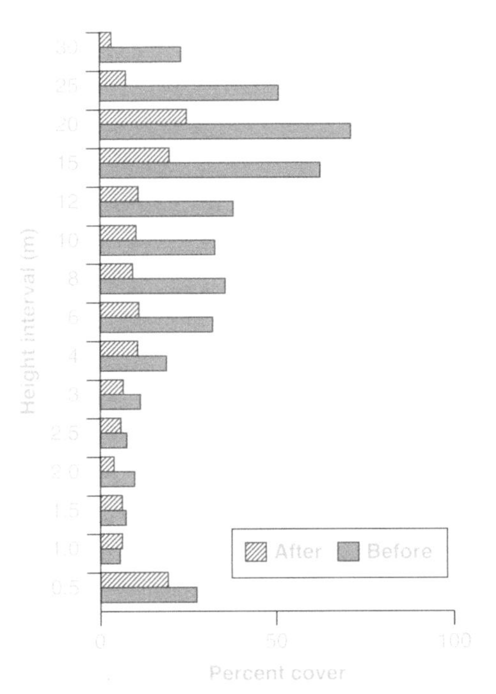

FIGURE 6-10
This figure shows vegetation height profiles for forest plots at El Verde, Puerto Rico, before and after Hurricane Hugo. Note how the hurricane reduced percent cover, opening a huge gap in the forest.

are dependent on gaps (Brokaw 1985; Hubbell and Foster 1986b; Murray 1988; Clark and Clark 1992). Of 105 canopy tree species studied as saplings at La Selva, about 75% are estimated to depend at least in part on gaps in order to complete their life cycle (Hartshorn 1978).

Gaps create a diverse array of microclimates, affecting light, moisture, and wind conditions (Brokaw 1985). Measurements made at La Selva in Costa Rica indicate that gaps of from 275 to 335 square meters experience 8.6% to 23.3% of full sunlight, compared with interior forest understory, which receives only 0.4% to 2.4% of full sunlight (Denslow and Hartshorn 1994). Thus a large gap provides up to 50 times as much solar radiation as interior forest. Further, it is "high-quality" sunlight, with wavelengths appropriate for photosynthesis. In contrast, the shaded forest understory is generally limited not only in total light intensity but also in wavelengths from 400 to 700 nanometers, the red and blue wavelengths most utilized in photosynthesis (Fetcher et al. 1994). Most high-quality solar radiation (61% to 77%) within a shady rain forest understory is received in the form of short-duration sun flecks (Chazdon and Fetcher 1984). The total amount and quality of solar radiation is probably the single largest limiting factor to plant growth inside tropical forests, which accounts for the importance of gaps. This restriction may be evident in the fact that many understory herbs have leaves that are unusually colored, with blue iridescence, velvety surface sheen, variegation, and red or purple undersides (Fetcher et al. 1994). *Abaxial anthocyanin*, the pigment responsible for the red

PLATE 6-22
Gaps open the forest to high levels of sunlight, stimulating both germination of seeds in the soil bank as well as growth spurts of saplings already present.

underside of some leaves, may be physiologically adaptive in aiding the plant in absorption of scarce light (Lee et al. 1979).

One hypothesis suggested that rain forest trees fall roughly into three categories, depending on how they respond to gaps and gap size (Denslow 1980):

- Large-gap specialists whose seeds require high temperatures of gaps to germinate and whose seedlings are not shade-tolerant
- Small-gap specialists whose seeds germinate in shade but whose seedlings require gaps to grow to mature size
- Understory specialists that seem not to require gaps at all

While the preceding categories are helpful, the reality is more complex. It is clear that pioneer species such as cecropias require gaps, but most shade-tolerant tree species show no gap association, demonstrating high levels of growth plasticity. They persist and slowly grow under the deeply shaded conditions of the forest interior, but grow more rapidly in gaps (Clark and Clark 1992; Denslow and Hartshorn 1994).

The leguminous tree *Pentaclethra macroloba,* common at La Selva Biological Station, is typical of many trees in that it is tolerant of deep shade but grows rapidly in high-light conditions provided by gaps (Fetcher et al. 1994). Only species that are completely shade-intolerant require gaps for growth and reproduction.

For many years, it has been known that sapling trees of some species persist in the understory, small but healthy, continuing their upward growth when adequate light becomes available. Understory specialists do not necessarily require gaps, but utilize them when the opportunity is presented.

Once in a gap, many tree and shrub species exhibit higher reproductive outputs, and thus larger fruit crops create more competition among plants for frugivores to disperse their seeds (Denslow and Hartshorn 1994).

The ecology of gap-dependent pioneer species is generally well understood. Brokaw (1982 and 1985) studied regeneration of trees in 30 varying-sized forest gaps on Barro Colorado Island. Pioneer species produced an abundance of small seeds, usually dispersed by birds or bats, with some capable of long dormancy periods. In another study, Brokaw focused on only three pioneer species and learned that the three make up a continuum of what he called *regeneration behavior* (Brokaw 1987):

PLATE 6-23
Trema micrantha, a rapid gap colonizer.

- One species, *Trema micrantha* (Plate 6-23), colonized and grew rapidly, growing up to 7 meters per year. This species colonized only during the first year of the gap. After that, it could not successfully invade, presumably due to competition with other individuals.
- The second species, *Cecropia insignis,* invaded mostly during the first year of the gap, but a few managed to survive that entered large gaps during the second and third year. This species grew more slowly (4.9 meters per year) than *Trema.*
- The third colonist was *Miconia argentea,* which grew the most slowly of the three (2.5 meters per year) but was still successfully invading the gap up to 7 years following gap formation.

Brokaw's study reveals how the three species utilize different growth patterns to successfully reproduce within gaps. Such subtle distinctions may partially explain the coexistence of so many different species within rain forest ecosystems. Or do they?

DO GAPS HELP EXPLAIN HIGH TREE SPECIES RICHNESS?

Ecologists have investigated the relationship between gap dynamics and tree species richness. On Barro Colorado Island in Panama, Hubbell and his colleagues studied 1,200 gaps of varying sizes over a 13-year period (Hubbell et al. 1999).

The study focused on a 50-hectare plot of old-growth forest. All woody plants with a stem diameter equal to or greater than 1 centimeter DBH were tagged, measured, mapped, and identified to species (Figure 6-11). This effort included more than 300,000 stems and comprised 314 species. Censuses were conducted in 1982, 1985, 1990, and 1995. Species were divided into three categories, called *regeneration niche guilds*: gap-dependent pioneer species, shade-tolerant species, and intermediate species. Researchers monitored changes in both gap and nongap control areas. Gaps varied 46-fold in size, ranging from 25 square meters to 1,150 square meters.

As predicted, gaps had greater seedling establishment and higher sapling densities than control areas. Species richness was higher in gaps than in control areas, and there was much variation in species composition among gaps (Figure 6-12). But when species richness and composition was expressed as a function of gap size, there were more species present in the gap than in an equal area outside the gap. But this is because there were many more stems represented in gaps, so there were more species.

The representation of species was:

- 47 pioneer species
- 33 intermediate species
- 94 shade-tolerant species

One key finding of the study was that of the available species pool in various categories listed here, there was a low recruitment rate per gap (which accounts for why gaps differed in species composition). For example, area-normalized gap occupancy rates were

(a)

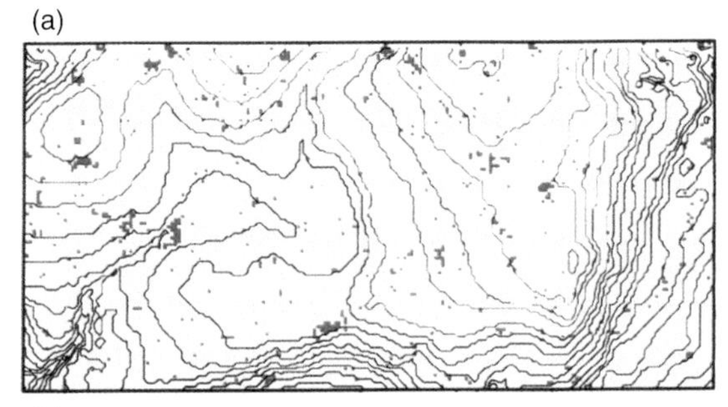

(b)

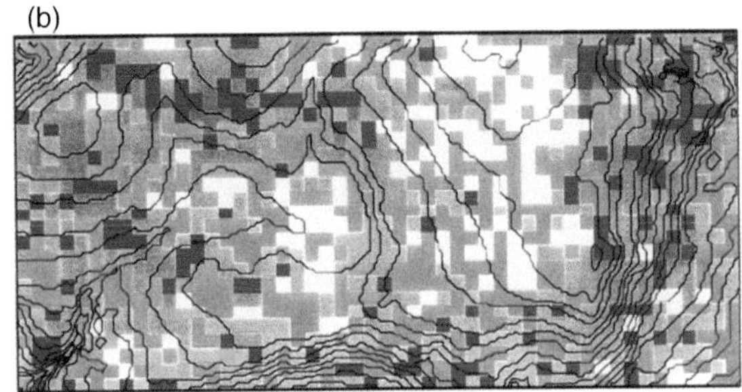

FIGURE 6-11
Distribution of light gaps and sapling species richness superimposed on a topographic map of the 50-hectare permanent forest plot on Barro Colorado Island (BCI), Panama. Contour intervals are 2 meters. (a) Distribution of light gaps in the 1983 canopy census. Each small square represents an area of 5 meters by 5 meters, the smallest gap censused. A c^2 analysis between habitat type and gap abundance showed no correlation between gaps and topographic features of the plot in 1983 or in any later year. (b) Distribution of sapling species richness in the 20-meter by 20-meter quadrats in the 50-hectare BCI plot (1,250 total quadrats), showing the relationship between topography and species richness. Yellow: <29 species per 400 square meters. Blue: 30 to 39 species per 400 square meters. Orange: 40 to 49 species per 400 square meters. Reddish orange: >50 species per 400 square meters. The central plateau and the small seasonal swamp (center and left, respectively) have 22% to 64% fewer species than slope areas to the east, south, and west. Note the lack of correlation with the 1983 gap sites in (a). A similar lack of correlation between species richness and gap disturbances also exists in the other years.

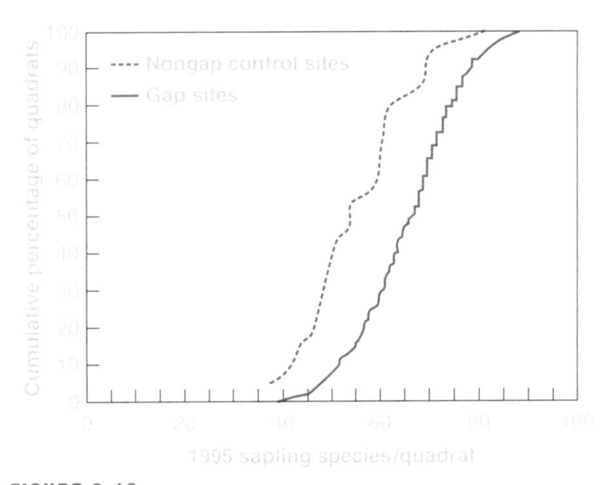

FIGURE 6-12
Distribution of species richness per quadrat for all 20- by 20-meter quadrats containing 1983 gaps after 13 years (solid line) and for quadrats in nongap control areas that remained in mature, high-canopy forest over the entire 13-year period of the study (dashed line).

2% to 3% for pioneer species and 3% to 6% for shade-tolerant and intermediate species (Figure 6-13). This suggests that most species in the forest could not take advantage of gaps because they could not get to them (via seed dispersal). Recall from the previous chapter that recruitment limitation may help account for why so many species of trees persist indefinitely. They cannot compete if they do not get to the arena, so to speak. If there is no strong competition, there will be no competitive exclusion.

A decade-long seed trap study and seedling census supports the argument that recruitment into gaps is limited. In excess of 1.3 million seeds from 260 census plots were collected and identified at BCI. Over 50 species of adult trees were not represented in any of the seeds collected. A mean of 88% failed to deliver even one seed to any given trap in a decade of continuous seed trapping. And only seven species dispersed more than one seed into more than 75 of the 200 traps. About 50% of the tree species seeds collected had seeds in 6 or fewer traps. Seedling censuses reflected a similar recruitment limitation. The most common species were found in only 14.9% of the quadrats, and 56% of the tree species on the plots were absent from the seedling census.

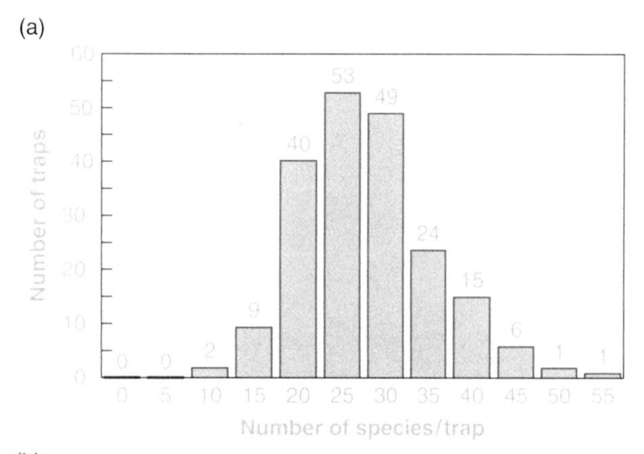

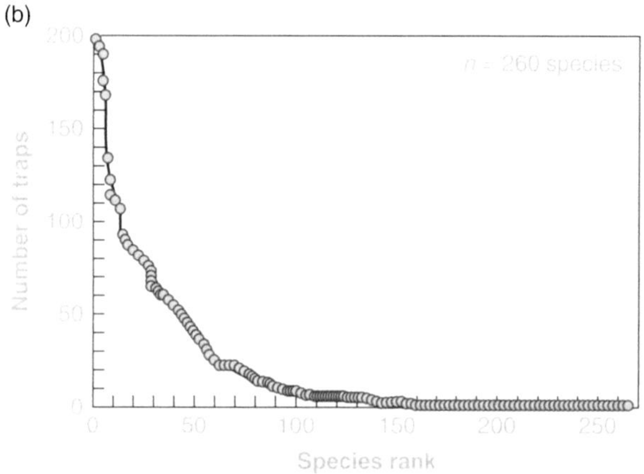

FIGURE 6-13
Evidence for dispersal limitation in BCI trees from a 10-year seed trap study using 200 traps in the 50-hectare plot. Seeds of a total of 260 species of the 314 species in the plot census were collected at least once. (a) Frequency distribution of the number of species captured per trap during the 10-year trapping period (1987–1996). The average number of species per trap was 30.8 ± 7.5 SD. (b) The total number of traps into which each species dispersed at least one seed during the 10-year trapping period.

The study concluded that a decoupling had occurred between gap disturbance regime and local tree species richness, essentially due to recruitment limitation. Nonetheless, the authors recognized that gaps promote tree diversity by increasing seedling establishment of whichever trees happen to occur near the gaps. Gaps are required for pioneer species, which collectively are an evolutionary result of the very existence of gaps in the first place.

A further study (Wright et al. 2003) also conducted at BCI further refined the role of gaps in tropical forest ecology and also suggested a role for gaps in promotion of tree species richness. This study looked at a suite of characteristics of tree species, including the following:

- Percentage of recruits located in tree-fall gaps
- Sapling growth
- Sapling mortality
- Wood density

The results of the study indicated that few tree species were to be found on either extreme of a continuum ranging from highly light-demanding species to shade-tolerant closed forest species (Figure 6-14). Most of the species were those that exhibit intermediate light requirements and life histories. Gaps provide conditions conducive to such a suite of species (Plate 6-24).

The role of gaps in tropical forest demographics is still under study. In some second-growth forests, for example, it has been documented that tree mortality may be sufficiently low that gaps are uncommon. These same forests show strong recruitment of trees into the largest size categories, suggesting that the recruitment of canopy trees is not linked with gap formation (Chazdon et al. 2005).

FOREST DEMOGRAPHICS

How long do rain forest trees survive? How long does it take for a canopy giant to grow from seedling to

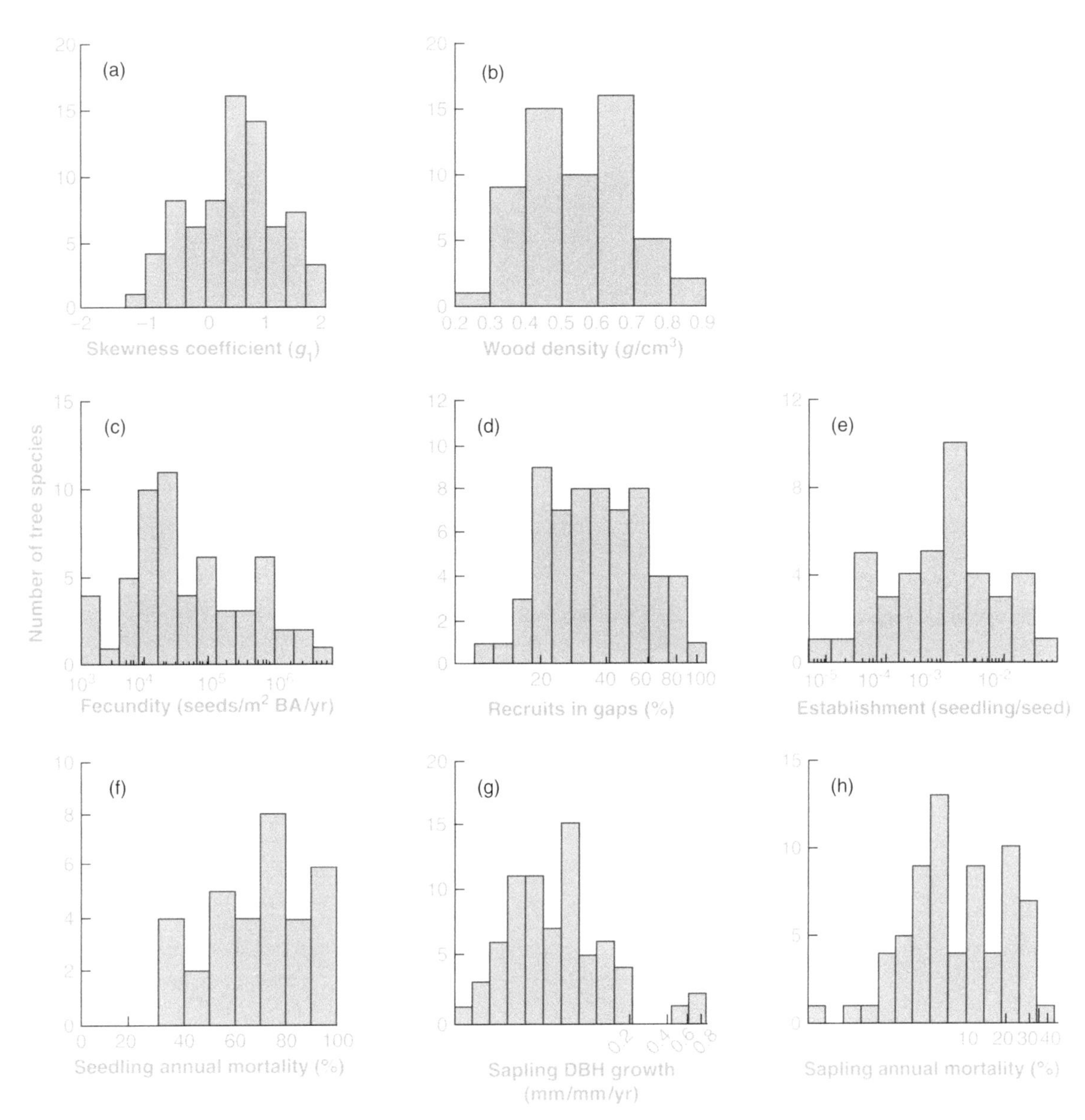

FIGURE 6-14
Frequency histograms for (a) the coefficient of skewness of the logarithm of DBH (*g*1); (b) wood density; (c) fecundity quantified as the number of seeds produced per square meter of basal area per year; (d) percentage of sapling recruits located in tree-fall gaps; (e) seed establishment probability; (f) first-year seedling mortality; (g) relative sapling DBH growth; and (h) sapling mortality for large canopy tree species from Barro Colorado Island, Panama. The distributions are approximately normal or log-normal. Note the *x*-axis log scales in panels c to e, g, and h.

adult? Does most growth occur in the rainy or dry season? How do short-term climatic fluctuations such as occasional droughts influence forest dynamics? What forces determine the probable survivorship of any given tree? These and many other questions compose the study of rain forest demographics. To answer these questions, it is necessary to initiate long-term, detailed studies of specific tracts of forest, literally monitoring the fate of each tree.

Studies of forest demographics have been ongoing at Barro Colorado Island (BCI) in Panama (Hubbell and Foster 1990 and 1992; Condit et al. 1992) and La Selva Biological Station in Costa Rica (Clark and Clark 1992; McDade et al. 1994).

PLATE 6-24
Palms are a common component of gaps in many tropical areas.

LA SELVA: THE LIFE HISTORIES OF TREES

The entire inventory of vascular plants at La Selva Biological Station totals 1,458 species. The majority of these are present in low population density.

Consider *Dipteryx panamensis* (Plate 6-25), a common emergent tree that favors alluvial soils. What might its longevity be? A Great Basin bristlecone pine (*Pinus longaeva*) atop the cold, windswept White Mountains of the western Great Basin Desert might live in excess of 4,000 years. Tropical trees show no comparable longevity. Ecologists study *forest turnover*, the average time that a given tree (defined within a certain size range) will survive in a particular spot. If you randomly select

PLATE 6-25
DIPTERYX PANAMENSIS

a *D. panamensis* that is at least 10 centimeters in diameter, how long before it is somehow destroyed or dies? The answer, for La Selva, is known: the rate of local disturbance is sufficiently high that the entire forest is estimated to turn over approximately every 118 plus or minus 27 years (Hartshorn 1978), and 6% of the primary forest is in young gaps at any one time (Clark 1994). One study, from 1970 to 1982, indicated an annual mortality rate of 2.03% for trees and lianas greater than 10 centimeters in diameter (Lieberman et al. 1985). Overall, adult survivorship of more than 100 to 200 years is uncommmon for subcanopy and canopy trees at La Selva (Clark 1994).

More recent work at La Selva using new techniques called *integral projection monitoring* and *age-from-stage analysis* refines the estimation of tree ages and forest dynamics (Metcalf et al. 2009). The researchers selected nine tree species for which there was an abundance of data from previous studies and that ranged from gap specialists such as *Cecropia obtusifolia* to the emergent canopy species *Dipteryx panamensis*. They calculated a *crown illumination* (CI) index for each tree. CI measures the amount of light received by each individual, as determined by the position of its crown with respect to openings in the surrounding canopy. CI values ranged from 1 to 6, where 1 indicates very low light and 6 indicates that the crown was completely exposed. Using a mathematical model, they calculated changes in tree diameter and changes in crown illumination with age. With the model, they could estimate life expectancy, time of passage to various sizes, and age patterns of mortality.

Pioneer species, unsurprisingly, thrived in high-light environments. Nonpioneer species exhibited high mortality when young, but the rate of mortality decreased as they aged, reaching a plateau. As individuals grew larger, their survivorship increased. Once trees were very large, survivorship then decreased. The pattern resembles that typical of human populations—namely, that infant mortality may be high, but if one survives infancy, mortality rate declines until old age, when it accelerates.

In comparison with previous work at La Selva, this study suggested somewhat different ages for various tree species and life expectancy patterns. For example, the genus *Lecythis,* in the Clark and Clark study (1992), was estimated to reach 30 centimeters diameter in as short as 59 years (maximum increment) or as long as 462 years (median increment). The Metcalf et al. study (2009) refined this estimate to between 96 years (when subject to high shade) and 76 years when grown with high light availability. The study concluded that it requires a minimum of 60 years to reach the canopy and possibly up to about 200 years for trees subjected to shady conditions. But once in the canopy, life expectancies were longer than those reported in the previous study, suggesting that canopy trees 300 years old or older may be relatively abundant. A study from Amazonia (Fichtler et al. 2003) that used carbon dating along with tree ring counts to age 12 fallen trees arrived at an age estimate of between 500 and 650 years for longevity of large canopy trees.

At Cocha Cashu Biological Station in Amazonian (southeastern) Peru, a forest on rich soils, mortality rate of adult trees (greater than 10 centimeters diameter) was 1.58% per year, implying an average life of 63.3 years (Gentry and Terborgh 1990). At San Carlos de Rio Negro in Amazonian Venezuela, mean annual mortality rates for trees greater than 10 centimeters DBH was 1.2% (Uhl et al. 1988b). Most tree deaths resulted in small gaps (large gaps were much rarer) such that approximately 4% to 6% of the forest area was in gap phase at any given time. At Manaus, Brazil, mortality was 1.13% for adult trees, with a turnover time of from 82 to 89 years (Rankin-De-Merona et al. 1990). These turnover rates are all for adult trees, with a minimum size of greater than 10 centimeters diameter. A tree often lives many years before attaining such a diameter, so the total age, from seedling to death, will be considerably longer (in the range of what was estimated by Fichtler et al. [2003]). In the Manaus study, it was learned that, in general, the larger a tree grew to be in diameter, the longer its probable life span from that point onward. For example, for trees greater than 55 centimeters DBH, turnover time increased to 204 years.

Mortality rates of seedlings or sapling trees exceed those of adults. Any recently germinated seedling stands a fairly high chance of being smashed by a falling branch, or a single fruit, or a whole tree, or perhaps being buried beneath a fallen palm frond or some other large-area leaf. Herbivores also occasionally consume seedling and sapling leaves to the point of killing the tree. For *Dipteryx panamensis*, seedlings ranging in age from 7 to 59 months experienced a 16% mortality rate from litter-fall alone (Clark and Clark 1987). Of course, many seeds never germinate because they are destroyed by a diversity of seed predators as well as fungal pathogens. Mortality rates are consistently highest in juvenile plants, sometimes very high indeed, declining steadily as the plants age (Denslow and Hartshorn 1994).

For example, in a study of six tree species, highest mortality rates, from 3% to 19% per year, occurred in the smallest saplings (Clark and Clark 1992). Mortality rates were much lower for intermediate to large juvenile sizes.

For most of a tree's life cycle, light acts as a significant limiting factor. Growth rates of trees in shaded interior forest are much lower than for those in more lighted, open areas. Trees such as *Dipteryx panamensis,* like many tree species, show extremely slow growth in low-light conditions but grow quickly taller and wider in a gap. For this reason, growth rates tend to fluctuate during the typical life cycle of a tree. Gaps open, close, and can reopen, so that any given tree might experience several periods of rapid growth (when in gaps) alternating with periods of extreme slow growth (under fully closed canopy). As described earlier, a substantial majority of tropical forest tree and shrub species show high levels of shade tolerance, with an accompanying high degree of *growth plasticity*—the ability to survive the very low light levels of the forest understory and grow rapidly in gaps (Denslow and Hartshorn 1994).

The existence of emergent trees has long been recognized as a characteristic of rain forests. Of what possible benefit is it to a tree to invest additional energy to grow above the majority of other trees in the canopy? Added light availability is certainly a possibility. But in a La Selva study of five emergent tree species, these species showed significantly lower adult mortality rates than nonemergent trees (Clark and Clark 1992). Perhaps emergents are more protected from being damaged by other falling trees, given that their crowns rise above the rest.

Peccaries and Palm Seeds

A study at Cocha Cashu Biological Station in Amazonian Peru demonstrated how a single species of seed predator affects the abundance and distribution of a tree species (Silman et al. 2003). White-lipped peccaries (*Tayassu pecari*) (Plate 6-26) disappeared from the region in 1978 and reappeared in 1990. Peccaries feed heavily on large fruits, including those of palms, in which case they act as seed predators, destroying the seed. Starting in 1978, researchers did transect counts of the number and spatial distribution of seedlings from the palm *Astrocaryum murumura*, a dominant tree species in the region. The counts were repeated in 1990, when peccaries had been absent for 12 years, and in 1999, when peccaries had reoccupied the region for 9 years. In the years of peccary absence, seedling density of *Astrocaryum murumura* increased 1.7-fold. Once peccaries returned, seedling density declined to what it had been when peccaries had been present before. The researchers realized that a single species of seed predator—in this case, white-lipped peccary—exerted a significant influence on the demography of *Astrocaryum murumura.* These results emphasize how species interactions may have major effects on forest species composition (see Chapter 7).

PLATE 6-26
WHITE-LIPPED PECCARIES

THE DYNAMICS OF DROUGHT

Drought was discussed earlier in this chapter, particularly when it results from El Niño effects. Beginning in 1980, a 50-hectare permanent plot was established at BCI. All woody plants at least 1 centimeter DBH were identified to species, measured, and mapped. Censuses were done in 1982, 1985, and 1990. Over the three censuses, 310 species were recorded in the plot, with data on 306,620 individual stems (Condit et al. 1992). In the brief time frame of this study, weather was an unexpectedly strong factor. An unusually protracted dry season coincident with a strong El Niño (see Chapter 1) brought a severe drought to BCI in 1983. Recall from earlier in this chapter that an ENSO year (1997–1998) had significant effects on patterns of secondary succession.

Tree mortality rates in Costa Rica were strongly elevated in the years immediately following the drought. From 1982 to 1985, trees with diameters greater than 8 centimeters experienced a mortality rate of 3.04% per year, a rate nearly three times higher than previously measured before the drought (Clark 1994). Compared with mortality during the interval 1985–1990, annualized forest-wide mortality from 1982–1985 was elevated as follows:

- 10.5% in shrubs
- 18.6% in understory trees
- 19.3% in subcanopy trees
- 31.8% in canopy trees

For trees with DBH greater than 16 centimeters, mortality was elevated fully 50% (Condit et al. 1992). The increased death rate among shrubs and trees was attributed to the drought. Approximately two-thirds of the species in the plot experienced elevated mortality from 1982 to 1985.

Those plants surviving the drought showed elevated growth rates. For example, growth from 16- to 32-centimeter trees was greater than 60% faster in 1982 to1985 than in 1985 to 1990. While this result might be surprising at first, it is really to be expected. The death of so many trees permitted much more light into the forest (the gap effect) as well as reduced root competition for water and nutrients among plants. Total gap area on the plot was increased after the drought but returned to its predrought level by 1991, an indication of how rapidly surviving plants responded to the added influx of light.

Many species' populations experienced changes in abundance during the period of the study, 40% of them changing by more than 10% in the first three years of the census (Hubbell and Foster 1992). Ten species were lost from the plot, and 9 species migrated into the plot from 1982 to 1990. Nonetheless, there was remarkable constancy in the number of species and the number of individuals within the plot at any given time:

- 1982 = 301 species, 4,032 individuals
- 1985 = 303 species, 4,021 individuals
- 1990 = 300 species, 4,107 individuals

The drought killed many trees but created opportunities for additional growth such that the deceased plants were quickly replaced. The speed of the replacement process was a surprise to the researchers (Condit et al. 1992).

The analysis of the BCI data suggests some important conclusions:

- The forest is responsive to short-term fluctuations caused by climate.
- The forest as a whole remains intact, though many species undergo population changes.
- The forest may be undergoing a long-term change in species composition.

This latter conclusion is based on the fact that there has been a decline of approximately 14% in annual precipitation at BCI over the past seven decades, dropping from 2.7 meters total in 1925 to 2.4 meters in 1995. The researchers hypothesize the future local extinctions of between 20 and 30 species, each of which requires a high level of moisture. Another reason for suggesting a long-term change in species composition is that after the 1983 drought, rare species declined more than commoner species, suggesting that plant species richness might be in decline (Condit et al. 1992). The BCI study demonstrated the dynamics of change as they relate to both a climatic drying period and short-term acute drought. But what of longer-term droughts, as might result from global climate change? Condit et al. (1995) looked at drought response of 205 tree and shrub species on BCI relative to the severe 1983 dry seson. Their conclusion is sobering: "The most important aspect of our hypothesis from the perspective of forest dynamics is that the effects of climate change have not been caused solely by the 1983 El Niño, but are due to a longer term pattern of drought" (p. 430). They conclude that drought-sensitive species will decline, suffering high mortality, should droughts become more severe and more frequent, as some models of climate change predict (see Chapter 15). This trend could reverse if wet conditions return. Clearly, climatic shifts alter forest demography and composition of the plant community.

In a more recent study (Engelbrecht et al. 2007) that examined plant distribution of 48 species of trees and shrubs on both regional and local scales and included 122 sites spanning a rainfall gradient across Panama, it was clear that differential drought sensitivity was an important variable influencing plant distributions. This suggests that changes in available moisture that may be influenced by fragmentation (see Chapter 14) and climate change (see Chapter 15) may cause significant changes in the plant community composition and diversity in the tropics.

Conclusion

The daily lives of plants ranging from newly sprouted seedlings to mature trees are subject to many vagaries, all of which combine to make tropical forests the complex mosaics that they are. Stochastic factors strongly influence whether a seedling will ever reach adulthood. Both biotic and abiotic "hazards" influence the likely success or failure of a seedling to endure and to attain adulthood. A study on BCI that looked at seedling risk showed that individual seedlings change over time in their exposure to risks from animals, disease, and litter-fall, adding even more complexity to forest dynamics (Alvarez-Clare and Kitajima 2009). From small gaps to large forest disturbances, the dynamics of tropical forests continually raise new questions for the tropical ecologist.

FURTHER READING

Gentry, A. H., ed. 1990. *Four Neotropical Rain Forests.* New Haven, CT: Yale University Press.

Leigh, E. G., Jr. 1999. *Tropical Forest Ecology: A View from Barro Colorado Island.* Oxford, UK: Oxford University Press.

Leigh, E. G., Jr., A. S. Rand, and D. M. Windsor. 1982. *The Ecology of a Tropical Forest: Seasonal Rhythms and Long-Term Changes.* Washington, DC: Smithsonian Institution Press.

McDade, L. A., K. S. Bawa, H. A. Hespendeide, and G. S. Hartshorn, eds. 1994. *La Selva: Ecology and Natural History of a Neotropical Rain Forest.* Chicago: University of Chicago Press.

7 Biotic Interactions and Coevolution in Tropical Rain Forests

Chapter Overview

The impressive species richness of tropical forests means that species are potential resources for one another, arguably to a more complex degree than is the case in areas outside the tropics. Consequently, species have evolved complex interdependencies. In some cases, competition and resource specialization lead to a process called *niche partitioning*. For many plant species, proximity to other individuals of the same species results in *negative density dependence*, when fitness is reduced as density increases. This leads to *Janzen-Connell effects* (see Chapter 4), where seeds farthest from the parent plant enjoy the highest probability of success. Some seeds are wind-dispersed, but animal dispersal is much more common. One estimate suggests that perhaps as many as 90% of tropical tree species require some sort of interaction with animals in order to reproduce (Jordano 2000). Fruit is an evolutionary adaptation to utilize animals as seed dispersers, and the ecology of animal-vectored seed dispersal is complex. Most tropical plants are also animal-pollinated. These interactions are sometimes examples of *coevolution*, a process in which two or more species exert strong selection pressures on one another, such that they evolve together. Coevolution may take the form of an "arms race" between predator and prey, or it may evolve to be a mutualistic relationship, with species locked in obligate interdependencies. Such a pattern of continuing coevolution among many species leads to what have been termed *mutualistic networks* (Bascompte 2009).

This chapter will provide examples of some intriguing forms of interactions evident in tropical ecosystems. The topics of fruiting ecology and coevolution of mutualisms will be developed in some detail. The following chapter will consider trophic interactions that characterize the complex food webs of tropical rain forest ecosystems.

Myriads of Interactions

In one Amazon rain forest, biomass of living vegetation was estimated at approximately 900 metric tons per hectare. The mass of animals—all the mammals, birds, reptiles, arthropods, and other creatures—totaled a mere 0.2 metric ton per hectare; thus the biomass of animals was only 0.02% of the plant biomass (Fittkau and Klinge 1973). Such an imbalance between plant and animal biomass is noticeable to anyone with experience in the rain forest. Green is everywhere, plants abound, but animals are far less obvious and represent dramatically less biomass.

The fact that plant biomass greatly exceeds animal biomass should not obscure a clear reality of forest ecology. Animals exert significant positive and negative influences on plants, and the complexity of interactions is a challenge for ecologists to understand. Plants also exert strong effects on animals. By some estimates, fully 50% of the world's terrestrial species inhabit the tropics. It should not be surprising that much of what fascinates us about rain forest ecology is the diverse interactions among the biota. Such interactions may be direct, subtle, obvious, or cryptic, but taken together, they form the essence of rain forest ecology. Interactions among organisms occur at time scales ranging from very brief (as when an invasive species is added to an ecosystem and begins interacting with resident species) to very long (as when one or more species evolve genetic adaptations in response to each other and become tightly locked in either an antagonistic or a mutualistic relationship). Because there are so many species in the tropics, interactions in tropical ecosystems, particularly rain forests, are especially complex.

Kinds of Interactions

An interaction between any two individual organisms may be positive, negative, or neutral for either party. Both may benefit from the interaction (+ +); one may benefit while the other loses (+ −); both may lose (− −), though to varying degrees; or one may benefit while the other experiences neither gain nor loss (+ 0).

Mutualism (+ +) occurs when two or more species engage in an interaction in which both species benefit. In evolutionary terms, the fitness of each is enhanced by the interaction. Animal-mediated pollination of plants is an outstanding example of a widespread form of mutualism. Mutualistic relationships are particularly common in tropical rain forests. Some are relatively casual, and some are obligate.

Commensalism (+ 0) occurs when one species benefits from the interaction and the other is not harmed or significantly compromised. Epiphytic plant species are generally commensal with their host plants. Normally, epiphytes do not reduce the fitness of host plants, and epiphytes obviously benefit from attachment on the host. It is true, however, that epiphytes may, on occasion, weaken a branch by their sheer weight, making it more susceptible to breakage, or they may be sufficiently dense to interfere with photosynthesis of the host tree (Plate 7-1).

There are two manifestations of interactions in which one species gains and one loses (+ −): *predation* and *parasitism*. Predators, parasites, and pathogens all gain while their prey or host species lose.

Last, there is *competition*, which occurs when two individuals both require the same resource and that resource is in limited supply. Competition may be intra- or interspecific. In general, one competitor obtains more of the resource and benefits, while the other loses. But the presence of a competitor usually results in some resource loss even to the "winner," and so competition may be viewed as a (− −) type

PLATE 7-1
Epiphytes on branch. Epiphyte load sometimes weakens the branch.

of interaction, where the minuses are not of equal magnitude such that the cost is not the same to each competitor. *Competitive exclusion*, which appears to be rare in nature, occurs only when one competitor completely excludes the other. More commonly, interspecific competition, to the extent that it occurs, results in *niche partitioning* (see the following section).

Niche Partitioning

While it has proven difficult to account for tree species richness by applying standard models of niche partitioning (see Chapter 5), such is not the case with many animal studies (see Chapter 4). Tropical moist forests and rain forests have a complex physiognomy (see Chapter 3) and thus provide many resources of potential space and food to animals. High species richness of various animal groups, particularly insects, is probably a result of the broad potential resource spectrum afforded by tropical forest structural complexity (see Chapter 4) (Plate 7-2).

Niche partitioning results from two processes that may both operate simultaneously. One is *interspecific competition*. When two or more species compete for a resource array, the force of competition is thought to result in selection acting to separate species such that they use somewhat different parts of the resource spectrum. The second process is *ecological specialization*. When a population exploits a particular resource array, selection may result in narrowly adapting the species to obtain the resource. An example is the long, sticky tongue of an anteater, which is an adaptation enabling it to capture ants and termites (Plate 7-3).

Niche partitioning is well described in numerous examples from temperate and tropical ecological studies and has led to the concept of the *ecological guild*. The term *guild* is derived from the medieval merchant and artisan guilds—groups of merchants and artisans offering similar goods or services. In ecology, *guild* refers to a group of species, not necessarily closely related, that share a common resource spectrum. An example of a guild is the bark-foraging guild, which can be observed in all tropical forests. In the Neotropics, for example, vertebrate animals including woodpeckers (Picidae), woodcreepers (Dendrocolaptidae) (Plates 7-4a and 7-4b), various other birds, and primates called marmosets are all bark foragers. Each obtains some resource by probing, drilling, or piercing bark. The various animals of the guild focus on a different aspect of bark foraging, resulting in foraging niche partitioning among the guild. At the same time, the reality that each species is dependent on bark in some way suggests that interspecific competition would be most intense

PLATE 7-2
The complex structure of a tropical forest provides ideal opportunities for niche partitioning among various animal groups.

PLATE 7-3
This tamandua (*Tamandua tetradactyla*), a Neotropical anteater, uses a specialized long tongue to extract ants and termites.

(a)

(b)

PLATE 7-4
Each of these two woodcreepers, (a) the small wedge-billed and (b) the much larger ivory-billed, is part of the diverse bark-foraging guild in Neotropical forests.

among the guild members. The guild concept has proven helpful in formulating comparisons among species, particularly regarding niche partitioning, as well as in developing models of interspecific competition and specialization.

AFRICAN RAIN FOREST SQUIRRELS

Globally, there are 22 living genera of tree squirrels, and 20 are either mostly or totally within the tropics. Louise H. Emmons (1980) performed a detailed study of the ecology of nine *sympatric* (overlapping ranges) species of tree squirrels inhabiting rain forest in Makokou, Gabon, in Africa (Plate 7-5). The study involved extensive fieldwork as well as experiments involving captive animals. Field data were collected over 31 months within a four-year period.

Nine sympatric species of squirrels represents a high species richness (*alpha diversity*), and it would be expected that there would be potential competition among the species. But at the same time, such competition, as well as other potential factors such as specialization to acquire resources, would be expected to lead to resource partitioning among the squirrel species. What was learned was that the squirrel species were indeed separated on a number of different niche dimensions, minimizing interspecific competition.

The Gabon study involved six genera that were divided into four groups based on degree of evolutionary relationship. The squirrels differed in body size

(a)

(b)

(c)

(d)

(e)

(f)

PLATE 7-5
AFRICAN TREE SQUIRRELS
(a) African giant (*Protoxerus stangeri*); (b) Lady Burton's rope (*Funisciurus isabella*); (c) red-legged sun (*Heliosciurus rufobrachium*); (d) ribboned rope (*Funisciurus lemniscatus*); (e) Thomas's rope (*Funisciurus anerythrus*); and (f) Western palm (*Epixerus ebii*).

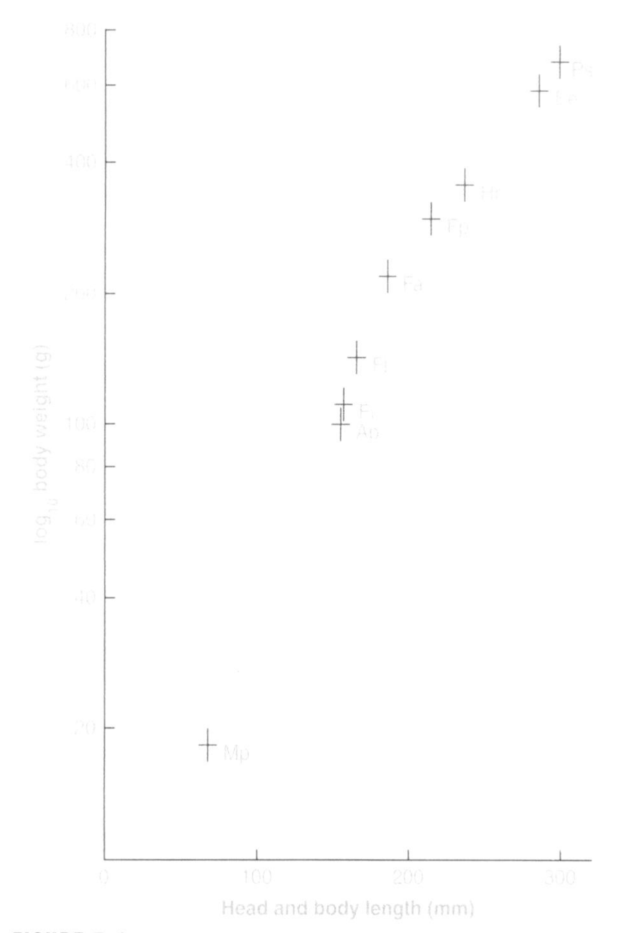

FIGURE 7-1
The size distribution of nine squirrel species from the Emmons study.

and weight, ranging from 16.5 grams to 691 grams (0.03 to 1.50 pounds) (Figure 7-1; Table 7-1). The study involved an analysis of food habits, feeding behavior, overall diet, the relationship of morphology to diet (as in characteristics of teeth, for example), food caching, temporal patterns of fruit exploitation, and seasonal variation in diet.

Two of the nine species were habitat specialists, one using dense growth at lower levels in the forest and the other using flooded forest at all vertical strata. The other seven species demonstrated niche partitioning along various niche axes within mature and disturbed forest. The results may be summarized as follows:

- The nine squirrel species differed from one another in diet because of differences of body size, habitat, and foraging height. This is a clear example of niche partitioning (Figure 7-2).
- The body size differences among the squirrel species were substantial such that two species within the same guild differed in weight by at least a factor of 1.3. This difference presumably means that each species feeds on somewhat different food resources, reducing potential interspecific competition among them.
- Few agonistic interactions were observed among the nine squirrel species, though one species, *Protoxerus stangeri*, defended trees with abundant fruit, driving away other squirrels. These trees typically contained abundant fruit, but only a few of the fruits were sufficiently ripe on any given day. The dominant squirrel had access to those fruits and could feed at an optimal time of day. This behavior is an example of both interspecific and intraspecific competition.
- All but one squirrel species were restricted in their vertical foraging range. The widest distinction was between arboreal and ground foraging species. This is another example of partitioning—in this case, along a vertical niche axis.
- Two species were in specialized habitats, but all others shared the same forests.
- The squirrels as a group exhibited a diverse diet that included hard nuts, arthropods, termites, leaves, ants, and bark scrapings. Various species focused on different foods, another example of niche partitioning. This is a much broader resource spectrum than is seen in temperate zone tree squirrels, most of which are essentially nut consumers.
- Competition for food resources was considered the main selective factor responsible for niche partitioning.
- Competition was not constant but was most severe during dry season (June to September). Large dietary shifts occurred such that squirrels became more specialized in diet during dry season compared with more generalized during rainy season.
- The greater diversity of tree squirrels in Gabon compared with typical areas in the temperate zone was explained by the following factors: mosaic nature of the environment permitting habitat specialization (two species); separation between arboreal and ground foraging groups; size differences among squirrels (small, medium, and large) at each vegetation level; use of arthropods including ants and termites and insect larva; and overall higher plant productivity, providing a broader resource base and greater food density.

TABLE 7-1 The comparative masses and body dimensions of four groups of squirrel species used in the Emmons study.

SPECIES	MASS (GRAMS)			HEAD AND BODY LENGTH (MILLIMETERS)			HINDFOOT (MILLIMETERS)			PAIRS OF MAMMAE	Pm_3[†]
	N	$\bar{x}$	SD	*N*	$\bar{x}$	SD	*N*	$\bar{x}$	SD		
Protoxerus stangeri ebivorous (Du Chaillu)	21	691	49.2	23	298	9.9	23	64.0	1.9	4	–
Epixerus ebii wilsoni (Du Chaillu)	7	592	34.1	5	284	11.2	5	68.0	2.3	4	–
Heliosciurus rufobrachium aubryi (Milne-Edwards)	37	363	24.2	17	238	8.9	17	51.0	2.9	3	–
Aethosciurus poensis (Smith)	8	100	6.9	6	152	5.3	6	33.0	2.1	3	+
Funisciurus pyrrhopus pyrrhopus (F. Cuvier)*	15	297	29.4	11	211	10.4	11	46.0	1.6	2	+
Funisciurus anerythrus mystax De Winton	15	218	20.5	20	182	8.9	11	44.0	1.1	2	+
Funisciurus lemniscatus lemniscatus (Leconte)	46	141	9.4	23	162	6.2	23	38.2	1.9	2	+
Funisciurus isabella (Gray)	17	110	11.9	23	154	6.4	22	36.0	1.3	2	+
Myosciurus pumilio (Leconte)*	6	16.5	3.6	6	66	5.0	6	18.8	0.7	2	–

*Includes measurements taken by G. Dubost from specimens collected in the same region.
†Pm_3: third premolar; + = present; – = absent.

SPECIES	HABITAT TYPE	VEGETATION HEIGHT	BODY SIZE	FOOD TYPE	ACTIVE PERIOD
M. pumilio			Tiny	Bark scrapings	Long
A. poensis		Arboreal	Small	Some diverse arthropods	Long
H. rufobrachium			Medium		Short
P. stangeri	Mature and disturbed forest		Large	Hard nuts	
F. lemniscatus			Small	Many termites	Long
F. pyrrhopus		Ground foraging	Medium		Short
E. ebii			Large	Hard nuts, few arthropods	
F. isabella	Dense growth	Lower levels	Small	Leaves, diverse arthropods	Long
F. anerythrus	Flooded forest	All levels	Medium	Many ants	Long

FIGURE 7-2
A summary of the niche partitioning of the African squirrels.

A GLOBAL COMPARISON OF PRIMATE NICHE SEPARATION

The Gabon study demonstrates how close analysis of various potential parameters of niche separation within a region accounts for coexistence among a guild of potentially competing species. It is also possible to perform a more-coarse-grain study comparing various wide-scale parameters that factor into niche partitioning. Such a study was done for primates, a pantropical mammalian order (Schreier et al. 2009).

Nonhuman primates are almost entirely tropical in distribution, and up to 10 sympatric species are present at some sites (Plate 7-6). Many have similar dietary requirements. Most are *frugivores* (fruit consumers) to varying degrees, and some are *folivores* (leaf consumers). Many have a generalized diet including leaves, fruits, and some animal food. Carnivory among primates occurs (chimpanzees, *Pan troglodytes*, have been observed pursuing, killing, and devouring monkeys), but not commonly. Small primate species are known to eat arthropods. What factors act to allow niche partitioning and coexistence among sympatric primate species, and do these factors weigh equally among various biogeographic areas of the tropics?

The study by Schreier et al. (2009) was *coarse-grained*, meaning that it compared species based on relatively few niche parameters. (The preceding study by Emmons was much more fine-grained.) Not only that, but it was restricted to *pair-wise comparisons*. That means that sympatric species were assessed as potential competitors based on a 30% or less biomass difference. This sometimes meant that two *congeneric* (species within the same genus) species were compared, but the study often involved two species of different genera. This methodology limits the extent to which potential interspecific competition is detectable. For example, a large ape such as a 35-kilogram orangutan (*Pongo*) (Plate 7-7) is not compared with a 3.5-kilogram monkey such as a macaque (*Macaca*), even though both might be competitors for fruits such as figs (*Ficus*). Allowing for these limitations, the study revealed some pattern differences in niche separation in various regions of the global tropics.

The researchers based their analysis on published studies from 43 sites (Figure 7-3). The parameters examined were:

- Forest type (open country, scrub, gallery forest, swamp forest, canopy forest, deciduous forest, montane forest)
- Forest location (edge versus core forest and primary versus secondary forest)
- Height in forest (different vertical strata within forest)
- Specific aspects of diet
- Prey capture methods (for those whose diets included arthropods)
- Ranging (distance traveled while foraging)
- Timing of activity (nocturnal, crepuscular, diurnal)
- Support diameters (small branches, trunk, large branches)

PLATE 7-6
This bald uakari monkey (*Cacajao calvus*) from South America is a specialist species, inhabiting seasonally flooded forests in the Amazon Basin.

PLATE 7-7
The orangutan (*Pongo pygmaeus*) is the largest of the Asian primates but nonetheless likely competes with smaller species for access to fruit, especially figs.

The analysis detected differences among global regions. Because comparisons were restricted to pair-wise comparisons based only on body weight similarity, the analysis was conservative in suggesting that only 7% of the primate species at the sites studied were putative competitors. The likelihood of species of greater body weight differences being potential competitors nonetheless remains strong.

Asian forests had the most potential competitors, with 17% of species pairs assumed to compete, compared with less than 5% for all other regions. Asia has many primate species, and their body mass differences are narrower than those of primate communities in other regions. Perhaps this is because of an abundance of resources, allowing minimal competition among similar species, but this has not been demonstrated. Perhaps they are more ecologically specialized and partition resources more finely than primates elsewhere. Another possibility is that competition among primates in Asia may be strong and they are evolving niche partitioning. This study points to those questions but does not provide data on any of them.

In Africa and Asia, the use of different tree strata was the primary mode of niche separation, but this was closely followed by dietary separation. In Madagascar, where lemurs prevail, differences in diet

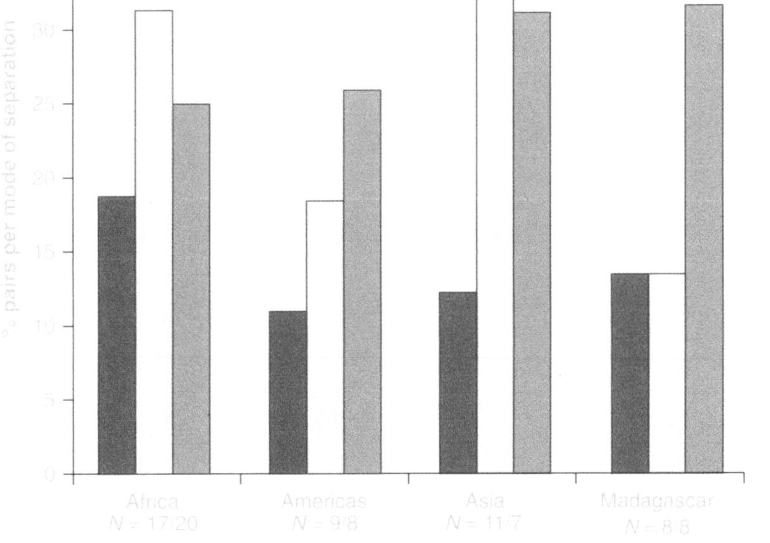

FIGURE 7-3
Comparison of continents for the three most common modes of niche separation, presented as percent of pairs of competing species (i.e., same body mass and same diet) that show the stated mode (more than one mode/pair possible). N = number of pairs of competing species/number of forests.

Competitive Irrelevance?

Animal ecologists have little difficulty inferring potential effects of interspecific competition within a guild of animals by use of niche partitioning analysis (though this is an inferential measure and not a clear demonstration that competition does or did exist). Plant ecologists also infer interspecific competition, particularly in examples such as secondary succession and the development of second-growth forest (see Chapter 6). Studies have clearly demonstrated negative shading effects of adult trees on seedlings and saplings of various species, and other studies have demonstrated that seedlings compete with root systems of adult trees. But that said, is it possible for competition among plants to sometimes be "irrelevant"?

An experimental study performed at Barro Colorado Island (BCI), Panama, and Cocha Cashu Biological Station in Peru focused on seedlings of three plant species (Paine et al. 2008). Researchers attempted to determine whether seedling competition was occurring. They manipulated the plots by randomly thinning plants in what were initially high-density plots. Plots were established with densities of 15, 30, and 45 to 60 individuals per square meter. In all plots, the survivorship declined over the 24-month period of the study. But there was no difference in survivorship based on plot density. In other words, there was not a shred of evidence for competition among the plants within the plots. Presumably, if the plants were competing, those in low-density plots would have grown more and survived better than those in medium- and high-density plots, but that did not happen. The researchers called this result *competitive irrelevance*. They suggested that competition may be strong at other phases of the life cycle of plants but not at the seedling stage, at least not in this study. They hypothesized that "if competition among seedlings were absent or weak, other processes would by necessity prevail as those structuring the forest understory, such as neutral dynamics, size-asymmetric competitive hierarchies, fundamental niche requirements, or interactions with consumers" (p. 439).

and activity periods formed the strongest basis for niche separation. In the Neotropics, diet and vertical foraging height were the most important parameters of niche separation (Plate 7-6).

The study did not consider seasonality. Recall how the dry season affected the niches of the African squirrel community, described in the previous example. This study also did not include such variables as resistance to plant defenses, an important potential selection pressure (see Chapter 8). And last, the study did not include potential competitors from other taxa such as marsupials in the Americas. But within these limits, the study does demonstrate real differences in primate niche separation among the various biogeographic regions.

Negative Density Dependence Revisited

Dipterocarps are among the most-species-rich trees of Southeast Asian tropical forests (see Chapter 8). One species, *Shorea quadrinervis*, was the focus of a study that revealed what is likely a significant characteristic of rain forest population dynamics among tree species (Blundell and Peart 2004).

Recall that negative density dependence was discussed in Chapter 5. Negative density dependence occurs when fitness of individuals is reduced at higher densities. If such a negative feedback is characteristic of rain forest tree species, it would be a potential stabilizing force that would prevent particular species from becoming inordinately dominant.

S. quadrinervis is considered a relatively dominant species of dipterocarp in Gunung Palung National Park, Indonesian Borneo. Using a natural distribution of adult trees, researchers studied the growth and survivorship of juvenile trees to ascertain whether negative density dependence was evident.

A total of 357 adult trees (equal to or less than 15 centimeters diameter breast height [DBH]) and 5,215 juvenile trees were mapped in eight areas with high adult densities (HAD areas) and eight areas with low adult densities (LAD areas). The growth and survival of juveniles was followed over a two-year period.

The results of the study were clear (Table 7-2):

- Juveniles in HAD areas were shorter and had fewer leaves than those in LAD areas (Figure 7-4).
- Juveniles survived significantly better in areas with lower adult densities (of conspecifics).
- Juveniles in HAD areas grew significantly less than those in LAD areas.
- HAD areas produced more seedings than LAD areas; thus seedling recruitment was positively related to adult density.

TABLE 7-2 Comparison of the condition of *Shorea quadrinervis* (Dipterocarpaceae) juveniles (less than or equal to 1 centimeter DBH) in areas of high and low density of conspecific adults at Gunung Palung National Park Kalimantan Barat (Indonesian West Borneo).

	HIGH			LOW				
CHARACTERISTIC	MEAN ± 1 SE	MEDIAN	UPPER QUARTILE	MEAN ± 1 SE	MEDIAN	UPPER QUARTILE	*F*	*P*
1995								
Height	34.0 ± 2.6	27	35	42.0 ± 2.7	31	46	8.6	0.011
No. leaves	12.8 ± 1.7	9	15	17.6 ± 1.8	12	21	12.2	0.001
+17 months								
Height	37.9 ± 2.8	29	40	47.0 ± 3.0	35	53	13	0.003
No. leaves	13.4 ± 1.9	9	15	18.5 ± 2.2	12	21	9.4	0.009
+24 months								
Height	51.4 ± 1.6	33	51	59.1 ± 1.4	39	65	19.8	<0.0001
No. leaves	14.9 ± 0.5	9	17	17.0 ± 0.5	11	20	7.8	<0.0001

Notes: Indices of juvenile condition (mean ±1 SE) were height to apical meristem (cm) and number of extant leaves. Juveniles were measured in October 1995, then 17 and 24 months later. *F* values are from the main effect of adult density in nested ANOVAs. Sample sizes: $n = 8$ (high-density areas); $n = 8$ (low-density areas at +17 months); $n = 6$ (low-density areas at +24 months).

- Though areas near adult trees recruited more seedlings, growth and survivorship among juvenile trees was greatest in LAD areas; thus growth and survivorship is negatively related to adult density.
- Juveniles in HAD areas showed more damage to meristems and greater tissue loss from herbivores than those in LAD areas. Thus it appears that greater concentrations of trees of the same species are subject to higher levels of attack by natural enemies.
- Results imply that HAD areas will recruit fewer adults than areas of LAD; thus *S. quadrinervis* will be limited in its capacity to attain greater dominance.

The negative density dependence observed in this study, like that seen in other rain forest studies, is believed to have been the result of natural enemies acting where plants of a particular species are most clustered. If

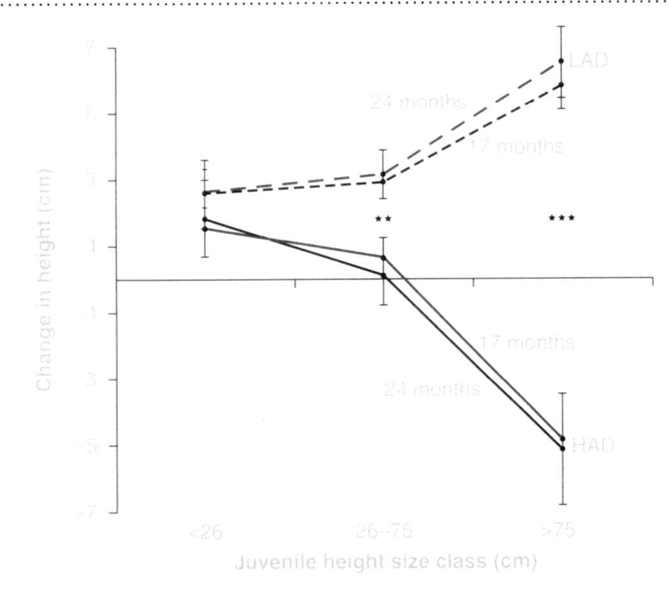

FIGURE 7-4
Effects of the density of conspecific adults (high versus low, HAD and LAD, respectively) and juvenile size on the growth of *Shorea quadrinervis* (Dipterocarpaceae) juveniles (≥1 centimeter DBH) at Gunung Palung National Park, Kalimantan Barat (Indonesian West Borneo) (means ± 1 SE). Growth (change in height to apical meristem) was measured between October 1995 and March 1997 (17 months, thin lines; 0.5-hectare areas of high [HAD, solid line; $n = 8$] and low [LAD, dashed line; $n = 8$] density of conspecific adults) and during the El Niño of March to October 1997 (thick lines; high-density areas [solid line; $n = 8$] and low-density areas [dashed line; $n = 6$]). Growth increased significantly with juvenile height in low-density areas but not in high-density areas. Asterisks refer to significant differences between high and low adult density (Tukey-Kramer hsd test): $^{**}P < 0.01$; $^{***}P < 0.001$.

negative density dependence caused by natural enemies is a general characteristic of tropical moist forest tree species, trophic interactions (see Chapter 8) are potentially responsible for maintaining high tree species richness.

Seed Shadows and Animal Dispersal

Seeds fall most densely near the parent tree. Seeds may be wind- or animal-dispersed, but in either case, the majority of seeds will accumulate beneath the parent tree. Even the most casual observation of seed dispersal in tropical forests reveals this pattern. Because of negative density dependence, seed dispersal, either by wind or by animal vectors, becomes an essential fitness characteristic of tropical trees. Somehow, some way, seeds must be dispersed at sufficient distances to permit eventual recruitment to adult size.

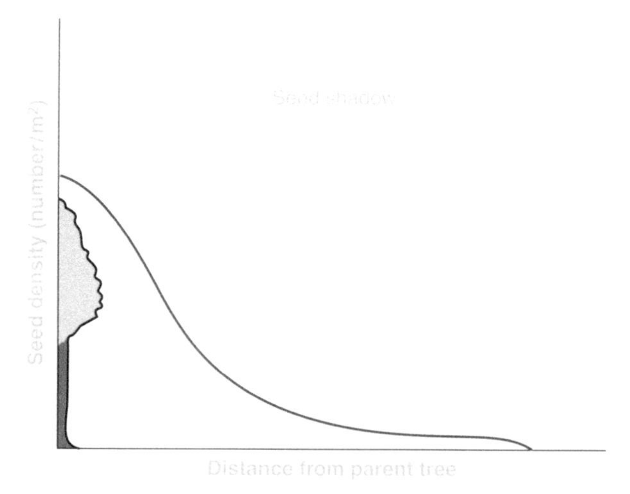

FIGURE 7-5
The seed shadow effect.

If seed distribution is examined at distances progressively away from the parent tree, a *seed shadow effect* is evident (Janzen 1970). Seeds are most dense near the parent tree and become much less dense the farther they are from the parent tree (Figure 7-5). The curve of density drops off sharply with distance from the parent tree. Proportionally few seeds make it far from the parent tree, but more distant seeds have greater likelihood of success. Seed shadows and seed dispersal are important, as they affect tree reproduction and recruitment not only in tropical forests but in all forests, including those in the temperate zone (Clark et al. 1999).

A study performed in Cameroon showed that differences exist in patterns of seed shadows from plants that are wind-dispersed compared with those dispersed by animals—in this case, large birds and monkeys (Clark et al. 2005). The Cameroon study showed that animal-dispersed tree species had longer mean dispersal distances than wind-dispersed trees but lower fecundities (Figure 7-6). This is viewed as an evolutionary

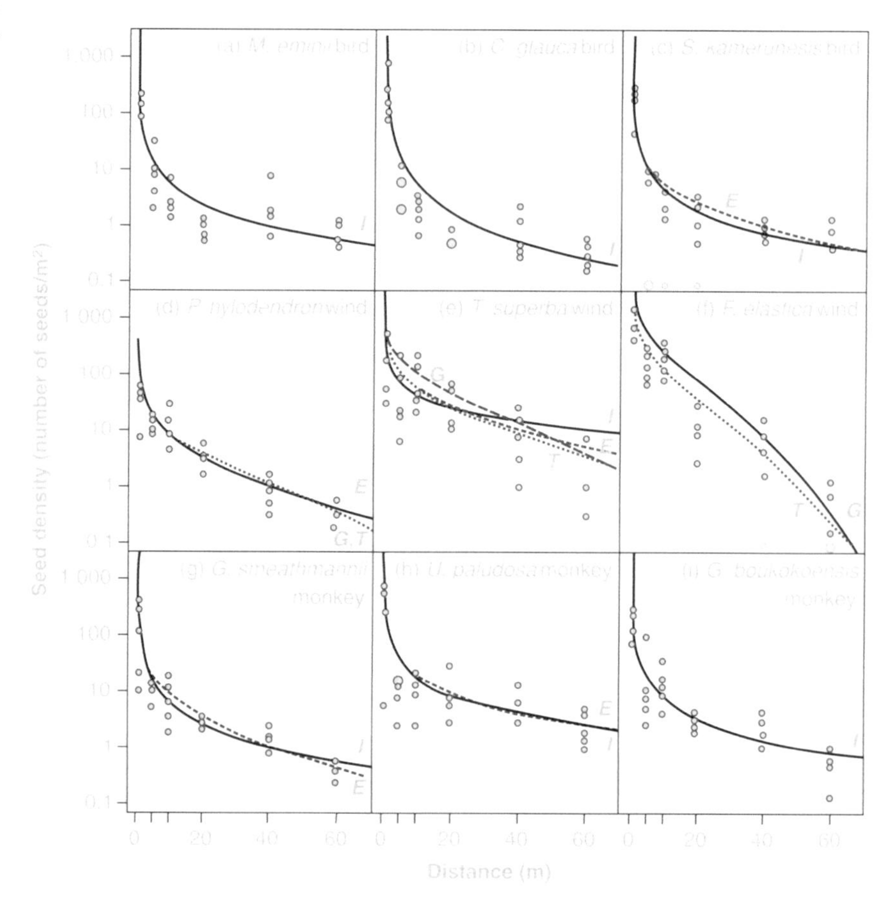

FIGURE 7-6
Seed shadows compared for various plant species used in the Clark et al. study. The curves are calculated using different statistical metrics (*G* is Gaussian; *I* is inverse power; *E* is negative exponential; *T* is Student *t*).

trade-off of sorts, in that animal-dispersed seeds are metabolically more expensive to make but animals move seeds greater distances than wind moves them, and thus each seed that attains such dispersal has a higher probability of success. Another constraint to wind dispersal is *wing-loading*, a concept familiar in aerodynamics. The heavier an object is, the greater the wing area needed to permit flight. (Look at the size of the wing on a Boeing 747, for example.) Therefore, wind-dispersed small seeds cost less per seed than large seeds, since the required amount of "seed wing area" is less. Plants with large seeds are under a selection pressure for animal dispersal, so it is no surprise that animal dispersal is an important factor in long-distance dispersal of most rain forest tree species (Holbrook and Smith 2000; Poulsen et al. 2001). Long-distance dispersal (greater than 60 meters) is uncommon to rare within any particular cohort of seeds. In other words, most seeds do not get dispersed very far. But some do, and long-distance dispersal is essential in that those seeds so dispersed have greater chances of germination, growth, and ultimately survivorship.

Because of the importance of seed dispersal, fruit has evolved in numerous species of plants. Because tropical climates are equitable, fruit is produced year-round in the tropics. Because it is nutritious, fruit serves to entice animals. The question remains, to what degree do the animals subsequently act as effective seed dispersers? What evolutionary costs and benefits are associated with producing fruit?

Howler Monkeys, Insect Larvae, and Seed Dispersal: Situation-Dependent Dispersal

Howler monkeys are both *folivores* (leaf consumers) and *frugivores* (fruit consumers) (Plate 7-8). A study that focused on black howler monkeys (*Alouatta caraya*) in forest in northern Argentina uncovered complexities in seed dispersal performed by the monkeys (Bravo 2008). Black howler monkeys are seed dispersers in flooded forests in the region of the study. Three plant species were studied, each of which produces seeds subject to infestation by insect larvae:

- Seeds from *Ocotea diospyrifolia* were present in monkey feces. Many of the seeds contained dead insect larvae, but most germinated. Only those where the larvae had reached an advanced stage of development failed to germinate.
- Seeds from fruit pulp of *Eugenia punicifolia* contained larvae that were killed by passage through the monkeys' digestive system, but only early in the fruiting season. The larvae remained alive (continuing to devour the seeds) later in the fruiting season, even after passage through the monkeys' guts.
- Fruits of *Banara arguta* contained larvae-infested and noninfested seeds. Infested seeds were destroyed by passage through the monkeys' intestinal system, but noninfested seeds were dispersed by the monkeys.

The study concluded that ingestion of larvae-infested fruits by howler monkeys may result in (1) dispersion of healthy seeds and killing insect larvae, (2) spread of the larvae, or (3) destruction of infested seeds. It all depends on which seeds and what larvae, and in what stage of development the larvae happen to be.

PLATE 7-8
This mantled howler monkey (*Alouatta palliata*) lounging on a branch is a different species from the one used in the study of black howler monkeys, but both are very similar.

The Ecology of Seed Dispersal and the Importance of Fruit in the Tropics

Fruit is both abundant and (relatively) constantly available throughout the year in tropical forests, making it an important resource for birds, mammals, certain reptiles, many fish, and all manner of arthropods. Fruiting trees are known to attract numerous animals of many species. Many are essential to dispersing seeds (Plate 7-9). Some of these are frugivores, creatures whose diet includes more than 50% fruit (Levey et al. 1994).

In the temperate zone, fruit is a seasonal resource, occurring from midsummer through autumn, with some fruits lingering through winter. Many birds migrating to winter in the tropics feed heavily on fruit in the fall, but because fruit is ephemeral in the temperate zone, no bird families have specialized as frugivores (Stiles 1980 and 1984). In the tropics, entire families of birds such as the manakins, cotingas, birds-of-paradise, bowerbirds, toucans, hornbills, parrots, figbirds, bulbuls, doves and pigeons, and tanagers depend heavily on fruit, and some species are almost exclusively frugivores (Snow 1976; Moermond and Denslow 1985). Large ground-dwelling birds such as cassowaries, pheasants, and tinamous are important consumers of fruits. In addition, numerous mammal species (many rodents, pigs, peccaries, tapirs, elephants, deer, and primates) utilize fruit as a major component of their diets. Fruits provide a calorie-rich, nontoxic, and easily acquired resource. But there are downsides to a diet of fruit. Protein is usually lacking, and thus an all-fruit diet, while rich in calories, is typically nutritionally deficient and needs to be augmented in some way by animal food. Also, fruiting patterns vary, often significantly, both in time and space. In other words, two fruiting fig (*Ficus*) trees may be widely separated, necessitating extensive searching by frugivorous animals. For example, the great green macaw (*Ara ambigua*), a large parrot, must make extensive movements throughout its range in Costa Rica, searching for satisfactory fruiting plants (Loiselle and Blake 1992). Seasonal changes in abundance of most fruits occur throughout the tropics (Fleming et al. 1987). Some montane frugivores must undergo regular seasonal migration to lower elevations in search of favored fruits (Wheelwright 1983; Loiselle and Blake 1992).

The evolutionary function of fruit is to advertise itself to some sort of animal so that it will be eaten. The seed(s) contained within the fruit then passes through the alimentary system of the animal (or is regurgitated), and because the animal is mobile, seeds are deposited away from the parent plant (Plate 7-10). Fruits evolved as adaptations for seed dispersal. The animal derives nutrition from the fruit but also disperses seeds; thus, the

PLATE 7-9
The unique ocellated turkey (*Meleagris ocellata*) is one of the larger seed dispersers in Central American humid forests.

(a)

(b)

PLATE 7-10
(a) This African hornbill and (b) this Neotropical tapir are both consumers of fruits and act as seed dispersers.

relationship between animal and plant is fundamentally mutualistic—the animal is bribed in return for its mobility (but see Wheelwright and Orians 1982), a relationship not unlike animal-vectored pollination. It is to the plant's ultimate advantage to invest energy to make fruit as well as to the animal's immediate advantage to eat the fruit. Some parrot and pigeon species, however, digest the seed as well as the pulp of the fruit (or else they injure the seed, and it does not germinate). These seed consumers are not useful as dispersers and are considered *seed predators* rather than mutualists (Plate 7-11).

Not all tropical plants produce expensive animal-dispersed fruit. Some rain forest canopy trees, vines, and epiphytes utilize wind or water dispersal of seeds (Kubitzki 1985). Wind dispersal is most common at the canopy level or in deciduous forests, where leaf drop can help facilitate wind movement of seeds (Janzen 1983). A study in Costa Rica surveyed 105 tree species in a deciduous forest and found that 31% were wind-dispersed (Baker and Frankie, cited in Janzen 1983). In dense interior rain forests, animals are far more important than wind for seed dispersal.

PLATE 7-11
This pair of Brazilian hyacinth macaws (*Anodorhynchus hyacinthinus*) crush and eat palm nuts and, like many parrots, are seed predators.

The High Cost of Fig Dispersal

Fig (*Ficus*) trees are pantropical, and wherever they occur, they attract numerous frugivores, ranging from apes to birds. One study from farmland in Kenya documented 70 bird species utilizing 27 fig trees (Eshiamwata et al. 2006). Precise measurements of fruit volume and seed dispersal were reported from a single fig tree in a lowland deciduous forest in Costa Rica (Jordano 1983). The estimated total crop was approximately 100,000 figs, all of which, because they were produced synchronously, were taken within five days either directly from the plant or after they had fallen to the ground. During the first three days alone, 95,000 were consumed. Birds were the principal feeders, eating 20,828 figs each day, about 65% of the daily loss. Mammals, many of which ate fruits falling to the ground, were the other source of loss. Parrots, which are fundamentally seed predators, accounted for just over 50% of the daily total of figs. The most efficient seed dispersers (those that flew away from the tree and therefore dropped seeds outside of the seed shadow) were orioles, tanagers, trogons, and certain flycatchers (Plate 7-12). However, these birds took only about 4,600 figs per day. Approximately 4,420,000 seeds were destroyed each day, mostly from predation by wasps and other invertebrates. Parrots were estimated to account for 36% of the seed loss. Only 6.3% of the seeds taken from the tree each day were actually dispersed and undamaged, indicating the high cost of seed dispersal.

PLATE 7-12
The slaty-tailed trogon (*Trogon messina*) is an important seed disperser in Neotropical forests.

HOW FRUIT INFLUENCES EVOLUTIONARY PATTERNS IN BIRDS

Ornithologists David Snow (1976) and Eugene Morton (1973) have described the evolutionary consequences to birds of a diet mainly of fruit. Fruit typically is temporally and spatially a patchy resource, meaning that it may be abundant on a given tree but trees laden with mature edible fruits may be widely spaced in the forest. For much of the year, a fruiting tree may be barren of fruit. Such a resource distribution selects for social behavior rather than individual territoriality. Flocks of avian frugivores can locate fruiting trees more efficiently than solitary birds, and there is little disadvantage to being part of a flock once the fruit is located since there is usually more than enough fruit for each individual. Even if not, it is extremely difficult for one individual to defend the resource, excluding all others. Therefore, it is not surprising (depending on where in the global tropics you

How Fruit Promotes Sociality in Purple-Throated Fruit-Crows

The purple-throated fruit-crow (*Querula purpurata*) feeds on both insects and fruits, but its frugivorous habits may have been instrumental in the evolution of its social behavior (Snow 1971). Not a true crow, this species is a member of the highly frugivorous Neotropical cotinga family (Plate 7-13). Purple-throated fruit-crows live in small, closely related communal groups of three or four individuals that roam the forest together in search of preferred species of fruiting trees. Within the social group, there is virtually no aggression, and all members of the group appear to cooperate in feeding the nestling bird. (They have clutches of only one.) Given the degree of relationship among the group, this form of cooperative nesting behavior has the potential to benefit even those individuals that are not parents of the nestling bird. The nest is in the open and is vigorously defended by the entire group.

The group organization of purple-throated fruit-crows likely assists them in both locating and defending trees laden with fruit. Because the fruit that the birds locate is not limited, there is little or no competition among the birds, a factor that may have been important in the initial evolution of their social organization.

PLATE 7-13
PURPLE-THROATED FRUIT-CROW

are) to encounter flocks of figbirds, bulbuls, tanagers, parrots, or toucans. Fruit-eating mammals also tend toward sociality. Pacas, coatimundis, peccaries, pigs, and elephants are organized in bands and herds.

Because fruits are generally easy to locate and require virtually no "capturing" time and effort, frugivorous birds tend to have much "free time." The male bearded bellbird (*Procnias averano*), a South American species that is entirely frugivorous, spends an average of 87% of its time on a perch in the forest subcanopy or canopy calling females to mate (see the section "Are Fruits the Evolutionary Stimulus for Sexual Selection" later in this chapter). Another South American frugivorous bird, the male white-bearded manakin (*Manacus manacus*), spends 90% of its day courting females! This pattern is profound in evolutionary terms because it strongly promotes sexual selection among various groups of conspecific males in tropical groups such as birds-of-paradise, bowerbirds, cotingas, and manakins (Plate 7-14).

(a)

(b)

PLATE 7-14
(a) Bearded bellbird. (b) Like the white-bearded manakin, the male red-capped manakin (*Pipra mentalis*) spends the majority of its time courting females.

THE ABUNDANCE OF FRUGIVORES

Frugivorous birds are more abundant per species compared with insectivorous species, again because fruit biomass may be high, therefore supporting high populations of frugivorous species. Insects, on the other hand, are widely dispersed, are often difficult to find and capture, and represent far less overall biomass. In one area of rain forest in Trinidad, Snow (1976) netted 471 golden-headed manakins (*Pipra erythrocephala*) and 246 white-bearded manakins, a total of 717 birds. In this same area, he caught 11 species of tyrant flycatchers (most of which are strongly if not entirely insectivorous), but their combined abundance did not equal that of the two frugivorous manakins. Flycatchers must focus on a more narrow resource spectrum that requires them to search, locate, and capture their prey.

A study of the ochre-bellied flycatcher (*Pipromorpha oleaginea*) (Plate 7-15) on Trinidad showed that the species is undergoing an evolutionary diet shift (Snow and Snow 1979). Though it is a tyrant flycatcher, it now feeds almost exclusively on fruit. The ochre-bellied flycatcher is the most abundant forest flycatcher in Trinidad, and its numerical success was attributed to its diet of fruit. Snow and Snow hypothesized that as the resource base for the ochre-bellied flycatcher expanded, its population increased.

In Belize, my colleague William E. Davis Jr. and I found the ochre-bellied flycatcher to be by far the most abundant of the 23 flycatcher species we encountered during a mark-and-release study. We netted 102 ochre-bellies compared with 14 sulphur-rumped

PLATE 7-15
OCHRE-BELLIED FLYCATCHER

PLATE 7-16
SULPHUR-RUMPED FLYCATCHER

flycatchers (*Myiobius sulphureipygius*) (Plate 7-16), the next most frequently netted flycatcher species.

Studies throughout the tropics confirm the high biomass of frugivorous bird and mammal species in lowland humid forest (Fleming et al. 1987; Gentry 1990). This is not surprising given that between 50% and 90% of tropical trees and shrubs (depending on site) have seeds primarily dispersed by frugivorous vertebrates (Fleming et al. 1987). Frugivores are less species-rich than are insectivores, but estimates are that between 80 and 100 species of primarily frugivorous primates, bats, and birds typically occur in forest sites ranging from Central America to Amazonia (Fleming et al. 1987). A survey of global tropical frugivore richness would produce similar estimates. For example, Paleotropical flying foxes (which are large bats all within the suborder Megachiroptera) have 173 species within 42 genera distributed from Africa to the Australasian tropics (Plate 7-17). All of these species are frugivores and are important seed dispersers of various plant species.

A diet of fruit is not without potential problems. Interspecific competition may occur due to the localized nature of the resource attracting many potential feeders (Howe and Estabrook 1977). Nutritional balance may be lacking (see the next section, "The

PLATE 7-17
A roost of Australian flying fox bats.

Oilbird—A Unique Frugivore"). Seeds are usually indigestible, and many fruits tend to be watery, containing little protein compared with carbohydrate (Moermond and Denslow 1985). Small birds tend to eat small, carbohydrate-rich fruit, and many diversify their diets to include arthropods. Large frugivores, such as toucans, eat many different-sized fruits, including those rich in oil and fat (Moermond and Denslow 1985). Many of these species regularly supplement their diets to include animal food.

THE OILBIRD—A UNIQUE FRUGIVORE

The oilbird, *Steatornis caripensis*, often called *guacharo*, is the only species in the family Steatornithidae (Plate 7-18). It ranges from Trinidad and northern South America to Bolivia. The oilbird is a large nocturnal bird, its body measuring 46 centimeters in length, its wingspread nearly a meter. Its owl-like plumage is soft brown with black barring and scattered white spots. Its broad head features a large hooked bill and bulging, wide eyes. Oilbirds are fascinating enough as individuals, but they come in groups. Colonies are widely scattered throughout the range of the species, as the birds live in caves, venturing out only at night to feed on the fat-rich fruits of palms as well as plants belonging to the Lauraceae, often obtained only after flying long distances from the cave (Snow 1961, 1962, and 1976; Roca 1994). Fruits are plucked on the wing as the birds hover at the trees, picking off fruits with their sharply hooked beaks (Plate 7-19). Oilbirds probably locate palms by their distinct silhouettes, but they are thought to find Lauraceae through odor. Fruits from some of these trees are aromatic.

Enter an oilbird cave and be greeted by a cacophony of sound, a chaotic chorus of growls and screams. In the dark, dank cave, the flapping wings of the disturbed, protesting host conjures up thoughts of tropical demons awakened. Soon, however, the birds calm and flutter restlessly back to their nesting ledges, snarling as they resettle. Among the din, you hear some odd clicking noises. These vocalizations are one of the reasons why oilbirds are unique. They are among the very few birds capable of echolocation, the same technique by which most bats find their way in the dark. The clicks are their sonar signals, sent out to bounce off the dark cave walls and direct the birds' flight. Oilbird echolocation lacks the sophistication found in bats, but it serves adequately to keep the birds from crashing against the cave walls as they fly in total darkness.

Oilbirds are closely related to the diverse family of nightjars (Caprimulgidae), of which the whip-poor-will (*Caprimulgus vociferous*) is a common example. However, nightjars differ from oilbirds in that are all insectivorous, none are colonial, nor do any live in caves. How did the oilbird evolve frugivory, sociality, and its cave-dwelling habit?

David Snow (1976) has suggested a scenario for oilbird evolution that begins with a critical diet shift from insects to fruit (Figure 7-7). Snow hypothesizes that oilbirds were originally "normal" nightjars feeding on insects. However, large fruits offer a potentially exploitable high-calorie resource, especially to a night bird, since there are few large bats in the Neotropics. That is not the case in the Paleotropics, where many species of megachiropteran (flying fox) bats feed on large fruits. In the Neotropics, many microchiropteran bats are frugivores, but none are comparable in size to an oilbird, so they cannot eat the fruits that oilbirds devour. Oilbirds thus became oilbirds when their ancestors shifted to a diet of fruit.

PLATE 7-18
Oilbird cave in Trinidad.
Note the germinated spindly plants in the lower-left corner. They sprouted from seeds dropped by the oilbirds.

PLATE 7-19
Oilbirds use their hooked beaks to pluck fruits as they hover.

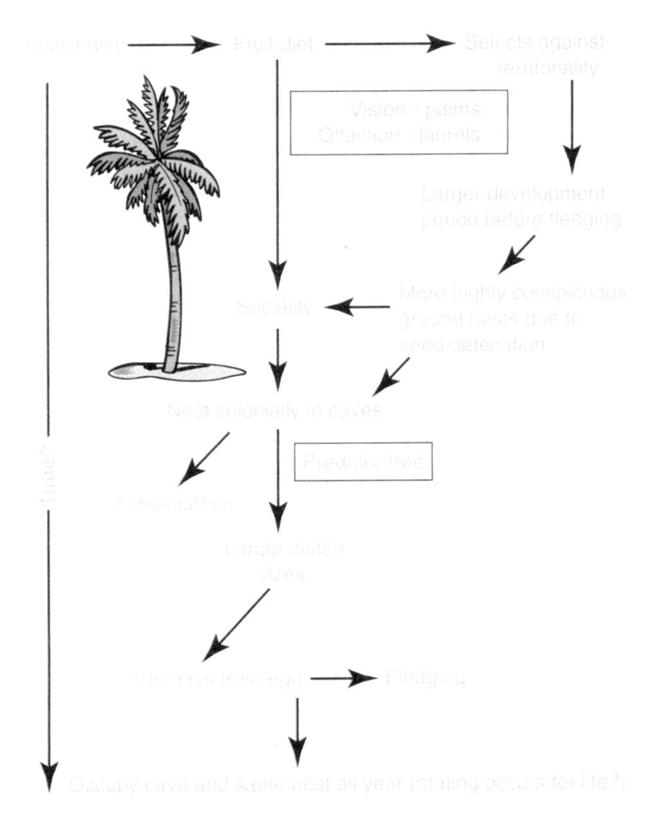

FIGURE 7-7
This diagram shows the hypothesized evolution of oilbirds.

Frugivory initiated a cascade of adaptations that profoundly affected the evolution of oilbirds. First, the birds' olfactory sense became enhanced, as it provided an advantage in locating aromatic fruits. Second, social behavior began to evolve because fruits are a patchy resource and birds would tend to come together at fruiting trees. More important, however, is that a diet of fruit, though rich in calories, is nutritionally unbalanced, and consequently, nesting time is prolonged. Incubation of each egg lasts just over one month. Once oilbirds hatch, they fatten up immensely in the nest due to a buildup of fat from the oily fruits, though they take a very long time to acquire sufficient and proper protein to grow bones, nerve, and muscle. By the time a juvenile is two months after hatching, it has still not left its nest but may weigh 1.5 times what either of its parents weighs. The name *oilbird* refers to the fact that juveniles put on so much fat that they can be boiled down to render the oil. Indigenous people also occasionally use them for torches, since the birds burn so well! The total time it takes an oilbird to go from newly hatched egg to fledging and independence is nearly 100 days, compared with about 30 days for insectivorous nightjar reproduction. It requires six months for a clutch of four oilbird eggs to develop. Such an extended development time makes it risky (even absurd) to nest on the forest floor, the traditional nightjar nesting site. Snow notes that the defecated seeds that would surround a ground nest would serve to bring attention to nesting birds. Cave dwelling offers much more protection for the nest, but caves are also very patchily distributed resources. Once oilbirds became colonial cave dwellers, it led to their being a social species.

Cave dwelling selected for the development of echolocation and also for a larger clutch size, one more typical of birds in the temperate zone. Most tropical birds lay only one to two eggs in a given nest, but birds in the temperate zone have clutches that often are four to five eggs or more. Predator pressure is likely to be a major reason for the small clutch sizes in the tropics, since nests can be more secretive if there are fewer mouths to feed. Given the safety of caves, however, oilbird nests are not under severe predator risk, and oilbird clutch size is normally four eggs. The nest is built up with droppings from the birds and located on a cave ledge. Thin, yellowish, light-starved seedlings sprout from defecated seeds around the nest. Birds are thought to pair for life.

Research using mitochondrial DNA (mtDNA) has suggested that the oilbird is an ancient species, which actually originated in North America perhaps as long as 50 million years ago (Gutierrez 1994), when North America was far more tropical than is the case today. Fruit would have been readily available throughout the year. Oilbirds may have invaded the present range by crossing the Pleistocene land bridge during the recent glacial periods. Such a migration could have seriously reduced the population, creating what evolutionists call a *genetic bottleneck*, where surviving individuals bear a strong genetic similarity, one to another. Oilbird mtDNA shows little variation among widely scattered colonies, evidence that the separated populations are not genetically distinct. Such similarity may also be due to young birds dispersing among colonies, promoting high gene flow (Gutierrez 1994).

The oilbird is an important seed-dispersing species. In an exhaustive study centered in Cueva del Guacharo (Guacharo Cave) near the town of Caripe, Monagas, Venezuela, a mountainous, heavily forested region in northeastern Venezuela, Roca (1994), using radio-tagged oilbirds, demonstrated that oilbirds have home ranges that encompass up to 96.3 square kilometers and may have to fly up to 150 kilometers (90 miles) between feeding sites. Indeed, dispersing individuals fly even farther in search of food, up to 240 kilometers (145 miles) in a single night. Given that an adult oilbird requires approximately 50 fruits daily, Roca calculated that the entire colony he studied collectively regurgitated approximately 15 million seeds each month, a biomass of about 21 tons of seeds. Roca estimated that about 60% of the seeds were dispersed in forest. Oilbirds are important species in maintaining the plant biodiversity of the forests in which they forage, and as such, they merit strict conservation measures, especially around their caves.

ANIMAL DIVERSITY AT FRUITING TREES

One of the best ways to see some tropical birds and mammals is to locate a tree laden with fruit and watch what comes by to feed. On January 14, 1982, over a 1-hour period in midmorning, I observed 17 species of birds and several agoutis (cavimorph rodents) at a single fruiting fig tree in Blue Creek Village, Belize (Table 7-3). I have repeated this experience numerous times in various tropical regions. Not all of the species fed directly on fruit, because insects are also attracted to fruit, so both frugivores and insectivores benefit from a visit to a fruit tree. At night, the same fruiting tree where I saw 17 bird species attracted mammals such as kinkajous and pacas. Leck (1969) observed 16 species of birds ranging over 11 families on a single tree species (*Trichilia cuneata*) in Costa Rica, of which 11 were observed directly feeding on fruit.

TABLE 7-3 Birds observed by the author at a fruiting fig tree in southern Belize.

1. Ruddy ground-dove (*Columbina talpacoti*)
2. Olive-throated parakeet (*Aratinga nana*)
3. Violaceous trogon (*Trogon violaceus*)
4. Blue-crowned motmot (*Momotus momota*)
5. Collared aracari (*Pteroglossus torquatus*)
6. Black-cheeked woodpecker (*Melanerpes pucherani*)
7. Lineated woodpecker (*Dryocopus lineatus*)
8. Lovely cotinga (*Cotinga amabilis*)
9. White-collared manakin (*Manacus candei*)
10. Common tody-flycatcher (*Todirostrum cinereum*)
11. Bright-rumped attila (*Attila spadiceus*)
12. Social flycatcher (*Myiozetetes similis*)
13. Tropical kingbird (*Tyrannus melancholicus*)
14. Magnolia warbler (*Dendroica magnolia*)
15. Black-and-white warbler (*Mniotilta varia*)
16. Olive-backed euphonia (*Euphonia gouldi*)
17. Red-legged honeycreeper (*Cyanerpes cyaneus*)

Fruit is an important resource in both montane and lowland forests. A comprehensive survey of fruiting trees and fruit-eating birds at Monteverde cloud forest in Costa Rica found that 171 plant species bore fruit that was fed upon by 70 bird species (Wheelwright et al. 1984). Some birds depended heavily on fruit, others casually. Among the birds were 3 woodpecker species, 9 tyrant flycatchers, 8 thrushes, 8 tanagers, and 9 finches, as well as toucans, pigeons, cotingas, and manakins. Though some birds were observed to feed on fruit only rarely, it was clear that fruit represents an important resource for a large component of the avian community. In comparison, at the lowland forest at La

PLATE 7-20
MASKED TITYRA (MALE)

Selva Biological Station, of 185 tree species studied, 90% produce fleshy fruits, of which approximately 50% of the species are primarily bird-dispersed, 13% bat-dispersed, and the remainder dispersed by other mammals such as monkeys and agoutis. Of 154 species of shrubs and treelets at La Selva, 95% bore small fleshy fruits, most dispersed by birds or bats (Levey et al. 1994).

A study of fruit dispersal of *Casearia corymbosa* in Costa Rica recorded 21 species feeding on its fruits, none of which really contributed to seed dispersal (Howe 1977). These species consumed the fruit at the tree and so were essentially useless as seed dispersers. Only one bird species, the masked tityra, *Tityra semifasciata*, was judged to be an efficient seed disperser (Plate 7-20). This thrush-sized black-and-white bird fed heavily on *Casearia* fruits and regurgitated viable seeds at considerable distance, well outside the seed shadow of the parent tree.

The tree *Tetragastris panamensis* on Barro Colorado Island attracted an assemblage of 23 birds and mammals that fed on the fruits, which were produced in "spectacular displays of superabundant but (nutritionally) mediocre fruit" (Howe 1982). But interestingly, actual seed dispersal of this species was less effective than another species, *Virola surinamensis*, which produced fewer but nutritionally better fruits. Fewer bird species fed on *Virola*, but those that did regurgitated seeds one at a time, well away from the plant where they were ingested (Howe 1982).

In a study performed in the cloud forest at Monteverde, Costa Rica, seeds of three species of gap-dependent plants were heavily consumed by six bird species (Figure 7-8). However, only three of the six were useful to the plants as seed dispersers. Each of

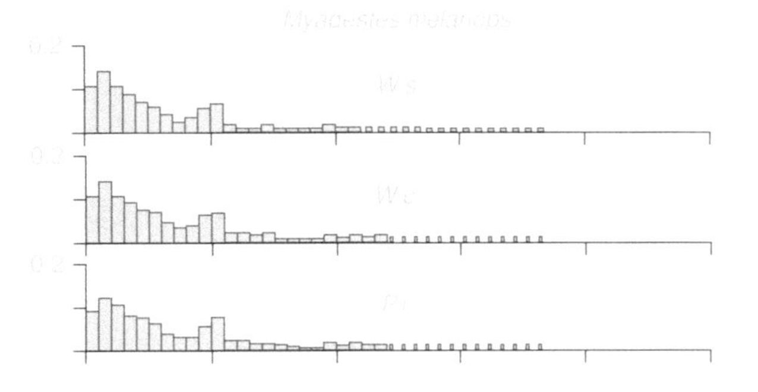

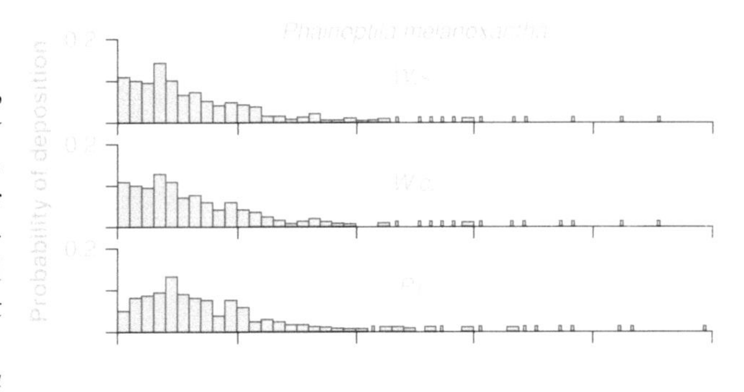

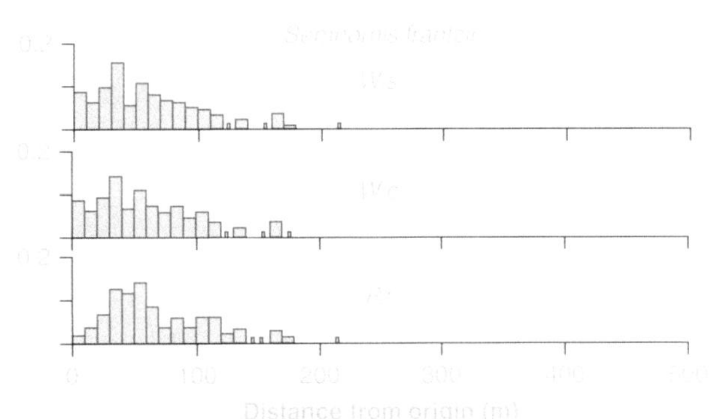

FIGURE 7-8
Estimated seed shadows produced by *Myadestes*, *Phainoptila*, and *Semnornis* around individuals of *Witheringia solanacea* (W.s.), *W. coccoloboides* (W.c.), and *Phytolacca rivinoides* (P.r.). Dots indicate probability of deposition greater than 0.005.

these three species produced seed shadows that extended beyond 100 meters from the plant, and in some cases, seeds were dispersed in excess of 200 meters. The other three bird species dropped seeds at the site of the parent plants or else destroyed the seeds in their guts (Murray 1988).

The efficacy of seed dispersal by frugivores is difficult to generalize, and thus there is less evidence of finely tuned mutualisms demonstrated in pollination systems (Wheelwright and Orians 1982; Levey et al. 1994). Different assemblages of frugivore species may exist throughout the range of a given plant species (Wheelwright 1988), so the important dispersal species in one location will not be the same as those at another locale. Some frugivore species that otherwise seem to be specialists on a given plant species may be ineffective seed dispersers. The resplendent quetzal (*Pharomachrus mocinno*) feeds largely on fruits from the laurel family (Lauraceae), but because of its habit of perching for long time periods as it feeds (thus defecating seeds beneath the parent plant), it is not a very efficient seed disperser (Wheelwright 1983).

Gapers, Gulpers, and Mashers

Birds are selective about the size of the fruits they eat (Wheelwright 1985; Levey et al. 1994). Species such as toucanets pluck fruit, juggle it in their bills, and then reject it by dropping it. Large fruits are particularly at risk of rejection, and may be found scarred by bill marks. Wheelwright hypothesized that plants are under strong selection pressure to produce small to medium-sized fruits, as larger ones are rejected by most bird species except those with the widest gapes. Thus large fruits will select for highly specialized, large birds (such as cassowaries in Australasia). Large fruits permit more energy to be stored in the seeds, an advantage once dispersal and germination have occurred.

In Australia and New Guinea, southern cassowaries (*Casuarius casuarius*) are known to feed preferentially on large fruits, and these nearly ostrich-sized birds seem to be effective seed dispersers (Plate 7-21). In a two-year study conducted in North Queensland, Australia, it was learned that 56 plant species were present as seeds in 198 cassowary droppings that were found and examined (Bradford et al. 2008). This is but a small fraction of the total plant species richness at the study site. A disproportionate number of seeds in the droppings of cassowaries came from tree species with large fruit and large seeds. The conclusion was that the cassowary is an essential disperser of these large-fruited, large-seeded trees over a substantial distance, in this case defined as greater than 2 kilometers. No other bird species or mammal in the forest is as effective a disperser for this group of tree species.

Studies in Costa Rica indicated two basic methods by which birds take fruit (Levey et al. 1984 and 1994; Levey 1985; Moermond and Denslow 1985). Calling them *mashers* versus *gulpers*, the researchers noted that some birds mash up the fruit, dropping seeds, while others gulp the fruit whole, subsequently either regurgitating or defecating seeds (Plate 7-22). In the Neotropics, mashers are finches and tanagers, and gulpers are toucans, trogons, and manakins. Mashers are more sensitive to taste

PLATE 7-21
SOUTHERN CASSOWARY

(a)

(b)

PLATE 7-22
(a) This palm tanager is an example of a masher; (b) this aracari (a type of toucan) is a gulper.

than gulpers. Levey (1987) found a distinct preference among masher species for 10% to 12% sugar solutions as opposed to lower concentrations. Gulpers swallow fruit whole and are taste-insensitive. Gulpers' wide wings aid them in hovering at the fruit tree, and they have wide, flat bills useful in plucking and swallowing the fruits.

Are Fruits the Evolutionary Stimulus for Sexual Selection in Tropical Birds?

Charles Darwin (1859 and 1871) developed his theory of sexual selection primarily to explain human racial diversity (Desmond and Moore 2009). But Darwin quickly extended his theory to account for why certain animal species ranging from beetles to birds and mammals show strong morphological differences between the sexes. Why would selection act differently between males and females? Among numerous bird species, males are typically brighter in plumage (a pattern termed *sexual dichromatism*) and often larger than females. Why females have cryptic plumage seemed an easy question to Darwin. Females have cryptic plumage because such coloration aids in reducing the risk of discovery by predators. But why are males so colorful? Adding to this mystery was the fact that elaborately colored males often augment their already gaudy selves by engaging in bizarre courtship displays.

Sexual selection may evolve through two pathways operating separately or simultaneously. One is male–male competition for access to females. This helps account for why males tend toward larger body size than females. The other is female choice of males that exhibit characteristics signaling high fitness. Characteristics of bright plumage in birds may reflect female choice.

In the tropics, sexual selection in birds is evident in numerous bird families, but the most dramatic examples are among the birds-of-paradise, bowerbirds, cotingas, and manakins, each of which depends heavily on diets of fruit.

AN EXAMPLE: THE GUIANAN COCK-OF-THE-ROCK

The Neotropical Guianan cock-of-the-rock (*Rupicola rupicola*), a large (nearly chicken-sized) cotinga (Cotingadae), provides an example of sexual selection. The elaborate courtship of this species has been studied in detail (Snow 1982; Trail 1985a and 1985b). Males are chunky, with short tails and bright golden orange plumage with black on the tail and wings (Plate 7-23a). In flight, they resemble winged, Day-Glo-orange footballs! Beaks, legs, eyes, and even the very skin are bright, vivid orange. The male's already striking plumage is further enhanced by delicate elongate orange wing plumes and a crescent-like thick fan of feathers extending from the base of the bill to the back of the

(a)

(b)

(c)

PLATE 7-23

(a) Three male Guianan cocks-of-the-rock (*Rupicola rupicola*) on their lek, facing off, preparing to display. (b) The female cock-of-the-rock is much less brightly colored than the male, a trait typical in sexually selected bird species. (c) The male cocks-of-the-rock have assembled on their lek, where they display and contest for females.

neck. Females are dull brown, with neither the wing plumes nor the head fan (Plate 7-23b).

Males gather in the rain forest understory in confined courtship areas called *concentrated leks* (Plate 7-23c). Each male clears an area of ground on which to display and defends perches in the vicinity of its display site. The lek becomes crowded, with males as close as 1.5 meters, with several dozen males occupying the lek. When a female approaches a lek, each male displays, first by landing on the ground and posturing to her. Each displaying cock strokes its wing plumes and turns its head fan sideways, presenting its profile to the female, staring at her with its intense orange eyes. The object of each cock's display is obviously to mate, presumably by suitably impressing the female. Females do not appear to be easily impressed. A hen will typically visit a lek several times before engaging in copulation. These visits, called *mating bouts*, always excite the males to display. Ultimately, only one male on a lek will get to mate with a visiting female, who may return to mate with him a second time before laying eggs (Trail 1985b). No extended pair bond is formed, only a brief coupling. The cock returns to the lek, continuing to court passing hens, while the newly fertilized hen attends to nest building, egg laying, incubation, and raising the young.

Darwin reasoned that in some species, female choice was the dominant factor in selecting male appearances. Put very simply, males are colorful (or musical or noisy or perform complex "dances") because females have tended through generations to mate mainly with males having these unique features. Since plumage color is heritable (as are behavioral rituals), gaudy coloration was selected for and continually enhanced. Recent work in sexual selection suggests that females may learn much about the evolutionary fitness of males by signals communicated both by plumage condition and male courtship behavior (Andersson 1994). For example, the appearance of males has been shown to be an honest signal of good health, lack of parasites, strong immune system, agility, and coordination. In other words, females are not being "frivolous" in driving male evolution toward more elaborate, gaudy plumage and exotic behavior but are looking intently for honest signals, expressed in plumage and behavior, of male fitness.

The other facet of sexual selection recognized by Darwin is that males must compete among themselves for access to females. Male–male competition takes numerous forms from one species to another: dominance behavior, guarding females, active interference with other males' attempts to mate, injury to other males, or merely being "sneaky" and mating before other males can react. Bright, conspicuous plumage may contribute to a male's success by intimidating other males and thus make it easier to gain the attentions of a female.

Sexual selection has costs for both males and females. Though the hen exercises choice in the mating process, she is left solely responsible for the chores of nest building, incubation, and caring for the young. These are risky, energy-consuming tasks. Males may at first glance seem the luckiest, rewarded by a life of lust in nature's tropical "singles bar," the lek. The combination of male–male competition plus dependency on female choice makes life surprisingly difficult for most males, however. Though some cocks are quite successful, mating frequently, others, the losers, spend their entire lives displaying to no avail. They eventually may die genetic losers, never selected even once by a hen. In one study in Suriname (Trail 1985a and 1985b), 67% of territorial males failed to mate at all during an entire year. The most successful male performed an average of 30% of the total number of annual matings, and the lek contained 55 cock birds. Many never mated. Such is the cost of sexual selection for males. In reproductive terms, most females do mate, though success in fledging young may certainly vary considerably among females.

Trail (1985a) also discovered another component in the mating process of the Guianan cock-of-the-rock. Some males habitually disrupted the mating of others. Trail found that aggressive males that disrupted copulations by other males fared better in subsequent mating attempts. He learned that males that were confrontational "were significantly more likely to mate with females that they disrupted than were non-confrontational males." He hypothesized that only the cost of confrontation in terms of energy expenditure, loss of time from the aggressive bird's own lek territory, plus risk of actual retaliation kept direct confrontational behavior from becoming even more manifest among the birds. On the other hand, Trail (1985b) found adult fully plumaged males remarkably tolerant of juvenile males that were still plumaged in drab colors, resembling females. Yearling males would actually attempt to mount adult males as well as females in a crude attempt at mating. Adult males did not respond aggressively to these misguided efforts, possibly because yearling plumage, being drab, does not stimulate an aggressive response.

The elaborate plumage and courtship behavior of cock-of-the rock is not unique. Most members of the Neotropical cotinga (Cotingadae, 97 species) and manakin (Pipridae, 48 species) families have colorful, often gaudy males and perform elaborate courtship behaviors. In the Australasian tropics, the birds-of-paradise (Paradisaeidae, 40 species) and bowerbirds (Ptilonorhynchidae, 18 species), though distantly related to each other and quite distantly related to the Neotropical cotingas and manakins, also have evolved elaborate plumage, courtship, and in many cases lekking behavior. What these bird families have in common is that each relies heavily on a diet of fruit.

David Snow (1976) and Alan Lill (1974) have suggested possible scenarios for the "release" of males from postcopulatory reproductive chores (such as feeding young or assisting in nest building), thus initiating the male–male competition and pattern of female choice that resulted both in the gaudy plumages and elaborate courtship behaviors. Snow emphasizes the importance of a fruit diet. He points out that cotingas and manakins feed so heavily on fruit that they are easily able to secure adequate daily calories with only a small percentage of their time devoted to feeding. Fruit is abundant and easily collected. It does not have to be stalked or captured and subdued. This makes access to females the limiting resource, initiating an evolutionary trajectory of sexual selection. In addition, a diet of fruit aids in permitting the birds to metabolically synthesize colorful pigment that characterizes male plumage.

A fruit diet may be the primary driver of evolution of sexual selection in some groups of tropical birds. But nest predation is also a factor. A largely frugivorous diet has metabolic costs as well as benefits (Morton 1973). Incubation time is relatively long and nestling growth rates slow in frugivorous birds because fruit is nutritionally not well balanced for a baby bird (low in protein but high in fat and carbohydrates). Lill (1974) argues that because of the slow development time brought about by a diet of fruit (recall the discussion of oilbirds earlier in this chapter), nest secrecy is of paramount importance. Heavy egg and nestling predation are best minimized by having only one bird, the cryptically colored female, attend the nest. A male's presence at the nest could potentially be detrimental to raising young, since one bird can easily find sufficient food for the small brood (at most two nestlings), and a second bird might inadvertently reveal the presence of the nest to potential predators. Unlike insectivorous bird species that also have small clutch size in the tropics, frugivores require more time to fledge young because the diet is unbalanced for nestling growth. Male absence actually increases the probability of egg and nestling survival. Males are dispensable, not needed for raising young.

Bowerbirds

Bowerbirds inhabit much of Australia and all of New Guinea. Sexual selection has taken an additional evolutionary twist with this unique group. Many of the 18 bowerbird species show a typical pattern of sexual selection in that the male bird is brightly plumaged and the female far more cryptic. But males of some species are not particularly gaudy. Why is that?

Bowerbirds are so-named for a unique courtship pattern. Males build *bowers*, display areas that are in many cases elaborately constructed. Some species construct avenue *bowers*, an arch of sticks and grasses forming an avenue. At the center is a display perch, also arranged by the male bird. Some species construct garden bowers, some featuring elaborate mats, some featuring what are termed *maypoles*, where a column of sticks rises from the center of the garden. Male birds collect items that are used to decorate bowers. The male satin bowerbird (*Ptilonorhynchus violaceus*), for example, favors blue (Plate 7-24). It collects flower petals, bottle caps, broken glass, or any other manageable object, as long as each item is blue. These items are each carefully placed in the bower by the males. (If they are moved by humans, the male will place them "back where they belong.") When a female visits the bower, the male displays. Copulation, if it occurs, may occur in the bower or elsewhere. Those bowerbird species that construct the most elaborate bowers have the dullest plumage. Thus the construction of the bower itself is a manifestation of sexual selection, demonstrating the quality of male fitness.

(a)

(b)

PLATE 7-24
(a) MALE SATIN BOWERBIRD; (b) FEMALE SATIN BOWERBIRD

Does Fruit from Exotic Plants Assist in Reestablishment of Forest?

The guava tree (*Psidium guajava*) is a common species in much of the Neotropics, but it has been widely introduced in various places in Africa, Australasia, and the Pacific islands. In those places, it is considered an exotic, invasive species. Normally, such a designation suggests that the species is undesirable, at least from an ecological standpoint. However, a study performed in western Kenya tentatively concludes otherwise (Berens et al. 2008).

Researchers surveyed 29 fruiting guava trees scattered in farmland at various distances from the Kakamega Forest, a highland rain forest rich in species. From September to December 2004, researchers documented the bird species that fed on the fruits and examined the *seed rain* beneath the tree, which came from dropped guava fruits but also from seeds of other plant species

FIGURE 7-9

Satellite image of Kakamega Forest, western Kenya (see map in background), showing the location of 29 study trees in the farmland between the main forest of Kakamega (south) and the forest fragment Kisere (northeast). Light gray area represents secondary forest, and dark gray represents near-primary forest. Distances of trees to the forest were calculated from the secondary forest boundary.

contained in bird droppings and regurgitation. They counted the seedlings beneath the tree, most of which were guava seedlings, but, importantly, not all.

In 203 hours of observations, researchers recorded 61 bird species feeding in the guava trees. Of these, 40 were frugivores. Eleven of the species were from forest, 27 from shrubland, and two were open-country species. It was clear that scattered guava trees attract a wide diversity of birds from various habitats (Figure 7-9). Some forest bird species such as the black-and-white casqued hornbill (*Bycanistes subcylindricus*) flew long distances from forest to feed in the guavas.

The seed rain beneath the guava trees included 12 tree and shrub species whose seeds were bird-dispersed. Of these, 58% were late-successional species, while 42% were early-successional species. In terms of seed abundance, 24% of all seeds were from late-successional plant species, 76% from early-successional species.

Excluding guava seedlings, 34 seedling species were documented beneath guava trees, of which 28 were animal-dispersed species. A total of 57% of these were late-successional, 36% were early-successional, and 7% were exotics.

This study suggests that scattered guava trees might assist in reestablishment of forest on abandoned agricultural sites. Because the exotic guava trees act as food magnets for a diverse array of frugivorous birds, the input of seeds from areas distant to the guava trees may facilitate regrowth of forest, literally beginning beneath the guavas. The researchers took care to qualify this conclusion by stating that such factors as interaction with pathogens and herbivores, competition of seedling trees with seedlings from the guavas, soil-water relations, and mineral cycling might reduce the efficacy of forest regeneration around guavas. But they point out that guavas, though they persist in early secondary succession on abandoned lands, appear to be eventually outcompeted by native species. The conclusion, though qualified, is that guavas may prove to be a potentially useful management tool in forest restoration.

In many areas in the tropics, fences are made with live plantings, stakes that literally consist of living plants. These have been shown to act in ways similar to what was learned with the study of guava trees (Zahawi 2008).

Fish as Amazonian Seed Dispersers

Floodplain forests within the Amazon Basin cover an area of approximately 150,000 square kilometers, which is roughly equivalent to an area the size of the state of Florida (Goulding 1980, 1985, 1990, and 1993). Floodplain forests, called *varzea* in Amazonia, are inundated by the annual flood cycle. Depending on location, floodplain forests may be submerged anywhere from 2 to 10 months out of the year. For example, the Amazon forest itself (from Manaus eastward) is flooded for about 6 months, whereas the upper Rio Madeira is in flood for only 2 to 5 months annually.

There are more than 2,400 fish species inhabiting the waters of the Amazon and its tributaries, and up to 800 additional species may remain yet to be formally described (Lowe-McConnell 1987; Goulding 1980; Goulding et al. 2003). Approximately 40% of the species thus far described are members of two groups, characins and catfish (Goulding 1985). These multitudes include many favorites of the aquarist (Lowe-McConnell 1987).

During the flood cycle, fish have direct access to forest. Many become fruit and seed consumers, and some act as seed dispersers. Approximately 200 species of fish devour fruits and seeds in Amazonian waters, far more species than do so in tropical Africa or Asia (Goulding 1985 and 1993; Goulding et al. 2003). A frugivorous diet is facilitated by the flood cycle, which enables fish to swim well within the gallery forest, foraging for dropped fruits, many of which float at the surface, making them easy to find and consume. There is a seasonal shift in diet among characin species in which they move from an omnivorous diet that includes zooplankton and various plants and algae to essentially a diet of fruit when the forest is flooded.

THE TAMBAQUI

The tambaqui (*Colossoma macropomum*), an inhabitant of blackwater rivers and *igapo* forests (flooded forests on poor soil), is an important seed disperser, particularly for *Hevea spruceana*, a rubber tree, and *Astrocaryum jauari*, a palm species (Goulding 1980). Both of these tree species are widely distributed, are relatively abundant, produce large seed crops, and have fruits that are laden with fat and protein and that are encased within hard nuts that many animals are unable to break. In the case of the rubber tree, seeds are contained in large capsules that eventually pop open and effectively toss their seeds as far as 20 meters. The seeds float, and tambaquis gather around rubber trees where seeds are being released.

The tambaqui is bass-like, an oval-shaped characin, weighing as much as 30 kilograms, and with its specialized, rounded, molarlike teeth, it is capable of crushing and grinding very hard fruits (Plate 7-25). Tambaqui feed almost exclusively on fruits for the first five months of the flooding season. The fruits contain sufficient protein and fat that the fish is able to survive during periods of low water from fat stored from its flood-cycle fruit consumption. Seeds often contain toxins, and thus, though fruit pulp is digested, seeds are not. Seed toxicity enhances the probability of dispersal, as the seeds pass through the digestive system of the fish.

Juvenile tambaqui feed in nutrient-rich whitewaters (where soil is rich in sediment), along *varzea* floodplains and lakes. At the close of the flood cycle, when the waters drop, adult tambaqui migrate from nutrient-poor to nutrient-rich waters to spawn. Young tambaqui feed on zooplankton, not fruit.

PLATE 7-25
Tambaqui (*Colossoma macropomum*), one of the most important seed-dispersing fish in Amazonian flooded forests.

OTHER FRUIT-CONSUMING FISH

Even the notorious (and carnivorous) piranhas are known to consume seeds, removing the husk and masticating the soft seeds within (Goulding 1985). Piranhas belong to the family Characidae, the characins, and many characins are seed predators, possibly the most important seed predators in the flooded forests.

Catfish (Plate 7-26) are not as destructive to seeds as characin species because they gulp the fruit whole rather than macerate it (Goulding 1985). Thus catfish are somewhat the piscine equivalent of birds such as toucans, in which the fruit pulp is digested and the seed passes out of the alimentary canal unharmed.

A study in Brazil looked at the diets of two important frugivorous fish species, *Colossoma macropomum* and *Piaractus brachypomus* (Lucas 2008). It was learned that 78% to 98% of the fishes' diets comprised fruits during the peak flood months of April to June. This included 27 species of woody plants and 4 herbaceous species. The abundance of fruit correlated closely with the flood cycle, obviously facilitating access of the fish to the fruits. The study revealed considerable dietary plasticity among the fish with regard to volume, species composition, and species richness of fruits taken within the high flood season. The fishes' diets reflected differences in local forest diversity and clearly documented the importance of fish as seed dispersers.

PLATE 7-26
Species such as these Amazonian catfish are often important seed dispersers.

A study done on the pacu fish (*Piaractus mesopotamicus*) in the Brazilian Pantanal, an extensive area of wetlands (see Chapter 12), demonstrated that it is an essential seed disperser for the palm *Bactris glaucescens* (Galetti et al. 2008). The size of the fish, which ranges in length from about 20 to 40 centimeters, and its weight and gape width were positively correlated with the number of intact seeds present in the fish's gut. None of the other potential dispersers of *Bactris* seeds, including the greater rhea (*Rhea americana*), iguanas, howler monkeys, peccaries, or tapirs were as effective as dispersers as pacus. For example, peccaries were seed predators. The authors of the study pointed out that pacus are routinely overfished in the study area and their diminishment could have a deleterious effect on seed dispersal.

Coevolution

Species may affect each other in intricate ways. Herbivores are dependent on plants for food, but plants are often spiny, noxious, or toxic, an apparent defense against herbivores (see Chapter 8). On the other hand, herbivores aid plants, as they help recycle nutrients by producing organic waste products, serve as pollinators (which amounts to being a surrogate sex organ), and disperse seeds. Herbivores may feed selectively, influencing competitive interactions among plants. The relationship between herbivores and plants is not necessarily simple. Predators are adapted to capture prey, but prey species are adapted to avoid capture. The ecological "cat-and-mouse game" between predators and prey has given rise to elegant adaptations such as cryptic and warning coloration. Parasites and hosts also engage in an evolutionary arms race, a war of adaptation. Many parasites evolve increasing degrees of virulence, and hosts evolve greater resistance to parasites. The myriad interaction among species within an ecosystem has an important historical aspect. Biotic selection pressures, the influences of predators on prey, of herbivores on plants, parasites on hosts, and so on, are selection pressures that determine the evolutionary directions determined by natural selection. When one species exhibits a trait that acts as a selection pressure on another species, and the second species in turn possesses a trait that acts as a counterselection pressure on the first, the evolutionary fates of both species may become permanently interlocked. This situation is called *coevolution* (Ehrlich and Raven 1964; Janzen 1980; Futuyma and Slatkin 1983; Thompson 2005 and 2006).

Coevolutionary interactions may result from parasitic or predatory interactions, where each species engages in a reciprocal arms race, as when both predator and prey evolve to be increasingly swifter. But much focus on coevolution involves mutualistic relationships, when both species gain from the interaction. Some mutualisms may be facultative, others obligatory. Recent work suggests intricate networks of mutualistic relationships in which a given plant species may, for example, depend highly on a particular animal species that is not nearly as dependent on the plant species. Such networks quickly become complex (Figure 7-10) and may typify species-rich regions such as lowland tropical forests. Mutualistic networks may facilitate biodiversity maintenance (Bascompte et al. 2006; Bascompte 2009).

Many species interactions are not strictly coevolutionary (Janzen 1980). Assumptions about coevolution should be based on rigorous demonstration of the nature of the interaction (Bernays and Graham 1988). For example, some tree sloths commonly harbor tiny moths that reside within the recesses of the sloth's dense fur. These diminutive lepidopterans are well attuned to the sloth's behavior. They lay their eggs on sloth droppings, where the larvae eventually feed

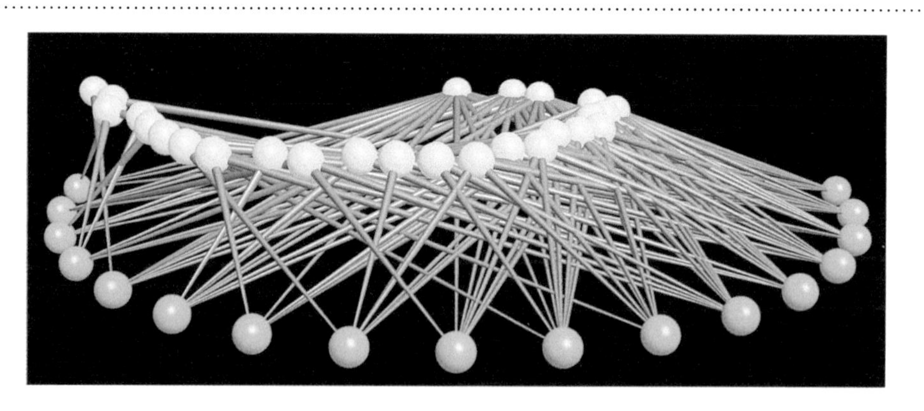

FIGURE 7-10
Plant–animal pollination network. Green and yellow nodes represent plant and animal species, respectively. A link between one plant and one animal indicates that the former is pollinated by the latter. Note how complex such a network becomes.

after hatching. These moths, while clearly adapted to life on a sloth, do not represent coevolution because there is no evidence that they have exerted any form of selection pressure on the sloth that has resulted in an adaptive response. These moths are *commensal* in that they benefit from being on the sloth but have no significant impact (at least none that has been demonstrated), positively or negatively, on the sluggish mammal. In addition to moths, some sloths may have an abundance of algae living within their fur, and the algae confer a degree of crypsis on the sloth, making the sloth somewhat more difficult to detect in the canopy. The sloth supplies habitat for the algae, and the algae aid the sloth's potential survival by adding an element of crypsis. Is this coevolution? If so, it is only in the most casual sense. But sloths have not been shown to have either a genetic propensity to harbor algae or to be algae-free. There is no evidence of any adaptive accommodation between the two participants. Neither appear to be exerting any selection pressures on the other (though one could argue that if sloths lacking algae are more susceptible to predation, the presence of algae is a selection pressure that affects sloth survival probability, but this would need to be demonstrated).

POLLINATION BY ANIMAL VECTORS

Pollination of plants by various animals is a widespread example of coevolution. Many flowering plants, particularly in the tropics, are dependent on insects, birds, or bats to facilitate fertilization. Animals use flowers as a food source, ingesting nectar and, in some cases, pollen. As they move from plant to plant, animals disperse pollen, making cross-pollination efficient and ensuring reproduction of the plants. This relationship is mutualistic because both plant and pollinating animal benefit.

DARWIN, POLLINATION, AND COEVOLUTION

Charles Darwin, in *Origin of Species*, wrote about the coevolved relationship between bees and the clover they pollinate: "The tubes of the corollas of the common red and incarnate clovers (*Trifolium pratense* and *incarnatum*) do not on a hasty glance appear to differ in length; yet the hive-bee can easily suck the nectar out of the incarnate clover, but not out of the common red clover, which is visited by humble-bees alone." Darwin discussed pollination and coevolution further in a monograph *On the Various Contrivances by Which British and Foreign Orchids Are Fertilized by Insects, and on the Good Effects of Intercrossing*, published in 1862.

Darwin's clear understanding of the interdependency between various plants and their specific pollinators, and how the behaviors of the insects as well as the morphologies of both insects and plants have evolved from the exertion of mutual selection pressures, makes him the person who discovered coevolution.

Animal pollination is widespread throughout all tropical regions, particularly in rain forests. In contrast to temperate forests, where wind pollination is common, rain forests are sufficiently dense that wind pollination would tend to be ineffective, except perhaps for emergent trees. It is not surprising that grasses, sedges, pines, and other open-area savanna species are the only tropical plant groups characterized by wind pollination.

Insects and vertebrates are major pollinators throughout the tropics. Among insects, pollination is accomplished by numerous species of bees, flies, beetles, butterflies, and moths (Prance 1985). Among vertebrates, several whole families of birds are adapted to feed on nectar. These include hummingbirds (Trochilidae, 332 species), sunbirds (Nectariniidae, 127 species), flowerpeckers (Dicaeidae, 44 species), sugarbirds (Promeropidae, 2 species), and honeyeaters (Meliphagidae, 174 species). Many other birds including honeycreepers, various tanagers and orioles, and parrots such as lorikeets (Plate 7-27) are highly nectarivorous, some with specialized brush-like tongues adapted to obtain nectar. Hummingbird-pollinated flowers take many shapes, but some have long tubes and are red, orange, purple, or yellow. In contrast, bat-pollinated flowers are often white (easy to locate in the dark) and may have a musky odor, an attractant to the bats. Many flowers are visited by a variety of vertebrates and insects, all potential pollinators.

Pollinators that fly long distances are most advantageous to plants. Such behavior helps ensure effective cross-pollination between widely separated plants (Janzen 1975). For example, fig wasps were shown to have a dispersal distance of between 5 and 14 kilometers (3 and 8.5 miles) (Nason et al. 1998). In a study conducted in Borneo, fig wasps appear to utilize wind dispersal, flying above the canopy where winds are strong (Harrison 2003). Euglossine bees are long-distance fliers, and males pollinate certain widely separated orchids. Compounds in the orchid flower that are absorbed by the male bees contribute to the longevity of the insects. Male euglossine bees live up to six months, a long life for a bee (Janzen 1971). Female euglossine bees, as well as carpenter bees, are important long-distance pollinators of many rain forest trees (Prance 1985).

PLATE 7-27
Rainbow lorikeet (*Trichoglossus haematodus*), a nectar-feeding Australian parrot.

Pollination of the Victoria Water-Lily

The huge Victoria (or royal) water-lily (*Victoria amazonica*) is found in quiet backwaters of Amazonian tributaries (Plate 7-28). Beetles pollinate this striking plant. Opening in synchrony, the large, conspicuous white flowers emit a strong odor and are warm, up to 11°C warmer than ambient temperature. These characteristics

PLATE 7-28
Victoria lily in a quiet lagoon in Amazonian Peru.

combine to attract beetles (*Cyclocephala* spp.) that enter the flower, only to become trapped inside at night, when the large petals tightly close. The imprisoned beetles feed on nectar-rich structures throughout the night, getting thoroughly sticky as they become covered with pollen. The next day the flowers open, having changed in petal color from white to red, as well as having lost their scent and cooled in temperature, all of which means that they are no longer an attractant to beetles. The formerly incarcerated pollen-bearing beetles leave the flower and fly off to seek out another white flower from another Victoria water-lily, where they will inadvertently deposit pollen as they feed (Prance 1985).

COEVOLUTION AND SEXUAL DIMORPHISM IN A HUMMINGBIRD

The purple-throated carib (*Eulampis jugularis*) is a hummingbird species in which males have conspicuously shorter bills than females but are larger in body size (Plate 7-29). Many hummingbird species show the same pattern. Hummingbirds have extraordinarily high metabolisms, even by bird standards. They hover to attain nectar—one of the most specialized forms of feeding behavior seen among birds. Much research conducted on the Caribbean islands of St. Lucia and Dominica (within the Lesser Antilles) has revealed a complex coevolution between the hummingbird sexes and the flowers on which they feed (Temeles et al. 2000, 2003, and 2009).

PLATE 7-29
MALE PURPLE-THROATED CARIB HUMMINGBIRD

On St. Lucia, the purple-throated carib is the only pollinator of two *Heliconia* species, *H. caribaea* and *H. bihai*. The two *Heliconia* species are easily separable, as *H. caribaea* flowers are within bright red bracts and *H. bihai* flowers are within green bracts (and birds have color vision). The males feed exclusively on *H. caribaea*, and the longer-billed, smaller females on either *H. caribaea* or *H. bihai* (Plate 7-30). The flowers of the *H. bihai* are deeper, with curved corollas, and thus the longer, curved bill of the female is ideal for obtaining nectar. The male's shorter bill requires much longer effort per flower (called *handling time*), consuming much energy, and thus males focus entirely on *H. caribaea*, which has a shorter, straight corolla ideally suited to fit the male's shorter bill. Though females are able to take nectar from the flowers of *H. caribaea*, the larger males forcibly exclude them. Thus male and female purple-throated caribs have coevolved with different species of *Heliconia*. It is important to note that the sexual dimorphism in bill length and body size is not the result of sexual selection in this case but rather of ecological selection for different food plants.

Some sites on St. Lucia lack *H. caribaea* but still have *H. bihai*. In these areas, *H. bihai* has evolved a second flower morph, one that has reddish-green bracts (similar to the bracts of *H. caribaea*) and shorter, straighter flowers. Male hummingbirds feed on this morph.

PLATE 7-30

Polymorphisms in bills of *E. jugularis* (a and b) and in flowers (c and d) and inflorescences (e and f) of *Heliconia* species on St. Lucia (e to g) and Dominica (c, d, h to j), West Indies. (a) *E. jugularis*, female bill. (b) *E. jugularis*, male bill. (c) *H. bihai*, flower. (d) *H. caribaea*, flower. (e) *H. bihai*, green inflorescence morph, St. Lucia. (f) *H. bihai*, red-and-yellow-striped inflorescence, Dominica. (i) *H. caribaea*, red inflorescence morph, Dominica. (j) *H. caribaea*, yellow inflorescence morph, Dominica.

On the island of Dominica, the pattern changes. In this case, *H. caribaea* evolved a yellow morph whose flower best fits the longer bill of the female purple-throated carib. Males continue to feed on the red morph, and females feed on the yellow morph.

The amount of nectar provided by those *Heliconias* that serve males is greater than that in those (regardless of species) that serve females. This accords with the greater energy requirements of males, because males are larger in body size.

Nectar Robbery

Nectar is a resource that functions to attract pollinators. It is easy to conclude that mutualism evolves in a direct manner between plant and pollinator. That is not necessarily the case. In the Neotropics, there are some bird species within the large family Thraupidae (tanagers, 271 species) that are called *flowerpiercers* (Plate 7-31). These birds eat nectar but do not act as pollen vectors. A flowerpiercer's bill has a small, hooked upper mandible. The bird typically perches at the base of a tubular flower and pierces the flower at the base, obtaining nectar without touching pollen. Flowerpiercers are thus parasitic on plants.

Flower traits have been shown to evolve not only to attract certain pollinating species but also to discourage species that are ineffective pollinators or pollen robbers (Irwin et al. 2004; Johnson et al. 2006). There is selection pressure to protect flowers from nectar robbers. It has been shown that nectar robbers have the capacity to reduce plant fitness in a way comparable to herbivory (Irwin and Brody 2000). Adaptations to reduce nectar robbery include adding toxins to nectar that discourage robbers but not pollinators, flowering at times when nectar robbers are inactive, growing near plants that offer better food sources for nectar robbers, or evolving flowers that are physically difficult for robbers to access. Some plants have evolved extrafloral nectaries that attract insects such as ants that defend the plant from nectar robbers (see Chapter 8).

It is important to understand that evolutionary patterns have resulted in two kinds of selection pressures on plants that utilize animals as pollen vectors. One is to evolve characteristics that attract and facilitate access by the pollinators, but the other is to evolve characteristics that defend against robbers.

PLATE 7-31
MASKED FLOWERPIERCER, *DIGLOSSA CAERULESCENS CYANEA*

CHIROPTEROPHILY

Pollination of flowers by bats (Plate 7-32) is common in the tropics, with more than 500 plant species wholly or partly dependent on bats as pollinators (Heithus et al. 1974). Plants adapted to host bats are termed *chiropterophilous*, meaning "bat-loving." (Bats are in the mammalian order Chiroptera.) Coevolution has occurred at behavioral, physiological, and anatomical levels in both bats and plants (Howell 1976).

Plants adapted to bats as primary pollinators usually have large white flowers, reportedly emitting a musky "bat-like" odor. These flowers open at night, when bats are active. Flowers may be shaped like a deep vase or may be flat and brushy, loading the bat's face with pollen as it laps up nectar. Many bat flowers are *cauliflorous*, growing directly from the tree trunks. Flowers may hang from long, whiplike branches (*flagelliflory*) or hang downward as streamers (*penduliflory*), a condition common in

PLATE 7-32
COMMON LONG-TONGUED BAT
(*GLOSSOPHAGA SORICINA*)

many vines. Cauliflory, flagelliflory, and penduliflory all have in common the fact that the flowers are positioned in such a manner that they are easily accessible to hovering bats.

In Guanacaste Province, Costa Rica, two bat species feeding on flowers of *Bauhinia pauletia* exhibited markedly differing feeding behaviors. *Phyllostomus discolor* visited high flowers, grasping the branch beneath the flower and pulling it down. *Glossophaga soricina* visited both high and low flowers and hovered as it fed. Both bat species aided in cross-pollinating *Bauhinia* (Heithus et al. 1974).

Nectar-feeding bats typically have large eyes and relatively strong vision, in contrast with insectivorous bats. The sonar sense on which insectivorous bats depend is often reduced in nectarivorous bats, but the olfactory sense is well developed. Nectarivorous bats have long muzzles and weak teeth, both advantageous in probing deeply into flowers. Last, they have long tongues covered with fleshy bristles that can extend well into the flower, and in some cases, their neck hairs project forward, acting as a "pollen scoop."

The pollen from bat-pollinated plants is significantly higher in protein than in non-bat-pollinated plants, and bats ingest pollen as well as sugary nectar. Pollen contains the amino acids proline and tyrosine, useless to the plant but important to the bats. Proline is used in making connective tissue such as is used in wing and tail membranes, and tyrosine is essential for milk production.

Once ingested, nectar helps dissolve the tough pollen coat, but bats aid this process because their stomachs secrete extraordinarily large amounts of hydrochloric acid. These bats also sometimes drink their own urine, which helps dissolve the pollen, liberating essential protein (Howell 1974).

The example of chiropterophily illustrates the following important components of coevolution:

- Coevolution may involve whole complexes of species, not merely two species evolving together.
- Anatomical characteristics may be most obvious, but behavioral and physiological characteristics are also readily evident in the coevolutionary process.

FIGS AND WASPS

An elaborate coevolution has been described between figs and fig wasps (Janzen 1979; Wiebes 1979; Herre et al. 1996; Anstett et al. 1997; Jousselin et al. 2003). Molecular studies have shown that the fig–wasp symbiosis stretches back through time to 60 million years ago (Rønsted et al. 2005). Figs (*Ficus* spp.) are pantropical, and approximately 700 species are known. Figs all produce a unique bulbous green flower that resembles a fruit more than a flower (Plate 7-33a). Old World figs are *monoecious*, meaning that male and female flowers occur on the same plant. Other species such as some found in the Neotropics are *dioecious*, with male and female flowers on different plants.

(a)

(b)

PLATE 7-33
(a) A cluster of fig syconiums. (b) Pollinator fig wasps (*Pegoscapus* sp.), mutualists found in the fruit of the strangler fig (*Ficus costaricana*), in the tropical cloud forest of Monteverde, Costa Rica.

Monoecious plants must avoid self-pollination, while dioecious plants require cross-pollination in all circumstances. In either case, wasps facilitate pollination (Plate 7-33b). Most figs are pollinated by only one or a few wasp species, so there are essentially as many fig wasp species as there are fig species, an example of an *interlocked speciation pattern.*

The fig inflorescence, called a *syconium*, is a cluster of enclosed flowers within a gourd-like covering. The wasp–fig relationship described here is for monoecious figs. Externally, the fig inflorescence reveals nothing of its unique inner structure. Internally, it forms a dense carpet of tiny male, female, and sterile flowers (which are sterile because they are parasitized by wasps, as described shortly). Though male and female flowers exist side by side in monoecious figs, male flowers cannot pollinate females in the same syconium because female flowers mature earlier than males. Since the syconium is enclosed, it is impossible for fig pollination to occur without the assistance of an animal pollinator. Enter (literally!) the diminutive fig wasps (family Agaonidae).

Some flowers contain fig wasp eggs laid by females of the previous generation. Other female flowers that contain seeds are shaped such that female wasps cannot oviposit. This is an adaptation to avoid sterility imposed by wasps laying eggs and is clearly a tradeoff in the evolutionary sense. Figs must commit some flowers to the role of wasp nurseries.

Male wasps hatch first and burrow into gall flowers to inseminate still unhatched female wasps. Each male commonly inseminates several females, a unique situation in which females are impregnated before they are born. Females hatch, already pregnant with the next wasp generation. Newly hatched females wander over the stamens of male flowers, which have reached maturity precisely when the females hatch. Laden with pollen, the female wasps either excavate an exit tunnel from the syconium or, in some fig species, the syconium opens sufficiently to permit females to exit.

Females, which are attracted to other figs by olfactory cues, have but a day or so to locate another flowering fig and tunnel into a flower via a narrow channel termed an *ostiole*. The female is often physically damaged from burrowing into the flower, but once inside, she has only to locate a suitable flower in which to deposit her eggs. After that, she dies. In her search for flowers, the female passes over many long-styled female flowers and inadvertently deposits the pollen she brought from the flower of her birth. (A style is the structure on the flower that contains the pollen.)

Dioecious figs contain two kinds of flower complexes, *gall figs* and *seed figs*. Wasps utilize gall figs in the manner described earlier. But seed figs contain only female flowers, and their dimensions are incompatible with wasp anatomy—so eggs cannot be laid, although pollination does occur. This is an example in which the wasp is essentially parasitized by the fig. It is not mutualistic. This is an interesting adaptation in that it becomes a form of frequency-dependent selection. Too many seed figs would reduce the local wasp population, which in turn would reduce fig reproduction.

Interactions among fig wasps become more complex when parasitic fig wasps also inhabit syconiums. These insects use the fig as a food source but do not pollinate it. Parasitic males roam about in the dark interior of the syconium, using their huge jaws to dismember other wasps as they all search for females with which to mate. To paraphrase William Hamilton (1979), who spent considerable time looking into fig syconiums, were a human to inhabit such a place it would be an uttely dark and crowded room filled with jostling people, some of whom would be homicidal maniacs wielding sharp knives. Quite an image.

The relationship between figs and fig wasps is complex, and not every aspect of it represents coevolution (Anstett et al. 1997). But it is clear that there really is no such thing as a fig tree without fig wasps. It would be a sterile plant. And there is functionally no such thing as a fig wasp without a fig tree. These species cannot exist in isolation from one another and thus have evolved into a single evolutionary entity, two united but separate genomes.

FUNGUS-GARDEN ANTS: THE LEAF-CUTTERS

Ants are abundant and ubiquitous inhabitants of the global tropics. Many ant species have evolved coevolutionary relationships with plants (see Chapter 8). One group, the Attini ants, has done so with fungus, a relationship that, like the figs and fig wasps, dates to about 50 million years ago (Mueller et al. 1998). Attini ants are largely restricted to the Neotropics, where they are obvious to even the most casual observer (Plates 7-34 to 7-38). Throughout rain forests, successional fields, and savannas, well-worn narrow trails are traversed by legions of ants of the genera *Acromyrmex* and *Atta* as they travel (both diurnally and nocturnally) to and from their underground colonies, bearing freshly clipped leaves (and sometimes buds and flowers). Their trails take them up into trees, shrubs, and vines, where they neatly clip off rounded pieces of leaves to be transported back to their underground colony. The ants live in colonies of up to eight million individuals, consisting of a single large queen and myriad worker ants, most of which remain subterranean (Plate 7-39). Workers are of several size classes: very small (minimas), medium-sized (medias), and large (maximas). Soldiers, the principle defense class, are large and well armed with formidable pincer jaws. Attini colonies are underground, but mounds of displaced soil and discarded leaves mark their multiple entrances on the ground surface.

Leaf-cutter ants have been considered a disproportionate force among tropical insects, but work from Barro Colorado Island has suggested that the

PLATE 7-34
Leaf-cutters (fungus-garden ants) marching across a forest log with their leaves, a common sight in the Neotropics.

PLATE 7-35
LEAF-CUTTER ANTS

PLATE 7-36
A prominent trail made by leaf-cutter ants as they cross a field. Look carefully and note the small fragments of leaves that the ants are carrying.

PLATE 7-37
Leaf-cutter ants cut large slices of leaves for transport to their subterranean colony. This leaf has already been largely cut away by the ants.

PLATE 7-38
This large leaf-cutter ant is transporting a leaf fragment to its nest, with some "help" from two minor individuals.

PLATE 7-39
The above-ground part of an Atta colony. The colony extends far below the surface.

impact of leaf-cutter ants is far less than some earlier published results suggest (Herz et al. 2007). This work will be discussed in Chapter 8. They have been shown to be effective in seed dispersal of plants such as *Miconia* (Dalling and Wirth 1998).

Leaf-cutters are generally selective as to which species they clip (Wetterer 1994). For example, in Guanacaste, Costa Rica, one *Atta* species clipped mature leaves from only 31.4% of the plant species available. Another species used leaves from only 22% of the available plant species, indicating a strong selectivity (Rockwood 1976). The commonness or rareness of a plant species has no correlation with *Atta* preference. The ants may travel relatively far from their colony to seek out a certain plant species. Rockwood hypothesized that internal plant chemistry strongly influences *Atta* diet, a suggestion borne out by subsequent research (Hubbell et al. 1984). Ants apparently concentrate on plants with minimal amounts of defense compounds in their leaves (see Chapter 8).

Leaf-cutter ants are part of a larger ant group called the *fungus-garden ants* (Myrmicinae: Attini), each of which, remarkably, cultivates a particular species of symbiotic fungus that makes up its principal food source (discussed shortly). Some fungus–ant relationships may be as old as 50 million years. There are approximately 200 fungus-garden species, of which 37 are leaf-cutters. The remaining species, most of which are inconspicuous, cultivate their fungus on some combination of decaying plant or animal organic matter (Hölldobler and Wilson 1990). Though most abundant in the tropics, fungus-garden ants also occur in warm temperate and subtropical grasslands. One species even occurs as far north as the New Jersey pinewoods (Wilson 1971).

Leaf-cutter ants do not consume leaves but rather clip and carry leaf fragments back to their colonies. There they convert the leaves to media to culture a specific fungus, called a *cultivar*. This fungus, which is never found free-living outside fungus-garden ant colonies, is the ants' food, though they taste and ingest the sap of the leaves they cut, using the sap as an additional food source (Wetterer 1994). Leaves brought to the colony are clipped into small pieces and chewed into a soft pulp. Before placing the pulpy mass on one of many fungus beds within the colony, a worker ant holds the bolus to its abdomen and defecates a fecal droplet of liquid on it. The chewed leaf is then added to the fungus growing bed, and small fungal tufts are placed atop it. Other ants sometimes add their fecal droplets to the newly established culture.

Worker ants collecting leaves avoid those that contain chemicals potentially toxic to the fungus (Hubbell et al. 1984). One tree, *Hymenaea courbaril*, a legume, has been shown to contain a terpenoid (see Chapter 8) that is antifungal (Hubbell et al. 1983). *Atta* ants avoid clipping leaves from this species. The tree has evolved a protection from *Atta*, not by defending against the ant, but by defending against its fungus.

The ants culture only a few fungal species, all of which are members the Basidiomycetes (family Lepiotaceae), a group whose free-living members include the familiar parasol mushrooms. The fungi are never found free-living outside the ant colony. The fungus garden is protected from contamination from other fungal species by constant "weeding" by ants (but see the following discussion). Without the attention of the ants, the fungus will be overtaken by other fungal species. Both ants and fungi are totally interdependent, an example of an *obligatory mutualism* (Weber 1972). Ant and fungus are coevolved, like the fig wasps and fig plants, to produce essentially two united genomes: only the queen reproduces, and when a queen ant founds a new colony, she takes some of the precious fungus with her inside her mouth. Fungus and ants disperse together.

Detailed studies of the fungus–ant relationship at the biochemical level have revealed the multiple roles that ants play in culturing the fungus (Martin 1970). The ants clean the leaves as they chew them to make the culture bed pure. Ant rectal fluid contains ammonia, allantoic acid, the enzyme allantoin, and all 21 common amino acids. These compounds are all low-molecular-weight nitrogen sources, and they are the key ingredients in making the culture optimal for the fungus. The fungus lacks certain enzymes that break down large proteins (all of which are made up of chains of amino acids). Thus it depends totally on the ant rectal fluid to supply its amino acids. Experiments attempting to grow the fungus in a rich protein medium failed. It can only grow in a medium of small polypeptides and amino acids. Ants also supply enzymes necessary to aid in breaking down protein chains. Martin (1970) summarized the functions of the ants as follows:

- Fungal dispersal
- Planting of the fungus
- Tending the fungus to protect it from competing species
- Supplying nitrogen in the form of amino acids
- Supplying enzymes to help generate additional nitrogen from the plant medium

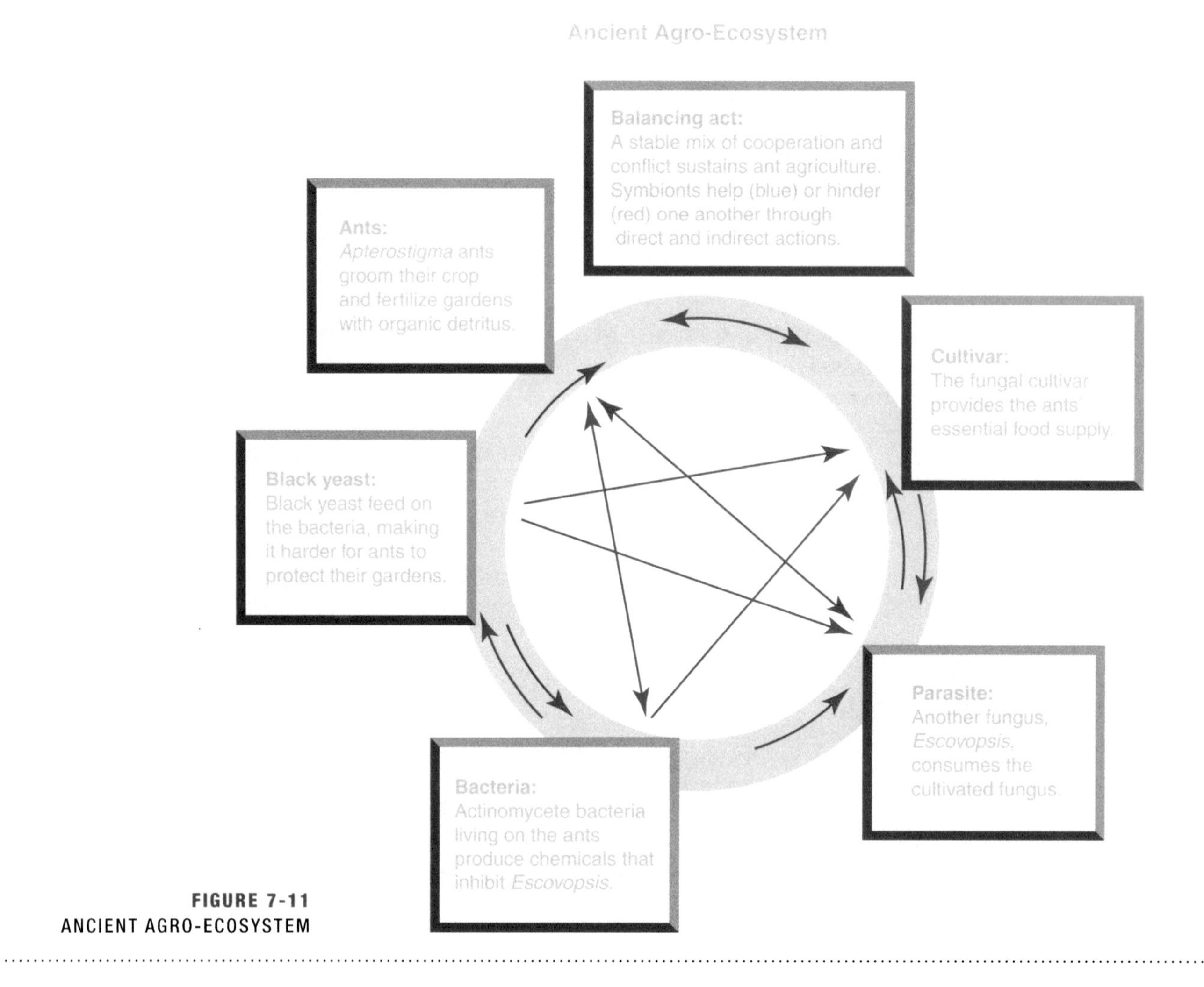

FIGURE 7-11
ANCIENT AGRO-ECOSYSTEM

While this seems complex, it gets even more complex, because the system actually is not confined to just ants and fungus (Figure 7-11). There is a second fungus, *Escovopsis*, that is antagonistic, invading and consuming the ants' cultivar. But ants are able to hold this fungus in check because Actinomycete bacteria that live on the ants manufacture antifungal chemicals that inhibit the *Escovopsis*. However, yet another organism, a black yeast, feeds on the Actinomycete bacteria, making it more difficult for the ants to protect their cultivars (Little and Currie 2008). Thus what appears to be a tightly linked two-species system (a very traditional way in which to view a mutualistic relationship) is, in fact, a far more complex five-species interaction of antagonistic and mutualistic interactions that has been called a "balancing act" (Youngsteadt 2008).

Note that each fungus garden is unique. The cultivar in any given colony is isolated from cultivars in other Attine gardens because queens that found new colonies transport only the fungi from their initial colony. Because of evolution, the fungi cultivars diverge genetically. Studies have shown that mycelial filaments from neighboring cultivars are rejected by the fungi in any given colony, and the rejection intensity is proportional to the genetic difference between the fungi from the two colonies (Poulsen and Boomsma 2005).

The fungus-garden ants are the expert agriculturalists of the insect world, and their labors pay off in evolutionary fitness. The fungus symbiont digests cellulose, an energy-rich compound that ants cannot digest. Not only that, but the fungus is unaffected by many, if not most, of the defense compounds contained within leaves of many plant species (see Chapter 8).

By eating the nutritionally rich cultivated fungi, ants circumvent many of the diverse defense compounds typical of Neotropical plants, while at the same time tapping into the immense abundance of energy in rain forest leaves.

IS NATURE COOPERATIVE?

The suggestion is often made that mutualistic associations such as that between ants and fungus represent a clear example of how nature evolves to be cooperative rather than competitive, a kind of anti-Darwinian view of nature. But cooperation, however apparent, is not at all contrary to predictions of natural selection theory. Any obligate or faculative mutualism is a case in which one species is exploiting another and being exploited in return (though to the ultimate benefit of each). Each species in the mutualism acts as a constant selection pressure on the other, so each mutualism is a balancing act of sorts played out in evolutionary time. There is nothing anti-Darwinian about that. Studies of the complex five-species interaction among fungus-garden ants, Actinomycete bacteria, black yeast, antagonistic fungus, and cultivated fungus have demonstrated that the stability of the ant–fungus relationship may be dependent on the presence of antagonistic species that, in essence, exert selection pressures that actually stabilize the mutualism, preventing either the ants or their mutualistic fungus from asymmetric exploitation of the other (Little and Currie 2008).

FURTHER READING

Burslem, D., M. A. Pinard, and S. A. Hartley. 2005. *Biotic Interactions in the Tropics: Their Role in the Maintenance of Species Diversity*. New York: Cambridge University Press.

Thompson, J. N. 2005. *The Geographic Mosaic of Coevolution*. Chicago: University of Chicago Press.

8 Trophic Dynamics in Evolutionary Context

Chapter Overview

Energy passes through ecosystems via a complex web of interacting species, a process termed *trophic dynamics*. Solar energy input is converted to chemical energy by plants. (This process, termed *primary productivity*, is a focus of Chapter 9.) Energy moves from plants through consumers and ultimately to decomposers before exiting the system as heat. Like goods and services in economic systems, energy flow structures ecosystems (in what has been termed the *economy of nature*), a reality reflected in diverse trophic-based interactions among species. The high species richness and interdependencies typical of tropical rain forest ecosystems make trophic relationships particularly complex. As with all ecosystems, influences on food web structure may be *top-down*, as when organisms strongly influence the dynamics and evolution of organisms at lower trophic levels, or *bottom-up*, as when organisms on lower trophic levels affect those on higher trophic levels. Bottom-up effects may also be *abiotic*, as when plants are affected by soil type, moisture level, and so on. Both top-down and bottom-up forces are important in tropical rain forest food webs. Add to this the reality of evolution. Trophic dynamics is strongly affected by phylogenetic histories and coevolutionary interactions such as plant–herbivore defenses and evolution of mimicry systems.

Trophic Dynamics 101: An Introduction

Given the hyperdiversity of humid tropical forests, it should be expected that food webs in the tropics would exhibit the greatest complexity of any kind of ecosystem. Add to that the existence of numerous mutualistic networks (see Chapter 7) and complex evolutionarily influenced top-down and bottom-up interactions, and it is clearly evident that trophic dynamics in the humid tropics is extremely complex, its level of interactions perhaps unique among the world's ecosystems.

It has long been recognized that energy moves through ecosystems by passing among organisms in such a way that the energy moves along at various "steps" from the sun (Figure 8-1). Plants, called *primary producers*, are one step away. They are *autotrophs*, capable of synthesizing complex energy-rich compounds using simple precursors and solar energy. All other species, all animals and fungi, for example, are *heterotrophs*, requiring complex organic matter. Herbivores, called *primary consumers*, are two steps from the sun. *Secondary consumers*, the carnivores, are at least three steps, and top carnivores such as harpy eagles or tigers, are four or five steps from the sun. The energy that a tiger uses has moved from the sun, to plants, to the herbivore that the tiger devoured, before its use by the tiger. These steps from

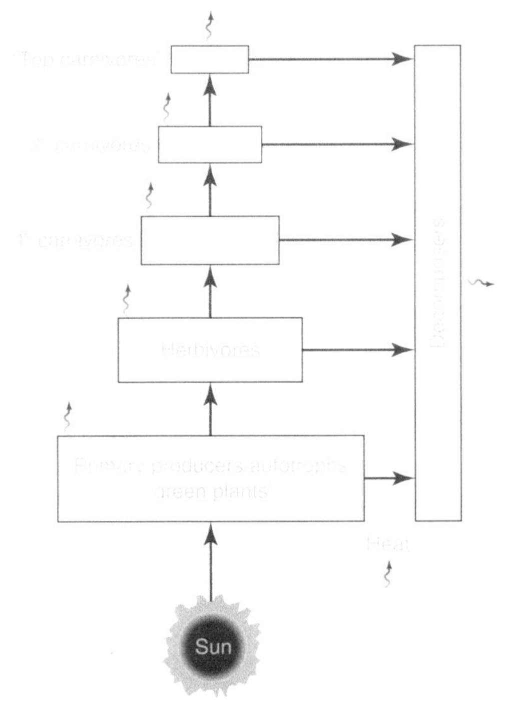

FIGURE 8-1
This simple compartment model illustrates the basic grazing food chain along with the decomposer compartment. Note that decomposers do not fit into the linear food chain, and note further that many animals may occupy more than one trophic level, something that is not evident in the diagram. Red arrows indicate direction of energy flow. Each compartment releases heat energy, and no energy is ever recycled.

the sun are referred to as *trophic levels*. This depiction is the classic *grazing food chain* that forms the conceptual backbone of trophic dynamics. Roughly 90% of energy present at a given trophic level is lost as energy moves to the next trophic level. This reality, which is largely derivative of the second law of thermodynamics (the law of entropy), means that there are far more producers (or producer biomass) than there are consumers, and there are far more herbivores (or herbivore biomass) than there are carnivores. This is why predators are far less numerous than their prey. The most accurate measure of movement of energy through a food chain is to measure the rate of passage, as in kilocalories per square meter per day. Standing crop biomass (as, for example, the weight of all dry plant biomass in a square meter) is a useful measure but does not take into account differences in metabolic rate among various organisms, differences that often affect energy flow.

When energy is measured as it flows through a typical grazing food chain, it may be depicted as an ecological pyramid, where plants, the primary producers, are at the base and top carnivores are at the apex. The pyramid shape is meant to show only that each trophic level contains much less energy than the level below. It is easy to understand that plants are ultimately responsible for the energy that is used by top carnivores just as the sun is ultimately the source of energy used by plants. But that reality obscures some important considerations.

The grazing food chain described earlier is simplistic. What really happens in ecosystems is that grazing and decomposer food chains are complexly interrelated into intricate, information-rich food webs. For example, some species, called *omnivores*, occupy positions at several levels of the grazing food chain, leading to numerous interactions with other species. Food webs are discussed in greater detail in the following sections.

The structure of the plant community may be strongly affected by the forces generated on it by higher trophic levels. This is called a *top-down effect*. Similarly, herbivores may be strongly affected by carnivores. Alternatively, the plant community may strongly affect the ecology and evolution of higher trophic levels, a *bottom-up effect*. Another bottom-up force is the effect that soil characteristics, or water availability, or energy input exerts on the plant community itself. The study of trophic dynamics has demonstrated that complex food webs result from a combination of both top-down and bottom-up forces. Changes in any of those influences produce *ripple effects* of various magnitudes throughout the food web.

Food webs are not sustainable unless the material comprising organisms, the atoms and molecules, are eventually recycled back to the primary producers. *Saprophytic* organisms (organisms that feed on nonliving organic matter) such as fungi and bacteria are termed *decomposers*. Decomposers are essential to all food webs (and will be discussed in greater detail in Chapter 10). They do not fit neatly on the energy pyramid (which is based on the traditional grazing food chain described earlier), but instead occupy their own compartment. Decomposers obtain energy from dead tissue and organic waste. In doing so, decomposers ultimately liberate low-energy atoms and molecules that are then recycled and used to rebuild organisms. Such recycling is not necessarily confined to the ecosystem in which the decomposers reside, as when they liberate carbon dioxide into the atmosphere, for example. The carbon dioxide released by a kapok tree in Venezuela may end up being dissolved into the Indian Ocean and then used by phytoplankton for photosynthesis.

Energy may be temporarily stored in an ecosystem in the form of living or dead biomass, or used kinetically or otherwise to drive metabolism. As described by the second law of thermodynamics, once energy is used, it ultimately becomes heat, a highly dispersed form of energy radiated to space. All energy entering an ecosystem is destined to degrade eventually to heat and will eventually be radiated away. Thus energy flows, it does not cycle, and Earth is an open system regarding energy. Once energy becomes heat, it may help keep organisms warm, but otherwise it is not reusable. Contrast this with recycling, where the material basis for organisms is constantly reassembled as organisms produce waste and eventually die and decompose, the decomposers releasing the atoms and molecules back to the abiotic pool (as in soil, for example), where such chemicals may be taken up, recycled, by primary producers. Decomposition will be discussed further in Chapter 10.

Food chains and trophic dynamics are often depicted as adversarial. Plants "don't want" to be grazed on by caterpillars, and a vervet monkey is presumably not anxious to become dinner for a leopard. Hostile interactions are obvious components of all food webs. But trophic dynamics, particularly in rain forest ecosystems, also include numerous mutualistic interactions such as animal-vectored pollination and seed dispersal. Carnivores that eat herbivores are indirectly adding to the fitness of plants, a top-down force. Complexities also exist within a given trophic level. Bird species in mixed foraging flocks are behaving as potential competitors for food within their trophic level, but also as mutualists to the extent that their flock organization aids each individual in food and/or predator detection (see the sidebar "Deception in a Mixed Foraging Flock of Birds").

All food webs consist of complex combinations of top-down and bottom-up forces (Pimm 2002). Top-down forces are exerted by species occupying a given trophic level on those species at lower trophic levels, as when predators eat herbivores, and thus reduce the effect of herbivores on the plants they consume. Bottom-up forces are exerted by organisms at lower trophic levels on those organisms occupying upper trophic levels, as when plants evolve toxic compounds that discourage herbivory. Examples will be described throughout this chapter. Food web complexity and dynamics has been extensively studied, particularly in relation to how it might contribute to ecosystem function and stability (de Ruiter et al. 2007). Modeling studies that have examined food webs in detail have suggested that food web stability is enhanced when species at a high trophic level feed on multiple prey species, and species at an intermediate trophic level are fed upon by multiple prey species (Gross et al. 2009). This implies that the number of food web connections (among the species in the ecosystem) is an essential component of food web stability.

Evolution adds a significant and often underappreciated component of complexity to trophic dynamics. This is because organisms evolve in response to each other, depending on the strength of the selection pressures they exert. This is evident in such ubiquitous patterns as cryptic coloration and warning (*aposematic*) coloration, each of which illustrates the evolutionary effects of antagonistic predator–prey interactions over evolutionary time.

THE EL VERDE RAIN FOREST OF PUERTO RICO

What does a food web of a tropical rain forest actually look like? Given the complex biodiversity of such an ecosystem, it might not come as a surprise if I said that no tropical rain forest food web has ever been completely diagramed. But, in fact, at least one tropical rain forest food web has been very well described (Reagan and Waide 1996). The food web of the El Verde rain forest ecosystem on the Caribbean island of Puerto Rico has been the subject of many studies that were combined into a now-classic book, *The Food Web of a Tropical Rain Forest.* The basic diagram of the food web appears in Figure 8-2.

The diagram of the food web of El Verde is a compartment model and does not attempt to list each species. Nonetheless, the complexity of the ecosystem's multiple energetic pathways is readily evident. El Verde's food web is not typical of rain forests in general. El Verde is an island rain forest, and as would be predicted from island biogeography theory (see Chapter 5), it has reduced numbers of species in comparison with mainland ecosystems. There are 468 vascular plant species and only 10 mammal, 40 bird, 14 reptile, and 13 amphibian species in El Verde. There are no large herbivorous or carnivorous mammals. The largest birds are quail-doves, pigeons, and a moderate-sized parrot species. The most important mammalian predator species is the Indian mongoose, an introduced, invasive species. This leads to some significant differences from food webs elsewhere in tropical forests. For example, the lack of large mammals and birds allows for potentially longer food chain lengths (Reagan et al. 1996). This is because large mammals and birds, being endothermic,

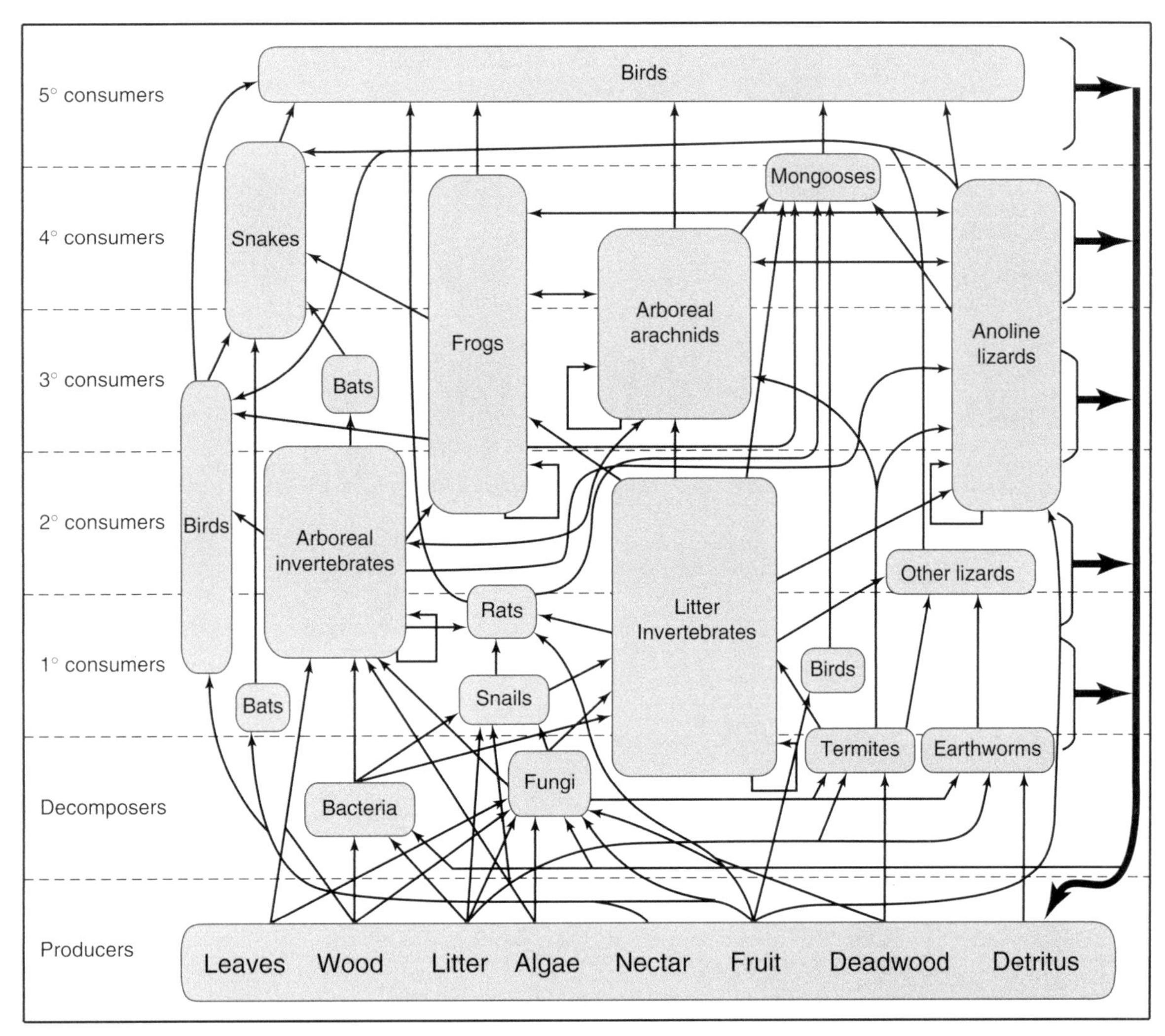

FIGURE 8-2
An aggregated food web of the El Verde community, presenting the groups discussed in this text.

require prodigious amounts of energy to support relatively little biomass. In the absence of large physiological endotherms, such constraints as they exert on food chain length are absent. The food web of El Verde therefore has longer food chain lengths than would be found in a tropical forest in, say, central Amazonia or Gabon, in Africa. The mean chain length for El Verde was 8.5 links, with a maximum of 19 links. Two predator groups dominate at El Verde, one diurnal and one nocturnal, and both small in body size. The diurnal predators are primarily anoline lizards, and the nocturnal predators are small leptodactylid frogs, numerically dominated by *Eleutherodactylus coqui*, the ubiquitous *coqui-frog* whose nocturnal vocalizations are heard throughout the island (Plate 8-1). The El Verde food web is therefore divided into two *subwebs*, one based on daytime predators and one based on nocturnal predators. The overlap in diet of frogs and lizards is only 13%. Omnivory among animals (which is defined as occupying more than one level on the food chain) was described as *pervasive*.

The El Verde study helps to make clear that food webs are very different from the simple linear conception of a grazing food chain. While it is insightful to appreciate the effect of thermodynamic law on structuring food chains (such that large top carnivores like jaguars and tigers are rare), it is essential to recognize that food chains exist clearly within the complexity of food webs.

PLATE 8-1
The ubiquitous coqui, a small tree frog.

Deception in a Mixed Foraging Flock of Birds

Studies have revealed high diversity and intriguing complexity of behavior within both canopy and understory mixed bird species flocks in the Peruvian Amazon (Munn 1984 and 1985; Munn and Terborgh 1979). These flocks move about in concert throughout the forest, in essence a form of *mobile food web* that travels as it feeds.

Each flock type consists of a core of 5 to 10 different species, each represented by a single bird, a mated pair, or a family group. Up to 80 other species join flocks from time to time, including 23 tanager, euphonia, and honeycreeper species, a remarkably high diversity. Mixed foraging flocks occupy specific territories, and when another flock is encountered, the same species from each flock engages in "singing bouts" and displays as boundary lines are established. Adult birds tend to remain flock members for at least two years. Nesting occurs in the general territory of the flock, the nesting pair commuting back and forth from nest to flock.

One long-held hypothesis about mixed flocks is that being part of a mixed flock serves to help protect against predation (Moynihan 1962). With so many eyes looking, predators have difficulty going undetected. Key species, called *sentinel species*, are part of every mixed flock. In one Amazonian study, two bird species, a shrike-tanager and an antshrike, aided the flock by giving general alarm calls when danger threatened. Both sentinel species, however, also gave false alarm calls, a behavior described as *deceitful*. False alarm calls are exactly that: alarm calls uttered when no danger is present. The hypothesized reason for the false alarms calls is that the alarmist has a better chance of capturing food that is also being sought by another bird. When a white-winged shrike-tanager (*Lanio versicolor*) (Plate 8-2) and another species were both chasing the same insect, the shrike-tanager's false alarm would cause momentary hesitation by the other bird, allowing the shrike-tanager to capture the insect.

PLATE 8-2
WHITE-WINGED SHRIKE-TANAGER (FEMALE)

PLATE 8-3
Boa constrictor on forest floor, demonstrating cryptic coloration. The photo was taken with flash, at night.

Predator–Prey Trophic Dynamics Reflects Evolution

CRYPTIC COLORATION

The Neotropical boa constrictor (*Boa constrictor*), if placed on a plain table, appears boldly patterned, complexly colored in browns and golds, with stripes, diamonds, and other markings. Once on the rain forest floor resting on decomposing leaves, the snake appears to blend into the background (Plate 8-3, and see Plate 1-12).

Place your hand on a tree trunk, and what you thought was bark might suddenly erupt in flight as a blue morpho butterfly (*Morpho* spp.), resting on the tree trunk, flies off at the disturbance. Once in flight, the iridescent blue wings make the large butterfly easy to see (Plate 8-4). At rest, however, the blue is concealed, making the insect largely indistinguishable from bark. Many insect species have wings patterned remarkably like leaves. Certain katydids mimic living

PLATE 8-4
BLUE MORPHO

(a) (b) (c)

PLATE 8-5
(a) to (c) These three images of katydids (Orthoptera) demonstrate how various species from the same insect group are adapted to be cryptic in the forest.

leaves (Plate 8-5), and many lepidopterans (and some katydids) mimic dead leaves, even to the point of including lighter areas simulating insect damage and leaf skeletal damage. An example is the butterfly species *Zaretis itys*, in the family Charaxinae, a species that feeds part of the time on the forest floor, on rotting fruits. The underside of the female butterfly bears an uncanny resemblance to a skeletonized leaf (DeVries 1987).

Cryptic coloration, or *crypsis*, is nature's camouflage. Thousands of species of tropical (and nontropical) insects, spiders (Plate 8-6), birds, mammals, and reptiles exhibit cryptic coloration to various degrees. It

PLATE 8-6
This large wolf spider exhibits crypsis as it searches for prey on the forest floor.

(a)

(b)

PLATE 8-7
(a) A great potoo (*Nyctibius grandis*) on a branch in Ecuador. Look very closely and notice the young bird in front of its parent. (b) A Papuan frogmouth (*Podargus papuensis*) at its nest (notice the sticks) in Australia.

is a prevalent reality of nature both on land and in the seas. Any cryptic animal, if removed from its normal background environment, would appear obvious. Each of these examples represents a case of evolutionary adaptation to trophic dynamics.

Cryptic coloration functions in one of two ways, and both may contribute to fitness. Crypsis may reduce detection by predators, and it may also reduce detection of a predator by its prey. In the tropics, what appears at first to be a green leaf may be a katydid, a twig may be a walkingstick or a mantis, a thorn may be a treehopper, dried leaves may be a butterfly, a bunch of dead leaves may be a coiled snake, bark may be a butterfly or moth, and a tree stump may be a bird.

The Neotropical potoos (Nyctibiidae, 7 species) are large nocturnal birds that feed on flying insects at night. By day, they sit utterly still atop a tree snag in plain sight (Plate 8-7). Any of these birds so much resembles a branch that it is easily overlooked. It even postures its body in such a manner as to more closely resemble a branch. In Australasia, a family of birds called frogmouths (Podargidae, 13 species) is the ecological equivalents of the potoos.

Tropical cats demonstrate cryptic coloration (Plate 8-8). The spotting and/or banding patterns of the Neotropical jaguar, the African leopard, or the Asian tiger, so obvious when the animals are observed in zoos, aid in concealing them when in nature. The coat patterns act to break up the animal's outline, rendering it less visible. Although cats are predators, cryptic coloration is no less an advantage to them, as it aids them in moving undetected toward their prey.

PLATE 8-8
Cheetahs (*Acinonyx jubatus*) inhabit savannas in Africa. Their spotting pattern breaks up their outline and renders them more difficult to detect. Note the second animal to the left, in shade and in front of the prominent one.

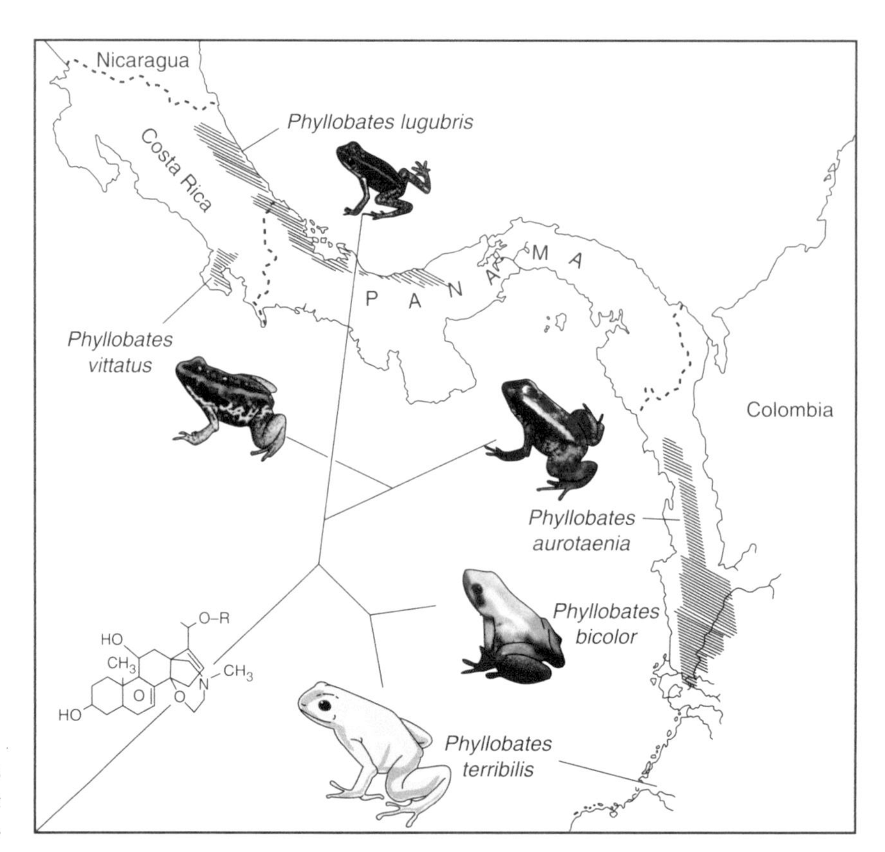

FIGURE 8-3
Ranges of species of poison-dart frog (*Phyllobates*).

WARNING (APOSEMATIC) COLORATION

Although many animals display crypsis, some send exactly the opposite message. Many groups of tropical butterflies and caterpillars as well as some snakes and frogs are brilliantly colored and stand out dramatically.

Consider the small frogs of the family Dendrobatidae, called *poison-dart frogs* (Figure 8-3). Within this group, there are about 30 species in the genus *Dendrobates* and five in the genus *Phyllobates*. All contain toxic alkaloids in their skin secretions. Each of these diminutive frogs of Central and northern South American rain forests is characterized by bold, striped patterns of orange, red, or yellow that glow like neon against a dark background (Plate 8-9). The Choco Indian tribes of western Colombia use toxic alkaloid

(a)

(b)

PLATE 8-9
(a) *Dendrobates pumilio*, sometimes called the strawberry poison-dart frog, sometimes called the blue jean frog, both for obvious reasons.
(b) Poison-dart frog (*Dendrobates auratus*).

compounds called *batrachotoxins* (a word that literally means "frog poison") extracted from the frogs' skins as potent arrow poison (Myers and Daly 1983; Maxson and Myers 1985). One species, aptly named *Phyllobates terribilis*, is reputed to be potentially lethal to the touch and may be the most toxic of all living creatures (Moffett 1995).

Skin toxicity and bright coloration are evolutionarily linked. These frogs represent a case of aposematic, or warning, coloration. Bright, bold patterning serves to signal potential predators to avoid the animal. Almost 300 noxious or toxic alkaloids have been isolated from various species of amphibians (Daly et al. 1993). Batrachotoxins are considered to be the most toxic of the various alkaloid compounds. (Alkaloids are discussed further later in this chapter.)

The precursors of batrachotoxin are obtained in the insect diets of frogs, particularly from ants. Frogs kept in captivity show reduced levels of toxicity, suggesting that certain components of diet are required to synthesize the skin toxins (Crump 1983; Daly et al. 1993).

The Toxic Birds of New Guinea

Alkaloids termed *homobatrachotoxins* have been found in the feathers of three brightly colored bird species called pitohouis (*Pitohui dichrous, P. kirkocephalus, P. ferrugineus)* of New Guinea rain forests (Dumbacher et al. 1992). These birds, members of a family called shrike-thrushes (Colluricinclidae, 14 species), represent the first and thus far only known examples of toxic bird feathers. Their bright colors suggest the possibility of warning coloration (Plate 8-10). The homobatrachotoxin in pitohouis evolved originally in the food the birds consume, a Melyrid beetle species, and it is essentially chemically identical to that found in Neotropical frogs. The bird, in its complex metabolism, must somehow biochemically sequester the beetle toxin into its feathers. How the beetle acquires the toxin is as yet unknown, as is how the biochemistry of the bird affects its ability to sequester the toxin and how that, in turn, relates to the bird's pigmentation pattern. Why should pitohuis evolve toxicity? One possible explanation is that species of bird-eating snakes inhabit New Guinea and the combination of neurotoxicity and possible aposematic coloration may adapt the birds to defend against serpent predation. If so, it has yet to be demonstrated.

It is curious how the discovery of pitohui toxicity was made. When a research team led by Bruce M. Beehler was netting birds in New Guinea, their indigenous assistants clearly knew that pitohuis were to be avoided, saying they "tasted bad." Years later, Jack Dumbacher, working with Beehler, discovered the toxic nature of pitohuis by inadvertently touching his mouth after handling several of the birds. His tongue and lips began to tingle and were soon numb. Further biochemical research on the birds' feathers confirmed the existence of the toxic chemical (Beehler 2008).

PLATE 8-10
HOODED PITOHUI (*PITOHUI DICHROUS*)

PLATE 8-11
CORAL SNAKE

Neotropical coral snakes (54 species) provide another example of boldly patterned and colorful animals that are extremely venomous (Plate 8-11). Coral snakes tend to be nonaggressive unless threatened. Even though potentially deadly, a coral snake could still be killed or suffer extensive harm if attacked. Various bird species routinely prey on snakes (see the sidebar "When Is a Coral Snake Not Protected?"). In the case of a coral snake, its well-defined red, black, and yellow pattern is presumably easy for birds to recognize, remember, and avoid. Avoidance of coral snakes may be innate in some bird species. Both the turquoise-browed motmot *(Eumomota superciliosa)* and the great kiskadee flycatcher *(Pitangus sulfuratus)* have been shown to instinctively avoid coral snake patterns (Smith 1975 and 1977).

Certain nonpoisonous snakes, including kingsnakes, closely resemble coral snakes and are thought to be coral snake Batesian mimics (Greene and McDiarmid 1981). *Batesian mimicry* will be discussed later in this chapter; it means that a nontoxic species evolves to look similar to a toxic species and therefore benefits from its resemblance. Experiments have demonstrated that the avoidance of ringed patterns similar to coral and king snakes is strong only where coral snakes are present (Pfennig et al. 2001). Bold patterning is common in tropical areas. Even caterpillars of the false sphinx moth *(Pseudosphinx tetrio)* have patterning suggestive of coral snakes (Janzen 1980) (Plate 8-12).

Warning coloration may be thought of as an example of how evolution influences bottom-up forces in

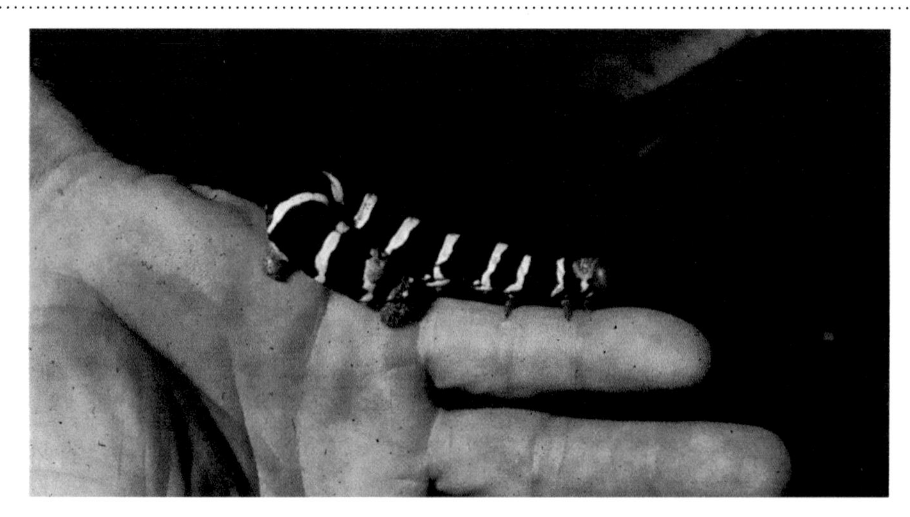

PLATE 8-12
PSEUDOSPHINX CATERPILLAR

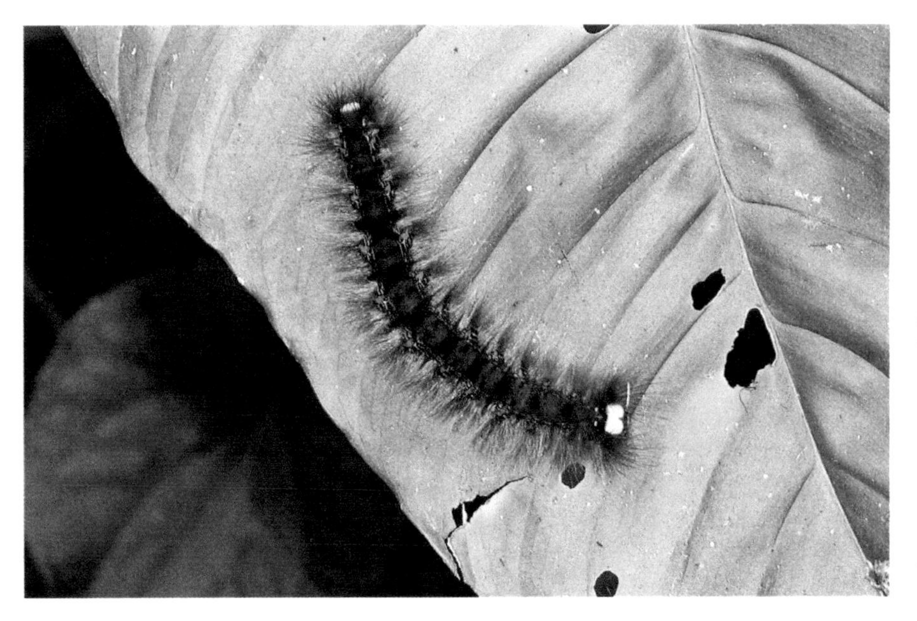

PLATE 8-13
This caterpillar also represents an example of aposematic coloration, and then some. Note its bright coloration (it appears to have "headlights"), but note also that it is covered with dense spines. These are *urticating hairs* that cause extreme itching and discomfort if touched. And if you do pick up this caterpillar, it will spasm such that you are likely to drop it immediately.

trophic dynamics. If an animal evolves a noxious chemical, the most visible and boldly patterned species present a signal that they are dangerous (Plate 8-13). The patterns and brilliant colors warn predators to stay away. An evolutionary adaptation at a lower trophic level affects would-be predator species at upper trophic levels.

When Is a Coral Snake Not Protected?

The warning coloration presumed to discourage birds from predation on coral snakes is not always effective. There is photographic proof that a bird of prey, the laughing falcon (*Herpetotheres cachinnans*), captures, kills, and eats at least one species of coral snake (*Micrurus nigrocinctus*). DuVal et al. (2006) observed two incidents of predation by a laughing falcon on a coral snake in Panama. In one instance, the falcon used its talons to hold the coral snake by its posterior end such that the serpent dangled from the branch. The snake lacked sufficient strength to raise its head to the point where it could strike the falcon. The snake eventually succumbed. In another incident, the observers reported seeing a laughing falcon with a decapitated coral snake draped over the leaning trunk of a tree (Plate 8-14). It is noteworthy that laughing falcons have apparently evolved at least two techniques by which they handle such a potentially lethal serpent as a coral snake. Quite a choice of diet.

PLATE 8-14
A perched laughing falcon (*Herpetotheres cachinnans*) holding a live coral snake (*Micrurus nigrocinctus*) by the posterior end.

Bottom-Up Plant Defenses

Looking at a sample of leaves in a tropical humid forest will usually reveal insect damage (Plate 8-15). Herbivore pressure on plants in the tropics is relatively constant because there is no cold winter (as in higher latitudes) when herbivores become inactive. But in general, tropical herbivores take a small percentage of what is potentially available to them (Coley et al. 1985). How do tropical plants defend against herbivorous hordes as well as against invasive pathogenic bacteria and fungi?

Leaves and stems may be covered with thorns of various sorts. Leaves tend to be thick and fibrous, of little nutritional value, and difficult for herbivores to digest. Also, leaves of both tropical and temperate zone plants are known to contain numerous chemicals that function in various ways that potentially defend the plant. Once known as *secondary metabolites*, they are now termed *defense compounds* and collectively called *allelochemicals* (Whittaker and Feeny 1971). Daniel Janzen (1975) suggested that the world of a herbivorous insect is not one of myriad green leaves, but rather of morphine, L-DOPA, and various other powerful chemicals such as strychnine and rotenone. Many familiar drugs ranging from curare to cocaine are derived from the allelochemicals of tropical plants.

Defense compounds likely originated as genetically accidental metabolic by-products or chemical wastes that, by chance, conveyed some measure of protection from attacks by microbes or herbivores. Such mutations would confer fitness, and natural selection would favor their rapid accumulation (Whittaker and Feeny 1971). Most plant species contain a diversity of secondary metabolites that could function as potential defense compounds. Some defense compounds function principally to protect against herbivores, some to protect against bacteria and fungi. The role of many allelochemicals remains uncertain (a reason why many researchers continue to refer to them as secondary metabolites rather than defense compounds). What follows is a brief introduction to some of the more prevalent groups of allelochemicals. It is not meant to be comprehensive.

ALKALOIDS

Alkaloids are familiar and often addictive drugs. Some, such as the frog toxins described earlier, are very toxic. Others are less so. Cocaine (from coca),

PLATE 8-15
Leaf damage by herbivorous insects is common in tropical leaves.

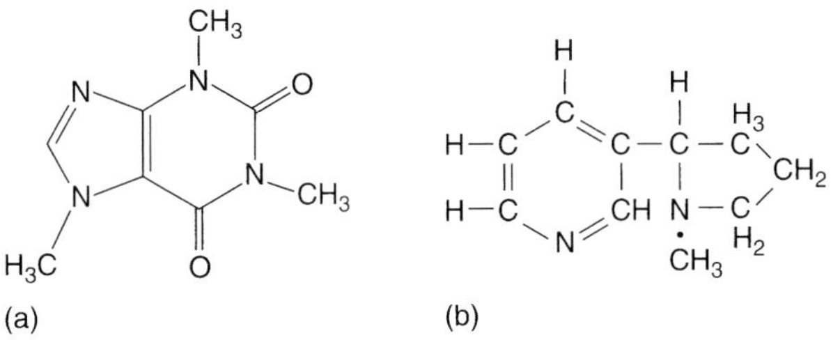

(a) (b)

FIGURE 8-4
(a) Caffeine molecule; (b) nicotine molecule.

morphine (from the opium poppy), cannabidiol (from hemp), caffeine (from teas and coffee), and nicotine (from tobacco) are alkaloids (Figure 8-4). Taken together, there are over 4,000 known alkaloids that are globally distributed among 300 plant families and over 7,500 species. A single plant species may contain nearly 50 different alkaloids (Levin 1976). Alkaloids are found not only in leaves but also almost anywhere in the plant, including seeds, roots, shoots, flowers, and fruits.

Most alkaloids taste bitter. In mammals, depending on dosage level, they may act as a stimulant or they may interfere with liver and cell membrane function. They may also cause lactation cessation, abortion, or birth defects. The bitter taste combined with the difficulties of digestion and liver function may discourage animals from consuming alkaloid-rich vegetation. Experimental evidence shows that caffeine discourages insect feeding (Nathanson 1984), as does nicotine (Stepphun et al. 2004). Alkaloids may act to toughen leaves (and that alone would discourage herbivory) as well as to store carbon and nitrogen (Futuyma 1983).

A Very Toxic Toad

The cane toad, *Bufo marinus*, is well known to visitors of Neotropical forests, where it is often a common resident on the forest floor (Plate 8-16). It has been introduced in many places outside of its normal range, including south Florida, much of the Caribbean, and as far away as Fiji and Australia. When it reaches adulthood, it is somewhere between the size of a baseball and a softball. It's a big toad, hard to miss, and it does not hop very quickly. It doesn't have to. Predators avoid it. The cane toad is chemically protected by a milky substance called *bufotoxin*, typically secreted from the animal's prominent paratoid glands should the creature be disturbed. One component of the toxin is an alkaloid called *bufotenin*, which has been shown to produce hallucinations in humans (who have been known to lick toads just to have this occur—although I would advise against it). Bufotoxin is present in tadpoles as well and has been known to cause death in animals that have consumed either tadpoles or adult toads.

PLATE 8-16
This adult cane toad is well protected by its chemical defenses. The bufotoxin is secreted by the large parotoid glands posterior to the head, shown as swellings in this image.

CHANGES IN LATITUDES

Levin (1976) made a detailed study of the geographic distribution of alkaloid-containing plants. Groups of herb, shrub, and tree species each contained a significantly greater percentage of alkaloid-bearing tropical species. In all, 27% of temperate species tested contained alkaloids, compared with 45% of the tropical species tested. Evolutionarily more ancient plants such as the magnolias, many of which are tropical species, contained more alkaloid-bearing representatives than species from more recently evolved families.

In Kenya, 40% of the plant species tested contained alkaloids compared with only 12.3% in Turkey and 13.7% in the United States. In Puerto Rico, a tropical but much smaller geographic area than the United States, 23.6% of the plants tested contained some alkaloids. Levin's analysis revealed a decrease in the percentage of plants bearing alkaloids from the equator northward, a trend termed a *latitudinal cline.* One explanation for this cline is that higher-latitude plant species are less subject to continuous pest pressure, and thus high alkaloid content has not been as strongly selected in these species. Another is that pest pressure may differ substantially between tropical and temperate areas, selecting for different defense chemicals. If there are more pest species in the tropics and they represent a greater range of genetic diversity, they will collectively exert greater qualitative evolutionary selection pressures on tropical plants.

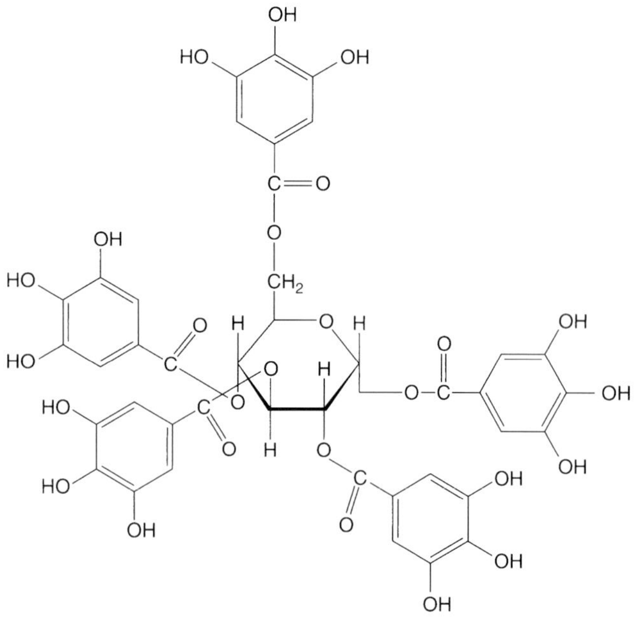

FIGURE 8-5
The basic molecular structure of a tannin molecule.

PHENOLICS AND TANNINS

Phenolic compounds are often abundant in plants (Levin 1971). One group adds the pungency to many of the most well-known spices, and another, known as the *tannins*, provides the basic compounds used in tanning leather (Figure 8-5). Tannins are complex polyphenols with abundant hydroxyl and carboxyl groups that bond with proteins. Tannins are abundant in temperate and tropical oak leaves as well as in many tropical plant species such as mangroves. Most people who enjoy wine know that tannin compounds are responsible for the astringency of certain red wines.

Research on Barro Colorado Island (BCI), Panama, that focused on the early successional tree *Cecropia peltata* showed that tannins are heavily concentrated in young trees but decline in concentration in older plants (Coley 1984), a characteristic that appears to be generally true for most tropical plants. Tannin levels are lower in plants grown in the shade, indicating that tannin production may be metabolically costly, requiring full sunlight. In field experiments, low tannin plants experienced twice the level of herbivory as those with high tannin levels. However, leaf production was inversely correlated with tannin levels. The more leaves on the tree, the lower the tannin per leaf, indicating that tannin production, though perhaps protective, is likely costly to the plant. Trees like *Cecropia,* experiencing intense competition for light, may have to limit tannin protection in favor of rapid growth, a situation termed an *evolutionary trade-off*.

Phenolics, less structurally complex than tannins, are compounds containing a hydroxyl group (—OH) directly bonded to an aromatic hydrocarbon group. The compound phenol (C_6H_5OH) is an example. Phenolics are stored in cell vacuoles that are broken when an insect or other herbivore bites the leaf. Upon release, the phenols combine with various proteins, including those enzymes necessary for splitting polypeptides (parts of proteins) in digestion, perhaps making it more difficult for a herbivore to digest protein. Leaf damage by insects or pathogens may stimulate production of phenolics (Ryan 1979).

Neotropical leaf-cutter ants (see Chapter 7) are apparently undeterred by phenolics (Hubbell et al. 1984). The role of phenolics and tannins as antiherbivore adaptations is unclear. It is uncertain as to what degree tannins serve as a general defense mechanism

(Zucker 1983). Some insects have evolved enzymes that detoxify specific defense compounds, an example of the evolutionary arms race to be described later in this chapter (Krieger et al. 1971). Many tannins retard microbes and pathogens and therefore affect rates of decomposition and nutrient cycling (Zucker 1983; Martin and Martin 1984; and see Chapter 10).

CYANOGENIC GLYCOSIDES

Many species of tropical plants contain compounds called cyanogenic glycosides consisting of cyanide (a potentially deadly compound) linked together with a sugar molecule. When combined with enzymes from either the plant or an herbivore's digestive system, the sugar is released, leaving hydrogen cyanide.

Cyanogenic glycosides as well as alkaloids, tannins, and other defense compounds are well represented in passionflowers (*Passiflora* spp.). Very few insect herbivores feed routinely on passionflower leaves or stems. The defense compounds apparently act to discourage most herbivores, but not all. *Passiflora* species are fed on by caterpillars of *Heliconius* butterflies. This group of lepidopterans has evolved resistance to the *Passiflora* defense compounds. *Passiflora* species vary in chemical profile and have been classified into five groups according to the presence or absence of various chemical classes (Smiley and Wisdom 1985). *Heliconius* larvae (caterpillars) of two species exposed to a range of *Passiflora* chemical profiles were able to consume the plants without difficulty. Only the level of nitrogen present in the *Passiflora* affected larval growth. *Heliconius* species are able to detoxify cyanogenic glycosides with specific hydrolytic enzymes. These remarkable butterflies will be discussed in more detail later in this chapter.

The Ethnobotany of Cassava

Manioc (*Manihot esculenta*), usually referred to as *cassava* or *yuca*, is a tuber crop whose root is highly rich in carbohydrate and has become a staple of the global tropics. A native of South America, the plant has long been grown throughout the tropical world. A perennial, manioc will grow annually without replanting. It is usually planted from cuttings taken from mature plants (Plate 8-17). The root is typically ground into a paste and made into hard bread, though it can also be fermented into a kind of beer. Whole recipe books have been written about use of cassava. The plant itself is a small, spindly tree, with palmate, compound leaves, most unpretentious in appearance. However, the thick root is long, often more than a meter (Plate 8-18).

Grown on nutrient-poor soils, manioc contains prussic acid, a powerful cyanogenic glycoside that defends

PLATE 8-17
Manioc cutting, recently planted, will grow into a full-size plant.

the root from herbivore attack. Indigenous people have developed various methods for removal of the prussic acid, an absolute necessity before further preparation. For example, in some cultures, toxic manioc root paste is soaked in water and repeatedly squeezed, compressed, literally wrung out, washing the water-soluble cyanogenic compounds from the paste, rendering the root safe to consume.

Many varieties of manioc exist, with variable levels of prussic acid concentration. There are *sweet manioc* varieties, with essentially no prussic acid, and *bitter manioc*, which has high concentrations of cyanide compounds. Sweet-tasting manioc varieties grow only in the most fertile soils, while the bitter strains are found in soils of low fertility, where herbivore damage would be more costly to the plant (Hansen 1983).

PLATE 8-18
Roots of cassava for sale at a market in Manaus, Brazil.

The Cyanogenic Millipede

Plants are not the only organisms to utilize cyanogenic compounds. Millipedes, in the huge phylum Arthropoda, are harmless ambling herbivores of the forest floor and should not be confused with swift-moving carnivorous centipedes (which inject toxin when they bite). Neotropical millipedes in the genus *Nyssodesmus* appear armored with a flattened shiny carapace protecting their delicate undersides (Plate 8-19). The creature may reach 10 centimeters in length and is readily visible as it moves about on the forest floor.

When threatened, millipedes roll up in a tight ball and tough it out (Plate 8-20). Some

PLATE 8-19
Millipede on forest floor.

also have an impressive array of chemicals at their command. The hindgut can squirt a volley of noxious liquid, containing both hydrogen cyanide and benzaldehyde (Heisler 1983). You'll notice this trait if you handle one of these creatures. Your hands will smell distinctly like almonds, from the cyanide. Wash your hands.

PLATE 8-20
Millipede in hand, curling into a ball.

TERPENOIDS

Terpenoids are a complex group of fat-soluble compounds. Some are used in the synthesis of compounds that may mimic insect growth hormones (preventing rather than promoting growth of the insect) or are modified into cardiac glycosides (Futuyma 1983). Some terpenoids discourage both insects and fungi. One terpenoid in particular, caryophylene epoxide, has been shown to repel the fungus-garden ant *Atta cephalotes* from clipping leaves of *Hymenaea courbaril*. This terpenoid was shown to be toxic to the fungus that the ants culture (Hubbell et al. 1983). In a survey of 42 plant species from a Costa Rican dry forest, 75% contained terpenoids, steroids, and waxes that repelled leaf-cutter ants (Hubbell et al. 1984).

TOXIC AMINO ACIDS

Some tropical plants, especially members of the bean and pea family (legumes, family Fabaceae), contain amino acids that do not build protein but instead interfere with normal protein synthesis. Canavanine, for example, mimics the essential amino acid argenine. Perhaps the best known of the toxic amino acids is L-DOPA, a strong hallucinogen. Both canavanine and L-DOPA are concentrated in the seeds of some tropical plants. In general, the major function of nonprotein amino acids, at least in legumes, seems to be to discourage herbivores (Harborne 1982).

OTHER TYPES OF PLANT DEFENSE

We drive on a defensive adaptation of tropical plants, namely rubber tires. The rubber tree *(Hevea brasiliensis)*, which can reach heights of 36.5 meters (about 120 feet), is one of many tropical trees that producc latex, resins, and gums, all of which act to render the trees less edible.

Rubber tree sap is a milky suspension in watery liquid contained in ducts just below the bark, external to the cambium and phloem (Janzen 1985). It congeals on exposure to air and may aid in closing wounds, protection against microbial invasion, and hindering herbivores. Latex is present in plants of many families (Euphorbiaceae, Moraceae, Apocynaceae, Caricaceae, Sapotaceae, and so on), a case of a convergent defense adaptation among distant species of tropical plants. The chicle tree, *Manilkara zapota,* which grows in rain forests throughout Central America, produces latex called *chicle*, from which chewing gum is made. Southeast Asian dipterocarp trees exude oily latex from bark that may serve as a defense against bacterial and fungal invasion.

In those trees whose leaves are chemically nonrepellent to leaf-cutter ants, sap adhesion to the insects' mouthparts and appendages discourages ants from clipping the leaves (Hubbell et al. 1984). Some insect herbivores have adapted to latex defense. One caterpillar species clips leaves of papaya in such a way as to cause the defensive latex to flow away from where the insect is feeding (Janzen 1985).

Tropical trees may be spiny, thorny, or have leaves coated with diminutive "beds of nails" called *trichomes* that literally impale caterpillars (Gilbert 1971). Experiments have shown that sharply toothed leaf edges reduce caterpillar grazing. When teeth are experimentally removed, caterpillars inflict much greater damage (Ehrlich and Raven 1967). Many palm species have spines lining the lower trunk. Some palms also have long, sharp spines on the undersides of leaf midribs. Wood of many tropical trees is hard, a possible adaptation to discourage termites and wood-decaying fungi.

Leaf toughness, nutrition value, and fiber content strongly contribute to herbivore resistance. Coley (1983) examined rates of herbivory and defense characteristics of 46 canopy tree species on Barro Colorado Island (BCI). She compared young leaves with mature leaves and gap-colonizing species with shade-tolerant species (Table 8-1). In general, young leaves were grazed much more than mature leaves, even though many contained phenols (indicating that phenols do not prevent herbivory, though some may discourage it). Researchers estimate that although leaves of shade-tolerant species may live for several years, 75% of the lifetime damage occurs during the early weeks, when the leaves have opened and are expanding

TABLE 8-1 Defensive characteristics of young and mature leaves of pioneer and persistent canopy tree species. Values within a row followed by different superscript letters are significantly different ($P < 0.05$, Mann-Whitney U).

	PIONEER (22 SPECIES)				PERSISTENT (24 SPECIES)			
	YOUNG		MATURE		YOUNG		MATURE	
	MEAN	RANGE	MEAN	RANGE	MEAN	RANGE	MEAN	RANGE
Chemical								
Total phenols (% dry mass)	12.7[b]	2.9–37.8	7.7[a]	2.6–19.1	19.2[c]	1.7–38.9	10.1[a]	1.7–22.6
Tannins—Vanil (% dry mass)	1.7[a]	0.04–8.4	0.8[a]	0.03–3.4	5.1[c]	0.03–15.8	2.7[b]	0.04–7.8
Tannins—Leuco (% dry mass)	3.4[a]	0–25.2	1.7[a]	0–7.5	9.7[c]	0.5–24.6	4.8[b]	0.07–13.2
Fiber—NDF (% dry mass)	38.1[a]	13.7–62.5	43.6[a]	27.2–64.2	40.4[a]	17.5–61.8	51.3[b]	32.8–67.5
Fiber—ADF (% dry mass)	27.5[a]	9.3–43.3	29.5[a]	19.7–47.7	31.5[a]	7.4–55.8	37.4[b]	15.6–49.9
Lignin (% dry mass)	9.6	0.9–17.7	10.5	3.3–18.5	11.7	1.4–21.5	12.5	3.3–20.8
Cellulose (% dry mass)	17.0[a]	7.4–30.4	17.6[a]	10.2–30.4	19.5[a]	5.9–41.6	23.4[b]	12.2–30.3
Physical								
Toughness (N)	2.20[a]	1.24–4.53	3.84[b]	2.46–7.09	2.45[a]	1.54–4.75	6.10[c]	2.73–11.56
Hairs (% of species)	65[b]		65[b]		38[a]		25[a]	
Hairs—upper (no./mm^2)	1.9[c]	0–7.0	0.4[b]	0–1.8	0.5[b]	0–6.7	0[a]	0–0
Hairs—lower (no./mm^2)	6.2[b]	0–18.0	5.1[b]	0–18.0	1.4[a]	0–10.0	0.5[a]	0–3.4
Nutritional								
Water (%)	74[c]	63–83	70[b]	57–82	76[c]	66–88	62[a]	49–77
Nitrogen (% dry mass)	3.2[b]	2.1–4.3	2.5[a]	1.7–3.1	3.3[b]	1.8–5.1	2.2[a]	1.2–3.1

(Coley et al. 2005). Nonetheless, some plants have adapted to reduce herbivory on young, expanding leaves (Kursar and Coley 2003).

Young leaves are richer in nutrients and defense compounds but lower in fiber and toughness. Mature leaves of gap-colonizing species were grazed six times more rapidly than leaves of shade-tolerant species, and overall, gap-colonizing trees had less tough leaves with lower concentrations of fiber and phenolics than shade-tolerant species. Gap tree leaves grew faster but had shorter lifetimes. Coley concluded that leaf toughness, fiber content, and nutritive value were more influential than defense compounds in affecting patterns of herbivory.

Studies on nutritional choices of howler monkeys (see the sidebar "Plants versus Howler Monkeys" later in this chapter) resulted in a similar conclusion (Milton 1979 and 1981).

Ecological Variability in Defense Compound Distribution

Defense compounds are particularly abundant in lowland forest occurring on nutrient-poor white sandy soils (Plate 8-21) such as are found in the northern Amazon region. (Chapter 10 discusses soil in more detail.) Leaves from the vegetation of white soil forests are metabolically costly to replace (Janzen 1985). These leaves are long-lived (several years), with high concentrations of defense compounds such that when a leaf finally drops, it must be leached of its defense compounds by rainfall before it can be broken down and its minerals recycled. Water from blackwater rivers characteristic of white sandy soil regions appears dark and "tea-like" because of the concentration of leached phenolics (e.g., tannins). Plants have greater fitness if they manufacture enduring leaves with concentrated defense compounds than replacing leaves ravaged by microbial pathogens, fungi, or herbivores (Janzen 1985). Given the shortage of minerals in the soil, the replacement of leaves is more costly than synthesis of defense compounds.

Most early successional plant species are adapted to maximize growth rates. They synthesize alkaloids, phenolic glycosides, and cyanogenic glycosides, all present in low concentration, collectively representing a relatively low metabolic cost. In contrast, plants on nutrient-poor soils invest in metabolically more costly defenses such as polyphenols and fiber (e.g., lignin) that are retained in leaves and bark. Trees grow more slowly but are better protected. These contrasting patterns in plant defenses are a function of resource availability (Coley et al. 1985). On sites where resources are poor, "expensive," long-lasting defense compounds are favored. On resource-rich sites, "cheaper," shorter-lasting defense compounds are favored because the

PLATE 8-21
White sandy soil near Manaus, Brazil.

PLATE 8-22
Cecropia leaves, showing extensive damage by herbivorous insects.

tree is able to both devote sufficient energy to rapid growth and replace defense compounds as needed.

Many successional species are subject to significant herbivore damage (Brown and Ewel 1987). Cecropias are fast-growing but routinely show extensively damaged leaves (Plate 8-22). Perhaps cecropias "trade off" protection by defense compounds for rapid growth (see the section "*Inga* Leaves: A Fork in the Evolutionary Road" later in this chapter). Vines and herbs typically show more damage in successional areas compared with adjacent mature forest. Janzen (1975) estimated that insect density is 5 to 10 times greater in successional areas than in the understory of a mature rain forest, presumably because successional species are more palatable to insects.

Ant-Plants and Plant Ants

Many tropical plant species possess nectar-secreting glands called *extrafloral nectaries* (EFNs) as well as other structures that attract ants. Extrafloral nectaries are found in 90 plant families (including ferns, epiphytes, vines, and trees) and 330 genera (Rudgers and Gardener 2004). These plants are called *ant-plants*, or *myrmecophytes,* because of their ant attractant properties (Benson 1985). Ant-plants occur widely in the Old World tropics, especially Southeast Asia, as well as elsewhere throughout the tropics. They are also present in areas such as the Sonoran Desert in southwestern North America. Ant-plants normally have some form of shelter for ants (*ant domatia*), as well as providing nutrition.

Domatia range from mere hollow stems to more sophisticated shelters such as specialized pouches or thorns. Extrafloral nectaries are present on leaf blades, leaf petioles, stems, or other locations on the plant. These glands manufacture various energy-rich sugary compounds as well as certain amino acids. In addition, some plants have *bead bodies*, which are modified hairs rich in oil (Benson 1985).

Extrafloral nectaries were initially puzzling, but it was quickly learned that plants with such bodies are populated by various aggressive ant species. This observation led to the *protectionist hypothesis*, that the relationship between plants and ants is mutualistic. The alternative idea, called the *exploitationist hypothesis*, argued that the ants fed on the sugary nectaries but provided no actual service to the plants. An early (1910) description of the exploitationist hypothesis suggested that the plants "have no more use of their ants than dogs do their fleas" (quoted in Bentley 1977). However, research has strongly supported the protectionist hypothesis.

Cecropias are common Neotropical trees with a unique form of defense similar to but distinct from extrafloral nectaries. Cecropia produces Müllerian bodies located at the base of the leaf petiole, where the large leaf attaches to the stem. Ants of the genus *Azteca* live in domatia within modified hollow pith of the stem and feed on the Müllerian bodies. I have frequently encountered these ants, and they are not nice ants. Cutting cecropias for use as net poles (to catch and band birds), I have experienced attacks by Aztec ants, and I have little sympathy for the exploitationist hypothesis. The ants of a cecropia are pugnacious and appear to behave protectively of their tree. The underside of the wide, palmate cecropia leaves is velvetlike, with a carpet of tiny hairs and hooks that allows ants to gain purchase and move easily across the leaf. Cecropia species that normally lack ants have leaves with smooth undersides (Benson 1985).

Daniel Janzen (1966) documented the protectionist activities of *Pseudomyrmex ferruginea*, an ant species found on five species of *Acacia* tree. Commonly called the bull's horn or swollen-thorn acacias, these trees have pairs of large hollow thorns on the stem that shelter the ants (Plate 8-23). A single queen ant burrows into a thorn of a sapling acacia to begin a colony that may grow to 12,000 ants by the time the tree matures. When the tree is 7 months old, 150 worker ants are "patrolling" the stem. The acacia ants attack other insects that land on or climb on the tree, including beetles, hemipterans (true bugs), caterpillars, and other ants. Ants also clip plants that begin to grow nearby or overtop and shade the acacia (thus taking its sunlight), and attack mammals, including people, if they should brush against the tree (another personal experience not easily forgotten). Ants appear agitated, swarming out of the thorns and over the foliage at any disturbance.

Ants obtain shelter within the thorns as well as nutrition from two kinds of extrafloral nectaries. One type is termed *Beltian bodies*, small orange globules growing from the tips of the leaflets of the compound leaves, the other type being *foliar nectaries*, located on the petioles.

Janzen performed a field experiment that discriminated between the protectionist and exploitationist hypotheses. He treated some acacias with the insecticide parathion, and he clipped thorns to remove all ants from the treated trees. The antless trees did not survive nearly as well as control trees, which were permitted to keep their ants. Janzen estimated that antless acacias were not likely to survive beyond one year, either falling

(a)

(b)

PLATE 8-23
(a) The branches of this understory acacia look innocent enough until you get close and meet the ants that guard the plant.
(b) The swollen, paired hollow thorns of the acacia tree house small but aggressive *Pseudomyrmex* ants, seen here swarming over the branch.

prey to herbivores or being overtopped by other competing species of plants. Janzen concluded that the ants and acacias are obligate symbionts, depending entirely on each other.

The swollen-thorn acacias attract protectionist ants in a biochemical manner. They synthesize the enzyme invertase that cleaves the dissacharide sucrose in the EFN, rendering it to glucose and fructose. Protectionist ants avoid sucrose but are attracted to glucose and fructose. And the contrary is also true. Those ants that would feed on acacia extrafloral nectaries without protecting the tree are attracted to sucrose and repelled by fructose and glucose (Heil et al. 2005). Researchers manipulated the sucrose concentrations in extrafloral nectaries of acacias and were able to attract or repel *Pseudomyrmex* ants on the basis of presence (repel) or absence (attract) of

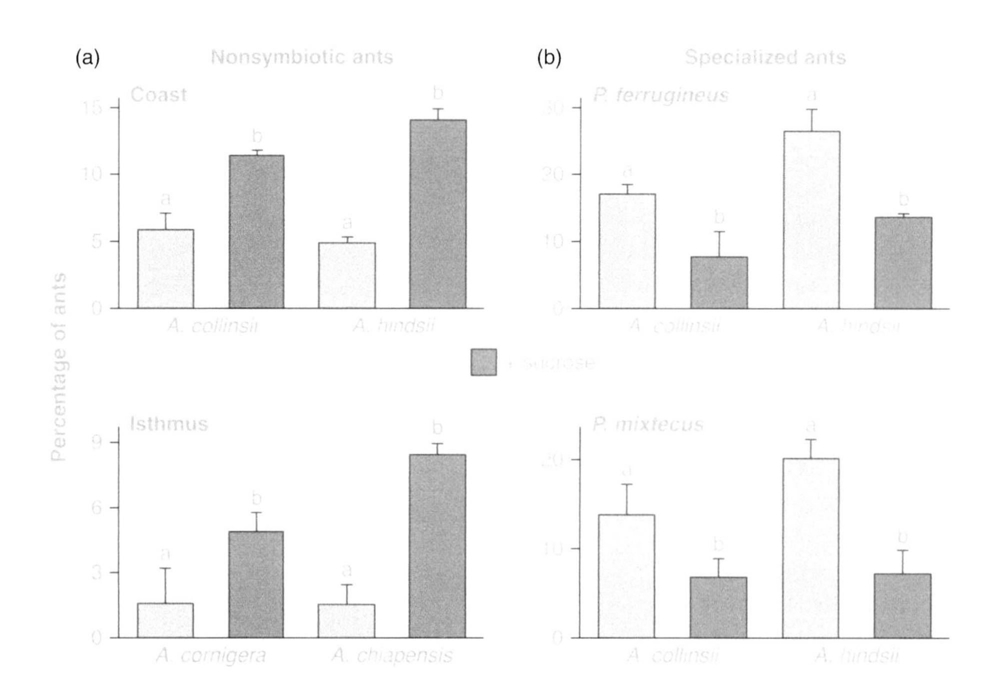

FIGURE 8-6

Responses of ants to EFN nectar of two Acacia myrmecophytes (*A. cornigera* and *A. hindsii*) supplemented with sucrose. Responses of nonsymbiotic (a) and specialized (b) ants to nectars without (tan bars) and with (brown bars) added sucrose (equal volumes of natural nectar and sucrose solution mixed, concentration 2% [weight/volume]) are given as percentages of ants attracted to these two solutions in cafeteria-style experiments. Nonsymbiotic ants were investigated at two sites (coast and isthmus), and the specialized *Acacia* inhabitants (*P. ferrugineus* and *P. mixtecus*) were investigated on *Acacia collinsii* shrubs. Bars represent means +SE. The number of ants attracted to nectar with and without added sucrose differed significantly (indicated by letters; $P < 0.05$ according to paired t test, $n = 7$ feeding sites [a] or colonies [b] in all cases]).

sucrose (Figure 8-6). This means that the acacia not only makes sucrose, it goes an additional step of synthesizing invertase to convert the sucrose to a chemical signal satisfactory to the protectionist ants.

Extrafloral nectaries are also known for some temperate-zone plants (Keeler 1980) but are far more abundantly represented among tropical species (Oliveira and Leitao-Filho 1987). In a survey of riparian and dry forests in Costa Rica, plants with extrafloral nectaries ranged in percentage cover from 10% to 80% (Bentley 1976). In the Brazilian cerrado, total cover by woody plants with extrafloral nectaries ranged from 7.6% to 20.3% (Oliveira and Leitao-Filho 1987). Many plants with extrafloral nectaries house ants, but the degree to which the ants act to protect their hosts varies (Bentley 1976 and 1977).

In the case of swollen-thorn acacias and some cecropia species, ants have assumed the function of defense compounds. Consider the trophic relationship. Plants exert a bottom-up but mutualistic evolutionary selection pressure on ants by providing food and shelter, at a clear metabolic cost to the plant. But ants protect the plant (which forms their exclusive territory), as it is in their evolutionary self-interest to be top-down predators on species that would reduce the fitness of the ant-plant. How delicate is this relationship?

Consider that ants, being animals, require abundant protein. The ant-plants offer the ants abundant carbohydrate in the form of EFNs, but nitrogen levels are low. This reality has led to the *deficit hypothesis*, which proposes that ants are aggressive defenders of plants because they need to kill other insects to acquire sufficient protein (Ness et al. 2009). In other words, selection pressures would favor plants that produce energy-rich carbon compounds for ants but not nitrogen-rich compounds, because in doing so, the ants would become more parasitic of the plant and less protectionist. By "forcing" ants to kill for protein, the plant gains from the relationship. The deficit hypothesis was tested for a guild of Sonoran desert-inhabiting ants and *Ferocactus wislizeni*, a species of barrel cactus, by manipulation of the C:N ratio available to ants (Ness et al. 2009). Researchers showed that the C : N ratio provided by the plant greatly exceeded that of the ants' insect food (meaning that the plant's EFN was deficient in nitrogen). Second, nitrogen enhancement of bait resulted in recruiting many more ants than did carbon enrichment. The research team cleverly titled their paper "For Ant-Protected Plants, the Best Defense Is a Hungry Offense."

In a sweeping meta-analysis of ant–plant relationships, Chamberlain and Holland (2009) showed that host plants virtually always benefited, in many cases very strongly, from the presence of protectionist ants. Plants that fared the best (that is, experienced the lowest levels of herbivory) were plants at tropical latitudes (more so than those in temperate latitudes) and plants that offered not only food bodies but also domatia to the ants.

EXTRAFLORAL NECTARIES PROMOTE MULTISPECIES INTERACTIONS

Because they represent an energy resource, extrafloral nectaries attract numerous arthropods, not merely ants. There is constant selection pressure operating to screen arthropods as antagonists or otherwise. Ant defenders are subject to selection pressures by organisms other than plants. In a remarkable example documented in Panama (DeVries 1990 and 1992; DeVries and Baker 1989), caterpillars of the butterfly *Thisbe irenea* entice ants to protect them, rather than the ants' host plant (*Croton*), and the caterpillars eat the leaves from the very plant the ants were once protecting. These caterpillars, termed *myrmecophilous*, for their "ant-loving" habits, have evolved at least three separate organs that act to attract and satisfy ants: nectary organs that produce protein-rich ant food; unique tentacles that release chemicals mimicking those of the ants themselves and signaling them to defend; and vibratory papilla that, when the caterpillar moves its head vigorously, make sounds that travel only through solid objects but that immediately attract ants. Ants appear to have a stronger preference for protein-rich caterpillar nectar droplets than for the carbohydrate-rich *Croton* nectaries. Ants are essential in protecting the otherwise vulnerable caterpillars from predatory wasps. Thus, by providing nectar for ants, the caterpillars have succeeded in averting the main protective adaptation of the plant as well as ensuring their own relative safety from their major predators, wasps.

The trophic dynamics of extrafloral nectaries remains poorly understood with regard to how these structures affect multispecies interactions. Because of the commonness of extrafloral nectaries in the tropics, this nectar source may be a widely utilized resource responsible for significant energy movement through tropical food webs. While some plant species attract specific ant species, it remains true that EFN-plants attract numerous species. One survey of 35 EFN-bearing plant species showed a presence of from 1 to 27 ants species, with a mean of 11.7 per plant species (Rudgers and Gardener 2004). Plants may compete both intra- and interspecifically for access to protectionist ants. This selection pressure would drive trophic dynamics to favor evolution of increasing numbers of extrafloral nectary–bearing plants. The authors of the study conclude: "EF nectar resources can affect the composition and dynamics of communities. In some systems, plants bearing EF nectaries contribute to a considerable fraction of the total flora and are visited by a large and diverse arthropod assemblage. Thus, attempts to understand the factors that influence arthropod diversity and abundance, as well as analyses of terrestrial food web dynamics, could benefit from considering EF nectar resources" (Rudgers and Gardener 2004, pp. 1500–1501).

THE PARADOX OF THE ANTS

Tropical ecologists who focus on animals soon realize how abundant and diverse ants are throughout tropical regions. Estimates based on samples collected after fogging studies have suggested that ants account for between 20% and 40% of arthropod biomass in tropical rain forest. But ants are thought to be primarily predators and scavengers. How is such high ant biomass supported? That is the paradox of the ants. One possibility should already be evident, that of extrafloral nectaries. But that is insufficient to fully explain the large ant biomass in forest canopies. In a study of herbivores in the canopy of tropical rain forest in Peru and Borneo, it was learned that some canopy ants essentially function as herbivores, at least in a manner of speaking, moving them lower on the food chain, "closer to the sun," allowing for greater biomass (Davidson et al. 2003). The ants were considered cryptic herbivores because rather than feeding directly on plants, they fed primarily on liquid exudates from a variety of other arthropods including leaf-chewing insects and sap-feeding insects, collectively called *trophobionts*. Ants were taking food from these herbivores, forcing them to continue to feed on the plants, exerting a greater impact on plants than would be the case if the ants were absent. Ants also fed

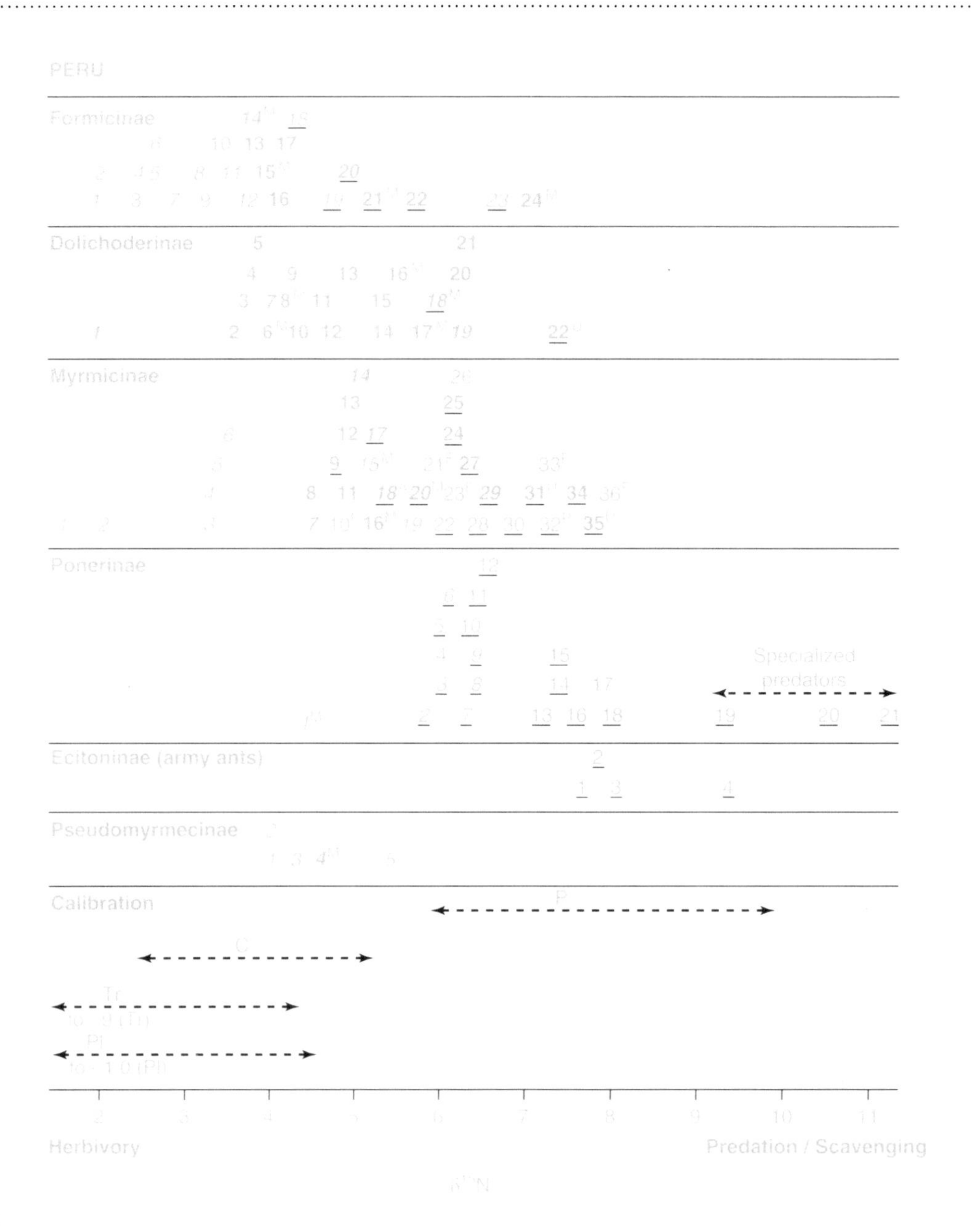

FIGURE 8-7
Mean $\delta^{15}N$ values (‰, by subfamily) for numbered Peruvian species. Predominant foraging modes are shown as follows: italics, leaf foraging; bold, trophobiont tending; underlined, predation; none of the above, poorly studied. Superscripts: A, ant-garden resident; F, attine fungus cultivators; M, specialized associate of myrmecophytes; P, brood parasite; U, unicolonial species. Calibration: Pl, 36 nonmyrmecophytic plant samples; Tr, 26 sap-feeding trophobionts, not from myrmecophytes; C, 13 chewing herbivores (Coleoptera, Orthoptera, and Diptera); P, 7 arthropod predators (Acari, Araneida, Hemiptera, and Pseudoscorpionida).

on pollen, fungal spores, fungal strands, and on tiny microflora that attaches to leaves (*epiphylls*).

Researchers examined the ratio of $^{15}N/^{14}N$, expressed as $\delta^{15}N$ [per million (0/00)]. This nitrogen isotope signature changes as energy moves up the food chain. Those organisms highest on the food chain have the highest $\delta^{15}N$ values. Thus by examining an organism with regard to its $\delta^{15}N$, it is possible to place it on a food web continuum (Figures 8-7 and 8-8).

Certain ant groups were clearly functioning mostly as herbivores even though most did not actually consume leaves. Predatory ants also inhabit rain forest canopy and may exert top-down effects on cryptic herbivorous ants. But the results of the study

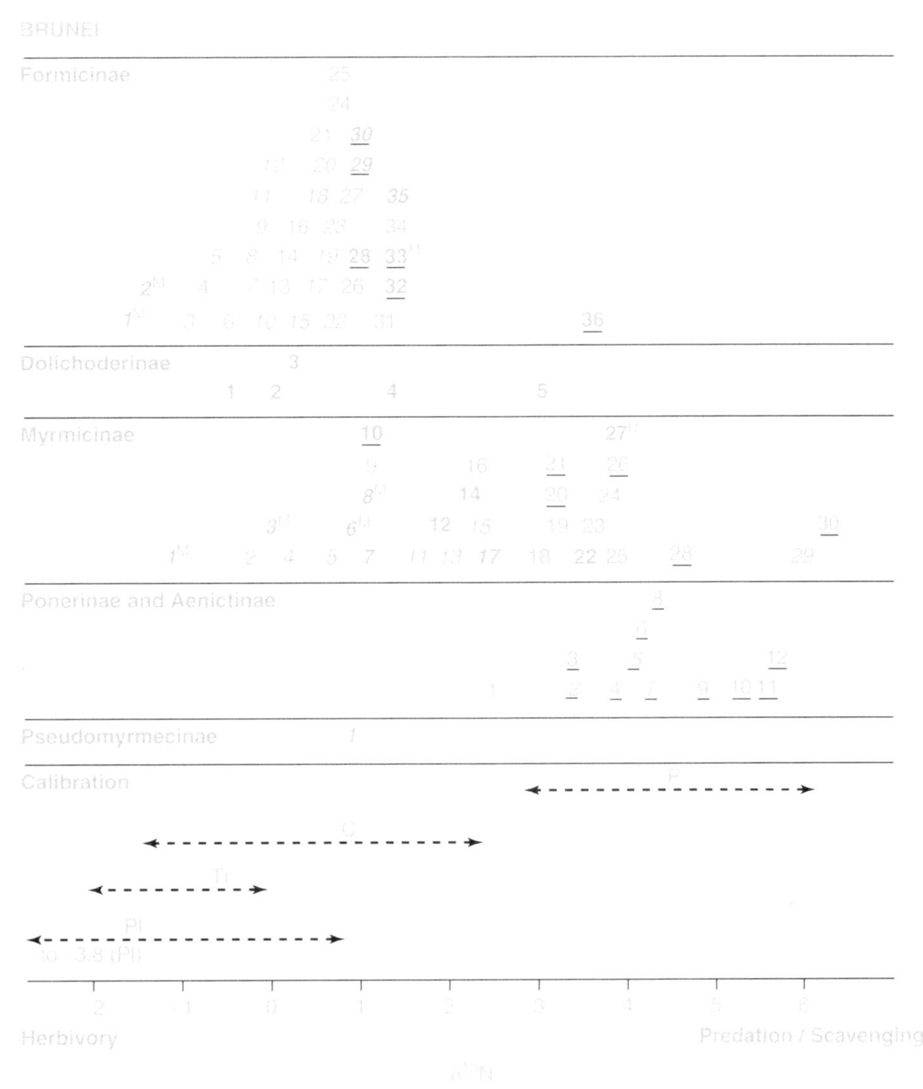

FIGURE 8-8
Mean $\delta^{15}N$ values (‰, by subfamily) for numbered Bornean species. Predominant foraging modes are shown as in Figure 8-7. Calibration: Pl, 20 nonmyrmecophytic plant samples; Tr, 9 sap-feeding trophobionts, not from myrmecophytes; C, 9 chewing herbivores (Coleoptera, Myrlapoda, Orthoptera, and Gastropoda); P, 9 arthropod predators (Acari, Araneida, Phalangida, Pseudoscorpionida, Chilopoda, and Coleoptera).

suggest that cryptic herbivory by ants adds significantly to plant biomass loss and, in effect, represents an adaptation to circumvent plant chemical defenses.

ARE LEAF-CUTTER ANTS THE DOMINANT HERBIVORES OF NEOTROPICAL RAIN FOREST?

In the previous chapter, it was noted that the coevolution between attine ants and the fungus they cultivate acts to help the ants circumvent chemical defenses of the plants they clip to use in preparing the cultivar. For this reason, leaf-cutter ants have often been assumed to be the major herbivores of Neotropical rain forest trees. A study on Barro Colorado Island quantified the impact of leaf-cutters in old secondary forest over a 15-month period (Herz et al. 2007b). A total of 49 leaf-cutter colonies were studied. The researchers measured refuse leaf fragments deposited monthly at ant colonies (refuse deposition rate) compared with harvesting rate (Herz et al. 2007a).

TABLE 8-2 Calculated consumption and herbivory rates of *Atta colombica* on BCI.

INTEGRATION LEVEL	RATE	±95% CONFIDENCE INTERVAL
Population		
Number of harvested fragments/yr	2.03×10^9	0.35×10^9
Harvested leaf area [ha/yr]	13.1	2.3
Harvested biomass [t/yr]	13.2	2.3
Deposited refuse [t/yr]	9.4	
Ecosystem		
Annual loss of		
leaf area [m²/ha/yr]	1,310.0	227
Biomass [kg/ha/yr]	132.4	23.0
Standing leaf area [%/yr]	2.1	0.4
Leaf area production [%/yr]	1.7	0.3
Leaf biomass production [%/yr]	1.5	0.3

The results showed that leaf-cutter ants were important herbivores (Table 8-2). The species studied, *Atta colombica*:

- Harvested 13.2 tons of biomass per year.
- Harvested 13.1 hectares of leaf area per year.
- Deposited 9.4 tons of refuse material per year.

At the ecosystem level, herbivory rates were 132 kilograms biomass/hectares/year and 1310 square meters foliage/hectares/year. These figures represent 2.1% of the foliage area in the forest or 1.7% of the annual leaf-area production. These figures, as impressive as they may seem, are considerably lower than previous published studies, which estimated that leaf-cutter ants consumed as much as 12% annual leaf area and 17% annual biomass. The impact of attines may vary considerably from site to site. Consumption rates among colonies varied sixfold, and thus these ants may have different impacts in different forests, making trophic dynamics ever more complex and generalization that much more problematic. The study concluded that on BCI, animals such as sloths, deer, and howler monkeys have more impact on leaf herbivory than leaf-cutter ants. Leaf-cutters appear not to be the *dominant herbivores* at BCI.

The Evolutionary Arms Race

Plants have evolved in concert with insects and other herbivores, pathogens, and seed predators for many millions of years. It should surprise no one that bottom-up forces by plants result in selecting for adaptations that are prevalent and diverse, particularly in the tropics. Given that many species of tropical plants possess numerous defense compounds as well as mechanical defenses, it may seem surprising that any kind of herbivore is able to consume them. But many do, because evolution works strongly in both directions. Herbivores, seed predators, and pathogens are considered by most plant ecologists to be the reason why negative-density dependence is widely seen among tropical plant species (see Chapters 5 and 7). Once a kind of defense compound evolves, it, in turn, acts as a counterselection pressure on herbivores and pathogens to evolve some form of adaptation that circumvents the plant defense. Once again vulnerable, the plant is then under enhanced selection pressures to evolve yet another defensive adaptation, which, should that occur, merely acts as yet another selection pressure on the herbivore, an evolutionary treadmill. One evolutionist (Van Valen 1973) has likened this process to the fabled Red Queen (of *Through the Looking Glass)*, always having to run to stay in place. Biological evolution has been called the ultimate existential game: you can never win, but only earn the right to keep playing. Because of coevolution, and because of the many millennia in which coevolution has been ongoing in the tropics, plants have accumulated a substantial reservoir of defensive adaptations as they "run through evolutionary time" to stay in place in the rain forest.

The arms race between insect herbivores and a plant species was documented for the understory species *Piper arieianum* at La Selva Biological Station in Costa Rica (Marquis 1984). Insect damage to this piper species varied greatly from plant to plant. Some plants were genetically more resistant to herbivores than others, and herbivory is a strong selective influence on the evolution of piper defenses. Some individual pipers appear to "risk" investing little in defense compounds but commit more energy to rapid growth and reproduction, while others grow more slowly but are better protected. Some *Piper* with few defenses may be partially protected by occurring among well-defended plants, where herbivores are scarcer. This example makes an important point. Even within a plant species, there is high variability in bottom-up defense adaptations. And such variability is the currency of natural selection, as the next example will demonstrate.

Inga Leaves: A Fork in the Evolutionary Road

Inga is a diverse and widespread genus throughout the Neotropics (see Chapter 4). By comparing two species of *Inga, I. goldmanii* and *I. umbellifera*, researchers showed that each species has evolved a different suite of adaptations in response to herbivore pressure (Coley et al. 2005). One species (*I. goldmanii*) has evolved what the researchers call a *defense* strategy and the other (*I. umbellifera*) has evolved an *escape* strategy. (Note that the use of the term *strategy* is evolutionary shorthand, widely applied even within professional literature. It in no way implies purpose or sentient decisions on the part of the plants.) In essence, the study, outlined here, shows that two closely related species reach a metaphorical "evolutionary fork in the road," where energy may be devoted to one set of traits versus another, both of which adapt the plant to withstand herbivory.

The study was conducted at Barro Colorado Island in Panama from March 2001 to November 2004. Research focused on 1- to 3-meter-tall saplings. Data for each *Inga* species were taken on herbivore–host associations, ant visitation to extrafloral nectaries, changes in leaf size and rate of leaf growth, and concentration and bioassays of leaf secondary metabolites (potential defense compounds).

The escape strategy hypothesizes rapid and synchronous opening and expansion of leaves. In other words, the growth rate of the leaves "outruns" the ability of herbivores to keep up with it. There is a delay in chloroplast development, and much energy, including use of nitrogen, is devoted to rapid, synchronous leaf growth. Defense by ants is low (little energy devoted to nectaries), and chemical defenses are predicted to be low. In contrast, the defense strategy hypothesizes slow expansion of leaves, nonsynchronous leaf growth, normal development of chloroplasts, and high levels of ant and chemical defense.

The study revealed clearly that each *Inga* species followed a different defense strategy but that the results of herbivory were essentially the same (Table 8-3). *I. goldmanii* lost 22.8% leaf area to herbivores, and *I. umbellifera* lost 21.4%. This result, specifically that each species experienced about the same degree of herbivory, may seem curious but should be unsurprising. If either species was clearly

TABLE 8-3 Observed and predicted associations among defensive traits of young leaves.

CHARACTERISTIC	ESCAPE	DEFENSE
Observed defense syndromes		
Expansion of young leaves	Fast	Slow
Nitrogen content	High	Low
Chloroplast development	Delayed	Normal
Synchrony of leaf production	High	Low
Ant defense	Low	High
Predicted chemical defenses		
Amount	Low	High
Bioactivity	Low	High
Diversity	Low	High
Biosynthetic complexity	Low	High

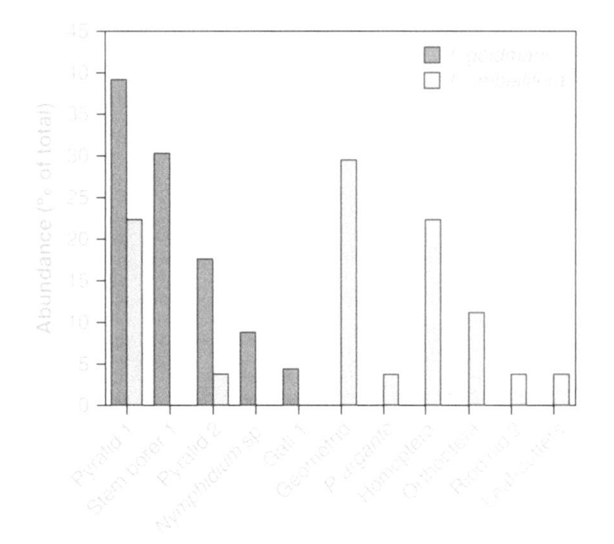

FIGURE 8-9
Herbivore communities on *Inga goldmanii* and *Inga umbellifera*. Data include only those herbivores that were identified to morphospecies, or to order in the case of Homoptera and Orthoptera ($n = 50$). Abundance is the percentage of the total observations assigned to a particular herbivore morphospecies (i.e., sums to 100% for each plant species).

superior to the other with regard to reduced herbivory, it should flourish far better than the other and eventually exclude it. Given the results, one could hypothesize that different herbivore groups are adapted to feed on each species, and that is the case. Only two herbivore species were found on both *Inga* species (Figure 8-9). Three herbivores were restricted to *I. goldmanii*, and six were confined to *I. umbellifera*.

Other results revealed:

- There were clear differences between the *Inga* species in two classes of secondary metabolites, nonprotein amino acids and flavenoids. *I. goldmanii* was considered to have more effective chemical defenses than *I. umbellifera*.
- *I. goldmanii* had higher levels of biotic defenses with more extrafloral nectaries per leaflet, attracting many more ants than *I. umbellifera*.
- Young leaves of *I. umbellifera* were produced in fewer but more synchronous flushes than was the case with *I. goldmanii*. Researchers noted that in each species, about 15% of the individual plants produced leaves in any given month but that the variability over time was far greater in *I. umbellifera*, indicating synchronous leaf production for this species.
- Young leaves of *I. umbellifera* expanded more quickly than leaves of *I. goldmanii*. Chlorophyll content was significantly lower during the growth spurt of *I. umbellifera* compared with *I. goldmanii*.

Researchers noted that while both *Inga* species are closely related (recall that they diverged as recently as 2 to 10 million years ago), they have diverged in defense adaptations. The conclusion of the study was that the two plant species are representative of an evolutionary trade-off in which energy is devoted either to strong defense or to rapid escape. The evolutionary arms race between bottom-up and top-down influence may take very different directions even within a closely related species pair, characteristic of evolution that is widespread, and not merely restricted to *Inga*.

Phylogenetic Approaches to Community Ecology

Community ecologists are increasingly focusing on integrating studies of evolutionary phylogenies with basic community ecology (Webb et al. 2006)). With the availability of molecular and cladistic technology capable of tracing phylogenetic histories linked with basic field research in community ecology, it is possible to look more deeply into the evolutionary context of trophic dynamics. By way of example, one such study examined a tropical herbivore community in relation to the phylogeny of its host trees (Weiblen et al. 2006).

Researchers looked at three insect groups, Lepidoptera (butterflies and moths), Coleoptera (beetles), and Orthoptera (grasshoppers and related species). The study was conducted on a site near Madang, on the north coast of Papua New Guinea in an area with about 150 tree species (greater than 5 centimeters DBH). About 12 genera of trees dominated at the study site. The study measured the relationship between the herbivorous insects and their host trees in relation to the phylogenetic distance between the hosts. The phylogenetic distance among host species (of trees) was based on DNA sequence divergence integrated across three genes. Two indices were measured, the *nearest taxon index* (NTI) and *nearest relatedness index* (NRI). Leaf-chewing insects were collected from 62 plant species representing 41 plant genera and 18 families.

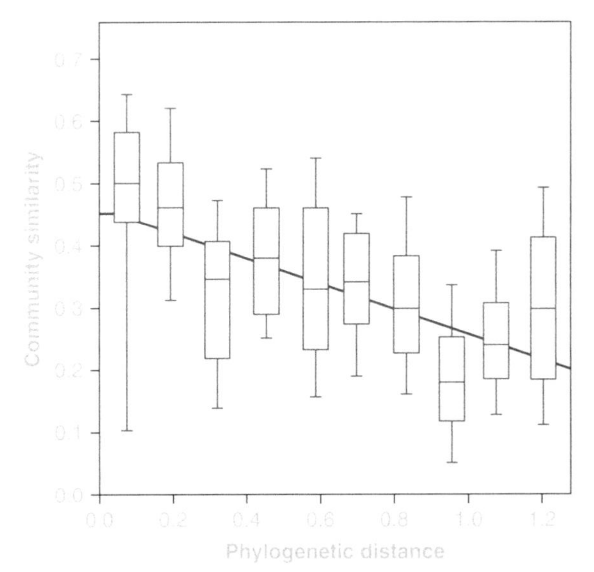

FIGURE 8-10
Herbivore community similarity as a function of the phylogenetic distance between host plants. Similarity is the fraction of the total fauna on two hosts that is shared between the hosts. Phylogenetic distance was derived from a penalized-likelihood ultrametric phylogram. Means, standard deviations, and ranges of community similarity are shown for selected distance intervals. The outgroup is excluded from the regression.

Feeding experiments were conducted on about 100,000 live insects collected from the host plant species. There was an overall negative correlation between faunal similarity and the phylogenetic distance between the hosts based on the DNA sequence divergence (Figure 8-10). This means that the greater the phylogenetic distance between two host plant species, the less similar are their herbivore communities. Two trees that are phylogenetically close share more herbivore species in common than two species of trees more phylogenetically distant. Herbivore community similarity among congeneric hosts was high, averaging 50% compared with only 20% to 30% with host families. Approximately 40% of caterpillar species showed significant phylogenetic clustering with regard to host plant associations, which was a bit more than for beetles or orthopterans. Overall, fully 25% of the variance among plants with regard to insect herbivore associations was explained by phylogenetic relationship with hosts, which the researchers considered to be very substantial considering the variability that environmental and population demographical heterogeneity must contribute to samples of herbivore associations.

This study, and others like it, demonstrates that integration of community structure and phylogeny add new insights into the long-term evolution of food webs.

Dipterocarps and Masting

Southeast Asia rain forests are unique because of the high diversity and numerical dominance of dipterocarp trees (see Chapter 2). Like most tropical trees, dipterocarps have an abundance of secondary compounds that defend them from herbivores. They also exude an oily bark resin that acts to protect them from invasion by pathogens such as bacteria and fungi.

Dipterocarps (Plate 8-24) are dependent on insects for pollination, but their seeds are not animal-dispersed. On the contrary, animals are major predators of dipterocarp seeds. Dipterocarps are so-named because each large seed is attached to two long wing-like sepals that literally spin when the seed is dropped, carrying it some distance (but normally not very far) from the parent tree. Seeds both on and off the tree attract an array of animals ranging from birds such as pigeons, pheasants, and hornbills to mammals such as wild pigs, orangutans, squirrels, rats, and flying fox bats. Many insects such as weevils also devour dipterocarp seeds.

Dipterocarps, like many other tropical tree species, have evolved extensive *synchronous masting*, a bottom-up adaptation to seed predation. At intervals that vary from 2 to 7 years, dipterocarp species over a large geographical area simultaneously flower and drop fruit (Sakai 2002). Masting is common among many species of temperate-zone trees such as various oaks and pines, but it is unusual in the tropics, especially as dipterocarp masting involves many species at the same time. The result is that many thousands of seeds drop at once, far more than the full array of seed predators is able to consume. Dipterocarps flood the market with seeds, becoming an important food source for animals during masting years, but in other years, animal populations are less numerous, as food is more limited. The result of dipterocarp masting is that without chemical protection of any sort, and with massive consumption by animals, some seeds nonetheless germinate and grow into seedlings. Like other tropical tree species, seedlings grow slowly in shade and persist for many years.

(a)

(b)

PLATE 8-24
DIPTEROCARP FOREST

For masting to occur, there must be some environmental trigger. One factor is likely related to energy buildup within the trees over time. Once they have sufficient energy stored up, masting happens. But the signal that seems to trigger it is thought to be a periodic brief drought that brings about cool night temperatures (Yasuda et al. 1999).

Keystone Plant Species in Rain Forest Food Webs

A *keystone species* is defined as a species with a uniquely strong influence on a food web, such that its removal creates a trophic cascade, altering and usually simplifying the food web. Keystone species, many of which are top carnivores, are usually not numerically abundant, but their influence on food web structure is disproportionate to their abundance.

Applying the concept of keystone species to complex tropical rain forest food webs presents challenges. The number of interactions is vast because of the high species richness. But it is to be expected that some species will exert very strong influences on food web stability.

A study performed at Cocha Cashu Biological Station (CCBS), within Manu National Park, southeastern Peru, showed that less than 1% of the plant species (amounting to 12 species of approximately 2,000) sustained the entire vertebrate frugivore community for three months of the year, when food is otherwise generally scarce (Terborgh 1986). The keystone plants were certain palms, figs, and others that produced nectar or fruit other than figs (Table 8-4).

Palm nuts (Plate 8-25) were a key resource to peccaries, agoutis, other kinds of rodents, and large parrots called macaws. Figs fed a diversity of birds, virtually all monkey species, and various marsupials and other mammals. Nectar was consumed by numerous

TABLE 8-4 Keystone plant resources in the flora of Cocha Cashu.

PLANT RESOURCE	PERIOD AVAILABLE	EATEN BY
Palm nuts		
Astrocaryum sp.	April–June	Capuchins, peccaries, squirrels, agoutis, other rodents, macaws
Iriartia ventricosa	May–July	Capuchins, titi monkey, spider monkey (exocarp only), peccaries (seed)
Scheelea sp.	All year	Capuchins, squirrel monkey (mesocarp), squirrels (seed)
Figs		
Ficus erythrosticta *Ficus killipii* *Ficus perforata*	Irregularly throughout season	Nearly all monkeys, marsupials, procyonids, guans, trumpeters, toucans, many passerines
Nectar sources		
Combretum assimile	July	7 species of primates, marsupials, procyonids, >20 species of birds
Erythrina ulei	July–August	Spider monkey, squirrel monkey, capuchins, many parrots and other birds
Quararibea cordata	Aug.–Sept.	Capuchins, squirrel monkey, tamarins, marsupials, procyonids, birds
Miscellaneous nonfig fruits		
Allophylus scrobiculatus	May–June	Spider monkey, capuchins, trumpeters
Calatola sp.	May–June	Titi monkey, trumpeters
Celtis iguanea	March–August	Titi monkey, tamarins, parrots

PLATE 8-25
Palm nuts form a keystone resource for many animal species.

PLATE 8-26
These two blue and yellow macaws (*Ara ararauna*) are attracted to a "macaw feeder" at this field station in Peru. But they are naturally attracted to feed on palm nuts and figs.

bird species, eight monkey species, various marsupials, and other mammals such as kinkajous. And three other fruit-producing tree species also were essential to certain monkey species and various birds, especially parrots (Plate 8-26). The study documented that fruit production from these keystone species is an essential resource at specific times of the year, when other resources are in low supply. The study also looked at rain forests in Asia and Africa and noted that figs are keystone species in those forests.

Keystone Animal Species in Rain Forest Food Webs

The mutualistic interdependencies between plants and animals that have evolved in rain forest food webs add a wider dimension to the concept of keystone animal species. It is not simply predatory species that exert top-down influences. Many animal species are essential to pollination and seed dispersal. Loss of clusters of such species would initiate a trophic cascade that would lead to loss of ecosystem species richness.

Cocha Cashu Biological Station formed a basis for a study that clearly documented how loss of certain large-animal seed dispersers initiates a potential cascade eventually leading to loss of plant species (Terborgh et al. 2008). Cocha Cashu was compared with a rain forest site named Boca Manu (BM) near a village of indigenous people who had extensively hunted throughout BM (Table 8-5). The two study sites, each 4 hectares, were 90 kilometers apart and represented similar forests on similar soils. The researchers hypothesized that the loss of large birds and mammals at BM would produce a cascade of effects on plants by altering seed shadows (see Chapter 5). Without the animals present to disperse the seeds away from the parent trees, seed shadows would be more constricted to the proximity of the parent tree. This constriction

TABLE 8-5 Densities of vertebrate seed dispersers and seed predators at Cocha Cashu Biological Station (CC) and Boca Manu (BM), Peru.

FUNCTIONAL GROUP	NUMBER PER SQUARE KILOMETER	
	CC	BM
Arboreal seed dispersers, diurnal		
Primates, large (spider monkeys, howler monkeys)	67.2	0.0
Primates, midsized (capuchins)	98.9	8.4
Primates, small (titis, squirrel monkeys, tamarins)	70.4	86.4
Birds, large (guans, trumpeter)	31.6	2.1
Arboreal seed dispersers, nocturnal		
Midsized (kinkajous, olingos, night monkeys)	20.7	33.5
Terrestrial seed predators, diurnal		
Large (collared and white-lipped peccaries)	98.2	16.7
Midsized (agouti, acouchi)	17.4	10.0
Small (squirrels)	41.1	35.3

TABLE 8-6 Comparison of stem densities at Cocha Cashu Biological Station (CC) and Boca Manu (BM), Peru.

STEM CATEGORY	NUMBER OF STEMS PER HECTARE		
	CC	BM	BM/CC (%)
No. adult trees (>10 cm DBH)	576	587	102
No. large saplings (≥1 m tall,)	5,129	3,087	60
No. small saplings (≥1 m tall, <1 cm DBH)	5,360	2,807	52
Total no. large and small saplings	10,489	5,894	56

would alter reproductive success among those species dependent on large-animal seed dispersers.

The sites differed greatly in diversity of large birds and mammals. The BM site was significantly depleted of most species due to persistent hunting. Cocha Cashu represented the normal mammal and bird fauna for the region and served as a control for comparison. The effect of indigenous hunting is severe in many tropical forests around the world and will be discussed further in Chapter 15. Hunting may be sufficiently substantial as to lead to what has been termed an *empty forest effect* (Redford 1992). This occurs when a tropical forest appears normal but where many plant species are experiencing reduced fitness because of lack of seed dispersal by keystone animal species. Once large animals are hunted to near or actual extinction, the recruitment of many tree species is severely reduced. The study comparing BM with CCBS showed the following:

- Both forests were similar in tree species density, average basal area, and approximate number of tree species. The two forests shared 183 tree species in common (out of 269 at CCBS and 252 at BM).
- Overall sapling densities were lower at BM. Large saplings were 40% lower and small saplings were 48% lower (Table 8-6).
- Sapling cohorts at BM contained a higher proportion of stems of species that were abiotically dispersed, such as by wind, and a lower proportion of animal-dispersed species.
- The relative abundance of stems in relation to adult trees was altered in the BM forest. There was a higher proportion of stems of abiotically dispersed species relative to adult trees of the same species. There was a lower proportion of stems of animal-dispersed species relative to adult trees of the same species. This result suggests that the BM forest is changing, losing tree diversity.
- Recruitment of tree species that depend on large animals for dispersal was depressed in the BM forest.
- The seed shadow of trees dependent on large animals for dispersal was reduced, with more recruitment occurring nearer the parent trees. This pattern has been shown to result in negative density dependence and reduced fitness.
- Large arboreal mammals and birds and large terrestrial mammals were reduced 80% to 100% in the BM forest, but nocturnal arboreal mammals and small (less than 1 kilogram) mammals were not depleted. This means that those trees dependent on this array of animals may increase in fitness relative to trees dependent on large-animal dispersers.

The overall result was that tree recruitment patterns are altered by the impact of indigenous hunting. The result of loss of large seed animal dispersers is that abundance patterns and ultimately diversity patterns begin to change. The end result is that tree species diversity will decline.

DOES SEED PREDATION ALTER THE PLAYING FIELD?

The loss of large mammalian seed dispersers clearly affected the diversity dynamics of plant species in the example described earlier. A study in a tropical forest in Mexico has demonstrated that within a forest that has experienced loss of large mammals, seed predation by small rodents also exerts a top-down effect on plant species, altering the diversity dynamics (Dirzo and Mendoza 2007).

The study was done at Los Tuxtlas (LT) Biology Field Station, a lowland rain forest in Veracruz, southeastern Mexico. Once a rich forest with 39 non-flying mammal species, the area was severely deforested and fragmented, resulting in loss of large mammal species. The result of the deforestation is that the LT site has an abundance of small rodents, all of which eat small seeds. Researchers hypothesized that the numerical dominance of small rodents would exert strong effects on small-seeded plant species, perhaps giving a fitness advantage to plants that produce large seeds. Even if large-seeded plant species lack large mammalian dispersers, they may not suffer the negative effects of constricted seed shadows if small rodents are consuming large numbers of small seeds. Researchers captured and confined small rodents (*Heteromys desmarestianus*) and showed that the animals preferentially consumed small seeds. In a field experiment, they showed that small seeds unprotected from rodents were consumed at 30 times the rate of small seeds that were protected in exclosures that kept out rodents. Large seeds, whether protected or not, were not affected by consumption by small rodents. Germination rates were four times greater among small-seeded plants protected from rodents than those that were left unprotected.

The study concluded that small-seeded species of plants are exposed to greater top-down herbivory brought about by an abundance of small rodents in areas where large mammals are absent. In contrast, large-seeded species essentially escape seed predation due to the lack of large mammals and thus enjoy a fitness advantage in such areas. The researchers noted that large-seeded species have unusually dense seedling carpets throughout the LT area.

Plants versus Insects

Ecologists debate the degree to which insect communities in the tropics exhibit greater host specificity and specialization compared with those at higher latitudes (Dyer et al. 2007; Novotny et al. 2007). Is increased insect specialization a general characteristic of tropical ecosystems, particularly rain forests? The answer is equivocal. It depends on how specialization is defined. Further complicating the question is that the full magnitude of insect species richness is still poorly known for rain forests.

Defense compounds, protective ants, and tough leaves are bottom-up selection pressures affecting the evolution of insect herbivores. Those insects that evolve enzyme systems that detoxify defense compounds or somehow sequester them are able to specialize on specific plant species. Other factors also select for diet specialization (Futuyma 1983). For instance, plant compounds may be repellent but not actually be toxic. Insects may overcome the repellency and adapt to recognize a host plant by its repellent compounds. Interspecific competition may, through niche partitioning, select for host specialization among insects, as might avoidance of certain parasitoids (attracted to certain plant-specific compounds).

Insects may evolve behaviors that minimize exposure to defense compounds. For example, the caterpillars of the butterflies of the genus *Melinaea* feed on plants in the Solanaceae (tomato) family, especially in the genus *Markea* and *Juanaloa*. The caterpillars cut the leaf veins of their host plants, preventing the defense compounds from reaching the leaf blade, where the caterpillars feed (DeVries 1987).

A research team performed a comprehensive study of host specificity of Lepidoptera (butterflies and moths) in temperate and tropical habitats (Dyer et al. 2007). Eight sites were chosen ranging from 15°S latitude to 55°N latitude. The diets of tropical caterpillars were more specialized than was the case at higher latitudes. (Larvae, the caterpillars, cause leaf damage; adults feed only on nectar.) The average tropical lepidopteran species fed on fewer species, genera, and families of plants compared with counterparts at higher latitudes. The study concluded by suggesting that greater specialization in the tropics was the evolutionary result of strong trophic interactions of the sort that result from the various arms races described earlier (Figure 8-11).

Other studies (Novotny et al. 2007) have concluded that host specificity is no greater in the tropics either with regard to leaf-devouring insects or those that focus on wood or fruit. The important point in these studies is that there needs not be dramatically greater host specificity to support greater insect species richness since the species richness of plants is so high to begin with. In other words, given the hyperdiversity of plant species, a relatively low host specificity of insect species will nonetheless compose a very large number in total. However, the overall arthropod diversity may be less staggering than some estimates suggest. Rather than 30 million species, there may be "only" 4 to 6 million (Novotny et al. 2007). That is still a lot of arthropod species.

Host specificity is low for plant fungal pathogens in tropical rain forests. In other words, pathogens of a given species are essentially *polyphagous*, attacking numerous plant species (the opposite of specialization). But most plant species in a local community have evolved resistance to various specific pathogens

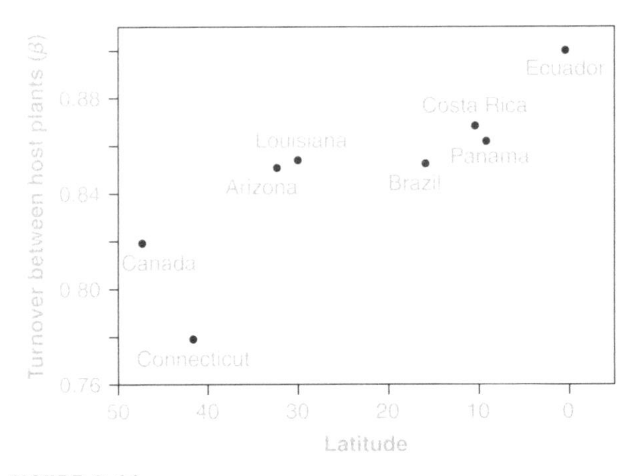

FIGURE 8-11
Caterpillar species turnover among host plant species for eight forest and woodland sites across a latitudinal gradient. Modified Whittaker's β is shown for two temperate (Canada and Connecticut), two subtropical (Arizona and Louisiana), and four tropical (Brazil, Costa Rica, Panama, and Ecuador) sites, ordered by an approximate modal latitude for each site. With this measure, insect species turnover across host plants is used as an index of host plant specialization, with higher values of β indicating higher levels of specialization. Symbol size was greater than the 95% CI for all sites except Connecticut.

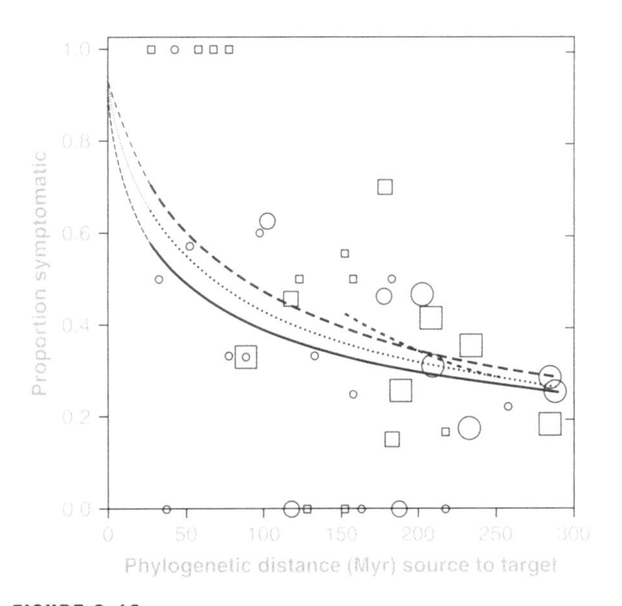

FIGURE 8-12
Proportion of target plant species that developed disease symptoms after inoculation with fungal pathogens from source plant species. Note how proportion drops with phylogenetic distance of source to target.

(Gilbert and Webb 2007). What is most noteworthy is that the likelihood that a pathogen can infect two plant species decreases continuously with phylogenetic distance between the plant species (Figure 8-12). This means that plants that are evolutionarily distant will not succumb to the same fungal pathogens. There is a *phylogenetic signal* in host range between pathogen and plant. The more evolutionarily similar two

Plants versus Howler Monkeys

Most primates are strongly herbivorous, and many eat mostly fruit. Some, however, like the howler monkeys of the Neotropics, are primarily leaf-eaters, or folivores. Thus they are exposed to defense compounds. In a study in Costa Rica, mantled howler monkeys (*Alouatta villosa*) were observed to occasionally appear disoriented and even fall from trees. Further observations showed that howlers must learn to be extremely selective in exactly which trees they dine on. Within one tree species (*Gliricidia sepium*), where 149 individual trees were available in the howlers' range, the troop fed on only three of the trees—always the same three (Glander 1977). The leaves of these trees were less difficult to digest. Mantled howlers favor young leaves relatively high in nutritional value but not yet concentrated with defense compounds. When only mature leaves are available, monkeys eat just a little and then move to a different tree, minimizing their exposure to various arrays of defense compounds. Sometimes they eat only the leaf stalk or petiole, ignoring the blade. The petiole has the lowest alkaloid content (Glander 1982).

Protein and fiber content may be even more important factors affecting leaf choice in howler monkeys (Milton 1979 and 1982). Fiber makes leaves difficult to digest and is least concentrated in the preferred younger leaves. Protein is proportionally higher in young leaves than in older leaves. The greater the ratio of protein to fiber, the more desirable the leaf is to howlers.

Howlers have long intestinal systems, especially the hindgut. It takes food up to 20 hours to pass through a howler's digestive system. Spider monkeys (*Ateles geoffroyi*) eat mostly fruits, much easier to digest because fruits contain more protein and fewer defense compounds than leaves. It takes food only about 4.4 hours to move through the shorter gut of a spider monkey. Howlers, with their long hindguts, are able to more efficiently digest leaves, coping to a reasonable degree with both high fiber and defense compounds (Milton 1981).

plants are, the more likely they will succumb to the same pathogen species. This suggests that "because close relatives are more likely to be susceptible to the same pathogens, pathogens may be most effective at maintaining diversity of higher taxonomic levels ('phylodiversity') rather than species diversity *per se*" (Gilbert and Webb 2007, p. 4982). The understanding of phylogenetic relationships has strong implications for interpreting complex trophic interactions and will be discussed further later in this chapter.

Coevolution of Plants and Lepidopterans

Lepidopterans are (to plant species) evolutionary examples of Dr. Jekyll and Mr. Hyde. As butterflies, they play the role of benevolent Dr. Jekyll, dispersing pollen essential to the plants' reproduction. As caterpillars, they are malevolent Mr. Hyde, devouring leaf tissue, reducing plant fitness. The trophic dynamics of lepidopterans and their plants is complex and made more so by evolution. Various lepidopterans in temperate and tropical areas are known for their affinities for feeding on specific plant families. Caterpillars are much more selective than adults because adult butterflies feed on nectar, aiding in pollen dispersion. Adult butterflies, whose interactions with plants are fundamentally mutualistic, often feed on a wider range of plants than larval lepidopterans that have evolved defenses against specific defense compounds. Caterpillars are voracious herbivores, and being folivores, they harm plants. Thus they encounter plant defenses, including secondary compounds, as they chew on leaves. Since different families of plants produce different combinations of defense compounds, natural selection has acted on caterpillars in such a way that various caterpillar species have evolved tolerance for defense compounds associated with different plant families.

HELICONID BUTTERFLIES AND PASSIONFLOWERS

Heliconid butterflies (Heliconiinae) (Plate 8-27) are a diverse and colorful group, almost all of which are Neotropical (DeVries 1987). They are usually considered part of the brush-footed butterflies (Nymphalidae), although some taxonomists place heliconids within their own distinct family. Nymphalids number nearly 3,000 species globally, and Heliconids are represented by about 50 species, with many local races throughout tropical America. Commonly called longwing butterflies, only two species, *Heliconius charitonius* and *H. erato*, regularly reach the United States, and both of these range through southern Texas and the Southeast.

PLATE 8-27
Heliconid butterflies (this one is *H. doris*) are found throughout the Neotropics and have been the object of much coevolutionary study.

PLATE 8-28
Passionflower vines (*Passiflora*) are fed upon by caterpillars (larvae) of *heliconius* butterflies. Photo shows the plant and its inflorescence.

Heliconid adults feed on a variety of plant species, but caterpillars feed almost exclusively on species of *Passiflora* or passionflower (Passifloraceae), a common vine numbering approximately 500 species (Plate 8-28). Like heliconids, passionflowers are largely Neotropical, and another name for heliconids is *passionflower butterflies* (DeVries 1987). Not many kinds of herbivores eat passionflower vines. Most passionflowers contain various cyanogenic glycosides and cyanohydrins. The high diversity of cyanogens among passionflower species may be an evolutionary response to herbivory by heliconids. Heliconids, however, seem able to adapt to the ever-changing cyanogen regime by evolutionary changes in their hydrolytic enzymes and by sequestration of cyanogens. Heliconid caterpillars (Plate 8-29) have coevolved with passionflowers. To passionflowers these caterpillars remain a significant top-down force.

Heliconid butterflies, like many other butterfly groups, are adapted to a life cycle on their host plant.

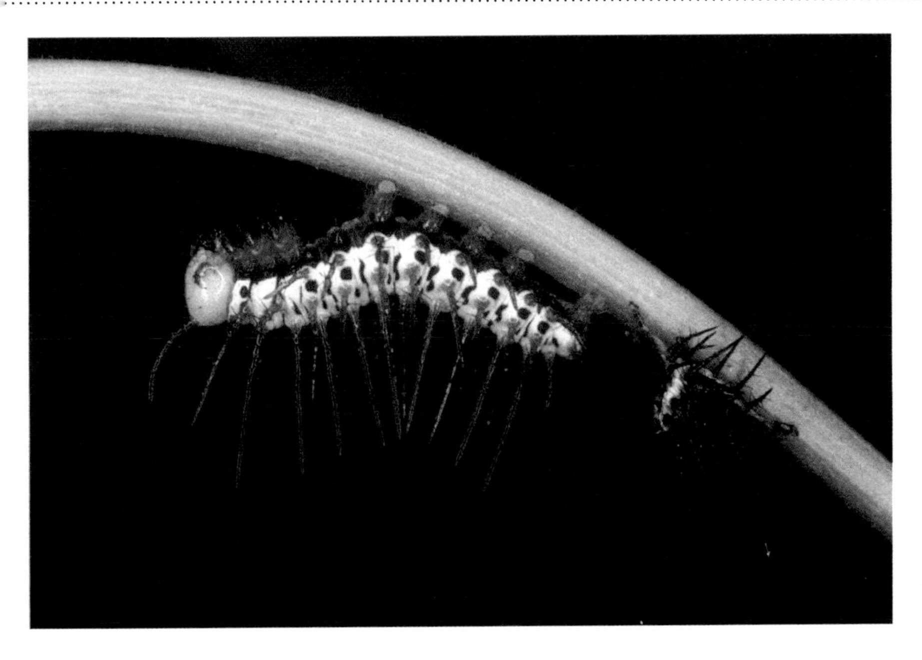

PLATE 8-29
HELICONID CATERPILLAR

Females lay small numbers of eggs in globular yellow clusters directly on passionflower leaves, favoring young shoots. When eggs hatch, caterpillars are sitting on their food source. The plant is therefore under selection pressure to somehow prevent the adult female butterfly from locating, selecting, and laying eggs on its leaves.

Detailed studies (Gilbert 1975 and 1982; Benson et al. 1976) of passionflowers versus heliconiines have demonstrated diverse passionflower defenses. Passionflowers produce extrafloral nectaries that attract various species of ants and wasps. These insects help repel heliconid caterpillars. Some passionflower species are protected exclusively by ants, some by wasps, some by both. At least one study has shown that caterpillar survival is much lower on *Passiflora* with attending ants (Smiley 1985). Caterpillar mortality rate was 70% on ant-attended plants compared with 45% on non-ant plants.

Some *Passiflora* extrafloral nectaries appear, to the human eye, to mimic heliconid egg clusters (Gilbert 1982). Passionflower vines typically have young leaves spotted with a few conspicuous yellow globs, the *egg mimics*. Female heliconids will not lay eggs on a leaf already containing egg masses. The mimic egg masses presumably prompt the female to continue searching. Gilbert believes the mimic eggs to be a recent evolutionary development in the plant–insect arms race because only 2% of passionflower species have mimic eggs.

Leaf shape varies within a species of passionflower, and passionflower leaves often resemble other common plant species growing nearby (Gilbert 1982). Perhaps heliconids may be tricked by the similarity of appearance (leaf mimicry) and thus overlook a passionflower. This speculation depends, of course, on the butterfly using visual cues to locate passionflower vines. If the insect depends principally on scent, such leaf mimicry would seem useless.

For the time being, at least one passionflower species, *Passiflora adenopoda,* may have a firm advantage in the coevolutionary arms race. Its leaves have a dense covering of minute, hooked spines, or *trichomes.* Resembling a bed of nails, trichomes impale the soft-skinned caterpillars. Once a caterpillar is stuck, it starves (Gilbert 1971). Trichomes occur on other plant species and at least one butterfly species, *Mechanitis isthmia* (an ithomiid, not a heliconid), has adapted to thwart trichome defense. *Mechanitis* caterpillars feed on plants of the tomato family and avoid impalement by spinning a fine web covering the trichomes, enabling the caterpillars to move over the leaf surface to feed on leaf edges (Rathcke and Poole 1975).

WHY ARE HELICONIUS BUTTERFLIES OBVIOUS?

Heliconid butterflies are not inconspicuous (Plates 8-30 and 8-33). In fact, they are among the most obvious and striking butterflies of the tropics. They fly slowly, almost delicately, and are very easy to see along forest edges as well as in interior rain forest.

PLATE 8-30
Even from a considerable distance, *heliconius* butterflies are obvious. This one is either *H. melpomene* or *H. erato*. They are very difficult to distinguish (see text).

PLATE 8-31
This caligo butterfly (*Caligo eurilochus*) shows substantial damage along the lower edge of its wings. The bite marks of one or more bird beaks are clearly visible.

PLATE 8-32
RUFOUS-TAILED JACAMAR

When atop a plant, the butterfly appears brilliantly iridescent. Why are they so conspicuous? Consider the potential risk to the insect. Most tropical bird species feed heavily on insects. It would seem suicidal for a butterfly group to be colored like neon signs, seeming to say "eat me."

But obviousness may serve as a warning. If heliconid butterflies are distasteful to would-be predators, a selection pressure favoring aposematic coloration becomes possible. Once a bird has eaten an unpalatable insect, bright coloration may facilitate learning avoidance behavior. It has been demonstrated that birds are capable of remembering butterfly warning coloration and will avoid unpalatable butterflies once they have experienced them (Brower and Brower 1964). Heliconid species showing warning coloration are likely to store plant defense compounds (or to use them as precursors to synthesize their own unique defense compounds) and thus to be able to sicken their predators (Brower et al. 1963; Brower and Brower 1964). In other words, the chemical defense of the plant becomes, in turn, the chemical defense of the insect.

In an experiment conducted in Costa Rica, the wing coloration pattern of the unpalatable butterfly *Heliconius erato,* was altered such that predatory birds would not have seen such patterning before (Benson 1972). A control group identical to specimens normally occurring was also established. Equal numbers of normal, "control" butterflies and altered, uniquely

PLATE 8-33
Heliconius doris adult on a milkweed flower.

patterned individuals were released. Significantly fewer of the altered individuals were recaptured, an indication that fewer survived. Some altered butterflies that were recaptured showed wing damage from bird attacks. Benson was even able to identify one bird species, the rufous-tailed jacamar *(Galbula ruficauda),* by the shape of wounds it left in a butterfly's wing (Plates 8-31 and 8-32). The experiment showed that wing pattern does confer protection, once predators learn the pattern and associate it with unpalatability.

There is an obvious cost associated with warning coloration. At each generation, some individuals are "sacrificed" in order to "educate" the predators. Why should a butterfly give up its life for the good of its species? The answer is that it probably doesn't but, rather, gives up its life only for the good of its own genes. Heliconid butterflies have a limited home range, and a single female may live for up to six months. Because the basically sedentary females lay many eggs over the course of their long lives, the local population in any given area is likely to consist of close relatives: brothers, sisters, cousins, second cousins, and so on. If a single individual is lost to a predator, but the predator subsequently avoids other members of the group, the other group members each benefit. And if most of the members of the group have many genes in common with the sacrificed individual, then in an evolutionary sense, the individual has indirectly promoted its own genetic fitness. It has acted to protect copies of its own genes. This process, called *kin selection*, may explain how warning coloration evolved in heliconids. Heliconid butterflies are among the only butterflies to have communal roosts, aggregations that may be *extended families*.

Unusual Speciation in *Heliconius* Butterflies

In an unusual case of speciation, two heliconid species have given rise to a new species by hybridization (Malvárez et al. 2006). The two parent species, *H. melpomene* and *H. cydno*, are found in overlapping ranges from Central through northern South America (Figure 8-13). The hybrid species, *H. heurippa*, occupies a small range in eastern Colombia, where it is sympatric with *H. melpomene* and parapatric with *H. cydno*. Researchers examined 12 microsatellite loci as well as two nuclear genes. In the case of microsatellite loci, the two parent species and putative hybrid species each fell into distinct genetic clusters. In the case of the nuclear genes, the hybrid shared genes with each of the parent species that they did not share with one another. Laboratory tests

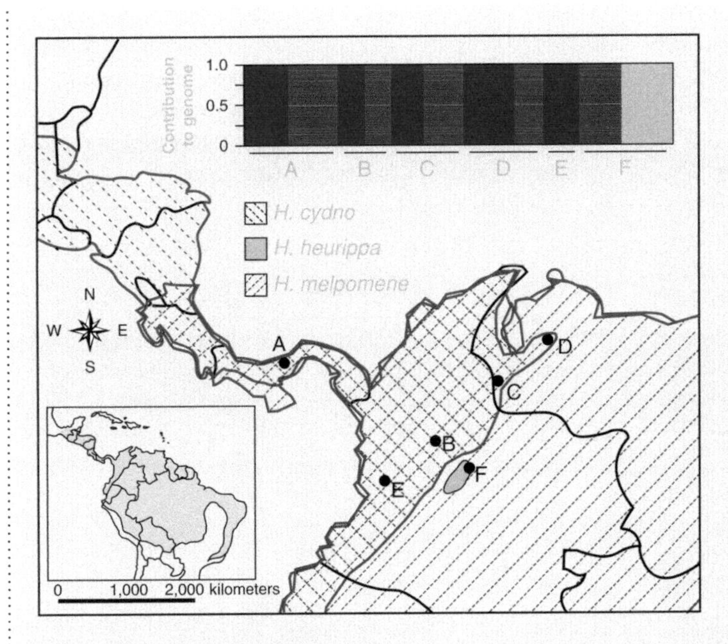

FIGURE 8-13
The geographic distributions and genetic differentiation among the three *heliconius* species in the study. The letters signify field areas where the animals were collected for the study. The upper inset shows the relative contributions of the three clusters to each individual's genome: blue—*H. cydno*; red—*H. melpomene*; green—*H. heurippa*.

confirmed that *H. heurippa* females were aversive to mating with males of either *melpomene* or *cydno*. The wing pattern of the butterflies is the cue that provides the isolating mechanism. The pattern of *H. heurippa* is intermediate between the two parent species, and each species, in general, will mate only with another butterfly with an identical wing pattern (Figure 8-14). This is the first documented example of a hybrid trait generating reproductive isolation by assortative mating within an animal species.

H. cydno cordula
bbN^NN^NBrBr

H. heurippa
BBN^NN^Nbrbr

H. m. melpomene
BBN^BN^Bbrbr

FIGURE 8-14
The three wing patterns of the species used in the study. Note that *H. heurippa* is intermediate in pattern between the two parent species.

Mimicry Systems

Although heliconids and many other tropical butterflies are striking in appearance, many are often difficult to identify because different species evolve to look alike: they mimic each other. Mimicry in tropical lepidopterans has been well documented and represents an evolutionary case of bottom-up trophic dynamics.

BATESIAN MIMICRY

Henry Walter Bates (1862), one of the first naturalists to explore Amazonia, was amazed to discover that some unrelated species of butterflies look alike. He suggested that a palatable species may gain protection from predators if it closely resembles a noxious unpalatable species, a phenomenon now termed *Batesian mimicry*.

A classic example of Batesian mimicry is the North American viceroy butterfly, *Limenitis archippus*. The viceroy, a member of the family Nymphalidae, bears a strong resemblance to the monarch (*Danaus plexippus*), in the family Danaidae. Both are striking orange with black wing veins (Plate 8-34). Monarch larvae feed on

PLATE 8-34
VICEROY (LEFT) AND MONARCH (RIGHT)

plants in the milkweed family (Asclepiadaceae), most containing cardiac glycosides. The glycosides are sequestered in monarch adults, making them unpalatable to birds. Viceroys feed mostly on plants of the willow family (Salicaceae) and are palatable to birds.

The calculus of Batesian mimicry is complex. The unpalatable species, the *model*, would seem to be parasitized by the palatable species, the *mimic*. Because it closely resembles an unpalatable species, the mimic enjoys the umbrella of protection provided by the presence of the model. For the model, the presence of the palatable mimics makes the education of predators more difficult. Suppose that a predator encounters one or even two palatable mimics as its first experience. It may be subsequently more difficult for the predator to "learn" that the noxious model is indeed, noxious. If the mimic were as abundant as its model, the entire system would be relatively unprotected, because predators would encounter palatable mimics as readily as unpalatable models. But there is an alternative possibility. By sheer numbers, even palatable mimics could act as mutualists with models. If there is more overall food available to predators because of the presence of mimics, top-down predation pressure on models will be relaxed. In simulation experiments using birds as predators and almonds as prey (some of which were made palatable, some partially palatable, some not palatable, but all of which appeared nearly alike), a mutualistic rather than parasitic relationship was shown to develop between mimic and model (Rowland et al. 2007).

Butterflies exhibit numerous examples of Batesian mimicry throughout the tropics (Gilbert 1983). However, they are not the only Batesian mimics. Many other insect species have evolved mimicry. For example, the insect species *Climaciella brunnea*, a predatory mantispid in the order Neuroptera (lacewings, alderflies, dobsonflies, and others), has evolved an uncanny resemblance to five different wasp species (Opler 1981). The stingless neuropterans mimic an array of models, and gain protection by appearing to be what they are not.

MÜLLERIAN MIMICRY

Although Batesian mimicry is well represented in the tropics, recall that most tropical plant species contain secondary metabolites, *defense compounds*. Therefore, any caterpillar species will likely have to cope

FIGURE 8-15
Comparison of distributions of *H. erato* and *H. melpomene*. Note the divergence within each species' convergence over its range but the convergence in pattern between the two species wherever they co-occur.

with defense compounds in adapting to a food source. Various degrees of unpalatability among caterpillars should be expected in the tropics, because so many of the food plants encountered by the larval insects have defense compounds that if stored or metabolically modified by the insect, could render the creature unpalatable. Such secondary metabolites represent strong selection pressures on the lepidopterans.

In 1879, Fritz Müller suggested that two or more unpalatable species would benefit in evolutionary fitness by close resemblance. If two unpalatable species look alike, the would-be predator needs to be educated only once, not twice. The greater the resemblance, the greater the advantage would be to both species. This concept of convergent patterns among unpalatable species is *Müllerian mimicry*. Müllerian mimicry is in theory mutualistic, because individuals of both species benefit from the mimicry. But there may be many cases in which species show varying degrees of palatability. Nonetheless, simulations have demonstrated that when predators have an array of prey from which to choose, any resemblance to one another among prey can induce intense selection for mimicry (Beatty et al. 2004).

Both *Heliconius erato* and *H. melpomene* are unpalatable, both are brilliantly colored, and they look remarkably alike (DeVries 1987). What is even more remarkable is that there are 11 distinct morphs of *H. melpomene* in the American tropics ranging from Mexico to southern Brazil (Figure 8-15). These morphs do not look the same. John R. G. Turner (1971, 1975, and 1981) learned that for every local morph (or race) of *H. melpomene,* there is a virtually identical local morph of *H. erato*. Both species have converged in morphological variation throughout their ranges. Only one morph of *H. erato*, which is restricted to a small range in northern South America, lacks an *H. melpomene* counterpart.

Field work in which wing patterns are manipulated has demonstrated selection for Müllerian mimicry. For example, in a study performed in western Ecuador using local morphs of *Heliconius erato*, *H. cydno*, and *H. eleuchia*, those butterfly morphs that were most distinct, whose wing coloration or pattern did not match the predominant model, consistently suffered higher rates of predation (Kapan 2001).

It has also been shown that divergent selection, driven by mimicry, may be sufficiently strong to result in assortative mating and eventually produce reproductive isolation by ecological specialization, resulting in speciation (Chamberlain et al. 2009).

(b)

FIGURE 8-15
(continued from page 314)

How to Catch a Marmoset: Vocal Mimicry by Margays

For some years, indigenous people in various locations throughout Amazonia have reported that some species of native cats emit sounds on occasion that appear to mimic the sounds made by certain primate and other mammal species. The obvious reason why doing so might be adaptive is that cats feed on mammals if they can catch them, but catching them is far from easy. Therefore, if a cat succeeds in making a primate or some other mammal come closer by mimicking its voice, the cat would stand a better chance of making a kill. Though primate specialists have been skeptical about the idea of cats mimicking primate vocalizations, apparently at least one species, the margay, does just that.

Margays (*Leopardus wiedii*) are small cats that are skilled tree climbers. Tamarins are small monkeys that are often found in the understory, where margays could conceivably capture them. A group of researchers (Oliveira et al. 2010) has observed an incident in which a margay vocalized such that it sounded like the voice of a pied tamarin (*Saguinus bicolor*), a case of vocal mimicry. Researchers were observing a group of eight pied tamarins feeding in a fig tree. A large vine tangle connected some of the surrounding trees to the fig tree. The male pied tamarin sentinel heard a vocalization from within the vines that to the researchers, and presumably the tamarin, sounded similar to a vocalization made by tamarin babies. The sentinel male then climbed up through the vines from where the call emanated, presumably to investigate. The sentinel male continued for about 15 minutes to move among the vines, and it eventually emitted an alarm call. Shortly thereafter, the researchers observed a margay descend from the vine tangle about 15 meters from where the tamarins were feeding. The sentinel tamarin immediately screamed, and the troop quickly fled.

No tamarins were captured by the margay in the incident, but researchers suggested that margays could save energy and effectively capture marmosets by frequent use of vocal mimicry. Perhaps they do.

EVOLUTION AND GENETICS OF MIMICRY

It may seem perplexing that two different species, with presumably different genetics, could somehow evolve to look essentially identical. But such an evolutionary concordance is not really unusual. There are many examples of convergence, or parallel evolution, where genetically distinct organisms evolve close resemblance under similar selection pressures. It should not be surprising that species evolve to be mimics. It has been suggested that the dramatic racial variation apparent in *Heliconius melpomene* and *H. erato* is an example of *punctuated equilibrium* (Gould and Eldredge 1977), but one in which dramatic racial variation over a species' range is not accompanied by subsequent speciation (Turner 1988).

Some mimicry patterns appear to be determined by one or a few genes, each with large effects on the organism's phenotype. For example, in *Heliconius melpomene* and *H. erato*, there are about 39 genes that affect color pattern, but the racial differences among the two species, as well as the racial concordances, appear to be influenced by only four or five genes (Nijhout 1994). Additional research on mitochondrial DNA (mtDNA) has shown that *Heliconius melpomene* and *H. erato* have not shared a recent common evolutionary history (A.V.Z. Brower 1996). The diverse races of both species appear to have evolved within the past 200,000 years, with convergent phenotypes having evolved independently both within as well as between species (A.V.Z. Brower 1996). However, given the possibility that color pattern could be determined by only a few regulatory genes, perhaps the wider evolutionary separation between the two heliconids, even in light of their uncanny similarity of wing patterns across their ranges, should not be so surprising.

Aposematic Insects as Guides for Bioprospecting

The presence of numerous allelochemical compounds in sweeping numbers of tropical plant species has led to the concept of *bioprospecting*, the attempt to discover compounds of potential medical use in plants. It has long been appreciated that tropical peoples are often keenly aware of the medicinal and other uses of plants (ethnobotany) and this topic will be discussed in greater detail in Chapter 13. Bioprospecting attempts to screen plant species for potential efficacy as drugs for use in treating a wide range of human diseases (Coley et al. 2003; Kursar et al. 2008). Thus far, bioprospecting has yielded some promising drugs for use in cancer treatment as well as treating severe parasitic agents such as those that cause leishmaniasis.

In a unique approach to bioprospecting, Helson et al. (2009) looked at the presence of aposematic insects as potential indicators of plants with active chemicals that could prove useful against disease. The logic was straightforward applied evolutionary theory. Plants with potent compounds exert a selection pressure such that only a few insect types evolve to utilize the plant (such as heliconid caterpillars gaining evolutionary access to *Passiflora*). In turn, these insect herbivores should be under selection pressure to evolve aposematic coloration. Thus the more aposematic insects associated with plants, the more likely the plants are to be biochemically active with regard to potential drugs. Plant species chosen for the study were taken from 1,380 species tested by the Panama International Cooperative Biodiversity Groups (ICBG) Program, which bioassays plant species for activity against various cancers and parasites (www.icbg.org). Plant species tested were taken from six plant families and were chosen based on consistent activity or inactivity in bioassays, accessibility in the field, and abundance in the field. *Active* (meaning that it had been determined that medically active compounds were present in the plant) and inactive plant species were paired and examined for presence of aposematic insect species. A total of 20 plant species (10 active, 10 inactive) was examined. The results showed that aposematic insect species were present on both active and nonactive plants but were proportionally more represented on active plants (Figure 8-16). The researchers suggested that, "The presence of aposematic insects can therefore indicate that a particular tropical plant may contain biologically active compounds. As non-aposematic insects are more equally associated with active and inactive plants, using all the insects collected on plants may not be as informative as using only aposematic insects" (p. 133). It was also observed that aposematic insects were common on plants with nonactive compounds, but the possibility exists that these insects are mimics of other species. Perhaps future bioprospecting research will begin with an insect survey.

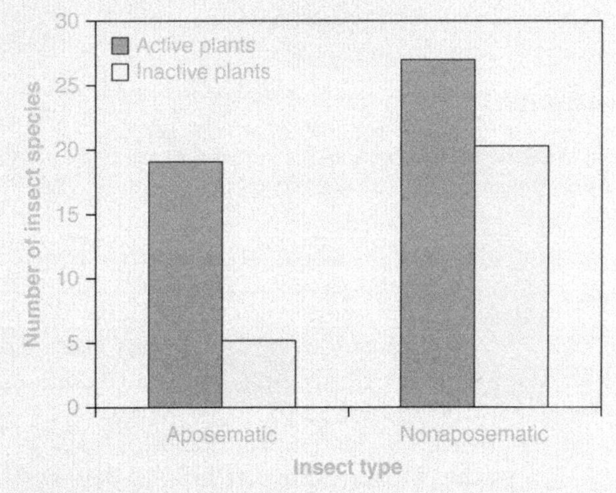

FIGURE 8-16
Number of aposematic and nonaposematic insect species found to feed on the 10 biologically active and 10 biologically inactive tropical plants studied.

Some Top-Down Influences in Rain Forests

BATS, BIRDS, AND ARTHROPOD HERBIVORY

Insectivorous bats and birds are prominent members of tropical rain forest communities all over the world and are often among the groups negatively affected by habitat changes such as fragmentation (see Chapter 14). Two studies demonstrate significant top-down effects exerted by bats and birds on herbivorous arthropods.

Both studies, one from Panama (Kalko et al. 2008) and the other from Mexico (Williams-Guillen et al. 2008), utilized exclosures to measure the impact of birds and bats on foliage arthropods (insects, spiders, mites, and so on). An *exclosure* is a mesh net placed over a tree or shrub that excludes large predators such as birds and bats (and large arthropods) while permitting access by smaller arthropods. Exclosures were established that (1) excluded birds but not bats, as the exclosure was in place only during the day; (2) excluded bats but not birds, as the exclosure was in place only at night; (3) excluded both bats and birds, with the exclosure in place day and night; and (4) did not exclude either birds or bats, a control.

The study in Panama was conducted in a lowland forest and covered a 10-week period. The study in Mexico occurred on an agro forest of shade-grown coffee (see Chapter 13) and covered a 7-week period during dry season and an 8-week period in wet season. In each study, arthropods were regularly censused.

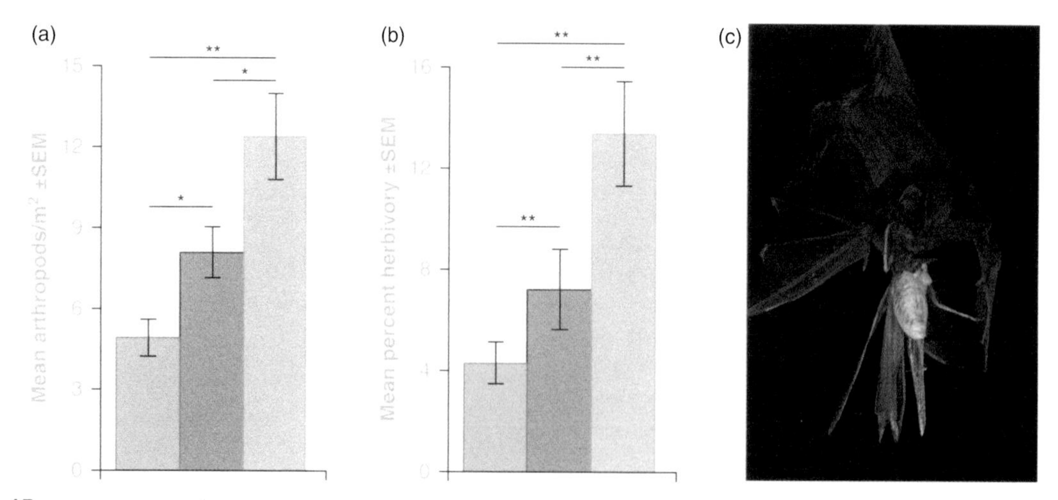

FIGURE 8-17
(a) Mean number of arthropods per square meter per census. (b) Mean herbivory as percent of total leaf area. Blue bars represent controls (birds and bats present); brown bars, diurnal exclosures (birds absent and bats present); green bars, nocturnal exclosures (bats absent and birds present); $^{*}P < 0.05$ and $^{**}P < 0.005$ according to Tukey's HSD. (c) A bat (*Micronycteris microtis*) consuming a katydid, Barro Colorado Island, Panama.

The results of the two studies were similar (Figure 8-17). In each case, both birds and bats exerted significant top-down effects in reducing arthropod abundances. And in both cases, bats exerted stronger effects than birds. The impact of birds and bats was high. In the Mexican study, arthropod densities averaged 46% higher on control plants where birds and bats were excluded. In the Panama study, control plants (allowing birds and bats access) averaged 4.3% leaf area lost to herbivory, bird-exclosed plants 7.2%, and bat-exclosed plants 13.3%. The studies demonstrate that birds and bats reduce arthropod densities, limiting herbivory, a top-down effect. The studies likely underestimated the impact of birds and bats, as they failed to measure predation on arthropods away from the exclosed plants.

BATS AS TOP-DOWN SEED DISPERSERS

A study in French Guiana looked at the effect of bats in dispersing seeds (Henry and Jouard 2007). By employing fine-mesh nets to capture bats, researchers artificially limited bat activity over a 60-night period in tropical mature forest. They set out seed traps and examined 39 plant species likely to be bat-dispersed. Bat exclusion resulted in a 30.5% reduction in seed species richness and, in addition, increased seed limitation (failure of seeds to reach all suitable sites for germination). The study showed that bat-dispersed plants may decline rapidly if bat abundance were to decline.

TOP-DOWN EFFECTS OF MAMMALIAN HERBIVORES

Carefully controlled studies in Mexico have documented the impact of mammalian herbivory on plant species diversity. These studies, like that of Terborgh et al. (2008) described earlier, compared an area in which mammals are at their normal diversity with an area that experienced loss of key mammals species (Dirzo and Miranda 1991).

The two study areas were each historically inhabited by approximately 39 nonflying mammal species, including tapirs and peccaries. One of these areas, Montes Azules (MA), still had its full complement of mammal species, but the second area, Los Tuxlas (LT), mentioned in a study described earlier, had lost many species, while others were much rarer than at MA. Herbivory by tapirs and peccaries was virtually nonexistent in the LT forest but represented about 27% of the total herbivory in MA. Plant density was higher at LT, and some species were numerically dominant, suggesting that the absence of mammalian herbivores in the LT forest was allowing a few plant species to thrive and outcompete other species. Two plant species in particular, *Nectandra ambigens* and *Brosimum alicastrum*, grew thickly throughout the forest floor.

The total number of plant species was low in the LT forest, much lower than in the MA forest, but the undergrowth was denser. The two forests were structurally different, with thick understory of fewer species in the defaunated site (LT) compared with the site where diverse mammals were present (MA).

To test the impact of mammals, researchers constructed selective exclosures at both sites. Exclosures excluded mammals from grazing on the plants within the exclosures. Each exclosure was 2 square meters, and each was located near five canopy trees.

After two years, the exclosure at the MA site showed a loss of plant species. A total of 91 species was found in the exclosures, compared with 112 plant species in the plots where mammals could visit. There was no change at the LT site. The implication is that plant interspecific competition in the absence of mammalian herbivory was the cause of the depletion of plant species richness at the MA site.

One additional manipulation was performed. Researchers planted seedling carpets using the species *Brosimun*. Half of the seedling carpet was kept in an exclosure, while the other half was outside the exclosure, a control, available to mammalian grazers. Pairs of seedling carpets were placed at both the LT and MA sites. The exclosures resulted in high survival rate of the *Brosimum* seedlings, while in the control plot, they were much reduced, a clear indication that mammalian grazers had consumed them.

Researchers concluded that mammals have a strong top-down effect on regeneration and maintenance of plant diversity in the rain forest. They hypothesized further that if medium and large mammals are eliminated, there is increased survivorship among large-seeded plant species, seeds that normally are consumed by large mammals. In the absence of medium and large mammals, smaller mammals such as rodents become much more important as seed predators, but these small animals shy away from plants with toxic seeds or large seeds. Therefore, these large-seeded plants "win" in competition with smaller plants, and a few large-seeded plants will eventually dominate the forest, at the expense of a diversity of smaller-seeded species.

What is complex about tropical food webs such as those just described is the opposing forces of mammalian seed predation and mammalian seed dispersal. (Recall the study in Cocha Cashu showing constricted seed shadows.) Mammals exhibit both a top-down negative force (seed predation) and top-down positive force (seed dispersal).

Top-Down Superorganism: Army Ants

Army ants represent a top-down force on small animals inhabiting rain forest floor. These remarkable insects may represent the most significant predator in some rain forests.

Army ants trace their history back to Gondwana, approximately 105 million years ago in the Cretaceous (Brady 2003). Dinosaurs presumably stepped on army ants. It was once believed that Neotropical army ants (subfamily Ecitoninae) represent a case of convergent evolution with Paleotropical driver ants (subfamily Dorylinae), but molecular and other analyses suggest a close relationship. These ants evolved before the separation of Gondwana.

Army ants are prevalent in the Neotropics, with two widely distributed species in particular, *Eciton burchelli* and *Labidus praedator*. In all, there are 5 genera and about 150 species of Neotropical army ants. Neotropical army ants likely represent more biomass per unit area than all vertebrate predators combined (such as cats, weasels, coatis, raccoons). In the Paleotropics, army ants, in the subfamily Dorylinae, genus *Dorylus*, are predominant in Africa but are less represented in tropical Asia. A third subfamily of army ants, Aenictinae, is found from Africa through Australasia (Gotwald 1982 and 1995). Army ants, like other social insects, have been treated as a form of superorganism because of their uniquely complex social order (Hölldobler and Wilson 1990).

Eciton burchelli (Plate 8-35) is one of the best studied of Neotropical army ants, and its natural history is the focus of this section. *Eciton* varies in size and color. The largest individuals are soldiers, with extremely large menacing mandibles, sharply hooked inward. The smallest workers are only about one-fifth as large as the soldiers. Color varies from orange and yellowish to dark red, brown, or black, depending on subspecies. *Eciton* armies are immense, often containing in excess of a million ants. Armies are nomadic, moving through the forest, stopping at temporary bivouacs during their reproductive cycles. They may bivouac only for a night or for several weeks in the same area. When the entire mass moves, it is usually a nocturnal migration. Bivouacs are either underground or in hollow logs or trees and consist of massive clusters of the ants themselves. There is a single queen per colony who remains in the bivouac except when the entire army is on the move,

PLATE 8-35
***ECITON* SOLDIER**

at which time an entourage of workers and soldiers transports her.

Eciton raiders stream from the bivouac, making a dense column in search of prey. Soon the column begins to fan out widely, raiding parties moving in different directions. Ants move over and through the forest litter. Though virtually blind (Doryline ants are totally blind), the ants effectively communicate by chemical signals, and once prey is discovered, ants quickly converge on it. Small prey are killed and taken to the nest; large prey are killed, dismembered, and carried to the nest by a specific worker caste. The raiders do not restrict their plundering to animals on the ground. They sometimes climb trees, even ascending into the canopy. They will enter human dwellings and attack cockroaches and other insects.

Eciton burchelli is a generalist predator. Prey consists of anything alive and small enough to subdue, most commonly arthropods such as caterpillars, spiders, millipedes, and other animals among litter and leaves. Small vertebrates such as tree frogs, salamanders, lizards, and snakes are routinely attacked, and baby birds in the nest are frequent army ant victims.

Raiding parties attract attention from groups of birds specialized in following army ants. Birds capture prey exposed or attempting to flee from the horde of army ants. Both Neotropical and African army ants have birds that typically follow swarms, but this behavior is much more developed in Neotropical birds.

In contrast with *E. burchelli,* another closely related species, *E. hamatum* has adopted a very different foraging pattern. This species never fans out across the forest floor but remains in tight columns, usually climbing trees to raid broods of social insects such as vespid wasps and other genera of ants. Both species are found together over most of their ranges, but they obviously have evolved very different patterns of predation.

ARMY ANTS AND ANTBIRDS: MUTUALISM OR PARASITISM?

Neotropical antbirds are members of two families, Thamnophilidae (206 species) and Formicariidae (62 species). The vast majority of these bird species do not follow army ants or attend ant swarms. There are 28 species (combined from the two families mentioned earlier) that are considered "professional" ant-following species. These birds are strongly associated with army ant swarms. In addition, other species of birds including various woodcreepers, tanagers, motmots, and ground-cuckoos join antbirds in attending ant swarms (Plate 8-36). The complex behavior of professional ant-following birds has been well documented (Willis and Oniki 1978; Willson 2004).

The ant-following birds devour arthropod and small vertebrate prey (depending on the bird species) exposed by the ants. The relationship between army ants and ant swarm birds has been puzzling. There are three possibilities: birds may be mutualists with ants, aiding ants by flushing prey; they may be commensal, taking prey of no consequence to the ants; they may be parasitic, taking prey away from the ants. One study performed at Soberania National Forest in Panama (near Barro Colorado Island) demonstrated that antbirds are significant parasites of army ants (Wrege et al. 2005).

Researchers, through a variety of creative techniques (one involved squirting persistent antbirds with a squirt gun to force them to flee), excluded ant-following birds from attending swarms, leaving other swarms with their attendant birds. They censused the arthropods and other animals captured by the ants. At swarms where birds were permitted to remain, they counted the prey items taken by birds. The results (Figure 8-18) revealed the following:

- The birds are highly dependent on ants to make prey available. Without the ants, their foraging success would be greatly diminished.

(a)

(b)

(c)

PLATE 8-36
(a) OCELLATED ANTBIRD (*PHAENOSTICTUS MCLEANNANI*)
(b) BICOLORED ANTBIRD (*GYMNOPITHYS LEUCASPIS*)
(c) RUFOUS-VENTED GROUND-CUCKOO (*NEOMORPHUS GEOFFREYI*)

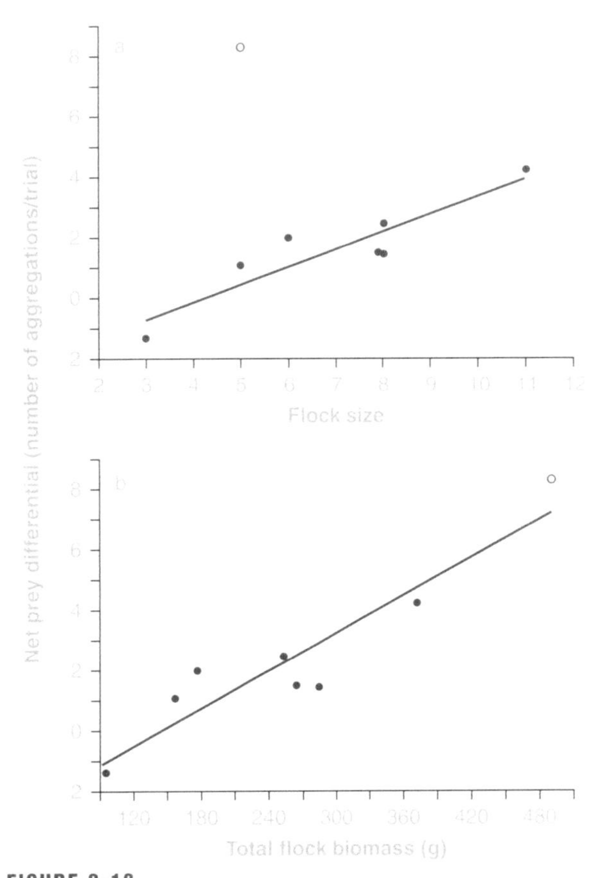

FIGURE 8-18
Costs to *Eciton burchelli* increase with number and species composition of ant-following birds. The open symbol shows the disproportionate effect of participation by the large rufous-vented ground-cuckoo.

- The average nomadic ant colony consumes about 22 grams of leaf-litter arthropods daily, plus an additional 22 grams from social insects (other than the army ants).
- The larger the flock size, the higher the cost of lost prey to the ants.
- Each day antbirds take more than 200 prey items that otherwise would have gone to the ants. This represents about 30% of the ants' daily leaf-litter arthropod intake, or 15% of the entire daily food requirement of a migrating ant colony.

The relationship between army ants and ant-following birds is, at least if this study is typical, significantly parasitic. The combined force of army ants and birds exerts a top-down effect on forest floor arthropods.

One More, Rather Awesome, Top-Down Predator

The humid tropics are often associated with the belief that poisonous snakes abound, and truth be told, they are not uncommon. In Central America, one of the most abundant and most frequently encountered poisonous snakes is *Bothrops asper*, commonly known as the *fer-de-lance* (Plate 8-37). This snake, a pit viper with an extremely powerful venom and long hypodermic fangs with which to deliver it, is believed responsible for the majority of human envenomations and snakebite mortalities in Central America (Wasko and Sasa 2009). In this author's experience in the Neotropics, this is the only poisonous snake species that I know to have struck humans, each of whom fortunately survived the experience. (Many apparently do not.)

Bothrops asper reaches lengths of up to 2.5 meters (8.25 feet) and has been reported to act aggressively when encountered, though such aggression is not common. Anyone who spends much time in Central American tropical forests will sooner or later encounter this species. A fer-de-lance is a sit-and-wait predator in that it does little searching, preferring to sit in striking position until prey ventures sufficiently near to strike. Pit vipers are sensitive to body heat, so birds and mammals (which are endothermic and give off detectable body heat) make up most of their prey.

In a somewhat courageous study performed at La Selva Biological Station in Costa Rica, two researchers used radiotelemetry to study the spatial ecology, activity patterns, and habitat selection of 11 female and 5 male fer-de-lances (Wasko and Sasa 2009). They brought captured snakes to the lab, anesthetized them, and implanted radio transmitters. They released the snakes back where they were initially taken, and they monitored their movements, attempting to relocate each animal at least once a day, either during the day or at night. They monitored the snakes' activity patterns (*inactive* if lying coiled; *active* if moving; *ambushing* if lying coiled and alert, head raised) over a two-year period. They calculated the average home range of the snakes (about 6 hectares), which was generally smaller than is the case with viper species in temperate regions (which helps account for their apparent abundance in many areas). They noted habitat preferences and learned that fer-de-lances have a strong focus on swamps, although they are also commonly present in primary and secondary forest. They avoid developed areas (much to the relief of humans who occupy such areas). The study showed that the animals were primarily nocturnal in their activity pattern, spending most of the daytime hours coiled and inactive.

This study is noteworthy in that it was the first of its kind to document the habitat selection, demographics, and activity pattern of a species that absolutely everyone who becomes involved with Central America ecology gets to know and to fear encountering. And unlike other top carnivores, such as jaguars, fer-de-lances are not rare.

PLATE 8-37
COILED FER-DE-LANCE
Notice the poison-dart frog atop the snake. This frog appears to be very well protected.

9 Carbon Flux and Climate Change in Tropical Ecosystems

Chapter Overview

Primary productivity is the total amount of carbon fixed as organic matter in the process of photosynthesis. It is usually measured as *net primary productivity* (NPP), the amount of carbon devoted to growth and reproduction. Tropical humid forests have the highest annual NPPs of any terrestrial ecosystem. *Carbon flux* is the rate at which carbon enters and leaves ecosystems. If carbon is stored in plant biomass such that more carbon enters (by photosynthesis) than leaves (by metabolic respiration), the plant community is functioning as a *carbon sink* and is adding biomass. If the opposite situation ensues, it is a *carbon source*, emitting more carbon than it stores. Tropical moist forests have potential to serve as carbon sinks, and some data suggest that some tropical lowland forests are adding biomass as atmospheric carbon dioxide concentration increases. Because tropical forests have such high biomass, their potential to act as carbon sinks could help offset the rise in atmospheric carbon dioxide associated with fossil fuel burning and global climate change. But forests consist of more than plants. What must also be considered is *net ecosystem productivity* (NEP), the amount of carbon added after carbon losses from respiration by plants, all consumer animals, and all decomposers are considered. It is unclear whether tropical forests do act as carbon sinks. Data vary among sites. Tropical forests are sensitive to changes in mean annual temperature, drought, and enhanced precipitation. These climatic changes could bring about shifting patterns of biomass accumulation, tree recruitment, species composition, and patterns of decomposition, all affecting carbon flux. Deforestation caused by anthropogenic activity also results in carbon loss from tropical ecosystems and, because it reduces area covered by high-biomass forests, compromises the ability of the global tropics to collectively act as a carbon sink.

This chapter weaves together two important concepts: (1) tropical humid forests are normally high in net primary productivity and have potential to gain additional carbon that has been added to the atmosphere by human activities; but (2) multiple studies show different degrees of carbon sequestration in different regions, and it is by no means clear just how tropical forests will respond to climate change with regard to carbon flux or other ecosystem characteristics. Please be aware that this area of research, as with all science, is a work in progress, but this is one of the more difficult, complex topics now confronting tropical ecologists. Studies show contradictory conclusions. This reality will become evident as you read this chapter.

Primary Productivity: An Introduction

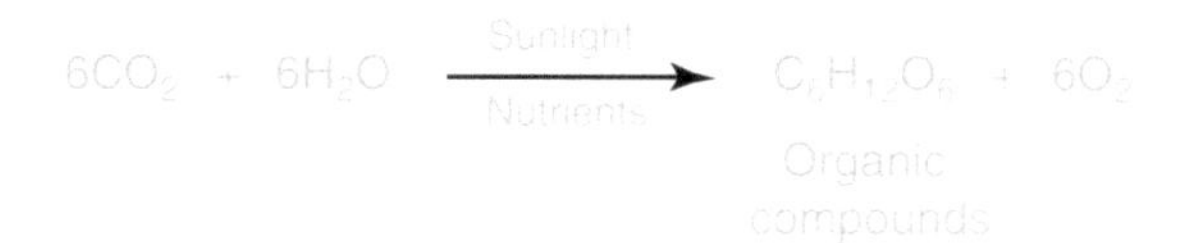

FIGURE 9-1
This is the basic empirical equation for photosynthesis.

Ecologists use the term *primary productivity* to describe the total amount of solar radiation (sunlight) converted by plants to high-energy molecules such as carbohydrates. Photosynthesis (Figure 9-1) is the complex biochemical process by which this energy transformation is accomplished. Plants capture red and blue wavelengths of sunlight, what is called the *photosynthetically active radiation*, from 400 to 700 nanometers. Plants use the energy to split water molecules into their component atoms, hydrogen and oxygen. To accomplish this, plants utilize the green pigment chlorophyll. The reason plants look green is because chlorophyll reflects light at green wavelengths while absorbing light in the blue and red portions of the spectrum. The essence of photosynthesis is that energy-enriched hydrogen from water is combined with the simple, low-energy compound carbon dioxide (CO_2, an atmospheric gas) to form high-energy carbohydrates and related compounds. This evolutionarily ancient and fundamental process provides the basis on which virtually all life on earth ultimately depends. Oxygen from water is released as a by-product. Photosynthesis, occurring over the past 3 billion years, has been responsible for changing the Earth's atmosphere from one of virtually no free oxygen to its present 21% oxygen.

Of all natural terrestrial ecosystems on Earth, none accomplishes more photosynthesis than tropical rain forest (Plate 9-1). On an annual basis, a hectare (10,000 square meters, or 2.47 acres) of rain forest is more than twice as productive as a hectare of northern coniferous forest, half again as productive as a temperate forest, and between three and five times as productive as savanna and grassland (Whittaker 1975; Whittaker and Likens 1975).

Ecologists distinguish between *gross primary productivity* (GPP) and *net primary productivity* (NPP). The former refers to the total amount of photosynthesis accomplished, while the latter refers to the amount of carbon fixed in excess of the respiratory (metabolic) needs of the plant—in other words, the amount of carbon (as plant tissue) added to the plant for growth and reproduction. The equation GPP = NPP + R describes the relationship. Most published data on productivity are expressed as NPP.

By way of example, if you measure the growth in biomass of a field of corn from seed to harvest, you are computing net primary productivity. You do not know how much energy the corn has used to maintain itself during its growing season. Such respiratory energy, essential to the metabolism of the plants, has been radiated back to the atmosphere as heat energy (along with a certain amount of carbon dioxide, oxygen, and water vapor from transpiration). A growing cornfield photographed from above with an infrared camera would reveal the deep red image indicating the heat that is continually emitted from the corn. This is energy of respiration. Net primary productivity is easier to calculate than gross primary productivity, because NPP can be measured by weighing biomass change over a period of time (keeping in mind that some biomass will be consumed by herbivores and pathogens).

PLATE 9-1
Tropical forests have the highest annual NPPs of any terrestrial ecosystem.

PLATE 9-2
El Verde rain forest in Puerto Rico.

Tropical forests currently cover about 7% to 10% of the global land area, and because they exhibit high net productivities, they store about 40% to 50% of carbon present within terrestrial ecosystems (Malhi and Grace 2000). It is estimated that tropical forests process (via photosynthesis and respiration) about six times as much carbon as is emitted from anthropogenic use of fossil fuels (Lewis 2006). Tropical humid forests, particularly rain forests, are the most productive. And keep in mind that tropical humid forests store carbon in the form of wood for long time periods, in contrast to carbon in aquatic as well as other terrestrial ecosystems. For example, at El Verde rain forest in Puerto Rico (Plate 9-2), leaves represent between 3.7% to 4.8%, palms from 0.3% to 2.2%, branches from 18.3% to 18.5%, and boles from 75.6% to 77.6% of the above-ground biomass (Lawrence 1996). Three of four grams of above-ground biomass is in the form of wood. Estimates from Brazilian grasslands and rain forest suggest that rain forests are about three times more productive than grasslands (Smil 1979). But it is also true that rain forests have rates of respiration that exceed those of other ecosystems, presumably due to temperature stress resulting in high rates of transpiration (Kormondy 1996). Rain forests expend as much as 50% to 60% of gross primary productivity in metabolic needs. What this means, of course, is that gross primary productivity, the total rate of photosynthesis (net productivity plus energy used for respiration) is very high in rain forests.

Many climatic and edaphic factors influence GPP, which is why productivity varies significantly among ecosystems. Solar radiation, the most fundamental ingredient, varies with season and cloud cover. The component molecules necessary for photosynthesis, CO_2 and H_2O, are equally essential. Water is limited in many ecosystems, including some tropical forests during dry season; thus water availability has a significant influence on global patterns of productivity. In addition, GPP is dependent on atoms such as nitrogen, phosphorus, calcium, and potassium (an incomplete list). In tropical forest ecosystems, phosphorus tends to be a common limiting factor. Atmospheric carbon dioxide has been increasing since the onset of the Industrial Revolution and continues to rise, potentially contributing to GPP as well as climate change (Figure 9-2). If water or other factors are not limiting, the addition of carbon availability, as is now occurring, could stimulate additional primary productivity, one of the reasons for the focus of research on forests as carbon sinks.

Carbon flux measures the rate of carbon passage through an ecosystem and the amount of carbon storage (sequestration) within ecosystems, acting as carbon sinks. Tropical rain forests are of particular interest because they have the potential to store a great amount of carbon in biomass, a factor that could potentially mitigate carbon dioxide increase in the atmosphere (Figure 9-3). But tropical forests also are metabolically active, and it is essential to understand how much carbon is released from tropical forests during normal metabolism not only of the trees themselves but also of the entire forest, especially the role of decomposers.

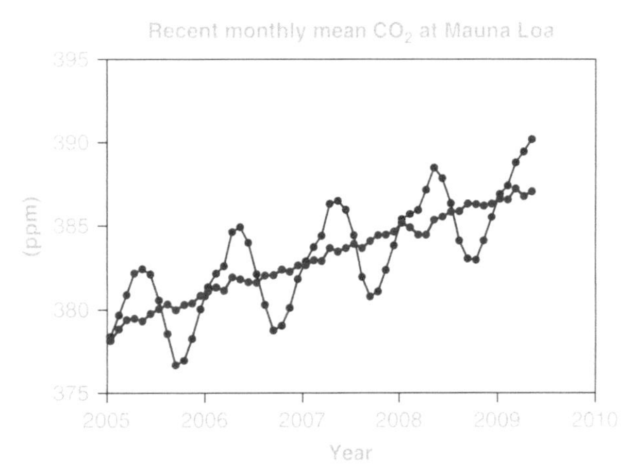

FIGURE 9-2
The pattern of carbon dioxide accumulation in Earth's atmosphere.

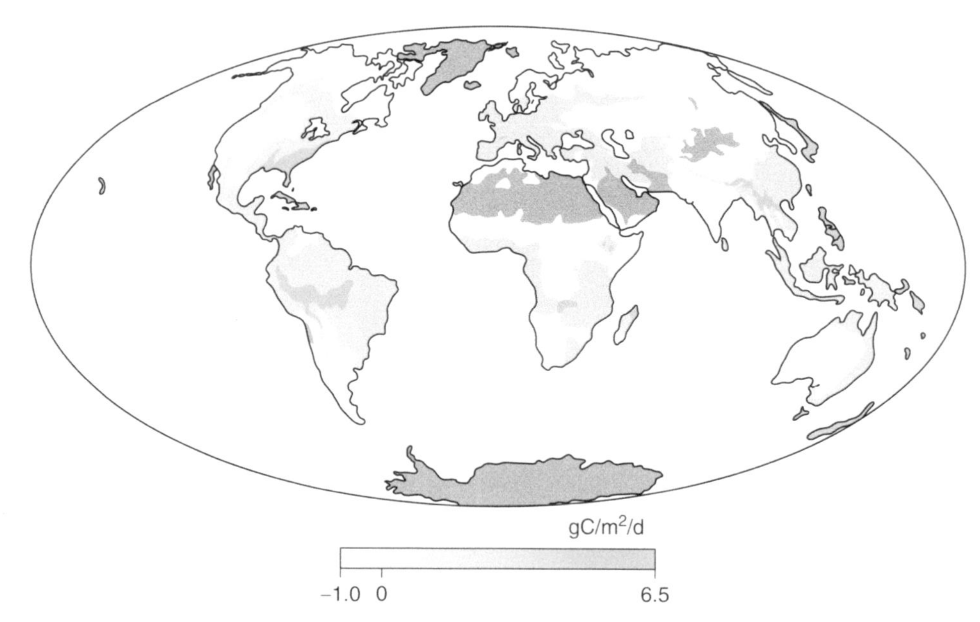

FIGURE 9-3
Map of global net primary productivity. Note the concentration of NPP in equatorial regions, specifically forested areas.

Net Ecosystem Productivity

Tropical humid forests demonstrate the highest net primary productivities of any terrestrial ecosystem, but one must realize that these forests are more than just trees. Ecologists understand that net primary productivity and respiration of the tree community do not, in themselves, explain carbon flux in any ecosystem, including tropical forests. Consider that all animals of the forest also emit carbon dioxide as they respire. Further, all decomposers emit carbon dioxide in their collective respiration. What this means is that carbon flux in the ecosystem is dependent on the net primary productivity of all plants as it relates to the total respiration of all plants, consumers, and decomposers. This is what ecologists term *net ecosystem productivity*. The following simple equation describes this reality:

$$\text{NEP} = \text{NPP} - R_{\text{plants}} - R_{\text{consumers}} - R_{\text{decomposers}}$$

The importance of NEP becomes readily apparent when disturbances, either natural or anthropogenic, result in losses of trees that are then decomposed, such that the ecosystem as a whole emits large amounts of carbon dioxide. The overall metabolism of a tropical forest, with its complex decomposer food web (see Chapter 10), makes it quite possible for tropical forests to release more carbon (in some areas) than they capture, making them carbon sources rather than carbon sinks. This reality will become more apparent in the description of the studies that follows.

Rain Forest NPP in Global Context

In the course of one year, a square meter of rain forest captures about 28,140 kilocalories of sunlight (GPP). Of this total, the plants convert a minimum of 8,400 kilocalories (about 35%) to new growth and reproduction (NPP), using the remainder for metabolic energy (Whittaker 1975; Whittaker and Likens 1975).

Considering the total global area covered by rain forests, these ecosystems have been estimated to produce 49.4 billion tons of dry organic matter annually, compared with 14.9 billion tons for temperate forests (Whittaker 1975; Whittaker and Likens 1975). However, this estimate is now over three decades old, and rain forest cover has been reduced by cutting and clearance, so the figure is lower today (Figure 9-4). Tropical forest is estimated to account for one-third of potential terrestrial net primary productivity (Mellilo et al. 1993). Tropical forests in Amazonia, Australasia, and Africa all have comparably high net primary productivities (Clark et al. 2001).

As rain forests are cut and replaced by *anthropogenic* (human created and controlled) ecosystems (see Chapter 13), much more NPP is directed specifically toward humans (in the form of agriculture or pasturage), and some is lost altogether (fields and pastures

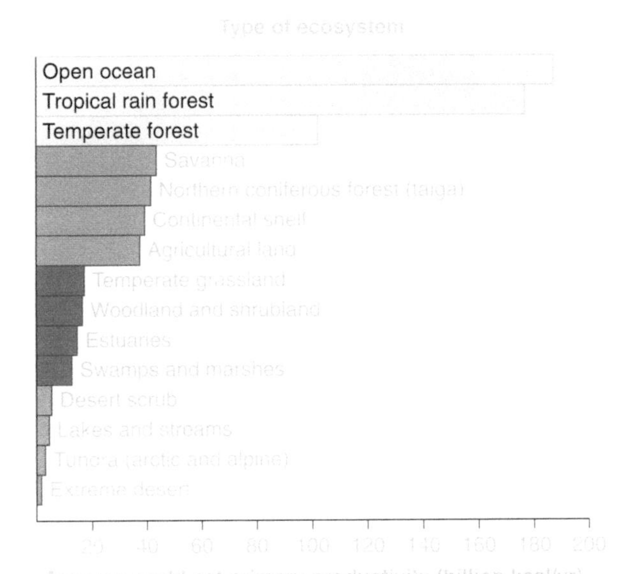

FIGURE 9-4
This graph shows the NPP of ecosystems as reflected in their total area. What is of interest is that the open ocean, a vastly larger area than tropical rain forest, barely exceeds tropical rain forest in total global NPP.

are less productive than forests), making less energy available for supporting overall global biodiversity and reducing the efficacy of tropical forests as carbon sinks. For example, when a forest is cleared, the slash burned, and the site converted to agriculture or pasture, there is far less biomass available for carbon absorption (Plate 9-3). It is simply a fact that agricultural ecosystems (and pastures) do not store as much carbon as forests. One research team has estimated that almost 40% of the world's NPP has either been co-opted by humans or lost due to human activities of habitat conversion (Vitousek et al. 1986). It is estimated that tropical forests store 46% of the world's living terrestrial carbon and 11% of the world's soil carbon (Brown and Lugo 1982). No other terrestrial ecosystem in the world stores so much carbon as living biomass (Plate 9-4).

PLATE 9-3
This old pasture contains far less high biomass than the tropical forest it replaced.

PLATE 9-4
In contrast with a pasture, interior tropical humid forest, with its massive tree boles, stores a great deal of carbon in the form of wood. Photo of the author next to a mature tree in the Peruvian Amazon.

The high productivity of broadleaf tropical rain forests is facilitated by a growing season much longer than in the temperate zone. At the height of the growing season, the daily NPP of a temperate forest will be similar to that of a tropical humid forest. Growth in the tropics is, however, fairly continuous throughout the year, never interrupted by a cold winter. Temperature hardly varies, and because the year is frost-free, there is no time at which all plants must become dormant, as they do in winter throughout much of the temperate zone. Water availability is an important variable in tropical primary productivity. The dry season may slow growth, and where it is severe, most trees are deciduous, dropping leaves at the onset of dry season and growing new leaves with the onset of rainy season. Droughts also exert severe effects on forest productivity (see the section "Drought Sensitivity of Tropical Forests" later in this chapter), often killing large numbers of trees.

Measuring Productivity in the Field

As discussed in Chapter 6, one widely used measure of productivity is called the *leaf area index* (LAI), the leaf area above a square meter of forest floor. In a mature temperate forest such as Hubbard Brook in New

Hampshire, at the height of growing season, LAI is nearly 6, meaning that the equivalent of 6 square meters of leaves covers 1 square meter of forest floor. Values of LAI in temperate forests range widely, from 0.5 to 8 (Le Dantec et al. 2000). Maximum values for LAI at El Verde rain forest in Puerto Rico range from 6.2 to 6.6 (Lawrence 1996), similar to Hubbard Brook and many other temperate forests. But, of course, El Verde is in the tropics and so maintains its LAI value throughout the year, making it annually far more productive than Hubbard Brook. For rain forest at Barro Colorado Island (BCI) in Panama, LAI is about 8 (Leigh 1975). Typically, LAI in the humid tropics ranges from about 5.1 (a forest on poor soil, Amazon Caatinga, at San Carlos, Venezuela) to a high of 22.4 (a lush forest on rich soil at Darien, Panama) (Jordan 1985). In forests with extremely high LAI, the intensity of shading is so great that many, if not most, understory leaves do not approach optimum NPP because they are severely light-limited.

Tropical leaves have greater biomass than temperate zone leaves because many tropical leaves are thick and waxy. When they drop from the tree, they contribute to the litter layer of the forest floor. In the tropics, one hectare of dried leaves weighs approximately one ton, about twice that of temperate zone leaves (Leigh 1975). Litter-fall was measured at over 9,000 kilograms/hectare/year for tropical broadleaf forest compared with just over 4,000 for a warm temperate broadleaf forest, and 3,100 for a cold temperate needleleaf forest (Vogt et al. 1986). Because tropical forests vary in productivity, so must leaf litter amounts. Leaf litter production on rich tropical soils can exceed twice that on nutrient poor soils (Jordan 1985). Nonetheless, tropical forests often have thinner litter layers than temperate forests because decomposer activity is high throughout the year. This topic will be discussed in Chapter 10.

Carbon Gain in Pioneer and Later-Successional Tree Species

Early-successional tree species (for example, cecropias) grow quickly, adding biomass (see Chapter 6). Since the essence of NPP is adding biomass, one might assume that the rate of NPP is fastest in early-successional ecosystems, progressively slowing as more-slow-growing trees eventually grow on the site (Plate 9-5). Thus more carbon would be gained early in succession compared with later. A study conducted in the Bolivian Amazon showed that the pattern of carbon gain over the course of succession is more complex (Selaya and Anten 2010). The study examined three species of trees defined as short-lived pioneers, four later-successional species, and three liana species. The study sites were 0.5, 2, and 3 years old. Researchers used a canopy model to measure canopy structure, mass distribution, and leaf photosynthesis to calculate *whole-plant daily photosynthesis*. This was done per unit leaf mass and above-ground mass (of the plant). The model integrated leaf longevities with the values for whole-plant daily photosynthesis in order to calculate lifetime carbon gain per unit of leaf mass. Whole-plant daily photosynthesis per unit leaf mass declined with age of the stand, but declines were far more pronounced in later-successional species. But later-successional species had much greater leaf longevities than short-lived pioneer species, which more or less balanced out the differences between the two groups of plants. In effect, lifetime carbon gain per unit leaf mass was relatively similar between the various species used in the study. This is because there is a strong negative correlation between daily leaf productivity and longevity. In essence, short-lived pioneer species gain carbon quickly (per unit time), but longer-lived species gain just as much by merely persisting over a longer time frame. It is important to keep in mind that this study looked at only three years of succession, but nonetheless revealed apparent trade-offs between carbon gain per unit leaf mass and leaf longevity. The study introduces the idea that gap dynamics and ecological succession are important considerations in thinking about carbon flux in tropical ecosystems.

PLATE 9-5
Early successional areas add biomass quickly with short-lived plant species that grow rapidly.

The Terrestrial Ecosystem Model

Using a complex, mechanistically based ecosystem simulation model called the *terrestrial ecosystem model* (TEM) (Figure 9-5), a team of researchers estimated the range in NPP among various major ecosystem types in South America (Raich et al. 1991). The model uses five compartments (state variables): C_V (carbon in vegetation); C_S (carbon in soils and detritus); N_V (nitrogen in vegetation); N_S (nitrogen in soils and detritus); and N_{av} (available soil inorganic nitrogen). The model was calibrated using data from 12 sites around the world, where NPP values were known. Additional variables included air temperature, cloudiness, elevation, precipitation, soil texture, and vegetation type. The model proved robust in that its estimates accorded well with other previous studies.

Unsurprisingly, the TEM showed that of the total NPP of South America, more than half occurs in tropical and subtropical broadleaf evergreen forest. Mean annual NPP estimates for tropical evergreen forest ranged from 900 to 1,510 grams/square meter/year, with an overall average of 1,170 grams/square meter/year. Lowest NPPs were for the coastal Atacama Desert of Peru that had an NPP of only 40 grams/square meter/year (Figure 9-6).

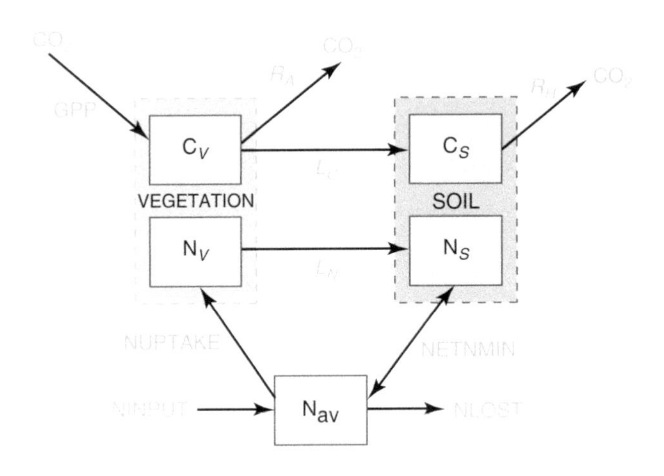

FIGURE 9-5
The terrestrial ecosystem model (TEM). The state variables are carbon in vegetation (C_V); nitrogen in vegetation (N_V); organic carbon in soils and detritus (C_S); organic nitrogen in soils and detritus (N_S); and available soil inorganic N (N_{av}). Arrows show carbon and nitrogen fluxes: GPP, gross primary productivity; R_A, autotrophic respiration; R_H, heterotrophic respiration; L_C, litter-fall C; L_N, litter-fall N; NUPTAKE, N uptake by vegetation; NETNMIN, net N mineralization of soil organic N; NINPUT, N inputs from outside the ecosystem; and NLOST, N losses from the ecosystem.

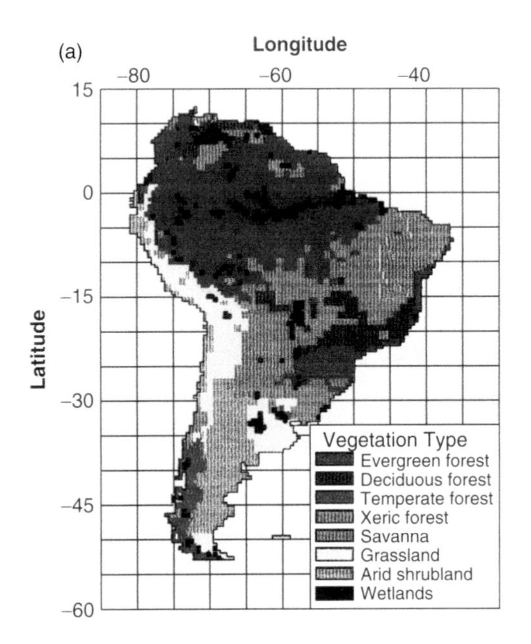

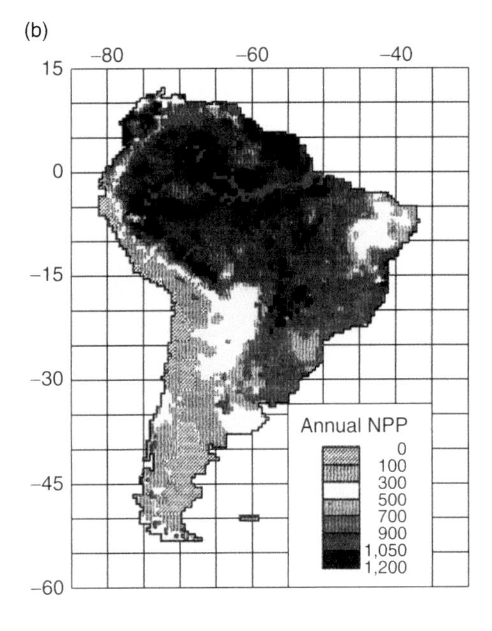

FIGURE 9-6
(a) Potential natural vegetation types of South America as defined for the terrestrial ecosystem model. Each vegetation type is presumed to have unique properties with respect to carbon and nitrogen cycling processes and is presumed to interact with the environment in a unique fashion. (b) Potential annual net primary productivity (NPP, as carbon) in South America as determined by the terrestrial ecosystem model. All vegetation is assumed to be mature and undisturbed by human land-use activities. Units are grams of carbon per square meter per year.

TABLE 9-1 Potential annual net primary productivity (NPP, as carbon) in grid cells containing the Ducke and San Carlos tropical evergreen forest sites, as predicted by the terrestrial ecosystem model. Each estimate is based on an independent model run with the parameter set and the initial conditions defined from the calibration site.

CALIBRATION SITE	PREDICTED NPP (g/m²/yr)	
	DUCKE	SAN CARLOS
Banco (Ivory Coast)	1,610	1,650
Ducke (Brazil)	*	1,090
El Verde (Puerto Rico)	920	990
Kade (Ghana)	1,400	1,410
Pasoh (peninsular Malaysia)	1,300	1,360
San Carlos (Venezuela)	1,180	*
Five-site mean ± SD	1,280 ± 250	1,300 ± 260

*The model correctly estimates NPP (as carbon) at the site for which it was calibrated. The field-based estimates of NPP at Ducke and San Carlos are 1,060 and 1,240 g/m²/yr, respectively.

The most productive ecosystems were lowland forests in Amazonian south-central Brazil, with NPP of 1,190 grams/square meter/year. This represents a nearly 30× difference between the least and most productive South American ecosystems. There was high variation in NPP values among ecosystems within the continent, a result attributed to the heterogeneity of environmental conditions. South American shrublands had an NPP estimate of 95 grams/square meter/year, and savannas averaged 930 grams/square meter/year. Broadleaf tropical forests are far more productive than either savannas or shrublands (Tables 9-1 and 9-2; Figure 9-7). The model

TABLE 9-2 Potential annual net primary productivity (NPP) in the major vegetation types in South America, as predicted by the terrestrial ecosystem model (TEM), in comparison with summaries of measured NPP rates in similar vegetation types. All values in this table are expressed in terms of organic matter (OM) to eliminate differences in the C:OM ratio used by different authors.

VEGETATION TYPE*	NET PRIMARY PRODUCTIVITY (g/m²/yr)				
	TEM†	LIETH (1973)	WHITTAKER (1975)	AJTAY ET AL. (1979)	OLSON ET AL. (1983)‡
Tropical evergreen forest	2040	2000	2000	2300	1680
Tropical deciduous forest	1780	1500	—§	1600	1200
Temperate forest	1200	1000	1250	1400‖	1260
Xeromorphic forest	1000	800¶	700	800¶	770
Savanna	1950	600#	900	1750**	1030
Grassland	620	625‖	600††	780††,‖	790
Arid shrubland	200	70	40‡‡	200	160

*Definitions of vegetation type differ among authors. The TEM estimates are for South America only, whereas the others represent global means.
†TEM estimates are not weighted by area.
‡Area-weighted means calculated from Table 2 in Olson et al. (1983) for comparable vegetation types.
§No value was reported for this vegetation.
‖Area-weighted mean for two vegetation types combined in this table.
¶Value for chaparral, maquis, and brushland.
#Value for woodlands.
**Area-weighted mean for four savanna types.
††Refers to temperate grasslands only.
‡‡Refers to deserts and semideserts.

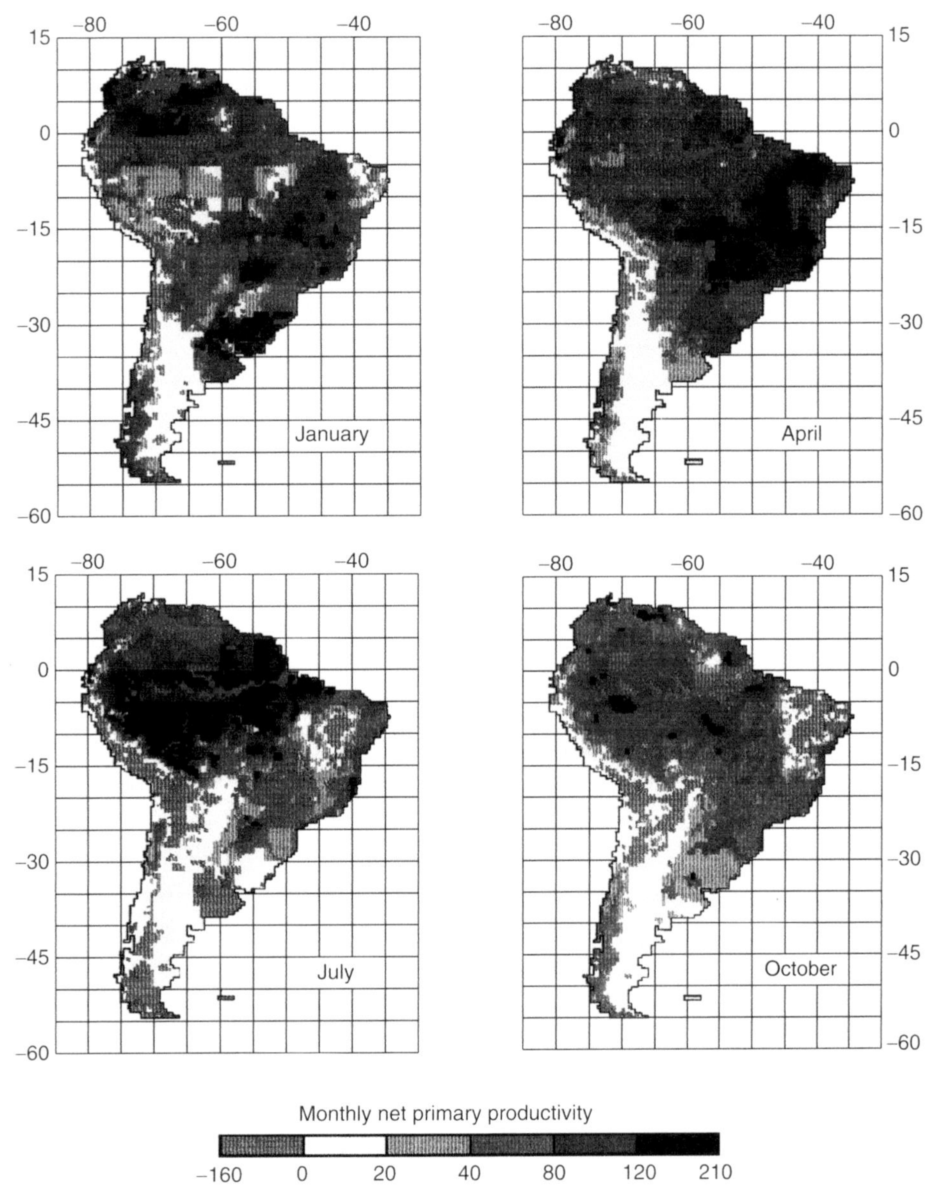

FIGURE 9-7
Potential net primary productivity (NPP) in South America (as carbon) for the months of January, April, July, and October. Values are grams per square meter per month of carbon. Negative NPP values (in red) indicate that autotrophic respiration exceeded gross primary production during that month. The blockiness in these figures is due to the poor spatial resolution of the cloudiness data set used to estimate the monthly irradiance of photosynthetically active radiation.

estimated that a total NPP of 12.5 petagrams per year of carbon, or 26.3 petagrams per year of organic matter. (A petagram, or Pg, equals 1×10^{15} grams.)

NPP showed strong seasonal variation. For moist forests, it correlated most closely with estimated annual *evapotranspiration*, a measure that combines temperature and moisture availability. Strong correlations were also shown to exist between predicted annual rainfall and mean monthly soil moisture (Figure 9-8). This is particularly interesting given that Amazonia is subject to periodic severe drought.

Solar radiation, or *irradiance*, also had a strong influence. Cloud cover reduced solar radiation input, for example, and cloud cover varied seasonally, being most severe during wet season. This means that during the height of wet season, NPP was possibly reduced

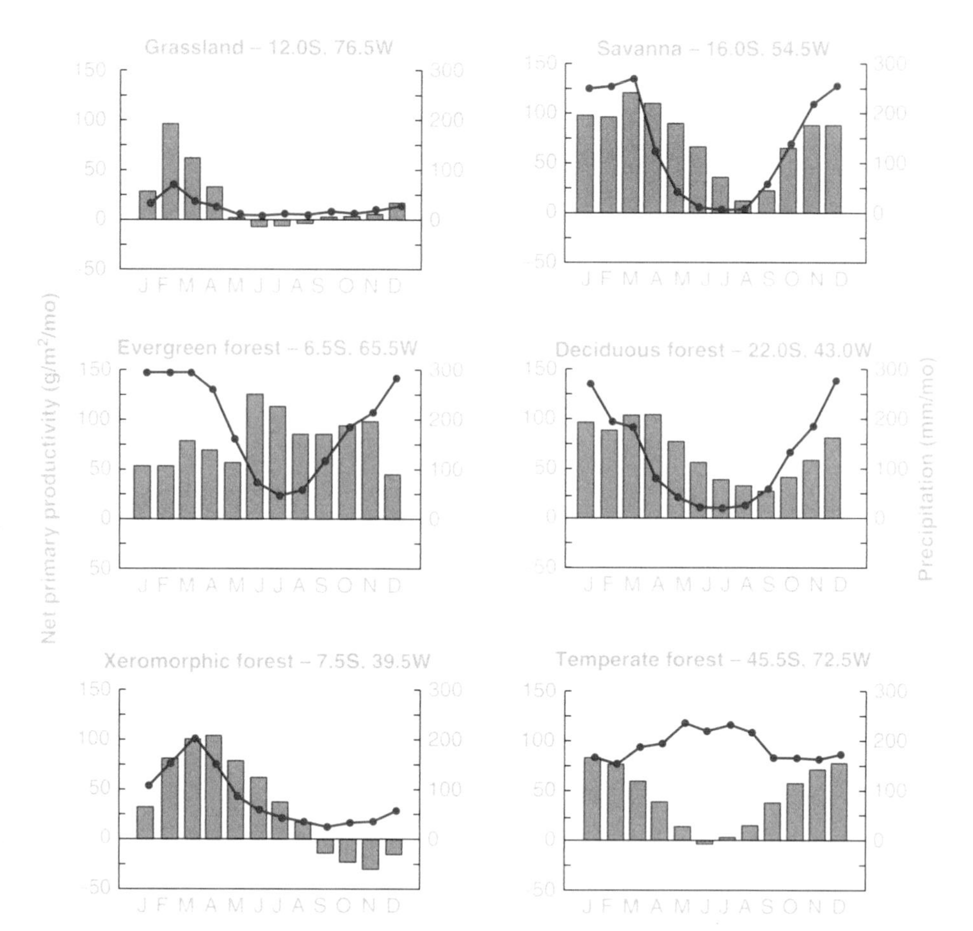

FIGURE 9-8
Estimated monthly net primary productivity for six locations in South America. Monthly rainfall (•—•) is also shown for comparison. Cartographic coordinates refer to the southwest corner of 0.5° latitude × 0.5° longitude grid cells.

because of cloud cover. The model showed that because of variance in cloud cover, rain forest NPP in the Amazon Basin is strongly seasonal (Plate 9-6).

The NPP of savannas, tropical deciduous forests, and dry forests correlated most closely with mean annual soil moisture. This is because cloud cover is less of a factor, and in these drier areas, rainfall becomes very important to NPP. For South American temperate forests (as in places such as southern Argentina and Chile), NPP correlated best with mean air temperature because these forests are located in high southern latitudes, where temperature variation is far greater than in equatorial regions, often becoming quite cold. For shrubland, NPP correlated best with annual rainfall because shrublands occur where it is too dry for forests, and thus rainfall is the prevalent variable affecting NPP.

PLATE 9-6
Cloud cover during rainy season may reduce NPP in lowland tropical forests.

Carbohydrate Storage Adaptations

Tropical plants face two obstacles to growth. If they begin growth in gaps or other areas of abundant solar radiation, they are at selective advantage if they accomplish rapid growth. But most tropical plants exist for much of their lives under shade (Plate 9-7). How are these plants able to have sufficient carbon to grow when the opportunity arises?

There is a hypothesized evolutionary trade-off between (1) utilizing carbon for growth and (2) allocation of carbon to defense and storage. The first, in terms of evolutionary economics, could be considered "investing carbon capital now," the second, "saving up carbon capital until really needed." A study performed in Bolivia compared moist forest with dry forest tree species with regard to use of *nonstructural carbohydrates* (NSC), which are starches and sugars contained within the plant that can be mobilized to support growth or other functions (Poorter and Kitajima 2007).

The study was performed in lowland tropical moist semievergreen forest and dry deciduous forest in Bolivia. A total of 49 tree species was studied in the moist forest, and 38 tree and shrub species in the dry forest. Species were studied in undisturbed (shaded) and disturbed (high-sunlight) plots within both forests. The study focused on NSC in stems from saplings between 0.5 and 2 meters tall. Field measurements included sapling height, leaf number, and diameter at 5 and 50 centimeters in height, and crown exposure (to light) and survival rates were measured at intervals. Samples were taken to measure NSC in stems of each plant.

PLATE 9-7
Deep shade affects patterns of carbon use by plants in rain forests.

Saplings of moist forest species had higher NSC concentrations than those of dry forests. These results suggest the importance of carbohydrate storage within trees that occupy persistently shady habitats such as interior moist forest. Among the moist forest species, sapling survival rates increased and growth rates declined as a function of NSC concentration. These trees were saving up their carbon. Carbohydrate concentrations did not show such a trend among the dry forest plants (Figure 9-9).

A comparison of partially shade-tolerant species to fully shade-tolerant species (in the moist forest) showed that partially shade-tolerant species have higher NSC than totally shade-tolerant species, a result that surprised the researchers. But the partially shade-tolerant species may be storing NSC that will be utilized when a gap opens, allowing them to grow rapidly. The overall results showed that the survival rate of moist forest species was positively correlated with concentration of NSC. These plants were storing carbon to sustain respiration and replace lost tissue.

This study demonstrates significant evolutionary adaptation that has major potential significance. Most tropical tree species in moist forests are shade-tolerant and grow slowly. They are sequestering carbon as they do so. And they grow large, so the total carbon sequestration represented in a tropical moist forest tree is not small. Do tropical forests act as carbon sinks?

What Is a Carbon Sink?

Carbon moves into and out of ecosystems. If the input equals the output, the system is in equilibrium. But that is almost never the case. Growing ecosystems add carbon in the form of biomass. That means that more carbon is removed from the atmosphere by photosynthesis than is released back into it by respiration. This is what *secondary succession* does (see Chapter 6). A successional ecosystem is a form of carbon sink, adding carbon through time.

There is a second way that a carbon sink may occur, and that is when more resources in the form of nutrients or water or carbon dioxide permit a higher rate of primary production in excess of increases in mortality and decomposition. With the ongoing increase in global carbon dioxide level, estimates (from 1980 through 2000) suggest that approximately 1.5 petagrams of carbon have been absorbed by a land carbon sink (Stephens et al. 2007). The question is just where has that carbon gone? One likely place would be that some of it might be going into tropical forests.

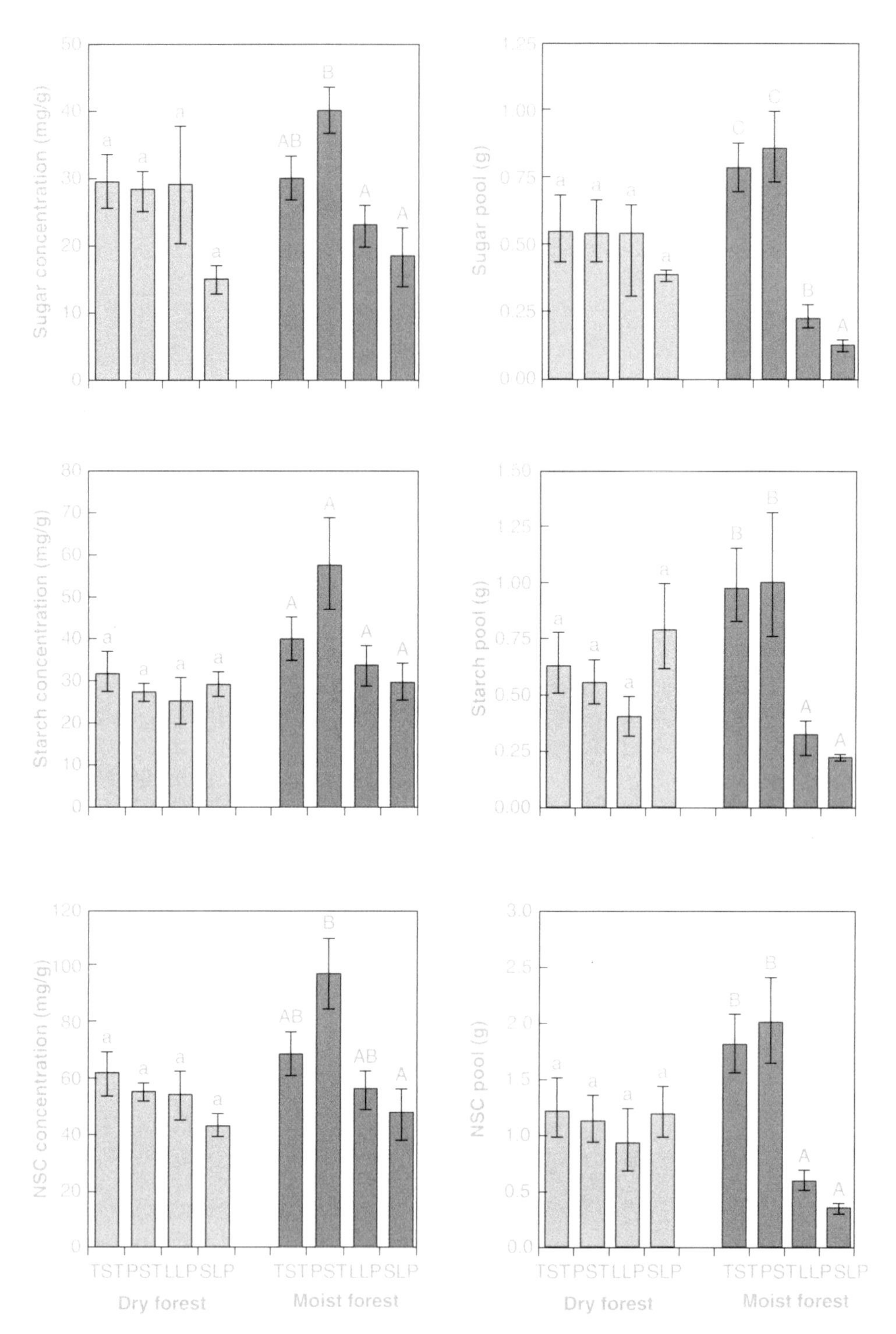

FIGURE 9-9

Carbohydrate concentrations and pool sizes (mean ±SE) of dry (light brown bars) and moist (dark brown bars) tree species belonging to functional groups differing in shade tolerance (TST, total shade tolerants; PST, partial shade-tolerants; LLP, long-lived pioneers; SLP, short-lived pioneers). Bars within forest type accompanied by a different letter are significantly different (Student-Newman-Keuls test, $P < 0.05$). The total carbohydrate pool sizes are estimated for stems of 1-meter-tall saplings.

Tropical Forests as Carbon Sinks

Because of atmospheric rise in carbon dioxide, it has been proposed that ecosystems such as ocean and lowland tropical rain forest act as carbon sinks, taking up and storing some of the additional carbon released by fossil fuel burning and deforestation. Approximately 10 petagrams of carbon (PgC) are released annually by fossil fuel burning and deforestation, an input that has tripled over the past half-century (Baker 2007). Estimates suggest that terrestrial ecosystems sequester between 20% and 30% of the carbon emitted annually as carbon dioxide (Saleska et al. 2003).

The question is how much of this additional carbon has become sequestered in rain forests? That is not an easy question to answer (Clark 2002).

One reason for the potential of tropical forest to store carbon is the very size of the trees (Plate 9-8).

PLATE 9-8
Carbon sink? The immense amount of wood in tropical forests is why they seem to be ideal as potential carbon sinks.

Carbon stored in wood, in trunks and branches of over 10 centimeters diameter and in roots in excess of 5 centimeters diameter, represents about 45% of the total carbon storage in the tree and 30% of above-ground production, with a mean residence time of 80 years (Chambers et al. 2001). One model that took into account such variables as maximum tree size, wood density, wood decomposition, recruitment, growth, and mortality of trees demonstrated that if net primary production increased (due to added atmospheric carbon dioxide), the production of wood would continue even after net primary productivity ceased increasing (Chambers et al. 2001). The model strongly suggests that Amazonian forests could act as an important carbon sink.

Tropical forests account for between 30% and 50% of terrestrial productivity and may store as much as 40% of the carbon present in terrestrial biomass (Phillips et al. 1998). A long-term monitoring study examined multiple permanent forest plots, most of which were in the Neotropics. Biomass gain by tree growth exceeded losses from tree death at most of the sites, particularly in the Neotropics. The plots showed an average increase of 0.71 tonne (+ or – 0.34 tonne) of carbon/hectare/year (Phillips et al. 1998). (Note, a *tonne* refers to a metric ton, which is a unit of mass equal to 1,000 kilograms or 2,204.62 pounds. Another term for metric tonne is megagram, or Mg.) The researchers measured basal area of over 600,000 trees from 153 plots ranging throughout the tropics. They measured mean rate of change in tree basal area across sites based on an initial and final census. They also estimated basal area change as a function of calendar year and were able to derive an estimate of regional accumulated biomass over time. All sites in the Neotropics showed gradual increases in biomass, though that trend was not shown in sites in the Paleotropics (which were fewer in number, only 18 sites) (Figure 9-10). The results showed much variability both in space and time in rate of biomass change among sites, though on average the tropics are gaining biomass. Biomass gain was particularly noteworthy in lowland Neotropical sites.

The question then becomes how to explain the biomass increase. There are several possibilities:

- Biomass is increasing in concert with continental-scale cyclical climate change, such as El Niño/ Southern Oscillation (ENSO) events.

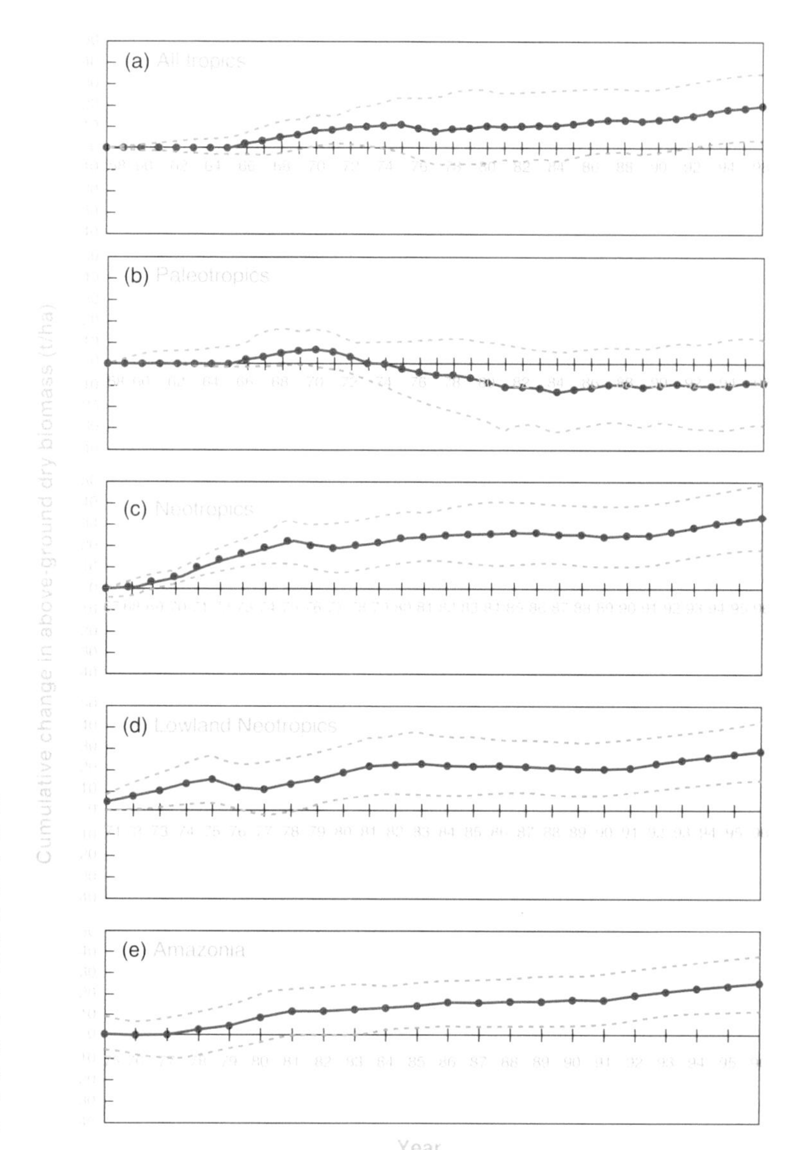

FIGURE 9-10
Cumulative above-ground net biomass change (tons per hectare per year) in humid forests in (a) the tropics since 1958; (b) the Paleotropics (tropical Africa, Asia, Australia) since 1958; (c) the Neotropics (tropical Central and South America) since 1967; (d) the lowland Neotropics since 1971; and (e) Amazonia since 1975. Annual mean (solid line) and 95% confidence interval (dotted lines) values are based on the cumulative changes in individual sites since the first year and are scaled by a/b, where a = the cumulative time elapsed since the first year and b = the mean monitoring period per site up to each year end.

- The ecosystems are recovering from widespread historical disturbance, either natural or anthropogenic, and are increasing biomass because, in essence, the forests are still growing back to a former state.
- Biomass is increasing due to global climate change or additional resource availability, as would be the case with added carbon dioxide.

The study was not able to suggest which of these possibilities is most likely. Amazonian forests have the potential to store a great deal of carbon for a long time period. Amazonian trees grow slowly and are not thought to be in carbon equilibrium (when loss of carbon = gain of carbon) until they are at full adulthood (mean age = 175 years). They may live over 500 years. Studies on Amazonian forests on *terre firme* have indicated that net carbon uptake is about 500 grams of carbon (gC)/square meter/year. Of this total, 70 grams are added to biomass, 70 wind up in the soil, 50 are lost as gases, and 60 are exported to the surrounding rivers (Grace and Malhi 2002). These amounts, each obtained with a different methodology, leave 250 gC unaccounted for. So questions about carbon sequestration remain.

African Forests as Carbon Sinks

The study described earlier focused mostly on the Neotropics, with relatively few sites in the Paleotropics. But another study, focused entirely on Africa, indicated that just as is the case with the Neotropics, African tropical forests may also act as a net carbon sink, where carbon is accumulating over time (Lewis et al. 2009).

The study was conducted on a network of permanent plots called the African Tropical Rain Forest Observation Network (Afritron). Data were taken on 79 plots (163 hectares) that had been sampled for 40 years (1968 to 2007). The plots are located in closed-canopy moist forest that includes West, Central, and East Africa. Trees (equal to or greater than 100 centimeters in diameter) in each area were measured at least twice, and allometric equations were used to determine carbon content based on tree dimensions.

The results of the study showed:

- Above-ground carbon storage in live trees increased by 0.63 MgC/hectare/year. (Recall that MgC refers to megagram, 1×10^6 grams, or one metric tonne.) African tropical forests are in general acting as a carbon sink (Figure 9-11; Table 9-3).
- Extrapolation to include such components as live roots and small trees and scaling up to include the entire African continent indicated that all African tropical forest trees combined show an increase in carbon storage of 0.34 PgC/year.
- When the researchers combined all known data from Africa, the Neotropics, and Tropical Asia, the estimate of global carbon storage by tropical forests was calculated to be 0.49 MgC/hectare/year, supporting a carbon sink of 1.3 PgC/year for all tropical forests over recent decades. But keep in mind that the research did not include potential losses in areas due to harvesting of forest and burning.

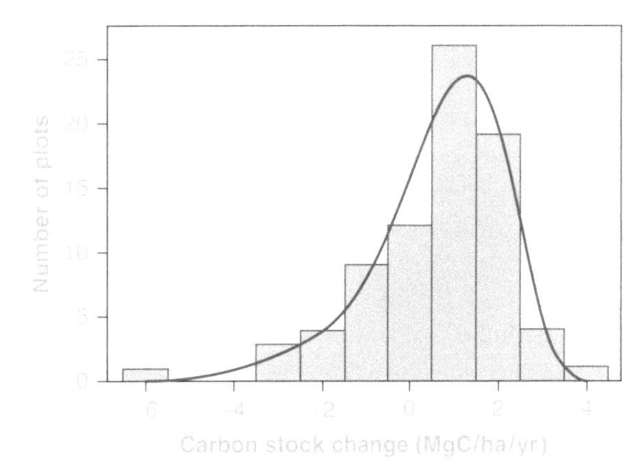

FIGURE 9-11
Histogram of annualized change in carbon stock from 79 long-term monitoring plots across 10 countries in Africa. Results are weighted by sampling effort. Note that most plots are accumulating carbon, but not all are.

TABLE 9-3 Estimated carbon stocks and their annual increase for African tropical forest.

STUDY	CATEGORY	AREA (106 ha)	ALTC (Pg)	TTC (Pg)	ΔALTC ≥100 MM (Pg/yr)	ΔALTC ≥10 MM (Pg/yr)	ΔATC (Pg/yr)	ΔTTC (Pg/yr)
GLC2000	Humid tropical forest	232.7	46.9 (40.5–56.8)	69.5 (60.9–80.7)	0.15 (0.05–0.22)	0.15 (0.06–0.23)	0.17 (0.08–0.25)	0.21 (0.09–0.27)
FRA CS	Closed forest	352.7	71.1 (61.4–86.1)	105.3 (92.3–122.3)	0.22 (0.08–0.33)	0.23 (0.09–0.34)	0.26 (0.12–0.37)	0.32 (0.14–0.41)
FRA RS	Tropical forest	518.5	104.5 (90.2–126.5)	154.8 (135.6–179.8)	0.33 (0.11–0.49)	0.34 (0.13–0.50)	0.39 (0.17–0.55)	0.47 (0.21–0.60)
WCMC	Tropical forest	401.0	80.8 (69.8–97.8)	119.7 (104.9–139.1)	0.25 (0.08–0.38)	0.27 (0.10–0.39)	0.30 (0.13–0.42)	0.36 (0.16–0.46)
Mean		376.2	75.8 (65.5–91.8)	112.3 (98.4–130.5)	0.24 (0.08–0.35)	0.25 (0.09–0.37)	0.28 (0.12–0.40)	0.34 (0.15–0.43)

Succession or Fertilization?

The researchers in the Africa study attempted to separate between two possible explanations for the annual increase in carbon. One explanation, as noted earlier, is that the forests are not in equilibrium due to a past history of disturbance (from drought, fire, or human use). They may be undergoing slow, continuing ecological secondary succession and thus accumulating biomass (carbon) as they continue to grow. The second explanation is fertilization by added carbon dioxide, a response to the changing concentration of atmospheric carbon dioxide. More CO_2 may be the essential nutrient that permits greater carbon sequestration. To complicate matters further, these explanations are not mutually exclusive.

In order to assess which of these possibilities is more likely, the researchers performed a taxon-level analysis. This involved comparing representation of lighter-wood tree species (lower wood mass density) typical of early and mid-succession with heavier-wood species typical of late succession. The assumption was that if succession were the cause of the carbon sequestration, there would be a clear pattern: stands accumulating carbon would show a predominance of lighter-wood, successional species. But this did not happen. Stands showed no trend between lighter- and heavier-wood species (Figure 9-12). This suggests that all stands were experiencing carbon increase regardless of their successional histories. Such a pattern supports the *carbon fertilization hypothesis*, suggesting that forests are acting as a carbon sink in response to greater concentrations of atmospheric carbon dioxide.

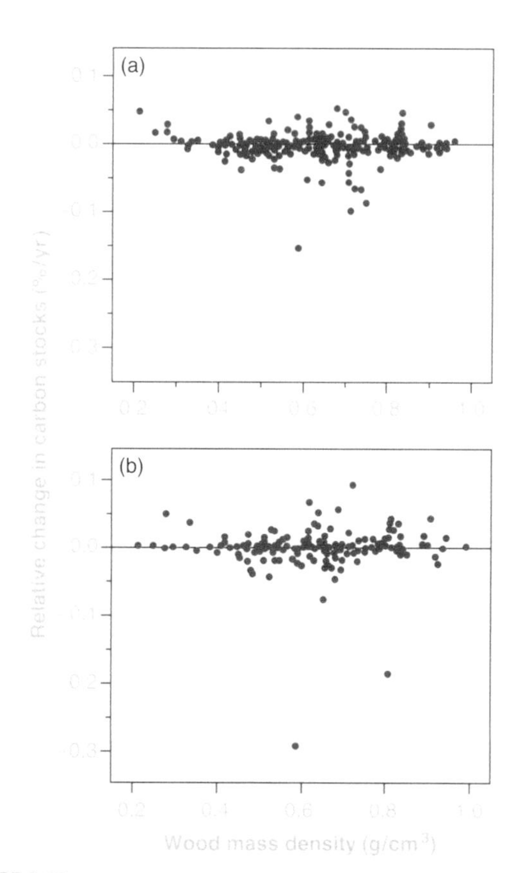

FIGURE 9-12
Relative change in carbon stocks and corresponding wood mass density values: (a) all 612 fully identified species; (b) 200 fully identified genera.

Seasonal Flux and Carbon Loss

Not all tropical rain forests are gaining carbon. A study conducted in two old-growth forest sites in Tapajos National Forest near Santarem, Brazil, showed a pattern of carbon gain and loss that was related to seasonality (Saleska et al. 2003). The annual rainfall is 1,920 millimeters/year, and the wet season (greater than 100 millimeters of rainfall per month) lasts for 7 months. Researchers examined changes in forest structure and monitored carbon stored in live wood and dead wood. Using a methodology termed *eddy covariance*, researchers measured above-canopy fluxes and carbon dioxide uptake within the canopy. Data were expressed as *net ecosystem exchange* (NEE).

The results were surprising. Net carbon gains occurred during the dry season, and carbon loss occurred in wet season. This is the opposite pattern seen in tree growth, where growth is usually greatest in wet season. Gross primary production was about the same in wet and dry season (contrast this with the results reported for the TEM study by Raich et al. 1991 described earlier), but respiration (loss of CO_2 from metabolic activity) was 40% higher at the peak of wet season (March) compared with the peak of dry season (November). Overall, release of CO_2 from dead wood (through decomposer activities) exceeded the uptake of carbon by live biomass. Carbon was taken up mostly by growth of small trees in forest gaps. There was no net change in above-ground biomass over 16 years in a nearby forest that served as a long-term study site. Researchers concluded that when dead trees dropped, carbon was released from them, their loss opened gaps, and that stimulated growth of young trees. The forest was not in equilibrium and may be showing the effects of an earlier drought and natural disturbance that resulted in killing trees. The results also suggest that carbon loss would accelerate if trees were killed during periods of drought, such as are typical of ENSO events.

Disturbance, Plot Size, and Landscape Effects on Carbon Sinks: The Plot Thickens

The overall hypothesis that global tropical forests act as carbon sinks and gain biomass has found mixed support. Some forests appear to be gaining carbon, some not. Add to this the complexities added by gap-phase dynamics and the shifting mosaic created by disturbance in tropical humid forests (see Chapter 6), and it becomes apparent that carbon flux in tropical forests varies from site to site. Add to this the fact that many tropical forests are being replaced by pastures and agriculture such that the tropics, taken as a large landscape, are quite possibly a carbon source in some regions. All of this makes generalizations about carbon sinks much more difficult.

A study by Feeley et al. (2007) looked at gap-phase processes in 50-hectare plots located at BCI (Panama), Pasoh and Lambir (both in Malaysia), and Huai Kha Khaeng (HKK, in Thailand). The objective was to evaluate the role of gap-phase processes in carbon flux. Overall, they learned that biomass increases were gradual and concentrated in earlier-phase forest, while biomass losses (which were of greater magnitude) were found in later-phase forest patches, which were less abundant. The researchers concluded that gap-phase processes strongly affect biomass changes in tropical forest, an important consideration in looking at the concept of forests as carbon sinks. In other words, areas with lots of early gaps could be adding biomass steadily and acting as carbon sinks. But some less-abundant areas of older patches might be losing carbon. Further, while above-ground biomass was stable at BCI and Lambir forests, it increased significantly at Pasoh and decreased at HKK. In both cases, Pasoh and HKK, researchers were unable to offer a clear reason for the biomass changes and they concluded that "idiosyncratic or regional factors" such as changes in temperature, cloudiness, precipitation, or anthropogenic action may take precedence over global increase in carbon dioxide availability.

Another study by Fisher et al. (2008) began with the observation that most data on forests as carbon sinks come from field plot studies, as evidenced by examples cited throughout this chapter. They asserted that such a focus introduces a bias in that inherently rare mortality events that characterize larger-scale landscapes are omitted and that such events may cause areas to become carbon sources rather than sinks. They developed a computer simulation model that showed that plots as large as 50 hectares displayed consistent bias toward growth, favoring the view of the forest as carbon sink. The researchers urged caution in extrapolating plot-based studies to larger landscape areas given that their model showed disturbance events that occur in various places throughout the landscape as well as such variables as size, number, and distribution of field plots, all of which may affect estimates of carbon balance (Fisher et al. 2008).

Another analysis (Sierra et al. 2007) looked at the range of net ecosystem production (NEP) in 33 permanent plots in tropical humid forests over a two-year period. Recall that NEP is the net amount of primary production after the costs of respiration by plants, consumer animals, and decomposers are included. The results showed a range of between −4.03 to +2.22 MgC/hectare/year. The researchers concluded that such a range in NEP did not provide sufficient evidence to reject the null hypothesis that these forests were, in fact, in carbon balance. In other words, they were, taken as a whole, neither sinks nor sources. They concluded by stating that ecosystems are behaving so as to reflect natural variations, such as gap dynamics, and that only in La Niña years do tropical forests tend to really gain carbon. Their blunt conclusion was that tropical forests are often not carbon sinks.

Last, in a massive study that involved measurement of over two million trees in 10 large (16- to 52-hectare) plots over three continents, there was substantial plot-to-plot variability in carbon accumulation (Chave et al. 2008). Above-ground biomass increased at 7 of the 10 plots, but significantly so at only 4 of the 10. One plot showed a large decrease. Carbon accumulation pooled across sites was statistically significant and measured ±0.24 MgC/hectare/year. Three sites showed no increase in biomass gain. Over all 10 sites, the fastest growing quartile of plants added disproportionately more biomass than the tree community as a whole (+0.33 compared with +0.15 MgC/hectare/year), but

PLATE 9-9
Is this tropical forest gaining or losing carbon? It is not possible to know with certainty without careful measurements.

this trend was largely due to the effect of a single plot that differed from the others. The study concluded that the plots studied may be recovering from past disturbances and responding to different levels of resource availability. The variability among plots strongly suggests that more studies are necessary before concluding anything about the general nature of tropical forests as carbon sinks (Plate 9-9).

Carbon Loss from Deforestation and Fire

Because most deforestation (see Chapters 14 and 15) is now occurring in tropical forests, it is reasonable to wonder what effect such activity is having on the carbon cycle. The answer, like so much else surrounding such a complex system, is simply not known with precision. Intact tropical forests act as carbon sinks. But if these forests are cut and burned (Plate 9-10), an immense amount of carbon is quickly released back into the atmosphere. One study concluded that 336 million tons of carbon emissions enter the atmosphere each year from deforestation in Brazil alone (Moran et al. 1994). Another study concluded that about 220 tons of carbon is released from soil and woody biomass to the atmosphere for every hectare of tropical forest that is cleared and burned (Holloway 1993). It is unavoidable that as tropical forests are cut and converted to ecosystems with less biomass and less overall primary productivity, carbon will be lost to the atmosphere.

PLATE 9-10
Fires severely reduce the potential for forests to act as carbon sinks.

Deforestation increases the potential for fire (Cochrane 2003), and fire results in the release of stored carbon. It is typical that deforested tracts are burned (to clear brush and slash), but the occurrence and effect of fire extends beyond this (Figures 9-13 and 9-14). Deforested areas and degraded ecosystems are

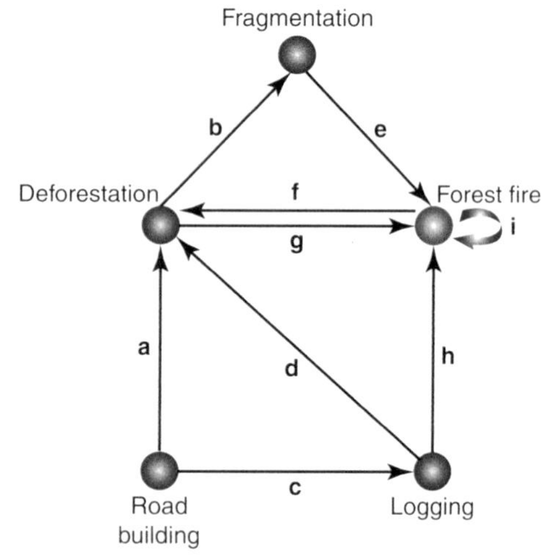

FIGURE 9-13
Diagram of interrelationships between tropical land-cover changes and forest fires. Arrows indicate forcing of each node on others in the system. Blue arrows directly affect forest fire occurrence; black arrows indirectly influence forest fire occurrence. Events a to i are: (a) Road building results in forest access that is strongly associated with deforestation. (b) Deforestation fragments the remaining forests, increasing amounts of edge. (c) Road building and paving directly affect transportation costs and area of economic accessibility. (d) Logging results in limited amounts of deforestation for roads and log landings. Post-logging colonization can increase deforestation. (e) Forest edges suffer biomass collapse and microclimate changes, making them susceptible to frequent fires. (f) Repeated forest fires can lead to unintentional deforestation. (g) Deforestation and pasture land maintenance fires result in many accidental forest fires. (h) Logging degrades forests, increasing fire susceptibility. (i) Forest fires can create a positive feedback cycle where recurrent fires become more likely and more severe with each occurrence.

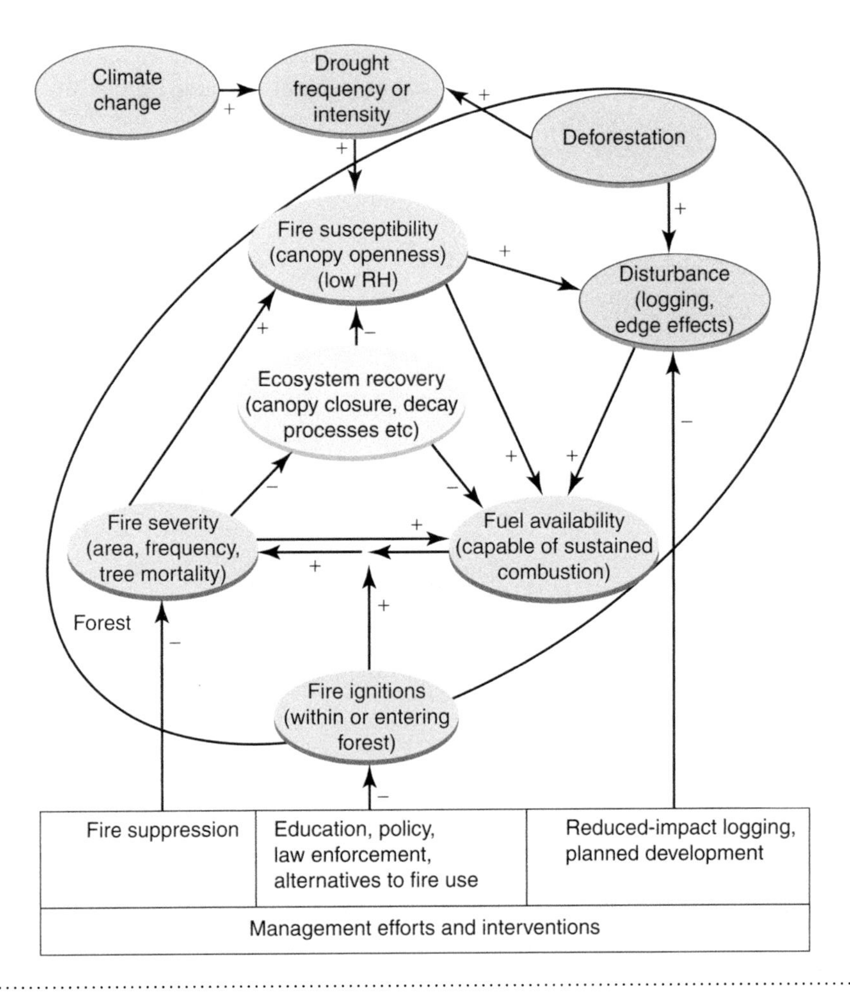

FIGURE 9-14
Positive and negative feedbacks controlling fire processes in tropical forests. Positioning indicates whether they occur within or outside the forest (beige shading). Items in gold control fire occurrence or behavior. Items in brown modify the potential fire environment. Green indicates ecosystem processes acting in opposition to fire, specifically regrowth, canopy closure, and decay of fuels. The management efforts and interventions box indicates how and where human actions can diminish tropical forest fires. Climate change encompasses effects of increased CO_2, land-cover change, and aerosol loading that result in regional drying in the tropics. RH, relative humidity.

much more likely to experience fire. It is estimated that net forest fire emissions may have released carbon equivalent to 41% of the world-wide fossil fuel use in 1997–1998 (Cochrane 2003). This represents a huge loss of carbon from tropical ecosystems. The year 1997–1998 was a year of a significant ENSO event. A study conducted in Central Kalimantan (Borneo, Indonesia) focused on widespread fires that characterized the region, particularly affecting tropical peatlands (Page et al. 2002). A *peatland* consists of dense swamp forest that sits atop deposits of peat that can be as much as 20 meters thick. Peat is partially decomposed plant material and represents a major carbon sink. During 1997, 32% of the area (representing 0.79 megahectare [Mha]) was burned, 91.5% of it peatlands. Measurements indicated that from 0.19 to 0.23 gigatonne (Gt) of carbon was released into the atmosphere from peat combustion and an added 0.05 Gt from burning of overlying vegetation. Considering all of Indonesia, estimates were that between 0.81 and 2.57 Gt of carbon were released due to peat and forest burning in 1997. Though 1997 was clearly an exceptional year for fire in the tropics, the data are sobering in that continuing deforestation, logging, and burning coupled with periodic ENSO events will have multiple effects, among them the accelerated loss of carbon to the atmosphere, reducing the efficacy of the tropics as a carbon sink. This example will be revisited in Chapter 15.

Carbon Loss from River Outgassing

The influx and efflux of carbon into and out of ecosystems may be thought of as a kind of equilibrium. Carbon enters as CO_2 via photosynthesis and ultimately exits as CO_2 via respiration. But is it in equilibrium? Studies suggest that carbon is increasing via added biomass throughout the global tropics, a net gain of carbon. But those studies are based on terrestrial measurements of biomass accumulation. Rivers and marshes are also prevalent ecosystems throughout the tropics, and the metabolism of these systems is closely associated with inputs from bordering terrestrial ecosystems (Raymond 2005).

The metabolism of riverine ecosystems is essential to understand in order to fully comprehend carbon flux in forests (Figure 9-15). This is because organic matter from forests is transported by flood and normal precipitation (as well as natural leaf and branch fall) from terrestrial to riverine ecosystems. This *allochthonous* (organic material washed into a river) matter forms an energy base for decomposers ranging from fish to bacteria. The result is to liberate carbon. In addition, rivers contain a diversity of primary producers that fix carbon (*autochthonous* [organic matter produced within the river] input). The carbon budget in

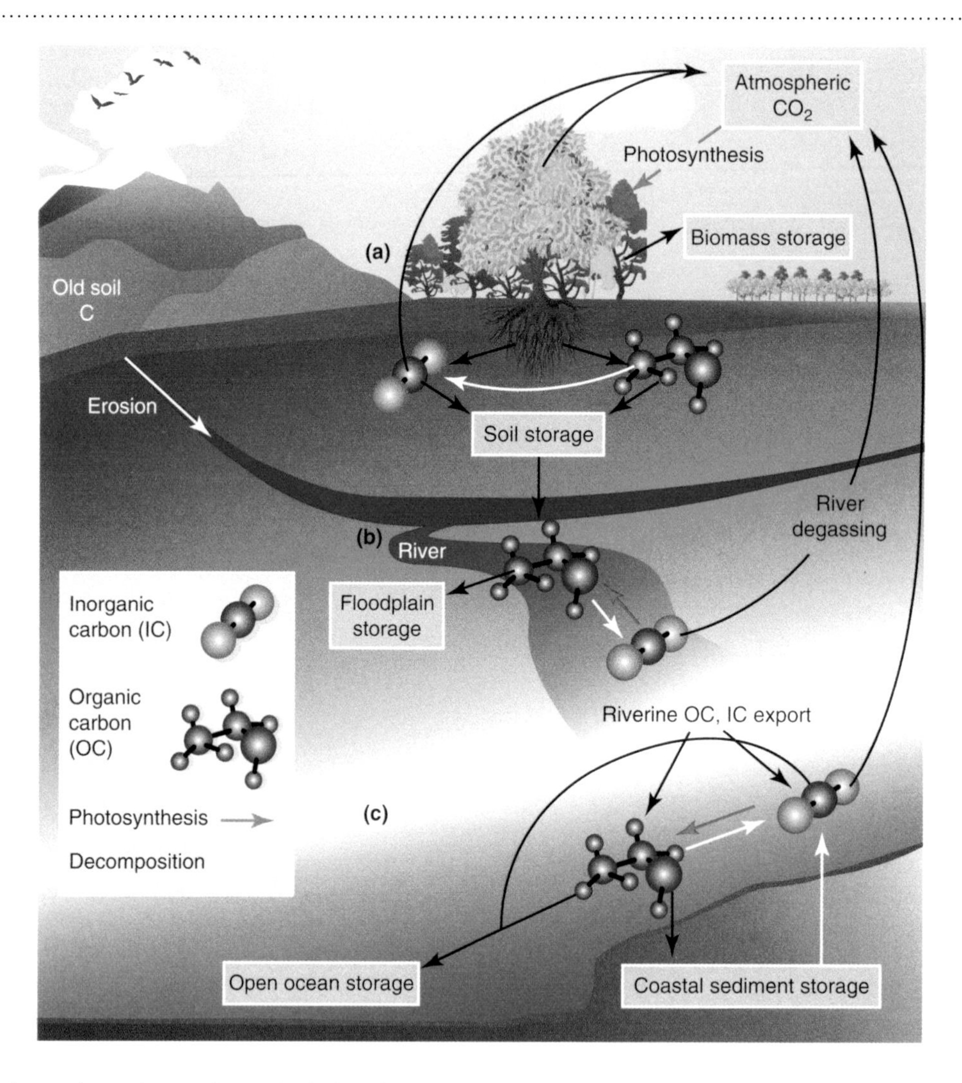

FIGURE 9-15

Rivers and the carbon cycle. (a) Photosynthesis in land plants fixes atmospheric CO_2 (inorganic carbon) as organic carbon, which is either stored as plant biomass or in soil, or is decomposed back to CO_2 through plant and soil respiration. This CO_2 can return to the atmosphere or enter rivers; alternatively, it can react with soil minerals to form inorganic dissolved carbonates that remain stored in soils or are exported to rivers. (b) The transformations of organic to inorganic carbon through decomposition and photosynthesis continue in rivers; here, CO_2 will re-exchange with the atmosphere ("degassing") or be converted to dissolved carbonates. These carbonates do not exchange with the atmosphere and are mainly exported to the coastal ocean. Organic carbon is also exported to the ocean or stored in floodplains. (c) In the coastal ocean, photosynthesis, decomposition, and re-exchanging of CO_2 with the atmosphere still continue. Solid organic carbon (such as soil particles and phytoplankton cells) is buried in coastal sediments, where it is stored or decomposes to inorganic carbon and diffuses back into coastal waters. Dissolved inorganic and organic carbon is also exported to the open ocean, and possible deep-ocean waters, where they are stored for many centuries.

rivers is measured in both organic carbon (OC) and inorganic carbon (IC) and is complex because rivers flow, moving both OC and IC from upstream to downstream. Gaseous carbon dioxide may enter rivers at the interface of air and water, and it may exit via that interface. Ecologists have investigated the net gain and loss of carbon from tropical rivers, as it relates to the carbon flux of surrounding terrestrial ecosystems.

Studies that measure the amount of carbon dioxide emitted from Amazonian rivers and wetlands suggest that outgassing (also known as *degassing* or *evasion*) may be important in returning carbon to the atmosphere (Richey et al. 2002). Outgassing appears to represent a net loss of carbon via riverine metabolism. Carbon loss by outgassing is far in excess of what could have been synthesized within the rivers and wetlands themselves. This means that allochthonous material from surrounding forest has washed into and accumulated in the riverine ecosystems and that metabolic activities of decomposers within the rivers and wetlands are the source of the carbon dioxide ultimately released to the atmosphere. The possibility exists that the overall carbon budget in Amazonia is roughly in balance, the amount of carbon sequestered by the trees being compensated for by the amount of carbon dioxide released from outgassing from aquatic ecosystems. This would mean that Amazonian lowland forests are not as effective as potential carbon sinks as initially believed. But is this the case?

A study in central Amazonia (Figure 9-16) looked at carbon flux in rivers throughout the year (Richey et al. 2002). The research showed that a large net amount of carbon is lost due to outgassing. Rivers are not in carbon equilibrium. They release far more carbon than they acquire. The amount lost, 1.2 Mg/hectare/year, is thought to have originated mostly from upland and flooded forests upstream. The outgassing is the result of metabolic activities within the river ecosystems.

Researchers estimated that within the entire Amazon Basin, a total of 470 TgC/year is lost by outgassing. Carbon loss was seasonal and varied among mainstem floodplain (MF), mainstem channels (MC), tributaries over 100 meters wide (T), and streams less than 100 meters wide (S), and varied seasonally (Figure 9-17). The high level of river outgassing observed in this study could reduce the efficacy of tropical forests as carbon sinks, moving them closer to being in carbon equilibrium.

Another study measured carbon isotope composition (^{13}C and ^{14}C) in dissolved inorganic carbon and three size-fractions of organic carbon in Amazonian river systems (Mayorga et al. 2005). Carbon-14 is radioactive, with a half-life of 5,730 years, and is used as a means of time dating. Carbon-13 allows researchers to learn the identity of a carbon source. This methodology is useful in determining the age

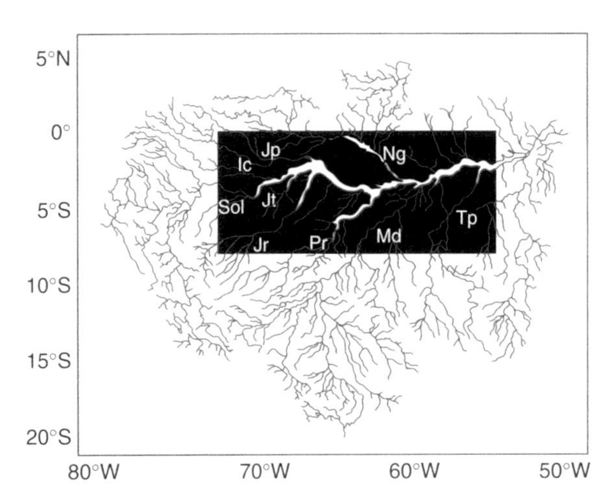

FIGURE 9-16
Flooded area of the central Amazon Basin at high water, as mapped from the Japanese Earth Resources Satellite radar data (May to June 1996). The flooded area is shown as light areas in dark inset (the study quadrant). Underlying the inundation image is a digital river network (derived from the Digital Chart of the World, the GTOPO30 digital elevation model, and ancillary cartographic information). Major tributaries are labeled: Negro (Ng), Japurá (Jp), Içá (Ic), Solimões (Sol, the Amazon mainstem exiting Peru), Jutaí (Jt), Juruá (Jr), Purus (Pr), Madeira (Md), and Tapajós (Tp).

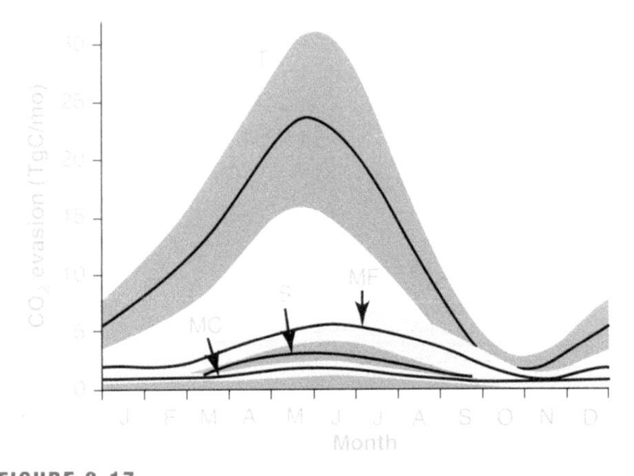

FIGURE 9-17
Spatially integrated sequences of monthly carbon dioxide evasion for the respective hydrographic environments. Lines represent the best estimate of long-term means, whereas shaded regions represent the 67% confidence interval for the range of values likely in a particular year. The upper confidence limit for streams, hidden from view, extends nearly to the upper limit for the mainstem floodplain.

and nature of the organic matter from which the carbon originated. Doing so establishes whether the carbon is recent or represents older sequestered carbon. In essence, it calculates *turnover time*, the time it takes carbon to go from plant tissue to being outgassed by riverine decomposer metabolism.

The dominant source of outgassed carbon was organic matter less than five years old. This carbon was from organic matter that originated on land and near rivers. The study concluded that the cycling of organic carbon (between land and river) is rapid in Amazonia and is essentially confined to recent plant growth such as leaves. Outgassing of Amazonian rivers (Plate 9-11) does not negate the notion that tropical forests serve as carbon sinks, nor does it support it.

PLATE 9-11
Outgassing by some tropical rivers reduces the efficacy of forests as carbon sinks.

Riverine Carbon Flux and a Fish

Many Amazonian fish species feed on *detritus*, dead organic matter that washes into rivers or originates from within the riverine ecosystem itself. One species, the flannelmouth characin (*Prochilodus mariae*, family Characidae), has been shown to be a keystone species of unique importance in its influence on organic carbon transport and primary productivity (Taylor et al. 2006). The study was performed in a river within the Orinoco Basin, an area of high fish diversity. Detritus-eating fish in the region represent between 50% to 80% of the fish biomass. The flannelmouth characin is heavily harvested by local people and is in decline in the region. Its loss could have significant effects on riverine productivity.

Flannelmouth characins (Plate 9-12) migrate into the piedmont rivers during the dry season (January to April) and spawn in floodplains in the wet season (May to December). While in the piedmont rivers, they feed heavily on various forms of detritus. This results in *bioturbation* (disturbance by organisms), consumption, and *egestion* (poop) of large amounts of detrital biomass. Experiments performed on caged fish showed that they grind up the detritus sufficiently to alter the microbial film. In essence, they grind the detritus, increasing overall surface area, and add microbial elements to it as it passes through their digestive system. The impact of the fish is measured as the flux of particulate organic carbon (POC) as it moves downstream.

Flannelmouth characin biomass was directly related to organic carbon flux. Researchers tested the effect of loss of flannelmouth characin by removing it from a region of river divided by a 210-meter plastic divider, thus establishing a control and an experimental comparison. They measured whole-stream primary production and respiration of organic carbon.

The removal of flannelmouth characin resulted in an increase in benthic particulate matter on the stream bottom (Figures 9-18 and 9-19). The biomass of POC on the streambed increased 450% because the fish were not there to consume it. But the downstream flux of suspended POC decreased by 60%. In the treatment area,

PLATE 9-12
FLANNELMOUTH CHARACIN (*PROCHILODUS MARIAE*)

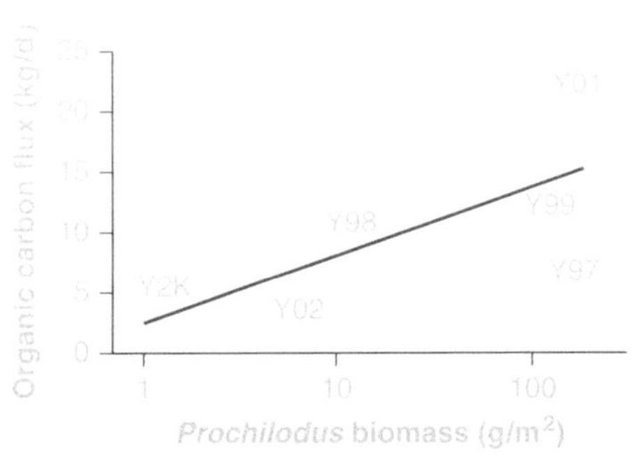

FIGURE 9-18
Interannual variation in organic carbon flux. Whole-stream flux of suspended particulate organic carbon increased as a function of *Prochilodus* biomass (as wet mass). Y97 indicates 1997 data; Y98, 1998 data; and so on, to Y2K, 2000 data.

(a)

(b)

FIGURE 9-19
Photographs of the split-stream removal experiment. (a) The plastic divider and 210-meter section of Rio las Marías. (b) Visual differences in benthic particulate matter after removing *P. mariae* (right) compared with the intact fish assemblage (left).

where fish were removed, primary productivity and respiration each increased, but respiration increased far more. The ratio of productivity to respiration decreased by 150% after fish removal. This was considered a highly significant alteration of the river metabolism in that far more carbon dioxide was now being dispersed into the atmosphere by the river.

Overall, the loss of flannelmouth characin had the following results:

- It altered the metabolism of the river, changing the proportion of primary productivity relative to community respiration.
- It demonstrated a lack of functional redundancy in that no other fish species substitute functionally for flannelmouth characin.
- The POC unused by the flannelmouth characin simply accumulates on the streambed and, during flood season, may be washed quickly downstream, perhaps unavailable to organisms because it is so rapidly pulsed. This represents an energy loss to the ecosystem.
- The flannelmouth characin acts to reduce spatial and temporal variability in organic carbon flow. This makes energy and atoms more constantly available.
- The flannelmouth characin is being overharvested, its average body size decreasing. This may result in significant alterations in ecosystem function in the river.

Drought Sensitivity of Tropical Forests

It may seem surprising to invoke drought as an important variable in ecosystem function in tropical lowland forests. This is, after all, the ecosystem known for its abundance of moisture. But it is also clear that tropical moist forests are seasonal, experiencing dry seasons of varying intensity. ENSO events substantially add to that variability. In addition, historical long-term droughts are known throughout tropical regions. Given that tropical lowland forest plant species are adapted for high-moisture regimes, it should be expected that drought would exert a strong effect on these ecosystems. To what degree do tropical plant species vary in drought sensitivity? If climate change and increasingly frequent ENSO events (see Chapter 15)

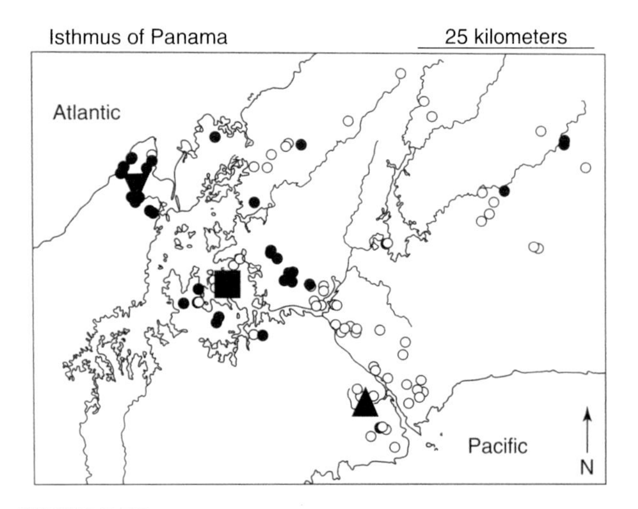

FIGURE 9-20
Map of the study areas and distribution of one tree species, as an example. Location of the study plots along the Panama Canal. The Cocoli plot on the dry Pacific side (upright solid triangle), the Sherman plot on the wet Atlantic side (inverted solid triangle), the 50-hectare forest dynamics plot on BCI (solid square), and the 119 additional inventory sites (circles) are shown.

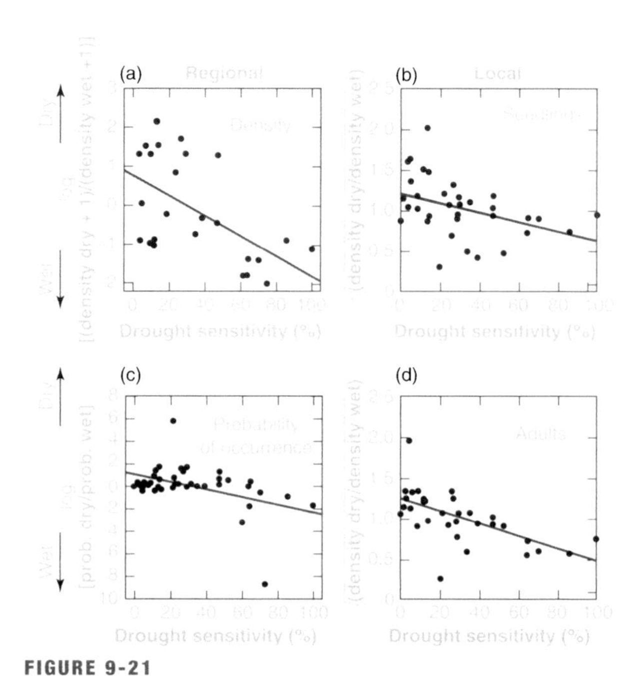

FIGURE 9-21
Significant relationships between drought sensitivity of seedlings and regional and local distributions of species.
(a) and (c) Regional distributions of tree and shrub species were assessed as (a) the density in a plot at the dry Pacific relative to the wet Atlantic side of the Isthmus of Panama and (c) as the probability of occurrence on the dry side relative to the wet side of the isthmus based on data from 122 inventory sites. (Note that the results were similar when omitting the two outlying species in [c].) (b) and (d) Local habitat affinities with topographic units within the 50-hectare forest dynamics plot on BCI were assessed for seedlings (b) and adults (d) as the density of a species in dry plateau sites relative to wet slope and streamside habitats. Drought sensitivity was experimentally assessed as the relative survival difference between dry and irrigated conditions.

result in more frequent and severe droughts, what sorts of changes in community composition might result?

An experimental study conducted on 48 tree and shrub species as they occur over 122 inventory sites in Panama (including Barro Colorado Island) showed strong variation in drought sensitivity among species (Engelbrecht et al. 2007). The inventory sites used in the study spanned a rainfall gradient across the Isthmus of Panama (Figure 9-20). The Atlantic side of the isthmus is wet, while the Pacific side is drier. Researchers tested transplanted seedlings to ascertain the difference in survival under dry and irrigated conditions. The results showed much variability in drought sensitivity among the 48 tree and shrub species examined. Researchers than examined the distribution of drought-sensitive species across the Panamanian isthmus. They found that experimental drought sensitivity was a significant predictor of the probability of occurrence of a species on the dry relative to wet side of the isthmus. Species most sensitive to drought in the experimental conditions were more common on the Atlantic side, the wet end of the climatic gradient (Figure 9-21).

The results of the study are clear. If drought becomes more frequent or more severe, those species on the Pacific side, the dry side of the isthmus, will likely spread toward the Atlantic side. Many species may decline due to drought effects.

The Amazonian Drought of 2005

A severe and widespread drought occurred over Amazonia in 2005. The worst of the drought occurred in dry season, July through September, and was focused in southwest and central Amazonia. The expectation would be that such an event would alter the pattern of carbon flux and primary productivity throughout the region. This happened, but not as predicted. A study conducted using Terra satellite's Moderate Resolution Imaging Spectroradiometer (MODIS) produced surprising results (Saleska et al. 2007). The MODIS program, which calculates an *enhanced vegetation index* (EVI), allows measurement of leaf area and chlorophyll content and is useful in tracking patterns of primary productivity (Figure 9-22).

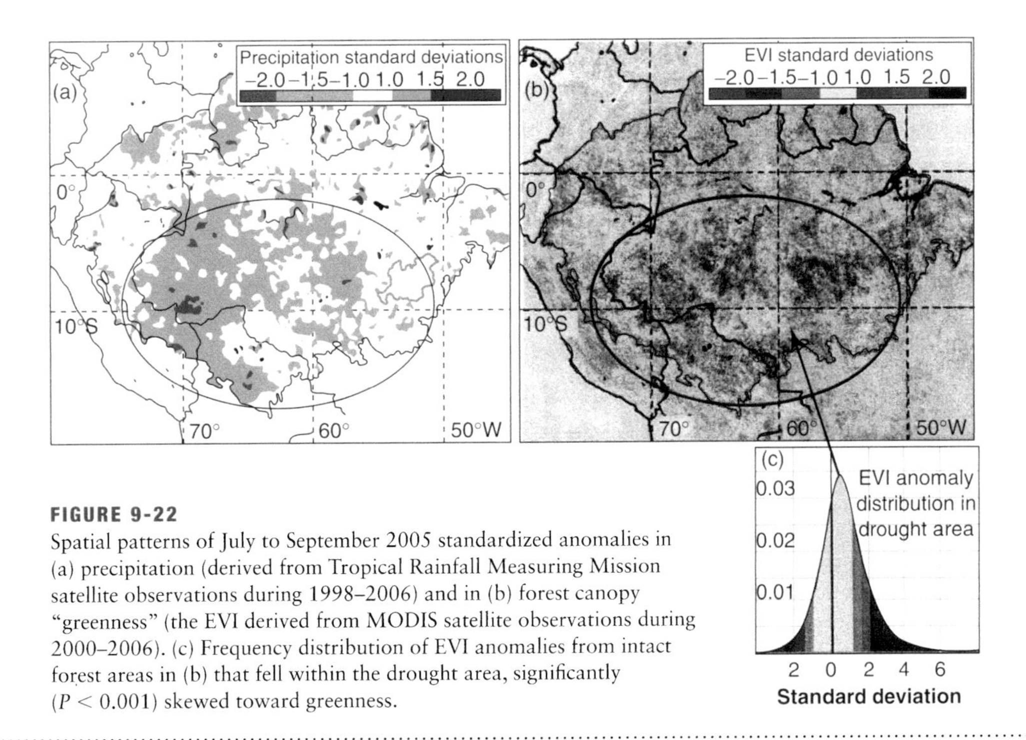

FIGURE 9-22
Spatial patterns of July to September 2005 standardized anomalies in (a) precipitation (derived from Tropical Rainfall Measuring Mission satellite observations during 1998–2006) and in (b) forest canopy "greenness" (the EVI derived from MODIS satellite observations during 2000–2006). (c) Frequency distribution of EVI anomalies from intact forest areas in (b) that fell within the drought area, significantly ($P < 0.001$) skewed toward greenness.

The satellite data showed an increasing greening of the region during the drought. The expectation was that drought would limit water availability and reduce photosynthesis. But that did not happen, and increased greening was a surprise. But maybe it should not have been. Trees may have sufficient root systems to tap into ground water and move it effectively where it is needed. The drought reduced rainfall that, in turn, reduced cloud cover. Increased irradiance may have been the principal cause of the enhanced greening, and thus rain forest function may be relatively resilient to short-term droughts, severe though they may be. But satellite data show only some of what was happening during the drought. The forest may have been greener, but it lost carbon.

The 2005 Amazonian drought was considered one of the most intensive in a century. A team of 67 researchers (Phillips et al. 2009) made it their goal to document the effect of the drought. Though droughts are often caused by ENSO events, this one was not. It resulted from an elevation of sea surface temperatures in the tropical Atlantic north of the equator that ultimately caused higher air temperatures and significantly reduced precipitation in the southern two-thirds of Amazonia.

The study was conducted on permanent plots established as part of a long-term monitoring network, RAINFOR. These plots have been monitored for 25 years. RAINFOR consists of 136 permanent plots located in old-growth forest across 44 landscape types. Data are periodically collected on tree diameter and wood density. Allometric models are used to calculate biomass and rate of change in biomass. For the 2005 drought, researchers focused on 55 plots, and they measured net biomass change, growth, and mortality, comparing these data with those of earlier years.

Prior to the 2005 drought, the 55 plots considered together showed a net increase in above-ground (dry-weight) biomass of 0.89 Mg/hectare/year. The biomass increase was consistent throughout years that included several ENSO events. The biomass increase was indicative of increasing productivity of Amazonian forests, acting as carbon sinks.

During the drought year, there was no net gain in biomass, but rather an overall loss (net rate of change = −0.71 Mg/hectare/year). Analysis indicated that the biomass loss coincided closely with drought conditions and represented the first time since the plots were monitored that there was a net biomass decline. Among 28 plots with the most severe water deficits, the rate of above-ground woody biomass accumulation declined by 2.39 Mg/hectare/year. Mortality rates were also elevated in forests that experienced the most severe drought conditions (Figure 9-23). The sample size of the study confirmed that the biomass loss and mortality increase was widespread, not confined to just a few areas. Overall, the drought resulted in a loss of 5.3 Mg of above-ground biomass of carbon per hectare, and the drought had an estimated total impact of 1.2 to 1.6 petagrams lost.

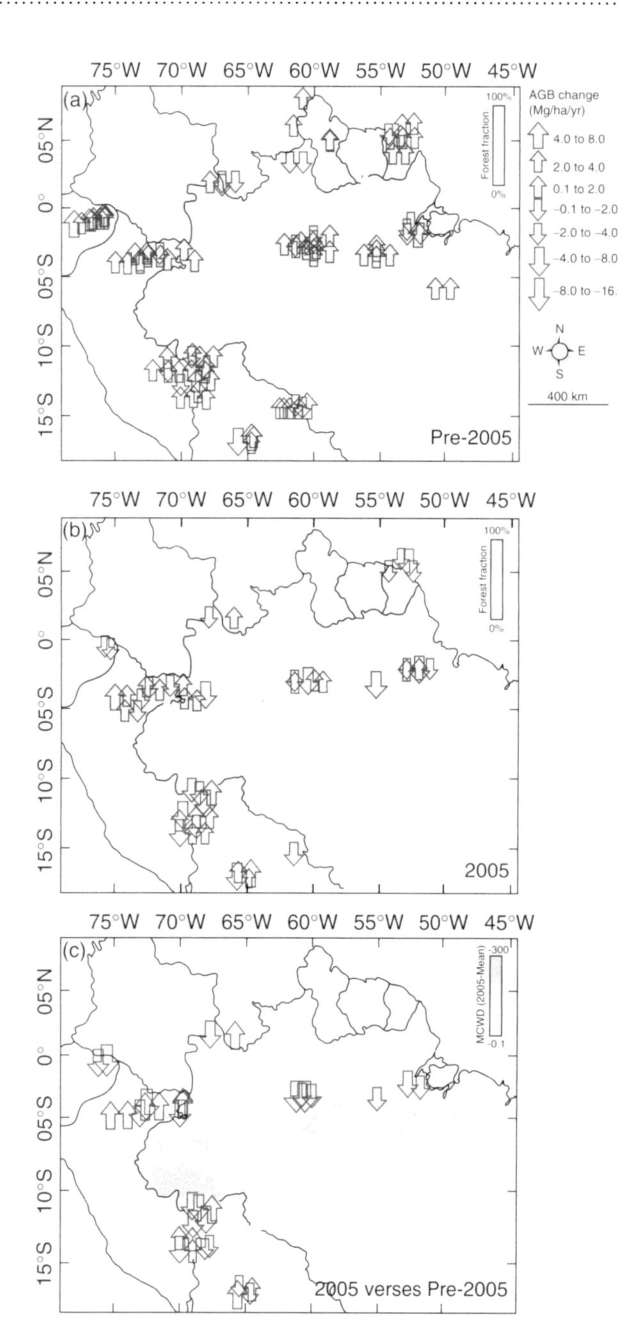

FIGURE 9-23
Above-ground biomass change in the Amazon Basin and contiguous lowland moist forests. The 2005 drought reversed a multidecadal biomass carbon sink across Amazonia. Symbols represent magnitude and direction of measured change and approximate location of each plot. (a) Annual above-ground biomass, 2005 versus pre-2005, for those plots monitored throughout. Grayscale shading in (a) and (b) represents proportion of area covered by forests. Colored shading in (c) indicates the intensity of the 2005 drought relative to the 1998–2004 mean as measured from space using radar-derived rainfall data (Tropical Rainfall Measuring Mission [TRMM]).

Should droughts such as that of 2005 become more frequent or more severe, it is clear that such events would bring about changes in forest composition throughout Amazonia and, presumably, the rest of the tropical world.

Will Too Much Rain Reduce NPP?

Drought is at one end of a spectrum of factors that affect levels of NPP in tropical forests. At the other end is precipitation. Models of climate change suggest increases in mean annual temperature (MAT) and mean annual precipitation (MAP) in the coming decades. The response of tropical forests to drought or added rainfall will determine whether these forests act as carbon sinks or as carbon sources. It is clear that tropical forests are sensitive to drought. How sensitive are they to increased precipitation?

To help answer this question, a study was performed using data from a source called the International Biological Program (IBP) combined with a recent survey of tropical forest NPP (Schuur 2003). The IBP was a global-level attempt to measure physical and biological ecosystem parameters that was conducted several decades ago. It served as a useful baseline for the present study. In addition, a climate gradient from Maui, Hawaii, was used to develop NPP–climate relationships. This model was designed to include both above- and below-ground plant production expressed as MgC/hectare/year.

The results (Figure 9-24) showed that as mean annual temperature increased, so did productivity. But more importantly, as mean annual precipitation increased, productivity peaked and then declined fairly steeply. The maximum NPP was attained at a precipitation level of 2,445 millimeters MAP. Above that, NPP declined. This result was a surprise, as it redefined the assumed relationship between global climate and NPP. The greatest effects of higher moisture level (reducing NPP) would be felt in tropical forests across Southeast Asia, western Amazonia, and to a lesser degree, coastal Africa and coastal South America.

Why should increased precipitation result in declining NPP? One reason might be that solar radiation, irradiance, is reduced. It has been shown in one study that added light (in a controlled study on a single tree species) resulted in higher CO_2 uptake (Graham et al. 2003). Another study showed no correlation between light availability and NPP (Clark and Clark 1994). The TEM model, cited earlier, suggested

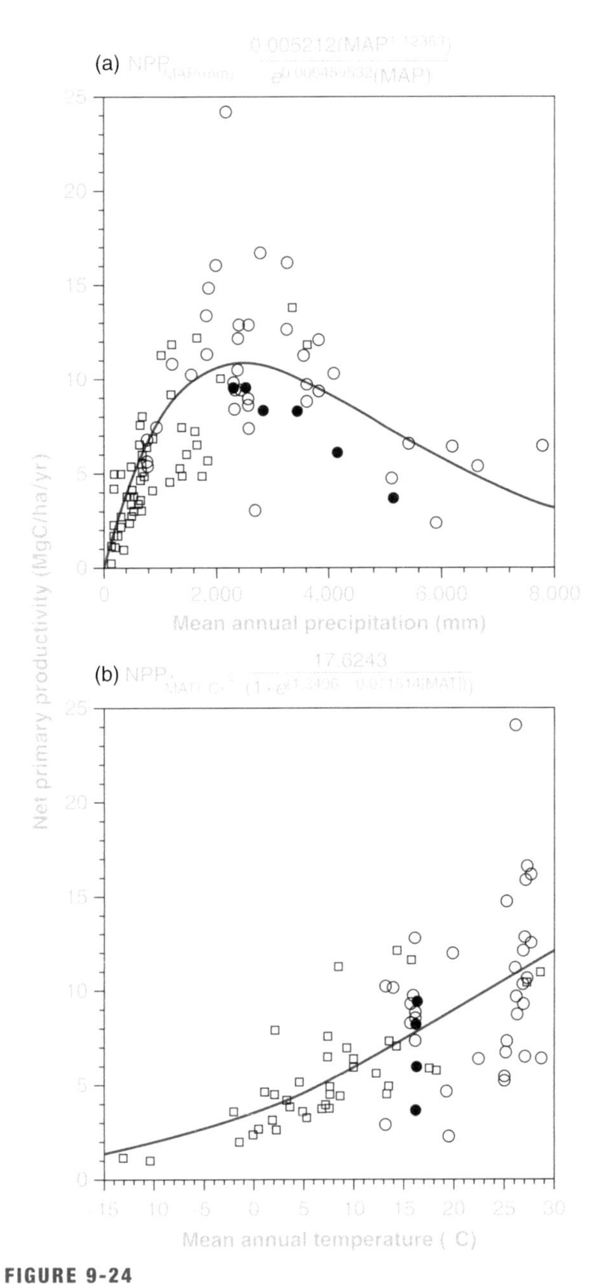

FIGURE 9-24
The relationships between net primary productivity and (a) mean annual precipitation and (b) mean annual temperature. Open squares are International Biological Program sites, open circles represent the tropical forests survey, and dark circles are sites on the Maui moisture gradient that varied in precipitation (MAP = 2,000 to 5,050 millimeters) but not in temperature (MAT = 16).

that added cloud cover would act to reduce NPP. Another effect of added moisture might be to decrease the efficacy of nutrient cycling (see Chapter 10) by leaching nutrients out of the soil and by decreasing decomposition rates. This could result from rainfall saturating soil, interfering with the aerobic demands of root systems and microbial organisms such as fungi and bacteria, the principal decomposers. There are thus two possible avenues, reduced irradiance and interrupted biogeochemical cycling, that might drive NPP rates lower with increased precipitation. Such an occurrence would compromise the function of tropical forests as carbon sinks (Plate 9-13).

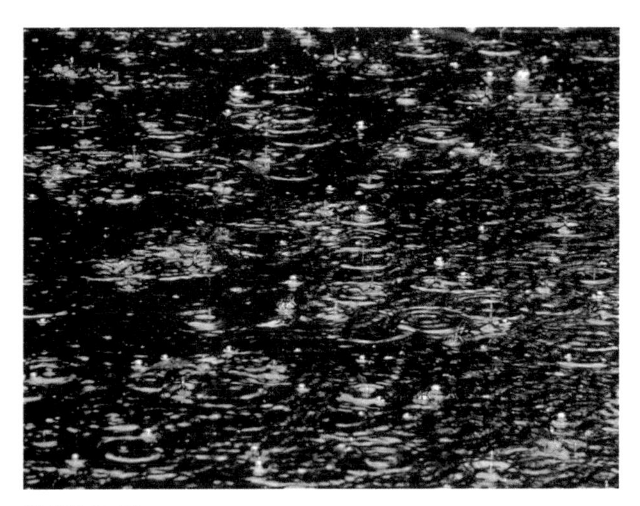

PLATE 9-13
In some tropical forests, precipitation may reduce NPP.

Climate Change and Tropical Forests

Data show that global temperature is rising. Over the past century, global average temperature has increased by approximately 0.6°C, and the expectation is that this trend will continue, if not intensify (Cubash et al. 2001; Houghton et al. 2001). At the same time that the world is gradually warming, geologists and ecologists have identified patterns that appear to be closely associated with climate change. These include the following:

- *Geological changes:* Shrinking glaciers, melting permafrost, earlier snowmelt, lake and river warming, increases in coastal erosion.
- *Biologic changes:* Shifts in leaf-out patterns, blooming dates, avian migration arrival, reproduction times, species distribution, ecological communities, including in the tropics.

Today, few scientists dispute the reality of climate change. The cause of the change is more debated, but the evidence is strong that it results from human activity, referred to as *anthropogenic climate change* (Rosenzweig et al. 2008) (Figure 9-25). All global regions are experiencing

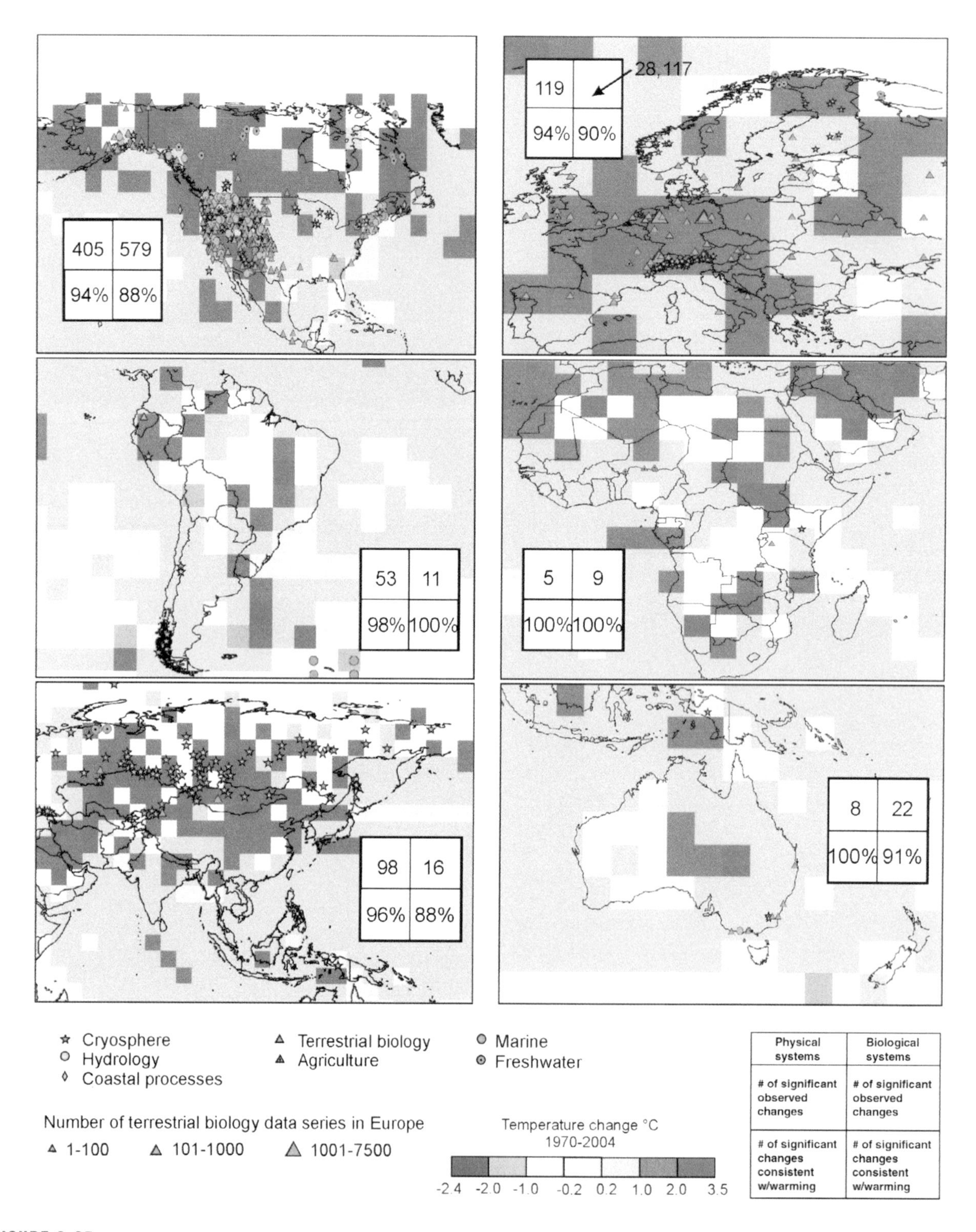

FIGURE 9-25

Location and consistency of observed changes with warming. Locations of significant changes in physical systems (snow, ice, and frozen ground as well as hydrology and coastal processes) and biological systems (terrestrial, marine, and freshwater biological systems), and linear trends of surface air temperature (HadCRUT3) between 1970 and 2004. Note that there are overlapping symbols in some locations. Africa includes parts of the Middle East.

effects of changing climate, and climate change is expected to have continuing multiple effects on biodiversity (Lovejoy and Hannah 2005).

The forcing of climate change is believed to be due to increasing levels of greenhouse gases in the atmosphere. Such gases include carbon dioxide, the principal gas emitted from fossil fuel burning, as well as nitrous oxides and methane. Of those, carbon dioxide is the most important. Levels of carbon dioxide began to increase with the onset of the Industrial Revolution in the mid-nineteenth century and continue to rise. In 2009, the carbon dioxide atmospheric concentration was 387 parts per million (ppm). It was only about 250 ppm at the start of the Industrial Revolution. Carbon dioxide is released into the atmosphere by burning of fossil fuel and by deforestation.

Methane levels have also more than doubled over the past two centuries (Schiermeier 2006), and the reasons for the increase are not clear. But some possibilities are that increase in rice cultivation and increasing numbers of cattle may be responsible (Beerling 2008). During rice cultivation, methane is released into the atmosphere by anaerobic decomposition or organic matter in flooded soils. Cattle ranches add methane for several reasons. First, the animals' complex digestion process releases methane from anaerobic bacterial action. Second, manure decomposition adds methane. And third, conversion of forest to pasture often stimulates termite activity, adding yet more methane.

Climate change has occurred throughout Earth's history, but natural climate change should not be confused with anthropogenic climate change. Recognition that climate changes naturally is not a valid reason for ignoring the effects of anthropogenic climate change. When climate changes, either from natural causes or from anthropogenic forcing, ecosystems are altered.

Research has demonstrated that Amazonian rain forest–savanna boundaries move in response to climate change. Over the past 3,000 years, for example, humid forests in Bolivia have been expanding southward, an expansion attributed to increased seasonal latitudinal migration of the *Intertropical Convergence Zone* (ITCZ), a broad equatorial oceanic zone where trade winds converge (Mayle et al. 2000). The ITCZ migration is ultimately explained by *Milankovitch cycles*, which are large-scale astronomical oscillations caused by slight cyclic changes in Earth's orbit (obliquity cycles, precession cycles, and eccentricity cycles). This is an example of natural climatic change. Ecosystem composition has been shown to change rapidly with climate shifts. Core sediments from parts of South America have shown that there were rapid changes in tropical vegetation following within decades after the last deglaciation from 15,000 to 10,000 years ago (Hughen et al. 2004).

Greenhouse Effect

The glass that covers a greenhouse admits photons of light energy, and some of this energy is converted to heat as it passes through the glass. The heat is a weaker and more dispersed form of energy than visible radiation, and as a result, the glass of the greenhouse allows the build-up of heat that keeps plants warm. Of course, heat does escape from a greenhouse, but at a rate sufficiently low as to allow a net increase in heat within the structure. A similar effect occurs by closing up an automobile tightly on a sunny day. If you leave the car in the sun all day, the light passing through its windows will cause a heat build-up within the interior, and when you finally open the door, it will be stuffy inside.

Some atmospheric gases have an analogous effect to that of the glass in a greenhouse. Gases such as water vapor, methane, nitrous oxide, and carbon dioxide act to block heat energy from passing easily through the atmosphere. Rather than a quick, easy transit from Earth to space, the heat is "trapped," retained (or, more accurately, "detained") within the atmosphere for a relatively long period. The more greenhouse gases there are, the more this effect of trapping heat is manifest. This phenomenon, the retention of heat energy by certain atmospheric gases, has come to be termed the *greenhouse effect*. It is tremendously important in mitigating rapid temperature fluctuations on Earth, and it has contributed in an essential way to making Earth a habitable planet.

Earth is situated at precisely the right distance from the sun for water to exist in liquid form. One profound benefit of the presence of oceans of liquid water is that greenhouse gases, and in particular, carbon dioxide, are absorbed into the oceans and, in the case of CO_2, by a series of physical reactions, eventually converted to insoluble carbonate and taken out of circulation. The importance of this cannot be overemphasized. Without the oceans, any buildup in carbon dioxide, such as from volcanic emissions, for example, would not be correctable, and CO_2 concentration would continually increase, trapping more and more heat. Eventually, this process would "run away," and the amount of heat trapped would be sufficient to raise the temperature of the planet beyond that which life could endure. Such is apparently the case with the planet Venus. Terrestrial ecosystems such as high-biomass forests are also potential carbon sinks, but the degree to which net carbon flux is positive in such ecosystems is uncertain.

Although oceans and high-biomass forests absorb CO_2, it is clear that since the onset of the Industrial Revolution, carbon dioxide concentration has increased steadily. This increase has correlated with the growing use of fossil fuel and, particularly in the latter part of the twentieth century, with increased global deforestation. These two factors have combined to release a significant amount of carbon dioxide, a process that is ongoing and that is altering the atmosphere to the degree that Earth is warming and climate is changing.

How Will Rising Temperature Affect Moist Tropical Forests?

There is evidence that mean annual temperature (MAT) is slowly rising in tropical regions (Malhi and Wright 2004). What effect will such a rise have on the carbon flux of moist forest ecosystems? A study employing the technique of *meta-analysis* (an analysis of multiple data sets) investigated how a rise in MAT will affect various parameters of carbon flux in tropical moist forests (Raich et al. 2006). Researchers examined mature forest sites with no histories of recent disturbance or human-caused modification. They confined the study to forests having fewer than three consecutive months of less than 50 millimeters per month of precipitation.

The results of the meta-analysis (Figure 9-26) suggested that significant changes would result from rising MAT:

- Above-ground litter production and above-ground tree biomass increment both increased significantly with rising MAT. Above-ground forest biomass increased by 7.5 to 13.2 MgC/hectare/°C.
- Soil organic carbon decreased with increasing temperature, indicating greater decomposer activity with rising MAT. The researchers noted that this provides a potential negative feedback, compensating for the greater carbon added in biomass. In other words, as NPP rises, so does decomposition, releasing some of the added carbon back to the atmosphere.
- The study indicated no net impact of changing environmental temperature on total carbon storage.
- The study suggested that a 1°C increase in MAT could result in the eventual transfer of about 8 petagrams of soil organic carbon from moist evergreen forests to the atmosphere through enhanced decomposition. That could be offset by added forest biomass.

Tropical deforestation could prevent offset of carbon loss by biomass increase. Deforestation rates are estimated to have averaged 14.2×10^6 hectares/year from

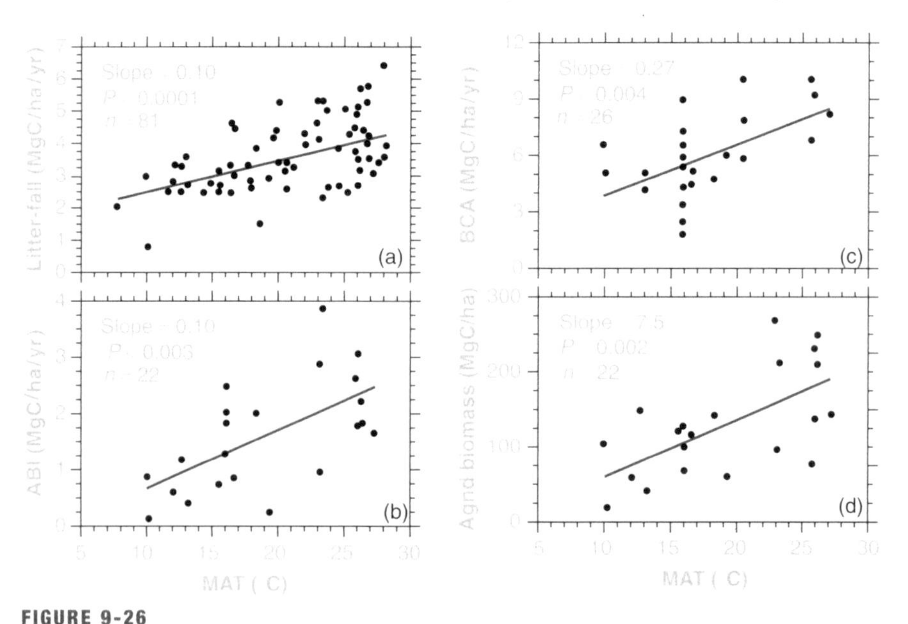

FIGURE 9-26
Ecosystem attributes in relation to mean annual temperature (MAT) in mature moist tropical evergreen forests: (a) fine-litter production increased significantly with MAT; (b) mean annual above-ground biomass increment (ABI) increased significantly with MAT; (c) below-ground carbon allocation (BCA) increased significantly with MAT; and (d) above-ground (Agnd) plant biomass increased significantly with MAT.

1990 to 2000 (Kitayama and Aiba 2002). With an estimated 60% of pantropical lowland tropical forest now disturbed by humans, the existence of sufficient forest area capable of significant carbon sequestration is an open question. Loss of carbon through soil respiration will continue regardless of whether high-biomass forest is present. This study concluded that it seems unlikely that carbon flux will result in net uptake of carbon. The opposite seems more likely.

Changes in Tree Communities in Amazonia

Within recent decades, there have been significant changes in tree species composition and forest dynamics of central Amazonian forests (Laurance et al. 2004). A study of 18 permanent 1-hectare plots (that had not been affected by any form of disturbance) spanning an area of 300 square kilometers in Amazonia demonstrated increases in rates of tree mortality, growth, and recruitment. Twenty-seven of 115 relatively abundant tree genera changed significantly in population density or basal area. Of particular interest is that genera of faster-growing species are increasing in dominance or density, whereas slower-growing species are declining (Table 9-4).

The researchers divided the census data for each plot into two intervals, 1984–1991 and 1992–1999. This allowed them to contrast rates of annual mortality and annual recruitment. Both rates rose significantly from the first to the second interval (Figure 9-27). Trunk growth accelerated for 87% of the genera between the first and second intervals.

The forest dynamics and composition changed between the two intervals. Such changes could result because the region was not in equilibrium due to a past history of disturbance. There was no evidence of any significant disturbance history. Nor did analyses support the possibility that

Table 9.4 Increasing or decreasing tree genera in undisturbed Amazonian rain forests.

GENUS	FAMILY	NET CHANGE (%)
Tree density increases over time		
Corythophora	Lecythidaceae	+9.8
Eschweilera	Lecythidaceae	+4.0
Tree density decreases over time		
Aspidosperma	Apocynaceae	−13.3
Brosimum	Moraceae	−8.1
Couepia	Chrysobalanaceae	−8.9
Croton	Euphorbiaceae	−35.0
Heisteria	Olacaceae	−25.0
Hirtella	Chrysobalanaceae	−13.0
Iryanthera	Myristicaceae	−16.3
Licania	Chrysobalanaceae	−11.0
Naucleopsis	Moraceae	−17.8
Oenocarpus	Arecaceae	−32.3
Quiina	Quiinaceae	−29.0
Tetragastris	Burseraceae	−15.0
Unonopsis	Annonaceae	−15.3
Virola	Myristicaceae	−14.0
Tree basal area increases over time		
Corythophora	Lecythidaceae	+12.0
Couepia	Chrysobalanaceae	+10.8
Couma	Apocynaceae	+14.4
Dipteryx	Leguminosae	+7.2
Ecclinusa	Sapotaceae	+13.8
Eschweilera	Lecythidaceae	+7.0
Licaria	Lauraceae	+17.2
Maquira	Moraceae	+9.9
Parkia	Leguminosae	+22.0
Peltogyne	Leguminosae	+15.9
Sarcaulus	Sapotaceae	+14.4
Sclerobium	Leguminosae	+76.6
Sterculia	Sterculiaceae	+23.4
Trattinnickia	Burseraceae	+13.6
Tree basal area decreases over time		
Oenocarpus	Arecaceae	−29.1

All increases or decreases in tree genera based on population density and basal-area data are significant ($P < 0.01$)

(a)

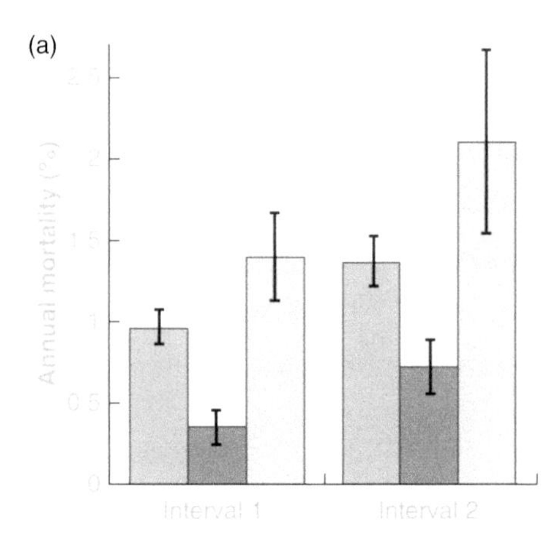

(b)

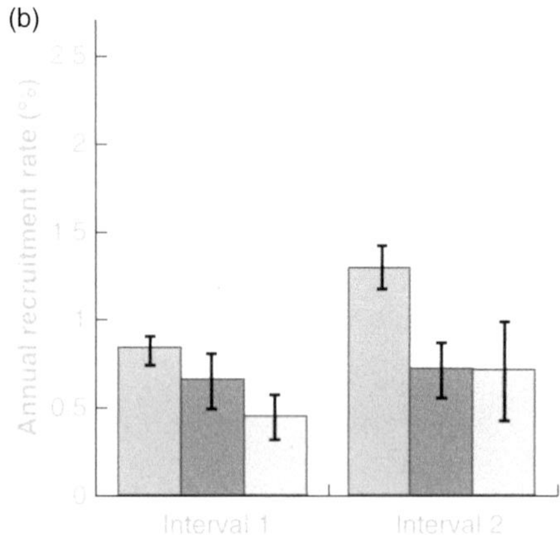

FIGURE 9-27
(a) Mortality and (b) recruitment rates (±SEM) for increasing and decreasing genera. Data are shown for all trees (blue bars), for 13 genera that increased in basal area (brown bars), and for 13 genera that declined in density (green bars). Overall mortality ($t = -2.38$, d.f. = 17, $P = 0.03$) and recruitment ($t = -4.45$, d.f. 17, $P = 0.0003$) accelerated from interval 1 (around 1984–1991) to interval 2 (around 1992–1999). However, there was no significant change over time ($P > 0.11$) in mortality or recruitment for the increasing and decreasing genera (paired *t*-tests).

ENSO events were driving the changes. No shifts in rainfall patterns were observed. The researchers suggested that accelerated forest productivity was most likely caused by increasing levels of carbon dioxide, or enhanced soil fertilization by transported ash, or by reduced cloudiness. Of these possible causes, they believed that rising carbon dioxide was probably the most important.

Bruce W. Nelson (2005) challenged the conclusion that the changes seen in the Amazonian study were likely the result of enhanced fertilization by carbon dioxide. Taking issue with Laurance et al. (2004), Nelson suggested that fire history, disturbance history, recent rainfall, and even the botanical collection of specimens could possibly account for the various patterns observed in the study. In a rebuttal, Laurance et al. (2005) reviewed each of Nelson's arguments (some of which were also addressed in their previous paper) and continued to argue that the most parsimonious explanation of the pattern observed was due to some combination of added carbon dioxide, elevated nutrient deposition from ash (transported atmospherically from deforested areas), and reduced cloudiness, all factors attributed to global change. They called for controlled physiological studies of rain forest trees to further investigate the results of CO_2 enhancement in tropical rain forests. Careful reading of both opinions (Nelson and Laurance et al.) shows that their views are not mutually exclusive. The polarization of views underscores the multiple variables that must be considered in research about such a complex topic as how tropical forests will respond to realities such as climate change. This topic will be discussed further in Chapter 15.

Do Lianas Portend a Changing Forest?

Liana cover has been increasing in Amazonian lowland forests. Recall that *lianas* are woody vines that begin growth as shrubs and soon climb into trees, draping themselves on tree branches, often interconnecting trees (Plate 9-14). Lianas are considered to be structural parasites of trees because they have the potential to suppress tree growth, add to tree mortality, and affect competitive relationships among trees because they tend strongly to infest certain tree species more than others (Snitzer and Bongers 2002). Lianas represent 5% or less of forest biomass but represent up to 40% of leaf productivity. A study of liana density and basal area at 47 interior-forest sites in four Amazonian regions (North Peru, South Peru, Bolivia, Ecuador), plus additional data from 37 other Neotropical sites showed major liana increase from 1980 to 2000 (Phillips et al. 2002). The researchers found that the dominance of large lianas relative to trees has increased by 1.7% to 4.6% annually (Figure 9-28). This represents a rapid, major shift in liana abundance. Researchers noted that tree basal area increased by an average 0.34% among sites but that the relative importance of large lianas approximately doubled over the same period. Dominance, basal area, and size of individual lianas increased throughout the study region.

Why have lianas increased so much? Researchers suggested that added atmospheric carbon dioxide is the likely cause. Lianas are known to react positively

PLATE 9-14
Changes in the abundances of lianas, many of which grow in deep shade, have characterized various tropical forests in recent years.

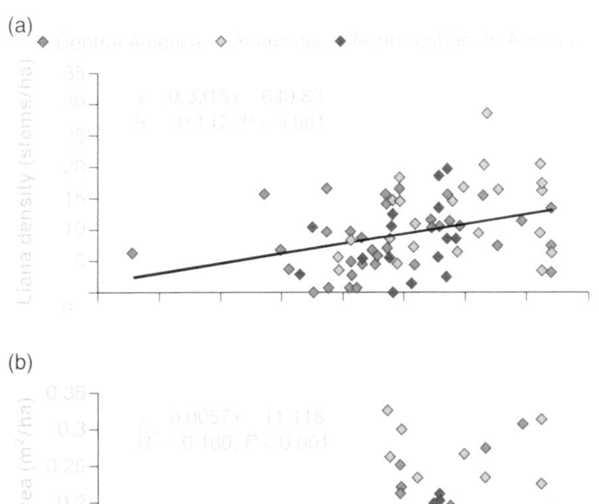

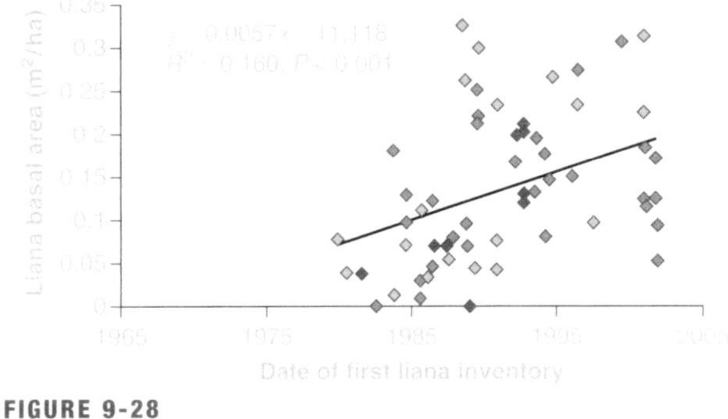

FIGURE 9-28
Structural importance of lianas over 10 centimeters in diameter in each Neotropical site as a function of date of first inventory. (a) Liana stem density in stems per hectare. (b) Liana basal area in square meters per hectare. Central America is Panama and tropical countries to the north; northwest South America is the Chocó bioregion, west of the Andes; Amazonia is the Amazon River Basin and contiguous forested zones of Guiana and eastern Brazil. Linear regressions are fitted to the Amazonian data.

to CO_2 fertilization. Lianas in deep shade are stimulated to grow if CO_2 increases (Phillips et al. 2002). If lianas continue to increase, there will be added tree mortality and possibly a shift in tree species composition because lianas are more frequent on certain tree species. They suggest that the terrestrial carbon sink may "shut down" sooner than expected if lianas continue to increase disproportionately.

A study at Barro Colorado Island, Panama, also looked at liana increase, but this study was somewhat equivocal (Wright et al. 2004). Liana leaf litter and proportion of forest-wide leaf litter composed of lianas increased over a 17-year period from 1986 to 2002. But liana seed production and seedling densities varied, apparently affected by ENSO years. The shift in liana dominance at BCI was therefore less pronounced than in Amazonia, though the studies used different measures and are not strictly comparable.

Lowland Biotic Attrition and Elevational Shifts in the Tropics

In the temperate zone, there have been numerous documented cases of recent range shifts among plants and animals, with the strong implication that climate change is driving these changes (Lovejoy and Hannah 2005). But such documentation is considerably more limited for tropical ecosystems. Consider that within the latitudinal span that encompasses the lowland tropics, there is generally less latitudinal variation than at higher latitudes. As the world warms, so will the tropics. How sensitive are extant species to increasing global temperature? And how will patterns of precipitation, cloud cover, and so on associated with climate change affect species' ranges?

What effects might climate change, and specifically global warming, have on ranges of tropical species? A study by Colwell et al. (2008) performed on an elevational transect in Costa Rica suggests that climate change will result in many range shifts and even extinctions among lowland species. Consider that as global warming occurs in the temperate zone, projections are that more southern-ranging species will gradually move northward. This pattern is already evident in numerous bird species. But what will happen in the tropics? As temperature warms in Amazonia and Borneo and the Congo, where will lowland tropical species go? Will they adapt to higher annual temperatures, or will they migrate elsewhere, and if so, to where? Another aspect of climate change is that montane species will migrate elevationally as climate warms, occupying higher elevations on mountains. Those species that require the highest elevations may be subject to *mountaintop extinction*, being replaced by species from lower elevations.

If lowland tropical species begin to move elevationally in response to warming temperature, there is really no species pool to replace them and, as a result, *lowland biotic attrition* will occur, an overall loss in species richness. Current data suggest that tropical lowlands have been warming by 0.25°C per decade since 1975 (Colwell et al. 2008) and that tropical climates are warmer now than they have been in the past two million years. It has been argued that Amazonian forests, as well as lowland forests elsewhere, are approaching a threshold beyond which many plant species will be unable to survive or to adapt to quickly enough to tolerate increased annual temperature. These species will either become extinct or will move elevationally. Looking at global geography, it is difficult to imagine them moving latitudinally. Amazonian trees will not migrate to Louisiana, for example, because they have no way of getting there. But some species might move upslope in the Andes.

Colwell et al. (2008) looked at 1,902 species of epiphytes, plants in the family Rubiaceae, geometrid moths, and ants, assessing species as to their potential for upslope range shifts and lowland biotic attrition. They studied a 2,900-meter transect from La Selva Biological Station to Vulcan Barvá. They also examined *range-shift gaps*, which are gaps between current and projected ranges. Many tropical species occupy narrow elevational ranges, and thus movement in response to temperature change could be highly problematic.

The researchers used a model assuming a 3.2°C increase in temperature over the next century. For a 600-meter upslope shift, based on such an increase, 53% of the 1,902 species were potentially subject to lowland attrition, a dramatic potential loss in species richness in the lowlands. In addition, 51% were concluded to face range-shift gaps that would make elevational shifts more difficult (Figure 9-29).

Studies such as that of Colwell et al. (2008) are necessary to assess the potential impact of global climate

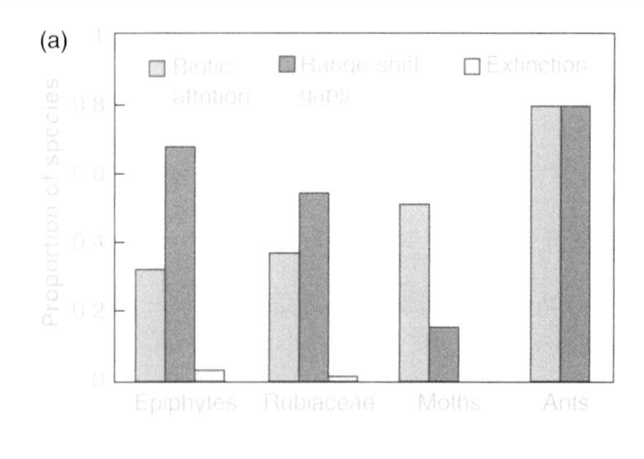

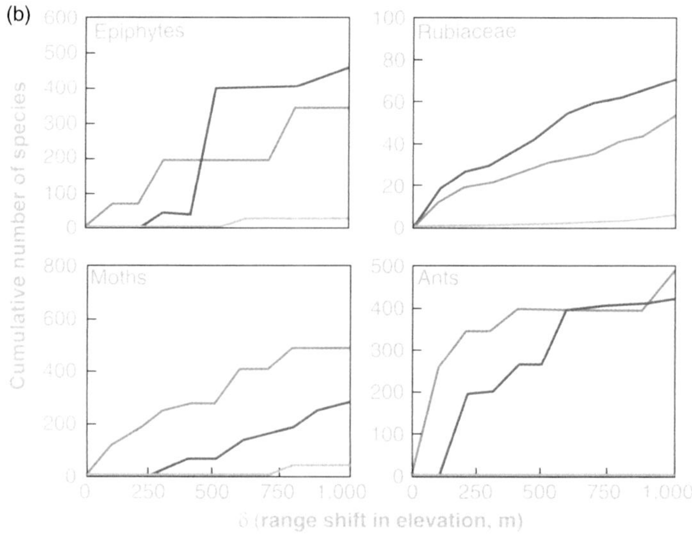

FIGURE 9-29

(a) Proportion of species in each of the four groups subject to decline or disappearance in the lowlands (biotic attrition), faced by gaps between current and projected elevational range (range-shift gaps), and exposed to mountaintop extinction, given a 600-meter upslope shift in all ranges. Proportions sum to greater than 1 because some species belong in two categories. (b) Cumulative number of species facing each of these three challenges as a function of warming-driven range shifts. The x axis represents model parameter δ, measured in meters of elevation range shift, on a continuous scale of warming-driven isotherm shifts of up to 5°C (nearly 1,000 meters), the upper range of projections for Central America for this century. The stairstep patterns are a consequence of sampling at discrete sites on the gradient.

Pheasants and Global Warming

A study of two pheasant species in Khao Yai National Park in Thailand shows that one species, the Siamese fireback (*Lophura diardi*), is increasing, while a similar species, the silver pheasant (*L. nycthemera*) (Plate 9-15) is not (Round and Gale 2008). The fireback is a lowland species whose typical elevational range reaches to about 700 to 800 meters. The silver pheasant occupies somewhat higher elevations, 700 meters and above. Observational records of encounters with each of these unique and colorful bird species have been kept at the national park since 1978.

The proportions of sightings of each of the two species have changed considerably. Until 1993, the fireback made up only 18.6% of the sightings between the two species. Since then, the number of fireback sightings has increased to 60.1% of the total pheasant sightings. Silver pheasants are being encountered at the same rate they have been throughout the period of the study, but firebacks are becoming more numerous. The change in fireback sightings is in part related to the species moving upward in elevation, more into the range traditionally associated with the silver pheasant. In addition, researchers suggest that enhanced NPP in the lowland habitat of fireback pheasants may add resources for the birds in the form of increased forest floor litter depth. (The birds feed like chickens, scratching in leaf litter.) The combination of enhanced NPP and expanding elevational range is consistent with warming temperatures. The study concludes that the change in fireback abundance was best explained by changes in an environmental variable, though other explanations are possible. Perhaps the pheasants may be early indicators of changes that will ensue as climate change continues.

PLATE 9-15
SILVER PHEASANT (*LOPHURE NYCTHEMERA*)

change in the tropics. What is important to keep in mind is that should it be true that lowland species will begin to move elevationally in response to warming temperature, corridors connecting habitats are essential. But throughout much of the tropics, deforestation and other human activities result in moderate to severe fragmentation of habitats, making the potential for range-shift gaps much greater (see Chapter 14).

Effects of Tree Species Loss on Carbon Storage: A Simulation

Climate change historically results in some species loss. The severity of such extinction depends on many factors, but the most important are the degree to which climate is changing and the rate of change. A study using data from a 50-hectare Forest Dynamics Plot (containing more than 300 tree species, of which 227 were the focus of this study) on Barro Colorado Island attempted to simulate various possible scenarios of tree species loss and determine how such losses would affect carbon storage (Bunker et al. 2005). The simulations used many combinations of species under three classes of extinction scenarios:

- Extinction associated with trees that have low population growth rates, low densities, and endemism.
- Extinction related to management or harvest strategies such as selective harvest or harvesting large trees.
- Extinction based on environmental change (changes in precipitation, disturbance rate, or elevated carbon dioxide).

The various extinction scenarios produced a diversity of outcomes. For example, extinction of species with the lowest wood densities led to an increase in carbon storage of 75%. Loss of species that attain large stature caused a significant decline in carbon stocks. Selective logging for species with high wood density, large diameter, high basal area, or maximal wood volume led to overall declines in carbon storage of 70%, 29%, 17%, and 21%. A scenario that simulated conversion of the forest to a plantation of tree species with high wood density increased above-ground carbon storage by up to 75%. Another scenario based on increased turnover and more liana abundance resulted in a shift to more rapidly growing tree species and resulted in a decrease of 34% carbon storage. The model also predicted a shift toward more drought-tolerant tree species accompanied by a modest (10%) increase in carbon stocks. Though a simulation, the study is insightful in demonstrating the multiple outcomes possible when species composition in tropical forests is altered.

Carbon Sink or Source?

It remains far from clear whether global tropical forests will ultimately prove to be carbon sinks or carbon sources (Clark 2004). The studies described here show that tropical forests have the potential to act as carbon sinks, but rising global temperature and more severe and frequent droughts, to say nothing of increasing anthropogenic effects, may collectively outweigh enhanced NPP, making tropical forests carbon sources, not sinks. Superimposed on this question is continuing deforestation resulting in liberation of large carbon stocks. This topic will be further explored in Chapter 15.

10 Nutrient Cycling and Tropical Soils

Chapter Overview

The process of *biogeochemical cycling* is essential to all ecosystems. In terrestrial ecosystems, the cycling of chemical nutrients is a complex process strongly affected by climatic properties as they influence soil characteristics that support a diverse community of decomposer species. In the tropics, where it is warm and wet, soils are often nutrient-poor. However, the rate of decomposition and subsequent recycling is high in the tropics, as is the efficiency of nutrient uptake and retention. Nutrient-rich soils support lush humid forest, but so do old, nutrient-poor tropical soils. Many organisms ranging from vultures to microbes are essential in the decomposition and recycling process. Those of particular importance include a diverse bacterial and fungal community—most notably symbiotic mycorrhizal fungi (mycorrhizae)—as well as animals such as termites. Bottom-up factors such as soil type interact with top-down factors such as herbivore effects in providing complex and highly heterogeneous nutrient flux that is characteristic of the tropics (Plate 10-1).

PLATE 10-1
A fallen leaf in a tropical humid forest undergoes a complex process of decomposition involving multiple organisms as the minerals contained in the leaf are eventually released back into the abiotic pool, where they are again taken up by plants.

Nutrient Cycling and the Soil Community

Primary productivity brings solar energy into ecosystems, some of which is converted to potential energy incorporated into organic compounds (see Chapter 9). As energy passes through food webs, it does so in material form, as high-energy, structurally complex compounds in organisms, organic waste, and eventually detritus. Thus energy and materials are coupled as they move through food webs. But recall that energy is ultimately lost as heat, never recycled. In stark contrast, materials, the atoms and molecules of life, are recycled. The material basis for nature comprises the key elements (in various proportions) essential to life, and those elements are "shared" by life forms in the process of biogeochemical cycling. Because Earth receives no significant input of matter from space (a year's worth of meteorites adds up to very little), atoms present in waste products and dead tissue must be reacquired, recycled back to living tissue. Decomposition and subsequent recycling is the process by which materials move between the living and nonliving components of an ecosystem. *Recycling* is a byproduct of decomposition, and *decomposition* is the means by which decomposer organisms acquire energy and nutrients.

In a rain forest, a unit of energy fixed during *net primary productivity* (NPP) (the amount of carbon fixed in excess of metabolic needs of the plant) will ultimately move in one of two paths. It may be consumed as part of living tissue, as when a caterpillar chews a leaf, in which case it will begin moving through the food web, passing through several heterotrophic organisms, the *grazing food chain* (see Chapter 8). Or it may remain within the leaf structure until the leaf eventually drops from the tree, at which time the energy becomes available to the soil community. This latter direction moves energy directly into what is termed the *decomposer food web*. The decomposer food web is a rich and diverse array of heterotrophic organisms ranging from scavenger animals such as vultures to minute bacteria and fungi that rely on dead material and waste products as their energy source (Dance 2008). A glance at a lush, green rain forest plus a dash of pure logic is enough to confirm that the vast majority of the energy fixed during photosynthesis is destined to directly enter the decomposer food web. If it were otherwise, trees, shrubs, and other green plants would show far more leaf damage than they usually do. Of course even when trees lack obvious leaf damage, they may still be affected by pathogens and predators that are taking some of the net primary productivity. Nonetheless, most net primary productivity remains as potential energy in the structural tissue of leaf, bark, stem, and root. This potential energy will eventually be released by a host of soil community organisms as they unpretentiously make their livings below your muddy boots in the forest litter and soil.

Numerous species of microbes—in particular, fungi and bacteria—are the principal organisms in this ongoing and essential process of decomposition, one of nature's most fundamental processes (Falkowski et al. 2008). Microbes, using a series of reduction–oxidation reactions, ultimately convert complex, high-energy organic tissue back to simple,

S#!t Happens—I'm Not Making This Up

Ecologists investigate all manner of topics. In southeastern Sri Lanka, ecologist Ahimsa Campos-Arceiz carefully examined 290 dung piles of Asian elephants (*Elephas maximus*), looking for seeds. Surprisingly, 5 of those 290 dung piles were the habitation of a frog. In fact, there were three frog species found together residing in dung. The environment in the region of this study is very dry, and the frogs were thought to be using the dung as a daytime refuge, much as they would utilize leaf litter in a forest. Campos-Arceiz also inspected 180 dung piles of free-ranging cows and buffaloes and found not a single frog in any of that dung. The reason may have been because of the structure of the dung. Bovid dung is soft and fine-grained, reflecting the efficient ruminant digestive system of the animals. Elephant dung is much more coarse, with undigested plant material and a complex physical structure. In addition to the frogs, elephant dung harbored many invertebrates including beetles, crickets, ants, spiders, centipedes, mites, and scorpions. Indeed, as Campos-Arceiz suggests, "a dung pile can become a small ecosystem on its own." The title of Campos-Arceiz's paper was "Shit Happens (To Be Useful)! Use of Elephant Dung as Habitat by Amphibians." I told you I wasn't making this up. In nature, little is wasted.

low-energy inorganic compounds, making them available for uptake by the root systems of plants. Many other organisms also significantly contribute to decomposition: slime molds, actinomycetes, protozoans, and hordes of animals ranging from vultures to arthropods, earthworms, and other invertebrates. Termites are uniquely important decomposers in tropical ecosystems (see the section "Termites" later in this chapter). All influence the complex process of converting a dead fig leaf, a dead sloth, or an elephant's feces back to basic chemical elements.

The soil community is a diverse, complex food web. The tropical soil community may rival the biodiversity found in the leafy canopy. There are relatively few detailed studies that make estimates of such parameters as bacterial diversity or fungal biomass or pathways of energy movement among the constituent flora and fauna of the decomposer community. Those that exist demonstrate the complexity of the microbial components of the decomposer community (Lodge 1996).

Fungi are immensely abundant in the tropics (and are also far from scarce in other regions). An individual fungal strand is called a *hypha* (plural *hyphae*), and a network of hyphae is called a *mycelium*. In some tropical forests, the net-like mycelium is sufficiently dense as to be visible on the forest floor. Fungi, like bacteria, are essential decomposers. They ultimately liberate atoms back to the soil. In addition, many fungal species, collectively termed *mycorrhizae* (see the section "Mycorrhizae" later in this chapter) are essential to trees and other plants in aiding the uptake of atoms from the soil (Whitfied 2007).

Organisms facilitate a process called *humification*, in which complex organic matter called humus is maintained at the interface between the tree roots and soil. Humus forms *colloids* (substances composed of particulate matter larger than molecules but not visible to the naked eye) that cement soil particles, helping aerate the soil. Humus particles are negatively charged, and by electrostatic attraction, they act to retain mineral nutrients such as potassium and calcium, which carry a positive charge, in the soil (Lavelle et al. 1993).

Soil represents a temporary repository for essential mineral nutrients such as phosphorus, nitrogen, calcium, sodium, magnesium, and potassium. Each of these minerals, as well as others, is necessary for biochemical reactions in organisms. A shortage of any one of them can significantly limit productivity. For example, phosphorus and nitrogen are both important in

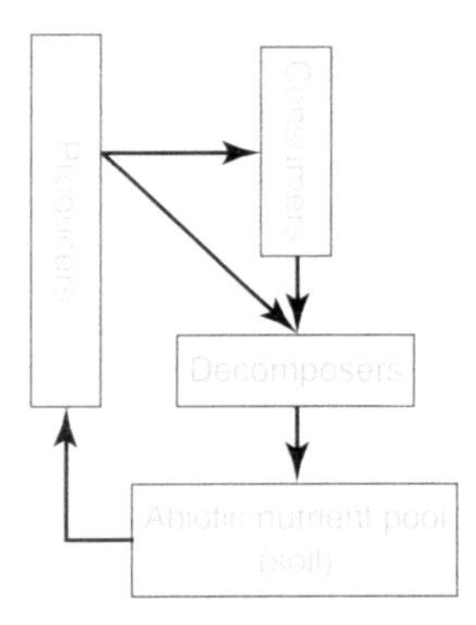

FIGURE 10-1
Simplified compartment model showing the recycling system of an ecosystem such as a tropical forest. Compartments are not to scale. Note that decomposers ultimately make mineral nutrients such as phosphorus available again to primary producers, which take them up during the process of photosynthesis. See text for details.

the structure of nucleic acids (DNA and RNA) as well as proteins and other necessary molecules. Magnesium is an essential part of the chlorophyll molecule, without which photosynthesis could not occur. Sodium is essential for the functioning of nervous and muscular systems in animals

Consider how an atom is cycled (Figure 10-1). A leaf drops to the ground. Inside the leaf are billions of atoms, but we will select a single atom of phosphorus. This phosphorus atom may initially pass through a termite or other invertebrate that consumes the dead leaf tissue, only to be returned to the litter through elimination of waste or the death and subsequent decomposition of the creature itself. Indeed, the atom may move through numerous organisms before becoming part of the humus. Or the atom may be taken up directly by a fungus. This same atom eventually may pass through several dozen fungal and bacterial species, each of which gains a modicum of energy by ingesting, digesting, and thus decomposing the deceased leaf (or termite). Within days, the phosphorus atom becomes part of the inorganic components of the soil. At that point, tree roots, aided by mutualistic fungi called mycorrhizae (discussed in detail later in this chapter) that penetrate tree roots or grow atop the root take up the element and pass it along to the tree. The cycle is complete.

Nutrient cycling is also termed *biogeochemical cycling* to describe the fundamental process of chemicals moving continuously between the *bios* (living) and the *geos* (nonliving) parts of an ecosystem. The movement of minerals in an ecosystem is strongly influenced by temperature and rainfall, the major features of climate. In the tropics, both high temperature and abundant rainfall exert profound effects on the patterns of biogeochemical cycling (Golley et al. 1969; Golley 1975 and 1983).

Heat stimulates evaporation. As plants warm, they evaporate water, cooling the plants and returning a great deal of water to the atmosphere in this heat-driven pumping process called *transpiration*. Transpiration is an essential process in plant physiology. It brings water and minerals up from the soil, helps cool the plant, and supplies essential water needed for metabolism. Water from rainfall is taken up by plants and transpired, returned to the atmosphere, under the stress of tropical heat. Nowhere is this continuous process of transpiration more obvious than along the wider stretches of the Amazon River. At midday, skies immediately above the wide river tend to be clear and blue. But should you look over distant forest beyond either of the riverbanks, you will likely see big, puffy, white clouds, formed by the condensing moisture transpired by the forest—you are watching the forest "breathe" (Plate 10-2). Approximately 50% of the precipitation falling on the Amazon Basin is directly recycled via transpiration from the vegetation (Salati and Vose 1984).

Roots acquire needed minerals by uptake of water from soil. But transpiration can be a mixed blessing. Plants may lose too much water when subjected to constant high temperature. Many tropical plants retard evaporative water loss both by closing their *stomata* (pore-like openings on the leaves for gas exchange) and by producing waxy leaves that inhibit the evaporation of water.

PLATE 10-2
The Amazon River at midday at a wide point. Notice how the clouds have formed over the forest but not over the river. The clouds reflect the transpiration process vital to the physiology of the forest trees.

Factors Affecting Nutrient Cycling

Nutrients cycle within ecosystems as well as enter and leave ecosystems. There are four primary components of nutrient cycling (Feeley and Terborgh 2005):

- The size of the nutrient pools as distributed among biomass, litter, and soil. In tropical forests, for example, many studies indicate that the total pool of phosphorus is small, and thus phosphorus is a primary potential limiting factor of plant productivity. Nitrogen, on the other hand, tends to be relatively abundant in tropical forests, as much nitrogen is taken from the atmosphere, or *fixed*, by the action of microbial organisms.
- The flux of nutrients into, out of, and through the ecosystem as reflected in rates of litter-fall, decomposition, mineralization, and plant uptake.
- The flux of nutrients into and out of the ecosystem by *volatilization* (loss of material through vaporization), *denitrification* (loss of nitrogen through the action of denitrifying bacteria), *leaching* (see the next section), and loss or input of nutrients due to herbivore movements.
- Environmental variables such as temperature and precipitation amounts.

LEACHING

High rainfall washes essential minerals and other chemicals from leaves, a process called *leaching*. Leaching is especially severe in areas subject to frequent heavy downpours. The protective waxy coating typical of tropical leaves contains lipid-soluble (but water insoluble) secondary compounds such as terpenoids that act to retard water loss and discourage both herbivores and fungi (Hubbell et al. 1983 and 1984). *Drip-tips* (the sharply pointed tips of many tropical leaves) probably reduce leaching by speeding water runoff. Such adaptations enable a typical tropical leaf to retain both its essential nutrients and adequate moisture (Plate 10-3).

Rainfall also leaches minerals from soil, washing them downward into the deeper soil layers. The degree of leaching in any area is in part related to the composition of soil with regard to particle size. Soil particles vary in size. The largest is rocky gravel, but gravel is not a prominent part of tropical soils except

PLATE 10-3
This array of tropical leaves demonstrates the waxy quality that they typically exhibit, as well as drip-tips.

in regions with recently formed volcanic soils. More typically, soil is a mixture of sand, silt, and clay. Sand particle size ranges from 2.0 to 0.5 millimeters. Silt particles are dust-like and range from 0.5 to 0.002 millimeter. Clay particles are so small as to be fundamentally microscopic. Coarse clay ranges from 0.002 to 0.2 micrometer, and fine clay is less than 0.2 micrometer. Tropical soils are typically rich in clay, a characteristic that strongly affects leaching and other characteristics of soils in the tropics.

Clay particles, like humus, have negative electrostatic charges that attract minerals with positive charges (called *cations*) such as sodium, calcium, and potassium. Clay also attracts alkaloids because they too carry positive charges. Because water adds positively charged hydrogen atoms to the soil, these abundant H^+ ions exchange with those of elements such as calcium or potassium, which, when free, then leach to a deeper part of the soil or may leach out of the soil entirely by washing into streams and rivers. This process reduces soil fertility and lowers soil pH, making the soil more acidic. Rainfall therefore influences soil acidity because the accumulation of hydrogen ions, either on humus or clay, lowers the pH (recall that the definition of pH is a measure of hydrogen ion concentration), thus increasing the acidity of the soil. In the tropics, the combination of high temperatures and heavy rainfall may result in much leaching and strongly acidic soils. The pH of tropical soils typically ranges between 4.0 and 6.0 but can be less than 4.0 to as much as 6.7 (Motavalli et al. 1995).

Because of high temperature and high precipitation amounts, tropical soils throughout the world are subject to accelerated leaching. For example, Amazon soils are typically mineral-poor, high in clay, acidic, and low in available phosphorus (Jordan 1982 and 1985b) and the nutrient-poor nature of the soil is a limiting factor to plant productivity (Uhl et al. 1990). One estimate suggests that nearly 75% of the soils in the Amazon Basin are acidic and generally infertile (Nicholaides et al. 1985). Age also affects soil mineral content. Older soils are more leached of minerals than younger soils.

One of the major differences between tropical and temperate forests is that in tropical forests, most of the rapidly cycling minerals are in the living plants, the *biomass*. Most of the calcium, magnesium, and potassium in a rain forest is located not in the soil, but in the living plant tissue (Richards 1952; Jordan 1982; Salati and Vose 1984). For example, in a study performed near San Carlos de Rio Negro in Venezuela, the distribution of calcium was as follows: 3.3% in leaves; 62.2% in wood; 14.0% in roots; 3.1% in litter and humus; and only 17.4% in soil (Herrera 1985). Another study concluded that 66% to 80% of potassium, sodium, calcium, and magnesium are in aerial parts of plants, not in soil (Salati and Vose 1984). However, this same study concluded that most nitrogen and phosphorus (somewhere around 70%) are in soil, roots, and litter. In the temperate zone, minerals are more equally distributed between the vegetation and soil bank.

TERMITES

Termites (insect order Isoptera) are pantropical social insects, occurring in great abundance. Termites are essential to biogeochemical cycling in the tropics. Like

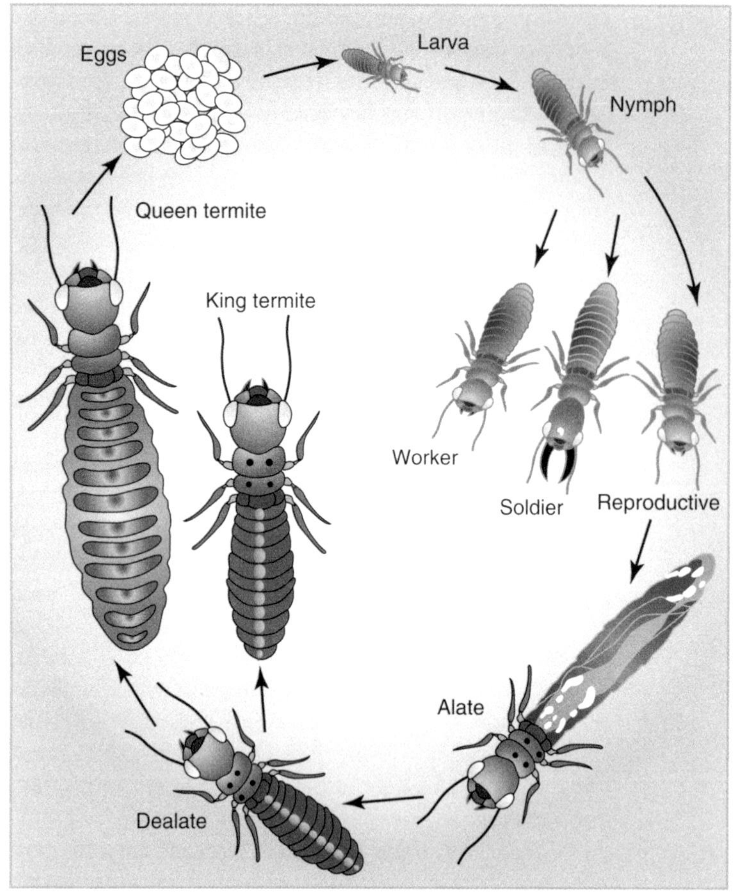

FIGURE 10-2
Termites are abundant throughout the tropics and have an immense influence on nutrient cycling. They also have a complex life cycle in which all reproduction is done by a queen termite.

ants, termites are complex social insects that have distinct castes (workers, soldiers, king, and queen). They resemble ants but are easily differentiated by the fact that termites lack sharp constriction between the thorax and abdomen. Another important difference is that termites do not undergo a complete metamorphosis (larvae, pupae, adults), as do hymenopterans (bees, ants, wasps) (Figure 10-2).

There are about 2,650 species of termites, and the majority of them occur only within tropical and subtropical latitudes (Abe and Higashi 2001). Termites represent the ancient insect order Isoptera, their closest relatives being cockroaches and mantises (also abundant in the tropics). Unlike other social insects such as bees and ants, termites are not *haplo-diploid* (females diploid [each individual containing a set of duplicate chromosomes], males haploid [each with a set of single chromosomes]). All individuals, like most other animals, are diploid (Wilson 1971). Thus termites evolved their complex societal structure independently of other social insect groups. Termites occur in all habitats, from rain forest to grassland, savanna, and mangrove forest.

No one can visit the tropics without seeing termite nests and/or mounds. Basketball-sized termite nests are typically attached to tree trunks and branches. From the rounded, blackish brown nests radiate termite-constructed tunnels through which the workers pass to and from the colony. In addition, especially in grasslands, dry forests, and savannas, cone-shaped termite mounds, some rising to heights of 3 meters (about 10 feet) or more, erupt directly from the ground. In Africa, some termites construct large, rounded mounds. Many termite species also nest underground in vast subterranean colonies that are not obvious from above ground (Plates 10-4 to 10-6).

Termites, though generally well protected within their colonies, are preyed upon by anteaters in the Neotropics, pangolins in Africa and Asia, and the odd *monotreme* (egg-laying) echidnas of Australia and New Guinea (Plate 10-7). Each of these mammals has developed unique adaptations, enabling them to break open termite mounds and, with long tongues, extract the termites. Chimpanzees (*Pan troglodytes*) are known to fashion sticks to poke into termite mounds and extract the termites for food. Emerging termites (that are flying to initiate new colonies) are fed upon by numerous bird species.

Among the most abundant of the termites are the many species of *Nasutitermes* (family Termitidae, subfamily Nasutitermitinae) that range throughout the tropical world and are particularly diverse and abundant in the Neotropics (Lubin 1983). The life history of *Nasutitermes* is typical of many termite species. There are four castes: worker, soldier, king, and queen. If you want to see workers and soldiers, locate a nest (usually on a tree), and cut into it. Nests are made of paper-like material called *carton*, a glue-like combination of digested wood and termite fecal matter. When a nest is breached, scores of workers and soldiers swarm out in reaction to the disturbance (Plate 10-4b). Workers are pale whitish, with dark mahogany-colored heads. Soldiers are similarly colored but larger than workers, with prominent heads and long snouts. Soldiers eject a

(a)

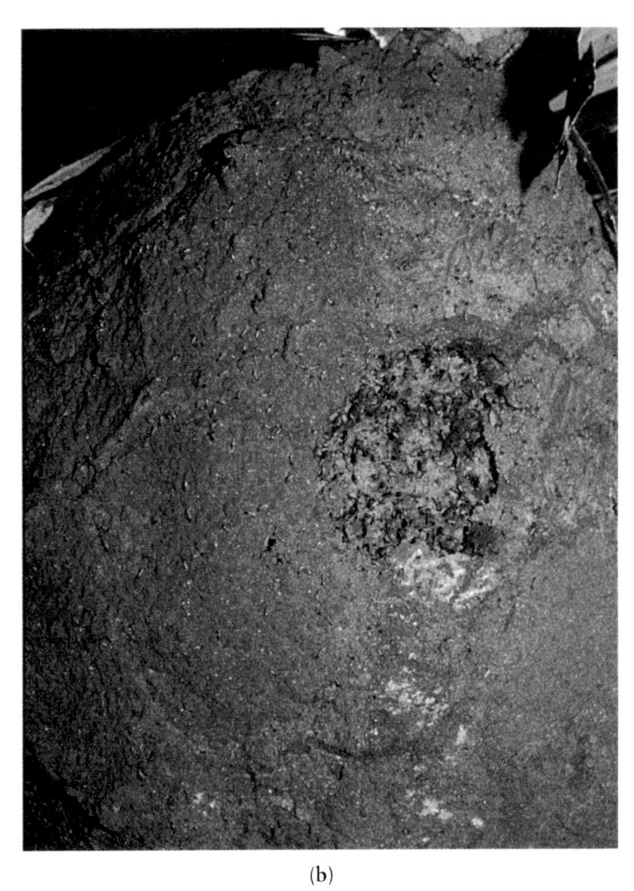

(b)

PLATE 10-4
(a) Basketball-size termite mounds such as this are common sights throughout much of the Neotropics. (b) This termite nest has been cut open and is swarming with worker termites repairing the damaged nest.

sticky substance with the odor of turpentine that apparently irritates would-be predators, including the anteater *Tamandua*. Termites quickly repair an injured nest. They rush about the surface, laying down new material to repair the damaged area. Another way to see termites is to break open their tunnels. Workers are blind and follow chemical trails laid down by other workers. They will continue to pass en masse along an opened tunnel, though some will eventually repair it. The termite queen (or queens—some species have multiple queens) is located deep within the nest and cannot be seen unless the entire nest is dissected. Queens are immense compared to workers and soldiers. Virtually immobile, weighed down by an enlarged gourd-like abdomen, their needs are attended to by workers as they pass their lives producing the colony's eggs.

PLATE 10-5
Look closely among the yellow flowering plants and see the large number of conical termite mounds in this dry forest area in southern Brazil.

(a)

(b)

PLATE 10-6

(a) Large termite mounds such as this are common throughout much of the Australian outback. The nearby trees likely benefit in nutrient uptake from proximity to the termite colony. (b) The author (third from left) and three colleagues are dwarfed by the size of the termite "skyscraper" typical of termite mounds throughout much of the Australian outback.

Many termite species ingest wood and digest it with the aid of mutualistic protozoans and bacteria inhabiting the hindgut of the insects. Cellulose is digested by the collective metabolism of various flagellate protozoans and bacteria. Once cellulose is reduced to simple sugar molecules, termites gain nutrition from the process of gut faunal and floral metabolism (much as ruminant mammals such as cattle do). Removal of protozoa will prevent a termite from digesting cellulose and other large molecules of wood. The termite, bacteria, and protozoans are *obligate symbionts*, another example of a complex coevolutionary mutualism (see Chapter 7). The flagellate protozoans gain fitness by inhabiting the termite in that it provides continuous food, shelter, and a means of dispersal. The hard woods of many tropical trees are perhaps an evolutionary response to selection pressures posed by termite herbivory.

PLATE 10-7

This ball of spines is a sleeping echidna. Eight species of echidna are found in Australia. They are *monotremes*, egg-laying mammals, and they all have a fondness for devouring termites and ants.

Termites do not feed only on wood. Many species are essential decomposers of forest floor litter. In Amazonia, they are believed to feed on somewhere between 20% and 50% of the fallen leaves, to collectively have a biomass of 2,000 kilograms per hectare (about 4,409 pounds per hectare), and number in excess of 1,000 individuals per square meter of forest floor (Martius 1994). Termites chew up leaves as well as pass them through their digestive systems. This action increases leaf surface area and enhances the microbial film, facilitating continued decomposition.

Termitaria, the termite nests, form patches of high nutrient concentrations in otherwise nutrient-poor tropical soils. A study conducted along the Rio Negro in Venezuela showed that termites consumed between 3% and 5% of the annual litter production, transporting it to their termitaria. Termitaria contained more nutrients than litter, and litter was more nutrient-rich than the soil. When termites abandon termitaria, which they do regularly (nest rate abandonment averaged 165 nests per hectare per year), these sites form patches of high nutrient level, ideal for many tree species. Termite activity has a potentially important influence on nutrient cycling and tree establishment in this area of very poor soils (Salick et al. 1983).

Many termites feed on soil, including various soil-inhabiting fungi. In one study from Amazonian Brazil, 57% of the 52 termite species were soil feeders (Ackerman et al. 2009). The actions of soil-feeding termites are believed to help release nitrogen and phosphorus, contribute to humification, improve soil drainage, and help aerate soil, as well as to increase exchangeable cations such as calcium and potassium. There is evidence that soil-feeding termites could be essential in supporting soil fertility, not only in forests but also in agroecosystems (Ackerman et al. 2009).

In a study conducted on the savannas of Kenya, it has been shown that termite mounds enhance the growth rates of plants and are what the authors of the study termed *hotspots* for plant productivity (Pringle et al. 2010). But beyond plant productivity, there was higher biomass and diversity of animals as well. There was increased biomass, abundance, and reproductive output of consumer species near the mounds. The termites were credited with having increased the nutrient-richness of soil as well as soil moisture. Mounds were relatively evenly spaced so that the region in general enjoyed a higher net primary productivity, essentially due to termite presence.

One Paleotropical subfamily of termite, Macrotermitinae, has evolved in a similar manner to Neotropical fungus-garden ants (see Chapter 7). These termites, found only in Africa, Madagascar, and parts of Asia, maintain subterranean fungus gardens in which they cultivate a genus of Basidiomycete fungus. Using their feces as fertilizer, they utilize the fungus to digest the plant cell walls. The termites then consume the remaining plant material. Some species also eat the fungus (Hyodo et al. 2003).

Another termite genus, *Hospitalitermes*, is called the *processional termite*. These termites, all of which occur only in Southeast Asia, are obvious on the forest floor, where they move in columns similar to those of various ant species. They ascend into the forest canopy and clip small epiphytic plants such as algae, lichens, and bryophytes that are converted to a ball-like mass and carried to a subterranean nest, to be used as food.

Termites are so abundant in the world's tropical areas that they may contribute to global climatic warming by enhancing the greenhouse effect. Their combined digestive abilities produce significant quantities of atmospheric methane, carbon dioxide, and molecular hydrogen. Forest clearance and the conversion of forests to agricultural ecosystems often result in increased termite abundance, thus accelerating the production of the atmospheric gases (Zimmerman et al. 1982). But, on the other hand, some areas that have experienced forest clearance or have been converted to uses such as banana plantations appear to result in loss of termite diversity and biomass (Ackerman et al. 2009).

Leaf-Cutter Ants and Tropical Soil Properties

Termites are not the only social insects whose colonies act to alter and enhance soil properties in the tropics. Leaf-cutter ant colonies (Attini; see Chapter 7), restricted to the Neotropics, also concentrate certain nutrients and alter the soil such that plant productivity may be stimulated (Moutinho et al. 2003; Sternberg et al. 2007). A study performed in Amazonia examined colony nests of *Atta sexdens*, a common leaf-cutter ant species. The work was remarkable in that the researchers constructed tunnels directly into the leaf-cutter ant nest to a depth of 3 meters (about 10 feet) (Figure 10-3). The study examined how soil properties and nutrient distributions differed between ant nest soil and soil away from the nests. The site of the study was a 17-year-old, 60-hectare (about 148 acres) secondary forest in eastern Brazil. The site had been cleared for cattle ranching and was abandoned and allowed to undergo secondary succession.

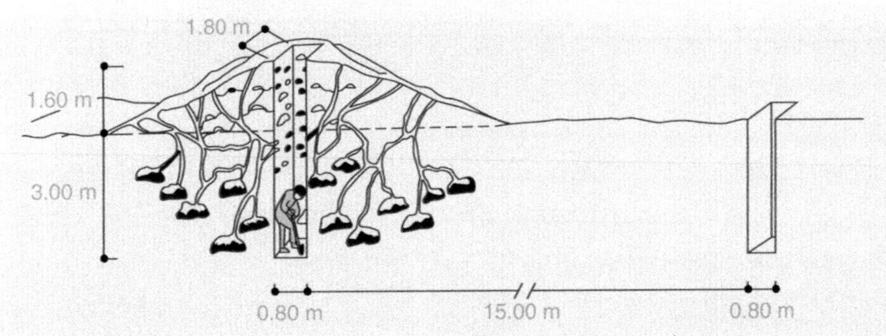

FIGURE 10-3
This is a diagram illustrating the methodology employed in excavation of shafts in the nests of *Atta* ants. Chambers with black dots are refuse chambers, and the others are fungus chambers. Note the overall size of the colony relative to the human figure.

TABLE 10-1 Extractable nutrient stocks (in kilograms per hectare) in nest soil of *Atta sexdens* and in soil without nest influence ($n = 3$) in a secondary forest.

ELEMENT	NEST SOIL			NON-NEST SOIL (DEPTH 0–4 METERS)	*P*
	NEST MOUND (HEIGHT 1.6 METERS)	(DEPTH 0–4 METERS)	TOTAL		
Nitrogen (N)	9,960 ± 3,200	15,000 ± 1,200	25,700 ± 3,000	21,650 ± 1,300	0.30
Carbon (C)	11,200 ± 230	101,500 ± 34,000	216,000 ± 31,000	200,000 ± 43,000	0.78
Calcium (Ca)	2,400 ± 600	4,000 ± 1,000	6,400 ± 700	3,500 ± 780	0.05
Magnesium (Mg)	600 ± 180	2,500 ± 450	3,000 ± 200	1,800 ± 460	0.06
Phosphorus (P)	2.8 ± 0.6	5.3 ± 3.6	8.2 ± 2.0	6.0 ± 3.0	0.59
Potassium (K)	230 ± 12	1,850 ± 80	2,700 ± 400	400 ± 200	0.02

Notes: The nutrient stocks were integrated to 4-m soil depth and nest mound (mean ± 1SE). *P* values from the one-way ANOVA comparisons between total nutrient stocks for nests and non-nest soil are given. Total nutrient stocks were integrated to 4 m using the soil bulk density by depth increment (0–10, 10–50, 50–100, 100–200, 200–300, and 300–400 m).

Leaf-cutter ants construct complex, deep nests extending to 6 meters (about 19.7 feet) below the surface and ascending a meter or more above the ground surface. The nest is divided into multiple chambers, some for use as fungus gardens (see Chapter 7) others to deposit organic debris.

Researchers measured stem and root growth of nine tree species seedlings near and away from ant nests. They also compared potted *Cecropia* seedlings grown in soil taken from ant nests and in normal forest soil. Because of the continuous activities of the ants, soil beneath ant nests was much more aerated and easier to penetrate than surrounding soil, encouraging the penetration of fine roots from surrounding trees. Both coarse root biomass and fine root biomass were highest in ant nest soils. Cations such as calcium, potassium, and magnesium were significantly more concentrated in ant nest soil (because of the high cation exchange capacity of soil that is rich in organic material). *Cecropia* seedling growth was stimulated by organic matter from the refuse chambers of ants contained within ant-soil (Table 10-1). Although tree diameter growth was not higher at ant nests, that may have been due to greater energy devoted to root growth.

Another study by Sternberg et al. (2007) used ^{15}N to trace whether plants growing near leaf-cutter colonies benefited in uptake of nutrients. The study was performed in tropical forest in Panama (Barro Colorado Island [BCI]) and savanna in Brazil. Two leaf-cutter ant species (*Atta colombica* and *A. laevigata*) were studied. Researchers introduced ^{15}N into the ant nests and subsequently measured plant leaves from trees near nests as well as distant from nests to examine nitrogen isotope ratios of leaves. In both Panama and Brazil, it was learned that plants near *Atta* nests differentially took up ^{15}N, demonstrating that plants do access the nutrients contained within the ant nests. Further, the calcium level within the leaves was greater in those plants that took up the ^{15}N. Given that calcium is sometimes a limiting nutrient in both tropical forests and savannas, plants near *Atta* colonies are in a unique position to benefit from such proximity.

Leaf-cutter ants, by their activities in altering soil properties, have a potentially beneficial effect in restoring abandoned pasture (such as the site of the study described earlier) to full fertility. Such a beneficial impact might somewhat offset the perception that Attini ants generally exert negative effects on plant productivity by their defoliating activities (Plate 10-8).

PLATE 10-8
These trees are presumably benefiting from growing directly out of an *Atta* nest.

MYCORRHIZAE

Throughout the tropics as well as most of the temperate zone, there is an intimate mutualistic association between tree roots and a diverse group of fungi collectively termed *mycorrhizae* (Plate 10-9). Up to 80% of all land plants contain mycorrhizae either on or inside their roots, and a single gram of soil may contain 100 meters of mycorrhizal filaments (Whitfield 2007). Mycorrhizae are ubiquitous components of soils throughout most terrestrial ecosystems.

Part of the mycorrhizal mycelium is inside the plant root, and part extends out into the soil. Mycorrhizae use some of the plant's photosynthate as food. In this regard, the fungi would seem to be parasitic. But though they take food from the plant, mycorrhizae are essential to the plant's welfare, as they greatly facilitate the uptake of minerals from the forest litter. When the soil is nutrient-poor, the benefit to the plant outweighs the energy cost it pays to the fungus. Many trees dependent on mycorrhizae have poorly developed root hairs; the fungal strands may substitute for the missing root hairs (St. John 1985).

Most mycorrhizal fungi are from the phylum Glomeromycota and are termed *vesicular–arbuscular mycorrhizae*, so-called because of the arbuscular (shrub-like) hyphae they produce within tree root cells for nutrient exchange. The other major group of mycorrhizae associated with trees is mushroom-forming Basidiomycota, found primarily in poor soils (such as white sandy soils), in monodominant stands, and in montane forests (especially where trees within

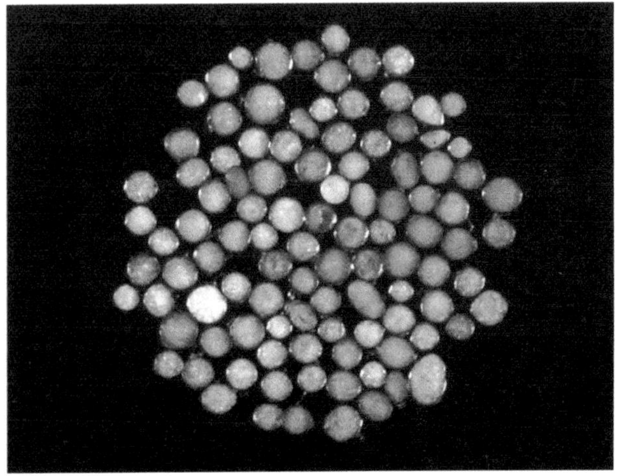

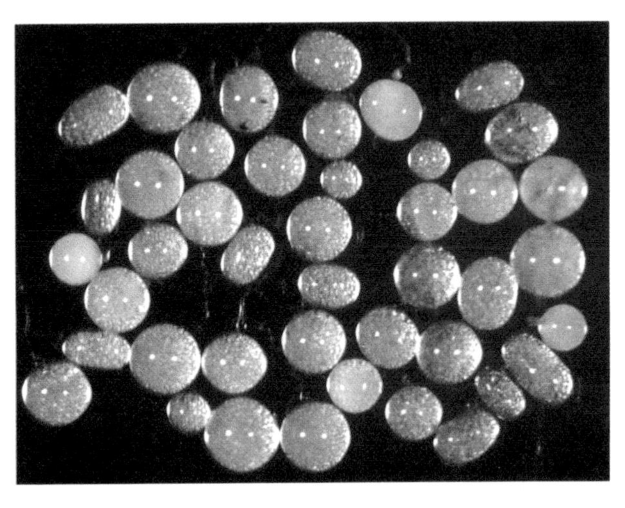

PLATE 10-9
These are images of spores from four species of mycorrhizal fungi.

the Fagaceae family are dominant, as well as in Southeast Asian dipterocarp forests). These ectomycorrhizae form a dense, visible mantle of mycelium on the surface of the tree root. Ectomycorrhizae are essential to many dominant tree species of temperate zone forests (Langley and Hungate 2003) but are much less common in tropical forests, which is partly why you see many fewer mushrooms on the floor of a tropical forest than in the temperate zone. Another group of mycorrhizae are essential to the roots of orchids, which are common epiphytes in many tropical forests.

The extensive surface area of the mycorrhizal mycelium is efficient in uptake of both minerals and water, as experiments have demonstrated (Janos 1980 and 1983). Mycorrhizae are essential in uptake of phosphorus, a nutrient that tends to be of limited availability in rain forest soils (Vitousek 1984, and see the section "Is It All about Phosphorus?" later in this chapter). Mycorrhizae may also have a role in direct decomposition and cycling, moving minerals from dead leaves into living trees without first releasing them to the soil (Janos 1983; St. John 1985), and they may affect competitive interactions among plants, thus influencing the biodiversity of a given forest (Janos 1983).

In early successional ecosystems, waterlogged areas, and high-elevation regions, mycorrhizae may be less ubiquitous (Parker 1994). It has been suggested that ectomycorrhizal fungi may provide their host plants with an initial competitive advantage over arbuscular mycorrhizae host plants (Lodge 1996).

Arbuscular mycorrhizae spores may be widely distributed by certain rodent species such as spiny rats (*Proechimys* spp.) and rice rats (*Oryzomys* spp.). A study performed in rain forest at Cocha Cashu, in Manu National Park in Peru, demonstrated that arbuscular mycorrhizae spores are well represented in the feces of spiny and rice rats (Janos et al. 1995). Though most mycorrhizae spread by direct infection from root to root, the researchers suggest that long-distance dispersal of arbuscular mycorrhizae fungi is facilitated by mammalian spore transport. Given that arbuscular mycorrhizae are essential in the uptake of minerals by the majority of rain forest tree species, the health and species richness of rain forest may be, at least in part, dependent on the wanderings of some unpretentious small rodents.

Diversity of Fungi and Negative Tree Density Dependence

Though mycorrhizae abound in the tropics, it should not be forgotten that not all fungi are mutualistic with vascular plants. Recall from Chapter 5 that the Janzen-Connell model suggests that seeds dropped near the parent tree experience reduced fitness due to attracting a concentration of predators and pathogens. Up to half of the seeds produced by 90% of tree species in tropical forests may be consumed by a combination of herbivores and fungi (Janzen and Vasquez-Yanes 1991).

For example, in a study done on Barro Colorado Island, Dalling et al. (1998) studied the seed bank dynamics of two pioneer tree species, *Miconia argentea* and *Cecropia insignis*. They learned that pathogenic fungi were by far the most important cause of seed mortality for both species. In the case of *Miconia*, seeds survived somewhat better at a distance of more than 5 meters from the crown (an example of the seed-shadow effect). For *Cecropia*, it did not matter. Ninety percent of seeds were destroyed, most by soil fungi. When seeds in an experimental plot were treated with fungicide, survivorship was much greater.

Another study performed on Barro Colorado Island in Panama compared seed-infecting fungi from four local *Cecropia* species with those of seeds brought from *C. insignis* from La Selva Biological Station in Costa Rica (Gallery et al. 2007). Seed survival varied significantly among species, and the seeds from La Selva fared poorly in comparison with those of the same *Cecropia* species on BCI (Figure 10-4). Fresh seeds were largely free of fungi, but once planted, that changed. In excess of 80% of soil-incubated seeds became infected by diverse Ascomycota fungi, some of which could be pathogens or saprophytes (that feed on dead material). Using molecular techniques, researchers demonstrated host affinity of fungi, high fungal diversity, and fine-scale spatial heterogeneity. Twenty-six fungal genotypes were isolated from more than one seed, and the remainder (64.3%) were recovered only once. This suggests a high fungal diversity in tropical soils. Although diverse, the ecological influence of fungi was unclear from this study and continues to be a promising area of ecological research in the tropics.

A study from Guyana that focused on ectomycorrhizal fungi challenged the Janzen-Connell hypothesis, at least for trees that demonstrate unique dominance in tropical forests (McGuire 2007). It is relatively uncommon, but some rain forest sites do contain tree species that comprise more than 50% of the canopy trees, a situation termed *monodominance*. Forests showing monodominance typically have ectomycorrhizal fungi networks associated with

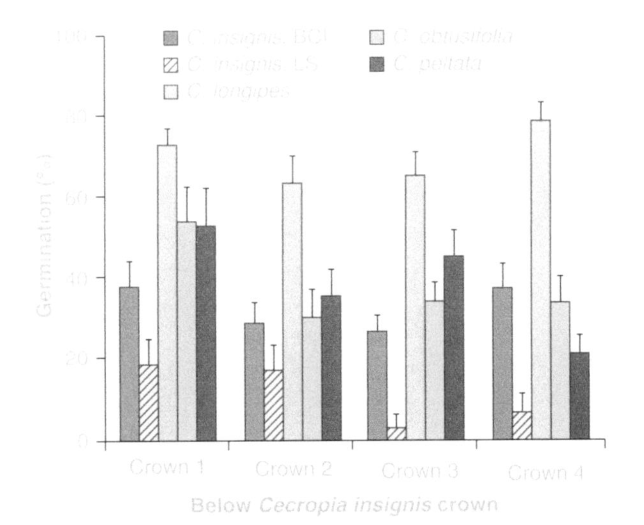

FIGURE 10-4
Differences in seed germination among *Cecropia* species at sites below four crowns of *C. insignis* trees at Barro Colorado Nature Monument, Panama. Bars represent the mean percentage (+SE) of germination success of 20 samples from each species at a given site. Abbreviations are: BCI, Barro Colorado Island; LS, La Selva.

the trees. The study involved manipulation of the seedlings of the monodominant species *Dicymbe corymbosa,* a species that makes up 70% to 90% of the canopy at the site of the study. Seedlings placed near a mature tree have easy access to a network of ectomycorrhizal fungi. But some seedlings were manipulated such that they could not access ectomycorrhizae. Seedlings with access to ectomycorrhizae networks did well. Those that didn't have access fared poorly. Seedlings with ectomycorrhizae experienced 73% greater growth, 55% more leaves, and 47% greater survivorship. In other words, because of ECM, there was a positive density dependence relationship associated with proximity to mature trees (Figure 10-5).

The benefit of growing near conspecifics is that it is possible to tap into beneficial host-specific microbes. At the same time, there is also a risk of increased predation and disease, as is predicted by the Janzen-Connell hypothesis. Hence, there is a cost/benefit issue dependent in part on the degree of host specificity of fungi.

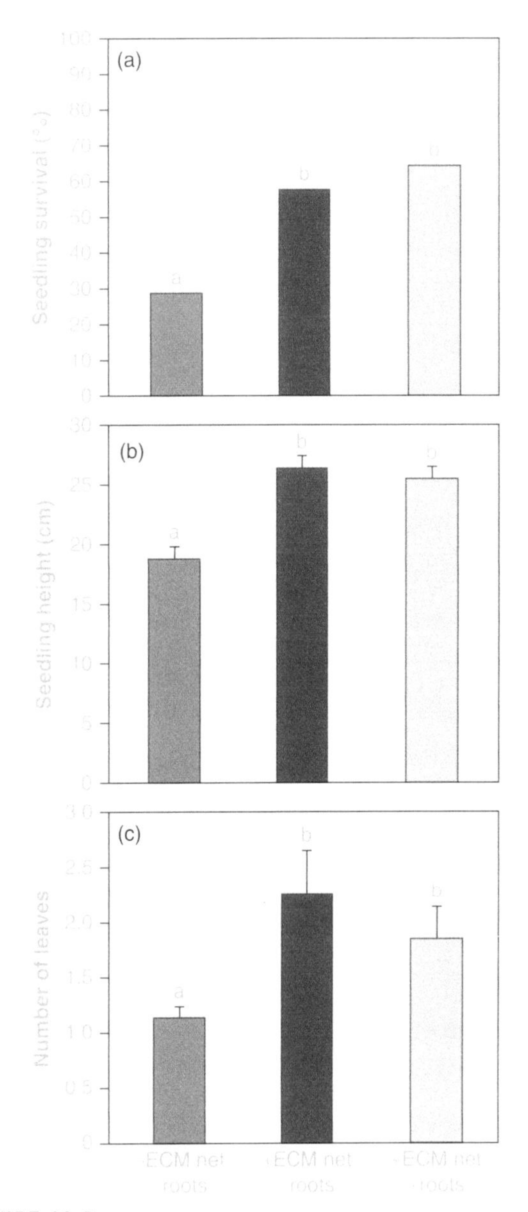

FIGURE 10-5
(a) Survival, (b) height, and (c) number of leaves across treatments for *Dicymbe corymbosa* seedlings after one year. Light brown bars (far left) represent the fine mesh net treatment where common ECM net access and root passage were restricted. Dark brown bars (center) represent the coarse mesh treatment where access to the common ECM net was permitted but root passage was restricted. Tan bars (far right) represent control treatments (no mesh) where ECM net and root passage were both permitted. Bars with different letters indicate significant differences between means of seedlings across treatments ($P \leq 0.05$). Each treatment was replicated 16 times for a total of 48 experimental units.

FUNGAL ENDOPHYTES

Fungi are not confined to soil and litter in the tropics. A vast assemblage is found within the photosynthetic tissue of leaves, a group collectively called *leaf endophytes.* These fungi, like mycorrhizae, appear to cause no harm and are thought to be symbiotic with leaves, though their exact influence largely remains to be discovered. Endophytic fungi occur at high latitudes as well as in the tropics, but patterns differ. A sweeping

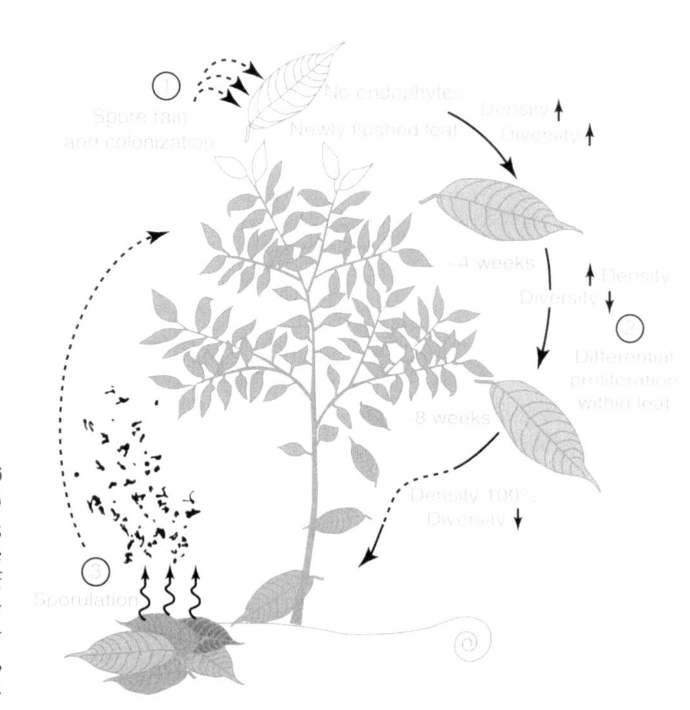

FIGURE 10-6
Proposed life cycle for tropical foliar endophytic fungi (FEF) and their host plants. Leaves are flushed, essentially free of FEF; spores land on the leaf surfaces and, upon wetting, germinate and penetrate the leaf cuticle. After a few weeks, the density of FEF infection within the leaf appears to saturate with a very high FEF diversity. Over several months, FEF diversity usually declines. After leaf senescence and abscission, FEF sporulate, and the cycle begins anew.

survey of 1,403 endophyte strains showed an increase in incidence, diversity, and host breadth of endophytes from arctic to tropical sites (Arnold and Lutzoni 2007). Leaves of tropical plants had such high endophyte diversity that they were described as *hotspots* (see Chapter 14) of fungal species diversity. But what do endophytes do for plants?

A study from Panama (Herre et al. 2007) looked at the effects of fungal endophytes and mycorrhizae on one species, *Theobroma cacao*, the cacao plant from which chocolate is derived. The study showed that the presence of foliar endophytes on leaves of cacao plants enhanced the cacao's defense against a common pathogen. In addition, root inoculations of arbuscular mycorrhizal fungi also aided in host resistance to the same pathogen. This study showed that endophytic fungi do have a strong symbiotic function when present on the leaves of plants such as cacao. The exact mechanism of protection extended by the fungi network to the plant awaits discovery (Figures 10-6 and 10-7 and Plates 10-10a to 10-10c). The widespread presence of fungal endophytes has been well established, and it is clear that the fungi affect the host plant's growth and physiology. The question remains as to how profoundly plant properties are affected by their abundant and diverse fungal communities (Herre et al. 2007).

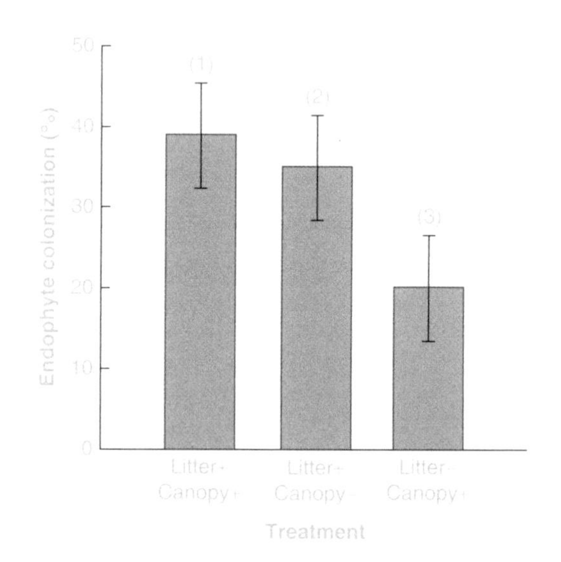

FIGURE 10-7
Local leaf litter is a more important source of foliar endophytic fungal inoculum than intact canopy cover. Mean percentage (±SE) of leaf tissue colonized by endophytes in endophyte-free seedlings of *Theobroma cacao* after a one-week exposure to each of three habitats: (1) intact forest (closed canopy with intact litter, ++; $n = 15$); (2) forest gap (open canopy but with leaf litter intact, −+; $n = 16$); (3) intact forest (closed canopy, with ~90% leaf litter removed within >20 meters of the seedlings, +−; $n = 16$). One leaf from each seedling was sampled, with 64 2-millimeter-square fragments per leaf assayed for endophyte infection.

(a)

(b)

(c)

PLATE 10-10

(a) to (c) Three examples of the fruiting (spore-producing) bodies of the many kinds of fungi found in Neotropical forests. (c) Note the flies on and around the stinkhorn mushroom (*Dictyphora*). It emits an odor that attracts them, and they aid in dispersing spores.

Rapid Recycling

There is often surprisingly little accumulation of dead leaves and wood on the rain forest floor, making for a generally thin litter layer. Unlike the northern coniferous forests, for example, which typically have a thick spongy carpet of soft fallen needles, or the broadleaf temperate forests, where layer after layer of fallen oak and maple leaves have accumulated, a rain forest floor is often sparsely covered by fallen leaves. This becomes particularly noteworthy when you keep in mind that more and heavier leaves occur in rain forest. The solution to this seeming paradox is that decomposition and recycling of fallen plant parts occur with much greater speed in rain forests than in temperate forests. Just as productivity is continuous, uninterrupted by the frozen soils of a northern winter, biogeochemical cycling continues unabated throughout the year. In tropical wet forests, particularly those on richer soils, litter is totally decomposed in less than one year, and minerals are efficiently conserved (Jordan 1985a) (Plate 10-11).

Ecologists measure the rate of decomposition of organic matter as a value k. The calculation is $X_0/X = e^{-kt}$, where X_0 is the original amount of litter on the forest floor, X is the amount remaining at a later time, e is the base of natural logarithms, and t is the time elapsed between X_0 and X (Montagnini and Jordan 2005). The k value in a tundra ecosystem is about 0.03 per year. In a temperate deciduous forest, it is 0.77 per year, and in a tropical moist forest, it is 6.0 per year. However, there is high variation from site to site, and therefore, while it is generally true that decomposition

PLATE 10-11

Leaf litter is abundant in tropical moist forests, but decomposition is sufficiently rapid that in many areas the litter layer is relatively thin.

rates are higher in the tropics, one must be careful not to overgeneralize (Montagnini and Jordan 2005).

Litter accumulation may be high, ranging from 12 to 18 kilograms per cubic meter (about 26 to 40 pounds per cubic meter) in lowland moist forests, higher in montane forests (Zinke et al. 1984, cited in Montagnini and Jordan 2005). At La Selva Field Station in Costa Rica, leaf litter-fall (all branches, leaves, and other plant material) ranged from 9.5 to 12.4 megagrams per hectare per year (about 21,000 pounds to 27,000 pounds) and averaged 10.9 megagrams per hectare per year (about 24,000 pounds) over six study plots (Parker 1994). Leaf decomposition is distinctly slower on poor soils than on richer soils (see the next section).

Key variables such as phosphorus level and amount of precipitation along with high temperature influence the rate at which litter decomposes. A study performed in a forest in Costa Rica that receives more than 5,000 millimeters (about 197 inches) of precipitation per year examined the rate at which litter from 11 canopy tree species decomposed (Wieder et al. 2009). Decomposition rate was high, typical of what has been observed in other humid tropical forests. There was a nearly fourfold variation in decomposition rate among species tested. The most important variable in rate of litter decomposition was deemed to be climate, a warm and very wet climate. But beyond that, the "quality" of the litter strongly affects decomposition rate. Litter rich in phosphorus decomposed more rapidly than litter with less phosphorus (Plate 10-12 and Figure 10-8).

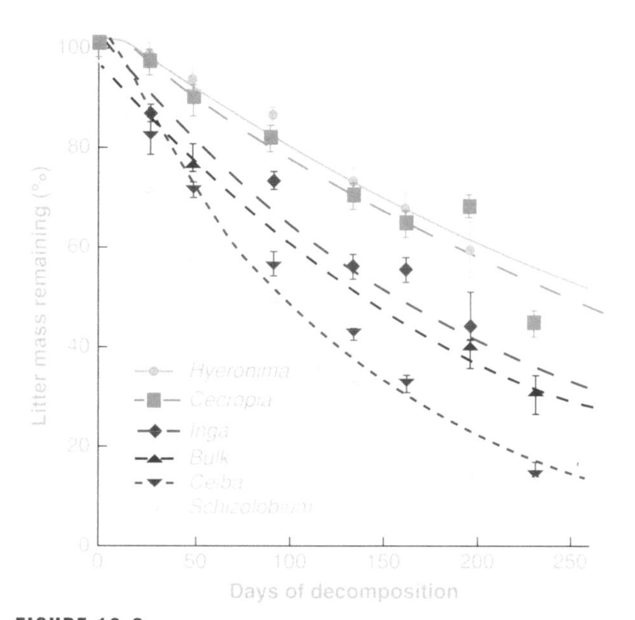

FIGURE 10-8
Decomposition curves illustrating how different genera of trees have different rates of decomposition. All trees were on control plots. Points are means ± standard error and are significantly different from one another.

Forests on poor soils show reduced rates of decomposition (Lavelle et al. 1993). Tropical moist forests also cycle minerals very "tightly." The resident time of an atom in the nonliving component of the ecosystem is typically brief (Jordan and Herrera 1981; Jordan 1982, 1985a, and 1985b). One study estimated that approximately 80% of the total leaf matter in an Amazon rain forest is annually returned to the soil (Klinge et al. 1975).

PLATE 10-12
The species of leaf as well as the overall quality of leaf litter, such as available phosphorus, are important in determining how rapidly the leaf will be decomposed.

Rain Forest Soil Types and Nutrient Cycling

Though most tropical soils are highly leached and of poor nutrient quality, there are some areas of fertile soils in the tropics (Vitousek and Sanford 1986). In some regions, such as the eastern and central Amazon Basin, soils are very old and mineral-poor (*oligotrophic*), while in other regions, such as volcanic areas of Costa Rica or much of the Andes, soils are young and mineral-rich (*eutrophic*) (Jordan and Herrera 1981). Soil characteristics, referred to as *edaphic characteristics*, vary regionally because soil is the product of several variables: climate, vegetation, topographic position, parent material, and soil age (Sollins et al. 1994). Because these factors vary substantially throughout the global tropics, so do soil types. Even within a relatively limited region, there may be high variability among soil types. For instance, a single day's ride in southern Belize will take you from orange-red iron-rich soils to gray-brown clayey soil. The gray-brown soil is calcium-rich, weathered from limestone, common throughout much of Belize.

Many of the soils throughout the humid tropics fall into one of three classifications: *Ultisols*, *Oxisols*, or *Alfisols*. Ultisols are generally well-weathered, meaning that minerals have been washed (leached) from the upper parts of the soils. Oxisols, also called *Ferralsols* or *Latosols*, are deeply weathered, old, and acidic and are found on well-drained soils of humid regions; typically, these soils occur on old geological formations such as the ancient Guianan Shield in northeastern South America. These soils have high iron content and are typically reddish (Plate 10-13). Oxisols are common throughout the global tropics and are typically heavily leached of minerals as well as quite acidic (Lucas et al. 1993). Alfisols are common in the subhumid and semiarid tropics and are closer to a neutral pH (though still acidic), with less overall leaching than typical Oxisols. It is estimated that Ultisols, Oxisols, and Alfisols, taken together, comprise about 71% of the land surface in the humid tropics world-wide (Lal 1990). This is generally similar to estimates made by Vitousek and Sanford (1986) suggesting that 63% of moist tropical forests are atop soils of moderate to very low fertility. In the Amazon Basin, about 75% of the area is classified as having Oxisols and Ultisols (Nicholaides et al. 1985).

PLATE 10-13
Red Oxisols such as are revealed in this roadcut in Belize are common throughout much of the tropics.

Not all tropical soils are old or heavily weathered or infertile. Vitousek and Sanford (1986) estimated that 15% of moist tropical forests are situated on soils of at least moderate fertility. Inceptisols and Entisols are young soils of recent origin, rich in minerals near the surface, with higher pH (still acid, but closer to neutral). Soils generated from deposits during the flood cycle (*alluvial soils*) or from recent volcanic activity typify these categories (Lal 1990; Sollins et al. 1994) (Plate 10-14).

Soil types are not absolute; soil types grade into one another along a continuum. For example, at La Selva Biological Station in Costa Rica, it is estimated that approximately one-third of the soils are fertile Inceptisols (some of recent volcanic origin) and some Entisols of alluvial origin, while the remainder of the soils are older, more acidic, and less fertile Ultisols (Parker 1994). Even within a 1-hectare (2.47 acres) tract of forest, soil characteristics may vary based on small differences in elevation (Poulsen et al. 2006). Such differences are reflected in the distribution patterns of vegetation. In one study of a 1-hectare plot in Amazonian Ecuador, the distribution of *pteridophytes* (ferns, club mosses, and so on), ground herbs, and palms correlated with edaphic variables, mostly based on subtle elevational differences (Poulsen et al. 2006).

Plants that grow on nutrient-poor soils such as white sandy soils exhibit different trade-offs than plants that grow on more-nutrient-rich soil. Experiments performed near Iquitos, Peru, near the Amazon River, have demonstrated that plants growing on white sandy soil typically invest more in defense (such as additional terpenes, phenolics, enhanced leaf toughness, and foliar protein) than these same species do if grown on nutrient-rich clay soil (Fine et al. 2006).

Another study using large forest plots in Colombia (La Planada), Ecuador (Yasuni), and Panama (BCI) employed a statistical analysis known as *Monte Carlo simulation* to examine species distributions of plants to determine whether plant species are distributed nonrandomly with relation to soil characteristics (John et al. 2007). A Monte Carlo simulation essentially tests for a null distribution, meaning that tree

PLATE 10-14
This farm along the Amazon River is sustained by the rich soil (ultimately from the Andes Mountains) that is annually renewed in the flooding cycle. Note the grove of bananas.

distribution is unrelated to other variables. Researchers used spatial distribution maps of more than 0.5 million individual trees of 1,400 species. (These sorts of distribution maps have become essential in the study of tropical plant ecology. Just imagine the effort that is involved in making them.) It was learned that 36% to 51% of tree species showed strong associations with distributions of 10 essential plant nutrients. The researchers concluded that below-ground resource availability is extremely important in affecting the assembly of tropical tree communities. The study strongly suggests that competition for nutrients may be an important component of tropical tree ecology.

Semiarid and arid regions in the tropics, because of climatic differences, have somewhat different soil types from those of humid and semihumid regions. Some of these soils are dark, heavily textured, and calcareous, sometimes subject to salt accumulation (Lal 1990). Because of frequent occurrences of burning, and sometimes animal grazing, litter is thin and poorly developed on savanna soils, and the decomposer ecosystem is more limited. Termites, however, can be particularly abundant in arid, grassy areas (Lavelle et al. 1993).

The general pattern throughout much of the humid tropics is that heat and heavy moisture input result in formation of oxides of iron and aluminum (which are not taken up by plants), giving the soil its characteristic reddish color. Clay content is normally high, evident as you slip and slide your way over a wet trail. Mountain roads become more dangerous and often impassable during rainy season because wet clay makes the roads slippery. Clay also has reduced porosity, impeding penetration by water. Thus clay soil enhances flooding potential.

In the Amazon Basin, sediments eroded from highland areas during the late Tertiary period were deposited in the western end of the basin, forming a flat surface about 250 meters above sea level. Much of this surface, called the Amazon Planalto, is made up of *kaolinitic clay*, a substance devoid of most essential minerals but rich in silicon, aluminum, hydrogen, and oxygen (Jordan 1985b). In the eastern part of Amazonia, soils are sandy, acidic, and nutrient-poor.

LATERIZATION

A process called *laterization* is associated with some tropical soils. It results from the combined effects of intensive erosion and heat acting on soil. If vegetation cover is removed and bare soil exposed to heavy downpours and heat, soil may be converted to a brick-like substance, preventing future productivity. Laterization has long been utilized by tropical peoples around the world for making bricks used in buildings as impressive and as venerable as some of the ancient temples in Cambodia. Though laterization has been widely reported as demonstrating the extreme delicacy of tropical soils and thus the futility of farming such soils, such a generalization is unfounded. Laterization occurs only with repeated wetting and drying of the soil in the absence of any vegetative cover. The loss of roots (which utilize aeration channels in the soil) plus repeated wetting and drying act to break up soil aggregates of bound clay particles. Only when these aggregates are eliminated and broken up, and the soil thus subject to extreme compaction, does laterization ensue (Jordan 1985b). In Amazonia, only about 4% of the soils are at risk of laterization (Nicholaides et al. 1985).

TROPICAL SOILS AND AGRICULTURE

Even without the extreme of laterization, attempts to farm the tropics by applying intensive agriculture have often caused rapid loss of soil fertility. This need not be the case. For example, much of the soil composition in Amazonia is surprisingly similar to that found in the southeastern United States, where successful agriculture is routinely practiced (Nicholaides et al. 1985). Soil infertility, though generally common throughout the Amazon Basin (Irion 1978; Nicholaides et al. 1985; Uhl et al. 1990), does not preclude agriculture. Where Amazonian soils are most fertile, as along floodplains, they will support continuous cultivation by small-scale family units (subsistence agriculture), with crops such as maize, bananas, sweet potatoes, as well as small herds of cattle (Irion 1978). Various approaches have been shown to be successful in achieving continuous farming of low-fertility Amazonian soils (Nicholaides et al. 1985; Dale et al. 1994) (Plate 10-14). Tropical agriculture will be discussed further in Chapter 13.

MINERAL CYCLING ON OLIGOTROPHIC SOIL

In parts of the Amazon Basin, white and sandy soils predominate, most of which are derived from the Brazilian and Guianan Shields, both ancient, eroded mountain ranges. Because these soils have eroded for hundreds of millions of years, they have lost their fertility and are extremely poor in mineral content. The paradox is that lush broadleaf rain forests grow on these infertile soils. I stress *on* and not *in* the soil because recycling occurs on the soil surface.

The word *oligotrophic* refers to "nutrient deprived." Poor-soil forests are found on *terre firme* or on *igapo* floodplain (see the following section). Forests on oligotrophic soils are less lush and of smaller stature than forests on rich soils. Henry Walter Bates (1862) commented about forest on poor soil *igapo* (which he spelled *Ygapo*) floodplain, comparing it with the forest on the rich soil delta: "The low-lying areas of forest or Ygapos, which alternate everywhere with the more elevated districts, did not furnish the same luxuriant vegetation as they do in the Delta region of the Amazons (p. 143)."

In forests with oligotrophic soil, up to 26% of the roots can be on the surface rather than buried within the soil (Jordan 1985a), and root mats as thick as several centimeters sometimes develop (Lavelle et al. 1993). This obvious mat (you can actually trip over it) of surface roots, intimately associated with the litter ecosystem, is much reduced or entirely absent from forests on nutrient-rich, eutrophic soil, where subsurface root mats occur. Surface roots are obvious as they radiate from the many boles across forest floor. A thin humus layer of decomposing material also covers the forest floor, and thus the root mat of surface roots, aided by mycorrhizal fungi, directly adsorbs available minerals (Lavelle et al. 1993).

RAPID RECYCLING OF ESSENTIAL NUTRIENTS

Carl F. Jordan and colleagues have made extensive studies of Amazon forests' nutrient conservation (Jordan and Kline 1972; Jordan 1982 and 1985a). Using radioactive calcium and phosphorus to trace mineral uptake by vegetation, they found that 99.9% of all calcium and phosphorus was adsorbed (attached) to the root mat by mycorrhizal fungi plus root tissue. The root mat, which grows quickly, literally grabs and holds the minerals. For example, in one study from Venezuela, the decomposition of fallen trees did not result in any substantial increase in nutrient concentration of leachate water, suggesting strongly that nutrients leached from fallen vegetation move immediately back into living vegetation (Uhl et al. 1990).

Phosphorus is usually limited in tropical soils because it complexes tightly with iron and aluminum and, due to high acidity, is held in stable compounds that make it unavailable for uptake by plants (Jordon 1985b; Parker 1994). It is thus the nutrient most difficult for plants to procure (Vitousek 1984; Cuevas 2001). However, arbuscular mycorrhizae enhance uptake of phosphorus, a fact considered the most essential nutritional benefit of mycorrhizae in the tropics (St. John 1985; Moyersoen et al. 1998; Lloyd et al. 2001; Ramos-Zapata et al. 2009).

A study performed in two neighboring rain forests in southwest Costa Rica demonstrated the rapid

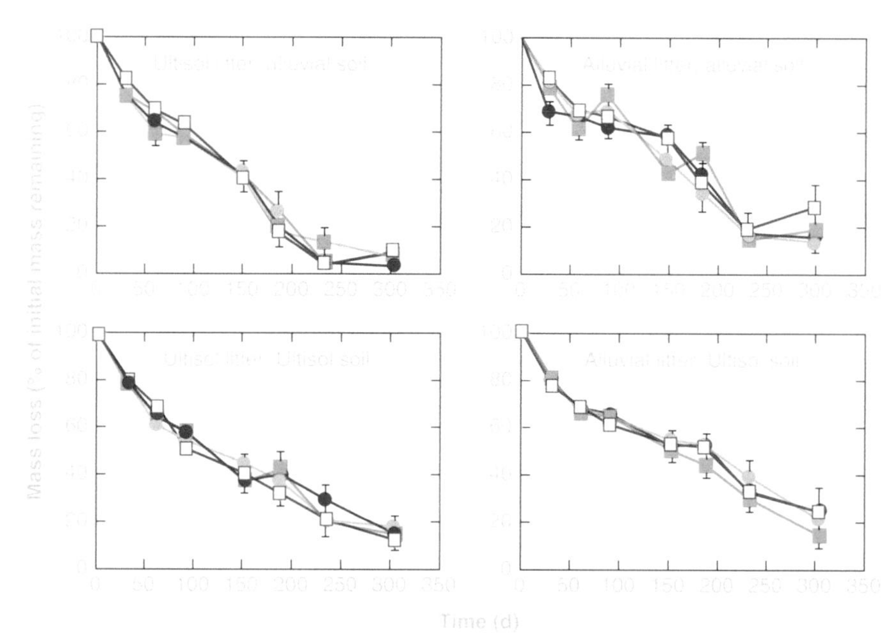

FIGURE 10-9
Effects of nutrient fertilization on loss of organic matter mass shown as percentage of mass remaining (mean ±SE). Treatment types are represented by purple circles (control), yellow circles (N added), solid squares (P added), and open squares (N + P added).

decomposition of tropical leaves as well as the effects of nitrogen and phosphorus fertilization on decomposition rate (Cleveland et al. 2006). One forest was located on rich alluvial soil, the other on poor Ultisol. One tree species, *Brosimum utile*, was studied in both forests. In each forest, leaves lost more than 80% of their initial mass in fewer than 300 days. Researchers showed that the rapid mass loss was associated with leaching of dissolved organic matter (DOM) (Figure 10-9).

Fertilization with P and N did not enhance decomposition rate of leaves, but adding P did have the effect of immobilizing P in decomposing material, lowering the ratio of carbon to phosphorus (C : P) in soluble DOM. The added P content of DOM enhanced the rate of microbial mineralization of DOM, meaning that more DOM was converted to CO_2. This result means that changes in P availability could alter carbon flux. Nitrogen fertilization produced a different result, appearing to inhibit soluble carbon decomposition (Figure 10-10).

Microorganisms living within the root mat and litter layer are essential in aiding the uptake of available minerals. The forest floor microbial community holds tightly to nutrients such as phosphorus, preventing minerals from being washed from the system (Jordan and Kline 1972; Jordan and Herrera 1981; Jordan 1982).

Such a uniquely rapid recycling system may be one reason for the presence of *buttresses*. The buttress allows the root to spread widely at the surface, where it can reclaim minerals, without significantly reducing the anchorage of the tree. This is probably the tightest recycling system in nature. If the thin layer of forest humus with its mycorrhizal fungi is destroyed, this recycling system is stopped, and the fertility lost. Removal of forest from white sandy soils may result in the regrowth of savanna rather than rain forest because of the destruction of the tight nutrient cycling system.

Though a dense surface root mat seems to be an obvious adaptation to the need for rapid and efficient

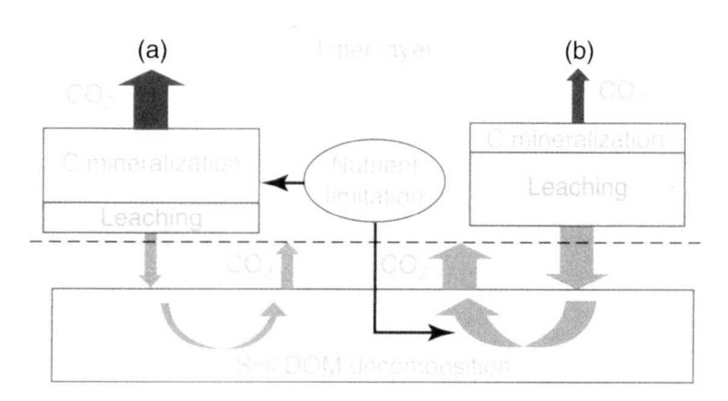

FIGURE 10-10
Conceptual model of the effects of nutrients on decomposition in systems where mass loss from the litter layer is dominated by (a) C mineralization or (b) leaching losses of dissolved organic matter (DOM) to the soil. Nutrient availability may constrain mass loss directly if most C mineralization occurs in the litter layer (a). However, in systems with high precipitation and/or highly water-soluble litter, nutrient availability may not limit mass loss (b) but may ultimately regulate DOM mineralization in the soil. The size of the solid arrows represents the relative flux of either CO_2 (dark brown arrows) or leached DOM (light brown arrows).

recycling on highly oligotrophic soils, the generalization is not universal. A research team working on Maraca Island, a lowland evergreen rain forest site in Roraima, Brazil, found that in spite of low nutrient concentration in the dry, sandy soil, the vegetation did not exhibit the types of adaptations described earlier. There was no surface root mat, nor was the root biomass unusually high, but the leaves were nonetheless relatively rich in nutrient content. Trees grew rapidly, and litter-fall was high. The rate of leaf decomposition was also high, indicating a rapid recycling mechanism. What intrigued and baffled the researchers was that this island forest showed none of the presumed adaptations of rain forests elsewhere on highly oligotrophic soils, yet it seemed to be functioning efficiently and without nutrient limitation (Thompson et al. 1992; Scott et al. 1992). The work is an example of the need for caution in generalizing about rain forest ecology and adaptations because different adaptations may evolve in response to a similar array of selection pressures.

OTHER NUTRIENT RETENTION ADAPTATIONS

Some tropical plants have root systems that grow vertically upward, from the soil onto the stems of neighboring trees. Termed *apogeotropic roots,* these roots grow as fast as 5.6 centimeters in 72 hours (Sanford 1987). The advantage of growing on the stems of other trees may be that the roots can quickly and directly absorb nutrients leached from the trees, as precipitation flows down the stem. This unique system, thus far described only for some plants growing on poor-quality Amazon soils, results in recycling without the minerals ever entering the soil!

A somewhat similar process, called *arrested litter*, has been documented at La Selva in Costa Rica (Parker 1994). Both epiphytes and understory plants, especially the wide crowns of certain palms (nicknamed *wastebasket plants*), catch litter as it falls from the canopy. The litter subsequently decomposes above ground, enriching the mineral content of stemflow and thus having the effect of fertilizing the soil in the immediate vicinity of the wastebasket plant.

Canopy leaves play a direct role in taking up nutrients. Algae and lichens on the surface of leaves, called *nutrient scavengers*, adsorb nutrients from rainfall, trapping the nutrients on the leaf surface. When the leaf dies and decomposes, these nutrients are taken up by the root mat and returned to the canopy trees (Jordan et al. 1979).

Some trees both in temperate and tropical regions have what are termed *canopy roots*. These adventitious roots, similar in structure to subterranean roots, grow into the thick litter and epiphyte layer that accumulates on the surface of thick branches far from the forest floor (Nadkarni 1981). This adaptation enables trees to tap into nutrients far above the forest floor and obviously is confined to forests where epiphyte density is high, such as temperate rain forests of the Olympic Peninsula of the Pacific Coast as well as montane tropical forests.

Nitrogen Fixation in the Tropics

Approximately 79% of Earth's atmosphere is *gaseous nitrogen*, a form of nitrogen that is not used in routine metabolism. But some bacteria convert gaseous nitrogen to forms such as nitrate, readily usable not only by themselves but also by other organisms, a process termed *biological nitrogen fixation*. This process is familiar in the temperate zone and is prominent throughout the tropics (Parker 1994; Cleveland et al. 1999). It is the likely reason why nitrogen is generally not a limiting nutrient in tropical soils. In some parts of the tropics, the availability of nitrogen may exceed the demand for it (Martinelli et al. 1999). Two kinds of biological nitrogen fixation are recognized, *symbiotic* and *free-living*, and both kinds are accomplished by various bacteria. Symbiotic nitrogen fixation is most closely associated with plants of the legume family (Fabaceae). Free-living nitrogen fixation occurs in soil with nitrogen-fixing bacterial genera such as *Azotobacter* and is also associated with epiphyllic microbes and lichens. Because of the distributions of legumes and free-living microbial nitrogen fixers, nitrogen fixation in the tropics extends vertically from canopy to soil.

One study estimated that about 20 kilograms (about 44 pounds) of nitrogen is fixed per hectare per year throughout the Amazon Basin (Salati and Vose 1984). In comparison with nitrogen fixation in crops such as soy, this is not a huge amount. As described in the following, however, the rate of nitrogen fixation in tropical forests varies widely.

Plants of the huge legume family (Fabaceae), abundantly represented in biomass and biodiversity throughout the global tropics, engage in symbiotic nitrogen fixation. Nitrogen is acquired through nodules in the root systems of the legumes (Plate 10-15). The nodules contain

PLATE 10-15
The nodules on the roots contain the symbiotic bacteria that facilitate uptake of nitrogen.

bacteria formerly called *Rhizobium* but now recognized to be four main genera and about nine others (bacterial classification is complex and changing due to DNA analysis). Both the plant and the bacteria benefit from their interaction, as the bacteria convert free nitrogen to nitrate and the bacteria are supplied with potential energy from the plant, an obligate mutualistic association.

Free-living nitrogen fixation also occurs abundantly in tropical forests. Certain epiphyllous lichens (*lichens* are intimate mutualisms between fungi and algae) convert gaseous nitrogen to usable form for plants in a manner similar to that of leguminous plants. In one study, it was determined that 1.5 to 8 kilograms of nitrogen per hectare per year (about 3.3 to 17.6 pounds) was supplied annually by canopy lichens (Forman 1975). Other studies have shown that leaf-surface microbes (*microbial film*) facilitate uptake of gaseous nitrogen (Bentley and Carpenter 1980 and 1984). Nitrogen fixation also occurs in termites because of the metabolic activities of bacteria in termite guts (Prestwich et al. 1980; Prestwich and Bentley 1981). Because of the abundance of termites throughout the tropics, these insects may contribute a substantial amount of nitrogen to the soil.

Rates of free-living nitrogen fixation differ among tree species and at different heights above the forest floor. A study performed in Golfo Dulce Forest Reserve in southwest Costa Rica looked at eight individuals each of six tree species, representing six common Neotropical plant families (Reed et al. 2008). These tree species are known to vary in canopy leaf content of nitrogen and phosphorus. Previous studies have documented significant interspecific variation in N and P content of live leaves as well as dissolved organic matter (DOM) leached from forest floor leaves. Researchers examined N and P content of live canopy leaves, recently dropped leaves, bulk leaf litter, and forest soil. Rate of nitrogen fixation was also measured. (The method by which researchers obtained canopy leaves is worth recounting. They shot them! Using a 12-gauge shotgun, sunlit leaves were literally shot from the canopy and collected as they dropped.) (Figure 10-11).

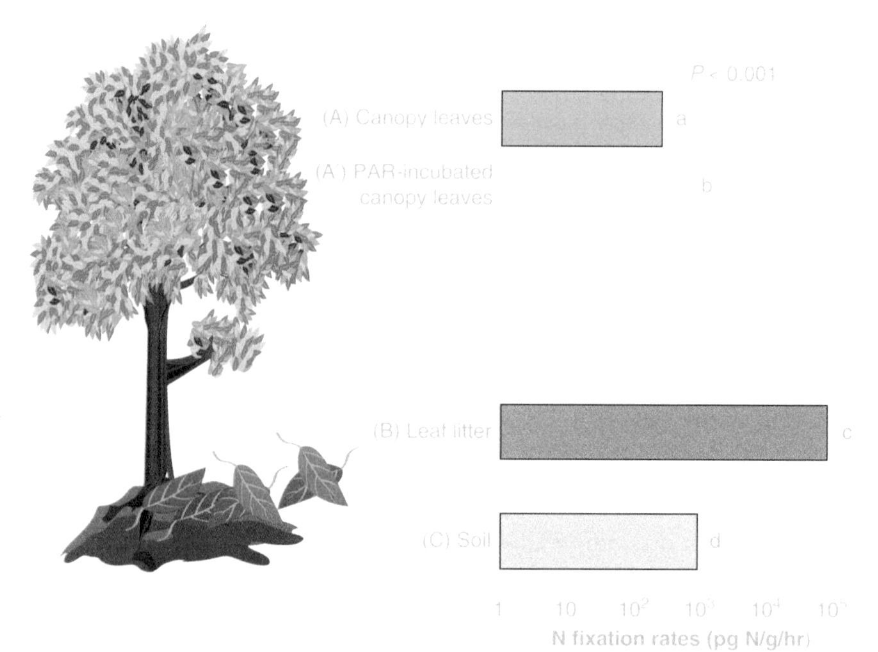

FIGURE 10-11
Rates of free-living N fixation along a rain forest vertical profile: (A) sunlit canopy leaves, (A′) sunlit canopy leaves incubated under elevated photosynthetically active radiation (PAR), (B) mixed-species leaf litter, and (C) topsoil (0 to 2 cm). Nitrogen fixation rates (pg/g/hr) varied along the vertical profile ($P < 0.001$), and means ($n \approx 48$ trees per layer) $\pm$SE are presented on a logarithmic scale (x axis). Significant differences ($P < 0.05$) in N fixation rates among groups are denoted by different lowercase letters.

Rates of free-living N fixation varied significantly along the vertical forest profile. There was also much variability among tree species not only with regard to nitrogen fixation of canopy leaves but also among bulk leaf litter and bulk soil (Figure 10-12). The variation in

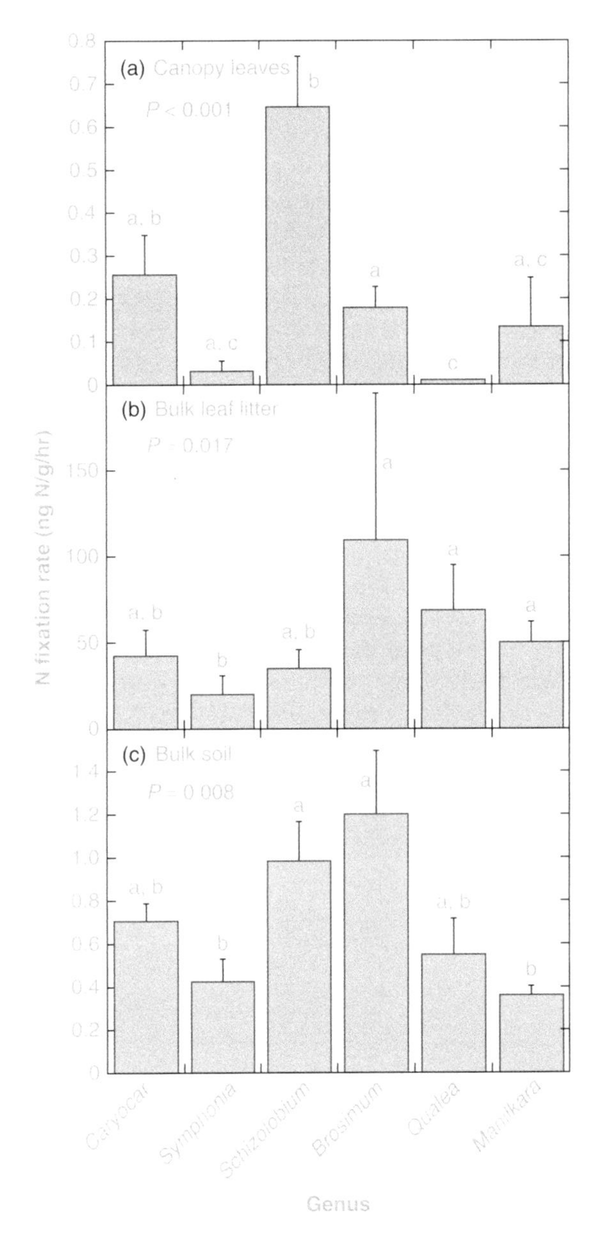

FIGURE 10-12
Comparisons of free-living N fixation rates among six tree species incubated on the forest floor for (a) sunlit canopy leaves, (b) mixed-species bulk leaf litter, and (c) surface soil (0 to 2 cm) collected within the crown radius of trees. Free-living N fixation rates were significantly different among the six tree species within each forest component ($P < 0.05$ for each), and significant differences among species within a component are denoted by different lowercase letters. N fixation rates (ng/g/hr) are means ($n \geq 7$ trees per species for each layer) ±SE. Note that the *y*-axis scale is different for each component.

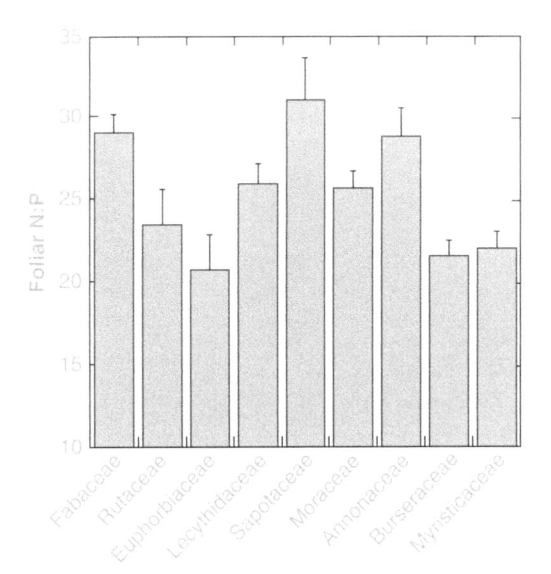

FIGURE 10-13
Variation in N:P among various plant families.

phosphorus was significantly related to the variation in nitrogen fixation rates for canopy leaves.

The study described earlier demonstrated fine-grain differences in nitrogen and phosphorus distribution in tropical forests. These differences are due to variations in rates of nitrogen fixation among tree species and are ultimately related to distribution and recycling of phosphorus. Previous work demonstrated that P fertilization significantly increased nitrogen fixation rates in both leaf litter and soil (Reed et al. 2007). Other work showed a wide variation in the ratio of nitrogen to phosphorus (N:P ratio) among 150 species of canopy trees sampled in Costa Rica and Brazil (Townsend et al. 2007) (Figure 10-13).

Researchers suggested that individual trees in essence create a *sphere of influence* or *canopy footprint* with regard to nutrient concentrations in their immediate vicinity. Because phosphorus availability correlated closely with nitrogen fixation rate, researchers concluded that differences in P concentration regulate N fixation rate (Figure 10-14).

Rates of nitrogen fixation in both soil and canopy vary widely. Some studies, as mentioned earlier, have suggested they are as high as 50 kilograms of nitrogen per hectare per year (about 110 pounds), but other studies suggest 2 to 5 kilograms of nitrogen per hectare per year (about 2.2 to 11 pounds). However, the study described earlier concluded that rates are far lower, averaging 0.035 kilogram of nitrogen per hectare per year (only about 0.08 pound). Reed et al. (2008) noted that estimates of rain forest canopy nitrogen fixation

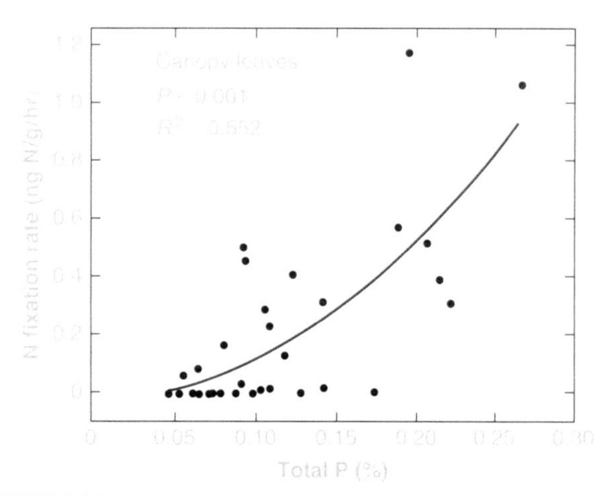

FIGURE 10-14
Total phosphorus as a function of nitrogen fixation rate for canopy leaves.

span 4 orders of magnitude! Seasonality strongly affects nitrogen fixation rate, adding to the complexity of the pattern. In a study done in Costa Rica, soil nitrogen fixation averaged 0.26 kilogram of nitrogen per hectare per year (about 0.06 pound) in dry season and 2.71 kilograms of nitrogen per hectare per year (about 6 pounds) in wet season (Reed et al. 2007).

Though nitrogen is considered not to be a primary limiting factor in forest net primary productivity, wet forests that experience high rates of leaching could conceivably experience high N loss from leaching and *denitrification* (a chemical reduction process by which certain bacteria liberate gaseous nitrogen into the atmosphere). Leaching and denitrification could create N deficits that offset nitrogen fixation. Studies performed in Amazonia demonstrate that secondary successional forests growing on poor soils of abandoned agricultural land quickly develop a tight nitrogen cycle during early succession, an indication that nitrogen is strongly conserved during the process (Davidson et al. 2007). As succession continues, nitrogen is less conserved (because it becomes proportionally more abundant), and the ecosystem shifts to greater conservation of phosphorus as the N : P ratio increases.

The complexity of nitrogen flux and the patterns by which basic elements are differently concentrated in tropical soil and vegetation is essential to understand. Generalities may be of limited value. Variability is apparent throughout the tropics. There is, to quote Townsend et al. (2007), "enormous environmental heterogeneity that exists within the tropics; such variation includes large ranges in soil fertility and climate (p. 107)."

Is It All about Phosphorus?

There is widespread agreement that phosphorus limits its primary productivity on many sites throughout the global tropics, particularly on soils of poor quality. The degree to which this occurs was demonstrated by a study performed at the Cabang Panti Research Station in Gunung Palung National Park, West Kalimantan (Borneo), Indonesia (Paoli et al. 2005). The study focused on the plant *nutrient use efficiency* (NUE) as it varied between rich and poor soils. NUE refers to the amount of carbon fixed per unit of nutrient taken up. To measure NUE, it is necessary to determine foliar nutrient concentration and photosynthetic rate. If phosphorus is a strong limiting factor, one would hypothesize that the efficiency of phosphorus uptake would be proportionally greater on poor soils because phosphorus would be in short supply.

The study in Borneo examined uptake of P over a 16-fold P-gradient in lowland rain forest. This gradient spanned the range of soil fertility from high to low. The results showed that phosphorus use efficiency increased as content of P in the litter declined (Figure 10-15). The ratios of above-ground net primary productivity to available soil P (ANPP/available soil P), uptake rate (P uptake/available soil P), and use (ANPP/P uptake) each increased with declining soil P. All of these ratios were significantly lower on P-rich soils compared with P-poor soils.

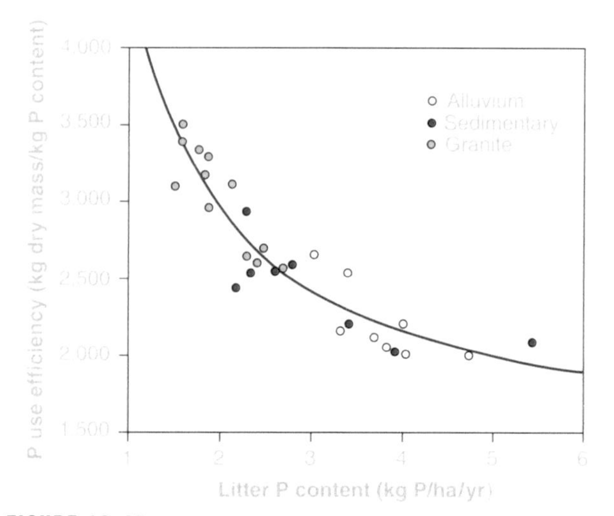

FIGURE 10-15
The P use efficiency of litter production along a P gradient in lowland rain forest at Gunung Palung National Park, Indonesia.

The researchers concluded that the higher nutrient use efficiency on P-poor soils enabled a greater above-ground net primary productivity.

Researchers identified different "strategies" associated with plants on P-rich soils compared with those on P-poor soils. On rich soils where P limitation is weak, trees grew rapidly and did not retain P for long. But on P-poor soils, growth was slower and plants retained P for longer periods. In both cases, the forests appear to thrive. It would not be possible to readily see the difference between the extremes (rich–poor) in the structure of forests, even though the efficiencies of nutrient cycling of phosphorus vary dramatically between them.

Numerous studies have converged in concluding that phosphorus availability is the key variable in litter decomposition in humid tropical forests (Cleveland et al. 2006; Reed et al. 2007; Wieder et al. 2009).

The Leaf Economics Spectrum and Leaf Decomposition

The evolution of leaf characteristics, not only in the tropics but in all terrestrial ecosystems, involves evolutionary trade-offs. A leaf may be adapted to grow quickly, devote prodigious energy to photosynthesis, drop and be replaced, or it may be adapted to grow slowly, acquiring chemical and structural protection, remaining for a long duration on the plant. The latter suite of characteristics would reduce the short-term rate of photosynthesis, "trading off" for longer leaf duration, and thus more long-term photosynthesis. The continuum between the two extremes noted here is the *leaf economics spectrum* (Wright et al. 2004).

A study in Panama examined leaf physiological traits as well as decomposition of leaves for 35 plant species, with the objective of investigating how the leaf economics spectrum related to rates of decomposition (Santiago 2007). The species chosen (trees, lianas, and understory plants) had variable leaf characteristics that placed them at different positions across the leaf economics spectrum. The decomposition rate of leaves (k) varied significantly among species but correlated closely with specific leaf area (SLA), leaf nitrogen (N), leaf phosphorus (P), and leaf potassium (K) in each of the species. Rate of photosynthesis also varied across the leaf economics spectrum, being lowest in those leaves that decomposed the most slowly. The close correlation of decomposition rate with leaf characteristics and leaf photosynthesis connects the decomposition process directly with adaptations of plants across the leaf economics spectrum (ranging from thin leaves with high nutrient concentrations that are easy to decompose, to thick leaves with more defense compounds), essentially extending the spectrum from living plant to decomposer dynamics and nutrient flux.

Neotropical Blackwaters and Whitewaters: The Poor and the Rich

White sandy soils are usually drained by rivers called *blackwaters*, best seen at areas such as the Rio Negro near Manaus, Brazil, or Canaima Falls in southeastern Venezuela. Water appears tea-like, dark and clear, and colored by tannins, phenolics, and related compounds—the humic matter. Blackwaters occur throughout the tropics and temperate zone, including North America, especially in such habitats as boreal peatlands and coniferous forests with mineral-poor, sandy soils (Meyer 1990).

Part of the humic matter in blackwaters consists of defense compounds (see Chapter 8) leached from fallen leaves. Leaves are believed costly to grow on poor soils because raw materials to replace a fallen or injured leaf are in limited supply. Therefore, leaves on plants growing on white sandy soils tend to concentrate defense compounds that help discourage herbivory. Leaf production may be less than half that in forests on richer soils, and leaf decomposition time can be in excess of two years (Jordan 1985a). When the old leaf finally does drop, rainfall and microbial activity eventually leaches the tannins and phenols, making the water dark, a kind of "tropical tannin-rich tea," the defining charateristic of blackwater. In South America, blackwater tributaries drain into the Rio Negro (which means "black river"). This water is dark and clear because there is little unbound sediment to drain into streams and rivers. Gallery forests, called *igapo* forests, bordering blackwater rivers, are subject to seasonal flooding, and their ecology is intimately tied to the flooding cycle.

PLATE 10-16
Thick deposits of sediment characterize this section of the Napo River in Ecuador. Such sediment is rich in nutrients.

Ecological relationships among species inhabiting blackwater forests are different in many ways from those of species in forests situated on richer soils (Janzen 1974).

In contrast, soils in places such as Puerto Rico, many parts of Central America such as much of Costa Rica, and throughout much of the Andes Mountains are not mineral-poor but mineral-rich. These eutrophic soils are much younger, mostly volcanic in origin, some up to 60 million years old, some much more recent. Though exposed to high rainfall and temperature, they can be farmed efficiently and will maintain their fertility if basic soil conservation practices are applied. Because so much sediment leaches by runoff from the land into the river, waters that drain rich soils are typically cloudy and are called *whitewaters*. Please do not let this terminology be confusing. Whitewater rivers do not drain white sandy soils; blackwaters do. Whitewaters drain nutrient- and sediment-rich Andean soils, and the term *white* refers to the cloudy appearance of the water, loaded as it is with sediment (Plates 10-16 and 10-17).

THE WEDDING OF THE WATERS

A dramatic example of the difference between blackwater and whitewater rivers occurs at the confluence of the Amazon River and the Rio Negro near Manaus, Brazil. The clear, dark Rio Negro, a major tributary draining some of the white sandy soils of the ancient Guianan Shield, meets the muddy, whitewater Amazon, rich in nutrient sediment load, draining mostly from the youthful though distant Andes. The result, locally called the "wedding of the waters," is a swirling maelstrom of soupy brown Amazonian water awkwardly mixing with clear black water from the Negro, a process that continues downriver for anywhere from 15 to 25 kilometers (9.3 to 15.3 miles), until the mixing is complete. The most remarkable feature is that both soil types support impressive rain forest—*igapo* in the blackwater areas, *varzea* in the whitewater areas (Plate 10-18).

PLATE 10-17
This photo shows sediment deposit on an island along the Amazon River approaching Manaus, Brazil. The river deposits sediment but also sweeps it away, depending on the flood cycle. This creates dynamic islands within the river.

PLATE 10-18
The "wedding of the waters," when the dark, mocha-colored, sediment-rich Amazon intersects with the clear, dark, sediment-poor Rio Negro. This confluence occurs near Manaus, Brazil.

PLATE 10-19
Parrots of several species at a *collpa* along the Napo River in Ecuador.

Collpas: Why Parrots Eat Dirt

Many species of animals throughout the world (including humans) have been observed to intentionally ingest soil, a behavior known as *geophagy*. Geophagy is common among many bird, mammal, and insect species throughout the tropics. It is also widely practiced by humans in many tropical areas. Several reasons for geophagy have been suggested, and they are not mutually exclusive. Geophagy typically involves ingestion of soils high in clay content. Clay, because of its negative charge, binds potential toxins such as alkaloids and phenolics. It has been thought to prevent diarrhea, help treat intestinal parasites, and supply vital minerals (Brightsmith et al. 2008). In humans geophagy is associated with pregnancy and may be an adaptation to eliminate plant toxins that could cause morning sickness (Plate 10-19).

Parrots throughout the tropical world are known to ingest clay. In western Amazonia, many parrot species (Psittacidae) gather along certain outcroppings of soil, usually along riverbanks, called *collpas*, or clay licks. The birds, which may number well over 100 at any given time in some places, are exposed to predation as they cling to an open embankment, but their repeated presence suggests that they are somehow enhancing their fitness by acquiring some sort of necessary chemical from ingesting the soil. In one clay lick (a 500-meter-long [1,640 feet], 25- to 30-meter-high [about 82 to 98 feet] dirt cliff) in southeastern Peru, a total of 28 bird species was observed at the clay lick, all ingesting soil. Up to 1,700 parrots of 17 species visited daily (Brightsmith et al. 2008). Research has shown that, in comparison with soils away from the clay lick, clay lick soil has the following characteristics:

- Greater percentage of clay
- Higher cation exchange capacity (CEC)
- Greater quinine binding (quinine is an alkaloid used to test for absorption of toxins)
- Greater sodium content
- Lower sand percentage

Use of *collpas* correlated closely with clay percentage, cation exchange capacity, sodium, and quinine binding (Brightsmith et al. 2008). The study showed that sodium was nearly 40 times more concentrated in the *collpa* soil compared with surrounding soils. Parrots feed on plant material, particularly seeds and fruits, vegetation that is poor in sodium content. Clay lick soil had sodium concentrations of greater than 245 parts per million, about six times the sodium available in natural foods used by parrots (Brightsmith et al. 2008).

A subsequent study convincingly demonstrated that sodium was the variable most likely to be the essential reason why parrots seek out specific *collpas* (Powell et al. 2009). A total of 18 *collpas* and 18 control sites was studied in southeastern Peru. At each of these, an analysis was done of soil particle size as well as concentration of 22 metals collected from soil. Soils were compared from used *collpas*, unused areas of the

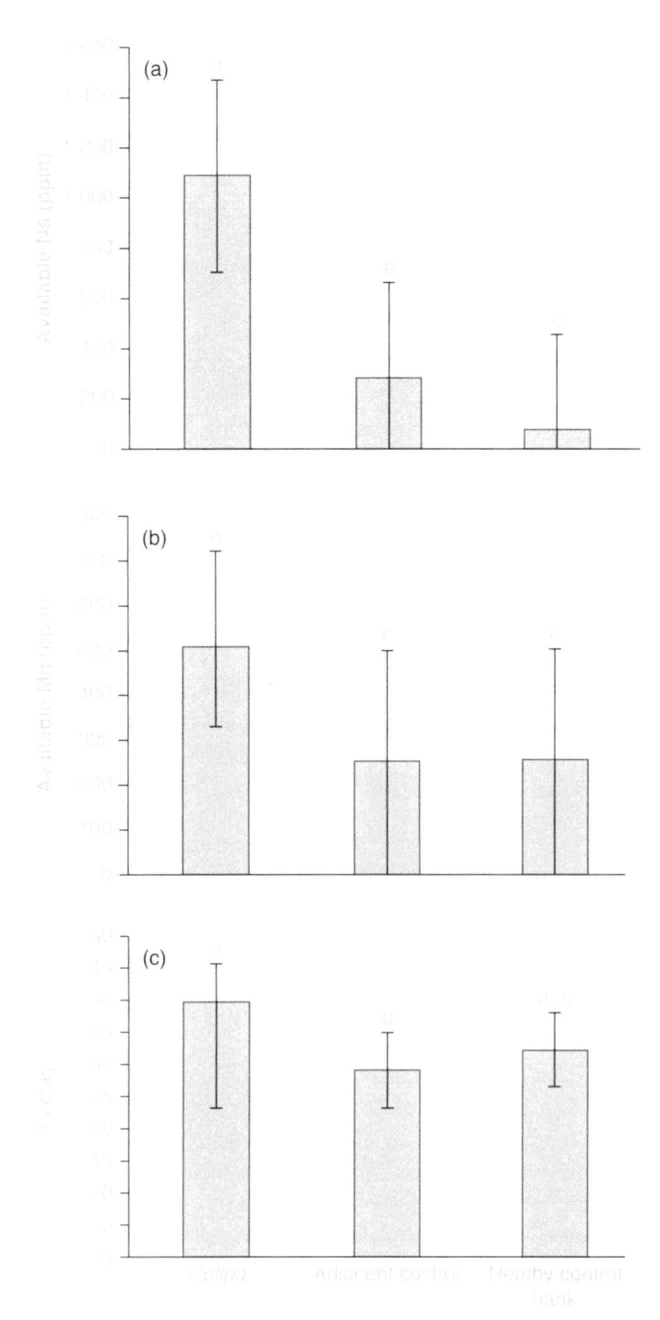

FIGURE 10-16
(a) Mean available sodium concentration, (b) mean available magnesium concentration, and (c) percent clay of soils in three treatments: *collpa,* adjacent controls, and nearby control banks ($n = 18$ for each). Error bars represent mean ±95% CI calculated using error terms in the ANOVA models. Treatment groups with the same lowercase letter indicate that rank-transformed values were not significantly different ($\alpha = 0.05$). Treatment group comparisons shown for percent clay despite marginal overall model significance ($P = 0.075$).

same riverbank, and unused soil from a different riverbank (Figure 10-16).

A total of 30 bird species (19 of which were parrots) was observed at the active clay licks. Of the 22 metals analyzed, only sodium and magnesium occurred at significantly higher levels in clay lick soil than in controls. The mean sodium concentration in *collpas* was 1,137 parts per million, 33 times the mean total concentration in the natural food sources used by parrots in the region. Other variables such as clay content did not correlate with parrot use as well as sodium levels. Magnesium is more readily available than sodium in natural foods, so sodium was concluded to be the essential nutrient sought by parrots. Because sodium is not an essential mineral for plants, it does not accumulate in plant tissues. This makes it more understandable as to why parrots, which are not carnivorous, seek soil, as they will not find sodium in their normal diets.

Some Bats Like *Collpas* Too

Parrots are perhaps the best known users of *collpas,* but many mammals, ranging from various monkeys to peccaries and rodents, visit clay licks and ingest the mineral substance. This mammalian host also includes bats of the family Phyllostomidae, the Neotropical fruit bats (Bravo et al. 2008). In a study performed in southeastern Peru, it was learned that Phyllostomid bats visit clay licks to drink water that has accumulated in depressions caused by the actions of other animals (such as large mammals) using the lick. Bats were sampled by capturing them with fine nets (called *mist nets*). Comparisons were made between bats captured at *collpas* and those captured away from *collpas.* Those bats that visited *collpas* all proved to be fruit bats. A total of 24 species were captured at *collpas.* Bat activity was also much higher at *collpas* than away from them. That fruit bats would seek out clay licks is perhaps not surprising, as other kinds of bats get significant minerals from their more-protein-rich animal food sources. But plants are poor in certain minerals essential to mammals, such as sodium and calcium. Most of the bats (70% of those captured) using the *collpas* were either pregnant or were lactating females. Away from *collpas,* only 44% of the captured bats were female (Figure 10-17).

Collpas appear to be an essential resource to a group of bats that are important seed dispersers of tropical plant species.

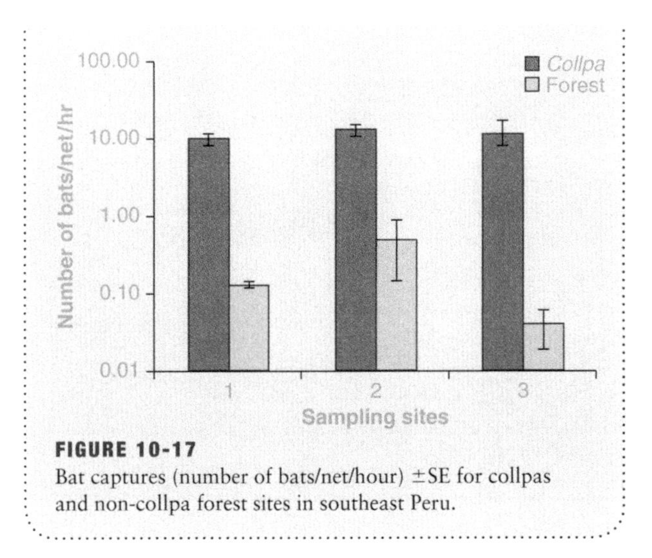

FIGURE 10-17
Bat captures (number of bats/net/hour) ±SE for collpas and non-collpa forest sites in southeast Peru.

Herbivores, Edaphic Variability, and Nutrient Flux

Herbivores, through their selectivity, may exert strong effects on soil-nutrient characteristics in a top-down manner. But herbivore patterns of abundance may be strongly affected by soil characteristics in a bottom-up manner. Several examples document the possibilities and complexity of the relationship between herbivory, soil characteristics, and nutrient flux.

THE INFLUENCE OF SOIL TYPE ON HERBIVORE DIVERSITY

A study done in Sabah, Malaysia, compared insect herbivores from alluvial (riverine) and sandstone soils (Eichhorn et al. 2008). The study focused on seedlings of five species of dipterocarps on both soil types in gaps and understory. The insects represented three families: Coleoptera (beetles), Orthoptera (katydids and allies), and Lepidoptera (butterflies and moths). The diversity of each insect group differed between soil types. Beetles and orthopterans were more diverse on alluvial soils, and lepidopterans were more diverse on sandstone soils.

Each of the three insect taxa had its highest species richness on plants of the alluvial forest, even though lepidopterans had greater diversity (because of higher evenness among species abundances; see Chapter 4) in the sandstone forest. The study shows that small-scale differences in edaphic characteristics may be reflected up the food chain, suggesting a bottom-up influence of soil characteristics on food web composition.

LEAF HERBIVORY AND DECOMPOSABILITY

There is a generality in tropical ecology (and temperate zone ecology as well) that the more strongly a leaf is defended from herbivory by combinations of defense compounds, lignin, and so on, the less quickly it will decompose. A study of 40 tree species in Lambir Hills National Park in Malaysia has cast some doubt on this generality (Kurokawa and Nakashizuka 2008). The correlation between leaf herbivory rate and leaf decomposition rate was quite weak. Those leaves that were least subject to herbivory in many cases tended to decompose equally rapidly as those much more subject to herbivory. In other words, leaves that appear well-defended did not necessarily decompose more slowly than those that appeared poorly defended. The one strong correlation that was noted was between herbivory and decomposition rates for continuously leafing species (Figure 10-18).

Plants that had synchronous leafing show no such correlation between herbivory and decomposition rate. In addition, the overall rate of herbivory was not significantly different in plants with continuous leafing (0.018% per day) and plants with synchronous leafing (0.009% per day). The researchers suggested that "continuous-leafing species may be more likely to have some condensed tannins that are effective in deterring herbivores, particularly generalists, because the plants do not escape temporally" (p. 2654), as tends to be the case with synchronous-leafing species.

Litter decomposition rate, apart from correlation (or lack of it) with herbivory, correlated (as expected) with traits such as lignin and nitrogen concentrations, ratios of carbon to tannins, and leaf mass per unit area (LMA). Researchers concluded that plant-decomposition interactions are much less specific than interactions with herbivores and, more importantly, have little if any effect on plant fitness (Figure 10-19). This is perhaps not surprising given that once a leaf is shed, it has no way of contributing to the fitness of the plant.

HOWLER MONKEYS ALTER PLANT SPECIES COMPOSITION, GROWTH RATE, AND LITTER CHARACTERISTICS

In a remarkable situation, described as a *found experiment*, the presence of a single herbivore species, the red howler monkey (*Alouatta seniculus*), was shown to alter

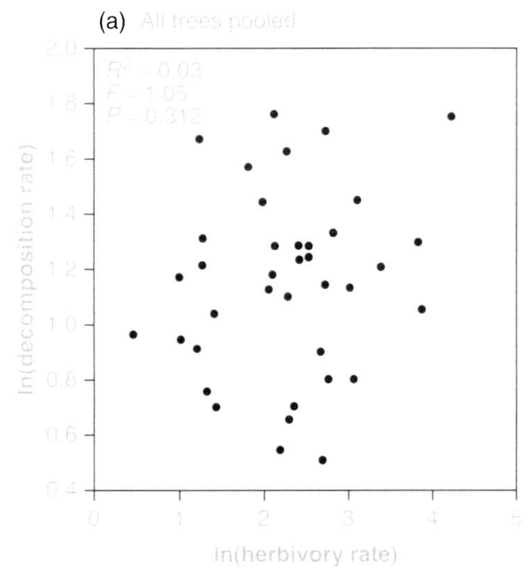

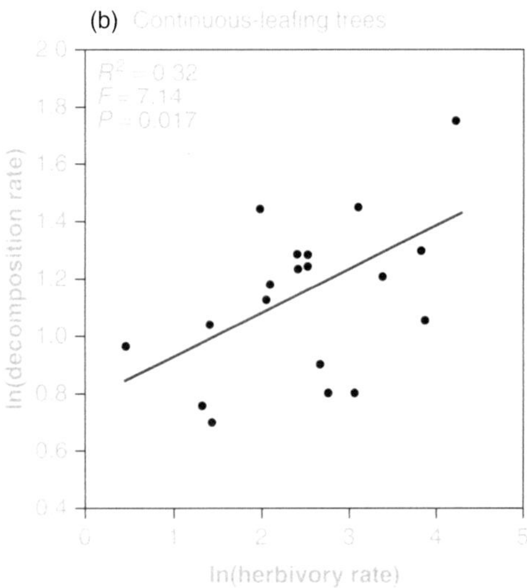

FIGURE 10-18
Comparison between all trees pooled (a) and continuous-leafing trees (b) with regard to the relationship between leaf decomposition rate and herbivory rate. Only continuously-leafing trees showed a strong correlation between the variables.

plant composition, productivity, and nutrient composition of the litter (Feeley and Terborgh 2005). Red howlers were exerting a strong top-down influence that was rapidly changing many characteristics of the forest. The study was performed at Lago Guri, a large hydroelectric reservoir in east-central Venezuela. When Lago Guri was formed in 1986, it caused the inundation of 4,300 square kilometers (1,660 square miles) of hilly terrain, resulting in the conversion of contiguous forest into hundreds of isolated hilltop islands. The islands ranged in size from

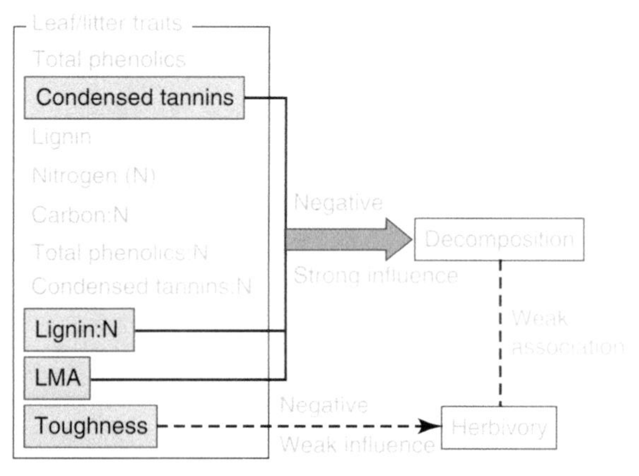

FIGURE 10-19
Relationships between herbivory and decomposition rates across tree species, and leaf and litter traits related to these rates in a Malaysian tropical rain forest. Solid, wide arrows indicate a strong influence, dashed arrows indicate a weak influence, and dashed lines indicate a weak association.

less than 0.1 hectare (about 0.25 acre) to greater than 700 hectares (about 1,730 acres). Some were separated from the mainland by as much as 10 kilometers (about 6 miles) of open water. Red howler monkeys, like other animals, were trapped on the newly formed islands. Some islands ended up with no monkeys, but on those islands where monkeys took refuge, their densities have increased to over 30 times that in normal mainland forest.

Researchers visited islands with and without red howler monkeys. They examined soil samples, measured total nitrogen and carbon concentrations (C:N ratio), measured relative availability of N and P for plant uptake, and measured nutrient concentrations in litter. They measured fine root biomass and calculated annual woody increment (AWI). The results were striking:

- Nitrogen concentration in the litter was negatively correlated with monkey density.
- Soil C:N ratio increased with monkey density.
- Annual woody increment increased with increasing monkey density.
- Fine root biomass was lowest on islands with high AWI.

It would appear that the monkey presence correlated with increasing AWI and decreasing soil fertility. The decrease in soil fertility was indicated by the changing C:N ratio. The less nitrogen there is relative to carbon,

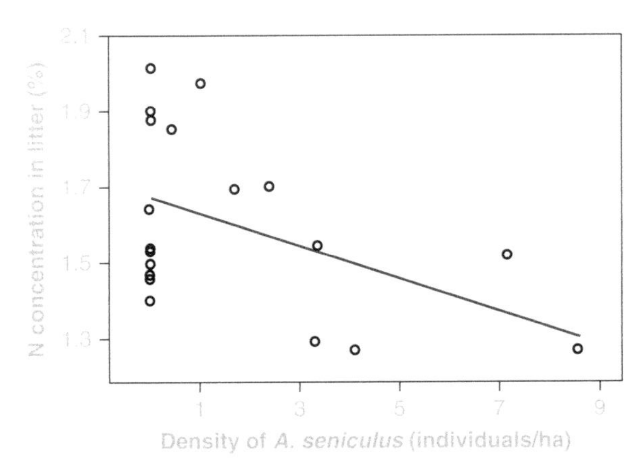

FIGURE 10-20
The concentration of nitrogen in leaf litter (in 2001) declines with increased density of *Alouatta seniculus* ($r^2 = 0.23, P < 0.05$).

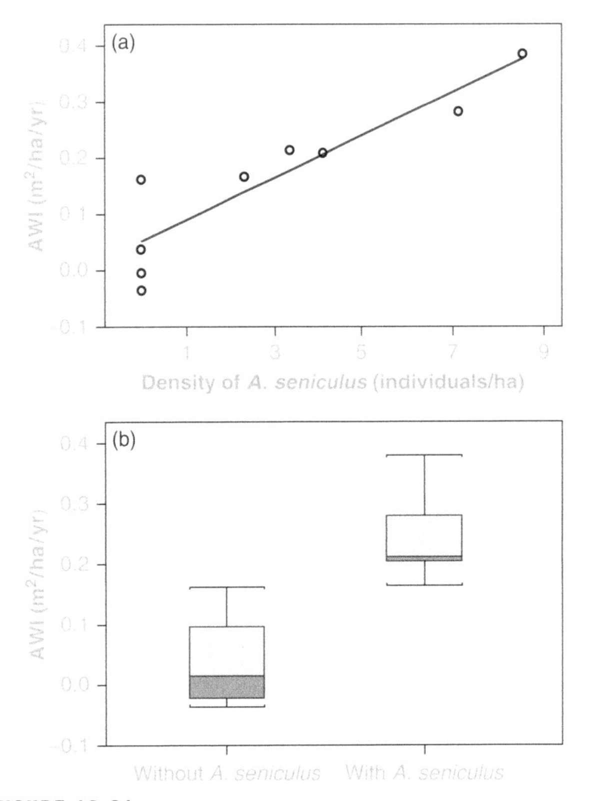

FIGURE 10-21
(a) The annual woody increment (AWI) of trees on the study islands increases with increased herbivore (*Alouatta seniculus*) density ($r^2 = 0.82, P = 0.001$). (b) AWI is, on average, 6.1 times greater on islands with *A. seniculus* than on islands without *A. seniculus* (*t* test, $P < 0.01$).

the poorer the soil. But why, on islands with monkeys, did AWI increase (Figures 10-20 and 10-21)?

Howler monkeys are folivores and are very selective about what sorts of leaves they eat (see Chapter 8). Researchers quickly learned that islands with howlers had significantly different tree communities with high abundances of tree species that monkeys avoid. For example, the tree species *Ocotea glomerata* was common on islands with monkeys. This plant (of the laurel family) is known to have powerful defense compounds and is rarely consumed by howler monkeys. The greater AWI typical of islands with monkeys was likely due to growth of species not consumed by monkeys, released from any competition with species that monkeys selectively dined upon. The loss of litter fertility (as shown in the greater C:N ratio) was likely due to slower rates of litter decomposition and nutrient mineralization due to the defense compounds of the plants. The low growth of fine root biomass on islands with monkeys was possibly the result of reallocation of net primary productivity to leaf and stem growth reflected in greater AWI.

The change in plant species composition on islands with monkeys occurred in only 17 years. Although litter and soil fertility declined, the increase in AWI of plants on monkey islands may reflect the greater nutrient use efficiencies typical of plant species adapted to low nutrient levels. The species composition of plants shifted dramatically, and the likely cause is the presence of the monkeys and their selective use of trees. The islands with monkeys are clearly not in equilibrium, as nutrient availability continues to decline. Ultimately, such a trend will likely result in reduced primary productivity.

11 Tropical Savannas and Dry Forests

Chapter Overview

Tropical savannas are ecosystems that range from mostly grasses to a mixture of grass and scattered trees. Where trees are dense, the ecosystem is dry forest. These widespread ecosystems are found in all regions of the tropics but are unique in Africa because of the presence of a diverse megafauna, large *ungulate* (hoofed) animals and their associated predators. Savannas are ecologically different from moist forests because seasonal effects and fire frequency are greater, and where there is a megafauna, grazing is an important influence. Savannas typically experience strong dry seasons and are subject to periodic droughts. Fire is an essential factor in savanna ecology, acting to maintain savanna against invasion by bordering forest. Savannas tend to interface with tropical dry forest. Many tree species, particularly acacias, are adapted to endure harsh dry seasons, and dry thorny woodlands are common in many tropical regions. Wherever savannas occur, there is a *zone of tension* between grassland-dominated savanna and woody dry forest. Various factors ranging from soil characteristics to fire intensity and grazing pressure strongly affect whether an area will be savanna or dry forest. This chapter will review the various factors that characterize savannas and dry forests. A major focus will be on African savannas, as they are widely studied and their ecology includes effects of large animals that are now absent from other savanna regions in the world.

What Is a Savanna?

Perhaps most people think of rain forest when they think of the tropics, but there are other terrestrial ecosystems in the tropics that are extensive and unique. Most prominent among these are extensive grasslands that grade into savannas (sometimes spelled "savannah") and dry forests. Tropical savanna is an area of grassland in which there is a rich scattering of trees. Ecologically, savanna forms part of a gradient from grassland to dry forest. Savanna ecosystems in temperate and tropical regions are estimated to cover approximately one-fifth of the land surface of Earth, including extensive areas within the tropics (Sankaran et al. 2005). J. S. Beard (1953) characterized savanna as having the following properties: it is a natural and stable tropical ecosystem; it typically contains a diversity of grasses and sedges that are adapted to tolerate hot and periodically dry tropical climate (termed *xeromorphic plants* because of their adaptations to withstand heat stress); and it usually has a discontinuous distribution of low trees and shrubs. Savanna plant species have both structural and physiological adaptations that enable them to survive high temperatures and periodic drought. For example, many savanna species, especially grasses, are C_4 plant species (see the sidebar "C_3 and C_4 Plants: Heat and Moisture Trade-offs"). In the Serengeti grasslands of East Africa, 90% of the total biomass of grasses consists of C_4 grass species (McNaughton 1983). Savannas appear stable, but the apparent stability is not representative of an equilibrium condition. Savannas undergo constant changes in amount of grass cover and woody vegetation. Such changes are related to effects of (1) varying rainfall amounts on soil, (2) grazing intensity, and (3) fire frequency, as will be discussed in the following section. There is a gradient evident: grasslands grading into savannas and savannas grading into dry forests. This gradient may shift direction from one end to the other. Where trees form a dominant component of the vegetation, the ecosystem is dry forest. In a survey conducted of 854 savanna sites across Africa, it was demonstrated that the amount of woody cover (relative to grass cover) is constrained by mean annual precipitation, providing it is less that 650 millimeters (about 26 inches). Such savannas are relatively stable. But above 650 millimeters mean annual precipitation, savannas become unstable, as there is sufficient precipitation for woody cover to proliferate, converting the ecosystem to dry forest (Sankaran et al. 2005). However, such an outcome tends to be prevented by periodic fire and herbivory, as will be discussed later.

Savannas, while adapted to dry conditions, may nonetheless experience periods of intense seasonal moisture. One familiar example is the Florida Everglades, a wet seasonal subtropical savanna in southern Florida.

C_3 and C_4 Plants: Heat and Moisture Trade-offs

The influence of temperature and precipitation acts as a significant selection pressure on the process of photosynthesis. Beginning in the late Mesozoic era and continuing during much of the Cenozoic (from the Oligocene epoch onward), world climate became more temperate, with many regions experiencing regular dry seasons of high heat and low rainfall. Carbon dioxide levels in the atmosphere also declined. Such a global change caused tropical forests to shrink, favoring the spread of numerous grass species. Many grasses, though by no means all, as well as other plants of hot, dry regions exhibit a distinctive form of photosynthesis.

Photosynthesis occurs by complex pathways of linked biochemical reactions. Its details are beyond the scope of this book. What is essential to understand is that CO_2 enters plants through leaves (though some plants, such as cacti, photosynthesize using stems), specifically through openings called *stomata*. Stomata can be kept open or shut. When open, they admit CO_2 that diffuses into the leaf from the atmosphere. (Recall that *diffusion* is a passive process, like the spread of perfume from an open bottle placed in a room. Energy from the random movement of the molecules is all that is required to make it happen, and the direction of diffusion is always from a region of high to low concentration.) Stomata represent an adaptation to plant life on land. They are not present on algae nor are they on flowering plants that live underwater, such as eelgrass.

Evolutionists presume that stomata evolved when Earth's early terrestrial ecosystems were much wetter. This is because stomata not only admit carbon dioxide, but they also release moisture from the leaf in the process of *transpiration*. In a warm, wet environment, having CO_2 enter and H_2O leave by the same "doorway" is not problematic. Transpiration cools the plant and CO_2 ultimately nourishes the plant, an efficient system, provided that there is plenty of water. And, in fact, the majority of the world's 300,000 species of flowering plants continue to utilize a cyclic biochemical system called *C_3 photosynthesis* because the principal compound, 3-phosphoglycerate, is a 3-carbon compound. C_3 photosynthesis is an ancient process, dating back 2.8 billion years (Edwards et al. 2010).

But consider that about 30 million years ago, climate became drier and warmer and there was a drop in atmospheric carbon dioxide to levels of between 350 to 550 parts per million (Edwards et al. 2010). Plants must keep stomata open to admit CO_2. Further, they have no way of forcing CO_2 in—no pump—and so they must rely on diffusion alone. But if their stomata are open, they lose water, which could desiccate the plant, even kill it, particularly in a dry, hot environment. Therefore, as carbon dioxide levels declined,

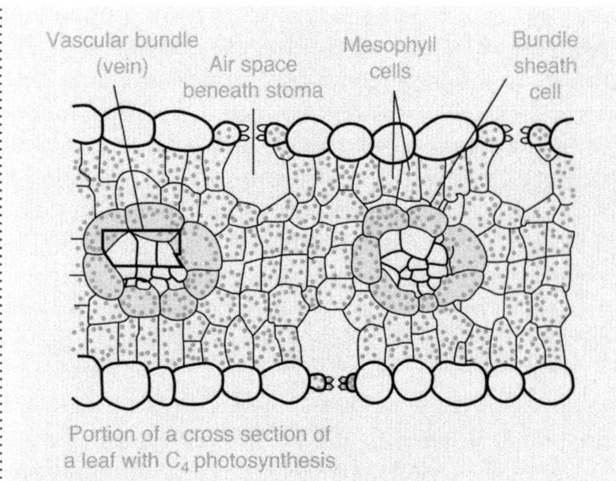

FIGURE 11-1
This is a cross section of a leaf that uses C_4 photosynthesis.

plants were more subject to desiccation because stomata would have had to remain open sufficiently long to admit adequate levels of CO_2. It is here that an important lesson in evolutionary biology can be learned: evolution cannot "go back." It does not re-evolve a system that already exits. Instead, it adapts, usually by altering or adding to what already exists. It is rather like what happens in human experience. High volumes of traffic, necessitating the addition of a superhighway, may render an old two-way road outmoded. The old road, still lined with houses, businesses, churches, libraries, and so on remains within the shadow of the new highway, which has ramps that provide access to the old "frontage" road for those who need it. The new highway is, in effect, an "add-on." So it is with much of photosynthesis.

Many plants, especially grasses, have evolved an additional component of photosynthesis, called the *C_4 pathway*. Today, about 60% of all grass species are C_4 plants, and they are found in warm-climate grasslands and savannas. Corn (*Zea mays*) is a C_4 plant. C_4 photosynthesis is not confined to grasses and has also evolved in more than 45 plant lineages (Edwards et al. 2010) (Figure 11-1). As you might guess, the critical compounds in the C_4 pathway are 4-carbon compounds, in this case malate and aspartate. The anatomy of the plant leaf is also modified. In both C_3 and C_4 plants, the cells involved with photosynthesis are called *mesophyll cells*. In C_3 plants, they surround the vascular bundles that act as "blood vessels" to transport water as well as the products of photosynthesis. But in C_4 plants, there is an additional group of cells, called *bundle sheath cells*, that surrounds the vascular bundles. In C_4 photosynthesis, there is an enzyme called *phosphoenolpyruvate* (PEP) in the mesophyll cells that catalyzes the formation of malate and aspartate. These compounds are then moved to the bundle sheath cells. There, a seemingly odd thing happens. Malate and aspartate are converted back to CO_2! Once this happens, another enzyme, called *RuBisCO* (whose full chemical name is ribulose-1,5-bisphosphate carboxylase oxygenase), completes the process of reducing CO_2 to sugar in exactly the same way as occurs in C_3 plants. In fact, RuBisCO is the critical enzyme in all C_3 plants and, by virtue of the abundance of plants in general, is the most abundant enzyme on Earth. Note that C_4 photosynthesis is essentially an add-on to C_3 photosynthesis. But why?

By sequestering CO_2 in the form of malate and aspartate in the bundle sheath cells, the concentration of CO_2 in these cells can rise to six times greater than in the air, an extremely important consideration when atmospheric CO_2 is declining, as occurred about 30 million years ago. Because they can concentrate CO_2 in this manner, C_4 plants can accomplish a higher rate of photosynthesis than C_3 plants. Perhaps even more importantly, C_4 plants have greater efficiency of water use. Relative to the time they leave their stomata open, they fix much more carbon than C_3 plants. This is a huge advantage in a hot, dry environment where water loss could mean loss of life. C_4 plants can "afford" to keep their stomata closed during high heat, thus avoiding desiccation.

There is a cost to C_4 photosynthesis, however. Since it requires PEP in addition to RuBisCO, it is metabolically expensive. However, this added expense is offset by the advantage gained by attaining greater rates of carbon fixation and more efficient water use, but only when water is in limited supply. The evolutionary "superhighway" of C_4 photosynthesis, constructed using the resources devoted to PEP, gives grasses and other C_4 plants an advantage in dry conditions that likely contributed to their rise and abundance throughout the Cenozoic era.

It's fairly hot and dry on the African savanna during the dry season, but what about in a desert? Dry, oppressively hot deserts pose an immense challenge to plants. How do they cope? Some families of widely distributed plants including orchids, bromeliads, and numerous others have evolved yet another unique kind of photosynthesis: *crassulacean acid metabolism* (CAM), named for the Crassulaceae, or stonecrops, one of the kinds of plants that uses CAM.

The key to CAM success is that these plants, which also include members of the cactus and euphorbia families, close their stomata during the day and keep the stomata open only at night. Doing so reduces water stress immensely, as it is both cool and humid at night. CO_2 is converted to 4-carbon malate and stored until daylight. During the day, malate is reconverted to CO_2, and photosynthesis is effected by the usual enzyme, RuBisCO. In CAM plants, there are no bundle sheath cells, because the biochemistry of photosynthesis occurs entirely in mesophyll cells.

Note that cacti and euphorbias are examples of convergent evolution. Though not closely related genetically, they have evolved to be anatomically similar. Their use of CAM is also likely an example of convergence at the metabolic level (Plates 11-1a and 11-b).

(a)

(b)

PLATE 11-1
Cactus such as this tree-sized Opuntia from the Galápagos Islands (a) and Old World euphorbias (b) are examples of convergent evolution.

Savanna Global Distribution

The continent of Africa contains expansive savanna forming a broad east–west belt across the central equatorial part of the continent and also forming the dominant biome in much of the southern part of the continent. All together, 27 African countries have large expanses of savanna. This includes the extensive and famous Serengeti ecosystem east of Lake Victoria covering parts of Tanzania and Kenya, one of the most intensively studied savanna ecosystems. African savannas occur between 15°N latitude and 30°S latitude. Dry forests, many dominated by various acacia species, typically border savanna regions. Savanna also occurs extensively in parts of Madagascar.

The African savanna is unique among savanna regions because it still has an intact megafauna typical of the large animal diversity that existed in many regions throughout much of the Pleistocene epoch. Large mammals such as antelopes, gazelles, giraffes, wildebeests, zebras, elephants, rhinoceroses, and others remain numerous in protected areas and are preyed upon by a diversity of predatory species such as lions, cheetahs, leopards, wild dogs, and crocodiles. Savanna areas elsewhere on Earth have lost their Pleistocene megafauna, perhaps due to pressures exerted by human hunters (Martin and Klein 1984; Diamond 2005), though other factors may have also contributed (Plate 11-2).

PLATE 11-2
Large and diverse mammals, the *megafauna,* still characterize the African savanna.

African grasslands and savannas range widely in plant species richness. Some African savannas are essentially monodominant with regard to tree species richness, where one species of *Acacia* predominates. For example, a study at Laikipia, Kenya, showed that the whistling-thorn acacia (*Acacia drepanolobium*) comprises 97% of the woody cover (Riginos and Grace 2008). Grass species richness is typically higher than tree species richness in African savannas, but it is also common for a single grass species to be numerically dominant in African grasslands. Grassland communities vary considerably in species composition and growth pattern. Some comprise short grasses, some (where moisture is greater), tall grasses. There is significant local heterogeneity in the composition of African savanna and grassland plant communities. In the Serengeti alone, a total of 16 different grassland communities is recognized (McNaughton 1983).

In South America, combinations of grassland, savanna, and dry forest are estimated to occupy about 250 million hectares, principally in Brazil (*Cerrado, Caatinga, Campo Rupestre*, and *Pantanal*), Colombia (*Llanos*), and Venezuela (*Llanos*). Large tracts are also found in eastern Bolivia (*Pantanal*), and in northern Argentina and Paraguay (*Chaco*) (Fisher et al. 1994). Broad expanses of savanna and dry forest also occur in Central America, particularly in the northern Yucatan Peninsula and in parts of Belize, Honduras, and

Baobab Tree

The baobab tree (*Adansomia digitata*, family Malvaceae) is one of the most distinctive trees in the world (Plate 11-3). Its uniquely thick trunk and spreading branch arrangement (often bare of leaves) make it appear as an "upside-down" tree, its roots where its crown ought to be. Baobabs occur throughout African savannas as well as Madagascar. There are several species, but *Adansomia digitata* is the most widely distributed. Only one species is found outside Africa and Madagascar, in the dry savanna of the Northern Territory of Australia. Baobabs are most common in savannas, but in some areas in Africa, they are part of the vegetation of dry forests. Baobabs thrive in dry savanna climates because they retain a great deal of water in the thick trunk. They are deciduous, and leafless during dry periods. Carbon dates suggest that baobabs may attain ages of 1,000 to 2,000 years, some even greater. The trunk is usually about 10 meters in diameter, but some trees attain a diameter of about 14 meters. The largest baobab measures 47 meters (about 150 feet) in circumference with a 15-meter (50 feet) diameter. The water storage capacity of the tree enables it to endure through severe dry seasons and protracted droughts. Flowers are large, white, and pendulous, and they emit a carrion-like odor, all characteristics typical of bat-pollinated flowers (see Chapter 7). Principal pollinators are various fruit bats in the family Pteropodinae.

PLATE 11-3
This baobab grows on Madagascar, but it is very similar to those that grow throughout African savannas.

Indigenous people utilize baobabs for many things. Leaves and bark are considered useful in combating maladies such as dysentery, inflammation, and even smallpox. Seeds are used for oil or eaten after cooking.

PLATE 11-4
Caribbean pine grows extensively in savannas throughout much of Central America and the West Indian islands. This stand is part of the savanna ecosystem on Abaco Island in the Bahamas.

PLATE 11-5
Spinifex grasses characterize dry forest in much of Australia.

Nicaragua. Savanna is common in many Caribbean areas too, particularly in the Bahamas and West Indies.

Numerous tree species populate savannas throughout Central America, the Caribbean Islands, and equatorial South America, including acacias, palmettos, palms, cecropias, and others, depending on location. Local plant species composition varies widely. Savanna plant species diversity is higher in South America compared with Central America and much of Africa. In much of Central America and the Caribbean, the most abundant savanna tree species is Caribbean pine *(Pinus caribaea)*, often adorned with bromeliads and orchids (Plate 11-4). Several species of oaks are common in Central American savannas, though no oaks are found in South American savannas. Fire-resistant tree species such as *Byrsonimia crassifolia*, *Casearia sylvestris*, and *Curatella americana* are abundant on South American savannas, as are large stands of moriche palm (*Mauritia flexuosa*). Grasses and (in wetter areas) sedges form much of the ground vegetation. Soil ranges widely from sandy to claylike, typically being described as *poor soil*.

Australia has vast savanna in its northern regions, particularly the Northern Territory. Australian savanna has a unique flora and fauna. As with most of the continent, savanna is dominated by various eucalyptus species, though some acacias also occur. Coarse, tall bunchgrasses such as *Spinifex* and *Triodia* also characterize much of Australian savannas (Plate 11-5). Cycads are often in the understory. The herbivore fauna consists mostly of marsupial animals such as wallabies and kangaroos (Plate 11-6).

Savanna also occurs throughout large regions of India and Nepal as well as a few parts of Southeast Asia, particularly in central Thailand and parts of Myanmar (Burma). In India, the most noteworthy savanna ecosystem is Terai-Duar. This area is a complex mosaic of lush, tall, and diverse grassland and wetland (supported by monsoon rains) and hosts a rich diversity of plants and large animals, including Indian rhinoceros (*Rhinoceros unicornis*), Indian elephant, five species of deer, and tiger (*Panthera tigris*).

PLATE 11-6
The wallaby (with a "joey" in her pouch) is one of many marsupial species extant in Australia today, but many larger species from the Pleistocene are now extinct.

The (Now Extinct) Megafauna Down Under

A drive across the Australian outback today will likely reveal several species of kangaroos and wallabies, flocks of emus, as well as various colorful parrots. But the large animals that occupy the wilds of Australia today are far fewer in species richness and generally quite a bit smaller than many species that lived there throughout much of the Cenozoic era, particularly in the Pleistocene. Australia, like North America, Europe, and South America, had a *megafauna*, an impressive assemblage of large animals that exerted strong ecological influences, just as creatures such as elephants and lions do in Africa today.

For example, one of the top carnivores of Australia during the Pleistocene was a varanid lizard named *Megalania prisca*. A close relative of the well-known Komodo dragon lizard, *Megalania* was substantially larger, estimated to be as long as 7 meters (about 23 feet) and weigh between 600 and 620 kilograms (about 1,300 pounds). A modern Komodo dragon would be dinner for such a creature. There was also a rather impressive serpent, *Wonambi narcoortensis*, a python that reached an estimated length comparable to the largest pythons and boas present today. The emu-like flightless bird *Genyornis newtoni* was about 2.2 meters tall (7.25 feet) and weighed up to 250 kilograms (about 550 pounds). Though superficially like an emu, *Genyornis* had a much more robust head with a more formidable bill and may have preyed on a diversity of animals. Many diverse marsupial species were also present in the Pleistocene, some much larger than any extant species. One was *Thylacoleo carnifex*, a marsupial lion. Another was *Diprotodon*, sometimes called a *giant wombat*, a unique animal somewhat similar in its ecology to a tapir or an elephant. It is considered to be the largest marsupial that ever lived. The animal reached a length of about 3 meters (about 10 feet), stood about 2 meters high at the shoulder (about 6 feet), and is estimated to have weighed up to 2,700 kilograms (about 6,000 pounds). There was even an animal called a *giant rat kangaroo* (*Propleopus oscillans*), which some paleontologists believe may have been a carnivorous kangaroo. (Extant kangaroos are all herbivorous.) For more on this amazing megafauna, none of which remain, see Rich et al. 1985; Vickers-Rich and Rich 1999.

What Factors Cause Savanna Formation?

Sarmiento and Monasterio (1975) summarized savanna causation, concluding that savannas exist as three basic savanna types:

- The *nonseasonal savanna* is largely the result of poor soils. It is the savanna of white sandy soils, where drainage is rapid and climate is wet most of the year.
- The *seasonal savanna* is the most widespread type in the tropics. It occurs on sites with a stressful dry season, where soils are sandy and nutrient-poor. Fire is an important component of these savannas.
- The *hyperseasonal savanna* is characterized by an annual period of water deficiency plus a period of saturation. In other words, there is either too much or too little water in the soil, making soils nutrient-poor and tending to waterlog. These savannas are typically all grass, with very few trees and shrubs.

There is no single environmental factor that determines that a given site will be savanna (Bouliere and Hadley 1970; Huber 1982 and 1987; Sarmiento 1983). Sarmiento and Monasterio (1975) note that savannas are somewhat paradoxical in nature because it has been very difficult for ecologists to identify exact causal factors leading to the formation of savannas. The existence of savannas once perplexed ecologists because it resists a simple encompassing explanation. Savannas occur on a wide variety of soil types and experience all extremes of tropical climate. Rainfall, while usually strongly seasonal, may be relatively nonseasonal, and water drainage may be rapid or slow (Huber 1987). Fire is an important influence, and savannas tolerate fire well, rebounding quickly after burning. Grazing by large animals is also typical of many savannas (particularly in Africa), and its influence is substantial. But the answer to what factors cause savanna formation resembles that on some multiple-choice exams: all of the above.

Tropical savannas occur at low elevations, less than 1,200 meters (3,960 feet) above sea level. Holdridge, in his life zone triangle (see Chapter 1), does not recognize savannas as specific tropical and subtropical life zones. Under the Holdridge schema, savanna falls within the broad range of tropical desert scrub, thorn woodland, very dry forest, and dry forest. These life zones are in moisture regimes ranging from superarid

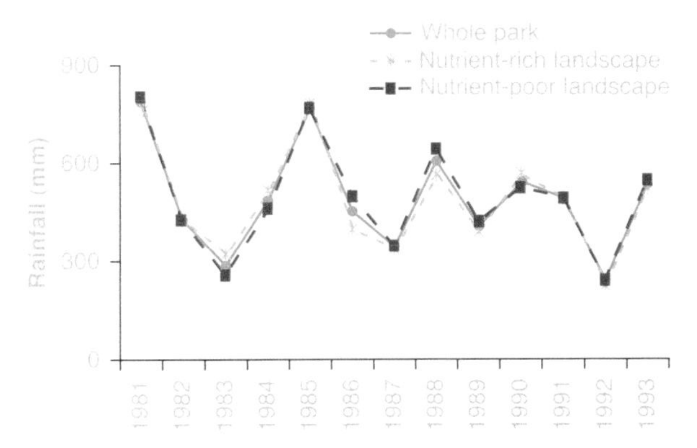

FIGURE 11-2
Average annual rainfall, calculated over the entire KNP and in the nutrient-rich versus nutrient-poor landscapes, plotted for the study period. Annual rainfall, calculated for the whole KNP, was 623 ± 124 millimeters (mean ± SD) in the six wettest years (1981, 1985, 1988, 1990, 1991, 1993) and 361 ± 86 millimeters in the six driest years (1982, 1983, 1986, 1987, 1989, 1992).

(desert scrub) to semiarid (dry forests), and they illustrate the gradient nature of savannas. Precipitation ranges from as low as 50 centimeters (about 20 inches) annually (thorn scrub) to as high as 250 centimeters (about 100 inches) or more (tropical seasonal savanna), an amount that also supports rain forest in many areas. What is important about savannas and related ecosystems is that the precipitation is usually highly seasonal, often with severe dry seasons lasting several months, followed by ample rainy seasons. Beyond that, annual precipitation is often highly variable, with "wet years" and "dry years." This is the case, for example, in the Serengeti ecosystem of East Africa and Kruger National Park (KNP) in South Africa (Figure 11-2).

Climate has a major influence on savanna formation, but it cannot be the only influence, because tracts of savanna frequently occur in the midst of otherwise wet forest areas, where rainfall is evenly distributed throughout the year. For this reason, local soil type (edaphic factors) as well as other factors must also influence savanna formation. Soil and climate strongly interact. In the central *Llanos* in northeastern South America, heavy rains result in soil forming a hardened crust of lateritic ferric hydroxide, usually at some depth in the soil, but occasionally on the surface. This crust, termed *Arecife*, is sufficiently hard to impede the growth of tree roots, except where the woody species encounter channels through the crust. Tree groves, or *matas*, occur only where roots have penetrated Arecife, resulting in the clustering of trees. Grass, with shallower root systems, usually thrives above the level of Arecife (Walter 1971 and 1973). In the African Serengeti grassland savanna, areas of lower rainfall have resulted in development of soil termed *Aridisol*. Calcium carbonate has been leached from the upper soil and deposited at a depth of about a meter, forming a hardpan termed *caliche*. The caliche pan retards water movement and keeps the pH well within the alkaline range (pH > 8). A hardpan also inhibits tree growth except along streambeds in the Serengeti Plains (Sinclair and Norton-Griffiths 1979). This is apparently the main reason that this area is covered with grassland rather than woodland. Where rainfall is higher in the Serengeti, soils weather more, leaching is greater, and the pH declines, forming *Vertisols*. Different grasses, including many primarily adapted to savannas, are adapted to grow on Aridisol and Vertisol (Aber and Melillo 2001).

Fire, normally set by lightning, occurs frequently on savannas and is of major importance in both savanna formation and propagation. Natural fires are most common during dry season. Some savannas have formed on sites where rain forest has been repeatedly cut and burned, suggesting that human activity can alter conditions on the site such that savanna takes over when the site is abandoned.

In the following sections, I review each of the major factors determined to be important influences in savanna formation and maintenance.

CLIMATE

Most areas of classic savanna (grassland with a rich scattering of trees) typically experience a prolonged and annually variable dry season. One prominent theory accounting for savanna formation is that wet forest species are physiologically unable to withstand the dry season, and thus savanna, rather than rain forest, is favored on the site. Savannas and moist forests share very few species, and moist forests are generally far more species-rich in comparison with savannas. Savannas typically experience an annual rainfall of between 500 and 2,000 millimeters (about 20 to 80 inches), most of it falling in a five- to eight-month wet season. Rainfall tends to be variable from one year to another, and periodic droughts occur, some of which are severe. Though annual precipitation on a savanna may be substantial, for at least part of the year there is drought stress ultimately favoring grasses and *xerophytic* (adapted to arid conditions) trees. In many areas, rainfall is the most critical variable in determining whether an area is essentially grassland (low rain), savanna, or dry woodland (moderate rain). Local

PLATE 11-7
Savanna is the dominant ecosystem of the Northern Territory of Australia.

variability in rainfall is typical in African and Indian savannas and results in patches of grassland, savanna, and thorn woodland all within the same general region. In the Serengeti of East Africa, the dry season extends from June through October. November brings light rain that continues through February. Only from March to May is there significant rainfall (Aber and Melillo 2001). In India, monsoon rains occur from June through September but vary in intensity from year to year.

In South America, savannas throughout Venezuela, Colombia, Bolivia, Surinam, Brazil, and Cuba all experience a significant dry season exceeding three months. However, many savannas in Central America (Nicaragua, Honduras, and Belize) as well as coastal areas of Brazil and the island of Trinidad do not have protracted dry seasons. There, for only three months at the most, is rainfall below 100 millimeters per month. Additional factors, likely edaphic in nature, contribute to savanna formation in these areas.

Australian savannas, which occur in the northernmost part of the continent, are strongly seasonal (Plate 11-7). Dry season is severe and typically lasts from May through October. The rainy season is most pronounced from December to March, when heavy monsoon rains may quickly flood some savanna areas.

FIRE

Fire is *the* variable on which savannas depend for their continued existence. Savannas typically experience frequent mild fires and major burns every few years or so. Many savanna and dry forest plant species are *pyrophytes*, meaning that they are adapted in various ways to withstand occasional burning. Grass is an outstanding example. Grasses have dense underground root systems that are protected from surface fire and allow rapid above-ground growth following fires. Fire burns above-ground vegetation, releasing minerals, fertilizing the upper soil layer, and enhancing regrowth. Tree species are less adaptable to fire, though ancient charcoal remains from Amazon forest soils dated prior to human invasion suggest that even moist forests occasionally burn.

Experiments in which fire is suppressed suggest that if fire did not occur in savannas, plant species composition would significantly change. When burning occurs, it prevents competition among plant species from progressing to the point where some species exclude others. Frequent fire favors grasses and selects against woody vegetation (van Langevelde et al. 2003). In experimental areas protected from fire, a few perennial grass species eventually come to dominate, outcompeting all others (Inchausti 1995). Evidence from other studies suggests that exclusion of fire results in markedly decreased plant species richness, often with an increase in tree density (Silva et al. 1991). There is little doubt that fire has a significant influence in maintaining savanna, certainly in most regions (Kauffman et al. 1994), and fire will be discussed in additional examples later.

Fire has been demonstrated to determine boundaries between grassy savanna and dry woodland (Hoffmann et al. 2009). A study performed in central Brazil demonstrated the effect of *fire traps*, where all tree biomass is killed above ground. The frequency of natural fires is such that trees have insufficient time to resprout, grow, and reproduce before the next fire. *Congeneric* (species within the same genus) tree species were compared between forest and savanna, and forest species typically had thinner bark, responsible for their greater susceptibility to fire. Rapid growth is key to escaping the fire trap in that larger trees resist the effects of fire better than small trees.

At the savanna–forest boundary, fire caused a mean mortality rate of 6.6% among savanna tree species, compared with 7.5% of forest species, which does not seem like much of a difference. But forest species are "caught in a fire trap," because after they resprout, there will be another fire in 1 to 3 years, insufficient time to grow to the point where they produce seed. The fire trap maintains a sharp boundary between savanna and forest (Hoffmann et al. 2009). The Brazilian study demonstrated a sharp transition in the abundance of savanna and forest tree species at the savanna–forest boundary and is an example of the concept of alternative stable states, a concept that will be reviewed later in this chapter.

SOIL CHARACTERISTICS

Many savanna soils, like many rain forest soils, are typically *Oxisols* and *Ultisols* (see Chapter 10) with a low pH (4 to 4.8) and notably low concentrations of such minerals as phosphorus, calcium, magnesium, and potassium, while aluminum levels are high. Other savannas, such as the Serengeti of Africa, have areas of calcium-rich soil that form caliche pans and maintain an alkaline pH. Some savannas occur on waterlogged soils, others on dry, sandy, well-drained soils.

Waterlogged soils occur in areas of flat topography or poor drainage. Because these soils usually contain large amounts of clay, they become water-saturated. Air cannot penetrate between the soil particles, making the soil oxygen-poor. In extreme cases, hardened pans form, as in the case of Arecife and caliche soils described earlier.

In contrast, dry soils are sandy and porous, their coarse textures permitting water to drain rapidly. Sandy soils are prone to the leaching of nutrients and minerals (see Chapter 10) and so tend to be nutritionally poor. Though most savannas are found on sites with poor soils (either because of moisture conditions or nutrient levels or both), poor soils can and do support lush rain forest. The white sandy soils of the upper Amazon (see Chapter 10) support such forests, unless the forest is cut and burned.

This wide range of soil characteristics may seem unusual, but it really means that extreme soil conditions, either too wet or too dry for forests, are satisfactory for savannas. More moderate soil conditions support moist forests. Indeed, soil degradation is blamed for promoting savanna formation in sites that once supported moist forest.

HERBIVORY

The savanna ecosystems most subject to significant levels of herbivory are those of Africa. Only Africa still retains its megafauna. The collective influence of over 40 species of hoofed mammals, some *grazers* (feeding on grass), some *browsers* (feeding on leaves from woody plants), and some *generalists* that both browse and graze, has a significant effect on savanna ecology. Savanna ecosystems in other regions are much less strongly influenced by grazing and browsing animals because of the loss of significant numbers of megafaunal species. Savannas of South America have no remaining large herbivorous mammals other than peccaries and tapirs (though once these areas supported megafaunas comparable to that of Africa today). The ecology of African ungulate (*ungulate* refers to hoofed) herbivores and their predators will be reviewed in the section "Foraging Niche Partitioning among Ungulates" later in this chapter.

HUMAN INFLUENCE

On certain sites, savanna formation correlates with frequent cutting and burning of moist forests by humans. Increase in pastureland and subsequent overgrazing has resulted in an expansion of savanna. Cutting and burning, if too frequent, destroys the thin, upper layer of humus necessary for rapid decomposition of leaves by bacteria and fungi and recycling by surface roots (see Chapter 10). Once the humus layer is lost, nutrients cannot be efficiently recycled and they more rapidly leach from the soil, converting soil from fertile to infertile and making it more suitable for savanna vegetation.

In some areas in South America, deep-rooted grasses imported from Africa to furnish fodder for cattle have come to dominate savannas, replacing native species. These grasses, in particular *Andropogon gayanus* and *Brachiaria humidicola* (often called *African elephant grasses*), now are estimated to cover about 35 million hectares of savanna (Fisher et al. 1994).

ARE ALL SAVANNAS "NATURAL"?

Savannas occur naturally because of impacts of the variables described earlier. There is no doubt, for example, that climate change in the middle part of the Cenozoic reduced forest cover and increased spread of savannas and dry forests. The emergence and subsequent evolution of bipedal hominids in Africa would be doubtful if savannas were not present. However, some ecologists have suggested that Neotropical savannas have essentially resulted from human activity rather than environmental causes. This claim is unsubstantiated by historical evidence, however. The fossil record of South American megafauna shows that savanna and open woodland were present in many areas throughout much of the Cenozoic era. Evidence exists that savanna vegetation grew in parts of the Amazon Basin as recently as 13,000 to 30,000 years ago (Sarmiento and Monasterio 1975). What remains controversial is just how much savanna was present (Salgado-Labouriau 1998).

Neotropical savannas demonstrate the highest plant species richness of any savanna ecosystems on the planet (Huber 1987). In numbers of both herbaceous and woody species, Neotropical savannas rank first. This high species richness suggests that evolution of savanna species has been occurring throughout much of the Cenozoic era, beginning in the Oligocene but particularly evident in the late Miocene. Savanna is as distinctive and intrinsic to the Neotropics as it is to Africa and other regions.

There is a dynamic, temporal interface among grasslands, savannas, and dry forests. One expands as the others contract in a climatically driven, edaphically influenced, long-term process that has produced and continues to produce far-reaching effects on evolutionary patterns of both plants and animals.

The African Megafauna and Savanna Ecology

INTRODUCTION TO THE MEGAFAUNA

Over 40 species of large herbivorous mammals are found inhabiting African savannas, grasslands, and dry forests. Some range widely throughout the continent, others are more restricted in range. Herbivorous mammals feed either on grasses (grazers), or on tree leaves (browsers), or on both (mixed foragers). In the Serengeti and other parts of East Africa, the most abundant grazing mammals are wildebeest (*Connochaetes taurinus*), zebra (*Equus burchelli*), Thomson's gazelle (*Gazella thomsonii*), buffalo (*Syncerus caffer*) (Plate 11-8), and topi (*Demaliscus korrigum*). Other widespread grazers include hartebeest (*Alcelaphus buselaphus*) and oryx (*Oryx gazella*).

PLATE 11-8
African buffalo are among the more formidable herbivores on the savanna. Even lions are wary of attacking them.

PLATE 11-9
The African lion is the top carnivore of the savanna ecosystem.

Browsers are less species-rich but include giraffe (*Giraffa camelopardalis*) and steinbok (*Raphicerus campestris*). Mixed feeders include Grant's gazelle (*Gazella granti*), eland (*Taurotragus oryx*), and savanna elephant (*Loxodonta africana*). In the Serengeti, the movements of many of these animals are migratory, following the seasons, as dry season forces them to move.

The predominant predators are lion (*Panthera leo*) (Plate 11-9), cheetah (*Acinonyx jubatus*), leopard (*Panthera pardus*), hyena (*Crocuta spp.*), and wild dog (*Lycaon pictus*). The lion accounts for over 50% of the predation on the Serengeti (Aber and Melillo 2001). Lions, hyenas, and wild dogs are pack hunters. Leopards and cheetahs are solitary hunters. Nile crocodiles (*Crocodylus niloticus*) are major predators of herbivorous mammals visiting watering holes and crossing rivers. Unlike herds of ungulates, predatory mammals do not migrate seasonally, but occupy local territories and home ranges throughout the year.

FORAGING NICHE PARTITIONING AMONG UNGULATES

Grazing mammals of the Serengeti in East Africa have been intensively studied, and much has been learned about their patterns of food consumption (Sinclair and Norton-Griffiths 1979; Sinclair and Arcese 1995).

Thomson's gazelle (Plate 11-10) is an abundant small ungulate throughout the East African savanna. Like other grazing mammals such as wildebeest, zebra, and various antelopes, Thomson's gazelles move continuously and widely on the Serengeti, a nomadic and migratory lifestyle that is ultimately related to the quality of their forage.

Thomson's gazelle is strongly affected by two characteristics of its food: the rate at which an animal

PLATE 11-10
THOMSON'S GAZELLE

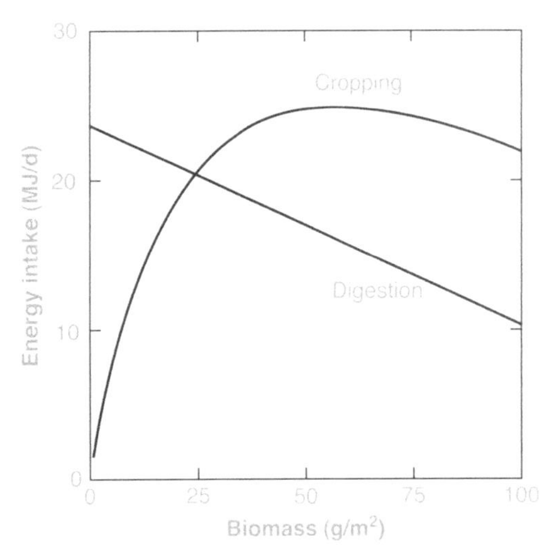

FIGURE 11-4
Grazing and digestion constraints as generated by the forage maturation model. Maximum energy intake occurs at the junction of the two constraints at a sward biomass of 25 grams per square meter.

can crop grass and the animal's physiological ability to digest it (Wilmshurst et al. 1999). On short swards of grass, those of low biomass but rapid growth, the grazing rate increases as biomass of grass increases (Figure 11-3). Thomson's gazelles thrive on short, rapidly growing grass, as it is rich in protein and easy to efficiently digest. The gazelles are limited on short grass only by the rate at which they crop and ingest the food. However, in tall swards of grass, no such correlation is observed. As grass grows, it becomes increasingly more challenging to digest, as it contains proportionately more tough, fibrous material. The digestibility of tall grasses is markedly less than that of short grasses.

Thomson's gazelles avoid the tallest grass, which is a better resource for larger grazing animals such as wildebeest, zebra, and hartebeest, whose physiology permits them to more easily digest the fibrous material. The larger ungulates facilitate grazing for the Thomson's gazelles by closely cropping the tall grass, stimulating regrowth, and then moving to other areas of tall grass. Thomson's gazelles, like domestic cattle, are *ruminants*, creatures with multichambered stomachs where bacteria do most of the actual digestion of the tough plant material. Small ruminant animals like Thomson's gazelles are able to detect small differences in forage quality, such that they frequent areas with low to intermediate forage biomass because of the quality rather than quantity of the grasses (Wilmhurst et al. 1999). The vast East African savanna is not a uniform ecosystem with regard to how grazing animals perceive its value as a food source. It is, instead, highly patchy and heterogeneous, with some areas optimal for certain species, other areas more energy-rich for other species.

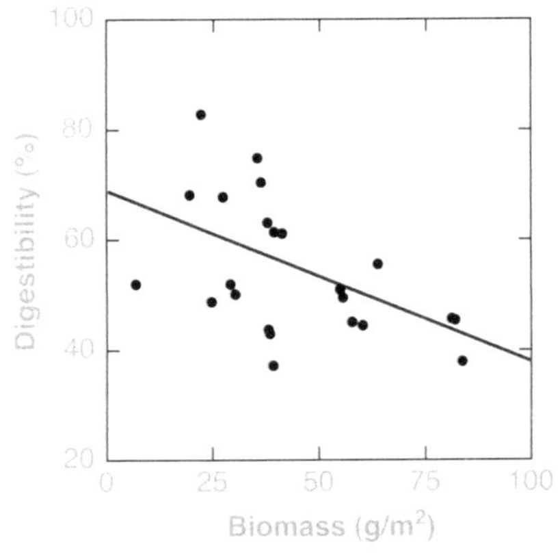

FIGURE 11-3
Digestibility of leaf and sheath tissue for Thomson's gazelles as a function of the biomass of leaf and sheath in the sward.

Given that Thomson's gazelles forage in short grasses, it seems reasonable to conclude that by doing so, they are maximizing the amount of energy they are gaining per unit time. The relationship between cropping rate and digestibility shows that grass biomass of around 25 grams per square meter (about 0.06 pound) is optimal for Thomson's gazelles (Figure 11-4). At that biomass, the digestibility is high, and the animals are able to adequately crop what they require.

Additional studies have shown that Thomson's gazelles closely track changes in the spatial distribution of short grass swards (Fryxell et al. 2004). Testing seven models to predict movements of grazing animals, researchers learned that gazelles abandoned short-grass patches when their local daily intake dropped below the expected intake as averaged across the landscape. The nomadic movements of Thomson's gazelles

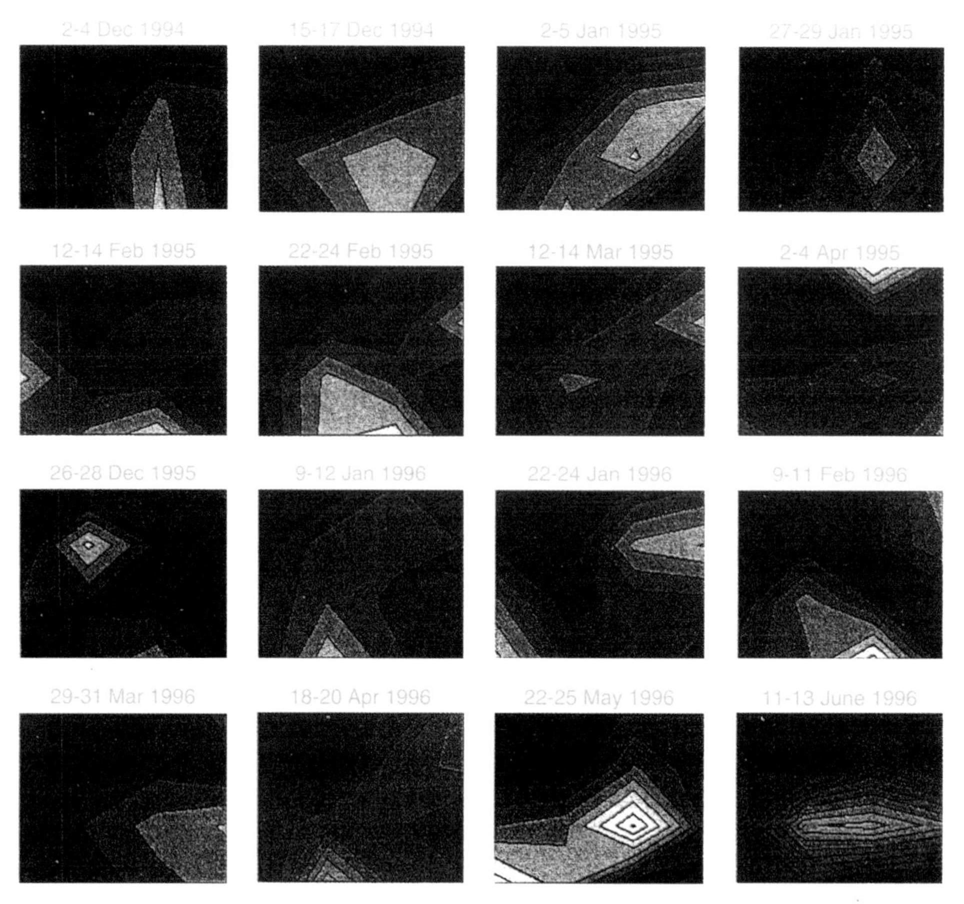

FIGURE 11-5
Sequential census (taken once every two weeks) of spatial distribution of Thomson's gazelles within a 40- × 40-kilometer grid centered on the northwestern Serengeti Plain (coordinates of northwestern corner: 2°32.1′S, 34°57′E). Shading corresponds to gazelle density, with the darkest shade being closest to 0 and the lightest shade closest to 300 animals per square kilometer.

were strongly predictable based on local energy gain. In other words, the gazelles recognize a *shifting mosaic* of food value among grass patches and switch accordingly, their daily energy gain being measured against that of neighboring grass patches (Figure 11-5). The highest energetic rewards to the gazelles are obtained from short grasses of low biomass (Figure 11-6). But this result is really not surprising, because short, rapidly growing grass is high in protein and low in tough, indigestible fiber.

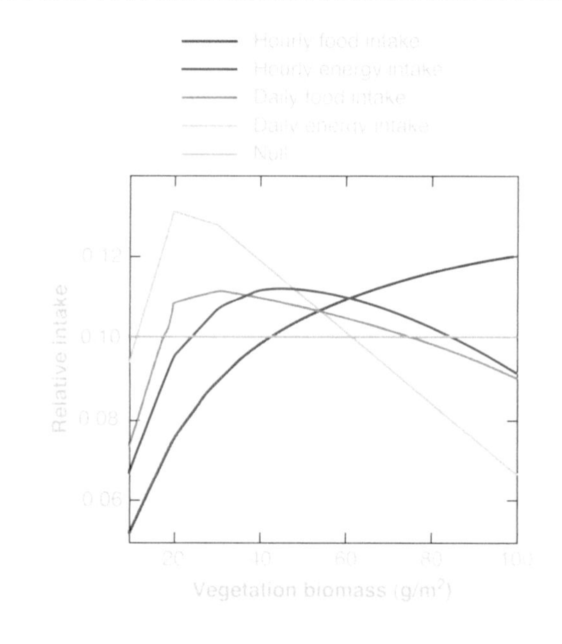

FIGURE 11-6
Relative gazelle fitness due to food or energy intake (over the grass biomass range 0 to 100 grams per square meter) predicted by the four foraging models and the null model. Relative ecological fitness was estimated by dividing intake at a given grass biomass by intake summed over the full range (10 to 100 g/m^2).

Browsing Lawns

The concept of a grazing lawn has been extended to browsing. A study performed on *Acacia nigrescens*, a palatable woody tree that covers much savanna in South Africa has shown that the trees develop strong responses to browsing by species such as elephants, kudus (*Tragelaphus strepsiceros*), steinboks (*Raphicerus campestris*), impalas, and giraffes (Plate 11-11) (Fornara and Du Toit 2007). Six *Acacia* stands were studied for traits that provide tolerance to damage. These include regrowth of shoots and leaves, branch growth rates, and branching patterns. Also investigated were resistance traits such as thorn size and density. Numerous savanna trees are *spinescent* (thorny), and researchers wanted to determine whether there was a difference in spinescence in lightly browsed trees compared with heavily browsed trees. Last, researchers also measured annual growth and seed production of heavily grazed and lightly grazed trees. Techniques included direct observation and measurement of ungulate grazing and artificial clipping of branches to examine rates and patterns of regrowth.

Mass allocated to regrowth on clipped branches was significantly higher on heavily browsed trees than on lightly browsed trees. Browsing, like grazing, appeared to stimulate the production of shoot mass and leaf production. Plant traits demonstrating tolerance of browsing were evident, but so were traits for resistance. Thorns were more numerous and dense on heavily browsed trees. (Thorns, of course, likely evolved to discourage browsing. In this case, it appears not to have worked.) Researchers found fewer trees with pods and fewer pods per tree on heavily browsed trees, suggesting that resistance traits remain essential to adapt the tree to browsing pressure, even when it is capable of rapid regrowth in the presence of browsers. Heavily browsed trees are typically reshaped by the browsing animals that consume them (Figure 11-7).

PLATE 11-11
Giraffes are among the browsers of the African savanna.

(a)

(b)

FIGURE 11-7
Heavily browsed *A. nigrescens* tree canopies showing different shapes due to the impact of different ungulate browsers. Note the clear height of giraffes' browsing limit in the tree in panel (b).

The interaction between herbivorous mammals and their forage has resulted in the creation of *grazing lawns* (McNaughton 1984). The grazing lawn results from close cropping of grasses by herbivorous mammals. The very act of grazing stimulates rapid plant regrowth, keeping the *net primary productivity* (NPP) high. For example, in a study on the Serengeti in which grazing was prevented by fencing vegetation, annual net primary productivity of grasses averaged 357 grams per square meter per year (about 0.79 pound), but in areas that were grazed, NPP averaged 664 grams per square meter per year (about 1.46 pounds) (McNaughton 1985). Grazers stimulate net primary productivity.

Grazers must continuously move for several reasons. First, net primary productivity is highly variable even among nearby sites. Short-term local net primary productivity may be extremely high, up to 30 grams per square meter per day (about 0.07 pound) in some cases. Such localized patches of high net primary productivity force animals to search for them. Local rainfall amounts vary, also forcing animals to move as productivity varies among sites. Diets of grazing animals change significantly from wet to dry season. Grazing is heaviest on grasslands where productivity is highest. This means that the rate of energy flow per unit of plant biomass is highest on grazed ecosystems (McNaughton 1976 and 1985). It is apparent that grasses and herbivores have evolved together and that plants are well adapted to continuous grazing. Superimposed on this complex mosaic are the effects of periodic fire and seasonal change. These factors lead to one of the most significant animal migrations on Earth, the annual migration of animal herds on the Serengeti.

THE SERENGETI MIGRATION

The three major migratory ungulates of the Serengeti are wildebeest, zebra, and Thomson's gazelle. Herds in excess of a million wildebeest move into the lush grasslands of the Serengeti during wet season, where they feed on the abundant growth of grasses (Plate 11-12). The collective effect of millions of wildebeest is to remove up to 85% of the grass biomass (Aber and Melillo 2001). They, along with herds of zebra and gazelle, move long distances into savanna and dry woodland during dry season, but this migration, necessitated by dry conditions, does not necessarily ensure ample food.

PLATE 11-12
Wildebeest are among the large, hoofed mammals that have a complex annual migration on the Serengeti.

The migration route is roughly circular but varies annually because of variation in weather patterns and food availability (Figure 11-8). In the dry season, the animals occupy the northern Serengeti (in Tanzania) and Masai Mara National Reserve (in Kenya). As rains begin and become more dependable, the animal herds drift to the southeast, into the southern parts of Serengeti National Park (in Tanzania).

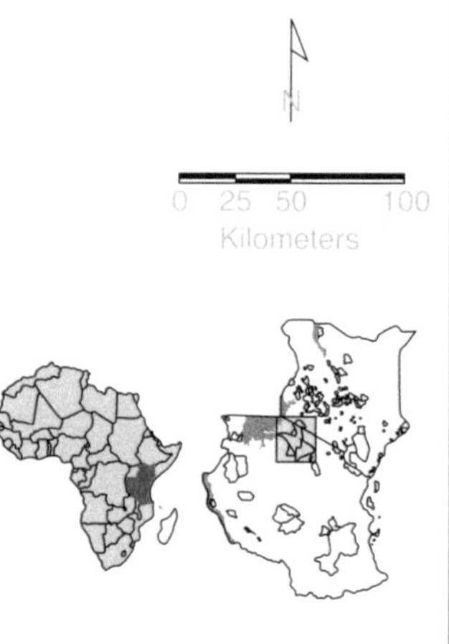

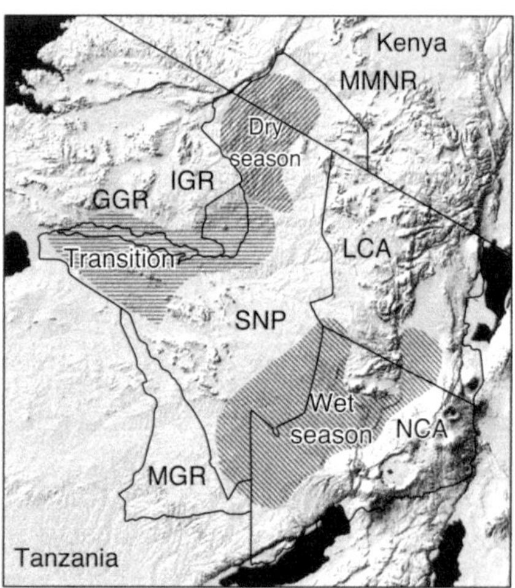

FIGURE 11-8
The Serengeti Mara ecosystem includes parts of northern Tanzania and southwestern Kenya (see insets). Areas include Serengeti National Park (SNP), Ngorongoro Conservation Area (NCA), Maswa Game Reserve (MGR), Masai Mara National Reserve (MMNR), Grumeti Game Reserve (GGR), Ikorongo Game Reserve (IGR), and Loliondo Conrolled Area (LCA). Wildebeest are in the northern portion of the ecosystem in the dry season, in the Western Corridor in transitional periods of some years, and in the southern part of the ecosystem in the wet season.

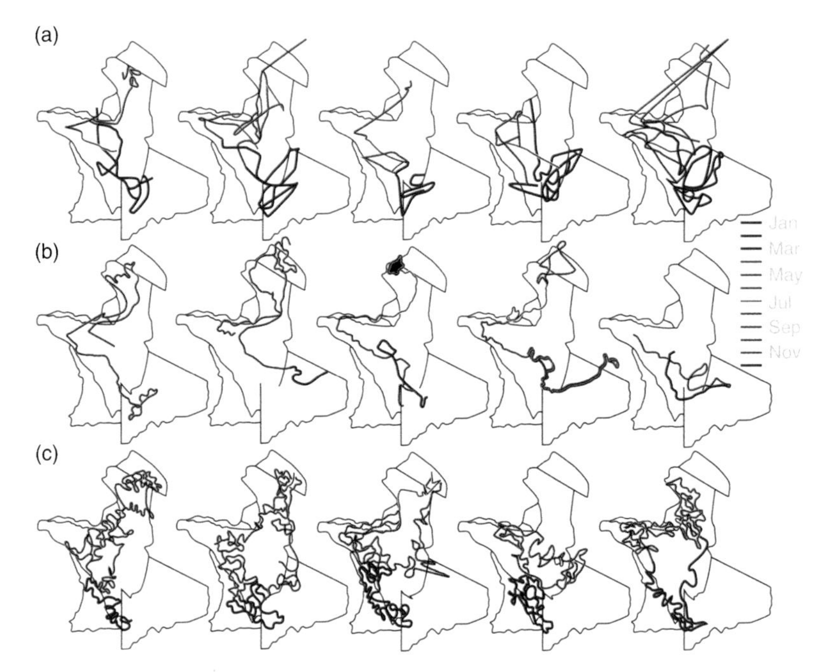

FIGURE 11-9
Pathways from (a) five VHF-tracked wildebeest, (b) five GPS-tracked wildebeest, and (c) five simulated wildebeest. Lines connect successive observations of tracked animals that are typically more than one day apart and do not indicate direct travel between points.

Studies utilizing a mathematical simulation derived from a methodology termed *evolutionary programming*, along with satellite data and GPS tracking of animals (Figure 11-9), have added support for the hypothesis that the animal herds move in response to rainfall amounts and vegetation growth (Boone et al. 2006). The simulation uses various parameters to "learn" how best to adapt to an ecological situation. In this study, a reasonable migratory pathway evolved in the simulation within merely 10 generations, and an optimum migratory route that is highly similar to what actually occurs evolved within 1,000 generations.

In the most accurate model, 25% of the model was influenced by rainfall and 75% by vegetation growth. The conclusion, based on the model, is that vegetation growth, or food availability, is the dominant parameter driving the migration.

While interspecific competition among ungulates is mild during wet season, the same does not apply during dry season. Many species concentrate in the more humid northwestern areas of the Serengeti, where interspecific competition is high. For example, impala (*Aepyceros melampu*) (Plate 11-13) switch from grazing on grasses to browsing leaves during dry season, perhaps in response to competition for grasses with other ungulates such as buffalo and wildebeest. It is during dry season that mortality peaks, a bottom-up force affecting ungulate population density.

Wildebeest herds give birth synchronously (Figure 11-10), most calves being born in February and March (Estes 1976). Food supply and seasonality are the

PLATE 11-13
Impala typically alternate from grazing to browsing.

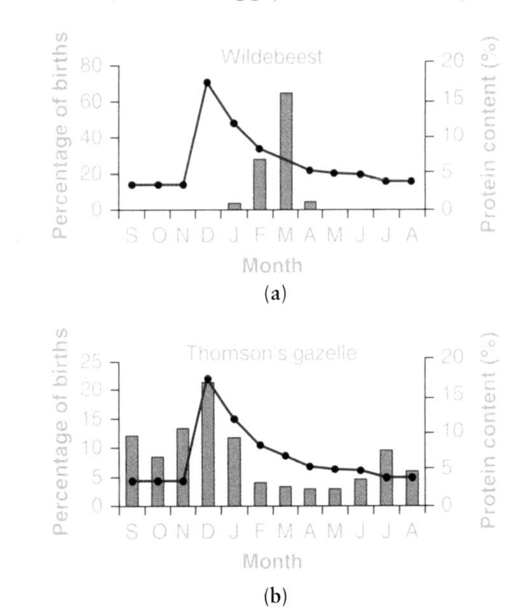

FIGURE 11-10
Comparison of (a) wildebeest and (b) Thomson's gazelle percentage of births over the course of a year. Dotted line is the percent crude protein of grass.

PLATE 11-14
Wildebeest cow giving birth on the Serengeti. Note the legs of the baby dangling from the mother.

PLATE 11-15
Wildebeest calves are precocial, able to run and keep up with the herd soon after birth.

major factors determining the timing of the birth season for various species of ungulates, including wildebeest (Sinclair et al. 2000). However, response to predators also exerts an influence that differs among species. Wildebeest are born precocial, and newborn animals are able to rise and follow their mothers after a short time (Plates 11-14 and 11-15). Such an adaptation allows for synchronous breeding. Because huge numbers of calves are born in a compressed time period, they benefit from the protection of the herd. Wildebeest give birth when rainy season has left the savanna green and lush. There is ample food for the pregnant females as well as other herd members. Because wildebeest form vast herds and newborns accompany the herd as it incessantly moves about the savanna, each newborn has a better chance of predator avoidance. Many newborn animals are preyed upon by lions, cheetahs, leopards, and hyenas, but most are not. Predators are "saturated" by the abundance of wildebeest and can only take so many.

In contrast, species such as Thomson's gazelle do not synchronize reproduction. Thomson's gazelles give birth to nonprecocial young that must hide in the grasses for days until sufficiently strong to accompany their mothers. Thus Thomson's gazelles practice *predator avoidance* rather than *predator saturation*, and consequently their calving season is much extended compared with that of wildebeest. Hence the timing of reproduction among the various species of the ungulate community of the Serengeti is based on a combination of food supply (which, in turn, is strongly affected by rainfall patterns) and adaptations to avoid predators.

Niche Segregation among Vultures

Savannas and dry forests are ecosystems where vultures tend to be a common sight. In the Neotropics, for example, the black vulture (*Coragyps atratus*) (Plate 11-16) is one of the most widely distributed birds of open habitats. The turkey vulture (*Cathartes aura*) (Plate 11-17) is also common throughout the Neotropics (and both species also occur in parts of North America), but it tends to favor forest habitats. This is because the turkey vulture has a strong olfactory sense and therefore locates its carrion by odor. Black vultures rely entirely on vision. Two other species, the greater yellow-headed vulture (*Cathartes melambrotus*) and the lesser yellow-headed vulture (*Cathartes borrovianus*), exhibit the same pattern. Greater yellow-headeds are forest vultures; lessers are savanna-dwellers.

In the Serengeti, there are as many as six vulture species that may feed on the same carcass (Lack 1971). They vary in size and morphology, and their modes of feeding differ (Plates 11-18 and 11-19). The white-backed vulture (*Gyps africanus*) and Rüppell's griffon (*Gyps reuppellii*) are both long-necked vultures, commonly called *griffons*. This anatomical characteristic permits them to insert their naked heads deep in the carcass, where they have access to organs and muscles. The griffons also have a slender skull, long beak, and a large tongue with horny and backward-pointing tooth-like denticles

PLATE 11-16
Black vulture at armadillo carcass.

PLATE 11-19
Rüppell's griffon, a long-necked vulture, at a carcass with other vultures. Its neck is stained with blood.

PLATE 11-17
Like other vultures, turkey vultures soar on thermal air currents as heat radiates from the ground to the air. This is how the birds search for prey.

that permit them to scrape the bones. In contrast, the even larger white-headed (*Trigonoceps occipitalis*) and lappet-faced vultures (*Torgos tracheliotus*) have short, powerful necks that are sufficiently strong to attack skin, sinews, and meat tightly stuck to bone. Last, there are the Egyptian vulture (*Neophron percnopterus*) and hooded vulture (*Necrosyrtes monachus*), both of which are small in comparison with the others. These vultures have short necks and slender beaks, and they feed on the scattered fragments remaining after all of the other scavengers have fed. When a carcass is abandoned by the predator(s) who killed it (lions or cheetahs, for example), the scavengers move in. Typically, hyenas are among the first, as well as jackals, but the vultures gather quickly, and as many as a dozen or more vultures may feed simultaneously on a single carcass. It is quite a spectacle (Plate 11-20).

PLATE 11-18
Lappet-faced vulture, the largest of the Serengeti vultures.

PLATE 11-20
Vultures massing at a carcass, many with blood-covered necks.

TOP-DOWN AND BOTTOM-UP INFLUENCES ON THE SERENGETI

Herbivores have been shown to enhance plant species richness on the Serengeti. An exclusion study (in which herbivores were prevented access) was conducted at Serengeti National Park in Tanzania on eight sites that varied in rainfall and soil nitrogen and phosphorus (Anderson et al. 2007). Excluding herbivores reduced plant species richness by an average of 5.4 species per plot. The most significant decreases in plant species richness occurred on sites of intermediate rainfall. There was greater growth of *forbs* (herbaceous plants) within exclosures, whereas sedges increased in richness outside the exclosures. The *compositional similarity* of plant communities (which refers to the number of species in common between two communities) on grazed compared with ungrazed sites was strongly correlated with precipitation. As precipitation increased, the compositional similarity decreased (Figure 11-11).

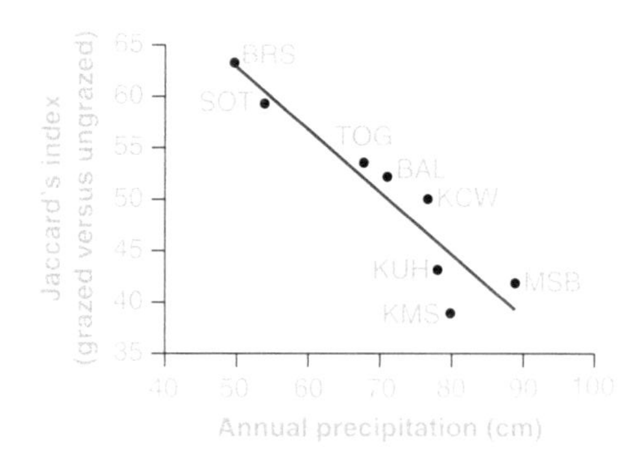

FIGURE 11-11
Relationship between Jaccard's index (a measure of similarity among plant communities, in this case grazed versus ungrazed) in relation to annual precipitation. The higher the precipitation, the less similarity between the grazed and ungrazed communities. The abbreviations are sites used in the study.

Grazing obviously exerts top-down effects on the composition of the plant communities, but precipitation is also an important bottom-up influence. The study also linked grazing to available soil phosphorus (P), which increased when grazing was prevented. On sites with low rainfall, herbivore exclusion led to a relatively greater loss of plant species

What's Good for the Olive-Hissing Snake?

The olive hissing snake (*Psammophis mossambicus*) is common on the Kenyan savanna. In a study in which snake abundance was compared between sites in which large herbivores are excluded and those where the large ungulates are present, it was learned that over a 19-month period, the abundance of olive hissing snakes increased significantly, but only where large herbivores were excluded (Figure 11-12).

Was this a top-down response or a bottom-up response? In other words, were the snakes increasing due to not being trampled by large animals or being exposed to predators that often stalk large animals, or were the snakes responding to greater food availability? The answer is that they seemed to increase because one of their major food sources, the pouched mouse (*Saccostomus mearnsi*), was increasing in the areas where large animals were excluded (McCauley et al. 2006). The pouched mouse, along with other small mammals, increased in abundance in the exclusion plots, providing more food for the snakes. Researchers were unable to determine whether the snake population increase was due to greater recruitment from surrounding areas or due to higher reproduction. There was no clear link between the absence of large animals and the greater abundance of the mice. But what seems clear is that the population density of the olive hissing snake is related to the number of large herbivores present, as they, in turn, appear to affect the populations of small mammals present.

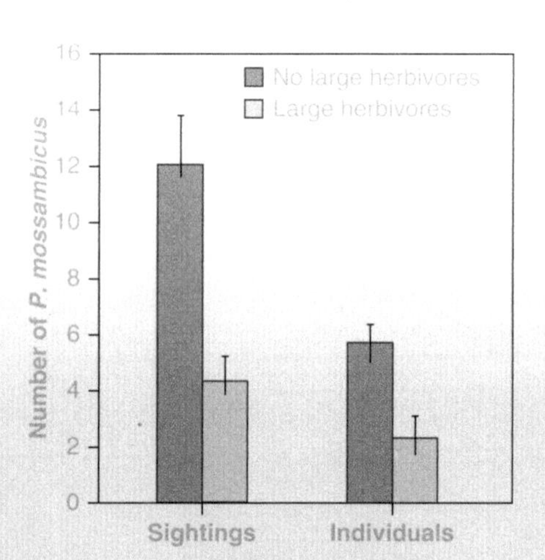

FIGURE 11-12
Abundance of olive hissing snakes (*P. mossambicus*) on plots with and without large herbivores.

(more local extinctions and fewer colonizations). These sites increased in P level. At sites with higher rainfall, herbivore exclusion led to reduced soil P, little change in species richness, but a large change in the composition of the plant community. Though soil phosphorus is a bottom-up influence on plants, its increase or decrease in the absence of grazing, coupled with effects of rainfall, suggests a complex top-down interaction with bottom-up forces. The researchers suggest that plant dynamics on the Serengeti result from multiple trade-offs between herbivory resistance and interspecific competition for resources such as soil N, soil P, water, and light (Anderson et al. 2007). Consider this statement in relation to the earlier discussions (Chapter 8) of factors that affect plant dynamics in tropical humid forests. There are compelling similarities.

Savannas vary in tree density. Some herbivores are browsers and depend directly on tree leaves as a food source. These animals must forage among trees. Grazers are also affected by tree density in two ways. Trees may affect the quality or quantity of herbivorous vegetation, a bottom-up effect. But trees also allow predators more cover, and herbivores that feed among trees may be at greater risk of predation, a top-down effect. Herbivorous mammals must be constantly aware of predation risk, and so the reality of their daily life is that they face a constant trade-off between predator avoidance and their biological need to maximize the quantity or quality of their food intake. This has been called a *climate of fear* (Brown and Kotler 2004). Large herds distant from trees may detect predators more easily, but such an advantage may be compromised if better forage is to be had near or among trees (Plate 11-21).

A study performed in an *Acacia*-dominated savanna at the Mpala Research Centre and Jessel Ranch in Laikipia, Kenya, examined a 10-fold range of natural

PLATE 11-21
Lion pride among acacias on the Serengeti.

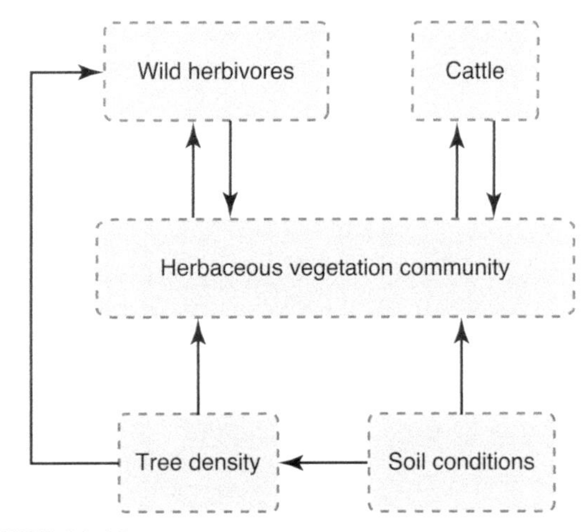

FIGURE 11-13
Schematic model guiding structural equation modeling (SEM) analyses. The boxes represent major conceptual categories of variables; the arrows represent the relationships and the directions of those relationships that were examined in SEM analyses.

tree densities, quantifying herbaceous vegetation, domestic (cattle) and wild herbivore use, and visibility among the sites (Riginos and Grace 2008). Eight large wild herbivore species were present, and predators included lions, cheetahs, and spotted hyenas. The study used structural modeling analysis (SEM) to examine the interactions and importance of variables (Figure 11-13).

Data were collected on animals by censusing dung piles (identified to species level) in areas that varied in tree density. The results indicated that the larger herbivores were found mostly in areas of low tree density, where visibility was greater (Figure 11-14). These animals were less influenced by vegetation characteristics (other than tree density) than avoiding predation. Herbaceous species composition did vary substantially with tree density, but that variation was not determining animal usage. While it is perhaps understandable that strict grazers such as zebra would avoid the risk associated with trees, the same tendency held true of browsers. For example, giraffe, which are strict browsers, preferred areas with the fewest trees (Plate 11-22).

The one large herbivore that was found in areas of high tree density was the elephant (Plate 11-23). Elephants are fairly immune from predation, even by lions. There was a positive correlation between tree density and percentage of trees damaged by elephants (Figure 11-14).

The study showed that top-down forces are more influential in affecting the distribution of herbivorous mammals than bottom-up forces. Apparently, the animals do exist in a landscape of fear, and they are behaviorally adapted to that reality.

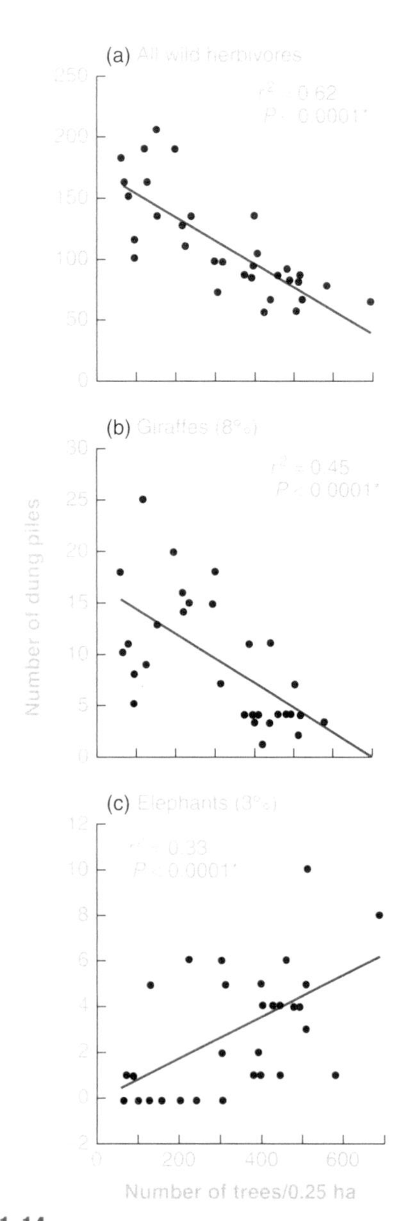

FIGURE 11-14
The relationship between tree and dung density for all (a) wild herbivores, (b) giraffes, and (c) elephants.

PLATE 11-22
Giraffes, though browsers, are usually found in open areas with relatively few trees.

Studies have shown that the collective effect of herbivores on the Serengeti is to maintain higher plant species richness than would occur in the absence of the large ungulate herbivores. The animals prevent competitive exclusion among the plants. At a finer grain, it has been shown that individual plant species enjoy greater seedling survival when large animals are present than when they are absent. A study performed at Mpala Research Centre in Laikipis in central Kenya (the same site used in the previous study) focused on the tree species *Acacia drepanolobium,* which accounts for > 98% of the overstory vegetation in the savanna (Goheen et al. 2004). The study was part of a broad program called the Kenya Long-Term Exclusion Experiment (KLEE), initiated in 1995 to evaluate the effects of ungulate herbivores (including cattle) on savanna ecology. The KLEE project consists of blocks measuring 400 × 600 meters (24 hectares, or about 59 acres), each consisting of six 200- × 200-meter (4 hectares, or about 10 acres) treatments. Each treatment plot excludes a particular combination of large mammals.

Another study, from the Ndoki Forest in northern Congo, examined the effects of forest elephants on tree diversity (Blake et al. 2009). Researchers examined 855 piles of elephant dung, looking for seeds. They learned that elephants were responsible for dispersing more intact seeds than any of the other herbivore species. Collectively, elephants dispersed seeds of 96 tree species. Using GPS telemetry to track elephants, it was

PLATE 11-23
Elephants have little or no aversion to being within areas of high tree density.

further learned that the ponderous mammals disperse seeds over longer distances than any other large hoofed mammal. Last, in examining the spatial distribution of 5,667 trees, it became clear that increasing amounts of elephant seed dispersal were associated with decreasing aggregation of trees. Elephants moved about 82% of the seeds more than 1 kilometer from the parent tree and frequently moved them farther than 5 kilometers. Seeds dispersed only by elephants were significantly less aggregated than would be predicted from random seed dispersal. If this pattern sounds familiar, it should. It is a clear example of the Janzen-Connell effect (see Chapter 5) that occurs in tropical moist forests with other seed dispersers. Seeds (and subsequently seedlings) dispersed far from the parent plant survive best. Further, researchers concluded that forest elephants at natural densities likely disperse more seeds than any other large herbivore. The efficacy of elephants as seed dispersers has strong conservation implications because there are some regions in Africa where elephant populations are in decline, even as there are areas where they are increasing dramatically (Marris 2007).

Goheen et al. (2004) performed a study involving *Acacia drepanolobium*. Seedlings were transplanted into study blocks to monitor their success in the presence and absence of large mammals. A total of 96 seedlings was planted in each of the three large-mammal-exclusion plots and in three control plots in which large mammals were not excluded. A total of 576 seedlings was planted across the plots. Plants were censused weekly and assessed for damage, death by predation, or death by desiccation.

A significant top-down effect was discovered. Approximately three times (15.3%) as many uncaged seedlings survived when large ungulate herbivores were present as when they were excluded (where only 5.6% survived) (Figure 11-15). The increased mortality of seedlings in exclosures was attributed to

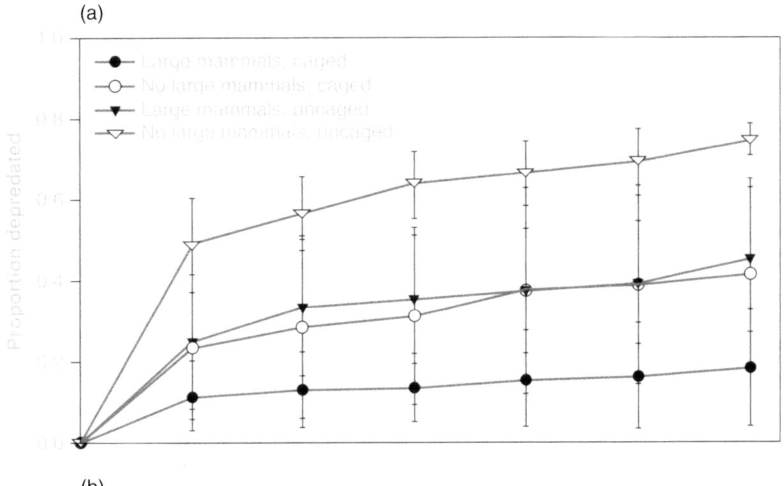

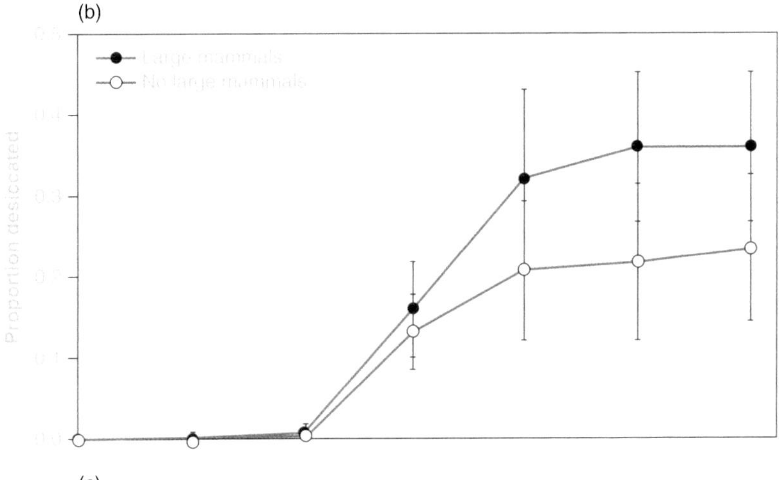

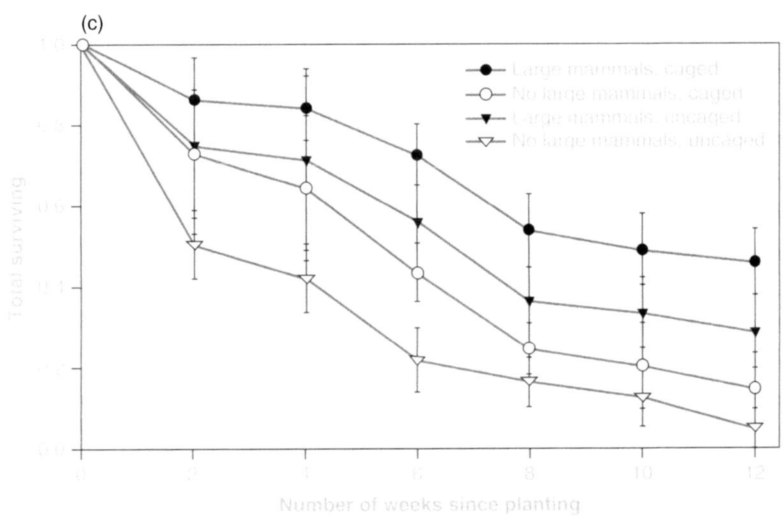

FIGURE 11-15
(a) Proportion of seedlings depredated, (b) proportion of seedlings desiccated, and (c) total proportion of seedlings surviving as a function of plot treatment and caged/uncaged status. Error bars represent ±1 SE.

Elephants as Ecosystem Engineers

The term *ecosystem engineer* has been suggested for organisms that uniquely control the availability of resources to other species by virtue of how they modify the physical environment (Jones et al. 1994). Elephants, as seen earlier, exert significant effects on tree diversity and forest structure in African savanna ecosystems and not just through long-distance seed dispersal. Their ponderous size, herding behavior, and continuous movement strongly facilitate the maintenance of open wooded grasslands (Shell and Salim 2004). In areas of elephant abundance, concerns have been expressed about elephants causing degradation of the savanna (van Aarde and Jackson 2007). But elephants have also been thought to enhance habitat complexity through their activities (Crooks 2002). Elephants turn over trees, split tree trunks, and act in other ways that add to habitat heterogeneity.

On a fine-grain scale, elephant ecological engineering was demonstrated in a study conducted at Mpala Research Centre (Pringle 2008). The broken tree limbs and stripped bark that were the work of browsing elephants created small crevices used by an arboreal gecko species (*Lygodactylus keniensis*). Through observations and transplant experiments, it was shown that geckos clearly depend on elephant-created damage for refuge sites. The geckos would vacate trees where elephant-engineered refuges were removed. Other studies (Pringle et al. 2007) have shown that on a broad landscape scale, the collective grazing activities of elephants and other large ungulates act to depress lizard densities by reducing tree density and (indirectly) arthropod density (lizards depend on arthropods as a food source). However, on a finer landscape scale, that of damaged, stripped bark, at least one lizard species is positively affected by top-down elephant activities Plate 11-24).

PLATE 11-24
(a) Elephant-damaged *Acacia mellifera* with multiple damaged stems. (b) Close-up of recent elephant damage illustrating crevices, with a pencil for scale. (c) Artificial "elephant damage" created for lizard transplant experiment, with pencil for scale. (d) Abandoned ant domatium lashed to *Acacia etbaica* to provide artificial refuge in lizard transplant experiment.

herbivory by small mammals such as rodents along with various insects. The most frequent cause of seedling death in plots with large mammals was plant desiccation. Researchers concluded that the movement of the ungulates kept the rodents and insects from unlimited access to the seedlings. For example, large herbivorous mammals are often accompanied by birds such as cattle egrets (*Bubulcus ibis*) that feed on insects stirred up by the moving ungulates. One caveat of the study noted by the researchers is that the overall effect of various consumers ranging from large herbivores to insects may be highly variable over time. But in this study, there was a clear top-down influence of large herbivorous mammals that resulted in higher seedling survival of *Acacia drepanolobium*.

INTERACTION OF FIRE AND HERBIVORY: ALTERNATIVE STABLE STATES

The view presented thus far has described a gradient from grassland to savanna to dry forest. There is another view to the gradient concept, called *alternative stable states*. Rather than a gradient approach, alternative stable state models argue for a given site to be either savanna or forest, depending on variables such as fire and herbivory (Figure 11-16).

Africa, because it has retained its megafauna, allows researchers to evaluate the impact of herbivory on overall savanna ecology, and this has led to an alternative stable state model. Grazing tends to reduce the *fuel load* of a savanna, which, in turn, reduces fire intensity. This effect acts to favor the spread of woody trees, as their bark is undamaged or less damaged by low-intensity fire. Browsing animals that feed on leaves have the opposite effect. Their impact on suppressing tree growth aids grass growth, adding to fuel load, resulting in more intense fires that reduce woody vegetation by directly killing seedling and mature trees. In a landscape of numerous browsers and grazers, the preceding scenario would tend to oscillate back and forth (Dublin et al. 1990; van Langevelde et al. 2003).

The interaction of fire and herbivory may produce alternative stable states. Researchers developed a model of tree–grass dynamics that allows for a more precise analysis of alternative stable states (van Langevelde et al. 2003). The model contains 24 parameters and variables, and it allows for multiple simulations of various interactions of grazing, browsing, fire frequency, and soil conditions. The study concluded that fire and herbivory produce strong interactive effects on the amount of grassy versus woody biomass. These two

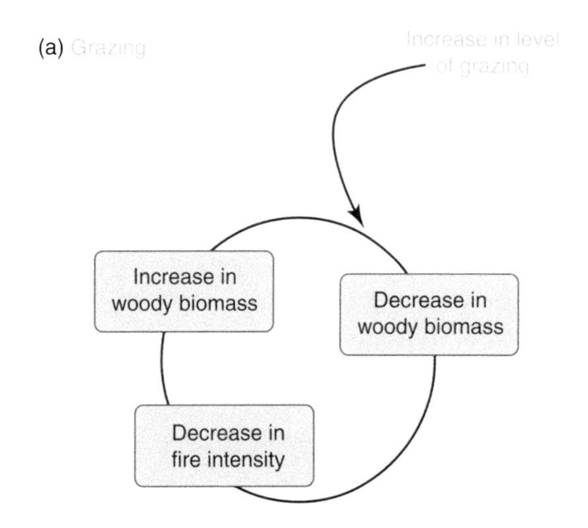

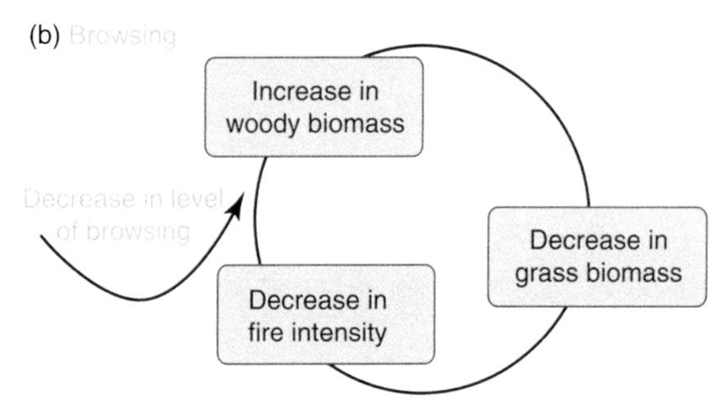

FIGURE 11-16
Positive feedback mechanism between grass biomass (fuel load) and fire intensity, triggered by (a) grazing and (b) browsing. A decrease in grass biomass leads to reduced fuel load that makes fire less intense and thus less damaging to trees, consequently resulting in an increase in woody vegetation.

factors both act such that savanna ecosystems oscillate from mostly woodland to trees and grass. The study also showed that a decline in grazing has less impact than a decline in browsing, which results in greater tree cover.

GRASS VERSUS TREES

Within any local savanna region that receives a given amount of rainfall, there is a zone of tension between grass cover and woody cover, related both to levels of herbivory (a top-down force) and frequency of fire (a bottom-up force). There is also the potential for competition among grasses and trees. In another study performed at Mpala Research Centre in Kenya, it was shown that grasses exert a strong negative effect on the tree species *Acacia drepanolobium* (Riginos 2009). The trees grow in an area dominated by five perennial

PLATE 11-25
Representative trees from the study: these two *Acacia drepanolobium* trees had the same height (1.4 meters) and similar basal diameters (grass-removal tree, 2.9 centimeters; control tree, 3.4 centimeters) when the experiment began in 2005. After two years, the grass-removal tree (right) had increased in height by 0.3 meter, whereas the control tree (left) had increased in height by only 0.07 meter.

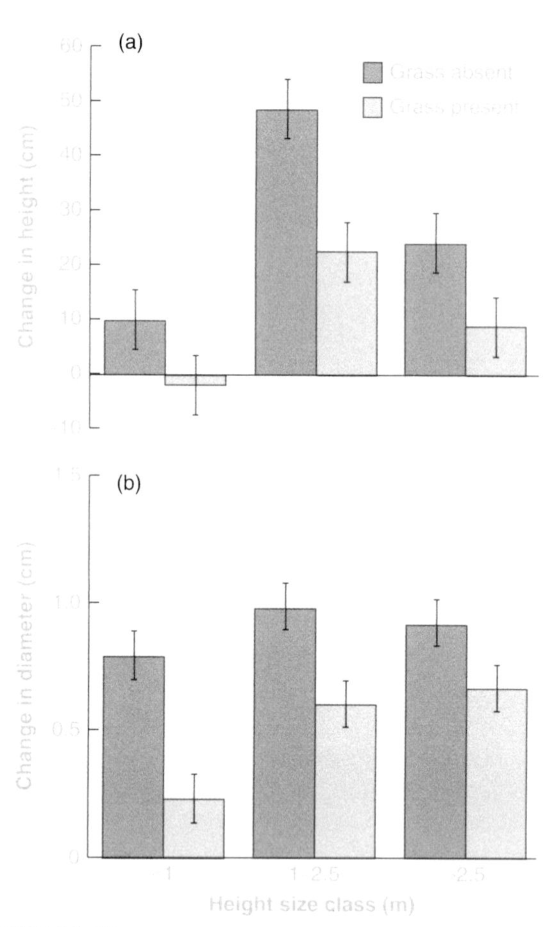

FIGURE 11-17
Growth of *Acacia drepanolobium* trees (mean ± SE) over two years with and without the subcanopy grass removed. Growth is measured as (a) change in height and (b) change in stem diameter.

bunchgrass species that together represent more than 90% of the herbivorous cover. A total of 180 trees was studied, ranging in height from 0.3 to 4.4 meters (about 1 foot to 14.5 feet). Grasses were killed (by applying a herbicide and clipping) in the immediate vicinity of half of the trees, while the others were used as controls. Tree growth was measured as changes in height, stem diameter, and total branch length. The study occurred over two years, and growth rates of trees differed because of differing rainfall between the two years. (The first year, 2005–2006, had 353 millimeters [about 14 inches], compared with 674 millimeters [about 26.5 inches] for 2006–2007.) However, the effect of grass removal was consistent in both years. Trees where there was grass removal grew in height more than twice as much as control trees. They also grew twice as much with regard to basal diameter and were more than twice as likely to grow enough to enter a different size class (Plate 11-25 and Figure 11-17). The experiment demonstrated that grass competition exerts a significant effect on *Acacia drepanolobium*. Therefore, it becomes clear that the dynamics of savanna ecosystems result from top-down, bottom-up, and within trophic-level forces.

The Asian sugarcane grass *Saccharum spontaneum*, which has become an invasive species in some areas of the Neotropics, provides a similar example (Hammond 1999; Wishnie et al. 2002). This grass grows in such dense stands that it successfully inhibits normal ecological succession. In addition, it responds to fire by regrowing far more rapidly than other species of plants. The presence of *S. spontaneum* produces what ecologists call *arrested succession*, where the normal process of vegetation development is severely inhibited. Application of herbicide treatments has been successful in retarding the severe effects of *S. spontaneum*, but the species is clearly problematic.

African Rinderpest: No More

Rinderpest is a viral affliction of ungulates, including domestic cattle, yaks, and other bovine animals. It is in the same group of viruses that cause measles, but it does not affect humans. It first occurred in Africa among domestic cattle in the late nineteenth century. By the approach of the twentieth century, it is estimated that 95% of the wildebeest and buffalo had perished from the virus. The mortality rate of domestic cattle was almost equally high, about 80%. The loss of the grazing animals set off an ecological cascade. The tsetse fly (*Glossina* spp.) disappeared from the region due to loss of ungulates, and lions began to frequently attack humans for the same reason. Tsetse flies feed on the blood of cattle and wild ungulates and serve as hosts for flagellate protozoans called *trypanosomes*, which cause the disease commonly known as *sleeping sickness*. They had little other prey to which to turn. As humans ceased farming and herding in the absence of cattle, areas cleared for pasture and agriculture returned to natural savanna. Natural selection resulted in some wildebeest becoming immune to rinderpest, and for a time, there was an equilibrium between the virus and the wildebeest, with the herd remaining at around 300,000 animals. The tsetse fly rebounded, as it depends on wildebeest for its blood food, and this, in turn, caused an increase in sleeping sickness. During the 1930s, a vaccine against rinderpest was developed, and the virus ceased to be a problem in the region, as domestic cattle all were vaccinated. The elimination of rinderpest resulted in stimulating the growth of wildebeest and buffalo populations. Zebras, unaffected by rinderpest, did not change in population (Plate 11-26) (Aber and Melillo 2001). Recently, the United Nations Food and Agricultural Organization declared that rinderpest has been eliminated. The last reported case was in Kenya in 2001 (McNeil 2010). Rinderpest is only the second viral disease to be eradicated; smallpox was the first.

PLATE 11-26
Zebras were unaffected by rinderpest.

The elimination of rinderpest virus was the equivalent of eliminating a strong top-down force on wildebeest and buffalo. Once decimated by the virus, then roughly stabilized by it, they were released from it and increased vastly in population, to the point where the impact of large predators such as lions is considered negligible.

WATER AVAILABILITY AND HERBIVORE FORAGING

Water is a bottom-up influence on herbivore distribution patterns, just as the availability of forage for grazing is. The availability of water and the quality and quantity of forage are each essential variables affecting the distribution of herbivores. In theory, grazing animals face a trade-off in how they distribute relative to surface-water availability compared with forage quality and quantity. Being near water may be more important to some animals than being among highest quality forage, for example. Other species may be more dependent on forage quantity than immediacy of surface water. At Amboseli, Kenya, during dry season, 99% of herbivore biomass occurred within 15 kilometers (about 9.3 miles) of surface water. This overall region represented only 52% of the total ecosystem area (Western 1975). Such a skewed pattern of herbivore distribution exerts different ecological effects on plants, dependent on whether the vegetation is near or distant from water. Water, a bottom-up influence, affects herbivore distribution, and herbivores, in turn, exert a top-down influence on plant biomass and community composition. A study performed at Kruger National Park (KNP), South Africa, demonstrated fine-grain differences among eight herbivore species in how water availability versus forage quantity and quality affected distribution patterns (Redfern et al. 2003). KNP experiences strong seasonality of rainfall, but its dry season is not as severe as in some other African regions. However, there is considerable variability from one year to another, some years being wetter and some drier (Figure 11-18).

The study consisted of a regression analysis based on aerial censusing of animal herds during dry season, and data on surface water distribution and for-

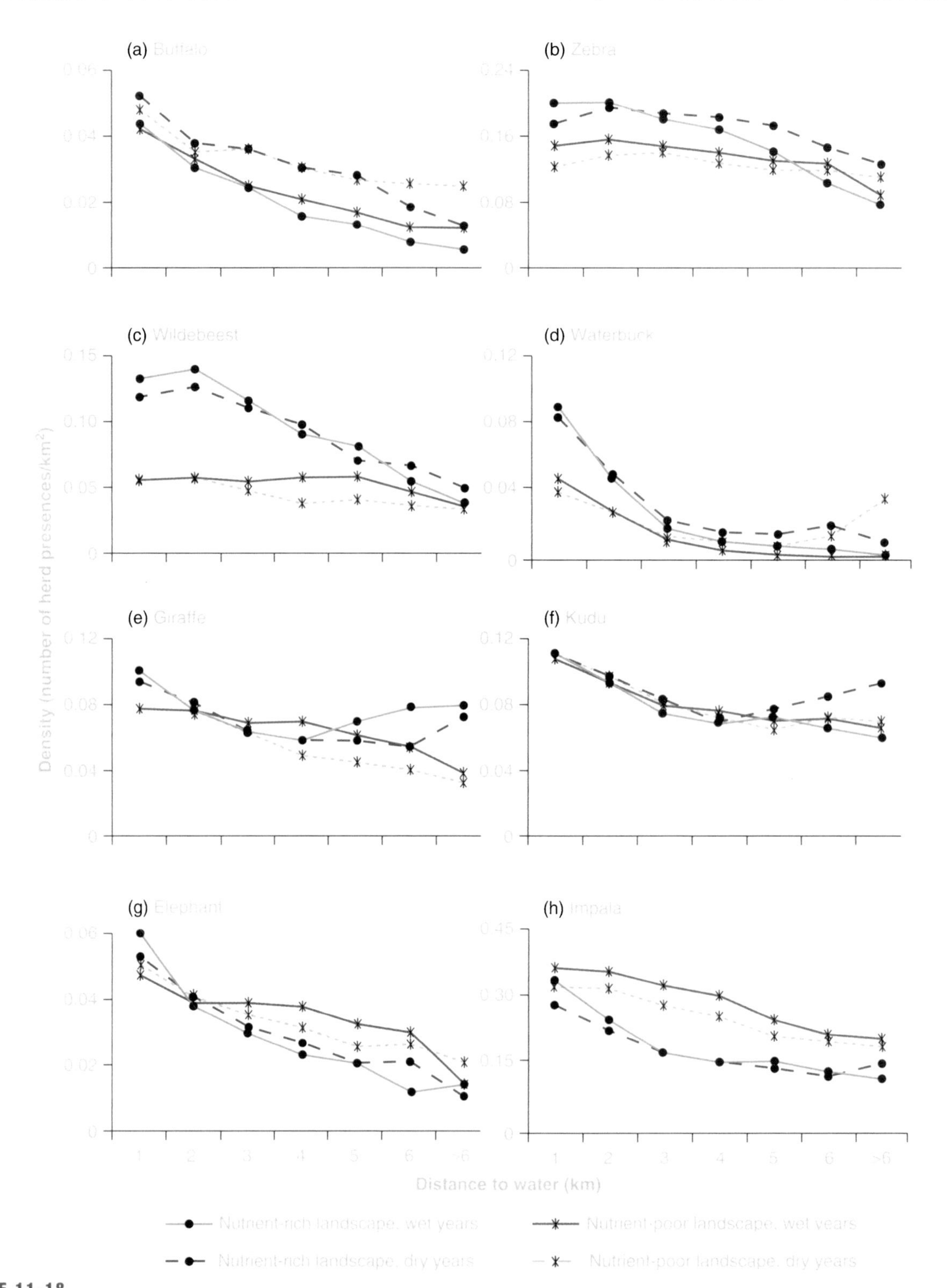

FIGURE 11-18
Observed densities of herd presences compared with distance to water. Note the differences between nutrient-rich and nutrient-poor landscapes in wet and dry years.

age quality and quantity as they vary across the landscape of the park. Animal numbers were expressed in 1-kilometer-square grids that could be correlated with water and forage variables.

Species varied in their proximity to surface water and in how they responded in wet and dry years. Browsers such as kudu and giraffe showed weak responses both to distance from water and to dry compared with wet years. These animals have been considered relatively water independent (Western 1975). Elephants showed a stronger response to water proximity, especially during dry years. Impala, which are both browsers and grazers, showed some response to surface-water proximity, particularly in dry years. Among the grazing species, waterbuck (*Kobus ellipsiprymnus*) showed the strongest response to water. The common name is suggestive of its need to be close to surface water. Both in wet and dry years, waterbuck herds were strongly tied to surface water. Buffalo, a large grazer, also showed a strong response to water proximity, with very little difference between wet and dry years. Wildebeest also had a strong response to water in nutrient-rich landscapes but a much weaker response in nutrient-poor landscapes. Both buffalo and wildebeest appear to face a trade-off between forage quality and quantity and proximity to water. Zebra showed a weaker response to water proximity than other grazers but also appeared to face a trade-off during dry years.

The overall conclusion of the study was that proximity to water is significantly correlated with all eight ungulate species during dry season. While this result may not seem surprising, it is also clear that substantial variation was evident among the species. These fine-grain differences likely exert local effects on the heterogeneity of the ecosystem.

Neotropical Savannas and Dry Forests: Some Examples

What follows is a brief overview of some of the most noteworthy savanna and dry forest areas in the Neotropics. It is not meant to be comprehensive.

THE DRY PINE SAVANNA OF BELIZE

The savannas of Nicaragua, Honduras, and Belize are populated abundantly by Caribbean pine. Riding through miles of savanna along the Southern Highway in Belize, one notices that many of the pines have darkened fire scars on their trunks. Lightning storms cause fires during dry season, and the effects of dryness and periodic fires combine to preserve savanna. Caribbean pines tolerate occasional mild fires better than other tree species in Belize. Grasses also thrive in an environment with periodic fire.

Throughout much of southern Belize east of the Maya Mountains, the dominant ecosystem type is savanna (Plate 11-27), abounding in Caribbean pine, but with many other species ranging from grasses, palms, and palmettos to *Cecropias* and *Miconias*. Compared with the nearby tropical moist forest nestled within the protective Maya Mountains, the pine savanna is an area of low species richness and a simpler, more arid, and rugged-looking ecosystem. During the dry season, which extends from about February through most of May, the pine savanna is subject to occasional fires, the evidence of which can be seen as charred stumps and burned bark on many of the pines throughout the region. In this area, fire is an important ecological influence, a factor that provides the key ingredient in maintaining dominance of savanna.

Wildlife is less diverse on savanna than in interior lowland moist forest, but many animal species typical of forest occasionally range onto savanna, including boa constrictors and jaguars. The pine savanna is inhabited by gray fox (*Urocyon cinereoargenteus*), tayra (*Eira barbara*), and white-tailed deer (*Odocoileus virginianus*), as well as numerous bird species.

The following two regions, *Llanos* and *Pantanal*, are each unique compared with other savannas because they are seasonally very wet. These areas typically flood in rainy season but become very dry during dry season.

PLATE 11-27
Pine savanna in southern Belize.

LLANOS—SEASONAL SAVANNAS OF VENEZUELA

The wide floodplain of the Orinoco River extends over an area of grassy savanna, interrupted by riparian forest and hammocks of woodland (raised areas in otherwise marshy areas). This habitat, which bears a strong physical resemblance to the Florida Everglades, is called the *Llanos*. Extending for an area of approximately 100,000 square kilometers, the vast *Llanos* habitat can be found throughout southern Venezuela into parts of Colombia. Grasses and sedges, especially those in the genera *Panicum, Leersia, Eleocharis*, and *Luziola,* and *Hymenachne* dominate much of the landscape. Trees and shrubs are widely scattered, often occurring as "island" woodlots called *matas* (Blydenstein 1967; Walter 1973).

The *Llanos* is a highly seasonal wet savanna, with pronounced dry season extending through most of the northern winter. Approximately 1,000 millimeters of rain is received over the seven-month rainy season (Solbrig and Young 1992). But for nearly five months, rainfall is quite low. It is then when natural fires are common. During dry season, vast flocks of wading birds such as ibises, storks, herons, and egrets are concentrated in the relatively limited remaining wet areas (Plate 11-28). These birds, plus the added presence of such species as capybara (*Hydrochoerus hydrochaeris*) (Plate 11-29), giant anteater (*Myrmecophaga tridactyla*), anaconda (*Eunectes murius*), spectacled caiman (*Caiman crocodilus*), and jaguar (*Panthera onca*), make the *Llanos* one of the best areas in the Neotropics for observing wildlife. Rainy season usually peaks in July. On average, rainfall exceeds 1,200 millimeters (about 47 inches) per year and can be over 1,500 millimeters (about 59 inches) annually. At peak rainy season, the *Llanos* is in full flood, though higher areas, hammocks (called *bancos*) of palms and other trees and shrubs, remain above water. Because of the strong degree of seasonality, plant and animal species must be generally adapted to endure both drought and flood (Kushlan et al. 1985). Many immense cattle ranches can be found scattered throughout the *Llanos*, some of which also serve to host ecotourist groups who want to see the birds and other animals.

PLATE 11-28
Large flocks of birds mass on the *Llanos*, including spoonbills, egrets, and ibis, shown here.

PLATE 11-29
Capybara with young, abundant on the *Llanos*.

The *Llanos* wet grassland savanna supports a diverse assemblage of waterbirds. Kushlan et al. (1985) studied the wading bird community of the Venezuelan *Llanos* and found 22 species of large wading birds, including 7 ibis species, 1 spoonbill, 11 herons and egrets, and 3 storks. Researchers compared the *Llanos* with the Florida Everglades, where only 15 species of large waders occur. Herons were the most diverse species in both ecosystems. Stork species, including the wood stork (*Mycteria americana*), the maguari stork (*Euxenura maguari*), and the huge jabirou (*Jabiru mycteria*), were richest on the *Llanos*, probably because large fish, their principal prey items, were more abundant than in the Everglades. Only the wood stork occurs in the Everglades. The researchers noted differences among species relative to their foraging behaviors, prey selectivity, and habitat use. They hypothesized that the greater diversity of waders on the *Llanos* is due in part to greater habitat diversity, increasing the types of feeding areas available. They noted that several species, including the buff-necked ibis (*Theristicus caudatus*) and sharp-tailed ibis (*Cercibis oxycerca*), fed in very shallow habitats on high ground. Such areas were less available in the Everglades. They also noted that during the dry season, many *Llanos* species fed together, using similar foraging behaviors and feeding sites. They speculated that prey availability is so high that competition among the wader species is minimal.

THE PANTANAL OF BRAZIL AND BOLIVIA

The vast *Pantanal*, a name that means "swamp," is a more southern ecological equivalent of the *Llanos*, sharing many of the same species. Larger in area than the *Llanos*, the *Pantanal* covers approximately 200,000 square kilometers (about 77,200 square miles), of which about 70% is within the state of Mato Grosso do Sul in Brazil, with the remaining area in eastern Bolivia. It is a region of low elevation, only about 150 meters (500 feet) above sea level, a vast, flattened basin created by deposited sediment eroded from the surrounding highlands. Eventually, all of the many *Pantanal* rivers flow into the Rio Paraguay, the *Pantanal* equivalent of the Orinoco. Dry season ranges from May through October (essentially the opposite pattern from the *Llanos*). During rainy season, at its peak from late January through mid-February, water levels can rise as much as 3 meters (about 10 feet), as much of the low-lying grasses and sedges are in full flood (Plate 11-30).

In general, the human population is low in this hyperseasonal wet savanna, with only a few large cattle ranches and scattered small towns and villages. Consequently, the wildlife diversity becomes a spectacle rivaling the African savanna. Riverbanks are lined with myriads of caiman kept well fed by the bountiful populations of piranha, tetras, catfish, and other fish. Giant otters (*Pteronura brasiliensis*) make dens along

PLATE 11-30
Large numbers of wading birds frequent *Pantanal* marshes.

PLATE 11-31
Jabirou nest on the *Pantanal*.

riverine embankments. Marsh deer (*Blastocerus dichotomus*) and red brocket deer (*Mazama americana*) can be seen among the tall *Pantanal* grasses, as well as giant anteater, Brazilian tapir (*Tapirus terrestris*), crab-eating fox (*Cerdocyon thous*), and crab-eating raccoon (*Procyon cancrivorus*). The *Pantanal* abounds with capybara and vast hosts of wading birds including three stork species (Plate 11-31), four ibis species, and a dozen species of herons and egrets, most of which can be found on any given day. The bulky southern screamer (*Chauna torquata*) can sometimes be seen by the dozens foraging like cattle in the wide grasslands. Among the copses of palms, the huge hyacinth macaw (*Anodorhynchus hyacinthinus*) (see Plate 7-11) can be found, the *Pantanal* being the final stronghold for this once-abundant and majestic parrot.

THE BRAZILIAN *CAMPOS CERRADOS*

The largest area of savanna vegetation occurs in central Brazil, the *Campos Cerrados*, forming a wide belt across the country from northeast to southwest (Sarmiento 1983). *Cerrados* are unique, occurring on acidic, deep, sandy soil. Vegetation is not simply grassland but rather ranges from open woodlands with a 4- to 7-meter canopy to dense scrub thicket. *Cerrado* soils are nutrient-poor, a factor probably partly responsible (see the following) for their existence, rather than richer forest. Crop yields are dramatically increased when soil is fertilized with trace elements (Walter 1973).

The *Campos Cerrados* is the preferred habitat of the red-legged seriema (*Cariama cristata*), a large ground-dwelling bird that resembles the secretarybird

PLATE 11-32
RED-LEGGED SERIEMA

(*Sagittarius serpentarius*) from the East African savannas (Plates 11-32 and 11-33). Both the secretarybird and the seriema are similar in morphology and ecology, examples of ecological equivalents. They each stalk insects, snakes, and rodents. Each bird walks as it stalks, often using its powerful legs and talons to subdue snakes and small mammals. The secretarybird is the only member of the family Sagittariidae, which is part of the large order of birds of prey, Falconiformes. There are two species of seriema in the family Cariamidae, part of the order Gruiformes that includes cranes, rails, and bustards, among others. Besides their unique ecology, perhaps the most distinctive aspect of the seriemas is their ancestry. Anatomical and genetic analyses strongly indicate that they are direct descendants of the so-called flightless "terror birds" (Phorusrhacidae) that were top carnivores in savanna ecosystems throughout much of South America during most of the Cenozoic era (see Chapter 2). Though closely

PLATE 11-33
SECRETARYBIRD

similar, there are marked differences between secretarybirds and seriemas, including the fact that secretarybirds are often seen soaring high on the warm thermals emanating from the savanna. This behavior reflects the fact that secretarybirds are members of the order containing hawks and eagles, known for their soaring abilities.

Woodland is part of *Cerrado*. It consists of shrubs and scattered semideciduous trees typically no taller than 8 meters (about 26 feet). The small, stocky trees have dense, twisted branching patterns and thick bark. *Cerrado* areas are highly seasonal and experience frequent natural fires, and the soil is typically very sandy (Walter 1971).

A somewhat similar type of ecosystem, *Caatinga*, scattered throughout parts of Brazil, consists of highly seasonal (with prolonged dry season) deciduous forest dominated by spiny trees and shrubs with thick leaves and thick bark, their branches covered with an abundance of lichens and mosses. The ground may also be rich in mosses and heath-like plants. *Caatingas* occur in climate that could support forest were it not for the nutrient-poor, sandy soils plus marked seasonality in precipitation. These ecosystems are not nearly as diverse as moist forests but nonetheless are characterized by a unique array of trees, grasses, and sedges (Pires and Prance 1985).

The *Cerrado* ecosystem, which is a combination of savanna and dry forest and includes some 4,000 endemic species, is one of the most threatened ecoregions in the Neotropics. It is being widely cut to make room for growing crops, particularly soybeans. In the past 35 years, approximately 2 million square kilometers (772,204 square miles) have been converted to agriculture, an area that represents about half of the original *Cerrado* (Marris 2005). At present 137 *Cerrado* species are now listed as threatened, including the maned wolf (*Chrysocyon brachyurus*).

THORNWOODS

Thornwoods occur in semidesert areas from Mexico through Patagonia. Dominant trees are usually *Acacia* species and other leguminous trees, of short stature, spaced well apart, and often interspersed with succulents such as cacti and agave. In many areas of thornwood, large herds of goats can be seen wandering about. Thornwood is very common along the Pan-American Highway throughout Peru as well as in central Mexico and many West Indian islands (Plate 11-34).

PLATE 11-34
Thornwoods are common in more arid areas in the Neotropics.

THE *PAMPAS*

In extreme southern Brazil continuing southward through Patagonia is an area termed *Pampas*. Mostly grassland, the *Pampas* comprise a vast grassland that is part of the southern temperate zone. The *Pampas* primarily consist of extensive stands of tussock grasses (*Stipa brachychaeta*, *S. trichotoma*). In areas of sandy soil and decreased rainfall, dry woodland occurs, consisting mostly of a single species, *Prosopis caldenia*. Some unique animal species inhabit *Pampas*.

The wind-swept, barren *Pampas* host a diversity of unique animals. The Pampas deer (*Ozotoceros bezoarticus*) is endemic to the *Pampas*. Also present is the maned wolf (*Chrysocyon brachyurus*). Rodents are abundant, including the mara (*Dolichotis patagonum*), sometimes called the Patagonian hare. Superficially resembling a hare (including long ears), this rodent, also called the Patagonian cavy, can leap 2 meters (about 6.5 feet).

Charles Darwin was fascinated by a group of burrowing rodents of the *Pampas* collectively called the *tucotucos (Ctenomys* spp.). Writing in *The Voyage of the Beagle*, Darwin said,

> This animal is universally known by a very peculiar noise which it makes when beneath the ground. A person, the first time he hears it, is much surprised; for it is not easy to tell whence it comes, nor is it possible to guess what kind of creature utters it. The noise consists in a short, but not rough, nasal grunt, which is monotonously repeated about four times in quick succession; the name Tucutuco is given in imitation of the sound.

PLATE 11-35
COMMON RHEA

The largest birds of the *Pampas* are the flightless rheas, relatives of the ostrich. Two species, the common rhea *(Rhea americana)* (Plate 11-35) and Darwin's rhea *(Pterocnemia pennata)* occur on both *Pampas* and more northern savannas. The common rhea, which can also be found on the *Campos Cerrados* and *Pantanal*, is the larger and more abundant of the two. Rheas have the unusual habit of laying eggs in a communal nest. Several females mate with one male, and each hen deposits two to three eggs in the same nest. Only the male incubates. Rheas run very swiftly but were successfully hunted by *gauchos*, the horsemen of the *Pampas*, who brought them down using the *bola*, a twine tied to three balls, which was skillfully tossed from horseback to entangle the bird's neck or legs.

12 Other Tropical Ecosystems: From the Mountains to the Rivers to the Sea

Chapter Overview

Many kinds of ecosystems are present in the tropics in addition to those already discussed. This chapter, with a strong focus on the Neotropics, describes some of these ecosystems, beginning with montane forests and high-elevation ecosystems. Increasing elevation along a mountain slope results in climatic changes that, in turn, result in changing ecosystems. Lowland forest grades into montane forest, much of which is cool, humid, and shrouded in mist and fog. Higher still, forests become stunted, elfin forests of small-stature trees densely laden with epiphytes. Above treeline are diverse areas of windswept grasses and shrubs (called *páramo* and *puna* in the Neotropics), in some places forming a heath-like ecosystem dense with peat. These areas, even though equatorial, are cold, and snow is often present. Nonetheless, many species of plants and animals have adapted to such conditions and thrive.

Riverine ecosystems are found throughout the tropics and have a major ecological impact in terrestrial ecosystem function and diversity patterns. Rivers interact with surrounding ecosystems through the annual flood cycle and also form a diversity of swamps, backwaters, oxbows, sandbars, and river islands. Each of these supports unique assemblages of species.

Mangrove forests are coastal ecosystems consisting of tree species uniquely tolerant of salt exposure. Mangrove forests are found throughout the coastal tropics. Ecologists have attempted to ascertain what factors are responsible for apparent horizontal zonation patterns among mangrove species. Mangroves, along with coral reefs, provide significant ecological protection from tropical storms such as hurricanes and monsoons. Mangroves are also essential in fertilizing coastal seas.

Montane Forests

Tropical regions contain many areas where there are mountains of various heights. With increasing elevation up a mountainside, temperature decreases and condensation and precipitation increase, supporting cooler but humid lower montane forests rich with trees laden with epiphytes. Temperature gradients are steep in tropical regions. There is a 5.2°C to 6.5°C decrease per 1,000-meter (this is roughly 10°F for every 3,300 feet) elevation (Colwell et al. 2008) (Figure 12-1). At greater elevation, upper montane forests prevail, and these typically have small-stature trees, termed *elfin forest*. Higher still, conditions become colder and increasingly windy, often too severe for trees to survive, so that unique high-elevation ecosystems of grass and shrubs prevail. Depending on various conditions, these alpine grassy and shrubby

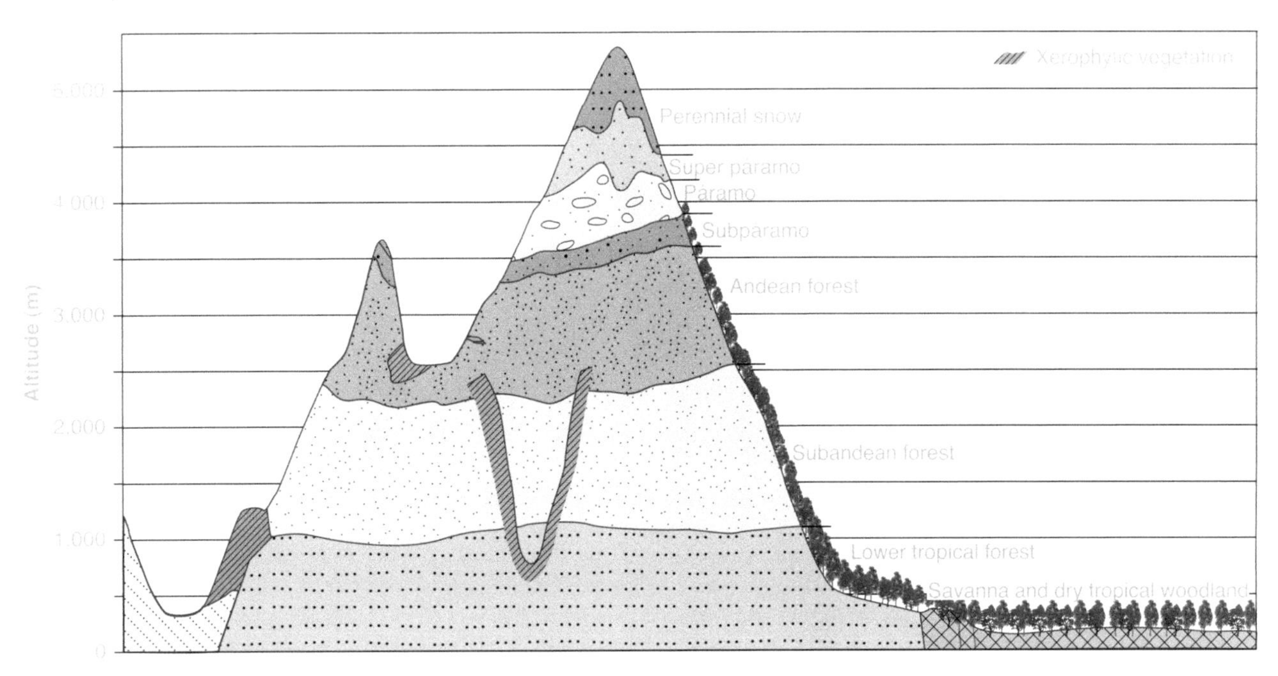

FIGURE 12-1
Vegetation belts in the Cordillera Oriental (Colombia) shown schematically.

areas are called either *páramo* or *puna* (see the following discussion). On rugged mountains such as the Andes, snow occurs at high elevations, even in equatorial regions. Conditions above treeline are frequently foggy and bog-like, with extensive areas of soft peat and mossy groundcover.

Typically, as moisture from the lowland tropical forest rises and cools, montane forest is enshrouded in heavy mist and fog for at least part of each day, giving rise to the term *cloud forest*. Clear morning skies yield to afternoon fog that persists through dark. For obvious reasons, cloud forests look and feel overcast and damp. Cloud forests are found throughout the global tropics and subtropics.

Alexander von Humboldt, who explored the Neotropics from 1799 to 1804, first described ecological changes that occur along tropical elevational gradients. Lowland moist forests grade into cloud (or fog) forests that, if elevation is sufficiently high, grade into elfin forests, páramo, or puna. The changes observed by Humboldt and his associate, Aimé Bonpland, form the backbone of what became the concepts of plant associations, life forms, and isotherms (Chazdon and Burslem 2002). Humboldt realized how changes in physical conditions resulted in changes in ecosystem characteristics.

Under the Holdridge life zone classification (see Figure 1-20), cloud forest (which is a generalized term) falls roughly within three elevational life zones: tropical premontane rain forest, tropical lower montane rain forest, and tropical montane rain forest. Holdridge also recognized transition zones between bordering life zones. Transition zones separating lowland from premontane and premontane from montane forests may be broad.

Tropical cloud forests vary, of course, from one site to another, but all share general characteristics (Hamilton et al. 1995):

- Species richness of most groups of plants and animals is comparable with lowland rain forests but gradually diminishes with elevation. Groups such as various epiphytes, particularly ferns and orchids, may be very species-rich, with higher diversities than in lowland rain forests.
- Groups such as mosses and ferns form abundant components of the vegetation community. Tree ferns are particularly characteristic of cloud forests.
- Most species of plants and animals found in cloud forests have strong taxonomic affinities with species found in low-elevation rain forests.
- The stature of trees diminishes at high elevation, sometimes resulting in dwarf trees. At Monteverde cloud forest in Costa Rica, canopy height varies from 20 to 40 meters (64 to 128 feet) in

sheltered sites but decreases to 5 to 10 meters (16.5 to 33 feet), becoming elfin forests, on exposed sites such as ridges and peaks (Haber 2000).

- Sunlight diminishes due to fog, limiting productivity.
- Precipitation is usually high. At Monteverde cloud forest, annual rainfall is between 2,500 to 3,500 millimeters (97 to 136 inches), and dry season is characterized by mist and cloud cover much of the day (Haber 2000). Other montane forests such as those in parts of Central America and the *Choco* (along the western coast of Colombia and Ecuador) experience much higher precipitation, up to 10,000 millimeters (390 inches) annually. These characteristics keep the forest moist, even during dry season.
- The saturated air inhibits evapotranspiration, making it more difficult for plants to obtain mineral nutrients from soil, and also limiting productivity. Recall that transpiration brings both water and soil nutrients from the substrate into the plant. If the air is cool and highly saturated with water, the physical conditions necessary for effective transpiration are limited.
- In addition to rainfall, precipitation occurs as *fog-drip*, where water from fog condenses on leaf and bark surfaces and drips onto the forest floor, saturating the soil. At Monteverde cloud forest, 22% of the annual hydrologic input is due to wind-driven cloud water from fog (Nadkarni and Wheelright 2000).
- Soil tends to become water-saturated and acidic, creating bog-like conditions in some areas, where decomposition is slowed by high soil acidity. Soils with such characteristics are termed *Histosols*.

Tropical montane cloud forests of some sort can be found in 45 countries and all tropical regions, including the United States (Hawaii and Puerto Rico).

Neotropical Montane Forests

Cloud forests are the dominant kind of ecosystem along a narrow altitudinal belt (from 1,400 to 3,500 meters or 4,620 to 11,550 feet) along the east slope of the Andes (Plate 12-1). Cloud forests occur in Venezuela, Colombia, Ecuador, Peru, and Bolivia, and also in parts of southeastern Brazil (De Barcellos Falkenberg and Voltilini 1995). Cloud forests also occur at higher elevations in the Greater and Lesser Antilles, and in parts of Nicaragua, Guatemala, Panama, and Costa Rica. In Central America, at Monteverde in Costa Rica, cloud forest occurs above 1,500 meters (4,950 feet) on the Pacific slope but extends to between 1,300 and 1,400 meters

PLATE 12-1
Clouds are beginning to cover this Ecuadorian cloud forest as afternoon approaches.

PLATE 12-2
This cloud forest in Venezuela contains numerous conifers (foreground). The fog has fully engulfed the forest.

(4,290 to 4,480 feet) along the Atlantic slope (Haber 2000). Cloud forests are also found in parts of Africa and Madagascar, as well as in Malaysia and New Guinea and other areas in Asia. In Uganda, for example, mountain gorillas still enjoy a tenuous existence in cloud forest habitat. Australia does not have cloud forests, though arguably the forests found on the island of Tasmania share characteristics in common with cloud forests.

Because of persistent cloud cover and fog, sunlight is reduced in cloud forests compared with lowland rain forests (Plate 12-2). Precipitation is abundant, both in the form of rainfall as well as interception of moisture directly from the foggy mist that prevails much of the day. Because the temperature is cool and moisture is abundant, evapotranspiration is reduced. Soil, not surprisingly, is wet, and may become waterlogged.

At Monteverde cloud forest in Costa Rica, three seasons are recognized: wet season from May to October; transition season from November to January; and dry season from February to April (Nadkarni and Wheelright 2000). Wet season typically has clear morning skies followed by cumulus cloud formation and precipitation in afternoon and early evening. Transition season features strong trade winds from the northeast, stratus and stratocumulus clouds, and wind-driven precipitation and mist throughout much of the day and night. Dry season features moderate trade winds and stratus clouds or clear sky part of the day, but also wind-driven mist and cloud water, particularly in the evening hours.

Neotropical cloud forests are lush, with high biomass and obvious abundance of epiphytic orchids, bromeliads, ferns, mosses, and lycopodia, densely covering and draping branches and tree trunks. Vines are also present but are relatively less represented than in many lowland forests. Shrubs are often abundant. Trees, which at high and exposed elevations exhibit gnarled trunks and branches, are from 25 to 30 meters (82.5 to 99 feet) in height, usually not as tall as in lowland rain forest. Two tree strata are often evident. Buttressing is common among trees at lower elevations. Bark characteristics are often difficult to discern, as bark is typically covered by epiphytic vegetation (Haber 2000). Tree crowns are usually compact, and most do not spread, parasol-like, as is typical of lowland forest trees. Leaves are *sclerophyllous*, meaning that they are small, hard, waxy, and usually thick, similar to temperate evergreen oaks. There are many montane palm species, and understory palms are sometimes abundant. Tree ferns (especially *Cyathea*), true ferns that grow to the size of small trees, are often common, adding an almost prehistoric look to the landscape (Plate 12-3). Small ferns, including may epiphytic species, are also often abundant. Bamboos thrive in humid montane forests, with one genus, *Chusquea*, often abundant.

Monteverde cloud forest, Costa Rica, has been intensively studied for many years (Nadkarni and Wheelright 2000). A total of 3,021 plant species, including 755 tree species, has been identified in the Monteverde cloud forest area (Haber 2000). The most species-rich

PLATE 12-3
Tree ferns are common in cloud forests.

component of the Monteverde plant community is epiphytes, with 878 species (Plate 12-4). More than 450 species of orchids occur there (Haber 2000). The five most abundant families of trees in Monteverde are Lauraceae, Rubiaceae, Fabaceae, Moraceae, and Euphorbiaceae (Haber 2000). At La Selva Biological Station in Costa Rica, a lowland rain forest site, the five most-abundant tree families are Fabaceae, Lauraceae, Rubiaceae, Annonaceae, and Euphorbiaceae (Hammel 1990). The overlap is obvious. Shrubs are well represented in cloud forests, particularly within the families Rubiaceae, Asteraceae, and Melastomataceae.

Cloud forests in southeastern Brazil are similar in structure to those along the east Andean slope, though with some differences (De Barcellos Falkenberg and Voltilini 1995). Central American cloud forests have a different species composition from those in South America. There are many oak species that tend to

PLATE 12-4
Cloud forests have an abundance of epiphytes, as is evident in this image.

dominate throughout the region. Various conifers are also common.

Montane cloud forests support many species of plants and animals found in lowland rain forest, but they also harbor numerous *endemic species* (species confined to just this area) (Gentry 1986; Leo 1995; Long 1995; Stotz et al. 1996). High endemism is a defining characteristic for tropical montane cloud forests (Leo 1995). For example, Stotz et al. (1996) point out that the tropical Andes and Amazon Basin each contain approximately the same number of bird species (791 and 788, respectively) but that the Andes have more than twice as many endemic species as the lowland area (318 compared with 152). In the tropical Andes, about 40% of the bird species are endemic, compared with 19% in Amazonia. Some cloud forests are so remote and difficult to reach that new species of birds, all endemics, have been discovered only recently. Ornithologists have described a new species of cotinga (Robbins et al. 1994), a wren (Parker and O'Neill 1985), two antpittas (Schulenberg and Williams 1982; Graves et al. 1983), and an owl (O'Neill and Graves 1977), all from northern and central Andean cloud forests. Monteverde cloud forest in Costa Rica has 5 endemic salamanders and 19 endemic frogs and toads (anuran), a total of 24 endemic amphibian species (but see the section "The Golden Toad and the Global Amphibian Crisis" later in this chapter). Monteverde also has 14 endemic reptiles, including 4 lizards and 10 snakes (Pounds 2000a and 2000b).

Though most Caribbean islands contain some endemic species, they are often not confined only to cloud forest; thus endemism of birds is less pronounced in Caribbean cloud forests than in those of the Andes (Long 1995). Nonetheless, it is interesting that the elfin-woods warbler (*Dendroica angelae*) (Plate 12-5), an endemic strictly confined to Puerto Rican cloud forest, was discovered only as recently as 1971. In Central America, Long (1995) identifies 16 bird species considered to have restricted ranges confined mostly to montane cloud forest. In the Caribbean, endemism is also common. For example, each of five species of todies (Todidae) is endemic to a specific island (Cuba, Hispaniola [2 tody species], Jamaica, and Puerto Rico.)

Other vertebrate groups show considerable endemism in cloud forests, though additional studies need to be done to fully document patterns. For example, there are no published surveys of the fauna in the southeastern Brazilian cloud forest (De Barcellos Falkenberg and Voltilini 1995). Leo (1995) lists 17

PLATE 12-5
ELFIN-WOODS WARBLER

mammal species endemic to Peru, all of which occur mostly in cloud forest. (Some occur in other types of montane forest.) These include 1 marsupial, 1 bat species, 2 primates, 1 sloth, and 12 rodents. Leo (1995) also lists 42 species of anurans (frogs, tree frogs, and toads) from five families that are endemic to Peruvian montane cloud forest.

SOME CREATURES OF THE NEOTROPICAL CLOUD FORESTS

The spectacled bear (*Tremarctos ornatus*) (Plate 12-6) is named for its facial pattern of beige lines surrounding its eyes and cheeks. Otherwise, the creature is black. It is a medium-sized bear, weighing about 200 kilograms (approximately 900 pounds). The only species of bear found in South America, it inhabits low-elevation montane cloud forests from Panama through Peru and Bolivia. The species is considered a *relict*, as it once ranged from the southern United States (California to the eastern seaboard) and throughout Central America (Eisenberg 1989). *Relict species* are those that once ranged widely but are today confined to very narrow ranges. Like most bear species, it is omnivorous, feeding on a wide variety of vegetation (including hearts of bromeliads) as well as small vertebrates and invertebrates. Mostly solitary, spectacled bears have been reduced in population in many areas by hunting, and they are among the most difficult large Neotropical mammals to observe.

PLATE 12-6
SPECTACLED BEAR

Cloud forests are habitat for numerous diverse bird species. One of the most spectacular Neotropical birds, the resplendent quetzal (*Pharomachrus mocinno*), nests in Central American cloud forests. Along with some other cloud forest bird species, quetzals migrate seasonally to lower elevations. Among the more elegant Andean cloud forest birds are the four species of mountain-toucans. Mostly blue-gray with yellow rumps and long, variously colored bills, these birds are restricted to the epiphyte-laden trees of cloud forests. Many of the most colorful tanagers and bush-tanagers are unique to cloud forests (Isler and Isler 1987). These include such gaudy species as the scarlet-billed mountain-tanager (*Aniognathus igniventris*) and grass-green tanager (*Chlorornis riefferii*) (Plate 12-7). Numerous hummingbird species occur in cloud forests, as well as high páramo and puna (see the following). Typical and ground antbirds also frequent cloud forests. The giant antpitta (*Grallaria gigantea*) (Plate 12-8), which at 22.5 centimeters (about 9 inches) is the largest and one of the most secretive of the ground antbirds, inhabits dense cloud forest and is more often heard than seen. The Andean cock-of-the-rock (*Rupicola peruviana*) (Plate 12-9), a large cotinga and close relative of the lowland Guianan cock-of-the-rock (*R. rupicola*) (see Plate 7-23), inhabits rocky ravines near streams at low- and

PLATE 12-7
GRASS-GREEN TANAGER

PLATE 12-8
GIANT ANTPITTA

PLATE 12-9
ANDEAN COCK-OF-THE-ROCK

mid-elevation montane forests. Males gather at subcanopy *leks* (areas where males gather to court females) at daybreak to court females.

THE GOLDEN TOAD AND THE GLOBAL AMPHIBIAN CRISIS

The golden toad (*Bufo periglenes*) (Plate 12-10) was first described for science in 1966 (Savage 1966). This amazing animal was found in the area of the Monteverde

PLATE 12-10
GOLDEN TOAD

cloud forest, on elfin-forest ridgetops. When discovered, the species did not appear endangered, but today it is considered to be extinct. Male golden toads were a brilliant, glowing orange color, highly conspicuous as they gathered for mating in small ponds within the forest. At one location, at least 200 males were observed in one pool with a radius of only 5 meters (16.5 feet) (Savage 2000). Female golden toads were colored differently from males: olive to black with large blotches of bright scarlet, outlined in yellow. It is extremely unusual to see such sexual dichromatism in *anurans* (frogs and toads). In 1987, it became apparent that golden toads as well as other amphibians such as the Monteverde salamander (*Bolitoglossa subpalmata*) had significantly declined. A survey conducted in November 1988 failed to find a single salamander (Pounds 2000a). In April to May 1987, more than 1,500 golden toads gathered at their traditional breeding site. In the following two years, only one golden toad male was observed each year (Pounds 2000b). The species has not been observed since, and its decline, as well as that of other amphibian species at Monteverde, is considered unprecedented (Pounds 1990; Crump et al. 1992). The apparent rapid extinction of the golden toad made worldwide news.

Scientists initially engaged in a strong debate about the apparent amphibian decline, some saying that it represented a global extinction event, others arguing that it reflected oscillations in amphibian abundance that were not particularly worrisome (Pounds 2000a).

It gradually became clear that the losses of amphibian species at Monteverde were not unique. Amphibians were experiencing global decline. What was happening went beyond any normal kind of population cycle. One study suggested that exposure to an amphibian pathogenic fungus (*Saprolegnia ferax*) was exacerbated by increased exposure to UVB radiation (ultraviolet light in strong concentration causes genetic mutations) ultimately brought about by climate change (Klesecker et al. 2001). The study linked climate change, including El Niño/Southern Oscillation (ENSO) events, with greater amphibian mortality ultimately caused by pathogens. By 2004, it was clear that about one-third of the amphibian species in the world faced potential extinction, and a Global Amphibian Assessment project was formed to survey the status of the world's amphibians (Stokstad 2004). The result of the global assessment revealed that amphibian decline exceeded that of birds and mammals (Stuart et al. 2004) (Figure 12-2).

In the Neotropics alone, significant declines in multiple amphibian species were reported for Brazil (Eterovick et al. 2005), Ecuador (Bustamante et al. 2005), Costa Rica (La Marca et al. 2005), and Mexico (Parra-Olea et al. 2005). Within only a few decades, there has been a 43% decline of global amphibian

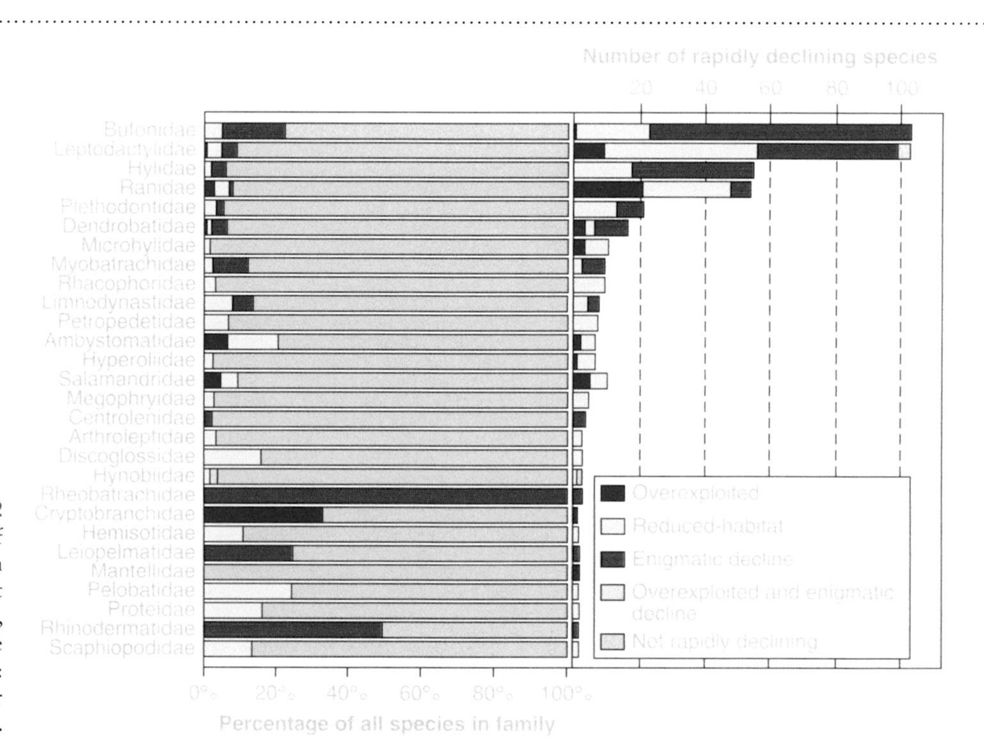

FIGURE 12-2
Percentages and numbers of rapidly declining species in amphibian families (with at least one rapidly declining species), broken into groups reflecting the dominant cause of rapid decline: overexploitation, habitat loss, or enigmatic decline.

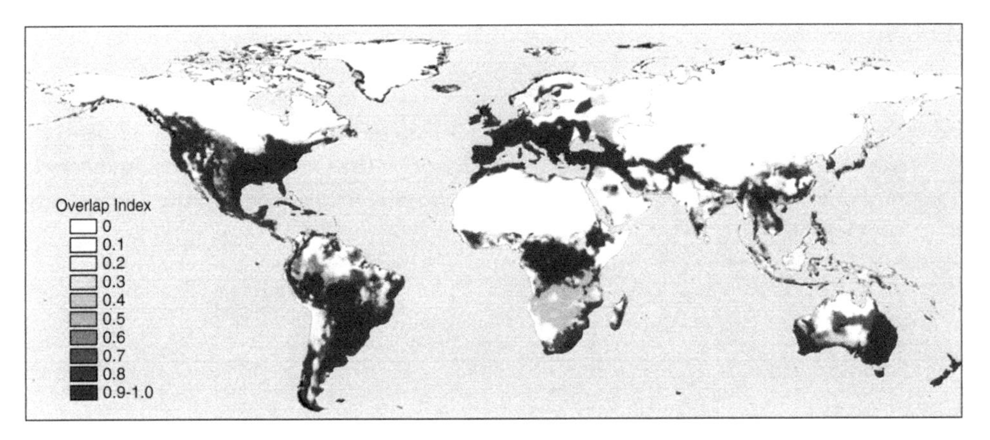

FIGURE 12-3
Predicted distribution of the fundamental niche of *Batrachochytrium dendrobatidis*. Darker regions are those where *B. dendrobatidis* niche presence was predicted by more models (i.e., overlap index 1 means that 10 out of 10 models predicted presence; overlap index 0 means that none of the 10 models did).

species, of which 32.5% are threatened, 34 species extinct, and 88 additional species possibly extinct (Lips et al. 2006).

The proposed causes of the global decline were multiple, ranging from severe habitat loss (not a factor in the case of the golden toad and other Monteverde amphibians) to climate change and pathogen spread (including the implication that climate change facilitates pathogen spread). In addition to *Saprolegnia ferax*, attention soon focused on the chytrid fungus *Batrachochytrium dendrobatidis* (Berger et al. 1998), now the leading suspect in global amphibian decline.

Batrachochytrium dendrobatidis is an emerging and deadly pathogen that invades keratinized areas of amphibian skin on juveniles and adults, as well as entering by mouth in tadpoles. The disease, called *chytridiomycosis*, has *low host specificity* (meaning that it jumps easily from species to species), and it infects only amphibians (Ron 2005). The fungus is able to survive outside the host. Modeling studies suggest that the fundamental niche of the fungus correlates closely with amphibian species richness (Ron 2005) (Figure 12-3). *Batrachochytrium dendrobatidis* is strongly implicated in the precipitous decline of Neotropical harlequin frogs (*Atelopus* spp.) (La Marca et al. 2005). Of a total of 113 species of harlequin frogs, 42 species have been reduced in population by at least 50%, and only 10 species are thought to have stable populations. A study conducted at El Copé, Panama, showed that several amphibian species have experienced precipitous declines, all due to infection by *B. dendrobatidis* (Lips et al. 2006).

Where did chytridiomycosis come from? Why did it spread so rapidly? One hypothesis traces the disease to Africa, where it infects African clawed frogs (*Xenopus* spp.). *Xenopus* frogs are widely used in research around the world and are also used in pregnancy testing. The possibility exists that chytridiomycosis was spread by introducing infected African clawed frogs to various places in the world (ranging from the Neotropics to Australia), inadvertently allowing the fungus to escape, infecting vulnerable populations with no immune defenses specific to chytridiomycosis (Weldon et al. 2004).

Frogs have delicate skin, and chytridiomycosis kills by disrupting skin functioning (Voyles et al. 2009). Studies of green tree frogs (*Litoria caerulea*) that have contracted chytridiomycosis show that electrolyte transport across the epidermis of the skin is impaired by more than 50%. The result is that concentrations of sodium and potassium, essential to normal physiology, are reduced by 20% (sodium) and 50% (potassium) in the blood plasma. The frogs eventually succumb to cardiac arrest.

An additional study suggested that *habitat split*, defined as human-induced disconnection between habitats used by different life history stages of a species, could also have strong negative effects on various amphibian species (Becker et al. 2007). Habitat split would occur if, for example, forest-dwelling amphibian species were forced to make potentially risky migrations over unsuitable habitat in order to find breeding pools. This hypothesis was tested for amphibians in the Brazilian Atlantic Forest, and those species with aquatic larva were found to be losing species richness more rapidly than those in which the life cycle (egg–tadpole–adult) is contained within the forest ecosystem.

At this juncture, there appear to be several causal factors affecting the global amphibian decline. In Latin America, and likely elsewhere, the causes are thought to be habitat loss, overexploitation, and declines caused by disease (chytridiomycosis) and climate change (Lips et al. 2005). The situation can be summarized as follows:

- Amphibian species found at mid- to high elevation, such as montane forests, are far more apt to be declining than those at low elevations, such as rain forest inhabitants. For example, all harlequin frog species occurring at higher than 1,000-meter elevation have declined and 75% have disappeared (La Marca et al. 2005). Some of the declines seem closely related to loss of habitat, but others do not.
- The virulent fungal pathogen *Batrachochytrium dendrobatidis* is clearly implicated, being found at sites that have shown strong amphibian declines. The fungus has been shown to grow best under cool, humid conditions, exactly what prevails in montane areas. This fungus, which causes chytridiomycosis, may be the single most important factor explaining global amphibian declines.
- Climate change is implicated because nine amphibian species (in Ecuador) have undergone upward shifts in their distribution (Bustamante et al. 2005).

Global amphibian declines continue to be monitored, and the situation continues to worsen for numerous species.

Conservation of Montane Forests

Globally, tropical montane cloud forests are among the most threatened ecosystems. In many regions, the loss of montane forest exceeds loss of rain forest (Nadkarni and Wheelright 2000). At the same time, montane forests are highly species-rich. In Panama and Costa Rica, for example, montane forests occupy only about 1.2% of the land area, but plant diversity is almost as high as lowland forest. In addition, about 50% of the cloud forest plant species are endemic (Henderson et al. 1991). Cloud forests in all tropical regions are being cut for many of the same uses as lowland forests: commercial logging, expansion of subsistence agriculture, hunting for sport or commercial trade, exploitation of non-wood forest products, and clearing to make cropland, sometimes for coffee, sometimes for coca. With the increased human activity has come an increase in human-caused fires, as well as threats from introduced plants and animals (Hamilton et al. 1995). Cloud forests are ecologically important both for their function as watersheds and for their unique biodiversity, and their value should be considered along with lowland rain forest in the development of comprehensive conservation and management plans for sustainable use (Doumenge et al. 1995).

High-Elevation Tropical Ecosystems: Focus on the Andes

You can't miss them, especially as you gaze out from a comfortable cruising altitude of 10,000 meters (about 33,000 feet) aboard an Aeroperú jetliner flying north from Lima, Peru, to Quito, Ecuador. Looking out from your window seat, you see beneath you the Andes Mountains (Plate 12-11), formed after dinosaurs became extinct and still growing and changing, geologically recent snow-covered peaks, an immense chain of granite stretching below as far as you can see in either direction. To the east, over the high peaks, lies the vastness of the Amazon Basin, while to the west is a narrow belt of coastline, much of it Atacama Desert, one of the most arid regions in the world. In between, within the mountains themselves, are the high puna, the páramo, and the flat altiplano of the high Andes. The puffy cumulus clouds that sometimes obscure the view of the mountains below are created by the effect of this massive chain forcing moisture-laden air over the mountain peaks. The Andes Mountains support a rich and evolutionarily active diversity of ecosystems and organisms (Morrison 1974 and 1976; Andrews 1982). The Andes have long been recognized for their impact on ecosystem diversity and climate.

The Andes are the dominant topographic feature throughout all of western South America (Figure 12-4). Approaching Quito, which itself is at an elevation of 2,858 meters (9,431 feet), you hope the clouds will lift sufficiently that you can see Cotopaxi, the currently quiescent volcano that looms above the city at an elevation of 5,897 meters (19,460 feet), one of many potentially turbulent mountains along the almost 10,000-kilometer-long (approximately 6,000 miles) Andes chain. The youthful mountain range stretches from Cape Horn, the southernmost tip of South America, all the way north to the Caribbean

(a)

(b)

PLATE 12-11
The Andes Mountains are extensive and active and include numerous snow-capped peaks. The top image shows terraced plots and a village.

Sea, finally terminating in the gentle, densely forested Northern Range on the island of Trinidad.

The Andes Mountains, known in South America as *Cordilleras de los Andes*, run from south to north, the major ridges bending northeast when they reach Colombia and continuing into Venezuela, one ridge continuing northward into Panama. For most of their extent, they are really a complex series of parallel chains divided by a flat tableland area, normally at about 4,000 meters (approximately 13,200 feet)

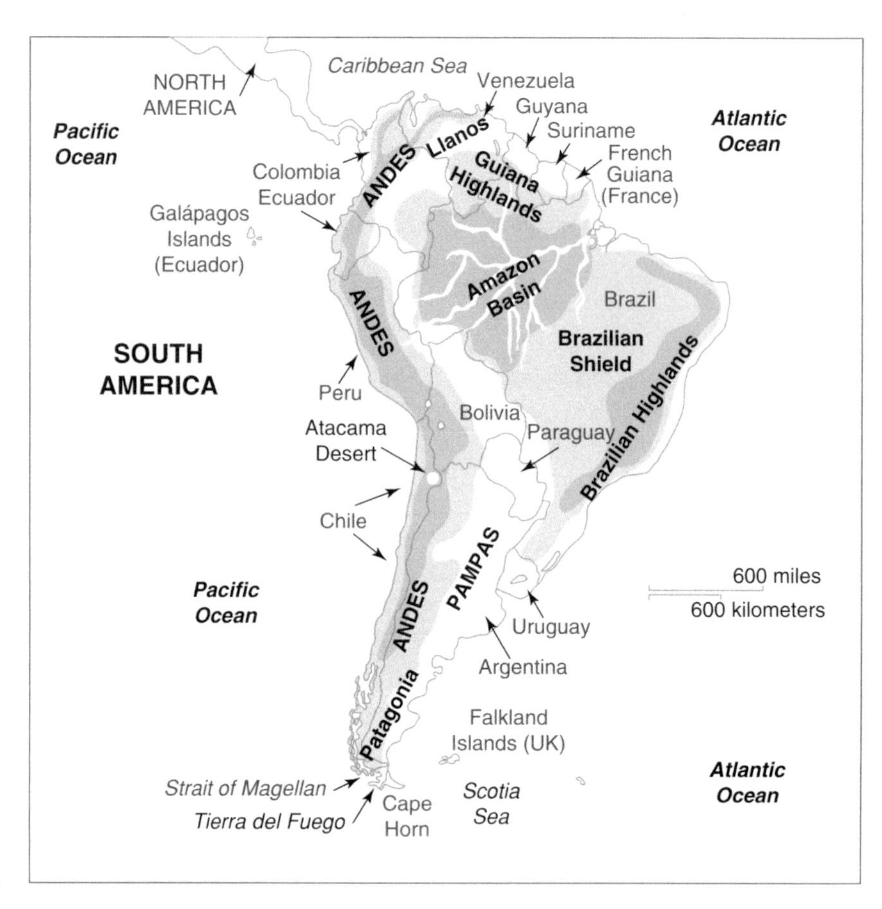

FIGURE 12-4
A general map showing the extensive range of the Andes Mountains.

elevation. This tableland is called *altiplano*, and it is the site of several unique, isolated high-elevation lakes, the largest being Lake Titicaca, which sits on the border between Peru and Bolivia.

Characteristic of young, geologically active mountains, the Andes are ruggedly tall peaks routinely ranging from 2,000 meters (6,600 feet) to 6,000 meters (19,800 feet). There are over a dozen peaks in excess of 6,100 meters (20,000 feet), the tallest being Mount Aconcagua in Argentina, at 6,962 meters (22,834 feet). You must go trekking in the Himalayas or Pamirs of Asia to find mountains of similar stature. (Everest is about a mile higher than Aconcagua.)

The Andes cross the equator, where snow falls at elevations between 4,500 and 5,000 meters (16,500 feet). Climb high enough, and you can in theory stand directly on the equator, face east, and toss a snowball off a very cold mountain. If you have exceptional throwing abilities, the snowball would melt in the hot, humid rain forest some 4,800 meters (15,840 feet) far below. As you move progressively north or south of the equator, the climate becomes more severe and snowline occurs at increasingly lower elevations. Approaching the southernmost part of the Andes, snowline is at only 1,000 meters (3,280 feet).

GEOGRAPHY OF THE CORDILLERAS: A CLOSER LOOK

The Andes Mountains extend well beyond the climatic zone of the Neotropics, beginning at the frigid southern tip of South America, just north of an area called *Tierra del Fuego* (Land of Fire). This land was once inhabited by the indigenous Yahgan (Yamana) Indians, a people living in an extremely harsh climate but who were physiologically capable of sleeping on snow, exposed in the open environment (Bridges 1949). Ships rounding Cape Horn continually face strong *westerlies*, gale-force winds that create some of the roughest seas known. It is here that Captain FitzRoy of the HMS *Beagle* discovered the Beagle Channel, and as he sailed through the channel, Fitzroy noted a 2,438-meter (8,045 feet) mountain that he named Mount Darwin, after the *Beagle*'s most distinguished passenger (Plate 12-12).

The Andes stretch northward, a relatively narrow ridge along the border between Chile and Argentina.

PLATE 12-12
The Andes Mountains near Tierra del Fuego.

Some of the tallest peaks occur east of the cities of Valparaiso and Santiago, near the mountain city of Mendoza in Chile, south of the Tropic of Capricorn. Here, in close proximity, one finds Mount Aconcagua (6,962 meters, or 22,834 feet), Mount Tupungato (6,802 meters, or 22,310 feet), and Mount Mercedario (6,772 meters, or 22,211 feet), and the Andes begin to widen into a series of ridges with extensive tracts of altiplano in between. The lower mountain slopes are temperate in climate, not yet tropical, and precipitation varies, depending on elevation, from between 25 and 102 centimeters (about 10 and 40 inches) annually. West of the mountain ridge, in northern Chile near the city of Copaipo, the Atacama Desert begins, an arid coastal region extending northward almost 3,218 kilometers (2,000 miles), finally becoming the Sechura Desert on the border between Peru and Ecuador.

Where the countries of Chile, Peru, and Bolivia meet, the topography of the Andes becomes increasingly complex, as the mountain range diversifies into a series of ridges with vast area of intervening high-elevation altiplano (Plate 12-13). It is here that there was once an extensive inland sea, the legacy of which remains as unique salt flats (*Salar de Uyuni* and *Salar de Coipasa*) as well as Lake Titicaca and a few other scattered, small lakes. The main chain of the Andes, the Western Cordillera, continues west of the salt lakes and Lake Titicaca toward Machu Picchu (the great city of the Incas) and Cusco, Peru. East of Titicaca, the Cordillera Real and Cordillera de Carabaya gradually descend on their eastern slopes through zones of humid montane forest, eventually terminating as tropical lowland rain forest in western Bolivia and eastern Peru.

In Ecuador and Colombia, the Andes diverge into three major ridges: the Western, Central, and Eastern cordilleras. The Western Cordillera extends north to Central America. The Central Cordillera extends roughly 800 kilometers (approximately 500 miles) northeastward through Colombia. The Eastern Cordillera passes through Bogota toward the northeast, dividing into two ridges: the Cordillera de Perija and the Cordillera de Merida. The former continues northeast in Colombia and terminates on the Guajira Peninsula bordering the

PLATE 12-13
ALTIPLANO

Caribbean Sea, while the latter bends farther northeast, passing into Venezuela, terminating finally in the Northern Range in Trinidad (an island today, but once part of Venezuela; it became isolated only when sea level rose after melting of the glaciers several thousand years ago).

The complex topography of the Andes has strong effects on the distributions and evolution of plants and animals (see Chapter 2). The immensity of the overall range coupled with the divisions of ridges and intervening altiplano, plus the elevational differences along the mountain slopes have provided ideal conditions for evolutionary divergence among many taxa. The fact that countries such as Colombia, Ecuador, and Peru have so many species is due in no small measure to the vicariance potential (see Chapter 2) constantly present because of the Andes.

SOME ANDEAN ECOSYSTEMS: ELFIN FORESTS

Trees and shrubs at high elevations (exact elevation varies with latitude) are noticeably shorter in stature, more gnarled, and more heavily laden with epiphytes, especially mosses, lichens, and bromeliads. Treeline varies depending on the location of the mountain relative to other mountains, a phenomenon called the *Massenerhebung effect* (Grubb 1971). Treeline is typically higher on mountains in close proximity with other mountains, whereas treeline is lower (meaning that the climate is more severe) on isolated mountains. The explanation for such distribution is the climatic influence of the mountains themselves. Mountains in close proximity moderate the influence of wind and retain heat better than isolated, more-exposed mountains.

In high montane ecosystems, there is an abundance of microphyllous epiphytes such as lichens, green algae, and bryophytes covering branches. Such a stunted elfin forest is typically made up of short, twisted trees barely 3 meters (10 feet) tall. These often-dense, diminutive forests exist in a climate of nearly perpetual mist. Tree growth is significantly slowed by a shortage of sunlight as well as low temperatures. Little energy is available for trees to invest in stems (nor is there much point to it since the light is so diffuse); hence stature is short (Grubb 1977). The atmosphere is saturated with moisture, so transpiration is difficult, and thus trees are limited in their nutrient uptake from the soil. Elfin forests have a lower species richness of trees than lower elevation forests, probably due to the greater rigor of the climate. Prominent genera include *Podocarpus*, *Clusia*, and *Gynoxys*. Elfin forests are common at high elevations in parts of Africa and New Guinea, as well as in South America.

POLYLEPIS WOODLANDS

At elevations between 3,500 and 4,500 meters (11,550 to 14,850 feet) in the Andes, above timberline, islands of gnarled trees dominated by the genus *Polylepis* (with about 20 species, family Rosaceae) can be found scattered in a landscape of wet páramo (Plate 12-14). *Polylepis* trees are found in wind-protected, rocky slopes. Though *Polylepis* can be found mixed among other species in lower-elevation cloud forests, the genus occurs in pure stands at higher elevations. *Polylepis* is evergreen, its leaves drought-resistant, with the largest trees reaching heights of about 18 meters (59 feet). Most are of smaller stature. Studies in central Ecuador

PLATE 12-14
Polylepis forest enshrouded in mist.

PLATE 12-15
GIANT CONEBILL

have shown that seedlings survive best when deep within the stand of *Polylepis*, where wind conditions are far less severe. However, vegetative propagation by shoots and ramets (a *ramet* is a stem that arises from an underground root, and thus is a form of asexual reproduction) is highest at the boundary between the woodland and surrounding puna (Clerjacks et al. 2007). This suggests that with protection, stands of *Polylepis* could increase in area. Several specialized bird species are found in *Polylepis* woods, including the unique giant conebill (*Oreomanes fraseri*) (Plate 12-15).

Though *Polylepis* woodlands occur from Venezuela to northern Argentina and Chile, this ecosystem type is considered to be highly threatened throughout most of its range (Stotz et al. 1996; Clerjacks et al. 2007).

TROPICAL ALPINE SHRUB/ GRASSLAND—PÁRAMO

Páramo is a shrub and grass ecosystem (Plate 12-16) occurring at altitudes above cloud and elfin forest from Costa Rica south to Bolivia, generally above 3,500 meters (11,550 feet). Climate is wet and cool (often cold), with frequent nightly frosts throughout the year. Large areas of wet grass are often interrupted by peat bogs.

Approximately 5,000 plant species are known from páramo ecosystems. Dominant vegetation consists of large, clumped grasses called *tussock grasses*, with sharp, yellowish blades, along with a scattering of terrestrial bromeliads and ferns. There are numerous endemic species. Shrubs grow among tussock grasses, some reaching heights of 4 to 5 meters (13.2 to 16.5 feet) so that they resemble small trees. Leaves grow from the base of the stem, surrounding it in a pattern termed a *rosette*.

Among the most distinctive shrubs are the *Espeletias*, perhaps the oddest members of the immense composite family (to which daisies, asters, and goldenrods belong). *Espeletias* have short, thickly-woolly trunks densely surrounded by withered dead leaves,

PLATE 12-16
This combination of grasses, shrubs, and forbs characterizes páramo at high elevations in the Ecuadorian Andes.

topped by a rosette of thick, elongate green leaves, each covered by soft hairs that help minimize evaporative water and heat loss (Plate 12-17). *Espeletias* are indicator species of South American páramo. Similar high-elevation ecosystems in Central America lack *Espeletias* and are called *pseudo-páramo*. *Espeletias*, scattered among the tussock grasses on the cold, windy Andean slopes, attract many hummingbirds and bees to feed on the nectar of the yellow flowers. Among them is the 23-centimeter (9-inch) giant hummingbird (*Patagonia gigas*), the largest of over 300 hummingbird species, and the unique bearded helmetcrest (*Oxypogon guerinii*) (see the sidebar "Some Unique Montane Hummingbirds").

Ecosystems similar to páramo, but with different species groups, are also found at high elevations in much of Africa (such as on the slopes of Mount Kilimanjaro) as well as in New Guinea and elsewhere in Indonesia.

(a)

(b)

PLATE 12-17
(a) *Espeletias* dominate the landscape high in the Venezuelan Andes. (b) *Espeletia* in flower. Note the thick leaves.

Some Unique Montane Hummingbirds

Hummingbirds (Trochilidae) are remarkable for many reasons. They include the world's smallest bird species, they hover and are able to fly backwards, and they attain a high species richness, with 332 species and 104 genera. Hummingbird plumages include much iridescence, making the birds literally glitter when seen in direct light. Hummingbirds feed mostly on nectar and supplement their diet with some arthropod food. Body sizes and bill lengths and shapes vary among species (recall the examples from Chapter 7), a reflection of coevolution between hummingbirds and their principal food plants.

Hummingbirds are entirely confined to the New World, and the vast majority of species are found in the tropics. This includes many lowland species, but perhaps surprisingly, hummingbirds have adapted to montane and high-elevation ecosystems. Many species are found throughout the Andes Mountains, some living under harsh climatic conditions. Some species endure the cold nights by entering a state of torpor, reducing their heart rate and body metabolism, lowering body temperature, to conserve energy.

The 13.5-centimeter-long (5.3 inches) swordbilled hummingbird (*Ensifera ensifera*) is remarkable for its extraordinarily long bill, 90 to 100 millimeters, approximately 4 inches (Plate 12-18). The bill appears adapted to obtain nectar from elongate tubular flowers, particularly

PLATE 12-18
SWORD-BILLED HUMMINGBIRD

PLATE 12-19
WHITE-TIPPED SICKLEBILL

in the genus *Datura*, though they also feed on other flowers with long corollas (Ridgely and Greenfield 2001). Swordbills are uncommon in most of their range because they are seeking widely spread flower species, and they inhabit successional woodlands and disturbed areas at temperate mid-elevations in the central Andes.

The white-tipped sicklebill (*Eutoxeres aquila*) inhabits lower montane forests from Panama to the central Andes (Plate 12-19). It is instantly recognized by its sharply down-curved sickle-shaped bill. Sicklebills inhabit shady understory in dense forest. The bird feeds by perching on *Heliconia* flowers and probing into the corollas, obtaining nectar and arthropods (Ridgely and Greenfield 2001). Sicklebills are among the hummingbirds that practice *trap-lining*, visiting a series of food plants in the course of a day, likely enabling cross-pollination among the plants.

The remarkable 11.5-centimeter (4.5-inch) Ecuadorian hillstar (*Oreotrochilus chimborazo*) is found between elevations of 3,600 and 4,600 meters (11,880 to 15,180 feet) in the windswept puna of the Ecuadorian Andes (Plate 12-20). On cold nights, the hillstar is known to roost in holes in rocks, where it is protected from wind (Ridgely and Greenfield 2001). Hillstars feed exclusively on orange flowers of *Chuquiragua* shrubs that are widely scattered among the tussock grasses of the puna.

The unique bearded helmetcrest (*Oxypogon guerinii*) is found among the tall, shrubby *Espeletias* that abound in the high páramo (3,600 to 4,500 meters, or between 11,880 to 14,850 feet) of Venezuela and Colombia (Plate 12-21). This 11.4-centimeter (4.5-inch) hummingbird conserves energy by actually walking on

PLATE 12-20
ECUADORIAN HILLSTAR

the ground, briefly flying up to capture an insect or to feed on low flowers. Unlike other hummingbirds, helmetcrests rarely hover at flowers. Instead, they perch, methodically plucking insects and obtaining nectar from within the *Espeletias* and other favored nectar plants (Hilty 2003).

PLATE 12-21
BEARDED HELMETCREST

TROPICAL ALPINE GRASSLAND—WET AND DRY PUNA

Puna is defined as cold alpine grasslands where conditions are severe (Plate 12-22). The essential difference between puna and páramo is that puna is more arid. Windswept and cold, puna at higher elevations is sometimes snow-covered. Wind seems to be constant, and billowing fog clouds are commonplace. Tussock grasses are abundant, as well as various succulents, such as cacti. Wet puna (overlapping with páramo) occurs in the northern and western

PLATE 12-22
Puna is windswept and dominated by tussock grass, as shown here.

PLATE 12-23
PUYA RAIMONDII

Andes (Colombia, Ecuador, Peru, and Bolivia). Dry puna predominates to the south (Chile and Argentina). Globally, puna, more generally described as *alpine grassland*, occurs in African mountains and in high elevations in New Guinea. Tussock grasses of higher elevations in New Zealand also form an alpine grassland ecosystem.

Perhaps the most striking of dry puna flora is *Puya raimondii*, the world's largest bromeliad, a huge and dense cluster of sword-like leaves with flowers on a stalk that may rise 8 to 9 meters (between 26.4 and 29.7 feet) (Plate 12-23). *Puya* flowers relatively infrequently. When it does, its hundreds of flowers cluster on a stalk that protrudes well beyond the dense basal cluster of thick leaves. *Puya*, as well as other puna plants, is visited by numerous hummingbird species, including such spectacular species as purple-collared woodstar (*Myrtis fanny*), Peruvian sheartail (*Thaumastura coa*) (Plate 12-24), and green-tailed trainbearer (*Lesbia nuna*).

PLATE 12-24
PERUVIAN SHEARTAIL

PLATE 12-25
ANDEAN CONDOR

Puna is habitat for many mammal species, including the vicuña (*Vicugna vicugna*), a South American member of the camel family. Dominant males group with up to 10 females in herds that roam about the barren puna. Another wild camel of the Andes, somewhat larger than the vicuna, the guanaco (*Lama guanicoe*) is perhaps ancestral to the llama (*Lama guanicoe glama*) and alpaca (*L.g. pacos*) whose origins date back to domestication by pre-Columbian peoples (Morrison 1974). Llamas are the beasts of burden for the mountain Indians, the modern Incas. The husky mountain viscacha (*Lagidium peruanum*), a member of the rodent family, is a close relative of the famed chinchilla (*Chinchilla sp.*), a species now quite local due to overtrapping.

High above puna and páramo, the immense Andean condor (*Vultur gryphus*) soars on heat currents rising from the surrounding valleys (Plate 12-25). The condor soars almost effortlessly, flying from mountaintop to seacoast. Twice the size of a turkey vulture, the condor has a 3-meter (about 10 feet) wingspread (surpassed, but only barely, by the largest albatrosses) and can weigh as much as 15 kilograms (about 33 pounds), making it one of the most massive birds. Vultures are scavengers, and historically the Andean condor was dependent on vicuña and other mammals for its food source. Condors are now known to rely on carcasses of sheep and cattle in some areas.

High Andean salt lakes of the altiplano are habitat for several flamingo species as well as other birds rarely encountered at lower elevations. The James's flamingo (*Phoenicoparrus jamesi*) was actually considered extinct until rediscovered on a lake 4,515 meters (14,900 feet) high in the Bolivian Andes in 1957 (Morrison 1974). More common are the Andean (P. *andinus*) and Chilean flamingos (P. *ruber*). All flamingos feed on brine shrimp and other small crustaceans, skimmed from the water using their peculiar hatchet-shaped bills (Plate 12-26).

Fast-flowing Andean rivers are habitat for the torrent duck (*Merganetta armata*) (Plate 12-27) and white-capped dipper (*Cinclus leucocephalus*). The sleek male torrent duck has a boldly patterned white head with black lines and sharply pointed tail. The female is rich brown. Both sexes have bright red bills. Six races of

PLATE 12-26
JAMES'S FLAMINGO

PLATE 12-27
TORRENT DUCK

torrent duck occur from the northern Andes to the extreme southern Andes. Torrent ducks brave the most rapid rivers, swimming submerged with only their heads above water. The white-capped dipper is a chunky bird, suggesting a large wren in shape. Like the torrent duck, it favors clear, cold mountain rivers, submerging itself in search of aquatic insects and crustaceans.

Elevational Migration and Migration Corridors

The resplendent quetzal (Plate 12-28) is one of many bird species to engage in elevational (also called altitudinal) migration (Stiles 1988). By attaching small radio transmitters (6 grams, less that 3% of the bird's body weight) to quetzals captured during nesting season, it has been possible to reveal a complex elevational migration (Powell et al. 2000). Nesting from January to June in high-elevation cloud forests such as Monteverde (1,500 to 1,800 meters, or between 4,950 and 5,940 feet), quetzals then move to lower elevations (1,100 to 1,300 meters, or between 3,630 and 4,290 feet) along the Pacific slope of Costa Rica, where they remain until October. At that time, they move back to the nesting areas, where they remain for a few weeks. Then they fly to the Caribbean slope (700 to 1,100 meters, or between 2,310 and 3,630 feet) until returning upslope to nest. This complex pattern describes

PLATE 12-28
RESPLENDENT QUETZAL (*PHARAMACHUS MOCINNO*)

their general movement, but there is much annual variability, suggesting that they are seeking fruiting plant species in the family Lauraceae, their principal food (Wheelright 1983; Young and McDonald 2000). The migration pattern of the quetzal demonstrates strong ecological connectivity between high- and low-elevation areas on both slopes. Other frugivorous bird species also have complex elevational migrations. Most hummingbird species at Monteverde move upslope to breed during wet season and migrate to lower elevations during dry season (Loiselle and Blake 1992). Insectivorous bird species do not engage in elevational migration, suggesting that the driving force is fruit availability, not factors such as seasonal changes in weather.

Butterflies are also elevational migrants. At Monteverde, more than half of the 658 butterfly species undergo seasonal elevational migration (Stevenson and Haber 2000). Butterflies' migration patterns vary among families and species, but in general, most migrating butterflies depart lowlands at the end of wet season, moving higher in elevation, into the cloud forest life zones. At one location, Windy Corner (the entrance to Monteverde Cloud Forest Preserve), on December 16, 1994, a total of about 6,000 migrating butterflies was observed over a 5-hour period (Plate 12-29). Other insects such as certain dragonflies, flies, beetles, bugs, parasitic wasps, and moths are also known to be elevational migrants (Stevenson and Haber 2000).

Some bat species as well as Baird's tapir and white-lipped peccary are also suspected to be elevational migrants, at least at Monteverde Cloud Forest Preserve (Timm and LaVal 2000).

The existence of elevational migration in such diverse groups of animals has strong ecological and conservation significance. Elevational migration is a clear example of ecological linkages among life zones. Many frugivorous birds, for example, are important seed dispersers (see Chapter 7), and their elevational migratory movements may prove to be an important influence on the distribution of various plant species whose seeds they disperse (Murray et al. 2000). It has been well established in the study of landscape ecology that ecological corridors serve essential functions in permitting the easy seasonal movement of animals among various life zones. It is essential to recognize the need to preserve contiguous habitats that form elevational migratory corridors. At Monteverde Cloud Forest Preserve, the reality of elevational migration helped in the creation of the International Children's Rain Forest and establishment of Arenal National Park to protect premontane rain forest essential to elevational migrants (as well as non-migrants).

Elsewhere in Costa Rica, a contiguous corridor has been established linking high-elevation Braulio Carrillo National Park with lowland La Selva Biological Station, a corridor that represents an elevational range of 2,871 meters (9,474 feet) (Pringle et al. 1984;

PLATE 12-29
Many butterfly species are elevational migrants.

Hartshorn and Peralta 1988; Pringle 1988). This corridor crosses six Holdridge life zones and transitional zones: tropical wet forest (< 250 meters [825 feet] elevation); wet forest—cool transition; premontane rain forest—perhumid transition; premontane rain forest; lower montane rain forest; and montane rain forest (2,500 to 2,906 meters [between 8,250 and 9,590 feet] elevation). In some places, the corridor is only 4 to 6 kilometers wide, rather narrow to ensure sufficient usage by migrant species. Areas outside the corridor are heavily used by humans and do not represent suitable habitats for elevational migrants and most other tropical forest species.

Ecosystems are strongly affected by climate (see Chapter 1), and current climate change models are expected to significantly alter the distribution of organisms along elevational gradients throughout the tropics. In addition, climate change could result in loss of some lowland species. For more on this important conservation topic, review the section "Elevational Range Shifts and Lowland Biotic Attrition" in Chapter 9.

Neotropical Riverine Ecosystems: Orinoco and Amazon

Rivers are essential habitat components throughout the tropical world. The Nile and its tributaries in Africa; the Mekong flowing through China, Laos, Cambodia, and Vietnam; and the Ganges in India, beginning high in the Himalayas and eventually emptying into the Bay of Bengal, could each merit detailed description. But I have chosen to focus on two major Neotropical rivers, one the largest river (by volume) on the planet.

Because of wet/dry seasonality, tropical rivers throughout the world experience a significant and often complex annual flood cycle that exerts a significant impact on bordering ecosystems, especially gallery (riverine) forests. In addition, rivers are bordered by numerous ecosystems: swamps, marshes, streams, oxbows, and river islands. Each of these habitats contains species that otherwise would not be present, adding significantly to regional diversity.

THE ORINOCO RIVER

The Orinoco River, nearly 2,560 kilometers (1,600 miles) long, flows northeast from the Rio Guaviare in eastern Colombia, bisecting Venezuela before exiting to the Atlantic Ocean (Figure 12-5). Considering average annual discharge, it ranks as the third largest river in the world. The Orinoco Basin, though large, is but one-sixth the area of the Amazon Basin, to which it connects via the Rio Casiquiare, which flows into the Rio Negro and is itself part of the Amazon Basin. Much of the Orinoco Basin drains the ancient Guianan Shield, located southeast of the main river (see the following discussion).

The Orinoco begins at an elevation of 1,074 meters (3,523 feet) in the Parima Mountains close to the border between Venezuela and Brazil. It soon bifurcates into the southern and northern streams, the former of which flows south, eventually joining the Rio Negro and flowing into the Amazon, the latter flowing north and east, joining major tributaries such as the Rio Meta, Rio Arauca, and Rio Apure. The major city located along the Orinoco is Ciudad Bolivar (in Venezuela), where the river is typically about 244 meters (800 feet) wide. Ships can navigate the Orinoco for about 1,120 kilometers (700 miles) from its mouth to the Cariben Rapids, 9.6 kilometers (6 miles) from the Rio Meta.

Like the Amazon and its tributaries, the Orinoco is strongly seasonal. At Ciudad Bolivar, the annual variation in water level is normally between 15 and 18 meters (approximately 50 and 60 feet). Rainy season in the Orinoco Basin is from April to October, dry season from November through March. Discharge rate varies with seasonality, with lowest flow during dry season, only 1/25 to 1/30 of the highest flow during wet season (Meade and Koehnken 1991).

The Orinoco River bisects two distinct geological areas. The right (south) bank of the Orinoco borders Precambrian bedrock from the Guianan Shield, among the oldest geological formations on the continent and, for that matter, on Earth. In contrast, the land bordering the left bank (north) of the main river is geologically recent, formed only a few centuries ago from sediments washed from the Andes and transported across the flattened *Llanos* (Meade and Koehnken 1991). The effect of these differing geological histories is reflected in the differing characteristics of the tributary rivers that drain into the main Orinoco. The right-side tributaries typically are stable, constrained by crystalline bedrock and, especially within the Guianan Shield, abundantly supplied with rapids and waterfalls. The left-side tributaries are unstable, with shifting channels formed by alluvial deposits from the river (Meade and Koehnken 1991).

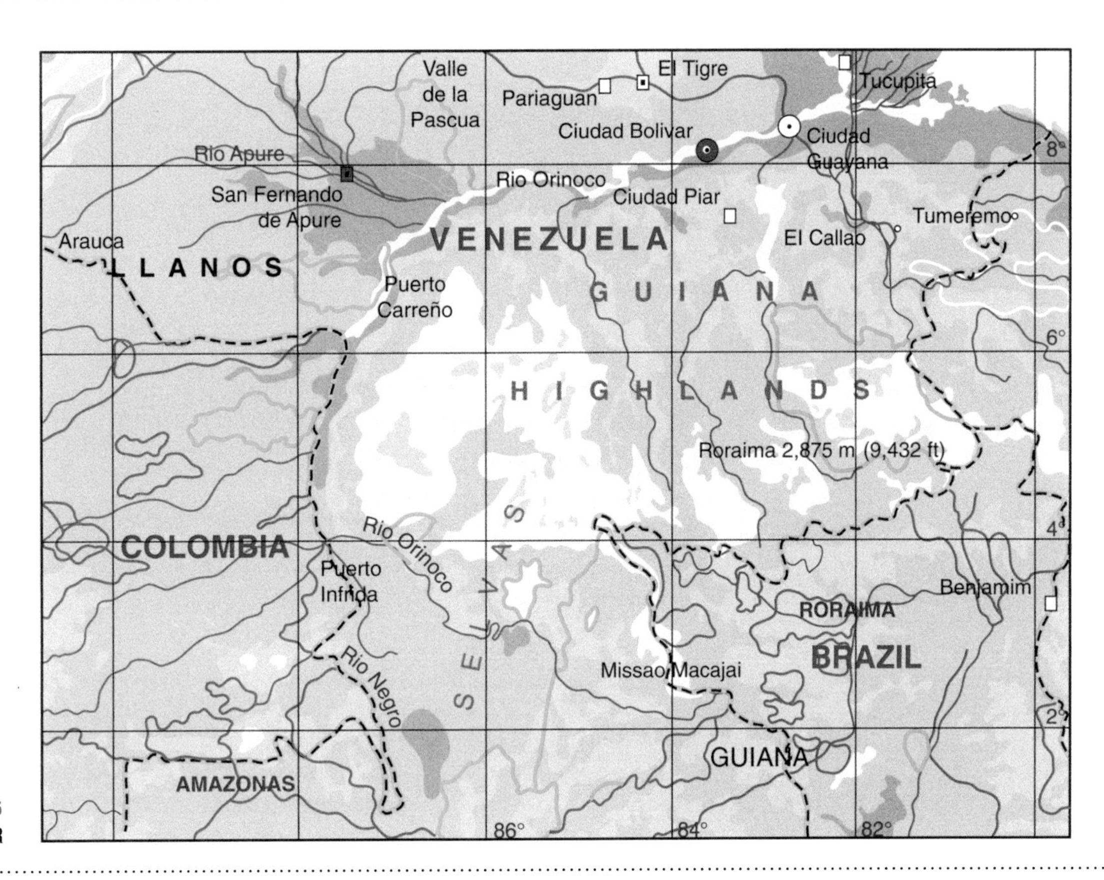

FIGURE 12-5
ORINOCO RIVER

The Orinoco flows west, then north, before beginning its major eastward flow. The river itself has had a strong influence in the geology of the region, having helped cut channels through parts of the Guianan Shield, and thus contributing to the isolation of a unique series of flat-topped table mountains called *tepuis*, some up to 1,500 meters (approximately 5,000 feet) tall (see the sidebar "The Unique Tepuis"). The highest waterfall on the planet, Angel Falls (also called *Kerepakupai merú*), drops about 1,000 meters (3,281 feet) from the top of one of these tepuis. Much of the Orinoco flows quietly and slowly through the vast marshy *Llanos* region of Venezuela (see Chapter 11), a region of relatively flat plains and marshes that supports abundant wildlife.

The Orinoco meets the Atlantic Ocean at the Amacuro Delta and Gulf of Paria, an area of extensive mangrove forests that Columbus explored in 1498, calling it a "gateway of the Celestial Paradise."

The Unique Tepuis

The smooth terrain that prevails in much of southeastern Venezuela, part of the geologically ancient Guianan Shield, is sharply punctuated by about 100 scattered, flat-topped mountains called tepuis, which together occupy an area of about 500,000 square kilometers (200,000 square miles). Tepuis are not part of the Andes, but are much older. Were they located in the United States, tepuis would be called *mesas* or *table mountains*, in reference to their characteristic flattened summits. The word *tepui*, taken from the Penon Indian language, means "mountain." Appearing abruptly from the flattened lowlands of the Gran Sabana and its surrounding tropical rain

PLATE 12-30
The "sky islands" that are the Venezuelan tepuis.

forests, tepuis rise from their forest-enshrouded bases as vertically steep, rocky escarpments to heights of over a mile (Plate 12-30). The tallest, Mount Roraima, is 2,810 meters (9,220 feet). The tepui region, located approximately 650 kilometers (400 miles) south of the coastal city of Caracas, is home to the world's highest waterfall, Angel Falls, which plummets 979 meters (3,281 feet) from atop Auyan-tepui. Angel Falls (named for the bush pilot Jimmy Angel, who discovered it in 1935) is one of hundreds of waterfalls spilling from various tepuis, continuing the ancient process of erosion.

Tepuis are of interest not only for their obvious stark beauty but also for their intriguing geological history. They represent some of the most ancient geological formations in all of South America. Indeed, the sandstone and quartzite rock of which the tepuis are essentially composed has been dated to be at least 1.8 billion years old (George 1989). If one could project oneself back through time, to somewhere between 400 and 250 million years ago, the tepui region would be in close proximity to what would eventually become the division between South America and Africa, an area of lowlands close to the sea. By somewhere between 180 and 70 million years ago, during the Mesozoic era, when dinosaurs were abundant, the future tepui region, known as the Roraima Plateau, was affected by the movement of the Earth's plates, as tectonic activity was separating the continents. At this time, the Andes were being uplifted to the west and the Roraima Plateau was being eroded by a combination of tectonic and meteorologic activity into what would become the tepuis. Evidence for continental drift is seen in that the sandstone of the tepuis is virtually identical to that found in the mountains of the western Sahara. Erosion continued throughout the Cenozoic era, and the flattened tops of today's tepuis are all that remain of the once-extensive Roraima Plateau. Most of the mass of sandstone that once comprised the plateau has long since found its way to the oceans through the continuous process of erosion. Today's tepuis represent but a fraction of that sandstone.

The flattened, eroded tops of tepuis represent *sky islands*, an archipelago of isolated mountaintops. The tepuis receive as much as 4,000 millimeters (about 157 inches) of rain annually, much of it in the form of deluges from thunderstorms, and the plants and animals that tenant the tepuis have evolved essentially in isolation from populations in the surrounding lowlands and, for that matter, on other tepuis. Sir Arthur Conan Doyle was so inspired by the splendid isolation of the cloud-enshrouded, wet tepuis that he chose the region as the setting for his 1912 science fiction novel *The Lost World*, a land where dinosaurs could still be found. No dinosaurs have as yet been located on any of the tepuis, nor are they likely to be. But the biota is nonetheless of great interest.

At least half of the approximately 10,000 plant species are endemic, a clear example of the effect of evolution on isolated populations. Orchids abound, with 61 species found on Auyan-tepui alone (George 1989). Also common are various plants that consume insects, such as pitcher plants and sundews. The soil atop the tepuis is poor, mostly eroded rock. Insectivorous plants are advantaged in such a soil-impoverished habitat because they can supply their need for such nutrients as nitrogen and phosphorus by digesting insect bodies.

THE AMAZON RIVER

The Amazon River, or *Rio Amazonas*, is a vast river that forms at the confluence of the Maranon and Ucayali rivers just west of Iquitos, Peru, flowing eastward 6,437 kilometers (about 4,000 miles) to the sea (Figure 12-6). In Brazil, the name *Amazon* is formally used from Manaus, Brazil, eastward, and west of Manaus, the river is called the *Rio Solimoes*. As mentioned, the main river first takes shape as a confluence of several major Andean tributaries, principally the Ucayali, the Maranon, and, to a lesser degree, the Tigre, all just west of Iquitos, Peru. The name *Amazon* is used in Peru from the confluence point eastward.

The headwaters of the Amazon were difficult to discover because of the severe conditions that prevail at high elevations in the Andes. But the river can be traced to a small, unremarkable tributary, the Carhuasanta, at an elevation of 5,598 meters (18,363 feet) in the cold, windswept Peruvian Andes only about 192 kilometers (120 miles) from the Pacific Ocean. The Carhuasanta flows into the Hornillos, which in turn joins the Apurimac, a major tributary that eventually joins the Ene, the Tambo, and finally the Ucayali. The Amazon system plunges in elevation initially but drops only about 5 centimeters per mile once outside the Andes, eventually flowing to the Atlantic Ocean.

The Nile is the world's longest river (about 6,650 kilometers [4,123 miles]), just exceeding the Amazon, but the Amazon carries by far the world's largest volume of water (Plate 12-31). About 16% of all river water in the world passes through the 320-kilometer-wide (200 miles) delta of the Amazon (Muller-Karger et al. 1988), which daily discharges about 4.5 trillion gallons, or about 200,000 cubic meters of water per second (7.1 million cubic feet per second). This

FIGURE 12-6
Amazon River, with all major tributaries labeled.

PLATE 12-31
Amazon River at one of its wider areas.

represents a discharge of about 4.4 times that of the Congo (Zaire) River, the next most voluminous river. The plume of sediment-laden water from the Amazon can be seen as far as 100 kilometers (about 60 miles) out to sea and has been traced by the NASA Coastal Zone Color Scanner (CZCS) as it moves toward Africa between June and January and toward the Caribbean from February through May (Muller-Karger et al. 1988). The river itself is over 10 kilometers (about 6 miles) wide as far as 1,600 kilometers (1,000 miles) upriver, and large ships can navigate for over 2,300 miles, eventually docking at Iquitos, Peru (Dyk 1995). Two Amazonian tributary rivers, the Negro and the Madeira, rank as the fifth and sixth largest rivers in the world with regard to annual discharge. In comparison, the Mississippi River ranks about tenth and has only about one-twelfth the annual discharge of the Amazon (Meade et al. 1991).

Recall that the Amazon River originally flowed in the opposite direction, draining into the Pacific Ocean near what is today the port city of Guayaquil, Ecuador. The river changed to its present west to east course as recently as 10 to 15 million years ago, when the Andean uplift profoundly altered the course of the river as well as altered patterns of biogeography, creating the Amazon Basin. Initially, the uplift of the Andes created a gigantic lake, bordered on the west by the newly-risen mountain chain and to the east by the extensive Guianan and Brazilian shields. The Amazon finally made its way to the Atlantic during the Pleistocene, cutting through its eastern barrier in the vicinity of Obidos, Brazil (Goulding 1980). Many widespread trees were probably dispersed eastward by the altered course of the river water (Goulding 1993).

The Amazon Basin, drained by the Amazon River and its gigantic tributaries, covers an area of about 6.92 million square kilometers (about 2.67 million square miles), essentially 40% of the total area of South America. Approximately 1,100 tributaries service the main river, and some of them, like the Rio Negro, Napo, Madeira, Tapajos, Tocantins, and Xingu, rank as major rivers.

Amazon tributaries vary in color from cloudy yellow to clear black, depending on where they originate and their geological and chemical properties (see Chapter 10). Examples of major blackwater rivers include the Rio Negro and Rio Urubu. Clearwater rivers include the Rio Tapajos, Rio Trombetas, Rio Xingu, and Rio Curua Una. Whitewater rivers include the Rio Jutai, Rio Jurua, Rio Madeira, Rio Purus, Rio Napo, and the upper Amazon itself.

Before the ever-increasing numbers of roads and airstrips, these tributaries served as the only access to the interior. Cities such as Iquitos, Peru, are, even now, approachable only by boat or airplane. There are no access roads. Where rivers flow, there is continual adequate soil moisture, and evergreen gallery forest lines the banks.

The Amazonian Flood Cycle

Standing in the downpour watching the trails turn into mud, I was at first surprised that the Amazon itself, near were I stood, was dropping nearly a meter a day, even though it was raining heavily throughout the region. But the river's depth was not closely related to the rain falling on this rain forest near Iquitos, Peru. The river was dropping because it had stopped snowing in the Andes, and the meltwater had already drained, absorbed by the Amazon. Now the huge river was receding from its peak flood.

The timing of floods and the distribution of floodwaters result from a complex pattern of seasonal precipitation, much of it in distant mountains. Because of the vast area of the Amazon Basin, at any given time some regions will be experiencing flood while others are at low water. This is because the equator divides Amazonia (though it is north of the river itself), and many of the major Amazonian tributary rivers are either partly north or entirely south of the equator. Rainy season generally occurs in southern Amazonia from October to April. Rainy season in Manaus, in the middle of the basin, is from November to May, and rainfall is highest in the northern part of the basin from April through June (Junk and Furch 1985). The wettest months in Iquitos, Peru (which receives between 3,000 and 4,000 millimeters of annual rainfall, or about 118 to 157 inches), are February through May, though there is much variability. In fact, according to meteorological records kept at Iquitos Airport, every month of the year except May has, at one year or another, been the low-water month (P. Jensen 1996, personal communication), so seasonality in Iquitos is, if anything, quite variable.

In general, flooding in the northern waters occurs as southern waters are low and vice versa. In areas fed by one major tributary, there is a single annual flood. But in those regions fed by both southern and northern tributaries, there are two annual flood periods, which may differ from one another in intensity. As rainy season proceeds, floodwaters build such that the peak of the flood cycle usually occurs at the onset of dry season. Because some parts of the Amazon are in flood while other parts are in low water, there is little difference between the minimum and maximum annual discharge rates, which vary by a factor of only two or three (Nordin and Meade 1982 and 1985). This is in marked contrast to the Orinoco River, all of which lies north of the equator (see the previous section).

The low, flat geomorphology of the Amazon Basin is conducive to flooding. Though sediment has a strong tendency to build up along riverbanks, forming levees, Amazonian rivers will routinely overflow their banks at full flood. The general floodplain is characterized by land that is not uniformly flat, creating habitats such as temporary lakes and swamps. Floodplain forest is estimated to occupy approximately 100,000 square kilometers within the total Amazon Basin (Goulding 1980).

The ecological importance of the flood cycle should not be underestimated. Amazonian rivers experience an annual fluctuation that averages between 7 and 13 meters (23 to 43 feet), which can result in a floodplain forest inundated up to 10 meters (about 33 feet) annually, a water fluctuation that can bring river water up to 20 kilometers (12.5 miles) into the neighboring forest (Goulding 1993). Flooding is essential in dispersing sediment, fertilizing *varzea* floodplain, and enabling fish and other organisms to make use of gallery forest during high water. Zooplankton reproduction peaks during high water, and this resource, which washes into neighboring rivers as the flood recedes, provides invaluable food for fish, especially, but not exclusively, during their juvenile life cycle stages. Until recently, the extensive damming activities routinely seen in countries such as the United States were unknown in South America. Damming changes the flood cycle, isolating previously flooded areas from the annual flood. Doing so throughout much of Amazonia would cause a substantial disruption of many ecological relationships and interdependencies (Goulding et al.1996).

Riverine Habitats

In an ecological sense, rivers do not exist apart from terrestrial ecosystems that border them. A river is dynamic: it varies seasonally and, with time, turns and twists within its floodplain, creating a diversity of habitats.

OPEN RIVER

From a ship sailing upriver on the Amazon, the width of the vast river can take on the appearance of a small inland sea (Plate 12-31). Strong currents and continuously changing underwater sediment bars make navigation challenging, and large ships sometimes run aground. There is not much wildlife to be seen on the open river (though there is much in it). In the central and lower Amazon, skies are typically clear above the widest stretches of the river, while neighboring forests have clouds above, the result of forest transpiration (Salati and Vose 1984). But along the upper Amazon, the humidity is often such that the river itself is cloud-covered and subject to intense cloudbursts.

Amazonian Fish Diversity

There are more than 2,400 species of fish known to inhabit the waters of the Amazon and its tributaries, and up to 800 additional species may remain yet to be formally described (Goulding 1980; Lowe-McConnell 1987). Approximately 40% of the species thus far described are members of two groups, the characins and the catfish (Goulding 1985). These multitudes include many such as the neon tetra (*Hyphessobrycon innesi*), cardinal tetra (*Cheirodon axelrodi*), pearl headstander (*Chilodus punctatus*), silver hatchetfish (*Gasteropelecus levis*), bronze corydoras (*Corydoras geneus*), and oscar (*Astronotus ocellatus*) that are favorites of the aquarist (Lowe-McConnell 1987) (Figure 12-7).

A large number of small species is a general characteristic of the Amazonian fish fauna. It has been hypothesized that small size evolved in response to a diet of tiny arthropods obtained during the flood cycle from within the flooded forest (Goulding 1993). Thus when the tiny neon tetras gather at the surface of the aquarium to grab up minuscule morsels of tropical fish food, they may be exhibiting a behavior originally evolved as they massed around flooded forest trees, gathering up the displaced insects and spiders.

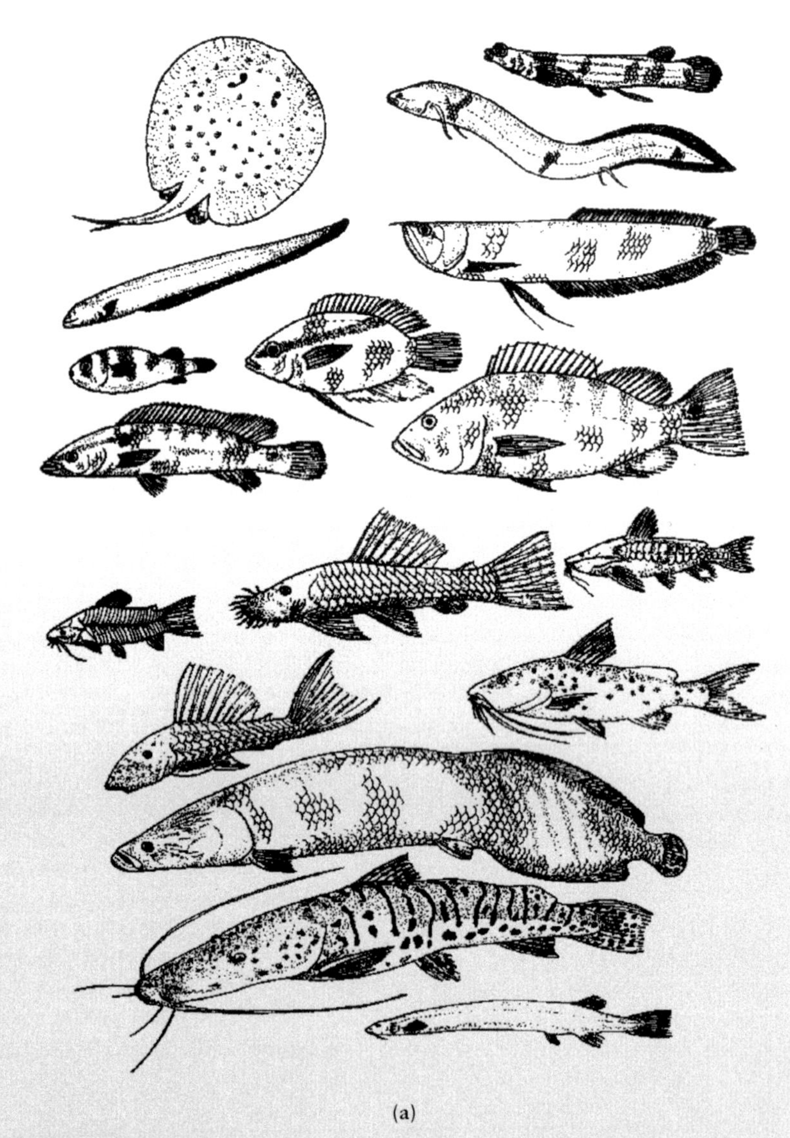

(a)

FIGURE 12-7
The Amazon has an amazing diversity of fish species.

The Amazon Basin is home for the infamous 35-centimeter (about 14-inch) red piranha (*Sarrasalmus nattereri*) and its relatives, a group of fish whose reputation for collective ferocity is rarely merited. Though they are widespread and abundant, piranha are only potentially dangerous when water levels are low and food supply is poor, concentrating the already hungry predatory fish and putting any potential protein source at risk of attack. Piranha do sometimes increase to an abundance where their collective predatory habits act to deplete some local fish species. Along parts of the Brazilian Amazon, a plant known locally as timbo (*Lonchocarpus utilis*) is used to reduce numbers of piranha and piranha eggs. The plant contains *rotenone*, a toxin often used throughout Amazonia by indigenous people to temporarily paralyze fish and thus make them easy to catch. The rotenone, at the concentration used, is lethal to piranha and piranha eggs but does little if any harm to other fish (Balick 1985).

Electric eels (*Electrophorus electris*), which can attain lengths of 1.8 meters (6 feet) and emit a jolt of 650 volts, are fairly common in Amazon waters. They are usually considered to be far more dangerous than piranhas.

(b)

FIGURE 12-7
Continued from page 452

The huge pirarucu (*Arapaima gigas*) (Plate 12-32) is an important protein source for people who live along the Amazon (Goulding et al. 1996). It reaches weights of 150 kilograms (about 300 pounds) and lengths of up to 3 meters (10 feet), a colossal size now rarely seen due to intense fishing pressure from humans. The pirarucu, which preys on many other fish species, occurs throughout Amazonia, especially in quiet lakes. A relative of the pirarucu, the smaller arawana (*Osteoglossum bicirrhosum*), is an elongate fish that always seems to be looking downward. It is one of many Amazonian fish that commonly shows up in the tanks of the aquarium fancier, but it is also an important food fish for Amazonian people.

Among the more unusual piscines is the South American lungfish (*Lepidosiren paradoxa*). This eel-like fish with large scales and thin ribbon-like fins can gulp air in the manner of its ancestors, which swam in stagnant lakes 350 million years ago. There is also a single lungfish species in Africa and one in Australia. The curious distribution of these three species of ancient freshwater fish on three widely-separated continents is almost surely the result of plate tectonics and the breakup of Gondwanaland (see Chapter 2).

PLATE 12-32
PIRARUCU

PLATE 12-33
GIANT OTTER

RIVER AND STREAM EDGE

For observing wildlife, it is far more useful to navigate moderate or narrow tributaries than the main river itself. It is in these areas that species such as giant otter (*Pteronura brasiliensis*) (Plate 12-33), capybara (*Hydrochaeris hydrochaeris*), various caiman (Plate 12-34), anaconda (*Eunectes murinus*), sunbittern (*Eurypyga helias*), and a selection of herons and egrets are best seen. Black-collared hawks (*Busarellus nigricollis*) (Plate 12-35) are common, perching by the riverside as they search for fish, as are several caracara species (yellow-headed [*Milvago chimachima*], black [*Daptrius ater*], and red-throated [*Daptrius americanus*]). Many other bird species are partial to gallery forest, including Amazonian umbrellabird (*Cephalopterus ornatus*), bare-necked fruit crow (*Gymnoderus foetidus*), and gray-necked wood-rail (*Aramides cajanea*). All five Neotropical kingfisher species may be encountered along a shady stream or tributary. At night, streamside trees make ideal perches for various potoo species (*Nyctibius spp.*). Where streams or small

PLATE 12-34
SPECTACLED CAIMAN (*CAIMAN CROCODILUS*)

PLATE 12-35
Black-collared hawk (*Busarellus nigricollis*), common along rivers and streams. It feeds primarily on fish.

tributaries meet the main river, both pink and gray river dolphins often congregate. Plant life along riversides and tributaries is successional (Chapter 6) due to constant disturbance from the dynamic river. Many small thin-boled trees as well as occasional large kapoks and other species may be evident on the floodplain. Cecropias and various leguminous trees are often abundant, and heliconias tend to dominate the understory. Other trees, such as rubber trees and Brazil nut trees, can be found on floodplain streamsides. Various arums are also common, as are palms, especially moriche palm.

Capybara: "Master of the Grasses"

The 1.3-meter-long (4 feet), 65 kilogram (about 140 pounds) capybara (*Hydrochaeris hydrochaeris*) is the world's largest rodent (Plate 12-36). Ranging throughout most of lowland South America, from Panama to northeastern Argentina, the capybara is aquatic, ecologically similar to the African hippopotamus. A cavimorph rodent (closely related to guinea pigs), stocky, and essentially tailless, the capybara has a light tan coat and short, thick legs. The toes are partly webbed, an adaptation to the animal's aquatic habits. The head is

PLATE 12-36
CAPYBARA

squarish, with eyes, ears, and nostrils located on the upper part of the head, an adaptation to an aquatic lifestyle. Being a cavimorph, capybaras' vocalizations sound strikingly like those of guinea pigs, though you must be lucky enough to approach a contented one very closely to hear its charming little squeaks, twitters, and grunts, sounds far more delicate than one might expect from 65 kilograms, or about 140 pounds, of rodent. Though usually found in small family groups, herds can grow to 50 or 100, especially in the Venezuelan *Llanos* or Brazilian *Pantanal*, where capybara remain abundant. Capybara groups typically spend most of their time in and along watercourses feeding on water lilies, water hyacinth, leaves, and sedges that line Amazonian rivers, lakes, and swamps.

The name *capybara* translates to "master of the grasses," a reference to their abundance on wet savannas. Their natural enemies, as might be expected, are caiman, jaguar, and anaconda. In many places humans have hunted them and populations have been drastically reduced. In other areas, however, such as the vast ranches on the Venezuelan *Llanos*, capybara populations, when properly managed, provide a sustained yield of meat and leather (Ojasti 1991).

BEACHES AND SANDBARS

The vast tonnage of sediment washed from the Andes and carried by the tributaries and Amazon itself is often deposited along the river edge or as bars in the river itself. Along the *varzea* forests of the upper Amazon, the sediment is deep blackish-gray, the rich volcanic soil transported from the high Andes. Altered annually by the flood cycle, sediment deposits form extensive beaches and sandbars (Plate 12-37) that provide habitat for birds such as plovers, black skimmers, and various herons and egrets, as well as good resting places for caiman. Swallows of various species, mostly seen in flight over the river, can often be seen perched on small snags and bushes along beaches and sandbars. Soon plants begin to colonize the newly-exposed rich sediment (Plate 12-38).

SANDBAR SCRUB

Andean sediment is rich in nutrients and does not go uninhabited for very long. Various plant species, including such familiar temperate genera as *Salix*, the

PLATE 12-37
Beaches forming from sediment deposit in the Napo River, Ecuador. Vegetation is beginning to colonize.

PLATE 12-38
Grasses quickly colonize exposed beach along the Napo River.

willows, quickly invade and often become sufficiently dense to stabilize the soil. Sandbar scrub is typically dense, composed of a low diversity of fast-growing colonizing plant species (Plate 12-39). Once sediment is deposited from the action of riverine dynamics, the area is subject to colonization by pioneer plant species typical of early succession (see Chapter 6). One comprehensive study of primary succession on western Amazonian floodplains recorded 125 species in the plant assemblage that initially colonized newly-deposited sediments. Of these colonizing species, 81% were dispersed by wind or water. Colonizing

PLATE 12-39
Sandbar scrub has stabilized this island in the Napo River.

PLATE 12-40
River island on the Napo River.

species included many perennial herbs as well as river margin trees, grasses, and various climbers. These pioneer species are typically replaced by more competitive plants as succession proceeds, but they continue to thrive by colonizing newly deposited sediment (Kalliola et al. 1991).

RIVER ISLANDS

Sediment deposition forms many varying-sized river islands (Plate 12-40) that are colonized by plants and that become diverse ecosystems. From the riverside walk at Iquitos, Peru, you cannot see across to the far side of the Amazon River, but that's not because the river is too wide. It is because the far bank is blocked from view by huge sediment islands. Padre Island is the main island visible from Iquitos, though just to the west of the city one finds Timarca and Tarapoto islands. River islands can be of all sizes, but the big ones are stable, composed of years of sediment deposit stabilized by vegetation invasion. Many humans inhabit and farm the river islands (as well as *varzea* floodplain bordering rivers), planting rice, corn, peppers, beans, and bananas. Whole towns can be found on the larger river islands. Forests grow on river islands and can be managed for sustained yield of various products. The riverine people, called *riberenos* in western Amazonia and *caboclos* in Brazil, have long inhabited riverine areas and actively alter the species composition of the forest in order to achieve economic gain. A study from the southeastern portion of the Amazon estuary in the vicinity of Bélem, Brazil, and the Rio Tocantins demonstrated that the local people employ several management techniques, including the active removal of undesirable plant species (such as certain spiny palms); removal of firewood species; cultivation of such species as cacao, avocado, and mango; and maintenance of potentially useful species such as rubber trees (Anderson et al. 1995). The result of such management activities is that the forest is altered in species composition but is nonetheless maintained as forest in a sustainable manner. *Riberenos* also are active agriculturalists, planting maize, rice, manioc, bananas, and other crops. Fish are the major protein source for most river inhabitants.

OXBOWS

Where the flow dynamics of the river become unstable (typically during the high-water period), the river may cut a new channel, effectively isolating a meander and creating what is called an *oxbow lake*, a habitat of essentially standing water (Plate 12-41). Oxbows are

PLATE 12-41
Oxbows are common along the tributaries of the Amazon, Napo, and Orinoco rivers.

common in rivers subject to a variable flood cycle, and they provide yet another kind of riverine habitat, where water stagnates rather than flows rapidly. Such still water supports vast growth of water hyacinth, as well as the giant Victoria waterlilies (see Chapter 7). It is here that the peculiar hoatzin can be found (see the sidebar "Hoatzin: Bizarre Bird of the Riverbanks").

Beaches and sandbar, sandbar scrub, and river islands are all related in a process that ecologists term *point bar ecological succession*. What happens is that when a sandbar forms, it provides habitat for plants, and through seed dispersal, plants quickly colonize (Plate 12-42). As the colonizers stabilize the sandbar, it builds, forming a point bar supporting a sandbar scrub community that replaces the original colonizers. But by then, more sand has been added, so the colonizers persist as the bar grows ever larger. It may, depending on the dynamics of the river, become sufficiently large to form a substantial river island. The dynamic nature of tropical rivers is evident in the ever-changing pattern of point bars and riverine island distribution.

PLATE 12-42
The point bar shown in this image is already being colonized by plants that will eventually stabilize it, contributing to its expansion.

Hoatzin: Bizarre Bird of the Riverbanks

Unique among riverine bird species, the hoatzin (*Opisthocomus hoatzin*) is found along slow meandering streams and oxbows within the Amazon and Orinoco basins (Plate 12-43). Hoatzins resemble chickens in size and shape. However, their overall appearance suggests a primitive, almost prehistoric, bird. A hoatzin is somewhat gangly, its body chunky, its neck slender, its head small. The face is not fully feathered but rather consists of bright blue bare skin surrounding brilliant red eyes. A conspicuous, ragged crest of feathers adorns the bird's head. Its plumage is a combination of soft browns with rich buff on breast and wings. Hoatzins are weak fliers, a feature that contributes to their primitive appearance. They live in noisy groups that occupy dense riverine vegetation. Their nonmusical, guttural vocalizations add to the auditory experience of Neotropical oxbow lakes.

Though originally considered taxonomically related to chickens, hoatzins are possibly most closely related to cuckoos, order Cuculiformes (Sibley and Ahlquist 1983 and 1990). The species is nonetheless sufficiently unique to be placed in its own family, Opisthocomidae. Hoatzins have an unusual diet, unusual breeding system, and unusual juvenile behavior.

Hoatzins are among the few avian folivores, feeding mostly on leaves (over 80% of the diet), often from plants that are typically loaded with secondary compounds, such as plants of the arum family (e.g., philodendrons). Leaves are bitten off, swallowed, and ground into a large bolus in the bird's oversized *crop* (the anterior of the digestive tract). With the aid of a diverse and abundant microflora housed within the expanded crop and esophagus, the bolus slowly ferments and is digested. The birds benefit from some of the digestive products of their microflora, and the bacteria, which are as concentrated in hoatzins as they are in bovines, also help detoxify secondary compounds. The odd amalgamation of partially decomposed leaves gives the bird an unpleasant odor (rather like cow manure), a beneficial characteristic since it renders the flesh distasteful to human hunters. Though a few other bird species are known to eat leaves, hoatzins represent the only known case of a bird species that exhibits foregut fermentation, a unique adaptation (resulting from coevolution with microorganisms) that enables the birds to survive on a diet of normally toxic plants (Grajal et al. 1989).

Hoatzins are communal breeders, and anywhere from two to seven birds cooperate in a single nesting. Nonbreeding birds called helpers typically assist the pair responsible for the eggs. Nests with *helpers* are considerably more successful at fledging young than nests lacking helpers (Strahl 1985; Strahl and Schmitz 1990). The helpers aid in incubation and feeding young, enabling the juvenile birds to grow more quickly and thus reduce their vulnerability to predators. The streamside nest is a cluster of thin sticks so loosely constructed that the eggs are usually visible from beneath.

Baby hoatzins bear a superficial resemblance to *Archaeopteryx*, one of the first birds, whose fossilized remains established that birds evolved approximately 120 million years ago during the Mesozoic era, when dinosaurs flourished. Young hoatzins possess claws on their first and second digits, enabling them to climb about in riverside vegetation. Juvenile hoatzins swim and dive efficiently. Should they be faced with danger, they escape by dropping from the vegetation into the water. When danger passes, they use their wing-claws to help in climbing back on to vegetation. Wing-claws were also present on *Archaeopteryx*, though no one suggests that the resemblance between the modern hoatzin and the first bird is other than coincidental. Young hoatzins lose their wing-claws as they attain adulthood.

PLATE 12-43
HOATZIN

FLOATING MEADOWS

Entire islands of floating grasses can be encountered along the Amazon and within Amazonian lakes (Junk 1970). Along the main rivers, some of these grassy islands occasionally reach a size where they can be a hazard to navigation. Two grasses, *Paspalum repens* and *Echinochloa polystachya*, are abundant components of the floating meadows, and together make up about 80% to 90% of all the floating grass species of Amazonia. *Paspalum* is adapted to float, forming dense, floating mats for the four- to five-month rainy season, when the river is high. The plants grow and spread asexually during this time, but also flower and make seeds so that during dry season multitudes of seeds fall on the newly exposed ground, to quickly germinate. Thus *Paspalum* is adapted to be both a floating and a terrestrial plant. Terrestrial *Paspalum* has a distinctly different morphology from the aquatic form, even though they are the same species. Unlike *Paspalum, Echinochloa* has no floating morph, but remains rooted throughout the flood cycle. This species is most common in lakes.

SWAMPS

A *swamp* is generally an area of woody vegetation that is inundated by standing water for a significant part of the year. Swamps are typically much lower in tree species richness than less wet sites, with some swamps mostly made up of a single species. Many swamp tree species have stilted root systems, and buttressing is extremely common as well. Along the coast, where saltwater incursion is normal, swamps are composed of various combinations of mangrove species (see the section "Coastal Hypersaline Forests: Mangroves" later in this chapter). Throughout interior Amazonia, the most characteristic swamp forests are composed of palms—notably the moriche palm, *Mauritia flexuosa*. Moriche palm is one of the most distinctive, abundant, and widespread Neotropical palms. Growing in swamps and along wet areas, often forming pure stands, a swamp forest of palms is termed an *aguajale*. Moriche palms are tall and slender, their fronds appearing as spike-tipped fans on elongate stalks that radiate from a common base atop the trunk (Plate 12-44).

FLOODED FORESTS

Floodplain forests within the Amazon Basin cover an area of approximately 150,000 square kilometers, which is roughly equivalent to an area the size of the state of Florida (Goulding 1985 and 1993). Overall, floodplain forest comprises only about 4% of the total Amazon Basin, the remaining forest all being *terre*

PLATE 12-44
MORICHE PALM SWAMP

Moriche Specialists: Birds and Moriche Palms

Stands of moriche palms are prime feeding areas for various large macaw species whose powerful bills are sufficiently strong to crack the hard palm nuts that occur in dense clusters below the fronds. Other bird species have become moriche specialists, rarely if ever found away from moriche stands. These species provide examples of an important component of Neotropical species diversity, the tendency toward extreme habitat specialization. The moriche oriole (*Icterus chrysocephalus*), a small, striking black oriole with bright yellow crown, shoulders, rump, and thighs, feeds and nests within moriche fronds (Plate 12-45). The sulphury flycatcher (*Tyrannopsis sulphurea*) is a gray-headed yellow-bellied bird, easily confused with the widespread tropical kingbird (*Tyrannus melancholicus*). Both species can be difficult to see well in the dense palm fronds. More obvious, the fork-tailed palm-swift (*Reinarda squamata*) is a common aerial feeder, streaking through the skies in the vicinity of moriche stands in search of insect prey. This pale swift with a deeply-forked tail builds its nest on the underside of a dead moriche palm leaf. Last, the point-tailed palmcreeper (*Berlepschia rikeri*) is a unique member of the Furnariinae, or ovenbird, family. It is entirely confined to moriche palm stands in Amazonia and is nowhere really common. It is difficult to spot among the dense fan-like palm fronds, but its presence can be known from its song, a loud series of ringing notes (Ridgely and Tudor 1994). The bird is bright cinnamon on the wings and tail and streaked boldly with black and white on its head and breast. See it really, really well, and you'll see its bright red eyes.

PLATE 12-45
MORICHE ORIOLE

firme (Goulding 1993). Floodplain forests are so named, of course, because they are inundated by the annual flood cycle. Depending on location, floodplain forests may be inundated anywhere from 2 to 10 months out of the year. For example, the Amazon forest itself (from Manaus eastward) is flooded for about 6 months, whereas the upper Rio Madeira is in flood for only 2 to 5 months annually (Goulding 1985).

Flooded forests have much less tree species richness than *terre firme* forests. Some are largely dominated by a single species. This was thought to be due to the greater physiological stress imposed on the roots of trees that are forced to remain submerged, the result being that fewer species have succeeded in adapting to such protracted inundation. One study showed that one tree species that is often dominant in flood forests is able to maintain its rate of photosynthesis while flooded, whereas three selected tree species from outside flooded areas were less able to do so (Lopez and Kursar 1999). However, it has also been shown that species routinely flooded do not respond differently (with regard to numerous anatomical and physiological parameters) to flooding than tree species that are never subject to flooding (Lopez and Kursar

2003). This poses a dilemma of sorts in explaining why flooded forest tree species persist. It is hypothesized that dry season effects may contribute strongly to survival of swamp-inhabiting species. In particular, the ability of root systems to grow deeply in flooded forests may be essential in surviving drought stress during dry season. In addition, efficient seed dispersal, particularly the ability of seeds to survive prolonged periods of immersion, may also contribute strongly to the success of flooded forest tree species.

Flooded forests may border whitewater, clearwater, or blackwater rivers. Because sediment load varies, forest productivity varies. Whitewater flooded forests are typically higher in stature and biomass (and probably species richness) than clearwater rivers. Flooded forests of blackwater rivers are typically low in stature, and species richness tends to be a bit lower and to vary less from site to site as compared with whitewater rivers (Goulding 1985 and 1993), and many species are important fruit and seed consumers as well as seed dispersers (see the following discussion). There is no simple explanation for how plant species that may be flooded for most of the year are able to survive such conditions. Their roots lack access to oxygen for much of the year, and yet there are no obvious physiological or anatomical adaptations that explain how they endure such immersion. Certain monkey species are restricted to floodplain forest, as are many species of birds and arthropods. Some plant species are unique to flooded forests, though many have closely related species in dry forests, suggesting a recent speciation between dry and floodplain species. For example, two species of closely related palms in the genus *Astrocaryum* are distributed such that one is abundant in flooded forests while the other is found only on *terre firme* (Goulding 1993). Floodplain forest is not an absolute term. Some forested areas are located immediately adjacent to the river and flood frequently, while areas farther from the river may be in flood infrequently. For Tambopata Reserve in Amazonian Peru, Phillips et al. (1994) recognize seven types of flooded forest:

- *Permanently waterlogged swamp forest*—Former oxbow lakes still flooded but covered in forest.
- *Seasonally waterlogged swamp forest*—Oxbow lakes in the process of filling in.
- *Lower floodplain forest*—Lowest floodplain locations with a recognizable forest.
- *Middle floodplain forest*—Tall forest, flooded occasionally.
- *Upper floodplain forest*—Tall forest, rarely flooded.
- *Old floodplain forest*—Subjected to flooding within the last two hundred years.
- *Previous floodplain*—Now *terra firme*, but historically ancient floodplain of Tambopata River.

Coastal Hypersaline Forests: Mangroves

Forests of mangroves line tropical coasts, lagoons, and offshore islands everywhere in the tropics. The ecological community dominated by mangrove trees of various species (see the following discussion) is sometimes called *mangal*. The term *mangrove* is not taxonomic, but rather refers to a series of characteristics that mangroves have in common. In other words, *mangrove* is based on physiological adaptations. Tomlinson (1995) defines mangroves as trees that are ecologically restricted to tropical tidal areas, where they tend to form pure or low species-rich stands, and where they reproduce often by making new plants (*viviparity*) rather than by seeds. In addition, they often exhibit aerial roots of some sort, and they are strongly adapted to tolerate high levels of immersion in saltwater. Last, many mangroves, though distinct from their nearest terrestrial relatives at the species level, have fairly close relatives on *terre firme*.

Mangrove forests are unique in several ways. One is that they form essential habitat for many kinds of animals. Frigatebirds (*Fregata* spp.) as well as many other species of birds commonly nest among mangroves (Plates 12-46 and 12-47). Migrant birds use mangroves as valuable wintering areas because mangrove forests are rich in arthropod food. Perhaps most importantly, mangroves capture a great deal of energy that supports a rich marine ecosystem and contributes to the health of offshore marine ecosystems such as coral reefs. Mangrove leaves that drop decompose slowly, with numerous invertebrates and microbial organisms using them as a food source. As microbial communities cover the decomposing mangrove leaf, the carbon-to-nitrogen (C:N) ratio declines because more nitrogen is added by the presence of the decomposer community. Small fish and invertebrates, ranging from larvae to immature animals, feed on the leaf fragments, now rich in both carbon and nitrogen. Thus the energy captured by the mangrove leaf is slowly released in little pulses, supporting a diverse animal and microbial community. Mangroves are

PLATE 12-46
This mangrove forest on a cay in Belize is a nesting colony of magnificent frigatebirds (*Fregata magnificens*). The red spots visible in the foliage are the expanded throat pouches of the male birds.

PLATE 12-47
This male frigatebird has its throat pouch fully expanded, an attempt to attract a female.

nurseries for many fish and invertebrate species that will enter the coral reef or *pelagic* (meaning open ocean) food web. Mangrove provides important habitat for many kinds of insects and other arthropods, nesting sites for birds, as well as shelter and food for many fish and invertebrates (Alongi 2009). Mangrove ecosystems are an essential component of coastal ecology. Mangrove ecology is summarized in Walsh (1974), Lugo and Snedaker (1974), Rodriguez (1987), Tomlinson (1995), and Alongi (2009).

The function of coastal mangrove forests in supporting marine diversity is destroyed if the mangroves are removed, and that is what is occurring in many areas throughout the tropics. Mangrove forests are being replaced by such things as shrimp aquaculture, which eliminates the energy pulses that mangroves otherwise supply. A tragic cyclone struck along the Irrawaddy Delta in Myanmar (formerly Burma) in May 2008, an event made more severe because many coastal mangrove forests, which would have buffered the interior from the brunt of the storm, were cut and replaced by rice paddies and shrimp farming. There is an increasing emphasis on the conservation value provided by mangroves, and efforts are under way to use *silviculture* (the cultivation and harvesting of trees for economic gain) techniques to enlarge mangrove ecosystems (Saenger 2009).

All mangroves are woody plants that tolerate high internal concentrations of salt, a consequence of frequent immersion of their root hairs in high-salinity

PLATE 12-48
Propagules of red mangrove attach to sediment, and the plant begins to grow quickly, forming prop roots.

tropical seas. These plants respond to physiological challenges of high temperature and concentrated salt exposure with a variety of adaptations. For example, some have salt glands on their leaves that effectively remove excess sodium and chloride. Others accumulate salt in special areas in their leaves, and still others filter salt from the root system. Mangroves are also tolerant of soils low in oxygen. The thick, odorous, muddy substrate that anchors them is virtually anaerobic, devoid of any gaseous oxygen. But some mangroves have roots that extend above the soil, modified to obtain oxygen from air. Mangrove leaves are similar to leaves of many rain forest tree species in that they are simple and unlobed and very thick, with a heavy waxy cuticle. This aids in storing water and preventing excess water loss through transpiration. Their major control of gas exchange, like all land plants, is through stomatal openings on the leaves.

About 34 mangrove species occur globally, all restricted to tropical waters where the average water temperature is never less than 23°C (73°F) (Rutzler and Feller 1987 and 1996). Southeast Asia has the greatest diversity of mangrove species, whereas the Neotropics has the fewest species of mangroves of any global region, with eight species, some of which are relatively uncommon (Spalding et al. 2010). Some mangroves are adapted to colonize shallow sandflats, trapping sediment and gradually building a dense, muddy, organic soil (Plate 12-48). As mangroves are far more salt-tolerant than other tree species, they tend to line tropical coasts, and their abundance extends inland along tidal rivers. Though subject to disturbance (especially from hurricanes and monsoons), they eventually rebound, though recovery is often protracted. Mangroves range in height from short and shrublike to anywhere from 10 to 20 meters (33 to 66 feet), occasionally taller.

TYPES OF MANGROVES

Red mangrove (*Rhizophora mangle*) is an abundant species that can grow as a bushy shrub, a stunted tree, or a full 20-meter-tall (66 feet) tree. It has reddish bark and numerous aerial *prop roots*, some of which are firmly anchored to the substrate and some of which grow downward toward it. Prop roots provide a firm anchor for the plant. The broadly spreading roots help assure stability against winds, tides, and shifting sands. Prop roots contain openings called *lenticels*, important in transporting air to the oxygen-starved deep roots. Leaves are oval and thick, dark green above and yellowish below. Flowers are pale yellow with four petals. Fruits are reddish brown and produce elongate green seeds that actually germinate while still attached to the parent plant. Seedlings resemble green pods and are about the length of a pencil. They drop from the plant and float horizontally in the sea, becoming flotsam in the tropical ocean. Dispersal is effective, and red mangrove has populated all tropical seas around the world. Seedlings eventually absorb sufficient water that they orient vertically as

PLATE 12-49
The interior of a mangrove swamp is dense with roots, stabilizing the sediment and helping expand the forest.

they float in the sea. Should the tide carry the vertical seedling to a shallow area, once it touches substrate, it will anchor and begin to put out roots (Plate 12-49).

Studies conducted on Twin Cays in Belize have shown that red mangroves that do not exceed a meter in height may, nonetheless, be several decades old. The reason for the stunted growth is the shortage of phosphorus (Rutzler and Feller 1987 and 1996). This study demonstrates that mangrove growth can be exceedingly slow, an important point when considering how mangrove recolonize after disturbance.

Black mangrove (*Avicennia germinans*) is abundant throughout the Neotropics, often forming pure stands in anoxic substrate. Black mangroves tend to grow in less-exposed areas than reds, but they are able to thrive in oxygen-starved sediment. They are bushy topped trees that can reach heights of up to 20 meters. Leaves are oval, leathery, and downy white underneath. The flower is yellow and tubular, the fruit green and oval. Seedlings float and, like adult trees, are tolerant of low oxygen levels. The most notable feature of black mangrove is its root system. Shallow horizontal roots anchor it in thick, smelly, anaerobic mud, but these roots send up vertical shoots, above ground, called *pneumatophores*. Lenticels on the pneumatophores feed into wide air passages connecting with underground roots, providing a means for air transport to the oxygen-starved root system.

Two other common mangroves are white mangrove (*Laguncularia racemosa*) and buttonwood (*Conocarpus erecta*). White mangrove tends to grow at slightly higher elevations than red and black mangrove. It is less tolerant of prolonged immersion in the sea. White mangrove has scaly reddish-brown bark and greenish-white flowers. It grows from 9 to 18 meters (30 to 60 feet) tall. Buttonwood resembles white mangrove but occurs only well away from daily flooding by saltwater. It is the least salt-tolerant of any of the four common mangroves and was once not considered to be a mangrove. As salinity declines, such as at higher elevations or with distance, pure stands of red or black mangrove give way to mixed stands of several species, not all of which are mangroves. The transition from mangal community to upland community is generally gradual.

HOW A MANGROVE CAY DEVELOPS

Red mangrove lines the outer edge of a cay, and in the sea just beyond the cay, red mangrove saplings grow. A careful look at the pattern of mangrove distribution in the cay suggests *zonation* among the various species. Outermost are *pioneer* red mangroves, followed by black mangroves, and innermost are white mangroves and an occasional buttonwood. This zonation pattern was once thought to correlate with the tolerance each species has for saltwater immersion, but in fact, black mangrove is the most salt-tolerant, so the pattern is not a simple case of response to salt exposure. The cay appears to be expanding outward. Red mangroves continuously colonize the outermost edges of the cay, but as sediment builds and the cay rises, black mangroves in turn expand their ranges outward, mixing with the reds. Whites and buttonwoods likewise expand their ranges as the sediment builds higher ground, as these species are most sensitive to immersion in saltwater. Such a pattern has also been described for mainland coastal mangroves. Reds are outermost, blacks intermediate, and whites and buttonwoods innermost.

Some authorities dispute the idea that mangroves are sharply zoned and represent a sort of successional sequence (Rodriguez 1987). Further, there is doubt about how routinely mangroves build up cays by trapping sediment. Experiments designed to test the efficacy of mangroves in cay reclamation following hurricane damage did not prove successful (Rutzler and Feller 1987). Geological evidence suggests that mangrove cays originate from coral deposition, not sediment accumulation by colonizing mangroves. Not all mangrove areas seem to accumulate sediment, and changing conditions caused by storms and tides certainly influence the pattern of mangrove distribution. Some mangrove cays have remained stable for many years, without significantly expanding or contracting.

Experimental work from Belize examining the response of both red and black mangroves to a tidal gradient has demonstrated the complexity of factors influencing zonation between these two abundant species (Ellison and Farnsworth 1993). Only red mangrove *propagules* (seedlings capable of colonization) were capable of becoming established in the lowest low-water zone, where the inundation of water is greatest and available oxygen lowest. The species not only could survive, but thrived, with high rates of root and stem growth, as it became firmly attached to the muddy substrate. Red mangrove propagules exposed to mean water depth grew less well, as a result of high rates of herbivory. Black mangrove could not become established in lowest low water, and thus red mangrove enjoys both freedom from competition and lower predation pressure in this zone. Both red and black mangrove can succeed at mean water (on the tidal gradient), and thus they may engage in interspecific competition in this zone. Insect herbivory is also intensive in this zone and could affect competition between mangrove species. More research is needed to know to what degree herbivory is a factor. It has been suggested that black mangrove may be a gap specialist species, dependent on chance disturbance in order to become established (Ellison and Farnsworth 1993). Differential propagule predation of mangrove species may also have a significant influence on zonation patterning (Smith 1987). Experimental work in Australian mangrove communities has suggested that propagule predation by such creatures as crabs can result in a loss of about 75% of propagules. Most important, significant differences exist among the rates of propagule consumption from one mangrove species to another. Should Neotropical mangrove communities be similarly affected by seed predation differences, distribution patterning could be strongly influenced. Like temperate intertidal communities, mangrove zonation at lowest elevation (maximum immersion) may be most affected by physiological tolerance differences among the various mangrove species, while at highest elevation biotic influences such as competition and herbivory are the predominant determinants.

It is also evident that disturbance history has a strong influence on zonation patterns. A study performed in Belize looked at Calabash Cay, a cay that had borne the brunt of a category five hurricane (Hurricane Hattie) in 1961 (Piou et al. 2006). Four sites were compared that had experienced different degrees of destruction from the hurricane, and four mangrove zones were examined at each site. The differences among the zones were best explained by disturbance intensity. Zones showed clear differences that could be traced back to the 1961 hurricane (Plate 12-50).

PLATE 12-50
This is Bird Cay off Dangriga, Belize, in 1978, some 17 years after Hurricane Hattie.

PLATE 12-51
Scarlet ibis flock near sunset at Caroni Swamp, a coastal mangrove forest in Trinidad, near Port-of-Spain.

Mangrove and coral reef ecosystems bear the brunt of tropical hurricanes and monsoons. But these ecosystems rebound. It is clear that coral reefs and mangrove forests provide a strong degree of protection from coastal erosion and damage from major storms. In addition, mangroves, by their high productivity, act as key species in nutrient flux in coastal tropical ecosystems. Mangrove forests are a unique resource that helps maintain the high diversity and productivity of tropical coastal regions, attracting colonies of unique nesting bird species such as the spectacular scarlet ibis (*Eudocimus ruber*) (Plate 12-51).

13 Humans as Part of Tropical Ecosystems: Focus on the Neotropics

Chapter Overview

Humans first evolved in the tropics and have been part of tropical ecosystems throughout the lineage of our species. People have exerted major impacts on tropical ecosystems extending back thousands of years in history. Archaeological discoveries confirm the intensive use of some tropical regions for sophisticated human civilizations and agriculture. In Amazonia, for example, there is currently a debate about just how much pre-Columbian human impact occurred throughout this vast region. Humanity began extracting food and fiber from tropical ecosystems through hunting and gathering, an activity that does not sustain dense human populations and that typically has limited impact at an ecosystem level. With the discovery of agriculture around 10,000 years ago, humans learned to simplify food webs and direct much more energy to themselves. This in turn stimulated human population growth, permitting the establishment of permanent villages and cities. The impact of agriculture on ecosystems is significant, involving widespread conversion of habitats, reducing normal biodiversity and replacing it with crops. Today, human populations continue to utilize tropical ecosystems in numerous ways. Few traditional hunter-gatherer groups remain, but many indigenous tropical peoples practice a combination of hunting, gathering, and rotational farming. This chapter reviews some of the various examples of how humans have used tropical forests in the past and continue to do so today, ranging from *slash-and-burn agriculture* (where trees are cut, burned, and the soil used for short-term crop growth) to *agroforestry* (the planting and cultivation of specific tree species to maximize and sustain a yield). Most examples are from the Neotropics.

Human Impact on Ecology

Today, it is obvious that human beings, *Homo sapiens,* are a force of unique, indeed extraordinary, importance in the influence we collectively exert on global ecosystems. It could fairly be stated that Earth has entered the "Anthropogenic era," a period in its history when a single species exerts a unique and inordinate influence over global ecology. Humans alter food webs, reducing biodiversity and simplifying energy movement to direct energy to people rather than dissipating it through a complex food web. Agriculture, invented approximately 10,000 years ago, is a process of simplifying the food web while amplifying net primary productivity (NPP). Today, approximately 40% of global NPP is channeled directly to humanity (Vitousek et al. 1986 and 1997).

PLATE 13-1
Olduvai Gorge, Tanzania, site of numerous hominid fossil discoveries.

Humanity evolved in and emerged from Africa, its history traced to the remarkable fossil *Ardipithecus*, estimated at 4.4 million years old (Gibbons 2009). Even earlier, approximately 6 million years ago, *hominins* (human-like apes) left trace fossils in Africa (Plate 13-1).

The species *H. sapiens* has been present for at most 200,000 years. Cultural advances such as use of fire as well as use of tools and metals and development of language and writing are more recent developments. The roots of our species are firmly planted in the tropical forests and savannas of Africa. Humans long ago colonized all regions within the global tropics and radiated as well into temperate and polar latitudes. In doing so, people have exerted significant influences on regional ecosystems. The history of agriculture, emergence of numerous cultivated plant species, and domestication of a select array of animals are all intimately involved with human ecology, much of it in the tropics. Even more noteworthy is the degree to which humans have altered tropical ecosystems, not merely today, but many hundreds of years in the past. As described in Chapter 6 with the example of Tikal in Guatemala, the legacies of ancient tropical civilizations are sometimes not obvious. Tropical forest undergoes ecological succession and returns after civilizations collapse, and anthropologists and archeologists are faced with the formidable task of discovering and uncovering what were once cities, villages, and roads.

The fossil record as well as genetic analysis of extant human populations leaves no doubt that *Homo sapiens* originated in Africa and radiated from that continent to other regions. Dispersal to Eurasia is estimated to have begun about 50,000 years ago. Humans reached Asia by about 40,000 years ago. A key colonization event was the influx of humans into North America via *Beringia* (the Pleistocene land connection between Siberia and Alaska), now estimated to have occurred sometime after 16,500 years ago (Goebel et al. 2008) (Figure 13-1). Analysis of mitochondrial and nuclear DNA firmly links indigenous peoples of the Americas with ancestral Asian populations. The conclusion is that all modern Native Americans descended from a single Asian source population (Goebel et al. 2008). Humans likely crossed Beringia both on foot and in boats. It is likely that they used boats to move southward along the Pacific Coast of North America, colonizing as they progressed, spreading rapidly as far south as Chile. Even later, there were other human colonization events, namely the influx of people into both Madagascar and New Zealand.

Human Ecology in the Tropics

As a generality, the culture of any people is deeply interwoven with their pragmatic knowledge of the land from which they extract food and fiber. Many cultural anthropologists now approach their studies from the perspective of human ecology, focusing on the ways

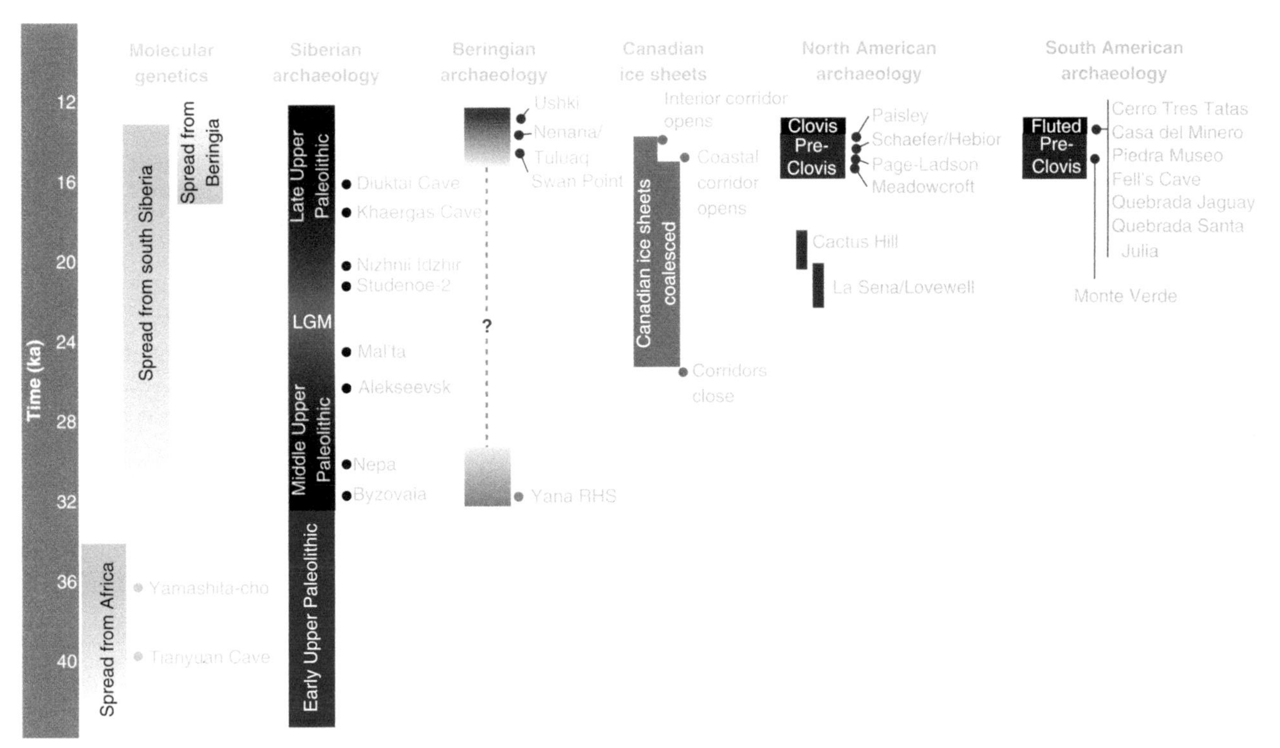

FIGURE 13-1
Combined, the molecular, genetic, and archeological records from Siberia, Beringia, and North and South America suggest that humans dispersed from southern Siberia shortly after the last glacial maximum (LGM), arriving in the Americas as the Canadian ice sheets receded and the Pacific coastal corridor opened, 15,000 years ago.

culture develops in response to environmental opportunities and challenges.

Successful exploitation of rain forest resources requires skill and local knowledge. For example, in northeastern South America, people plant between 17 and 48 varieties of manioc together in the same plot (Dufour 1990). Manioc is also planted close together to form a canopy of shade to keep the ground cooler, and some weeds are permitted to remain in crop rows to guard against rapid soil erosion (Plotkin 1993). In some areas (though not most areas where manioc is planted), the manioc root contains cyanide in the form of prussic acid. This dangerous compound is skillfully eliminated by various means, rendering the root edible.

Throughout its extensive range, Amerindian peoples make diverse use of moriche palm (*Mauritia flexuosa*). Named the *koi* in Suriname and the *buriti* in Brazil, this species, which has been called the *tree of life* (Carneiro 1988), provides wood for canoes and houses, as well as thatch and material for weaving. Its fruit is used for oil, and its unopened flowers are used to make wine or for flavoring (Plotkin 1993; Goulding et al. 1996). It is also used for making bowstaves, spears, arrow shafts, and manioc strainers (Carneiro 1988). Fruit from moriche palm is reported to be the third most important fruit, after bananas and plantains, sold at the markets in Iquitos, Peru (Goulding et al. 1996). Some other palm species are equally as important as moriche (Balick 1985).

Some aboriginal groups have learned to extract potent poisons, ranging from batrachotoxins (see Chapter 8) in frog skin to curare from various plants. The spiritual world is extremely important in many cultures—hardly surprising when one considers how many hallucinogenic drugs are extracted from tropical plants and mushrooms (Schultes and Hoffmann 1992; Schultes and von Reiss 2008). Consequently, one of the most important members of many societies is the village shaman, the person who holds the knowledge about the varied uses of local plants and animals and who, it is believed, is able to communicate with the spirit world. Beyond their local knowledge of plants, the hunting skills of tropical indigenous peoples have been widely documented, from the stealth and speed with which they move through the forest, to their accuracy with a bow and arrow or a blowgun.

Human Impact on Lowland Tropical Forests: How Much, How Little?

There are many areas in the global tropics where large tracts of lowland rain forest are lush and diverse. It is not unreasonable to believe that such ecosystems represent ancient virgin forest. But as discussed in Chapter 6, forests have historically all been subject to natural disturbances at various scales of time and area. The view of a forest primeval, pristine and undisturbed, is incorrect. But how much impact have people had throughout history on tropical humid forests? Evidence is strong that humans have exerted major influences on lowland rain forest, including, at least in some regions, extensive forest clearance for habitations and agriculture (Figure 13-2). Areas deep within Amazonia, the Congo Basin, Indo-Malaysia, and Papua New Guinea all suggest extensive human occupation with significant habitat alteration (Willis et al. 2004).

In the Congo Basin, for example, evidence of extensive human occupation includes subsoil layers of charcoal (indicating burning of forest; see the sidebar "*Terra Prieta del Indio*"), stone tools, palm nuts (cultivated by humans), pottery fragments, and other archeological evidence dating from 3,000 to 1,600 years ago. Other tropical regions show similar archeological patterns. For example, the domestication and subsequent cultivation of *maize* (*Zea mays*), now called *corn*, began with a plant called *teosinte* in Mexico (Iltis et al. 1979) and has been archeologically traced to southern Peru, dating to approximately 3,600 years before the present (Perry et al. 2006). There is an ongoing debate about just how much impact pre-Columbian indigenous people actually had

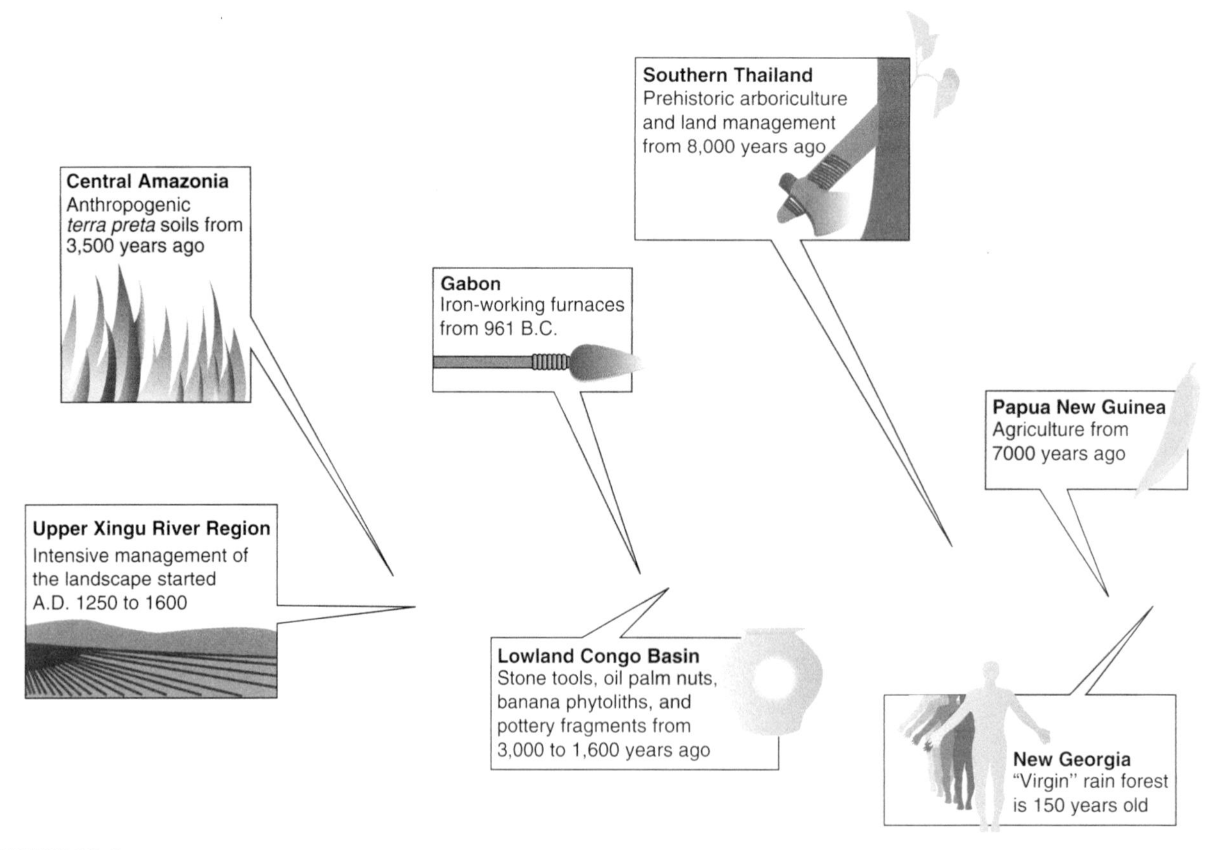

FIGURE 13-2
Anthropogenic modification of "virgin" tropical rain forest. The map indicates the three largest undisturbed rain forest blocks remaining worldwide. These are located in the Amazon Basin, the Congo Basin, and Southeast Asia (Indo-Malay region). Evidence is accumulating from archeological and paleoecological studies that each of these regions was disturbed by prehistoric human settlements but subsequently regenerated once the populations moved on or died out. This suggests that given sufficient time, tropical rain forests disturbed by modern human activities may be able to regenerate.

on Amazonia (Bush and Silman 2007) to be discussed later in this chapter. It is important to state here that, as in the case of Tikal (see Chapter 6), rain forest, at least in some regions, is resilient and capable of regrowth once humans abandon a region or once human populations significantly decline.

The indigenous peoples of tropical America inhabited Amazonia for several millennia before the Spaniards arrived (usually recognized as 1492, the first voyage of Columbus). Their total population density remains an open question, with estimates ranging from as low as about a million (in all of Amazonia) to as many as 11 million or more (Mann 2008). When Europeans arrived in Amazonia between A.D. 1500 and A.D. 1600, the indigenous population crash began, largely due to the introduction of such diseases as smallpox, diphtheria, and

Angkor: Cambodia's Ancient City in the Rain Forest

The dense rain forest covering a 1,000-square-kilometer (about 4,000 square mile) area within Cambodia was once largely cleared to support a thriving civilization of some 750,000 people, an area now called Angkor. In the late sixteenth century, Portuguese missionaries discovered the sacred temple, Angkor Wat (Plate 13-2), cloaked in strangler figs within a tropical moist forest, all of which had regrown from when Angkor suffered a catastrophic population decline beginning in the mid- to late 1300s, and complete by the mid-1400s. Angkor lasted from the ninth to the fifteenth century, and then was essentially abandoned. After thriving for nearly six centuries, in less than a single century, Angkor effectively collapsed.

Angkor was sustained by an intricately engineered network of canals and reservoirs (called *barays*). Rainwater brought annually by wet-season monsoons was captured and stored in the vast barays. During wet season, the excess water was easily dispersed, and during dry season, water was distributed to rice fields, allowing sustained agriculture throughout the year. Some barays were massive, one measuring 8 kilometers in length and 2.5 kilometers in width (about 5 miles in length and 1.5 miles in width). By diverting water at times of potential flooding and storing water for use during dry months, Angkor was able to support the large human population that farmed rice in extensive fields, many of them raised, surrounding the city. The collapse of the great Cambodian city is difficult to explain. As was the case with Tikal, many reasons have been suggested, most related to unstable politics and war. But a more fundamental reason for Angkor's collapse is likely ecological. At the time of Angkor's demise, Earth was in the grip of the Little Ice Age, a time characterized by much unpredictable weather, more severe winters than usual, and generally cool summers. Archeologists have suggested that the complexity of Angkor's water system led to its deterioration, exacerbated by the occurrence of the Little Ice Age. Evidence suggests that, like Tikal, severe and frequent droughts may have triggered the catastrophic decline of Angkor (Stone 2009).

PLATE 13-2
ANGKOR WAT

influenza (Denevan 2003). So, because it remains unclear as to how dense indigenous human populations were in pre-Columbian times, it is also not clear what their collective influence on the landscape was. Human prehistory in Amazonia remains uncertain (Meggars 1985 and 2003), but more archeological data are being discovered (Heckenberger et al. 2003 and 2008; Mann 2008).

Prior to the recent work at Santarém and Marajó and the Xingu region of Brazil, described in more detail shortly, the only known complex civilization to occur within a Neotropical lowland moist forest environment was that of the Classic Mayans. Some evidence now suggests that comparable urbanization, population concentration, and widespread agriculture existed within parts of Amazonia. *Varzea* regions, where soil fertility is annually renewed during the flood cycle, may have supported large and permanent settlements from about A.D. 500 until the European conquest (Meggars 1988 and 2003). Francesco de Orellana, European discoverer of the Amazon River (of course the river was "discovered" many years earlier by the first people to colonize the region), reported dense human populations along much of the river when he navigated it in 1542 (Goulding et al. 1996). However, many of Orellana's observations (actually reported by his scribe, Friar Gaspar de Carvajal) have been considered at best to be highly questionable as to their veracity. It was Carvajal, for example, who reported an aggressive tribe of warrior women many of whom allegedly removed one of their breasts to facilitate use of the bow and arrow. These women were, of course, the so-called *Amazons*, very likely a mythical tribe from which a sizable river obtained its name. Such is the nature of some early historical accounts.

Until recently, it was generally believed that South American Amerindian civilization originated in the Andes and slowly spread eastward into Amazonia. That view has been challenged by the discovery of artifacts that may predate Andean artifacts. Middens (*middens*, in the archeological sense, are ancient trash heaps where discarded objects of old civilizations are found) have been uncovered in the Santarém region of Brazil, near the confluence of the Tapajós and Amazon rivers, containing pottery and other artifacts that date from about 8,000 to 7,000 years before the present (B.P.) (Roosevelt et al. 1991). The pottery dates to about 1,000 years earlier than that found in northern South America and 3,000 years earlier than Andean and Mesoamerican pottery. Archeologists investigating this site suggest that by 2,000 years ago, a large and agriculturally sophisticated population could have been supported on the rich alluvial soils deposited by the annual flood cycle along Amazonian floodplain forests (*varzea*).

Evidence exists of complex societies from about A.D. 500 to 1400 from Marajó Island, a large island at the very mouth of the Amazon River (Bahn 1992; Gibbons 1990). On Marajó, mounds contain multiple house foundations, one upon another, plus an abundance of pottery, indicating long human occupation. The artifacts found at both Santarém and Marajó are elaborate, consisting of burial urns and other sophisticated pottery, finely carved jade, and large statues of presumed chiefs.

Archeological efforts have uncovered much evidence for human habitation of the Upper Xingu area within the state of Mato Grosso, in Brazil (Heckenberger et al. 2003 and 2008). These settlements date from about A.D. 1200 to 1600. The Upper Xingu is today an extensive tract of rain forest, but archeological work demonstrates that this forest was anthropogenically altered. Satellite imagery (Figure 13-3) reveals excavated ditches around ancient settlements, linear mounds at the margins of major roads, circular plazas, and features such as bridges, artificial ponds, and other structures (Heckenberger et al. 2003). This effort required massive alteration of forest, converting rain forest to a built environment that used sophisticated engineering to channel water and sustain agriculture.

The pattern of human occupancy of the Upper Xingu area has been described as *galactic* (Heckenberger et al. 2008). This means that rather than extensive city-states that form the nexus of a political and geographic region (such as ancient Rome, for example), the pattern in Amazonia was one of small independent political units (a *galaxy* of local independent communities), all within a relatively unified regional political system (Figure 13-4). In this sense, the Upper Xingu civilization does not represent a "lost city" such as Tikal or Angkor but, instead, represents a network of permanent plaza communities integrated within about 250 square kilometers (96.5 square miles) and sharing regional interactive politics.

Heckenberger et al. (2008) have discovered 28 prehistoric residential sites, most of which are associated with two galactic clusters. These clusters integrated large and medium-sized plaza towns and small villages ranging from greater than 40 hectares to less than 10 hectares (99 to 25 acres). The people inhabiting the region relied on fish as a protein source but also practiced agriculture. Using raised beds of soil to minimize damage to plants from flooding, mounds allowed for drainage and high agricultural productivity.

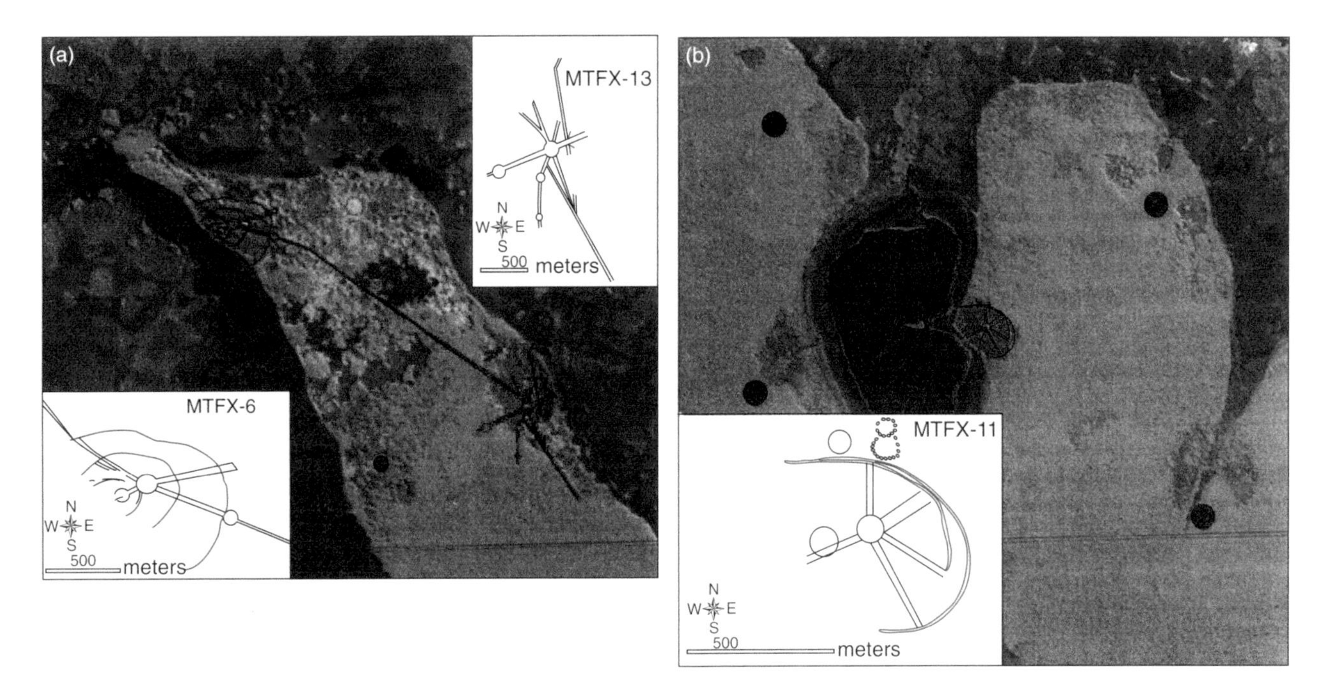

FIGURE 13-3
Satellite image [Landsat 4 Thematic Mapper, path 225, row 69; 21 June 1992; bands were assigned as 5(red)–4(green)–3(blue)] with global positioning system–mapped Ipatse cluster sites X6 and X13 (a, insets) linked by the north–south road and transit-mapped X11 (b, inset). Ditches are colored in red; road and plaza curbs are black. Fieldwork in 2003 demonstrated that roads extend fully from X13 to X18 and continue on to X19, X20, and beyond along the north–south road, as well as across high ground to X17 and X22; X11 roads also connect it to the four satellites. MTFX, Mato Grosso (the state), Formadores do Xingu (the archeological region). The number refers to the site number.

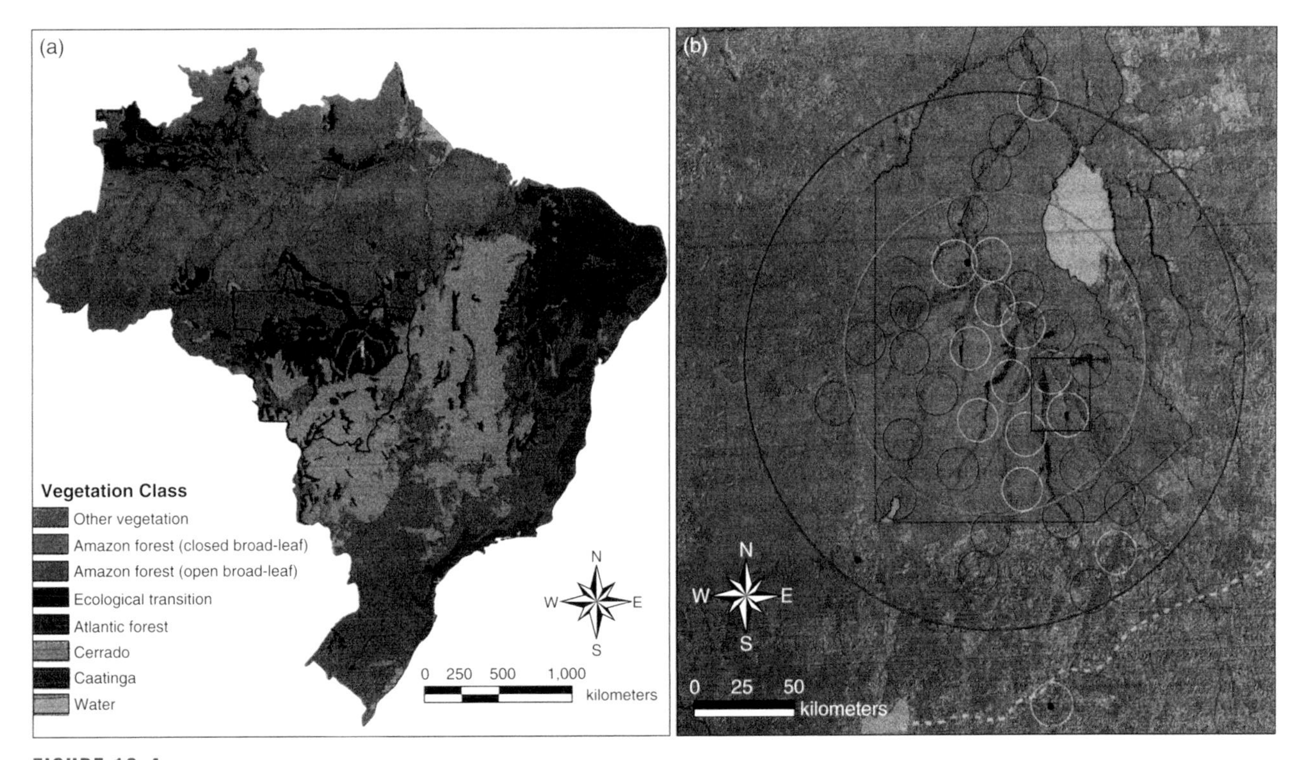

FIGURE 13-4
(a) Map of Brazil showing major vegetative regions and areas of Amazonian complex societies, including (1) Mato Grosso state outlined; (2) blue-green forest in eastern Brazil is open broad-leaf associated with the Atlantic forest biome. (b) Proposed cluster distributions in the Upper Xingu Basin. White circles are based on the presence of major (> 30 hectares) walled centers; red unnumbered circles are hypothetical. Area of acute anthropogenic influence is denoted by an orange circle (150 kilometers diameter, ~17,500 kilometers square) and moderate to minimal anthropogenic influence by a red circle (250 kilometers diameter, ~50,000 kilometers square).

Terra Preta del Indio

In parts of Amazonia, there are scattered patches of a unique Latosol-type soil (see Chapter 10) called *terra preta del Indio*. The name is from the Portuguese for "black soil" or "black earth" (Plate 13-3). Unlike most Latosols, *terra preta* is very dark, often black in color. The reason for its color is that it is rich in charcoal. It also contains pottery shards and organic material including plant remains and animal waste and bones from fish and other animals. Unlike most soils in Amazonia (particularly on *terre firme*), *terra preta* is nutrient-rich, with high concentrations of nitrogen, potassium, calcium, and other essential elements. In addition, *terra preta* soils have as much as 14% organic matter. *Terra preta* is a type of soil that was made by humans, an anthropogenic soil dated between A.D. 450 to about A.D. 950. Apparently, indigenous people developed the technology and knowledge to cut the forest and burn it at low temperature, allowing the accumulation of charcoal. *Terra preta* soils are sometimes 2 meters (6.6 feet) deep, indicating that people had been present in the same location over many years, steadily building up the nutrient-rich soil. The utility of *terra preta* was obviously to enhance local agriculture. These human-constructed soils were far greater in nutrient content than the normal Oxisols that make up most of the substrate in Amazonia. *Terra preta* remains are found along rivercourses scattered throughout the Amazon Basin, and likely reflect the locations of what were then permanent human settlements in pre-Columbian times. The high organic content and overall nutrient richness of *terra preta* shows a sophisticated understanding of how to enhance agricultural production such that it could support relatively large human populations. It has been suggested that *terra preta* could be used to enhance agriculture in the present century. For more on *terra preta*, see Glaser (2007).

PLATE 13-3
Terra preta soil (very dark) at the surface of this cut.

In Central America, there is much archeological evidence suggesting that Mayans supported their dense population through techniques of intensive agriculture and silviculture (Hammond 1982; Fedick and Ford 1990; Peters 2000). Such practices would result in large-scale landscape alterations, with much forest cutting. Intensive agriculture in Central America was accomplished largely by two methods: *hill terracing* and *raised fields* in swamps and marshland. Hill terracing involves the construction of stone walls along hillsides, the walls acting to retard erosion and trap soil washed by rains. (This method was also widely practiced by Incas of the Andes.) Hill terracing permitted Mayans to cultivate a given plot for much longer than ordinary slash-and-burn techniques because the soil fertility was preserved. Raised fields involve the excavation of drainage canals to reduce water levels and thus raise "dry" fields from what was previously swampland. Ancient Mayans not only used the raised fields for agriculture, but also probably used the canals for keeping fish and turtles, both important protein sources. Imagine the image from a small plane at low altitude, flying over the Mayan Yucatan during the height of the Classic period. It would resemble a flight over Midwestern North America, where vast acreages of agriculture characterize the landscape. This intensive and productive land usage permitted the construction of large population centers such as Tikal, with its impressive temples (Plate 13-4).

PLATE 13-4
Temple 2 at Tikal dates to the Mayan Classic period, when much of the surrounding forest was cleared for intensive agriculture.

Past raised-field agriculture is revealed in part by patterns evident in aerial photographs of the landscape today. A large area in northern Belize (Pulltrouser Swamp) was under intensive wetland cultivation by Mayans between 200 B.C. and A.D. 850 (Turner and Harrison 1981 and 1983). Mayan techniques succeeded in preserving soil fertility and may have permitted uninterrupted farming throughout the year. Using both aerial photography and remote-sensing side-looking airborne radar imagery, a pattern was exposed showing that vast areas of northern Guatemala and Belize contain canals that were probably constructed by early Mayans and used in connection with agriculture (Siemans 1982). Modern attempts to model early Mesoamerican intensive agriculture using raised fields and canal systems (called *chinampas*) produce high yields (Gomez-Pompa et al. 1982; Gomez-Pompa and Kaus 1990).

The Mayans of Tikal cultivated the *ramon* (or breadnut) tree (*Brosimum alicastrum*). This tree is today abundant throughout the Guatemalan Peten region, and a single tree has the potential to yield 1,000 kilograms, or about 2,200 pounds, of edible nutritious seeds (Hammond 1982). In addition, the breadnut tree is tolerant of many soil types and grows rapidly, an ideal tree for cultivation. Its fruits and seeds would have provided sources of nutrition for humans and domestic animals, its leaves would have been used for animal forage, and its wood would have been used for construction (Gomez-Pompa et al. 1982; Peters 2000). The present abundance of *ramon* throughout areas formerly densely populated by Mayan civilization is likely due to Mayan *silviculture*, the skillful cultivation of select trees for use by the people. Evidence cited by Hammond (1982) suggests that Mayans preserved *ramon* seeds in underground chambers called *chultunobs*. Breadnuts probably served as a "famine food," to be used when times were difficult. Because *ramon* grows on limestone outcrops, it is known to grow even on ancient Mayan ruins, which were built largely of limestone.

While there is no doubt whatsoever about the extent of Mayan influence over the Yucatan landscape during the Classic period of Mayan history, there are serious questions about just how much human occupation there was throughout Amazonia in pre-Columbian times. The discoveries at Upper Xingu, Santarém, and Marajó each suggest that dense aggregations of people settled permanently and practiced efficient and sustained agriculture, as well as relying on the river to supply needed animal protein. As these discoveries have mounted, the suggestion has been made that much of Amazonia may have been a *manufactured landscape* before European arrival at around A.D. 1500. Two views have emerged, one being that human population density was generally low throughout most of Amazonia (except in those few places that lent themselves to dense settlement), and the other that human influence was pervasive and widespread (Erickson 2000), transforming much of Amazonia into what has been called a *cultural parkland* (Heckenberger et al. 2003). Bush and Silman (2007) challenge each of these extremes and suggest a middle ground. They argue that dense human populations, urban centers (as evidenced by cultural artifacts), pollen remains (such as corn pollen, for example), and charcoal distribution (including *terra preta* soils) were essentially limited to areas near or along major rivers. Away from these urban centers, human population density was much less. Evidence from charcoal remains suggests widespread fires (and, by implication, human-set fires). But these fires may also have resulted from climatic influence, such as El Niño/Southern Oscillation (ENSO) years (where drought results in more fire) and were perhaps intensified because of human-set fires that "got away" during drought years. They note that the highest fire frequency thus far detected was around A.D. 700 to 800, during the peak of a severe ENSO event (Thompson 2000). Last, Bush and Silman (2007) caution that the view of Amazonia as having been largely converted to human settlement in pre-Columbian times coupled with the reality of what it is today, an area of dense rain forest, suggests to some that the forest is quite resilient, as appears to have been the case with much of the Yucatan following the fall of the Mayan civilization. Such resilience may not be the case for much of Amazonia, particularly in light of predicted effects of climate change (see Chapter 15).

Hunting and Gathering: The First Human Societies

From the time when hominins first evolved on the African savanna until approximately 10,000 years ago, all human populations everywhere on Earth were hunter-gatherers. For the vast majority of human history, humans survived by selective use of plants taken directly from their ecosystems as well as by hunting and scavenging various animals. Hunter-gatherer groups normally do little in the way of manipulating or altering the ecosystem. The impact of hunter-gatherers is generally low but not negligible (Odum 1971; Vickers 1988; Bettinger 2009). Humans often burn forest to facilitate nomadic movement or for enhancing game detection. The most significant impact of human hunting is that during the latter part of the Pleistocene, skilled human hunters may have contributed to the rapid extinction of an array of large mammals ranging from mammoths and mastodons to sloths and giant kangaroos (Martin and Klein 1984; Flannery 1994 and 2001). This view, termed the *Pleistocene overkill theory*, is based on strong correlations between the appearance of humans and the rapid loss of megafaunal species. The theory has been challenged by some anthropologists (see Grayson and Meltzer 2003, for example), and debate continues.

Hunter-gatherer ecology has been widely studied (Lee and Daly 2004), and lifestyles of hunter-gatherer people are variable (Panter-Brick et al. 2001; Kelly 2007). Because not all hunter-gatherer societies are the same, this account is meant only as an introduction to some of the more general characteristics of hunter-gatherer ecology.

The population density of hunter-gatherers is one person or fewer per square mile. This is a very low population density by modern standards, and to understand it you must consider the ecological context of hunting and gathering. Hunter-gatherer groups identify essential plant and animal resources, seek them out, and use them. If the people inhabit tropical rain forest, they might, for example, select certain palm species to use for thatching houses, making rope, and a variety of other functions. In much of the Neotropics, many species of palms (*Mauritia*, *Bactris* [Plate 13-5], *Astrocaryum*, *Attalea*, *Euterpe*, *and Iriartea*) have multiple uses for indigenous people.

PLATE 13-5
Palms in the genus *Bactris* are widely utilized by indigenous peoples in the Neotropics.

Hunter-gatherer groups are commonly egalitarian, sharing food and other resources among the group. Egalitarianism is a pragmatic social order in a society in which individual hunters experience variable biomass yields from one day to another. Should a hunter fail on any given day, he nonetheless eats. Should he succeed, he shares. As stated earlier, hunter-gatherer groups usually do not permanently alter ecosystems, though their impact may vary. Settlements are meant to be temporary. Hunter-gatherers are typically nomadic, moving after having depleted essential resources, which could be particular animal species or certain plant species. Because they are nomadic, there is normally wide birth-spacing in hunter-gatherer groups. Children must be able to walk to keep up. Consequently, hunter-gatherer peoples sometimes practice both birth control and infanticide (Diamond 1997). Hunter-gatherer people transport only the possessions they can carry along with them. Writing has never evolved in hunter-gatherer societies, perhaps because it is pointless to make tablets or books or anything else of bulk that must be transported. Accumulated knowledge is passed along by word of mouth, particularly from learned shamans.

The diets of hunter-gatherer peoples have been studied and found to be heavily dependent on animal food, usually between 45% to 65% of the total daily energy intake. Protein makes up 19% to 35% of the daily energy intake (Cordain et al. 2000). In this regard, hunter-gatherer societies take in considerably less carbohydrate in relation to protein than non-hunter-gatherer groups that rely more on agriculture.

In ecological terms, humans in hunter-gatherer societies harvest few calories relative to what is present in the ecosystem as a whole. In one study, for example, humans in tropical rain forest attained only 0.4 kilocalorie per square meter per year, although solar radiation into the ecosystem was approximately 1,500,000 kilocalories per square meter per year (Odum 1971). The low human productivity is because most of the plants and animals in the forest are not used. This should be no surprise. Walk into any forest, particularly a rain forest, and ask yourself just what could you eat? Could you eat the bark of the trees? Which leaves, if any, are edible? Recall that plants concentrate defense compounds throughout their tissues, leaving most inedible. There are perhaps millions of tiny insects in the forest, but how time-consuming would it be to capture a few thousand for dinner? You might identify some wild grapes that are sufficiently ripened to consume, and you might notice squirrels or other animals that, if you could kill them, could provide sufficient nutrition. But the total biomass of these potential food sources would be tiny in comparison to that of the total ecosystem. Humans have routinely starved to death surrounded by calories in forms that they cannot access or digest. It is for this reason that hunter-gatherer groups must acquire a deep pragmatic knowledge of the species within their ecosystems. Which fruits can be eaten? When are these fruits ripe? Which roots are safe to consume? Which leaves can be used for medicinal purposes?

Hunter-gatherer groups require large areas. A group of 50 people typically needs no less than 130 square kilometers, or about 50 square miles, to sustain it, though densities do vary among hunter-gatherer groups. And bear in mind that *sustain* does not mean indefinitely but only for some period of time, perhaps months to a few years. The group will eventually abandon that site and move to another area where resources are sufficiently plentiful.

Because hunter-gatherer groups remain small, inbreeding is a potential genetic problem. Human behaviors such as incest taboos and tribal raids that procure females who are then brought into the raiding group as wives are sociological adaptations to this reality. So is tribal warfare among neighboring groups. Since settlements are not permanent and because resources are ultimately limited, periodic warfare among hunter-gatherer groups is commonplace. In ecological terms, hunter-gatherers provide an example of *intraspecific territorial competition* among groups.

The use of fire, which was probably discovered by various early human populations, made it possible for humans to eat a wider variety of foods because difficult to digest items such as nuts could be cooked and rendered digestible. In a similar manner, the making of tools, such as the Clovis spear point (Plate 13-6) so widely distributed in North American archeological sites, made it possible for human hunters to procure larger animals. The first carefully constructed hand

PLATE 13-6
CLOVIS POINTS

axes date from about 1.5 million years ago, well before *Homo sapiens* evolved. The use of fire and the construction and use of tools represent technology, which is itself a reflection of human culture and ingenuity.

Because all hunter-gatherer groups possess various forms of technology, it is a mistake to assume that they exist in some sort of ecological balance with their environments. The concept of *ecological balance* is naïve (Kricher 2009). Hunter-gatherers are not intrinsically adapted to exert minimal impact within their various habitats nor do they attempt to limit their impact. The reason hunter-gatherers are nomadic is because they do exert significant but local impacts within their ecosystems. They routinely perturb and deplete certain resources and must eventually abandon an area for another where the affected species have had time to regenerate. The possibility that Pleistocene hunters may have brought about the extinctions of numerous large animals demonstrates that even a sharpened spear point represents a powerful technology.

There are few true hunter-gatherer populations in the world today (Cordain et al. 2000; Lee and Daly 2004). This should not be surprising. It is doubtful that any hunter-gatherer populations will persist for very much longer. Such populations require not only a large quantity of land but also isolation from other human populations with more complex technologies capable of major ecosystem alteration. Neither of those requirements is likely to be sustained in the future.

Examples of Hunter-Gatherers in the Neotropics

Because of the variety of hunter-gatherer tribes throughout the Neotropics as well as differences in habitat from one region to another, cultures vary. It is not possible to describe one culture as typifying all. Neotropical hunter-gatherers are relatively nomadic, living in a small temporary village or encampment for some time and eventually moving on when they have exhausted the game or essential plants from a given locality (though some villages are fairly permanent and people walk long distances to hunt and gather). Hunting is accomplished by careful, quiet stalking, using a blowgun, a bow and arrow, or a spear to bring down essential protein: large birds, monkeys, sloths, agoutis, pacas, tapirs, and others. Often, but not always, arrows or darts are tipped with poison. Rifles and shotguns are becoming more frequent. Protein is also supplied by certain arthropods, especially large grubs, and, where tribes are living along rivers, by hunting for fish (also sometimes using poison), turtles, capybara, and crocodilians.

Some Neotropical hunter-gatherer tribes are territorial and are known, at least in the past, for high levels of aggression. (Shrunken human heads are part of the cultural artifacts of some Amazonian aboriginal groups.) In Brazil, Mundurucu headhunters in the recent past made no distinction between people from different tribes and animals such as peccaries and tapirs—all were hunted. Tribal warfare was probably a response to the need to protect areas of forest for the exclusive use of a single tribe, thus increasing the forest's yield (Wilson 1978). Tribal raids were also conducted to procure women, ensuring genetic outbreeding through aggression, a custom inculcated within the culture. Given those explanations, it should not be lost on the reader that aggression in extreme forms has been a major component of many Amazonian cultures, including groups such as the Mundurucu and Yanomami.

There are few "pristine" hunter-gatherers in the Amazon Basin and essentially none in Central America. Most people now use agriculture of some sort to supplement their diets, and most have some periodic direct contact with the "modern world." Shotguns are rapidly replacing blowguns in many remote areas. When Europeans first began coming to South America around A.D. 1500, the estimated total population of aboriginal humans throughout Amazonia was about 6.8 million (Denevan 1976), though some estimates suggest a number half that size. Most people were settled along riverine areas, where *varzea* floodplains ensured annual renewal of soil fertility. Interior *terre firme* forests and savanna areas apparently were far less populated (Dufour 1990). By the early 1970s, the indigenous population was only a half million, and in Brazil alone, the number dropped precipitously from about a million to about 200,000 during the twentieth century (Dufour 1990). By 1988, the estimate for all of Amazonia is only about 250,000, a 24-fold decrease from an estimated six million (Carneiro 1988). Amerindian populations were reduced by a lethal combination of outright conquest and genocide, slavery, and, probably most significant, by the introduction of European diseases to which the Amerindians had little natural resistance (Diamond 1997). In Amazonia, tribes that inhabited *varzea*, who represented the largest Amerindian populations, fared worst, being essentially decimated by Europeans. Only those tribes such as the Yanomami, the Javari, the Xingu, and others who inhabited remote and inaccessible forest survived the conquest period, and even they suffered reductions in population whenever there was European contact.

Today, most Amerindian tribes live on anthropological reserves or *indigenous areas*, called *resguardos*, lands in which aboriginal groups are permitted to follow their traditional lifestyles. In Brazil, Indian lands are administered by the government agency Fundaçáo Nacional do Índio (FUNAI, or the National Indian Foundation). This agency governs a huge area representing 100.2 million hectares in 371 reservations in the Brazilian Amazon, representing roughly 20% of Brazilian Amazonia (Peres 1994). In the northern state of Roraima, about 42% of the land area is reserved for use by Indians, even though Indians represent only about 15% of the population of Roraima (Brooke 1993).

Numerous issues face indigenous tribes. Amerindian populations continue to be forced into retreat in some areas, as people with different cultural backgrounds, often from overpopulated, extremely poor urban areas, migrate to the new frontier of the rain forest, some to begin subsistence farming, some in search of gold. This trend has greatly accelerated in Amazonian Brazil due to the continually expanding Transamazonian highway system. In some areas, aboriginal tribes (for example, the Nambikwara tribe of Mato Grosso, Brazil) are deliberately exploiting their own lands for short-term profit, granting permission to outsiders for logging and gold mining (Peres 1994).

The Yanomami group of the state of Roraima in northern Amazonia, who inhabit the highland rain forest near the border between Venezuela and Brazil, represents an aboriginal population who combine hunting and gathering with simple agriculture (Plate 13-7).

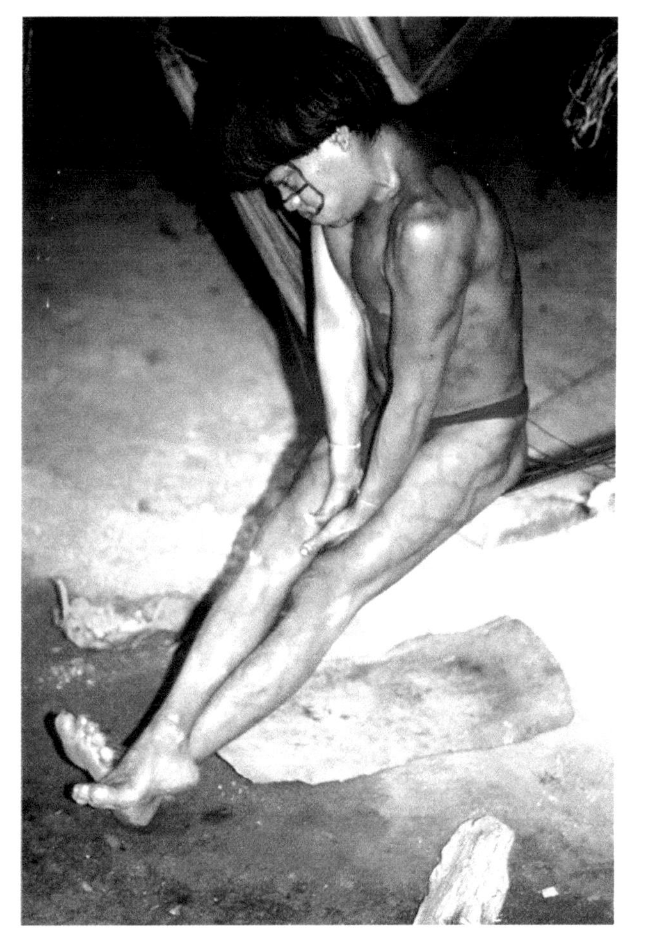

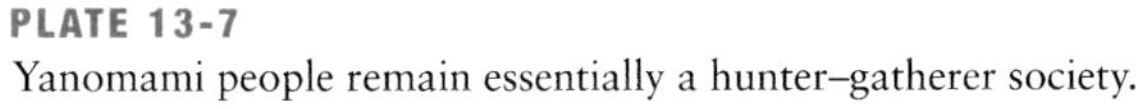

PLATE 13-7
Yanomami people remain essentially a hunter–gatherer society.

Approximately 24,000 Yanomami remain, the largest of the remaining forest-dwelling Amazonian tribes (though they are much reduced from previous times), and they are scattered in groups throughout an overall territory of about 8,400 square kilometers (32,400 square miles). In the past, the intense level of aggression manifested among local villages earned the Yanomami the description of being "fierce people" (Chagnon 1992), though it has been argued that the Yanomami level of aggression is not significantly different from that of modern societies (Plotkin 1993). A significant controversy has arisen among cultural anthropologists dealing with the techniques and ethics by which groups such as the Yanomami have been studied and their culture reported and interpreted (Borofsky 2005). Yanomami live in scattered villages, hunting local game, gathering foods and fiber from the forest, and clearing small plots for the cultivation of plantain. They continue to practice rituals such as the use of powerful narcotics to interact with the spirit world and the consumption of the ashes (mixed into a soup) of their dead. Today, the Yanomami are under threat from highway construction through their territory, bringing an influx of gold miners (Collins 1990; Brooke 1993).

Do hunter-gatherers eventually deplete local game populations? There is much documentation of wildlife depletion due to hunting (Robinson and Redford 1991). But this need not always be the case. A study of the Siona-Secoya Indian community in *terre firme* rain forest of northeastern Ecuador documented 1,300 kills, representing 48 species, including various mammals, large birds, and reptiles (Vickers 1991). The average number of kills per 100 man-hours of hunting was only 21.16 (with sample size of 802 man-days and 6,144 man-hours), and the mean number of kills per man-day of hunting was only 1.62. These figures do not suggest that hunting pressure was sufficient to deplete the animal populations and are consistent with other studies (Redford and Robinson 1987). In the total hunting area of 1,150 square kilometers (444 square miles), only one bird species, a curassow (*Mitu salvini*), was depleted due to hunting, though in the area immediate to the village, two species, a bird (trumpeter, *Psophia crepitans*) (Plate 13-8) and a monkey (woolly, *Lagothrix lagotricha*), were depleted. Other animals did not suffer population declines. The human population was low, approximately 0.2 person per square kilometer (about 0.08 person per square mile), which is obviously a low human population density. Hunters were selective, concentrating on large game such as peccaries, tapir, and woolly monkeys. It has even been argued that human modification of rain forest (through the planting of selected fruiting species) can increase animal populations such as tapir, deer, and peccary (Posey 1982). In Vickers's study, factors such as depletion of a mahogany species (*Cedrela odorata)* for canoes as well as cutting palms for thatch were more important than loss of game animals in forcing the settlement to move. The settlement remained in a given area for a decade and game was not significantly depleted (Vickers 1988).

In contrast, if one looks at the entire rural population of Amazonian Brazil, including colonists as well as aboriginals, subsistence hunting takes on much greater impact. Given a presumed population of nearly 3 million persons (living outside cities and the vast majority of whom are not aboriginal people living in tribes) in an area of about 3.6 million square kilometers (1,389,600 square miles, or a population of about 0.46 person per square mile), and considering the average per annum consumption rate, about 14 million mammals are killed each year, a very significant number. Adding birds and reptiles, the total number of animals killed by subsistence hunting jumps to about 19 million annually. Add to that figure the number of animals fatally wounded but not taken, and the estimate jumps to a staggering 57 million per year (Redford 1992). There are numerous examples of hunting pressure being responsible for the local depletion of animals (Robinson and Redford 1991). Hunting impact is a complex issue, as it is frequently done not only for subsistence but also for commercial profit.

PLATE 13-8
Gray-winged trumpeter, a bird often sought by Amazonian indigenous people.

The Ache: Hunter-Gatherers of Paraguay

The hunting habits and caloric intake of the Ache in eastern Paraguay have formed the subject of an exhaustive study that serves well to illustrate the adaptiveness of the hunter-gatherer lifestyle (Hill and Hurtado 1989). The Ache are a society of four independent groups, where a group typically comprises from 10 to 15 small bands (of about 48 people each) that roam throughout a territory of about 18,500 square kilometers (7,141 square miles), most of which is semideciduous evergreen forest. On average, an Ache forager consumes 3,700 calories per day, which, when compared with 2,700 calories per day for an active adult North American, is certainly a large intake. Further, an average of 56% of the Ache calories come from mammalian meat, with honey providing 18% and plants and insects together providing 26%. Of course, these averages vary from day to day and with season. A typical Ache foraging trip involves men leading, carrying bows and arrows, and women and children following, carrying supplies in baskets. The group *bushwhacks* through forest, without following existing trails. Eventually, the sexes separate, as the men move more rapidly in search of prey. A simple camp is made at day's end.

The Ache hunt approximately 50 species of vertebrates, including fish, a far smaller number than the total potentially available to them. They exploit only about 40 species of edible fruits and insects, though, again, there are many more potentially available. Only 17 different resources are estimated to account for an astonishing 98% of the total caloric intake, showing that the people are highly selective in their choices, ignoring most resources. Like numerous other hunter-gatherer societies that have been studied, the Ache practice extensive food sharing, a habit strongly encouraged by cultural reinforcement from childhood through adulthood. Hunters readily share their kills to the point where a hunter rarely eats any of his actual kill. Meat and honey are shared the most, with less sharing of plant and insect food. This is hardly surprising, and it is clearly adaptive to the individual as well as to the society as a whole, because meat and honey are highly scattered, rather unpredictable resources compared with insects and plants. A hunter cannot be successful every day, but he might be very successful on any given day. But with food sharing as the norm, a hunter who fails to make a kill does not go hungry, nor does his family. Food sharing, a clear example of *reciprocal altruism*, reduces the day-to-day caloric variability that would otherwise prevail from person to person if sharing did not occur.

The Ache selectivity with regard to prey choice is an example of what ecologists call *optimal foraging*. In other words, does a hunter kill whatever he finds, regardless of the amount of energy required to stalk and make the kill (relative to that received), or does he ignore certain prey, preferring to invest more energy searching for prey with a higher "payoff"? Is it better to stalk and shoot a small monkey than to ignore it and keep searching for a peccary, which will provide far more meat? In the Ache study, return rates for a whole day of foraging (the actual work of finding and procuring food) were calculated to be 1,250 calories per hour (of actual hunting effort) for men and about 1,090 calories per hour of gathering effort for women, but the average caloric value for food acquired per hour of effort was about 3,500 calories per hour for men and 2,800 calories per hour for women. So both men and women (and thus, ultimately, the group) profited from their respective efforts to obtain food.

One prevalent belief is that hunter-gatherer peoples have relatively large amounts of *down time*, when they are not hunting and gathering. They do not, after all, need to labor over clearing forest or planting and weeding crops. The Ache study showed that men spend about 6.7 hours per day searching, making a kill (or finding honey), and processing food, plus another 0.6 hour per day working on their hunting tools, rates of labor comparable to the average working day of a North American. Ache men spend about 4.5 hours per day socializing. For women, about 8 hours per day is spent in light work or childcare, 1.9 hours in subsistence activities that include not only the search for various plant foods and fiber but also the time-consuming preparations and associated work that occur after the material is obtained, and 1.9 hours moving camp. In total, men supply about 87% of the energy in the Ache diet and almost 100% of the protein and fat.

Birth and death rates among the Ache are generally reflective of most hunter-gatherer peoples, though there is certainly variability among groups. About one in every five Ache children die before reaching one year of age, and about two out of every five fail to survive to age 15, so the survivorship from birth to puberty is about 60%. Only about 32% of child mortality is due to illness; fully 31% is from homicide(!). Among Ache adults, most (73%) die from warfare or accidents (including snakebite and jaguar attacks, the occupational hazards of hunting). Illness accounts for only 17% of adult mortality. On average, Ache women give birth every 38 months (Hill and Hirtado 1989).

Emergence of Tropical Crops

A crop is a species of plant used for some purpose by humans, usually nutritional. Though there are many thousands of species of tropical plants, few species have been domesticated for use by humans (Diamond 1997). Many others, while not domesticated, are widely utilized for various purposes (see the following).

The Neotropics are the places of origin for:

- Moschata squash, about 10,000 B.P.
- Maize (corn), sometime between 9,000 to 8,000 years before the present (B.P.)
- Peanut, about 8,500 B.P
- Manioc, about 8,000 B.P.
- Potato, about 7,000 B.P.
- Chili pepper, about 6,000 B.P.
- Other important crops including cacao, tobacco, peach palm, and rubber

Tropical Africa is the place of origin for:

- Sorghum, about 4,000 B.P.
- Pearl millet, about 3,000 B.P.
- African rice, about 2,000 B.P.
- Coffee, about 1,500 B.P.

Tropical India and Australasia are the place of origin for:

- Rice, about 8,000 B.P.
- Yam, about 7,000 B.P.
- Banana, about 7,000 B.P.
- Taro, about 7,000 B.P.
- Mung bean, about 4,500 B.P.
- Various millets, about 4,500 B.P.

(Most dates are from Balter 2007.)

The cultural evolution of domestic crops has been widely discussed and debated (Bellwood 2004; Cowan and Watson 2006; White and Denham 2006). The next section provides a summary, taken from various accounts, of how agriculture might have emerged.

From Simple Beginnings: Discovery of Agriculture

Agriculture's beginnings are unclear, but the practice of cultivation was likely slow and complex (Pringle 1998). Agriculture, in its most primitive form, began approximately 11,000 years ago. By about 8500 B.C., it was flourishing within the area in Southwest Asia (the Middle East) known as the *Fertile Crescent*, the land between the Tigris and Euphrates rivers (Diamond 1997). Today this area is largely desert, and includes parts of the countries of Iraq, Syria, Jordan, and Turkey. The discovery of agriculture in combination with the domestication of animals acted to stimulate a rapid increase in human populations and, consequently, the collective impact of humans on ecosystems. Agriculture changed the world to one in which human beings exert a unique and inordinate ecological influence.

Agriculture is normally thought of in relation to cultivation of selected species of vegetation, but in its broader sense, it also includes the domestication of animals for use as beasts of burden or for food production. Agriculture, including animal domestication, required many human generations to fully develop (Mann 2008) (Figure 13-5).

Agriculture differs from simple gathering (finding plants useful as food or fiber and then collecting them) in that the plants in question are selectively chosen and cultivated, increasing their population densities at the expense of cohabiting species. Humans nurture and protect the plants chosen for agriculture until it is time to harvest them. The same is, of course, true of domestic animals such as cattle, sheep, goats, pigs, and chickens.

Agriculture requires work of a much different sort from that required for hunting and gathering. Like most of the earliest events in human history, the actual discovery of agriculture is obscure. It was likely accidental, and various scenarios have been offered. What is known is that people in the Middle Eastern region of the Fertile Crescent soon became capable of altering the local ecology to redirect a significantly greater portion of the sun's energy to themselves. This alteration or perturbation of the ecosystem, the conversion of a natural ecosystem to one containing only selected, harvestable species, forms the essence of agriculture.

In order to accomplish agriculture, a plot of land must be cleared of all competing species, and desired species must be planted and continuously protected as they grow (Plate 13-9). This requires human labor. People are concentrated around relatively small agricultural plots, tending crops.

Because land can be farmed repeatedly, and because agricultural labor requires constant human effort, permanent or semi-permanent villages form. For

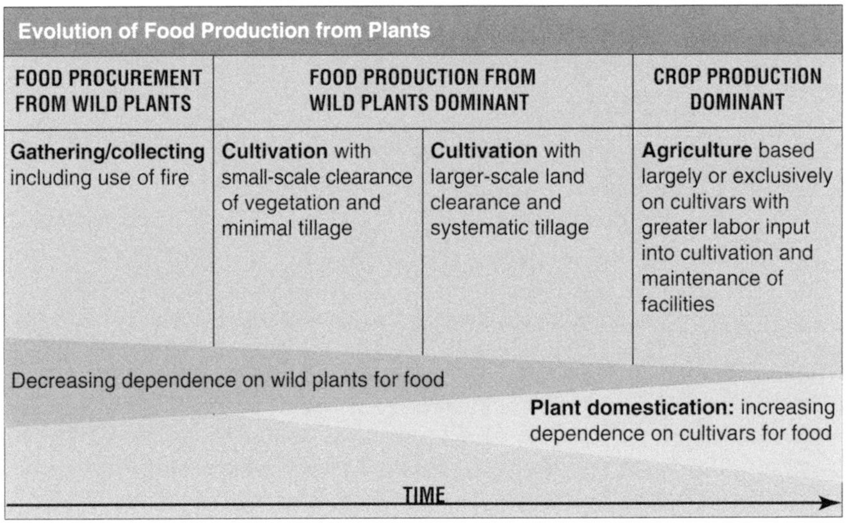

Evolution of Food Production from Plants

FOOD PROCUREMENT FROM WILD PLANTS	FOOD PRODUCTION FROM WILD PLANTS DOMINANT		CROP PRODUCTION DOMINANT
Gathering/collecting including use of fire	**Cultivation** with small-scale clearance of vegetation and minimal tillage	**Cultivation** with larger-scale land clearance and systematic tillage	**Agriculture** based largely or exclusively on cultivars with greater labor input into cultivation and maintenance of facilities

Decreasing dependence on wild plants for food

Plant domestication: increasing dependence on cultivars for food

TIME

FIGURE 13-5
A scenario for the evolution of food production from plants.

this reason, the need for wide birth-spacing (required when groups are nomadic) is essentially eliminated, and women can have more children in the course of their reproductive years (Diamond 1997). Without the need for birth-spacing, and with many additional calories provided by the net primary productivity (NPP) of the crops, human population growth is accelerated. Thus traditional agriculture supports somewhere between 10 and 100 persons per square mile, a one to two orders of magnitude increase from that of hunter-gatherer societies. Howard T. Odum, in a now-classic analysis (Odum 1971), showed that in an agricultural ecosystem in India, humans ingest about 200 kilocalories per square meter per year, far more than the 0.4 kilocalories per square meter per year attained by hunter-gatherers (these figures assume the same solar input in each region) in an equivalent area.

Agriculture routinely provides more food than is actually needed by the local population. There is typically a surplus that can then be used to trade for goods with other peoples. Thus agriculture ultimately supports the organization of more complex villages

PLATE 13-9
This plot of cleared land, called a *milpa*, is planted mostly with corn. It is typical of rotational agriculture in southern Belize (in this case) as well as much of the rest of the tropics.

and towns where barter and trade sustain many of the people. This creates a division of labor among people such that some percentage of the population who never farm or hunt nonetheless survive by doing business with those who do. Thus with the permanency of villages, such activities as technology and writing were promoted, and these in turn led to more effective means of cultivation, such as the invention of the plow, for example.

Odum (1971) has provided a concise analysis of the ecological energetics of agriculture (Figure 13-6). In its most basic form, agriculture first involves rechanneling and then capturing some of the sun's energy through photosynthesis of selected crops. For this to successfully happen, there must be several catalytic factors that act in concert, what Odum terms *multipliers*. These multipliers combine to greatly enhance the eventual net primary productivity of the crops. Rain is a multiplier. Without moisture input, crops will wither and die. With appropriate amounts of precipitation, crops will flourish, their net productivity "multiplied" many times. Human labor (and that supplied by beasts of burden) is a huge multiplier. Humans must carefully clear the land, fertilize the soil, and plant the crops. They must then protect crops from weed invasion and herbivore devastation. They must harvest in a timely manner. They must replant for the next harvest. Those activities require much in the way of work, but without that effort, the crops will surely fail.

Ecologically speaking, crop ecosystems are unstable, open to invasion by both competitors and herbivores. Thus energy input from human labor is fundamental to preserving the stability of crop ecosystems. What this means, of course, is that humans must *reinvest* some of the energy (in the form of human labor) that they derive from the present growing season's crops to ensure the success of crops in the forthcoming season. They must also devote some of the productivity of the crops to sustaining domestic animals. Domestic animals also represent important multipliers (Plate 13-10). Work provided by animals includes not only pulling objects such as plows but also supplying essential fertilizer. Animals are also used as food. Thus the sun's radiation, along with the multiplier effects of precipitation, human and animal labor, and fertilizer input, combine to maintain the dynamic equilibrium of the agricultural ecosystem and ensure high net primary productivity.

PLATE 13-10
Water buffalo are used for labor and food in many tropical areas.

Agriculture cannot be expected to simply appear anywhere that humans occur. Diamond (1997) argues that agriculture originated in only nine places on Earth: the Fertile Crescent, China, Mesoamerica, Andes and Amazonia, West Africa and Sahel, India, Ethiopia, eastern North America, and New Guinea. For agriculture to begin, it must be either discovered or imported. There are reasons why hunter-gatherers, even if aware of agriculture, might choose instead to remain hunter-gatherers

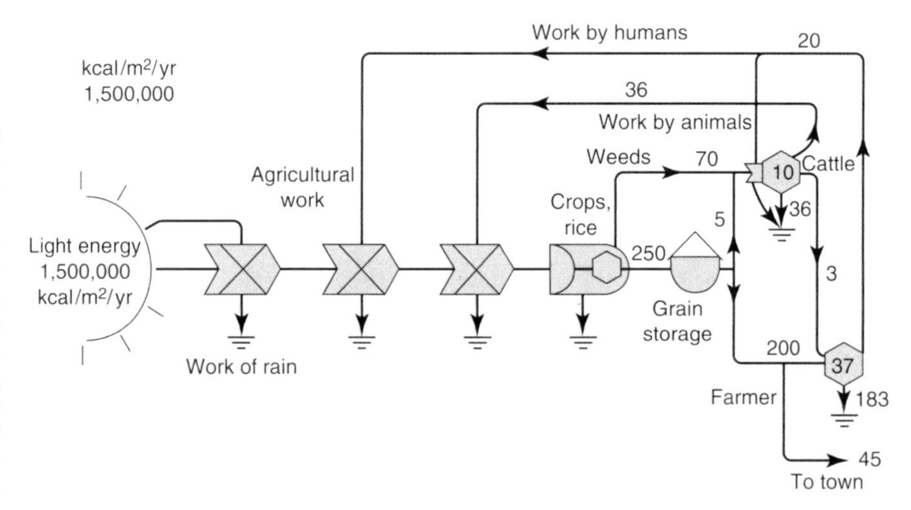

FIGURE 13-6
This diagram illustrates H. T. Odum's concept of the energetics of simple agriculture (in this case, in India). The arrow-shaped boxes with an *X* inside are called *multipliers* because they enhance the productivity of the land. Typical multipliers are abiotic factors such as rain and biotic factors such as the use of farm animals for labor, as well as a feedback loop involving human labor tending crops. Note that the system is sufficiently productive to allow some export.

PLATE 13-11
Fields of rice characterize much of the landscape of Bali, Indonesia, where rice has been successfully farmed for hundreds of generations.

(Diamond 1997). But once agriculture becomes firmly established, it clearly stimulates population growth as well as technological growth, and thus societies that perfect agriculture soon are potential threats to the existence of hunter-gatherers, who have far smaller population densities and are thus less able to resist ecological intrusion by expanding agricultural societies.

Though well over 200,000 species of flowering plants exist today, only a small fraction have been domesticated for agriculture. This should be less surprising than it first might seem once one considers the requirements necessary for a plant species to be suitable for domestication (Diamond 1997). Rapid growth, at least in the initial stages of the evolution of agriculture, would seem to be an obvious requirement. Slow-growing trees and other plants would hardly be suitable, whereas rapidly growing grasses offer far more potential for quickly attaining needed calories. Only after agriculture based on grasses and other fast-growing crops made it possible to establish permanent towns and cities did humans begin domesticating slower-growing species, such as fruit-bearing trees like olives. Only about 34 species of cereals, grasses, and pulses (legumes; see the following discussion) have been domesticated, plus an additional six species used to obtain fiber for such things as rope and cloth, plus eight species of root and tuber crops, and about seven melon species. These represent the key crops of humanity. Just four cereal crops—wheat, barley, corn, and rice—are responsible for 50% of the calories consumed by humans today (Plate 13-11).

The combination of cereal crops and pulse crops provides a generally adequate diet. Wheat, barley, rice, and corn as well as other cereal crops such as millet, sorghum, and sugar cane supply essential carbohydrates. *Pulses* are species of legumes such as peas, soybeans, lima beans, chickpeas, peanuts, and lentils, and all are rich in protein.

The essence of agriculture lies in two considerations. The first is the ability to identify, cultivate, and harvest specific species for use by humans. This is the process of *domestication.* The second is the ability to create technology such that productivity per unit area is enhanced, thus providing an ever-greater crop yield. To increase yield involves identifying and employing additional multipliers beyond those present in basic agriculture (Odum 1971). The Industrial Revolution did exactly that. With the discovery and subsequent usage of fossil fuel came the development of farm machinery and other forms of technology such as chemical fertilizers and pesticides, all of which act in concert as large agricultural multipliers. Thus today, much of the world is fed by the high productivity of industrialized agriculture.

Agriculture in the Neotropics

Agriculture basically represents a disturbance, an *anthropogenic gap*, created in the ecosystem. The farmer works to prevent the normal successional processes from closing that gap so that crops can be grown for exclusive use by humans and/or their domesticated animals. Just as successional areas are temporally unstable, eventually succeeding to forest, so agricultural systems are intrinsically unstable systems. The farmer's labor provides the stability against nature's tendency to diversify the system.

Today, as in the past, tropical peoples face a challenge in attempting to farm nutrient-poor rain forest soils. Most of the minerals and nutrients are not in the soil but in the biomass: the trees, lianas, and epiphytes. The problem is that to clear an area for farming, it is obviously necessary to remove the mass of vegetation. But to do so seems to doom the farming effort, because the often mineral-poor soil will not sustain very much in the way of crops. The way out of this dilemma is fire, applied in a practice that has come to be termed *slash-and-burn agriculture* (Beckerman 1987; Dufour 1990).

Agriculture has a long history in the Neotropics, dating back to about 7,000 years ago (Piperno and Pearsall 1998). As discussed earlier, there is now historical evidence suggesting that human populations made extensive use of fire in parts of Amazonia (Roosevelt 1989) and throughout Central America. For example, charcoal fragments contained in soil from lowland rain forest at La Selva, Costa Rica, dates to as long ago as 2,430 years before the present (Horn and Sanford 1992). Similarly, charcoal fragments suggest extensive burning activity by humans in the high-elevation páramo zone of Costa Rica (Horn and Sanford 1992; Horn 1993). While charcoal fragments could be due to natural fires, it is thought more likely that the fires were set by humans, probably in order to facilitate hunting, to open understory, or in preparation for agriculture.

Within Amazonia, researchers have discovered *terra preta*, numerous patches of soil (from less than 1 hectare to several hectares) that have been modified by the addition of large amounts of crumbled charcoal (Mann 2008). The unique use of charcoal is thought to have resulted in enhanced and prolonged soil fertility, thus promoting more permanent agriculture (see the sidebar "*Terra Preta del Indio*" earlier in this chapter).

Slash-and-burn agriculture follows a typical pattern: a small plot of land (usually between 0.4 to 0.6 hectare, or about 1 to 1.5 acres) is chosen, and machetes and axes are used to cut down all of the vegetation (Plate 13-12). Trees too large to be cut are girdled, killing them. The tangled mass of vegetation is then set on fire rather than removed. Fire eliminates the leaves and wood while at the same time releasing the nutrients and minerals contained within. The soil surface, fertilized by the ash from the biomass, tends to be alkaline, raising the pH, making the soil less acidic. The farmer plants crops for a few years on relatively fertile soil. Rainfall will still act to erode the now-exposed soil and leach minerals. The crops themselves are removed, of course, and with them go more of the minerals. The result is that fertility and yield decline steadily. Typically, staple crops include manioc (various varieties), plantains, bananas, sweet potato, pineapple, chili peppers, and others (Dufour 1990).

PLATE 13-12
The charred remains of trees are readily visible among the bananas growing on this slash-and-burn agricultural plot on Trinidad.

Plots are normally planted as polycultures rather than monocultures, a practice that helps with pest control and that slows the rate of natural succession. Crop losses from predation by such creatures as agoutis are actually anticipated, and "extra" sweet-manioc is often planted for rodent consumption. In a detailed study of the Tukanoan Indians of Colombia, it was learned that most intensive farming of the swidden plots occurs from the 12th through the 24th month after burning. At least two manioc crops are harvested over this time, and manioc harvesting continues until the 36th month after cutting. Individual households may have several swidden plots of differing ages, including some older plots that are used for other products such as fruit trees, medicinal plants, or plants that produce fish poisons (Dufour 1990).

In one Amazon region, the yield for a single village was 18 tons/hectare the first year, 13 tons/hectare the second year, and only 10 tons/hectare the third year, a reduction from the first year of 45% (Ayensu 1980). Within a few years, the plot will be abandoned, allowing natural succession to occur, closing the gap. The typical time sequence for swidden agriculture is to farm the plot for 2 to 5 years, sometimes only for 1 year, sometimes for as long as 7 years, and then abandon it for at least 20 years. Ideally (but rarely), an area just abandoned will not be recut for nearly a hundred years or so, permitting substantial recovery of the system. Slash-and-burn agriculture requires constant rotation of sites and often results in a nomadic population that must move around in the rain forest to find suitable plots to farm. Because of soil-nutrient limitations and therefore the need to allow forest regeneration, the human population density remains generally low.

In some areas, a practice known as *swidden fallowing* significantly extends the usefulness of the plot. A swidden fallow is somewhat akin to *agroforestry* (see the following) in that people plant longer-lived species such as peach palm, guava, coca, various tuber crops, breadfruit, and copal, along with encouraging the growth of various shade trees useful as firewood, thatch, or medicinal plants (Plate 13-13). This practice is also common around permanent campsites and trails, resulting in a significant alteration of the plant species richness of the local area, forming a complex mosaic of agricultural and agroforestry plots as well as natural but disturbed forest in various stages of succession. Given the extensive and varied use to which indigenous people have put the land, it is probably hard to know exactly what "natural forest" is throughout floodplain areas in Amazonia (Dufour 1990).

An experimental study in Costa Rica demonstrated that slash-and-burn does not, in the short run, degrade the soil (Ewel et al. 1981). Researchers cut, mulched, and burned a site that contained patches of 8- to 9-year-old forest and 70-year-old forest. Before the burn, there were approximately 8,000 seeds per square meter of soil, representing 67 species. After the burn, the figure dropped to 3,000 seeds per square meter, representing 37 species. Mycorrhiza

PLATE 13-13
Small farms such as this one commonly practice swidden-type agriculture.

fungi survived the burn, and large quantities of nutrients were released to the soil following burning. The remaining seeds sprouted, and vegetation regrew vigorously on the site. Studies in Amazonia also indicate that ecosystem function is in no way permanently impaired by agricultural practices (Uhl 1987). Nonetheless, throughout most of Amazonia, soils being what they are, it is possible to obtain only five harvests of corn within a period of 2 to 3 years. After that, soil nutrients need replenishment, and that requires a fallow period of from 20 to 30 years (Kellman and Tackberry 1997).

Amazonian Life on *Terre Firme* and *Varzea*

Because of the greater availability of fish and other aquatic game, plus the renewal of soil fertility by the annual flooding cycle, Amazonia *varzea* populations were estimated to be about 70 times greater than those of indigenous people on *terre firme* (38 persons per square kilometer, or 14.6 persons per square mile) compared with (0.52 persons per square kilometer, or 0.2 persons per square mile) (Denevan 1976). Meggars (1985) has compared cultural adaptations on *terre firme* and *varzea*, and there are some striking differences.

Peoples on *terre firme* rely heavily on sophisticated slash-and-burn shifting cultivation, with manioc being the principal crop. Crop yield drops annually, and thus plots are typically abandoned within about 3 years. For productivity to continue, new plots must be put into cultivation each year, and at least 20 years must pass after plot abandonment in order for soil fertility to be restored. Manioc is an ideal crop, as it grows well in poor soil, is easily harvested as needed, and is a fine source of continuous carbohydrate. Protein is obtained from animals. In addition, many kinds of naturally occurring plant materials (nuts, fruits, seeds, and so on) are consumed. Villages must move after about 5 to 7 years due to depletion of local resources. Interestingly, among some aboriginal cultures, the motivation for abandoning a village is based not on a direct measurement of resource availability but rather on the cultural belief that the death of an adult in the community is a sign that the people must move, a belief that ensures nomadism. Sorcery and warfare are common, which means that as populations increase, violence becomes more common, and populations are then reduced.

Varzea populations enjoy an uninterrupted protein source in the form of fish, caiman, birds, manatees, and turtles, as well as mammals such as capybara (though such creatures have often been seriously depleted by too much hunting pressure). Wild rice is harvested and crops (beans, peppers, cacao, and bananas) are planted to make maximal use of the annual flooding cycle. The variety of manioc that is planted matures in 6 rather than 12 months. Maize is grown as well. The period of most stress is the three months during full flood, when food from the previous harvest must be consumed, as none can be grown at that time. Because of variability from year to year in the flooding cycle, crop yields can vary dramatically. Cultures reflect this reality—practices such as infanticide, especially female infanticide, are common and, of course, act to lower the population.

Nonindigenous Farmers in Amazonia

Should you travel anywhere along the Orinoco, the Amazon, or the major river tributaries, you will notice immediately that areas along rivers are inhabited by people, particularly in *varzea* areas. When Europeans colonized Amazonia, they bred with Amerindians, and their descendants became the people who today make their living by farming and fishing the floodplains. The riverine peasantry are called *caboclos, riberenos, mestizos*, or *campesinos*, depending on region. In many ways, these people work the floodplain as described earlier for traditional *varzea* populations, with the exception that they make much more use of market economy rather than rely entirely on subsistence. They cash-crop rice, for instance, and sell fish at market. Indeed, the largest and most diverse fish market in Amazonia is at Manaus, Brazil, where between 30,000 and 50,000 tons of fish are marketed annually (Goulding et al. 1996). Riverine people also harvest such things as Brazil nuts, palm fruits, and rubber for commercial sale (Padoch 1988; Dufour 1990). Note that such usage is not necessarily environmentally damaging, unless overharvesting occurs (and it often does, with such creatures as tapir, manatee, turtles, capybara, and other game animals, though not so much with plants).

Mestizo Life along the Tambopata River in Southeastern Peru

A detailed study by Phillips et al. (1994) demonstrates the importance of diverse floodplain habitat to the local *mestizo* population along the Tambopata River in southeastern Peru. *Mestizos* are relatively recent colonists who live along riverine floodplain and who are of diverse heritage (Plate 13-14). Interviews conducted with men, women, and children ranging in age from 5 to 67 years indicated that the vast majority of plant species occurring in the region were of at least some use to the people (though some species much more so than others). This, of course, demonstrates that the people have a strong pragmatic knowledge of the ecology of the region as well as a dependency on it.

Mature forest on *terre firme* tended to have the broadest range of usage because it supplied both construction materials and food, but floodplain forest was most useful for medicinal purposes, and swamp forest was important for commercial harvesting. The people used an estimated 57 woody plant families and 87.2% of the tree and liana species found on the inventory plots. Remarkably, in one 6.1-hectare plot, 94% of the woody stems were regarded as useful to the *mestizos*. Not surprisingly, palms were of particular importance. Studies from other riverine areas indicate a generally similar broad usage of the biodiversity of the floodplain.

PLATE 13-14
Young *mestizo* boy skillfully paddling a dugout canoe.

Agroforestry: Focus on Coffee

Coffee (Plate 13-15) and cacao (Plate 13-16) are each widely cultivated in the tropics, and each is fundamentally an understory plant. Coffee is a shrub; cacao a small tree. *Agroforestry* is the practice of crop cultivation within a forest. Only the planting of the understory crop alters the forest. The canopy and other attributes of the forest remain more or less unaltered (but see the following discussion).

Ecologists have a strong interest in agroforestry for two reasons. One is that agroforests may be used as model systems to investigate various questions in tropical ecology (Greenberg et al. 2008). This is because these forests are floristically less complex than other forests, often with a dominant canopy species such as one of the species of *Inga*. Manipulations within agroforests are easier than within more complex

PLATE 13-15
Coffee beans ripening on the plant.

PLATE 13-16
Cacao fruits grow from cauliflorous flowers on understory trees.

forests, so ecological questions are more conveniently investigated. For example, one study manipulated epiphyte growth in a shade coffee forest in Veracruz, Mexico, in an attempt to determine how lack of epiphytes might affect select bird species. Of the two species investigated, one, the common bush-tanager (*Chlorospingus ophthalmicus*), was negatively affected by epiphyte removal, while the other, the golden-crowned warbler (*Basileuterus culicivorus*), was far less affected. They differed because the bush-tanager relied heavily on epiphytes as nesting sites (Cruz-Angon et al. 2008). Another study looked at the effect birds have on arthropod prey densities in agroforests (Van Bael et al. 2008). Using exclosures that prevented birds from accessing arthropods, it was learned that birds significantly reduce arthropod densities and thus reduce leaf damage, particularly when migrant birds (that winter in the tropics) are present (Figure 13-7).

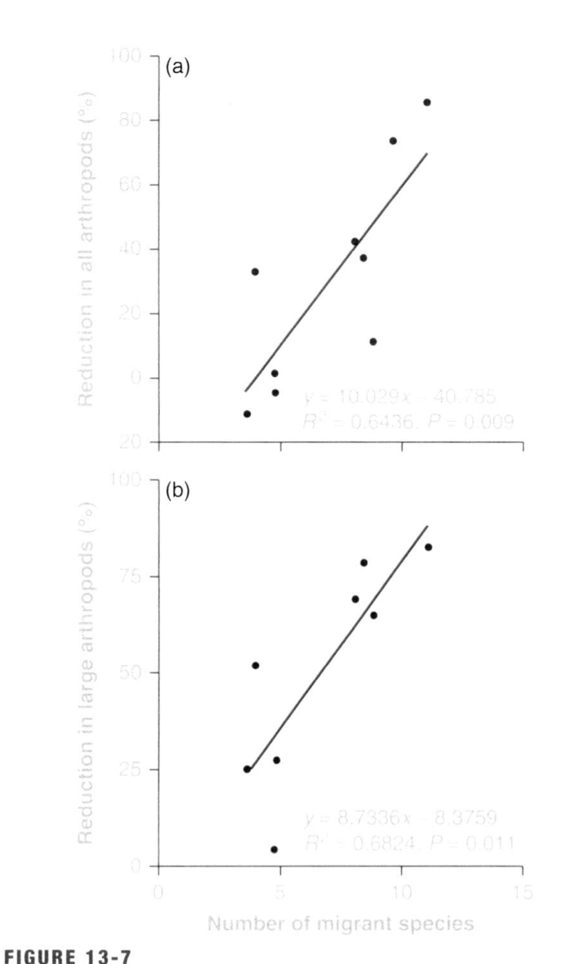

FIGURE 13-7
Bird effects on (a) all and (b) large (>5 millimeters) arthropods as a factor of migrant bird diversity. Each point represents bird data for only omnivorous and insectivorous birds that primarily forage in the strata in which bird exclosures were placed. Correlations were determined using linear regressions, with the percentage reduction of all or large arthropods as the dependent variable and species diversity as the independent variable.

Another study tested the so-called *insurance hypothesis*, which states that high diversity helps ecosystems resist environmental perturbations (Perfecto et al. 2004). In this study, two coffee farms in Tapachula, Chiapas, Mexico, were used, one with diverse shade, one with monodominant shade (using *Inga*). Researchers studied insect predation by birds by placing lepidopteran larva on coffee plants and measuring the rate at which the larva disappeared, presumably due to avian predation. Researchers placed exclosures on some coffee plants in each of the farms, preventing birds from having access. Significant differences in insect consumption were evident such that on those coffee plants where birds had access, lepidopteran larva disappeared much more quickly. The conclusion was that bird diversity may protect against potential outbreaks by insects. In other words, birds provide *biodiversity insurance* to maintain the stability of the forest. In addition, it was clear that the farm with diverse shade contained many more bird species than the forest with monodominant shade (Table 13-1). These kinds of studies are often easier to perform in agroforests than in more complex forests.

The study of agroforests also has the potential to contribute to better understanding of landscape heterogeneity and beta diversity as well as the importance of floristic complexity in ecological processes (Greenberg et al. 2008). For example, there is a gradient of shade coffee agroforests ranging from *rustic*, where there has been essentially no canopy alteration, to various degrees of *plantation* forests, where select species such as *Inga* are planted to varying densities. This range of structural complexity allows ecologists the opportunity to investigate numerous questions about the relationship between diversity, species richness, and structural complexity.

The second reason agroforests are of interest is that agroforestry represents a potentially sustainable use of tropical forest without significant loss of biodiversity. This second focus has led to efforts among conservation biologists to promote the use of what is termed *shade coffee*, coffee grown within traditional agroforests (Kricher 2000).

Coffee is part of the plant family Rubiaceae, a large family of some 600 genera and 13,500 species. Two species, *Coffea arabica* and *C. canephora* (also known as *robusta*) are cultivated. Caffeine averages from 0.8% to 1.4% in *arabica* and 1.7% to 4.0% in *canephora*. Many varieties, called *cultivars*, have been developed within

TABLE 13-1 Vegetation characteristics and bird densities and richness at two large coffee farms in Tapachula, Chiapas, Mexico.

TYPE OF FARM†	NUMBER OF TREE SPECIES	CANOPY COVER (%)	NUMBER OF BIRD SPECIES		NUMBER OF BIRD INDIVIDUALS	
			TOTAL	AVERAGE‡	TOTAL	AVERAGE‡
Diverse shade	7.3[a] (0.396)	64.0[a] (3.978)	35	11.66[a] (1.383)	679	56.58[a] (9.053)
Monodominant shade	3.0[b] (0.211)	30.4[b] (2.146)	22	5.58[b] (0.679)	219	18.25[b] (3.608)

Notes: Numbers in parentheses represent 1 SE of the mean. In a given column, means with different lowercase letters are significantly different at the $P < 0.001$ level.
† Diverse shade and monodominant shade correspond to commercial polyculture and shaded monoculture, respectively.
‡ Average number of bird species and individuals refers to the mean of 5-minute counts for a total of 30 minutes.

each species. *C. arabica* is endemic to Ethiopia, southeastern Sudan, and northern Kenya, but today it is planted on over 10 million hectares within more than 50 countries, representing about 70% of the world's coffee (Vega 2008). In the Neotropics, about 95% of the coffee grown is *C. arabica*. In 2006, the countries producing the most coffee were Brazil, Vietnam, and Colombia. While coffee is genetically adapted to grow in shade, cultivars have been produced that thrive in full sunlight. This has resulted in the progressive replacement of traditionally grown coffee with high-intensity agriculturally grown coffee, a trend that has caused concern among those interesting in conserving tropical biodiversity through more traditional agroforestry.

Traditional coffee production involves replacement of understory species with coffee plants in what are called *rustic coffee plantations* (Plate 13-17). What this means is that forests remain relatively unchanged aside from the dominance of coffee in the understory. But coffee plants grow more rapidly with more sun, so in many cases some canopy trees are taken out to open the forest and permit much more light, reducing the biodiversity value of the coffee plantation (Plates 13-18a and 13-18b). In many cases, the coffee planta-

PLATE 13-17
This coffee plantation near Gallon Jug, in western Belize, is representative of a rustic coffee approach. There is a dense canopy of indigenous trees. The biodiversity of this plantation is high.

(a)

(b)

PLATE 13-18
(a) This is the same area in Belize some years later, when some of the canopy trees were removed to permit more sunlight for coffee growth. The dense coffee understory is readily evident. (b) Added light enhances the productivity of the coffee crop but reduces the ecosystem's overall biodiversity.

tion is combined with other crop plants such as various fruits, vegetables, and medicinal plants to form a complex *polyculture*. A more focused approach, usually called *commercial polyculture*, requires additional manipulation and simplification of the ecosystem to permit greater sunlight and faster growth of coffee. This approach may be extended to create reduced or specialized shade by replacing canopy tree species with select species of *Inga*, *Erythrina*, *Gliricidia*, or other species. *Inga* is a canopy-level species, but the others are small trees that allow far more sunlight to penetrate. Last, there is *full-sun coffee*, a monoculture that appears identical to traditional large-scale agriculture. Sun-coffee plantations have been simplified to the degree that they no longer support forest biodiversity and thus represent a form of habitat loss, just as the cutting of forest for soybean production does.

It has been long recognized that rustic coffee production acts to preserve local biodiversity (Griscom 1932). As noted earlier, some plantations are not rustic but instead utilize a single species of canopy tree, in many cases *Inga*. Studies of bird species richness have demonstrated that rustic coffee plantations act to maintain a high species richness of birds (Greenberg et al. 1997b; Wunderle and Latta 1998 and 2000; Wunderle 1999). When compared with plantations that have an overstory of *Inga* or *Gliricidia*, somewhat more numerous bird species are found in *Inga* plantations, but considerably fewer than are recorded in mature forests (Greenberg et al. 1997a). In other words, as coffee plantations are structurally simplified, bird species richness declines, especially in the case of sun-coffee plantations (Greenberg et al. 1997a; Wunderle and Latta 1998).

The type of agroforestry used to produce coffee has strong conservation implications (Perfecto et al. 1996). As forests are converted to grow increasingly sun-adapted cultivars of coffee, biodiversity is lost. It is clearly possible to continue to grow coffee in a manner in which that outcome is prevented. Shade coffee is not the same as undisturbed forest, but it has clearly demonstrated the potential to maintain a reasonably high biodiversity. There are efforts by conservationists to promote the use of shade coffee, an attempt to develop a market for a coffee product that does not deplete biodiversity (McLean 2000). But problems exist with the pragmatics and economics of shade coffee. Typically, it is considered to be *premium coffee* and costs more (Perfecto et al. 2005). Profit from the higher cost should find its way to the farmers who produce the coffee as incentive to continue to do so. There is also the matter of certification. How does one know that the coffee is really shade coffee? Is it rustic or monodominant shade coffee? There must be some form of enforcement in the field to ensure that the coffee really is shade coffee. And certification programs differ (Mas and Dietsch 2004), so one brand of shade coffee may be certified differently from another. For example, some certification programs exclude shaded monocultures, while others do not. Because of various economic and social factors, shade coffee may not always ensure that biodiversity will be preserved (Tejeda-Cruz et al. 2010). Nonetheless, if the challenges of social and economic factors are met successfully, there is little question of the potential value of shade coffee to sustain much biodiversity.

Ethnobotany

Native peoples in the tropics have generations of experience with plant and animal defense compounds. For example, consider *cassava*, also known as *manioc* (Plate 13-19). Some manioc is *sweet*, and cultivars of sweet manioc are a staple source of carbohydrate in many areas. Sweet manioc has little in the way of defense compounds in its thick root, and thus it is safe to consume. But there are also bitter-tasting strains of manioc, whose meter-long, thick roots are protected from herbivores by cyanogenic glycosides. It is only in the poorest, least fertile soils that one finds bitter, high-cyanide manioc. Just as with most plants on infertile soils, manioc synthesizes powerful defense compounds because the cost of replacing herbivore-damaged tissue is too great relative to the potential for productivity.

PLATE 13-19
Cassava crop in Indonesia. Cassava is native to the Neotropics but has been planted throughout the tropical world.

Because of its cyanide protection, manioc is easy to grow, one of the few crops that people can successfully farm on poor soils. But what to do about the cyanide? After the root is harvested, it must be first grated and then thoroughly washed, soaking throughout the night so that the toxic compounds become soluble in water. The dough-like mash is then placed into an elongate, flexible cylinder, normally woven from palm fronds, that constricts when pulled and is used to "wring out" the dough. The cylinder is hung from a tree branch, and a horizontal pole it attached to the base. Usually, two women sit on the pole, one on either side of the cylinder, and their combined weight elongates, squeezes, and compresses the cylinder, which forces out the cyanide-containing liquid. The mash, now essentially free of cyanide contamination, is wrung out and baked on a flat stone (Schultes 1992). In western Amazonia, the cylinder is called a *tipi-tipi*. In Belize, it is a *whola*, the local name for the boa constrictor. For a broader discussion of detoxification of manioc, see Padmaja (1995). Cassava is an essential crop throughout much of the tropical world, and its biology has been well described by Hillocks et al. (2003).

Given the abundance and diversity of plants and animals in the tropics, coupled with the long history of human occupation, it is not surprising that indigenous people have found multiple uses for the diverse array of chemicals contained within the many species of native flora and fauna. Chemicals have been extracted for use in arrow (dart) poisons, hallucinogens, fish poisons, drugs for medical and related use, stimulants and spices, essential oils, and pigments (Gottlieb 1985).

Ethnobotany (Cox and Balick 1994; Balick and Cox 1997; Joyce 1992; Plotkin and Famolare 1992) is the study of how indigenous people have learned to use ambient vegetation for a diversity of pragmatic purposes. It is broader in scope than merely the extraction and subsequent usage of chemicals contained within the plants. Ethnobotany also includes a consideration of all uses of plants, including for food and fiber. It is an interdisciplinary field involving botany, anthropology, archeology, plant chemistry, pharmacology, history, and geography (Schultes 1992; Schultes and Raffauf 1990; Schultes and Hoffmann 1992; Schultes and von Reis 2008).

Ethnobotany is not confined to the tropics. A perusal of old herbal manuals will quickly reveal that numerous North American plant species were relied on for pharmaceutical applications in years past, until the advent of modern medicines. Some are still used. As one example, resin from the mayapple (*Podophyllum peltatum*), a common understory spring wildflower throughout eastern forests, was commonly employed by Native Americans to remove warts. It is still used today to treat venereal warts (Schultes 1992). In the Neotropics, many make the assumption that ethnobotany applies only to isolated indigenous tribes such as the Yanomami, for instance, but this is false. Modern populations of mixed-heritage humans such as the *mestizos* and *riberenos* of Peru or the *caboclos* of Brazil, all of whom are linked to native Amerindian cultures, but also under strong Western influence, make heavy use of ethnobotanical knowledge (Phillips et al. 1994).

Ethnobotanical insight is gained culturally through the generations, essentially by trial and error. Not all indigenous groups possess equally sophisticated ethnobotanical understanding. With regard to extraction and preparation of various drugs and drug combinations, the knowledge is often housed in the mind of but one revered individual, the shaman of the village. Nothing is ever written down but is instead passed on from one generation to the next by the shaman, who is both a teacher and a practitioner, a person of substantial power in the community. Illnesses are rarely if ever blamed on organic causes but are usually assigned to evil spirits or curses (Schultes 1992). It is the shaman who communicates with the spirit world—and who cures headaches, back pain, bug bites, and constipation. Unfortunately, shamans may die of old age before passing their knowledge to the next generation. There is widespread fear among ethnobotanists that much knowledge is currently being lost as traditional tribes experience the impact of Western culture, and fewer young people study to be shamans.

Ethnobotanist Mark Plotkin (1993), who studied with shamans in northeastern South America, describes many intriguing examples demonstrating the sophisticated knowledge of local people in the use of tropical defense compounds. Alkaloid-containing sap from a common liana is used to help cure fever in children. Rotenone, a potent vasoconstrictor, is extracted from another common liana and employed to kill fish, a critical protein source. Plants that even a skilled botanist has difficulty telling apart are easily recognized as separate species by the shaman. Equally intriguing is Plotkin's vivid descriptions of how he was tutored in this knowledge (including the use of hallucinogens) by various shamans whose trust and respect he patiently won.

I met Piwualli, a shaman, whose circuit included the villages along a section of the Napo River in Ecuador. Said to be 73 but looking considerably younger, Piwualli ushered my group to a dilapidated wooden table outside his house. The table was piled high with clumps of dried herbs, part of the bush doctor's pharmacopoeia. Piwualli spoke no English and only a little Spanish, but our guide knew Piwualli's language and acted as translator. Piwualli described the multiple uses of the various plants. One sounded remarkably like it was used for symptoms characteristic of certain severe nervous disorders such as Parkinson's disease. Interestingly, Schultes and Raffauf (1990) list three species of unrelated plants used by indigenous people in this region to treat "palsy-like trembling."

The suggestion that many serious ailments may be helped by potent compounds from the tropics is both intriguing and promising. For many years, the alkaloid quinine, from the Neotropical shrub/small tree genus *Cinchona*, has been reasonably effective in combating certain malarias. Resin extracted from plants of the genus *Virola*, used as a powerful hallucinogen (see the following discussion), may also prove to be very effective in controlling or even curing chronic fungal infections, which currently can be only suppressed by Western medicines (Schultes 1992). Plotkin (1993) notes that only about 5,000 of the world's 250,000 species of plants have been thoroughly investigated as to pharmacological properties and that the 120 plant-based prescription drugs are derived from only 95 species. Gottlieb (1985) points out that as of 1977, only 470 of the estimated 50,000 Brazilian flowering plant species, or about 1%, had been examined for the existence of chemical compounds. Cox and Balick (1994) assert that only a minuscule number of plant species (compared with total plant diversity) has been thoroughly studied to ascertain their chemical composition and potential medicinal uses. Surveys are being conducted in an attempt to evaluate the pharmacological potential of Neotropical plant species. Schultes (1992) describes how scientists aboard the research vessel *Alpha Helix*, normally an oceanography ship, sailed the Amazon for a year and collected 960 plants, representing 3,500 specimens, conducting biochemical analyses of these plants using the sophisticated shipboard laboratories.

Thomas Eisner of Cornell University applied an innovative approach. He was catalytic in convincing the Merck Pharmaceutical Company to enter into an agreement with the government of Costa Rica, such that Merck would provide one million dollars for inventory and conservation purposes in exchange for exclusive rights to survey the flora for compounds of potential medical use (Cox and Balick 1994). Eisner calls his approach *chemical prospecting*, and though the eventual outcome of such searching is unknown at present, it is undeniable that there is potential for finding medically useful drugs within tropical flora. Shaman Pharmaceuticals, which was created in 1988, is a pioneer company attempting to discover and apply ethnobotanical knowledge to the needs of Western medicine (Joyce 1992).

Plotkin (1993) emphasizes the obligation to share any benefits that may be derived from ethnobotanical studies with the indigenous people who, in fact, obtained the knowledge in the first place. Such a policy is not only morally compelling, it has strong conservation potential. For example, the Terra Nova Rain Forest Reserve in Belize was established in 1993 by a group named the Belize Association of Traditional Healers, an assemblage that includes people from most of the cultural and ethnic groups in Belize, a country in which about 75% of the people are estimated to be dependent on plant medicines for their primary health care needs (Balick et al. 1994). The reserve, a 2,400-hectare (5,928 acres) area of lowland forest, will be managed to accomplish the following: cultivation and documentation of medicinally useful plants and protection of the plants from overharvesting; conducting of ethnobotanical and ecological research; and encouragement of ecotourism, with walks and seminars designed to teach about the uses of the plants.

Schultes and Raffauf (1990), in their book *The Healing Forest*, discuss approximately 1,500 species and variants of plants from 596 genera and 145 families, all of which have medicinal or toxic uses by indigenous peoples in northwest Amazonia. It is fascinating to see the range of symptoms that are treated as well as to see the diversity of plants that are applied to certain common ailments or conditions. For example, there are 38 plants that can be used for diarrhea, 25 for headache, 18 for muscular aches and pains, and 38 for toothache. There are many plants that can be used for various insect bites (including 16 for ant bites), 36 for intestinal parasites, and 29 for snakebites. There are 26 plants listed for use as contraceptives. In addition, there are plants alleged to have use in treating such conditions as sinusitis, stiff neck, bleeding gums, stomach ulcers, cataracts, asthma, swollen breasts, epilepsy, testicular swelling, tumors, boils, blisters, mange, and baldness, a selection that is by no means comprehensive. Of course, you should bear in mind that the degree to which these diverse applications achieve success is debatable. Where available, many indigenous groups readily accept Western medicine (though it can be argued that part of the reason for such acceptance is that these peoples have been afflicted with various Western ailments from exposure to settlers). Still, the efficacy of ethnobotanical treatments for many afflictions seems undeniable, and as mentioned earlier, there is still so much to learn.

Besides medicinal uses, many plants are used to extract various poisons for hunting, and many other plants are used for hallucinogenic or narcotic purposes. I will close this chapter with a brief look at some of the better known of these.

PLATE 13-20
CURARE VINE AND LEAVES

CURARE

Charles Waterton, whose first journey to Amazonia was during the same year as the War of 1812, was undoubtedly a wonderfully entertaining dinner guest. What stories he must have told. This aristocratic, eccentric explorer of the Amazon demonstrated uncommon skill at taxidermy as well as an intrepid drive for exploration and discovery. And one of his discoveries was that indigenous people had found a very powerful drug, one now called *curare*. Waterton (1825) describes a vine, called *Wourali*, that supplies the primary ingredient for arrow poison and the "gloomy and mysterious operation" in which the poison is extracted and prepared, only by certain skilled individuals. He details how a large ox, estimated to weigh nearly 1,000 pounds, died within 25 minutes after being shot in the thigh with three poisoned arrows. The poison, said Waterton, produced "death resembling sleep."

Curare (Plate 13-20) has such a powerful effect of relaxing muscles that it induces paralysis. And that's the basic idea. Curare is added to the tips of arrows and darts and then used by skilled hunters to bring down various species of mammals and birds. If you look at the small darts that are the ammunition of blowguns, you will see immediately that these weapons would do little more than make a pin prick in their intended prey were it not for the presence of the poison. The arrow or dart does not bring the creature down—the curare does. Curare and its derivatives are well known by practitioners of Western medicine, as they are commonly employed during certain surgical procedures.

Curares are extracted from many different kinds of plants from an array of different families. Indeed, over 75 plant species are utilized for this purpose in the Colombian Amazon. Most curares are mixtures of several plant species (often prepared specifically for the kind of animal sought), with much variation not only from tribe to tribe but from one shaman to another (Schultes 1992). The art of preparing curare requires careful attention to detail. It is a dangerous substance. It is remarkable that so many different combinations of curare poisons have been discovered and utilized by the indigenous Amazonian peoples (Schultes 1992; Gottlieb 1985).

Curare takes its name from the genus *Curarea*, formerly *Chondrodendron*. *Curareas* are lianas, beginning as rooted shrubs that eventually become climbers. *Curarea toxicofera* is a species that is widely used by many tribes. The curare is extracted from the bark and wood of the stem and is often mixed with other species, particularly those in the genus *Strychnos* (Schultes and Raffauf 1990).

COCAINE

Cocaine is an addictive narcotic, a powerful alkaloid extracted principally from a small, unpretentious shrub, *Erythroxylum coca*, var. *ipadu*, commonly called *coca* (Balick 1985). A second species, *E. novogranatense*, is cultivated along the eastern slopes of the Andes and does not occur in lowland areas (Plate 13-21). Coca contains numerous alkaloids, but cocaine is the one in greatest concentration. Though cocaine is considered a scourge of society in North American culture, it has important traditional uses

PLATE 13-21
COCA LEAVES

by South American indigenous peoples: as medicine, in certain rituals, for chewing, and for nutrition (Balick 1985). Studies cited by Balick (1985) show that ingestion of 100 grams of coca leaves is sufficient to supply one's daily needs for calcium, iron, phosphorus, and vitamins A, B_2, and E. Chewing wads of coca leaves is also important in suppression of fatigue, providing added endurance for people in the rarefied air of the high Andes. It should be emphasized that a leaf contains only about 1% cocaine, and even those effects are modified by other compounds in the leaf (Boucher 1991), so chewing coca leaves is not the same as snorting crack cocaine (which affects the brain in as little as 5 to 10 seconds). Coca leaves are also applied to wounds or boiled to make a tea. I can assert from personal experience that coca tea helps to attenuate the unpleasant effects of high-altitude sickness common to visitors in the Andes.

Most coca that is grown to be used as a narcotic is from Peru and Bolivia (though it is purified and shipped from Colombia, which produces about 80% of the world's cocaine), especially the Upper Huallaga Valley in Peru (along the east slopes of the Andes), where it is estimated that 60% of the world's coca is grown (Boucher 1991). Unfortunately, it is very lucrative to farm coca for cocaine. For example, a hectare of coca in Bolivia can yield $6,400 compared with $1,500 for coffee, $600 for bananas, and $300 for corn, and these profit estimates are nearly two decades old (Boucher 1991). The profit for cocaine growing may be even larger now. Such profitability, coupled with the reality that coca has many traditional uses, suggests that the eradication of the cocaine trade is at best problematic.

One historic note of interest is that the soft drink Coca-Cola was at one time really *coca* cola. In 1903, based on the recommendations from a report by the U.S. Commission on the Acquisition of the Drug Habit, the producers of Coca-Cola eliminated the minute amount of cocaine that, up to that time, had been included in the recipe (Moeser, in Boucher 1991). The report asserted that cocaine was being used mostly by "bohemians, gamblers, prostitutes, burglars, racketeers, and pimps."

The unique history of coca continues. At the time of this writing, Evo Morales, president of Bolivia (and the nation's first indigenous president), has developed a drink using the extract of the coca leaf; the name of the drink is Coca Colla. According to the *Telegraph* (telegraph.co.uk), Bolivia is the third largest producer of coca, after Colombia and Peru. In 2008, the Bolivian coca crop was 30,500 hectares (75,370 acres). President Morales reportedly intends to increase that crop.

INTOXICANTS AND HALLUCINOGENS

Perhaps the best-known hallucinogen in the Neotropics is the genus *Virola* in the family Myristicaceae (nutmegs). There are between 62 and 65 species of these understory trees throughout the Neotropics, and a few of them are widely used throughout western Amazonia and much of the Orinoco Basin to achieve rapid and extreme intoxication with subsequent hallucinations. This practice serves multiple functions, ranging from spiritual divination to ritualistic diagnosis and treatment of disease (Schultes and Hoffmann 1992). In many tribes, only the shaman takes *epena*, *ebena*, or *nyakwana*, as the *Virola* preparation is known, while in others, such as the Yanomami, for example, all male members of the group participate. The drug itself is obtained from *cambial exudate* (a gummy substance secreted by breaching the tree's cambian layer just beneath the bark) on the inner bark of the tree, which is boiled and simmered and then refined into a reddish powder. In most cases, the drug is taken as a powdered snuff, blown with great force into the nostrils and sinuses, using an elongate pipe made from a plant stem. In some cases, however, the drug is administered orally, in the form of a pellet. Once administered, the drug, which is a combination of various strong alkaloids (Gottlieb 1985; Schultes and Hoffmann 1992), causes immediate eye tearing and mucus discharge followed soon by a restless sleep during which the person is subject to extreme visual hallucinations described as "nightmarish." Details of this experience can be found in Schultes and Hoffmann (1992), Schultes and Raffauf (1990), and Plotkin (1993). In addition to use as a hallucinogen, *Virola* is used for an array of medical problems (Schultes and Raffauf 1990).

Another well-known Amazonian hallucinogenic preparation is *ayahuasca*, also called *yagé*. Ayahuasca, which means "vine of the soul" in the Quechuan language, is derived from the vine *Banisteriopsis caapi* (Malpighiaceae). The active compounds are harmine, harmaline, and tetrahydroharmine (Calloway et al. 2005). Ayahuasca is used not only in religious ceremonies but also to purge parasites.

Aztecs and Mayans of Central America routinely used mushrooms and various "psychedelic fungi" in their religious rituals. One mushroom in particular was said to provide its users with "visions of Hell" (Furst and Coe 1977). *Peyote*, made from a cactus (*Lophophora* spp.), is a widely used alkaloid hallucinogen throughout Central and North America.

People living in tropical regions have developed rich and complex cultures and obviously learned how to extract a great many varied resources from the land they inhabit, resources that range from food and fiber to intoxicants and stimulants. A rich field remains for investigation, as this chapter has, by necessity, touched on only the highlights.

14 Forest Fragmentation and Biodiversity

Chapter Overview

This chapter discusses two essential elements of tropical ecology: the effects of habitat fragmentation and the importance of understanding and conserving biodiversity. These concepts are strongly related because fragmentation is an important force driving the loss of biodiversity. *Fragmentation* occurs when forests are cleared in varying-sized parcels. A patchwork of forest fragments results. Fragmentation represents loss of area, and area loss results in numerous ecological effects that may lead to eventual loss of species. Fragmentation increases *edge effects*, as the proportion of edge habitat changes in relation to total forest volume. Isolation of forest fragments interferes with the dispersal of many plant and animal species. Fragmentation has now been widely studied in the tropics. Knowledge of how organisms respond to fragmentation is helpful in designing areas intended to conserve species.

Biodiversity is a distinguishing characteristic of tropical ecosystems, particularly forests. Concerns have been widely raised by ecologists and others about ongoing and pending loss of biodiversity throughout the world, and particularly in the tropics. This chapter and the next discuss why biodiversity is threatened, why it is justified to act to preserve it, and what actions might be taken to minimize biodiversity loss.

Fragmentation

When an ecosystem is subdivided into patches (of varying areas), isolated to some degree from other such patches, that ecosystem is fragmented. *Fragmentation ecology* is the study of the ecological effects of reduced area, relative degree of fragment isolation, and modes of connectivity within the landscape matrix among fragments. Fragmentation is an outgrowth of both the theory of biogeography (see Chapter 5) and the field known as *landscape ecology*. Many years ago, particularly in Europe, ecologists began to study anthropogenic-generated landscapes as representing complex matrices of interactive ecological units. Agricultural fields divided by natural hedgerows, some with abutting forest, form an example of a landscape mosaic, the initial focus of landscape ecology. Fragmentation is a branch of landscape ecology, because fragments represent natural ecosystem patches isolated to varying degrees by anthropogenic ecosystems. Humans alter landscape by removing areas of natural ecosystems (forests, savannas, and grasslands) and replacing those areas with crops, pasture, housing, or something else. Fragmentation occurs throughout the world (Plate 14-1). Think of any suburban area in North America, for example. But the focus in this chapter is on fragmentation in tropical forests.

PLATE 14-1
This landscape mosaic in Ireland illustrates both fragmentation of forest as well as how narrow hedgerows provide natural corridors to connect forest fragments.

What ecological effects occur when rain forest is fragmented? Recall how species richness is related to area (see Chapter 5). The theory of island biogeography predicts that, with time, islands (and, by implication, nonisland ecosystems) attain what is termed a *dynamic equilibrium* of species richness such that immigration of new species is roughly offset by extinction of some resident species. Should island area increase, theory predicts that more species will be accommodated and the equilibrium point will shift upward. But should area decrease, the theory predicts that extinction rate will then exceed immigration rate and species richness will decline until reaching a lower equilibrium point (see Figure 5-18a). An island that suddenly experiences an area reduction will then exceed its equilibrium point simply due to area loss. Fragmentation is a form of area loss in that it breaks up a contiguous area such as a forest, and subsequently isolates fragments from other such fragments, creating an *archipelago effect*: where once there was uninterrupted ecosystem, there are now fragmented forest islands.

Millions of hectares of tropical forest (ranging from lowland rain forest to tropical dry forest) are cleared annually (Achard et al. 2002), converted to pasture, agriculture, or some other use. Deforestation does not necessarily result in fragmentation. If huge areas are cut, then no fragments remain. But many times, various-sized tracts of forest are cleared, leaving scattered forest remnant *islands*. That is fragmentation. Many of the world's tropical forests are increasingly becoming either fully cleared or forest fragments, a reality with significant ecological implications. Forest fragmentation has become a major topic of ecological and conservation focus in tropical landscapes (Schelhas and Greenberg 1996) (Plate 14-2).

PLATE 14-2
This photograph was taken near Alta Floresta, Brazil. The foreground shows essentially total deforestation. Forest remnants are barely visible in the distance. The dense haze is smoke from fires set to burn the slash.

The Biological Dynamics of Forest Fragments Project (BDFFP)

Brazilian law requires that each landowner maintain at least 50% of the land as forest (Holloway 1993). Given that most activities result in deforestation, the result of the Brazilian policy has been that landscapes have become increasingly fragmented, with varying-sized islands of forest remaining amid a landscape of pastures, croplands, and other human-created ecosystems.

Approximately 70 to 90 kilometers (43.5 to 56 miles) north of Manaus, Brazil, on *terre firme* forest, there is an ongoing study that was established specifically to evaluate the numerous ecological effects of forest fragmentation, with the associated objective of learning how best to structure biological preserves. The project was formerly known as the Minimum Critical Size of Ecosystems Project, but has since been renamed the Biological Dynamics of Forest Fragments Project (BDFFP). The project was initiated in 1979 with the support of the Brazil National Institute for Research in Amazonia (INPA) and the World Wildlife Fund. In 1989, the administration of the project was assumed by the National Museum of Natural History at the Smithsonian Institution, working in cooperation with INPA. The study has involved over 25 principal investigators working with such taxa as woody plants, birds, primates, bats, nonflying mammals, ants, butterflies, euglossine bees, and various beetles. As would be expected, there are numerous publications from such a comprehensive study, and general reviews can be found in Lovejoy et al. (1986); Lovejoy and Bierregaard (1990); Bierregaard et al. (1992); and Bierregaard et al. (2001a and 2001b).

The essence of the BDFFP is to document ecological effects created by differing patch size (area) and degree of isolation of forest fragments (Plate 14-3). This goal, as described earlier, follows directly from the theory of island biogeography. Because fragmented forests are, in a sense, islands (but see the following discussion), it is clear that the project has conservation implications with regard to the design of nature preserves and parks (Diamond 1976; Simberloff and Abele 1976). The Minimum Critical Size of Ecosystems Project was structured in essence as both a test and an application of the theory of island biogeography. One objective was to learn how best to design preserves amid an area of deforestation such that normal biodiversity is maintained. The original name of the project reflects this initial objective of determining the minimum area that must be preserved to maintain the biodiversity typical of intact forest (Plate 14-4).

PLATE 14-3
Aerial view of BDFFP sites.

Researchers worked with cattle ranchers in designing the project (who, you will recall, were required by Brazilian law to leave 50% of their lands forested). The ranchers were persuaded to clear forest in such a way as to create forest fragments of different size and distance from an undisturbed, protected 1,000-hectare (2,470-acre) forest area that served as a control (analogous to the "mainland," or *source*). Fragments varied in area as follows (Table 14-1):

- 1 hectare (2.47 acres) (5 fragments)
- 10 hectares (24.7 acres) (4 fragments)
- 100 hectares (247 acres) (3 fragments)
- 200 hectares (494 acres) (1 fragment)

TABLE 14-1 History of surrounding land use and isolation of BDFFP reserves.

RESERVE	SIZE	ISOLATION DISTANCE	HISTORY
Fazenda Esteio: Colosso, Florestal, Cidade Powell, and Gavião camps			
1104	1 hectare	150 meters	Reserve isolated in 1980. The surrounding area was poorly burned. High second growth was present in 1983 (approximately 4 meters). In August 1987 and August 1994, 100 meters of second growth was cleared and burned around the reserve to maintain its isolation. Cattle grazed around reserve.
1112	1 hectare	400 meters	Reserve isolated in 1983. Approximately 600 hectares were cleared but never burned around this reserve. As of 1985, this whole area was high second growth dominated by cecropia. No grazing around this reserve.
1202	10 hectares	700 meters	Reserve isolated in 1980. The burn was successful and intense. High second growth was present in 1983 (approximately 4 meters). When the ranchers burned the second growth around the reserve in November 1982, the fire got into the reserve and burned approximately 1 hectare in the reserve's southeast corner. Grasses were planted in 1983, and cattle grazed around the reserve. The second-growth area between Colosso camp and the reserve was cut and burned in August 1985. Second growth was cleared again around the reserve in 1987. In 1989 and 1994, a 100-meter band of second growth was again cleared and burned around the north, west, and south sides of the reserves.
1207	10 hectares	100 meters	Reserve isolated in 1983. Only a band of 100 to 150 meters was cleared around the isolate on the east, west, and north sides. This area was never burned. Second-growth forest dominated by cecropia rapidly reconnected this reserve to the adjacent forest. No grazing around reserve.
1301	100 hectares	Not isolated	A continuous forest reserve with one side exposed. East side of reserve isolated in 1983, but area was never burned. As of 1985, this whole area was high second growth dominated by cecropia.
1401	1,000 hectares	Not isolated	West side of reserve exposed in 1977–1978 by the first clear-cut area on the farm. This area was pasture until 1989. Presently, high second growth (>15 meters) delineates the west side of this continuous forest reserve.
Fazenda Porto Alegre: Porto Alegre camp			
3114	1 hectare	300 meters	Reserve isolated in 1983 by 1,000-hectare clear-cut that was never burned. Rapidly surrounded by exuberant, cecropia-dominated second growth. A band of 100 meters was cut and burned all around reserve in 1991 and 1994. Grass planted after cut and burn. No grazing around this reserve.

(continued)

TABLE 14-1 *(continued)*

RESERVE	SIZE	ISOLATION DISTANCE	HISTORY
3209	10 hectares	900 meters	Reserve isolated in 1983 and rapidly surrounded by cecropia-dominated second growth. In 1991 and 1994, 100-meter band around the reserve was cut and burned. No grazing.
3304	100 hectares	450 meters	Approximately 700 meters of the north and west margins were exposed in 1981 by pasture installation. Cattle grazed in this area. Reserve isolated in 1983 by 1,000-hectare clearing that also left a connection with continuous forest along the stream about 2 kilometers to the north. A 300-meter break in the north end of the corridor was cut and burned in 1984. Cecropia-dominated regrowth on the east, south, and west sides was 1 to 1.5 meters tall in late 1984 and nearly 10 meters tall by 1988. This reserve re-isolated with the 1- and 10-hectare reserves in 1991 and 1994.
Fazenda Dimona: Dimona camp			
2107	1 hectare	150 meters	Reserve isolated in 1984. Cut area was burned the same year. Pasture installed and cattle grazed. The area between camp and the reserve was cut in 1987 and burned two months later. In 1989, 1990, and 1994, a 100-meter band was cleared of second growth around the reserve. *Vismia* spp. dominates second growth around reserve.
2108	1 hectare	600 meters	Reserve isolated in 1984 with Reserve 2107. Cut area was burned the same year. A 100-meter strip around the reserve cleared in 1989 and 1994. Pasture planted and cattle grazed around reserve, *Vismia* spp. dominates second growth around reserve.
2206	10 hectares	225 meters	Reserve isolated in 1984. Cut area was burned the same year. A 100-meter strip cleared in October 1989. In 1994, a 100-meter strip around this reserve was again cut and burned. As with neighboring 2108, pasture installed and cattle grazed. *Vismia* spp. dominates second growth around reserve.
2303	100 hectares	150 meters	The north side of this reserve was exposed by pasture in 1980. In 1982, that area was solidly-established pasture. Abandoned as pasture in 1983 and returned to high second growth (*Cecropia* spp. and *Vismia* spp.). The west side was cut and burned in 1984. In 1990, the reserve was finally isolated by a 200-meter band cut and burned on the remaining east and south sides of the reserve. Currently, this reserve is surrounded by high second growth (*Cecropia* spp. and *Vismia* spp.).

PLATE 14-4
Forest fragment from BDFFP showing fragment isolation.

Fragments were separated by varying distances (100 meters to 900 meters, or 330 to 2,970 feet) from the 1,000-hectare (2,470-acre) control forest (Figure 14-1).

What, for instance, would be the differences in biodiversity between two 10-hectare (24.7-acre) plots, one of which was 500 meters (1,650 feet) from the source and one of which was 100 meters (330 feet) from it? What are the differences between a 1-hectare (2.47-acre) and a 10-hectare (24.7-acre) plot, both of which are 200 meters (660 feet) from the source forest? Does the tree community change in fragmented patches, and if so, how? Are understory bird species more sensitive to area effects than canopy species? Which species of monkeys are most sensitive to area and isolation? What

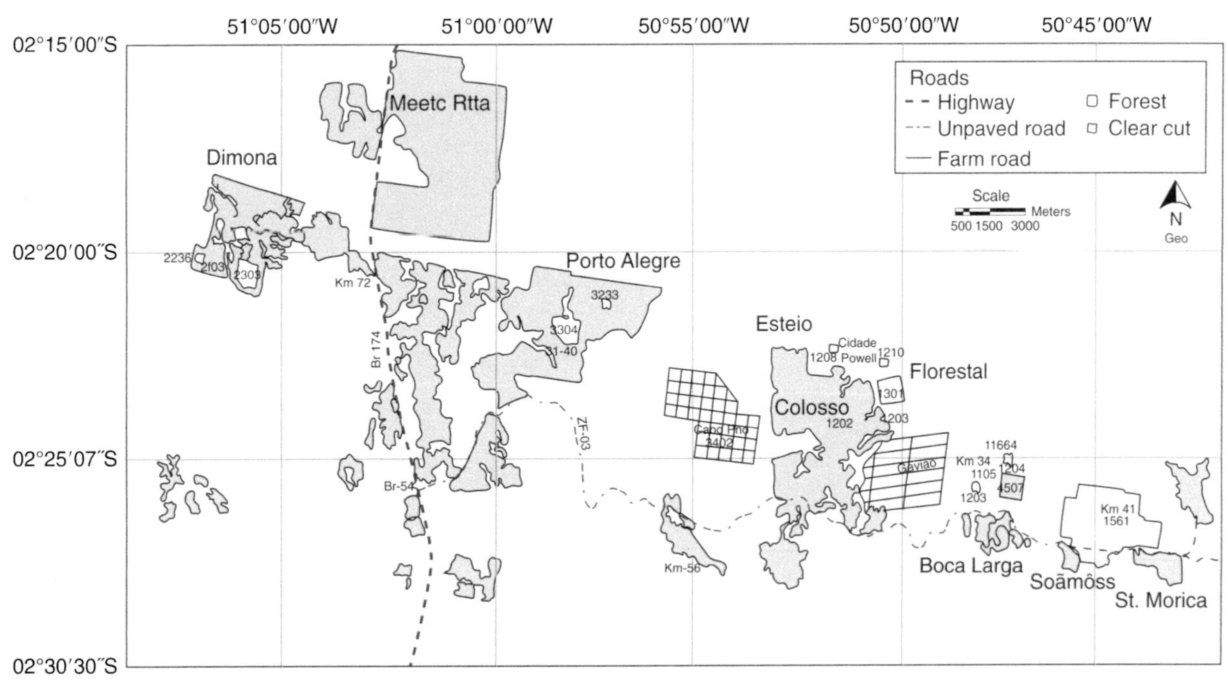

FIGURE 14-1
Map showing the location of the study area of the BDFFP, north of Manaus.

species increase in density with fragmentation? These are the kinds of questions that were posed.

Various researchers have shown that virtually all taxonomic groups studied are sensitive in varying ways to both area and distance effects. Some species decline; others increase. For example, after isolation:

- Isolated fragments experienced an influx of understory birds (presumably immigrating from the surrounding deforested area), but after about 200 days, the total number of birds dropped to below what it was before the forest fragment was isolated. In other words, biodiversity plummeted, what Bierregaard et al. (2001a) describe as a *faunal collapse*.
- Army-ant-following birds were negatively affected, declining and disappearing from fragments. This occurred because army ants require large areas in which to forage for prey. They are among the species most likely excluded in small-area fragments. A similar pattern was evident on Barro Colorado Island (BCI) after it became an island.
- Euglossine bees, which are important long-distance pollinators, were apparently reluctant to fly into fragments isolated by 80 meters (264 feet) or more from other forest.
- Primate species richness in four 10-hectare (24.7-acre) fragments combined was less than the number of species in a single 100-hectare (247-acre) fragment.
- Small nonflying mammals such as various rodents increased in species richness, biomass, and abundance in 1-hectare (2.47-acre) fragments compared with 10-hectare (24.7-acre) fragments and continuous forest.
- In areas where forest was cut but not burned, the rate of succession in the patch was faster, and the dense vegetation supported a larger community of butterflies than either isolated forest fragments or continuous forest.
- Three species of *Phyllomedusa* frogs were lost from small fragments, probably because peccaries had abandoned these areas, and peccary wallows formed the breeding pools for these small frogs. With the peccaries gone, the frogs could not breed. This example illustrates that many faunal changes are subtle and difficult to predict.

Fragmentation thus produces various ecological domino effects, with first-, second-, and third-order effects. But these effects need not be permanent.

Species of dung beetles declined in small fragments, presumably from the lack of excrement and carcasses that disappeared as various animals migrated out (Klein 1989). However, dung beetles made what was described as a "rapid recovery" (Quintero and Roslin 2005). Within little more than a decade, the initial loss of dung beetle numbers and species richness had recovered to its density prior to fragmentation (Table 14-2).

TABLE 14-2 Dung beetle diversity in various-sized plots and habitats in two years.

DIVERSITY MEASURE, BY YEAR	SECOND GROWTH	1 HECTARE	10 HECTARES	CONTINUOUS FOREST	TOTAL
Total number of genera observed					
1986	7	14	13	14	15
2000	17	17	17	16	17
Total number of species observed					
1986	12	38	39	44	55
2000	53	56	54	53	61
Total number of individuals observed					
1986	717	935	749	1,381	3,784
2000	2,584	3,983	4,002	3,047	14,657

Note: The sampling effort implemented in 2000 was three times higher than that of 1986. Hence, absolute numbers are not directly comparable across years.

This recovery was attributed to what happened not in the forest fragments themselves but in the surrounding areas where cutting had initially occurred. These areas went through ecological secondary succession (see Chapter 6), and though quite different in species composition from forest fragments, the successional areas provided suitable habitat for dung beetles as well as animals on which dung beetles depend. This connectivity of the surrounding habitat matrix with forest fragments permitted the beetles to recolonize the fragmented forest as they moved easily within the secondary vegetation and, along with other animals, recolonized the fragmented forest patches. This is an example of *fragmentation turnover*, *habitat connectivity*, and *matrix suitability*, concepts that will be discussed further in the following sections. It also shows that forest fragments are not, in the strict sense, islands, as they are affected strongly by the habitat matrix surrounding them. As it changes, the dynamics of fragments change too.

Edge effects were also evident in isolated fragments. Many injured and dead trees were found along edges, and the overall turnover rate of trees was highest on edges. Rates of litter accumulation accelerated near edges. Seedling recruitment patterns varied as well. Edge effects have been a focus of study in relation to forest fragmentation and will be discussed in more detail later.

Isolation is frequently problematic. Small, distantly isolated forest fragments become ecologically *depauperate* (meaning that they have lost essential species) and function differently from normal continuous forest, sometimes resulting in *ecological meltdowns* (see later in this chapter). This should come as little surprise given that so many species of Neotropical trees, for instance, are dependent on long-distance pollinators or seed dispersers or both. A tree isolated in a small fragment well over 100 meters (330 feet) from other forest could, in effect, be made sterile for want of seed dispersers (see the box below). But isolation due to the creation of forest fragments is occurring throughout Brazil and other tropical countries throughout the world. In general, fragmentation will exert its most severe impact on those species that require a large area but are reluctant to cross small gaps. There are many such species in the tropics. In the Neotropics, the small frog *Chiasmocleis shudikarensis* requires only 0.0001 hectare (or about 0.00025 acre) of forest, but because the species is very patchy in distribution, it is estimated that it requires a total of 500 hectares (1,235 acres) as the minimum area to sustain a viable population (Dale et al. 1994). Any forest fragment of less than 500 hectares will not sustain this species.

Fragmentation Reduces Seed Dispersal by Disrupting Mutualism

Tropical tree species are usually dependent on specific seed dispersers. Fragmentation affects species of plants and animals to varying degrees. If a tree species is dependent on a particular suite of pollinators and/or seed dispersers and these organisms are reduced or eliminated as a result of fragmentation, the future recruitment of that plant species is in jeopardy. A study of the emergent canopy tree species *Duckeodendron cestroides* conducted at the BDFFP demonstrated that seed dispersal is strongly reduced in fragments compared with continuous forest (Cramer et al. 2007). *Duckeodendron* is a rare and widely dispersed species, and thus its seed dispersal is particularly critical. It produces large fruits consumed by various primate species as well as peccaries, agoutis, and pacas. (*Agoutis* and *pacas* are moderate-sized rodents related to guinea pigs.) The study was done over a three-year period.

Trees in continuous forest had greater seed dispersal distances than did those in fragments. Recall that it is normal for seed germination and seedling survival to be higher as distance from the parent tree is increased (*negative density dependence*). Over the three-year period of the study, both the quantity (proportion) and quality (distance) of dispersed seeds in continuous forest were double those in fragments. The distance of the farthest dispersed seeds in continuous forest was triple that in fragments. Thus the future recruitment of *Duckeodendron* is higher in continuous forest than in fragments, presumably because seed-dispersing animals tend to be less abundant or in some cases absent from the fragments. The authors of the study noted that because organisms such as tropical trees are long-lived, factors such as fragmentation that disrupt their mutualistic relationships, particularly those related to reproduction, could be very difficult to detect. It would be many years before their populations would show obvious decline.

Tropical Fragmentation: Focus on *Heliconia*

Heliconias are widespread and common understory plants throughout much of Amazonia (Plate 14-5). *Heliconia acuminata* has been the subject of intensive studies to by E. M. Bruna to ascertain the effects of fragmentation on the plant's overall ecology (Bruna 1999 and 2003; Bruna and Oli 2005).

In one study, seeds of *H. acuminata* that were planted in forest fragments were shown to be less likely to germinate than those planted in continuous forest (Bruna 1999). The experiment, conducted in the BDFFP fragments, involved the collection of 1,668 seeds from 200 plants. The seeds were distributed among seven forest fragments, four of which were 1 hectare (2.47 acre), and three of which were 10 hectares (247 acres). Three continuous sites were selected as controls. Seeds were planted 10 meters (33 feet) apart on 100-meter (330-feet) transects. Some seeds were placed in a cup covered with fine mesh to prevent seed predation and accumulation of leaf litter. Seeds in continuous forest were 3 to 7 times more likely to germinate. Seeds in small fragments were subject to greater edge effect (see the following discussion) and

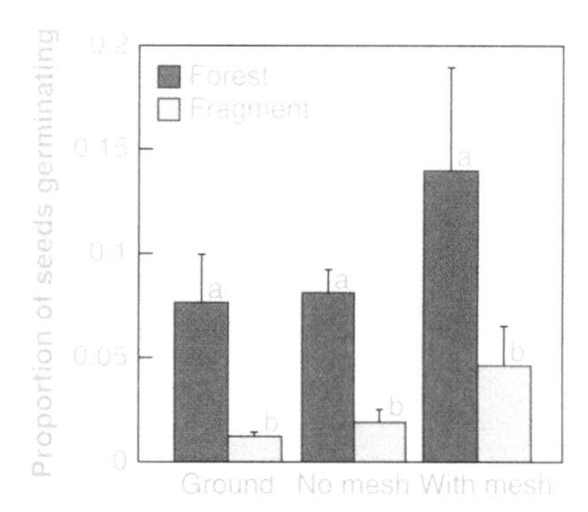

FIGURE 14-2
Mean proportion of seeds germinating in continuous forest and fragmented sites in different experimental treatments (± standard error of the mean). Bars with different letters (a/b) were statistically significantly different.

experienced hotter, drier conditions with greater light penetration (Figure 14-2). Demographic analysis (using a technique known as *life-table response experiment analysis*, or LTRE) allowed the calculation of growth rates in fragments of varying areas (Bruna and Oli 2005). The results showed that populations in 1-hectare (2.47-acre) fragments grew much less than those in continuous forest. One reason was poor seed germination, and another was poor seedling establishment in the small fragments. Recruitment rates of

PLATE 14-5
Heliconias, like this one in flower, are common throughout the Neotropics; some species are negatively affected by forest fragmentation.

Howler Monkeys Choose Their Fragments

A study focused on the Mexican mantled howler monkey (*Alouatta palliata mexicana*) showed that the simians are highly selective in choosing among forest fragments (Arroyo-Rodriguez et al. 2007). The study area at Los Tuxlas, Mexico, consisted of many small forest fragments (less than 10 hectares, or 247 acres) with differing tree species richness. The monkeys presumably had selected which fragments they would occupy, as they had access to all of the fragments. Researchers studied nine fragments with howlers and nine without the monkeys. Those fragments with monkeys were consistent in that they all had greater densities of large trees (diameter breast height [DBH] greater than 60 centimeters, or about 24 inches), greater total basal area, and greater basal area of top food species in comparison with those fragments avoided by the howlers. It is important to note that the appearance of the fragments with and without monkeys was similar. There were no obvious differences to the researchers (until they collected data), but there were apparently clear differences to the monkeys. The researchers noted that although howlers are known in general to feed on numerous tree species, in any one location they usually focus on a small, select number of species. This particular subspecies of howler (*mexicana*) is considered critically endangered, so the study was important in showing how best to conserve the monkeys. The researchers suggested that even small fragments are suitable habitat for these monkeys if the trees are sufficiently large and provide suitable food resources for the animals. Though population sizes of monkeys were not given for the study just described, it has been shown (see the section "An Ecological Meltdown: Lago Guri" later in this chapter) that some howler populations have large population densities in small fragments.

plants into larger-size classes were depressed in the small fragments, most likely due to microclimatic changes: elevated air temperature and lower relative humidity. These factors stimulate leaf loss as a short-term adaptation to minimize water loss.

A further study (Bruna 2003) that also focused on demographic analysis predicted that *H. acuminata* should decline by a rate of 1.0% to 1.5% per year in fragments but grow by about 2.3% to 4.0% per year in continuous forest. However, fieldwork did not support this prediction (which was based on a demographic matrix model). What seems to be the case is that although the plants fare poorly in the fragments (and thus should be in decline), there is high dispersal from continuous forest into fragments, largely offsetting the loss in fragments. This kind of situation is familiar to ecologists who study fragmentation and is called a *source-sink model*. It is frequently the case that populations fare poorly in fragments and thus fragments are *ecological sinks*, where extinction would be predicted to occur. However, in some cases, if dispersal is adequate, lost individuals from fragments are replaced by dispersal from growing populations, *ecological sources*. In the case of *H. acuminata*, the continuous forest is the source that allows the species to persist in fragments, the sinks. The study clearly suggests that proximity to continuous forest is essential in mitigating biodiversity loss from small fragments.

Tropical Fragmentation: Focus on Birds

Fragmentation of tropical forests has been shown to exert significant effects on bird species, but those effects vary among species (Ferraz et al. 2007) and are distinct from fragmentation effects on bird species in temperate forests (Stratford and Robinson 2005).

Tropical bird species typically occupy larger territories than species that nest in temperate forests. For example, one Amazonian study showed that the smallest territories were typically about 3 to 4 hectares (about 7.5 to 9.9 acres). Comparable species in temperate forests may occupy a territory of only a single hectare (2.47 acres) (Terborgh et al. 1990). Requirement of a large territory obviously makes species more vulnerable to fragmentation if the fragments are small. In addition, numerous species of tropical birds are reluctant to cross barriers such as rivers or nonforested habitat, and thus the habitat matrix surrounding the fragment is a significant parameter. One amazing example showed that birds in mixed-species foraging flocks in Amazonian Brazil tended to avoid crossing roads with a total width of about 30 meters (99 feet) (Develey and Stouffer 2001). This behavior makes these species highly vulnerable to negative effects of isolation (Plate 14-6).

PLATE 14-6
Narrow roads such as the one pictured here usually have no effect on patterns of bird behavior, but wider roads, at least in some Amazonian areas, apparently alter mixed-flock foraging behavior.

A study of 55 bird species inhabiting the forest and forest fragments of the BDFFP demonstrated that bird species are each sensitive (to varying degrees) to patch area and to isolation (Ferraz et al. 2007). Researchers applied a patch-occupancy model to examine the following parameters: initial occupancy, local extinction probability, local probability of colonization, and probability of detection given that the species is present. The data, collected by *mist-netting* (mist nets, widely used in field ornithology, are nets made of fine mesh that permit safe capture of birds) birds in fragments that varied in size and in degree of isolation, showed many species to be strongly affected by area, while the effect of isolation was considerably more variable among species. For example, the black-throated antshrike (*Frederickena viridis*) is a typical forest interior species (Plate 14-7a). It is a poor colonizer that virtually never leaves the interior forest understory and is thus highly sensitive to area loss. Another species, the white-chinned woodcreeper (*Dendrocincla merula*) is one of several species that typically follow army-ant swarms. This behavior makes it particularly sensitive to isolation (because army ants are confined to forest) but less sensitive to area loss.

Forest fragments are subject to increased variability in microclimate because of factors such as greater light penetration around edges. Comparisons of understory bird distribution during dry and wet seasons in lowland forest in Panama demonstrated subtle but important changes in bird communities that correlated with microclimate (Karr and Freemark 1983). When fragmentation occurs, microclimate changes, and such change could create stresses on understory birds that are physiologically adapted to exist in a highly narrow range of temperature and humidity (Stratford and Robinson 2005).

Some bird species are adapted to fragmentation. A study performed on 12 rain forest fragments in Kenya showed that mobile species that typically experience high dispersal as well as species described as *ecological generalists* are tolerant of habitat deterioration and thus persist in forest fragments (Lens et al. 2002). In the Neotropics, it has been shown that bird species such as the barred antshrike (*Thamnophilus doliatus*) that are typically found in edge habitat (Plate 14-7b) or in canopy gaps are tolerant of fragmentation (Stouffer and Bierregaard 1995).

Because of the varying responses among species to fragmentation, it is clear that fragmentation alters the community structure of birds. It not only generally reduces species richness, but it also selects for a predictable species assemblage in any given region.

It has been shown in at least one study that bird species richness may increase as a result of forest fragmentation (Feeley and Terborgh 2006). In Venezuela, studies performed on island forest fragments in Lago

(a)

(b)

PLATE 14-7
(a) The black-throated antshrike (*Frederickena viridis*) (female shown in photo) is an obligate forest interior species and poor colonizer. (b) The barred antshrike (*Thamnophilus doliatus*) is a widespread species that thrives in disturb ed successional areas and along edges of forests. It is not sensitive to fragmentation.

(Lake) Guri (to be described in more detail in the next section), showed that birds species richness on several small islands of forest (in the lake) correlated closely with annual woody increment (net primary productivity, or NPP). In other words, as plant productivity increased, so did bird species richness. The stimulus for greater plant productivity was that red howler monkeys (*Alouatta seniculus*) were up to eight times more abundant than normal on some of the fragmented, isolated islands in the lake. (These islands were once elevated areas of contiguous forest that are now isolated as a result of the creation of Lago Guri, a hydroelectric reservoir.) The monkeys devour leaves and are thought (because of their increased population density) to have greatly enhanced the nutrient-recycling system that in turn stimulated plant productivity, ultimately making more food available for birds (Figure 14-3).

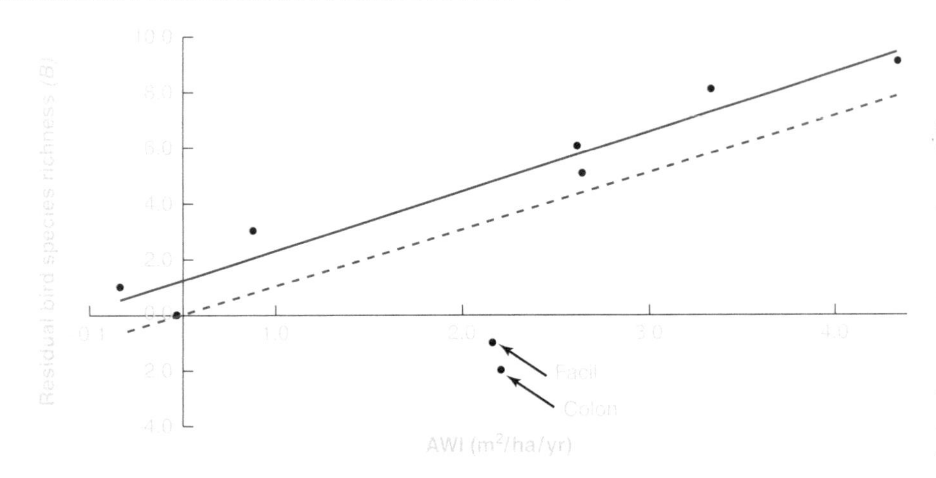

FIGURE 14-3
The residual bird species richness (B) increases significantly with the annual woody increment (AWI). If all of the islands where AWI was measured (n + 9 islands) are included, the relationship is described by the linear function $B + 0.2 + 20.1(\text{AWI})$ (solid line; $r^2 + 0.48$, $P < 0.05$). If Facil and Colon islands are excluded as outliers, the relationship is significantly improved (dotted line, $B = 1.3 + 20.9[\text{AWI}]$; $r^2 = 0.94$, $P < 0.0005$).

An Ecological Meltdown: Lago Guri

Lago Guri, with an area of 4,300 square kilometers (1,660 square miles), was formed in 1986 when Venezuela constructed a hydroelectric dam in Caroni Valley in the state of Bolivar. The lake flooded valleys of forest and created an archipelago of isolated forest islands, former hilltops of a once-contiguous forest. The forest islands range in area from 0.25 hectare (0.6 acre) to more than 150 hectares (370 acres)and range in distance from 0.2 to 4.9 kilometers (0.1 to 3 miles) from the nearest point on the mainland. The forest is classified as semideciduous tropical dry forest (Plate 14-8).

In 1993 and 1994, a study was performed to inventory the fauna on select forest islands and document how it had changed since isolation in comparison with control sites on the mainland (Terborgh et al. 2001). Researchers inventoried six small islands (0.25 to 0.9 hectare, or 0.6 to 2.2 acres), four medium islands (4 to 12 hectares, or 9.9 to 29.6 acres), and two large islands (> 150 hectares, or 370 acres) and compared them with two sites on the mainland. The results were dramatic, suggesting to the research team that they were witness to what they termed an *ecological meltdown* of species loss on the smaller islands. The two large islands were far less affected, losing almost no species. The results may be summarized as follows:

- The small and medium-sized islands lacked about 75% of the vertebrate species found on the mainland and two large islands. This effect can be partly explained by island biogeography theory that predicts loss of species with reduced area.
- Animals present on the small islands fell into three trophic categories: (1) small predators of invertebrates (spiders, frogs, lizards, birds); (2) seed predators (small rodents); and (3) herbivores (howler monkeys, common iguanas, leaf-cutter ants). Medium islands had these groups plus armadillos, agoutis (Plate 14-9), and, in a single case, capuchin monkeys. Two groups that were absent from small and medium islands were frugivores and predators. The researchers characterized the communities as a "suite of consumers without predators." Ecologists have noted such situations before and have used the term *ecological release*, meaning that one trophic level (in this case herbivores) is released from top-down effects of a higher trophic level.
- Those animals that persisted on small and medium islands became hyperabundant. The rodent population increased by a factor of 35 (compared with the mainland and large islands). Iguanas increased 10-fold. Howler monkeys increased on some of the islands to densities of 1,000 per square kilometer (386 per square mile) compared with a normal density of 20 to 40 per square kilometer (8 to 16 square miles) on the mainland. Leaf-cutter

PLATE 14-8
LAGO GURI ISLANDS

PLATE 14-9
Agoutis became more abundant on medium-sized islands in the absence of predators.

ant density increased by two orders of magnitude compared with the mainland. These abundance changes demonstrate the strong top-down effects now missing that would be exerted if the normal array of predatory species were present.

- Vegetation changes were also monitored. Small saplings were less than half as dense on the small islands in comparison with larger areas. Plants less than 1 meter (3.3 feet) tall were rarest on small islands, and more than half were lianas, the rest being shrubs. Few canopy species were represented among the small saplings on small islands, suggesting a reduced recruitment of canopy trees.

Researchers noted that the changes evident on the small and medium islands represent an *ecological cascade*, initiated by the loss of predators and subsequent hyperdiversity of herbivores. The changes had not yet resulted in establishment of an equilibrium on the small and medium islands. The researchers suggested that herbivores would continue to reduce the species richness of plants, setting up a strong selection pressure favoring herbivore-resistant plants. As this trend continues, further reductions in animal and plant species richness will occur. Because of the dearth of predator species, herbivore species populations increased dramatically. However, this reality will probably be relatively short-lived because selection pressures exerted on vegetation by the herbivores likely will initiate bottom-up responses that will reduce at least some herbivore populations. It should not be overlooked that once this bottom-up equilibrium is attained, the overall biodiversity of the ecosystem will be far less than what was.

Fragmentation, Edge Effects, and Matrix Suitability

Edge effects occur because at the border of an ecosystem such as a forest fragment, microclimate differs from the interior of the fragment and favors a different suite of plant species. Colonization of various species is also promoted by proximity to a different landscape matrix. This was demonstrated in studies done at the BDFFP sites (Laurance 2001; Rankin-De Merona and Hutchings 2001; Laurance et al. 2006).

W. F. Laurance and a team of researchers spent two decades monitoring densities of 52 tree species in genera comprising mostly successional species at the various BDFFP sites (Laurance et al. 2006). They noted that successional tree species were uncommon prior to fragmentation, comprising only 2% to 3% of

PLATE 14-10
Cecropia trees are among the most aggressive colonizers along edges.

all tree species. But this pattern rapidly changed following fragmentation. Successional tree species tripled in abundance along edges and in small and medium fragments. By 13 to 17 years following fragmentation, successional species represented about 25% of all trees on some of the fragmented plots. Among the 52 tree species studied, there was much variation in abundance patterns following fragmentation. Cecropia species are well-known edge colonizers (Plate 14-10). *Cecropia sciadophylla* increased density by more than 1,000% in edge plots (Figure 14-4). About one-third of the species remained the same or declined in density. Those species that showed the greatest increases in density were also those with the fastest growth rates. The study showed that the number of nearby edges increases tree mortality rate, tree density, and the species richness of successional tree species.

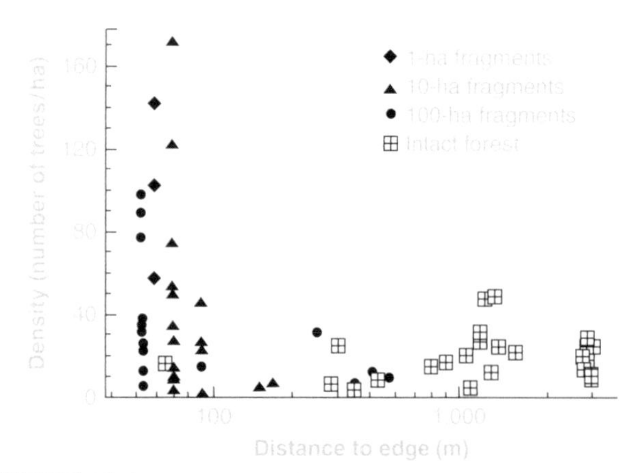

FIGURE 14-4
Density of successional trees as a function of distance from forest edge (note the log scale), in 66 1-hectare plots in intact forest and in 9 forest fragments ranging from 13 to 17 years old. The relationship is highly significant ($r_s = -0.393$, $n = 66$, $P = 0.001$; Spearman rank correlation).

A similar study performed in Alta Floresta in northern Mato Grosso, Brazil, resulted in similar conclusions (Michalski et al. 2007). Sixty quarter-hectare plots were sampled at the cores of 21 forest fragments (ranging from 2 to 14,480 hectares, or 5 to 35,766 acres) and two undisturbed continuous forest areas. A total of 8,248 trees, representing 130 genera, was tallied. There were significant differences in distributions among the fragmented areas and continuous forests. The small fragments had a much larger concentration of small-seeded softwood trees (typical successional species) than larger fragments or continuous forest. Tree genera composition varied from plot to plot and was related to time since fragmentation, distance to the nearest edge, and fire history. As in the previous study, this study showed a proliferation of successional, fast-growing tree species in disturbed fragments and near edges (Figure 14-5). The authors noted that the influx of pioneer tree species has potentially major effects on forest structure with regard to parameters such as carbon stocks, wood density, overall biomass, and vulnerability to surface fires.

Why do the successional tree species increase along edges and in small and medium fragments? One reason is that tree mortality of nonsuccessional species

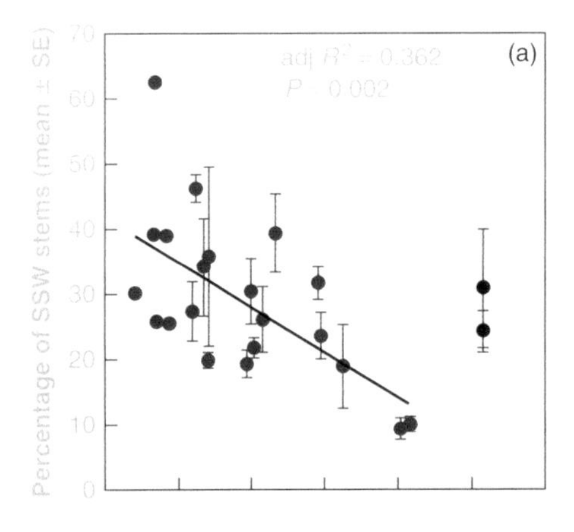

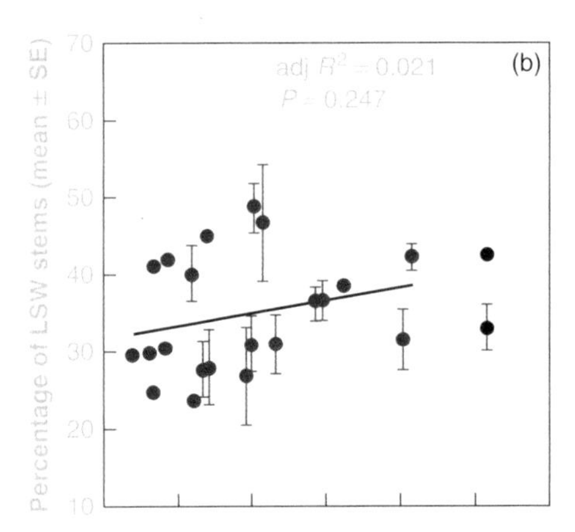

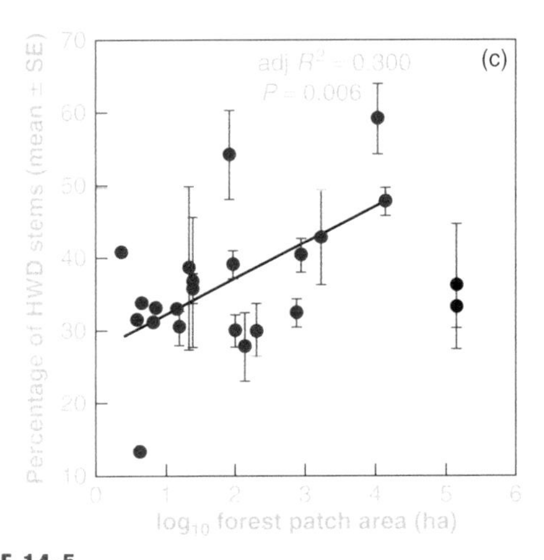

FIGURE 14-5
Relationships between forest patch area ($\log_{10}$) and the abundance of (a) small-seeded softwood stems (SSW), (b) large-seeded softwood stems (LSW), and (c) hardwood stems (HWD) in 21 forest fragments (brown circles) and two continuous forest sites (black circles). Multiple plots sampled in all but the smallest forest patches are represented by means ± SE.

increased along fragment edges, opening up the edge to colonization by successional species. Such mortality is part of edge effect in many tropical forests and is attributed to changes in microclimate and wind exposure, subsequently altering such factors as leaf fall, seedling recruitment patterns, and animal distribution (Gascon et al. 2000). Also, in a surrounding landscape matrix of successional habitat, the *seed rain* (the falling of mature seeds) from species inhabiting such ecosystems will penetrate the forest edge, allowing successional species to penetrate the forest.

Much variability in edge effect is attributable to the plant communities of the surrounding areas. Areas that are disturbed but then allowed to experience normal secondary succession, as noted earlier, will form a favorable matrix for recolonization of fragments by some species as well as provide a seed rain of successional species. The edge will close as successional species grow to canopy size. But frequent disturbance of the matrix (creating what is termed *high-matrix harshness*) surrounding a forest fragment leads to a different outcome. Forest regeneration along the edge is undermined by disturbance frequency (often resulting from frequent fires to maintain matrix grassland), allowing greater penetration of edge effects, in essence destroying the forest (Gascon et al. 2000) (Figure 14-6). It has been shown that edge effects and dominance of successional species may occur on fragments of up to 1,000 hectares (2,470 acres).

One of the most essential ecological factors to consider with regard to forest fragmentation is the influence of the vegetation matrix surrounding the fragments (Bierregaard et al. 2001b). If forest is permitted to regrow adjacent to forest fragments, the ecological effects of fragmentation are considerably mitigated. It has been demonstrated that tree mortality rates increase in fragments (and edge) adjacent to pasture compared with such areas adjacent to secondary forest.

Because varying landscape matrices affect forest fragment ecology differently, I reiterate that it is incorrect to perceive fragments as *islands*. They are not. Fragments are linked with the surrounding vegetation matrix in significant ways with implications for conservation biology.

The following three principles may be applied to minimize ecological deterioration of fragmented areas and edge (from Gascon et al. 2000):

- Protect large forest remnants rather than smaller ones in order to maximize the area/perimeter relationship (see the discussion of single large area or several small areas [SLOSS] in the next section).

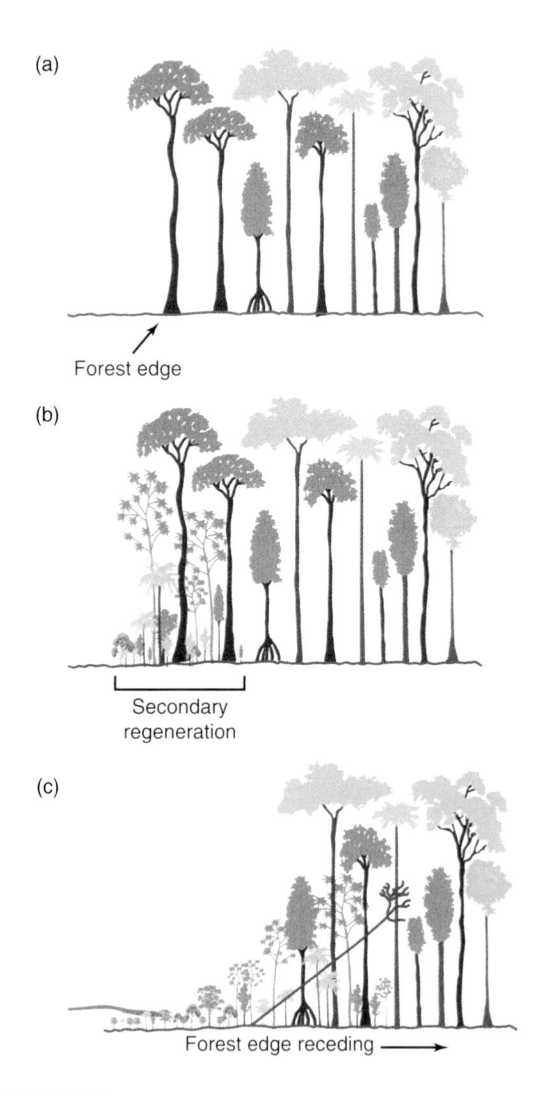

FIGURE 14-6
The death of a forest. Forest edges at three different stages after isolation. (a) Recent cutover area shows abrupt forest edge subjected to lateral winds and light penetration that allow for significant microclimatic changes inside the forest edge. (b) After several years of isolation, regenerating vegetation is found along the forest border and in the understory, closing the edge. In a landscape with low matrix harshness, this edge will be maintained (and could expand outward), buffering the forest interior from the severity of the initial edge effects. (c) In landscapes with high matrix harshness, the process of forest regeneration along the edge breaks down, resulting in greater penetration of edge effects and a gradually receding edge. The original edge will be replaced by scrubby vegetation, eventually leading to vanishing reserves.

- Protect forest edge from structural damage such as effects of fires or invasion of exotic species from the surrounding matrix.
- Minimize matrix harshness through careful land use, including careful control of fire, minimizing use of toxic compounds, controlling exotic species, and, most important, less-intensive types of land use, including less road building and restriction of hunting in forest fragments (see Chapter 15).

Connectivity, Corridors, and SLOSS

One possible way to mitigate the effects of fragmentation is to connect isolated fragments by corridors of uncut forest. Instead of islands of forest utterly isolated by pasture and other hostile anthropogenic ecosystems, corridors permit the movement of species within ecosystems in which they are adapted. Fragments thus connected represent an interconnected matrix where their respective areas are functionally additive, making for a much greater area and therefore sustaining a higher equilibrium point with regard to biodiversity (Figure 14-7).

Even an uncut forest of 1,000 hectares (2,470 acres) may not be sufficient to meet the ecological requirements of certain species. Terborgh (1974) argues that national parks must have between 100,000 and 1 million hectares (247,000 to 2,470,000 acres) to ensure that maximum biodiversity is maintained. This is particularly true for top carnivore species, each of which requires a large range.

Some years ago, there was a debate among conservation biologists, nicknamed the SLOSS debate (Quinn and Hastings 1987), short for "single large area or several small areas" (which would be of equal total area to one large reserve). Many ecologists favored protection for large-area tracts of land, arguing that such vast areas are necessary in order to ensure maximum species richness at all trophic levels. Large areas, for example, are best for supporting apex predators such as large cats and eagles. This remains the prevailing view. Some ecologists supported positioning smaller reserves in crucial areas, where certain species or certain unique ecosystems are most at risk (such as polylepis forests along high Andean slopes), or where corridors connect small reserves. This approach is not as effective as large-area preservation but does help in protecting certain specialized habitats.

The very shape of a fragment is critical with regard to its potential to retain biodiversity (Desouza et al. 2001). What is important to realize is that many tropical species of plants and animals are sensitive to

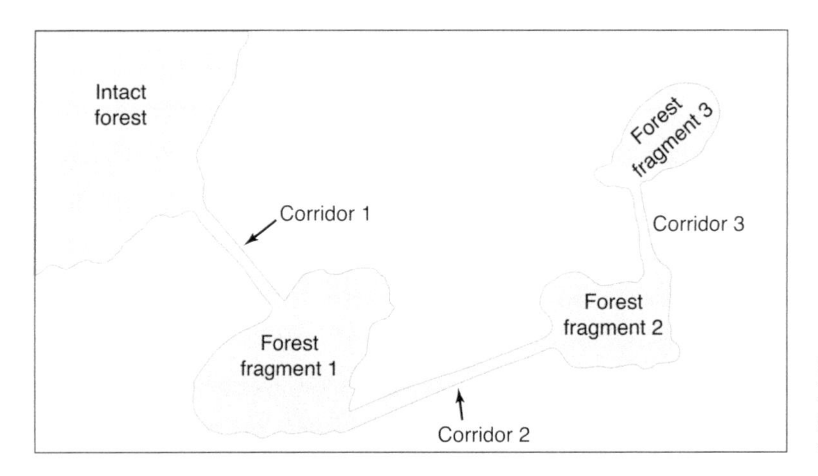

FIGURE 14-7
Natural corridors between isolated fragments provide connectivity, reducing isolation and helping maintain biodiversity.

fragmentation, that ecological processes are easily and substantially altered by fragmentation, that edge effects and surrounding habitat matrix exert much influence on how biodiversity responds in an isolated fragment, and that fragments are highly dynamic and changing (Bierregaard et al. 2001b).

What is clear is that area matters, and matters in a complex way. *Minimal critical size* varies among species. It will be a challenge to design bioreserves that are suitable for numerous species without compromising ecosystem function.

Approaches to Understanding Biodiversity

No one knows exactly how many species of plants, animals, and microbes inhabit the global tropics or, for that matter, other ecosystems on Earth. But to consider one example, there is little disagreement that the total Amazonian biodiversity probably exceeds that of any other ecosystem on the planet. Other tropical areas have uniquely high biodiversity as well. Regional biodiversity in the tropics encompasses more than lowland rain forest habitat. It also includes such ecosystems as savannas and dry forests. Although all tropical rain forests exhibit high species richness, Amazonia, probably because of its vastness and complex biogeographic history, achieves a total species richness that is higher than that of all other tropical forest regions. South America, including the Atlantic Forest and areas west of the Andes as well as central Amazonia, contains 26 areas of uniquely-high diversity and endemism (Prance 1982) (Figure 14-8). Though endemism is characteristic of rain forest, even higher levels of endemism, at least for birds, are found in dry forest, savanna, and grassland (Stotz et al. 1996). For example, 90% of all bird species restricted to dry forests are local endemics, and about 80% of grassland bird species are endemics. Endemism level among bird species of humid forests is 45%.

As forest and other ecosystems are converted to anthropogenic systems, it is difficult to estimate rates of species loss. No one really knows, nor is there any reliable way to predict rates of species loss considering the overall species richness in tropical rain forests and the complexity of interactions among species. Add to that the reality that many species in various taxa have yet to be discovered and described (May 1988 and 1992).

Birds, unlike most invertebrate groups, are well documented throughout the world, including the tropics (though species new to science are still being discovered). Birds represent a group where there are reasonable estimates of extinction probabilities in the coming century. A total of 128 bird species have become extinct over the past 500 years, with 103 of those extinctions having occurred since the year 1800. One estimate suggested that if current rates of deforestation continue, within a century (from 1984, when the study was published), 12% of Amazonian bird species would become extinct and 15% of all Neotropical plant species would become extinct (Simberloff 1984). This represents a loss of about 84 bird species and nearly 14,000 plant species. Predictions from BirdLife International are that up to 400 bird species face extinction within the twenty-first century, most of them in the tropics (Stattersfield and Capper 2000). Many, such as the Philippine eagle

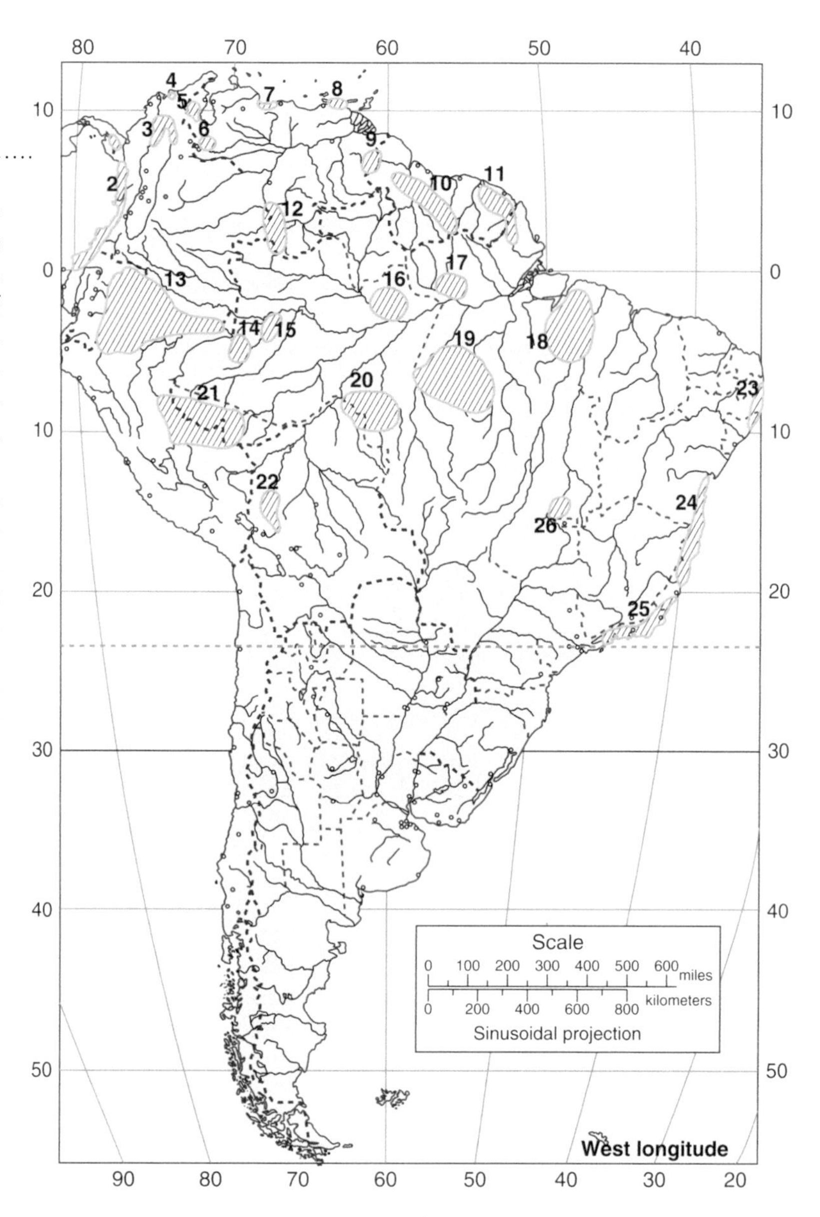

FIGURE 14-8
Areas of high endemism in South America. Note that these areas are small in relation to the total area of Amazonia.

(*Pithecophaga jefferyi*), are at risk due to loss of primary habitat, in this case the loss and fragmentation of dipterocarp forests. From an original population of about 6,000, only about 350 to 650 Philippine eagles remain, all within a shrinking area of dipterocarp forest now totaling a mere 9,220 square kilometers (3,559 square miles) (Plate 14-11).

Many Amazonian species of plants and animals may be susceptible to extinction because they have limited ranges, and even within those ranges, many species are present at low population densities. For example, at the Rio Palenque Biological Station in Ecuador, an area of only 0.8 square kilometer (0.31 square mile), there are 1,033 species of plants, and about one in every four is known only from coastal Ecuador, a relatively limited area (Gentry 1986a; Wilson 1988). Thus the loss of any particular tract of rain forest such as Rio Palenque increases the extinction risk to some of these range-limited species.

Some tropical forests contain a higher proportion of endemic species than are typically found in temperate forests (see the section "Hotspots" later in this chapter). It is estimated that about 20% of the plant species in the Choco Department of Colombia are endemic, and within certain genera, that figure rises to as high as 70% (Gentry 1986b). Endemic species are also common on Andean slopes, areas where deforestation has been active. Related to endemism is the tendency for habitat specialization. Gentry (1986a) noted that four very similar species of *Passiflora* occurred in the Iquitos area, each restricted to a different substrate: one on *terre firme*, one on seasonally inundated floodplain forest, one on white-sand forest, and one on noninundated rich alluvial soil. The combination of high endemism and extreme specialization makes many species obviously susceptible to extinction by loss of habitat.

Gentry (1986a) recognized two kinds of endemism, both of which are significant in the tropics. One kind, *paleoendemism*, describes a species with a restricted range today but a much wider range in the past. This is similar to the concept of a relict species, one whose current range is but a fragment of what it once was. In the Neotropics, there are many examples of paleoendemic species, especially in such places as the tepuis of the Guiana Highlands. Another form of endemism, *pseudoendemism* or *anthropogenic endemism*, occurs when human activities have so reduced suitable habitat that a once-widespread species is now confined to a few local populations where its habitat remains. Anthropogenic endemism is well documented for the region around Rio Palenque in Ecuador. The tree *Persea theobromifolia*,

PLATE 14-11
PHILIPPINE EAGLE

once a widespread species in western Ecuador, was overharvested for timber as well as suffered from loss of habitat to the extent that its total population is now fewer than a dozen trees.

In trying to ascertain the probable patterns of extinction resulting from deforestation, it is instructive to examine the Atlantic Forest. Though this forest once occupied 1.3 million square kilometers (501,800 square miles) in Brazil, Paraguay, and Argentina, less than 8% remains, most of which is highly fragmented (Galindo-Leal and de Gusmao Camera 2003). The Atlantic Forest is a region of high endemism among many taxonomic groups and thus would seem a region where extinction from habitat loss ought to be common (Figure 14-9). It is considered one of the world's *hotspots* (see the discussion later in this chapter) for biodiversity. Some of South America's most endangered species occur here: the golden lion tamarin (*Leontopithecus chrysomelas*) (Plate 14-12) is one of 13 endangered

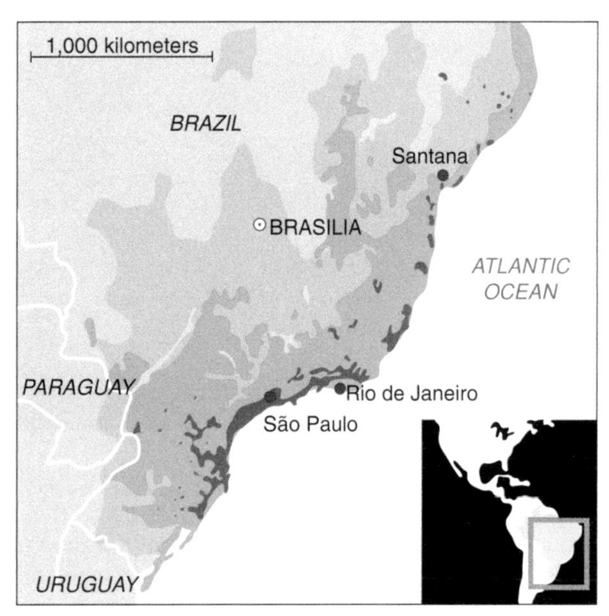

FIGURE 14-9
BRAZILIAN ATLANTIC FOREST (IN GREEN)

primates; the maned sloth (*Bradypus torquatus*) (Plate 14-13) is confined to the same forests as the tamarin; and the black-fronted piping-guan (*Pipile jacutinga)* is faced with extreme habitat loss and undue hunting pressure. But these species, while clearly endangered, are not yet extinct. A study by the Brazilian Society of Zoology listed 171 animal species from the Atlantic Coastal Forest considered to be vulnerable, endangered, or extinct. Only six of the 171 (two birds and four butterflies) were considered to be actually extinct.

Brown and Brown (1992) suggest why extinction rates in the Atlantic Forest are thus far perhaps less than expected. First, the region is physically heterogeneous, a patchwork of microhabitats. Brown and Brown suggest that most species have historically lived

PLATE 14-12
GOLDEN LION TAMARIN

PLATE 14-13
MANED SLOTH

in small, isolated populations and are thus adapted to remain so. This may be so, but small populations are normally vulnerable to extinction because they harbor less genetic diversity. Second, the area is subject to a high degree of natural disturbance (from heavy rains, cold spells, and varying seasonality), resulting in species-specific adaptive responses to such unpredictable and potentially catastrophic events, an adaptive buffer against extinction. There may be high levels of adaptive plasticity among endemic plants and animals, allowing rapid adaptive responses to sudden perturbations. None of these suggested explanations is really satisfactory, and species in the Atlantic Forest are generally regarded as highly vulnerable to potential extinction. It may be simply a matter of good luck that extinction rates have not been higher thus far.

Susceptibility to extinction clearly varies among species and there are *extinction-prone species* (Terborgh 1974). It has been well documented that bird species have become extinct on Barro Colorado Island, Panama, since the island was created (Willis 1974; Robinson 1999). The current total of bird species now extinct from BCI is 65 species (Robinson 1999). Some of these species were typically found only in early successional ecosystems, and with regrowth of forest at BCI, their habitats literally disappeared. But there has been a loss of both forest and edge species. Eighteen interior forest species (whose habitat has remained intact) have been lost from the island, and others continue to decline. For example, the slate-colored grosbeak (*Pitylus grossus*) has declined from an estimated population of about 500 in 1970 to only two pairs in Robinson's survey from 1994 to 1996 (Robinson 1999). Most of the now-extinct species are either large in size, ground nesters, or ground foragers. It is not clear why some of these species have disappeared from the island, but Terborgh (1974) suggests some characteristics that may typify species most prone to extinction. These include species that are large in body size and on the top of the food web, such as harpy eagle and jaguar. These animals require large areas in order to survive, and their populations cannot be sustained in fragmented areas, such as BCI. But Terborgh also notes that widespread species with poor dispersal and colonizing abilities are also likely to be extinction-prone, as are endemics. In Robinson's study (1999), he did not observe recolonizations of BCI by any of the interior forest species that had previously disappeared, even though many of these were numerous on the mainland. The 1,600-hectare (3,952-acre) area of BCI may be simply too small to support various bird species.

Ecologists recognize that in most ecosystems there are certain species that are essential in maintaining prevailing food webs; these uniquely important species are termed *keystone species*. Terborgh (1986) suggests that the abundance of fruit-dependent large animals in rain forests makes certain kinds of fruiting plants such as figs (*Ficus*) keystone species. For example, about three-fourths of the biomass of birds and mammals in Cocha Cashu rain forest in Peru is from animals that are in large part frugivorous. Most of these creatures are dependent on palm nuts, figs, and various miscellaneous other fruits. Indeed, Terborgh suggests that without figs in the ecosystem, it would be likely that species reductions would be severe, leading to a possible collapse of part of the food web. Adding nectarivorous species to frugivores, Terborgh concludes that at Cocha Cashu, only 12 plant species (out of a total of 2,000) are responsible for sustaining all of the diverse species of fruit- and nectar-consuming animals for three months out of the year. The lesson here is obviously that the loss of certain plant species, such as figs, would have much greater impact than random loss of plant species that are less essential resources to most animals. And bear in mind that the argument works both ways. The loss of a keystone pollinator or fruit disperser could bring about the loss of a keystone plant species.

Ehrlich and Wilson (1991) offer three broad reasons why there should be concern for loss of biodiversity (and efforts to prevent it):

- Biodiversity provides aesthetic satisfaction, and it can be argued that humans should be ethically bound to provide protection and stewardship for the rest of nature.

- The richness of biodiversity provides humans with many goods of direct economic benefit (foods, medicines, and industrial products), and there is future potential for even more pragmatic usage.
- Natural ecosystems, and thus biodiversity, provide humans with many essential services having to do with regulation of the atmosphere and climate, recycling of materials, and maintenance of soil fertility, to name but a few.

This final point, *ecosystem services*, has been overlooked, at least by the public, until relatively recently. While it is clear that the reason to develop ecosystems (such as cutting tropical forests to make pastures) is to use the land to yield valuable resources for people, it must be recognized that natural ecosystems are, in and of themselves, essential to people for the goods and services they provide. Forest degradation in tropical areas has severe consequences (Foley et al. 2007; Bradshaw et al. 2009). Ecosystem services include the pollination and dispersal of plants essential to the ecosystem, some of which supply resources for humans. Another service is flood control, another is carbon sequestration, and still another is potential global climate regulation (by forest evapotranspiration). Ecosystem services, only now being fully recognized and understood, will be discussed again in Chapter 15.

High-biodiversity forests (in the temperate zone as well as the tropics) appear better able to resist invasive plant species than simplified ecosystems, though there are clear indications that mature, complex forests are nonetheless subject to invasion by exotic species, provided such species are shade-tolerant (Martin et al. 2009). Last, complex tropical forests harbor numerous pathogen species, and the wholesale cutting of forest has the potential to deregulate some of these organisms. An example is the mosquito *Anopheles gambiae*, one of the species that is a vector for various forms of malaria. *A. gambiae* larvae survive better in deforested landscapes than in interior forest. Survivorship was 55% to 57% in sunlit areas compared with 1% to 2% in full forest. Large-scale deforestation may promote an increase in malaria frequency (Tuno et al. 2005).

Accelerated loss of biodiversity eventually results in disassembling food webs and forcing redirection of energy through ecosystems, as noted in the fragmentation examples described earlier. This affects ecosystem functions such as primary productivity as well as seed dispersal and other species interactions. Theoretical models as well as empirical data show that food webs tend to initially deteriorate at upper trophic levels by the loss of essential species that exert top-down effects (Dobson et al. 2006). This is largely because organisms at upper trophic levels require the largest area to sustain a viable population. Thus as biodiversity is lost, it may be lost far more in upper trophic levels in proportion to lower ones. This pattern leads to what is termed *trophic collapse*. Depending on how severe such an alteration may be, it could affect ecosystem services.

It is as yet unclear how tight the relationship is between ecosystem functioning and biodiversity (Loreau et al. 2001). Ecologists now understand that loss of predators leads to a trophic cascade that radically alters an ecosystem. But ecologists are also investigating how biodiversity relates to such variables as primary productivity, nutrient cycling, decomposition rates, and resistance and response to perturbations. For example, if a large tract of forest in the tropics experienced a loss of 2% of its tree species, would net primary productivity change? What if the species loss represented 50%? How resistant are high-biodiversity ecosystems to environmental change and biodiversity reduction (Figure 14-10)?

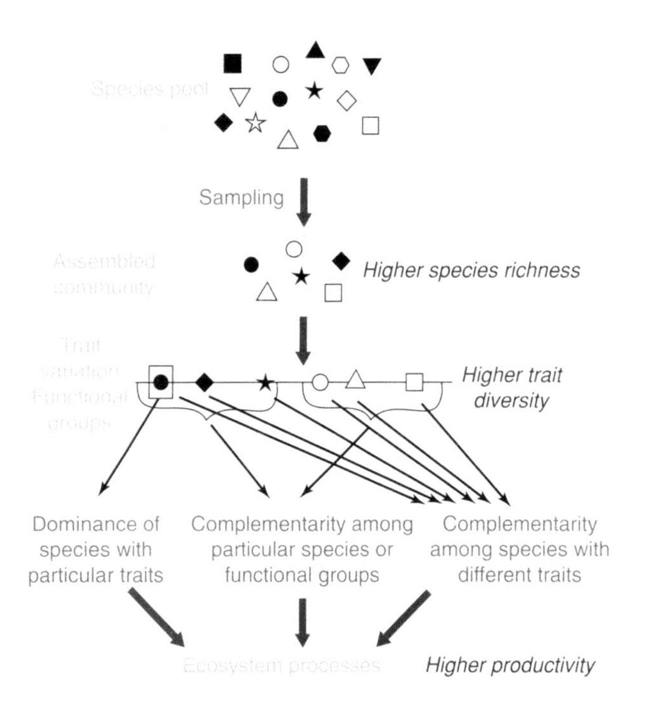

FIGURE 14-10
Hypothesized mechanisms involved in biodiversity experiments using synthetic communities. Sampling effects are involved in community assembly, such that communities that have more species have a greater probability of containing a higher phenotypic trait diversity. Phenotypic diversity then maps onto ecosystem processes through two main mechanisms: dominance of species with particular traits, and complementarity among species with different traits. Intermediate scenarios involve complementarity among particular species or functional groups or, equivalently, dominance of particular subsets of complementary species.

Flying Fox Bats: A Real Numbers Game for Plants

Flying fox bats (Megachiroptera) are essential seed dispersers in many areas of the Paleotropics (just as various Microchiropteran bats disperse seeds in the Neotropics). These bats are the principal seed dispersers throughout the tropical Pacific. A study performed in the Vava'u Islands in Tonga (in the tropical Pacific) focused on the bat species *Pteropus tonganus,* an important seed disperser for a variety of plant species. The study demonstrated that the efficacy of the bats as seed dispersers depended on the density of the bat population. Once the abundance of the flying foxes dropped to below a critical number, seed dispersal was severely reduced.

This threshold relationship between flying fox abundance and seed dispersal efficacy is believed to be largely due to the bats' behavior when in dense groups. The animals become aggressive when clustered in fruiting trees, and many (especially juveniles) limit their time in the fruiting tree, electing to ingest some fruit and fly away, thus enhancing the probability of efficient seed dispersal. Below threshold level, *Pteropus tonganus* dispersed less than 1% of the seeds they handled compared with 58% above threshold level (Figure 14-11).

This study demonstrates another essential facet of mutualistic relationships: density matters. The ecosystem-level function facilitated by the flying foxes is severely reduced long before these bats even become rare. This is another element to consider in an analysis of how biodiversity functions in maintaining ecosystem services

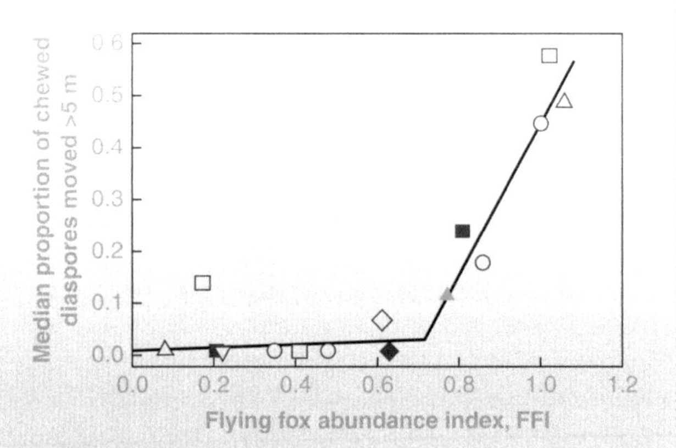

FIGURE 14-11
Relationship between flying fox abundance and median proportion of diaspores dispersed > 5 meters at each site in Vava'u (Tonga, Polynesia). Lines are fitted separately to the data points above and below the threshold flying fox abundance required for them to function as seed dispersers of large fruit. Points of the same shape and fill represent repeat visits to the same site.

Dealing with Potential Loss of Biodiversity

Global biodiversity is in decline. The main factor in biodiversity decline is outright loss or severe fragmentation of habitat (Ehrlich and Wilson 1991). Throughout the tropics, there has been cumulative loss of lowland and montane forests over many decades, and the loss continues. Tropical rain forest has declined globally by over 50% of its historic coverage. Consider that tropical forests are thought to contain at least 60% of all known species, while these forests cover only 7% of Earth's surface (Laurance 1999; Dirzo and Raven 2003). Biodiversity is *the* unique feature of these ecosystems. Given that numerous rain forest species are endemic and occupy limited areas, calculations based on island biogeography equations suggest that as many as 4,000 species may become extinct annually (Ehrlich and Wilson 1991). Deforestation is high throughout all equatorial areas, the Neotropics, Central Africa, and most of tropical Asia (Rudel 2005; Spray and Moran 2006). In parts of South America, there has been a rapid loss of Amazonian rain forest (Wood and Porro 2002) as well as other ecosystem types, principally dry savanna (called *Cerrado*) and dry shrubby desert (called *Chaco*). These ecosystems, once rich in endemic species, have been largely converted to soybean fields.

The ongoing loss of complex ecosystems such as forests has been suggested as a primary cause of species extinctions. A detailed study of species extinctions following deforestation in Singapore (Brook et al. 2003) supports that contention. Singapore, a tiny island country off the southernmost tip of the Malaysian Peninsula, is in the humid tropics and was originally heavily forested. In the Singapore study, habitat loss is estimated to be up to 95% over a period of 183 years, from the time when the British first established their presence there. Forest reserves now occupy a mere 0.25% of the island's total area of 540 square kilometers (208 square miles), but they hold an astonishing 50% of the biodiversity still extant on Singapore. Major extinctions have been recorded for all vertebrate groups as well as invertebrates such as butterflies. The highest percentages of extinctions were among butterflies, birds, fish, and mammals. The observed loss of biodiversity from Singapore over the 183-year period was 881 species (28%) out of a total of 3,196 species. These numbers will continue to escalate with any further loss of protected reserves. The authors calculated the total percentages likely to become extinct with loss of reserves by adding the number of species already extinct plus the number of species restricted to reserves and dividing this total by the original number of species. The percentages are sobering in that 78% of amphibians, 39% of birds, 69% of mammals, and 77% of butterflies would become extinct with loss of the tiny forest reserves that still remain. The authors of the study acknowledge that some taxonomic groups such as vascular plants have been difficult to document as to actual extinctions, but their best estimate is that the projected biodiversity loss in Singapore by the year 2100 could be between 12% and 44% of vascular plant species. Using a species-area model, the authors expanded their study to estimate extinction rates throughout Southeast Asia if current rates of habitat loss continue. Their estimate, based on the species-area model, predicted a loss of between 13% and 42% of species from regional populations by the year 2100.

The continent of Africa represents a land of rapidly increasing human population and rich biodiversity. Conservation of biodiversity will be a significant challenge in Africa's future. A study of the distribution of biodiversity and people in sub-Saharan Africa demonstrated that the areas with the highest intrinsic biodiversity are also the areas most populated by humans (Balmford et al. 2001). This is largely due to the fact that primary productivity is highest in these areas, making the regions desirable for people's uses but also rich in wildlife. Thus a conflict is bound to arise, as humanity wants to expand its claims on the land for anthropogenic use. The researchers looked at the distribution of 940 mammal species, 1,921 bird species, 406 snake species, and 618 amphibian species, many of which are unique to particular areas and some of which, due to habitat loss, are threatened species. The pessimistic conclusion was that conflict is bound to arise generated by human pressures to develop the land. Habitat fragmentation has become and will continue to be a pressing issue.

Extinction is an ongoing natural process, at times accelerated by rapid climate change or even extraterrestrial events affecting the planet (as was apparently the case at the close of the Mesozoic era). There have been times when speciation rate has outpaced extinction rate, as happened in the early Cenozoic with the expansion of mammalian, avian, vascular plant, and insect diversity. But presently, global extinction rate is thought to be well in excess of historical background extinction rate. E. O. Wilson (2002) estimates that about 27,000 species become extinct each year. That amounts to 74 per day and 3 per hour. Wilson's logic for such a claim rests on numerous case studies that demonstrate the cumulative negative impacts of anthropogenic activities. For example, Wilson cites data on freshwater fish species whose ranges occur in Canada, the United States, and Mexico. Of a total of 1,033 species, 27 became extinct within the twentieth century and another 265 are vulnerable. The threats are as follows: destruction of physical habitat (73% of species), displacement by introduced species (68%), alteration of habitat by chemical pollutants (38%), hybridization with other species and subspecies (38%), and overharvesting (15%). The reason percentages add to well over 100% is that many species face multiple threats. This concept, *multiple threats*, is particularly evident throughout the global tropics. Another estimate suggests that somewhere between 10,000 and 10 million species suffer extinction each decade, much of it due to deforestation (Pimm and Raven 2000). Added to these biodiversity threats is global climate change. Recall from Chapter 9 that many lowland tropical forest species may be facing lowland biotic attrition as temperatures rise, forcing some species higher along elevational slopes. It may not be possible for all lowland species to adapt or to migrate as climate warms. This topic will be discussed further in Chapter 15.

Jaguars Photograph Themselves in Belize

Top carnivores are often among the most threatened species because they require large areas. Not only that, but they can be difficult to study, as they range widely, and some, like the jaguar of the Neotropics, are active mostly at night. This reality makes monitoring population trends among some top carnivores very difficult. In a remarkable study conducted in the forest around Gallon Jug, in western Belize, Carolyn Miller of the Wildlife Conservation Society has established that jaguar density is 11.8 animals per 100 square kilometers (about 38.5 square miles). This is thought to be the highest jaguar density reported anywhere in the Neotropics (C. Miller

2010, personal communication). The area where Miller conducted the study is where Chan Chich Lodge, a major ecotourist destination, is located, and the vast surrounding forest is totally protected from hunting. There is an abundance of white-tailed deer and collared peccaries as well as numerous other prey species. Livestock is maintained in the area, but there have been few jaguar attacks, presumably because the jaguars are feeding in the forest. The attacks to livestock were from two jaguars injured by shotgun wounds from illegal hunters. The Gallon Jug jaguar population is thought to be an important source population for the surrounding area.

But how were jaguars at Gallon Jug counted? Miller placed numerous cameras along various trails throughout the Chan Chich Lodge forest and elsewhere. The cameras were tripped when an animal passed closely, so the jaguars basically documented themselves by taking their own photographs. By examining the spotting pattern on the animals, Miller was able to identify and keep track of individuals (Plates 14-14a and 14-14b).

(a)

(b)

PLATE 14-14
(a) Jaguar walking a trail, having just photographed itself. (b) This male jaguar, named Spot by the researchers, is walking along a road.

Hotspots

One approach to conserving biodiversity that has received much attention is to focus on what are termed *hotspots*. These are regions where there are many species not found elsewhere (i.e., a high level of endemism) and where anthropogenic activity threatens to further erode the biodiversity (Myers 1988). The reason it is called a *hotspot* is because in addition to having a high density of endemic species, the region is threatened by ongoing, often rapid habitat loss (Myers et al. 2000). By identifying hotspots, it is possible, in theory, to preserve the most species relative to per species cost of conservation. A key element in identifying a hotspot is imminent threat from anthropogenic activity. For example, the Galápagos Islands contain numerous endemic species but are collectively protected by being a national park. Thus the archipelago would not be considered a hotspot, as it currently receives a strong measure of protection.

The question of what is and what is not a hotspot is subject to interpretation and is clearly a matter of ecological scale. An evaluation based on the distribution of endemic terrestrial vascular plants identified 18 hotspots, most of them in equatorial regions (Myers 1988; Wilson 2002). A more detailed assessment by Myers et al. (2000) expanded the number to 25 hotspots (Figure 14-12). An additional study (published online by Conservation International, at www.biodiversityhotstops.org) shows 34 hotspots, of which 20 occur in the tropics (Bradshaw et al. 2009). Exactly how were these hotspots defined?

Hotspot boundaries are determined by what are described as *biological commonalities*, such that each hotspot contains a biota that can be considered as a biogeographic unit. While this criterion is obvious for such hotspots as Madagascar (see the following discussion) and the Philippines, it is less obvious for such hotspots as the tropical Andes Mountains of western South America. A second criterion is endemic

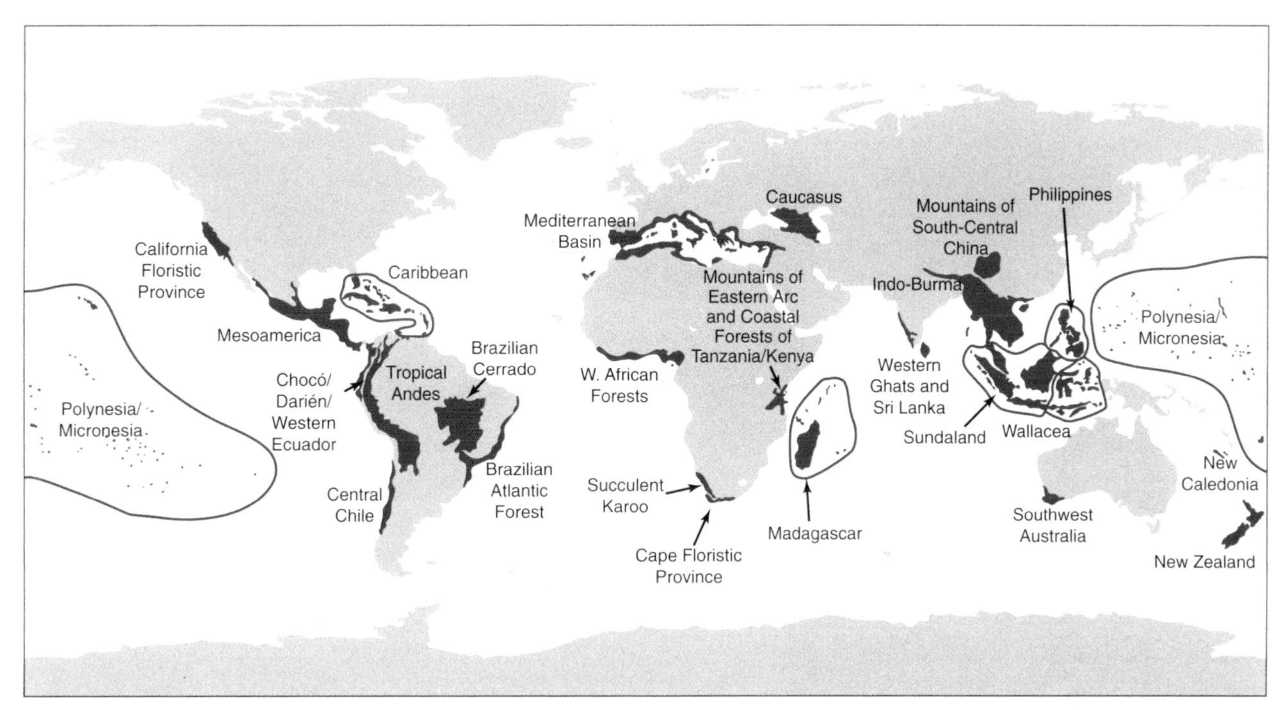

FIGURE 14-12
The 25 global hotspots, based on the distribution of endemic plant species.

species richness. A hotspot must contain at least 0.5%, or 1,500, of the world's 300,000 plant species as endemics. Of the 25 hotspots, 15 contain at least 2,500 endemic plant species. Terrestrial vertebrates do not directly factor into hotspot definitions other than as "back-ups." In other words, it is assumed that threats to vertebrates parallel threats to plants and thus plants will suffice as the taxonomic group to use in hotspot analysis. This is, of course, not strictly true. Large vertebrates may decline due to hunting or fragmentation, factors having little directly to do with plant diversity. Invertebrates, because their species richness is in general so poorly documented, do not factor at all into hotspot analysis. But many insect species are dependent on various families and genera of plants, and therefore there is a connection between insect and plant diversity. Last, to be a hotspot, the region must have lost 70% or more of its primary vegetation, a reflection of the degree of anthropogenic threat. Eleven of the 25 hotspots have already lost at least 90%, and 3 have lost 95%.

The authors of the hotspot study believe their analysis to be within the acceptable 5% range of error, though they are frank to also admit that the data contain much variability and are thus imprecise. They note that the 25 hotspots contain 133,149 plant species (44% of those worldwide) and 9,645 vertebrate species (35%) in a total area of 2.1 million square kilometers, or about 1.4% of Earth's land surface. These species formerly occupied 17.4 million square kilometers, or 11.8% of Earth's land surface; thus, combined, the hotspot areas have lost 88% of their primary vegetation (Myers et al. 2000). Sixteen hotspots are within the tropics, where human population growth is greatest and conservation resources the most limited (Cincotta et al. 2000). Indeed, there is a strong correlation between human population growth and hotspot location (Figure 14-13).

Madagascar, in the Indian Ocean off the east coast of Africa, is a critical hotspot. This island has been detached from Africa for millions of years, and numerous endemic species have evolved there, including many that are already extinct. For example, all of the world's 30 extant lemur species occur only on Madagascar. Approximately 80% of the 10,000 plant species of Madagascar are endemic and 90% of the reptile and amphibian species (including two-thirds of the world's chameleons) (Plate 14-15) are endemic

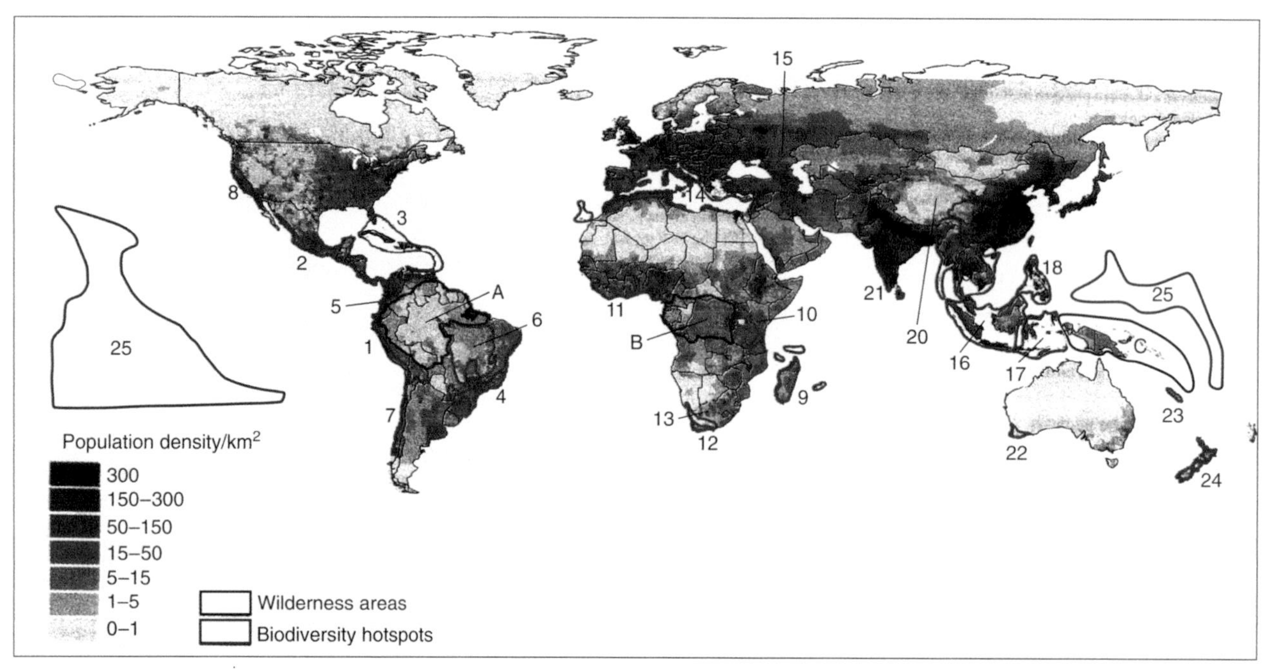

FIGURE 14-13
World population density (1995) and the 25 biodiversity hotspots (outlined in red, numbered) and three major tropical wilderness areas (outlined in black, lettered). Hotspots: (1) Tropical Andes; (2) Mesoamerica; (3) Caribbean; (4) Atlantic Forest Region; (5) Chocó-Darién-Western Ecuador; (6) Brazilian Cerrado; (7) Central Chile; (8) California Floristic Province; (9) Madagascar; (10) Eastern Arc Mountains and Coastal Forests of Tanzania and Kenya; (11) West African Forests; (12) Cape Floristic Region; (13) Succulent Karoo; (14) Mediterranean Basin; (15) Caucasus; (16) Sundaland; (17) Wallacea; (18) Philippines; (19) Indo-Burma; (20) Mountains of South-Central China; (21) Western Ghats and Sri Lanka; (22) Southwest Australia; (23) New Caledonia; (24) New Zealand; and (25) Polynesia and Micronesia. Major tropical wilderness areas: (A) Upper Amazonia and Guiana Shield; (B) Congo River Basin; and (C) New Guinea and Melanesian Islands.

PLATE 14-15
This chameleon is one of many endemic species on Madagascar.

to the island. At the same time, the human population on the island continues to grow, and as it does, more and more forest is felled simply to supply firewood for cooking. The rate of loss is accelerating rather than diminishing, and thus it is clear that some action is merited to protect what remains of the island's extraordinary biodiversity.

Unique ecosystems also qualify as hotspots. The Nullarbor Plain of southwestern Australia (Plate 14-16) is a heathland ecosystem of some 3,630 plant species, of which 78% are endemic (Wilson 2002). This area is rapidly being lost to agriculture and mining activities, and 25% of the species are now listed as either rare or threatened.

The "hottest" hotspots (those under the most threat) are currently considered to be Madagascar, the Philippines, Sundaland (islands of western Indonesia), Brazil's remaining Atlantic coastal forest, and the Caribbean region (Myers et al. 2000). Each of these hottest hotspots is in the tropics. It is noteworthy that the 25 hotspots are the sole remaining habitats of 44% of Earth's plant species and 35% of its vertebrate species, all of which are at risk.

PLATE 14-16
The Nullarbor Plain, in Australia.

Efforts are under way to conserve hotspots. That, after all, is what the objective is in identifying them as such. When Myers et al. published in 2000, approximately $400 million had been invested by the MacArthur Foundation, the W. Alton Jones Foundation, Conservation International, the World Wildlife Fund, and other nongovernmental organizations (NGOs). Governments in various hotspot areas have also made efforts to take firmer action in conservation policy. Myers et al. estimate that it would require about $20 million per hotspot per year over a five-year period to ensure the preservation of the regions, a total of about $500 million annually. They point out that this figure is twice the cost of a NASA Pathfinder mission to Mars that, ironically, is justified on the basis of a search for biodiversity—life on Mars!

While compelling, the hotspot approach to conservation has its critics. Hotspot definitions are based entirely on (1) endemic plants and (2) anthropogenic-generated loss of area. Not included in the definition are such considerations as ecosystem function, ecosystem services, or taxonomic groups other than plants.

Has the hotspot model resulted in leaving other essential ecological sites overlooked? Kareiva and Marvier (2003) argue that regions containing far fewer endemic plants nonetheless merit equal attention by conservation biologists, regions they refer to as *coldspots*. They provide a straightforward example by comparing the country of Ecuador with the state of Montana.

Both Ecuador and Montana are similar in total area, but Ecuador contains some 19,362 vascular plant species and 2,466 vertebrate species, while Montana has only about 12% of Ecuador's species richness. As a hypothetical goal, suppose that it is decided to conserve 20,000 species in total from the two areas. There are lots of combinations as to how such a goal could be met, but here are two: a total of 18,000 species could be saved in Ecuador and 2,000 in Montana—or a total of 19,000 species could be saved in Ecuador and 1,000 in Montana. Either of these approaches would allow Ecuador to retain the bulk of its species richness, either 82% in the first case, or 87% in the second case. Montana would not fare nearly as well. The second strategy would reduce the percentage of protected species from 74% to a meager 37%. Montana includes part of the region of the Yellowstone ecosystem, where the remnants of

PLATE 14-17
Coral reef ecosystems, such as this on the Great Barrier Reef of Australia, are uniquely threatened ecosystems.

the vertebrate megafauna (elk, bison, moose, deer, grizzly and black bears, and wolves) can be found, but the region is quite unremarkable with regard to numbers of endemic plant species. Ignoring Montana in conservation priorities misses some important considerations.

As a second example, Kareiva and Marvier (2003) cite the well-established nature of coastal ecosystems (such as salt marshes, tropical mangroves, and coral reefs) in providing essential ecosystem services. Salt marshes and coastal mangrove forests are vital for pollution control, for oceanic productivity, and as nurseries for juvenile fish and invertebrates, and their worth in dollars, estimated by Costanza et al. (1997) comes to about $2,000 per hectare per year. But salt marshes contain virtually no endemic plant species and, overall, have low plant species richness, with no more than about 20 to 30 species. A salt marsh, as ecologically essential as it is, cannot be a hotspot, but such ecosystems are certainly deserving of conservation action. Many of the world's coral reefs are in jeopardy, and taken as a whole, coral reefs could be considered as hotspots, though they contain no vascular plants. Coral reefs represent less than 1% of the ocean floor but contain about 25% of all marine species. As well, coral reefs are essential to marine productivity and fisheries success. They are currently threatened by a combination of human activities including overfishing, pollution from increased sedimentation, and physical damage from tourists. Beyond that, as coastal ecosystems such as mangrove swamps are cleared for various reasons, coral reefs suffer subsequent damage (Plate 14-17).

Last, Kareiva and Marvier (2003) suggest that more attention be paid to conservation of genera, which is a form of "evolutionary investment in the future." They examined the distribution of bird and mammal genera that are in serious jeopardy of extinction and mapped the regions just as was done with identifying hotspots for endemic plant species. While the genera map bears clear similarity to that containing the hotspots, there are major differences. For example, there is more attention placed on conservation in Alaska, Australia, and East Africa when bird and mammal genera are considered (Figure 14-14). Although Kenya, in East Africa, contains only 265 endemic plant species, it nonetheless hosts eight highly threatened mammalian

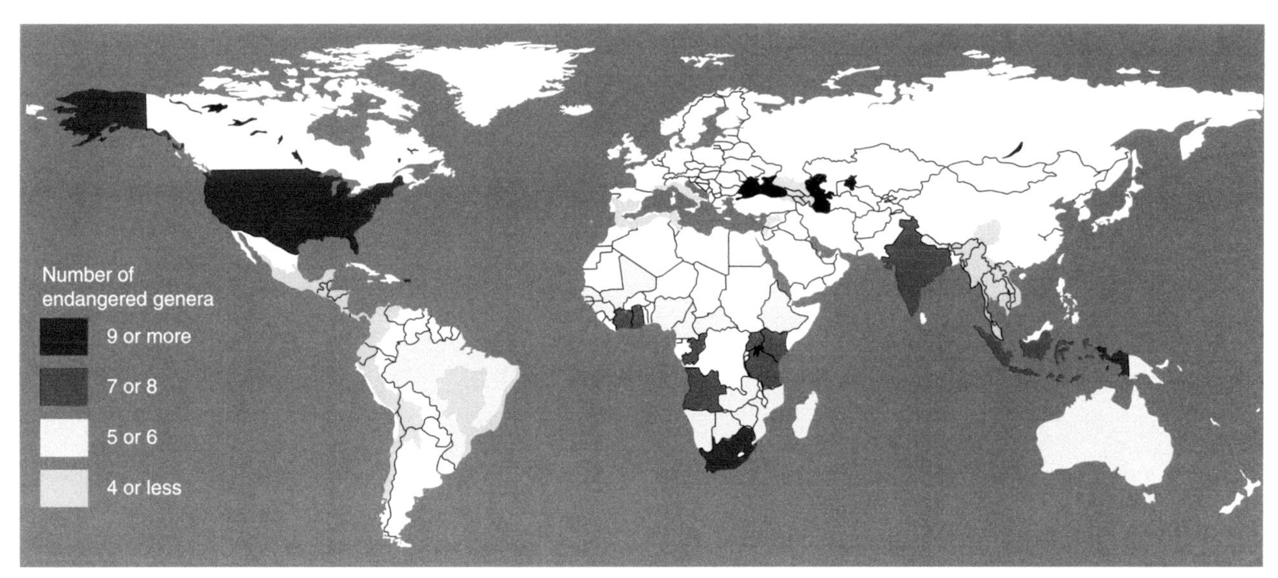

FIGURE 14-14
Conservation priorities can be set using any number of different strategies. Here, the distribution of bird and mammal genera that are in danger of being entirely wiped out is examined, and countries are scored according to how many of these genera are still found there. The resulting map shows some similarity with the hotspots of Myers (dark shading) but also many differences.

genera, mostly ungulates, making it an important hotspot for such species.

The final chapter will continue the discussion of biodiversity and conservation, and present an overview of the current status of conservation issues in the tropics, considering such issues as logging, invasive species, emergence of pathogens, and overall outlook for the future.

15 Conservation Outlook for the Tropics

Chapter Overview

The forests, savannas, and other natural ecosystems of the global tropics were once far more extensive than today. The loss of tropical habitat due to human economic activities and population growth brings with it the potential loss of species and ecosystem services. Habitat reduction and fragmentation continue throughout the tropics. Ecologists have raised concerns for the future of tropical ecosystems. The science of conservation biology attempts to apply scientific principles to analysis of the present and future state of the global tropics. Outlooks vary widely from cautiously optimistic to deeply pessimistic. This chapter will provide an overview of some of the most pervasive issues that affect the future of the tropics.

The Concept of Conservation

Somewhere between 39% and 50% of Earth's land surface has been transformed or degraded by humans (Uriarte et al. 2009). Never in the evolutionary history of this planet has a single species had such a profound ecological impact. We humans must recognize our impact and address it. It is not in our collective self-interest to do otherwise. As the previous chapter noted, the major issues of conservation concern in the tropics are habitat loss, fragmentation and degradation, and biodiversity decline. But why does loss or alteration of natural habitat matter? Why would a monoculture of soybeans be less ecologically valuable than the forest it replaced? Why is loss of natural habitat of concern to ecologists?

The answer ultimately focuses on two concepts: one is loss of species richness (the *biodiversity crisis*), and the second is reduction of ecosystem services. The public at large generally has some knowledge of the biodiversity crisis (see Chapter 14), but the concern about disruption of ecosystem services remains vague to most nonscientists.

Consider that a soybean field is a monoculture. It is a far simpler form of ecosystem than what preceded it because it contains very few species. It replaced an ecosystem with literally thousands of species of plants, animals, and microbes that collectively occupied that area before it was cleared for soybean production. The myriad interactions that formed the food web among the organisms and the various ecological functions (like recycling of minerals and potential carbon storage) that were present in natural ecosystems were eliminated, reduced, or altered. Conservation biologists attempt to evaluate the collective impacts of such habitat alterations, especially as such ecosystem changes multiply throughout the global tropics.

The main assumption behind modern conservation biology is that biodiversity has demonstrable pragmatic as well as aesthetic value (recall the previous chapter) and that biodiversity is at the core of ecosystem services. Further, fundamental ecosystem processes are essential to humans, so much so that from our anthropocentric viewpoint, we refer to them as *ecosystem services* (see the sidebar "Ecosystem Services"). These assumptions provide the basis for diverse research programs focused on conservation biology, not only in the tropics but also elsewhere. The main scientific body devoted to the study of tropical ecology in the Americas is named the Association for Tropical Biology and Conservation. Its journal bears the full title *Biotropica: The Journal of Tropical Biology and Conservation.* Conservation is no longer considered an elitist and arcane area devoted to preserving nature. The stakes are much greater today. Conservation biology is a rapidly developing science.

Ecosystem Services

Ecosystems are functional units, and how these units actually function is the very definition of what Earth is or will become with regard to climate, biodiversity, and other encompassing variables. Ecosystems, whether in the tropics or elsewhere, provide an essential range of important ecological functions on which humans ultimately depend. These include such things as plant and animal productivity for food and fiber; the generation and maintenance of biodiversity (including such things as pollination, seed dispersal, and natural pest control through trophic dynamics); climatic effects such as the mitigation of floods and droughts; the cycling of chemicals such as carbon, phosphorus, and nitrogen; and the cleansing of pollutants from air and water, to name but a few. Planet Earth is unique because of its ecosystems, making it the "ecological planet." Ecosystem services operate on such a grand scale and are so intricately complex that most ecosystem services could not be easily or inexpensively duplicated with any form of extant technology. Ecosystem services are economic externalities, provided without cost to humans. Costs become apparent only when human activities disrupt ecosystem services to the extent that habitat restoration (if it is possible) is required. Human activities are currently impairing various ecosystem services on a large scale.

A sad and dramatic example of the consequences of the loss of ecosystem services was evident when an immense tsunami from the Indian Ocean struck the Cuddalore District in Tamil Nadu, India, on 26 December 2004. The devastation in loss of life and property was severe elsewhere as well, but coastal sites with dense mangrove forest were significantly less damaged (Danielsen et al. 2005). Areas where mangroves prevailed as well as *shelterbelts* (protective buffers consisting of planted trees) composed of the Australian tree *Casuarina* suffered relatively little damage. Prior to the tsumani, human activities had reduced mangrove area by 26% in the five countries most affected by the tsunami (Danielsen et al. 2005). This was costly, as mangrove forests help protect against severe coastal storms, aid in fertilizing coastal fisheries, and may be used for timber production. Coastal mangrove forests are ideal examples of how ecosystems buffer against the vagaries of nature and add to the health and productivity of a region (Plate 15-1).

PLATE 15-1
A coastal mangrove forest such as this one in Australia provides essential ecosystem services such as storm protection.

The following is a list, compiled by the Ecological Society of America and in no particular order of importance, of basic services on which we humans depend, each of which is provided by ecosystems when such systems function normally:

- Moderate weather extremes and their impacts
- Disperse seeds
- Mitigate drought and floods
- Protect people from the sun's harmful ultraviolet rays
- Cycle and move nutrients
- Protect stream and river channels and coastal shores from erosion
- Detoxify and decompose wastes
- Control the vast majority of agricultural pests
- Maintain biodiversity
- Generate and preserve soils and renew their fertility
- Partially stabilize climate
- Purify the air and water
- Regulate disease-carrying organisms
- Pollinate crops and natural vegetation

The theoretical value of ecosystem services supplied by the world's remaining tropical forests has been estimated at somewhere around $2,000 per hectare per year, a substantial sum (Costanza et al. 1997).

For further discussion of ecosystem services, visit the Ecological Society of America's website, and read "Ecosystem Services: Benefits Supplied to Human Societies by Natural Ecosystems," *Issues in Ecology*, vol. 2 (1997). The document is available for download as a pdf at http://www.esa.org/science_resources/issues/FileEnglish/issue2.pdf. Also see Ninan (2009).

Threats to Tropical Forests: Overview

Ecologists have identified a suite of potential threats to the integrity of tropical forests (Laurance and Peres 2006). The most obvious of these threats is the loss of forest itself, the continuing process of deforestation, forest degradation, fragmentation, and increasing loss of biodiversity (see Chapter 14). Deforestation is a complex topic, because it is sometimes not a matter of simple forest clearance. Forests may not be fully cleared, but they may be significantly degraded. Forest degradation results from activities such as logging, agriculture, and prospecting for minerals. Local hunting for *bushmeat* (the term applied to wild animals killed by subsistence hunters) may leave a forest essentially intact but lacking in animal species essential for seed dispersal (Redford 1992). But other more subtle and complex threats are evident. One threat that is particularly poorly understood is the effect that climate change will bring about in the lowland tropics. How will climate change affect forest dynamics (see Chapter 6), biotic interactions (see Chapter 7), trophic dynamics (see Chapter 8), carbon sequestration (see Chapter 9), recycling of nutrients (see Chapter 10), and land-use patterns (see Chapter 13)? How will invasive species and emerging pathogens affect tropical ecosystems and their human inhabitants? Even something as seemingly innocuous as a road through a tropical forest may pose serious problems for conservation (S. G. W. Laurance 2006) by exposing forest to anthropogenic activities that degrade it (Plate 15-2).

PLATE 15-2
This road through a protected forest in Tanzania is a popular resting place for a troop of baboons. However, most roads through tropical forests result in significant local impacts to wildlife and may result in loss of local biodiversity.

What Is a Forest?

It is essential to understand something of the terminology that ecologists use in describing and categorizing forest types (Putz and Redford 2010). Forests may be defined ecologically or even politically. For example, *old-growth forest* has not been subject to cutting or clearance by humans, or it has recovered from previous anthropogenic disturbance for at least 200 to 300 years. Old-growth rain and moist forest characterizes the most diverse tropical forests. On the other hand, consider the United Nations Food and Agriculture Organization (FAO) definition of a forest: an area of greater than 0.5 hectare (1.23 acres) with more than 10% tree canopy cover, with trees defined as plants capable of growing 5 meters tall (16.5 feet). This definition would apply over much of the tropics and the rest of the world, but it is nothing remotely similar to old-growth rain forest. Scrublands, simple monocultures, fragmented tracts, and seriously degraded forests would all be considered *forests* under the FAO definition (Plate 15-3).

The definition of a degraded forest also varies. *Degraded* means forest that has been subject to activities such as logging or various degrees of forest clearance. Degraded forests are also considered to result from excess hunting, invasive species encroachment, and fire. Putz and Redford (2010) attempt to clarify various terms associated with forests:

> Forests that develop after complete deforestation should be referred to as *secondary forests*. Although some researchers do not distinguish between *secondary* and *degraded forests*, we believe that these types of forest are different enough in structure, composition, dynamics, and management options to warrant distinction. *Secondary forests* that develop on land used for intensive agriculture and pastures, for example, regenerate mostly from seeds that are either dispersed from nearby remnant forest or, to a lesser extent, germinate from buried seeds. In contrast, regeneration after logging in *managed* or *degraded forests* is typically dominated by vegetative recovery of plants that remained on site through the episode of damage. Furthermore, although *secondary forests* rapidly develop some of the structural, compositional, and functional characteristics of *old-growth forests* (Chazdon 2003), other features recover only after centuries if ever (Clark 1996). (pp. 13–14)

Figure 15-1 shows the relationship among the various types of ecosystems that are forest and are related to forest.

PLATE 15-3
This tract of mallee scrub in Australia fits the FAO definition of forest.

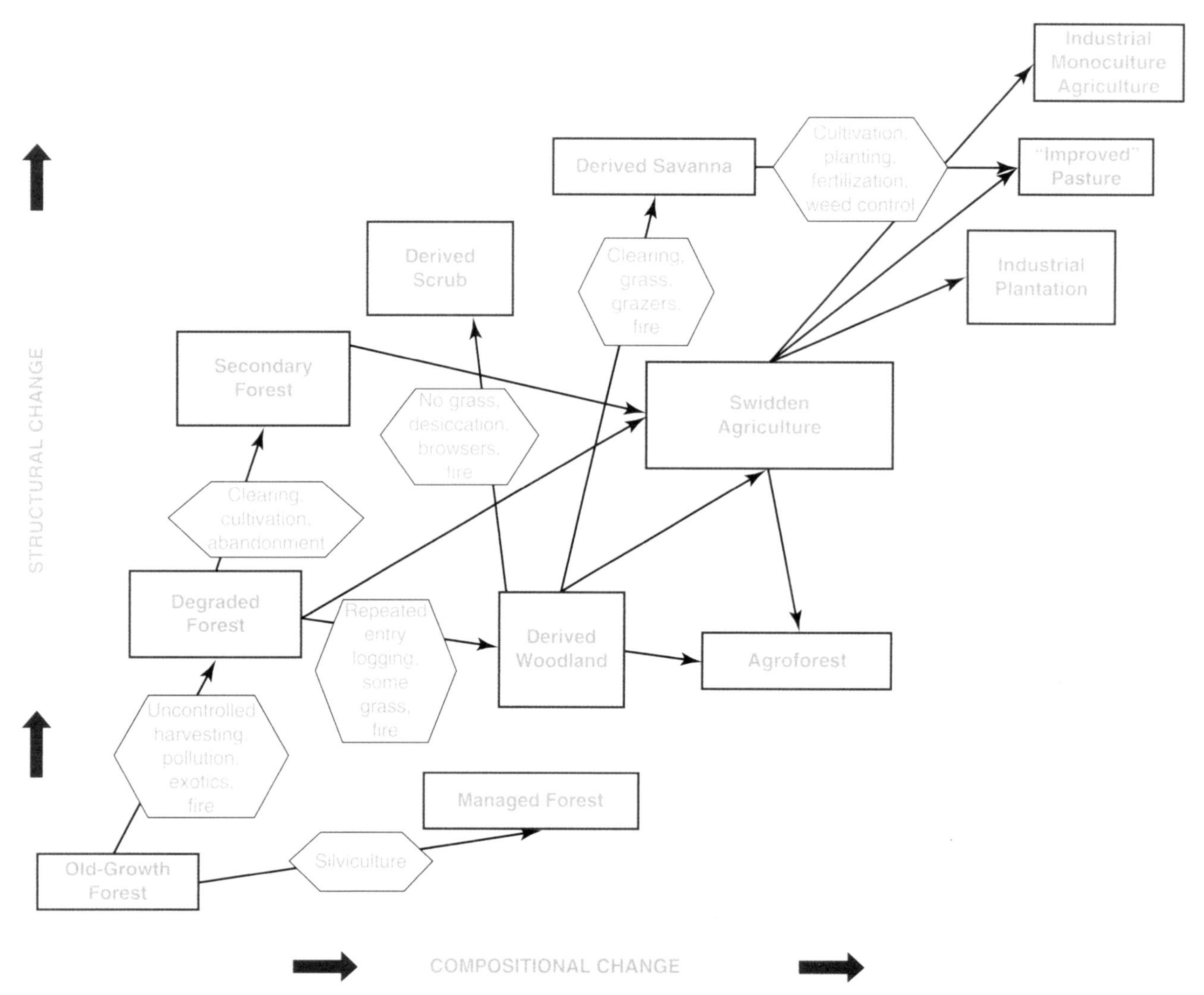

FIGURE 15-1
Ecological state changes starting from the reference condition of tropical old-growth forest. Only the principal drivers are included (in hexagons), and not all possible transitions away from the reference condition are depicted. Back transitions that result from natural succession as well as afforestation, reforestation, forest management, and other restoration activities are not indicated, to reduce figure clutter.

Deforestation: How Much, How Bad?

Since the dawn of agriculture, humans have cleared forests to make room for crops. But as human population has grown vastly over thousands of years (with the growth rate having dramatically accelerated in the twentieth century), the rate and impact of deforestation has also grown.

Recall that the once-extensive Atlantic Forest of southeastern Brazil (see Chapter 14) has been reduced and severely fragmented to a mere 10% of its original area (Galindo-Leal and de Gusmao Camara 2003). Approximately 70% of the animal species considered endangered in Brazil are confined to the Atlantic Forest. There are about 8,000 endemic plant species, 55 threatened endemic bird species and 21 threatened endemic mammal species. European colonists began cutting the forest immediately upon their arrival because, like European colonization of North America, the settlers believed that the forest had to be cleared to allow for social progress (Dean 1995).

At the core of concern for conservation in the tropics is the reality of continuing, indeed accelerating, deforestation (Wood and Porro 2002; Montagnini

PLATE 15-4
This degraded forest in Amazonian Brazil (in Mato Grosso) has been largely cleared for logging and for conversion to agriculture. Perhaps ironically, the words *Mato Grosso* come from the Portuguese for "thick wood." The color of the sky is due to thick smoke created by the burning of wood (during the dry season) from the cleared forest.

and Jordan 2005; Rudel 2005; Spray and Moran 2006). Estimates of deforestation vary (see Table 15-1), but one detailed study estimated that between 35% and 50% of original closed-canopy forests around the world (not just the tropics) have been removed (Wright and Muller-Landau 2006a and 2006b). This represents between 8 and 12 million square kilometers lost (Plate 15-4).

TABLE 15-1 Recent estimates of potential and extant closed-canopy forest area for tropical and extratropical continents. Ramankutty and Foley (1999) estimated potential forest area as the area that would be covered by closed-canopy forest in the absence of human intervention using 1-kilometer-resolution satellite imagery supplemented by a global vegetation model. The FAO (2000) estimated extant closed-canopy forest cover from national forest inventories supplemented by expert opinion. Achard et al. (2002) estimated extant closed-canopy forest cover from 30-meter-resolution satellite imagery using a random 6.5% sample stratified by forest cover and recent levels of deforestation. Hansen and DeFries (2004) estimated extant closed-canopy forest cover from 8-kilometer-resolution satellite imagery using a 100% sample of the tropics.

	POTENTIAL FOREST COVER[a] RAMANKUTTY AND FOLEY (1999)		EXTANT FOREST COVER FAO (2000)		EXTANT FOREST COVER ACHARD ET AL. (2002)		EXTANT FOREST COVER HANSEN AND DEFRIES (2004)	
CONTINENT	(10^4 KM^2)[b]	(10^4 KM^2)[c]	(10^4 KM^2)[b]	(%)	(10^4 KM^2)[c]	(%)	(10^4KM^2)[c]	(%)
Tropical								
Africa	545	548	344[d]	63.4[d]	193	35.2	172	31.3
Americas	1085	1056	891[d]	82.1[d]	653[e]	61.8	701	66.4
Asia	661	632	256	38.7	270[f]	42.7	199	31.5
Total	2291	2236	1496[d]	65.3[d]	1116	49.9	1072	47.9
Extratropical								
Americas[g]	510		231	45.3				
Asia[h]	456		186	40.8				
Europe[h]	421		171	40.5				
Russia	1219		834	68.4				
Total	2606		1422	54.6				

[a]Includes tropical evergreen, tropical deciduous, temperate broadleaf-evergreen, temperate needleleaf evergreen, temperate deciduous, boreal evergreen, boreal deciduous, and mixed evergreen/deciduous forests and woodlands (i.e., biomes 1 through 8 of Ramankutty and Foley 1999).
[b]Forest area in tropical and extratropical areas is summed over countries whose geographic centers are at latitudes $\leq 24°$ and latitudes $>24°$, respectively.
[c]Forest area in the tropics is for the area between the Tropics of Cancer and Capricorn.
[d]Comparisons among sources are problematical because the FAO (2000) "closed forest" includes an unknown proportion of the extensive Brazilian *Cerrado*, African *Miombo*, and other relatively open formations that the other sources exclude. Such open formations are of relatively limited extent in tropical Asia.
[e]Excludes Mexico and the Atlantic coastal forest of Brazil.
[f]Includes the "evergreen and seasonal forest of the tropical humid bioclimatic zone" for all continents and also the "dry biome of continental Southeast Asia."
[g]Excludes Canada due to its unique definition of forest cover.
[h]Excludes Russia, which is reported separately.

If agriculture on once-forested land is abandoned, forest will usually return. In eastern North America, forest clearance peaked in the eighteenth and nineteenth centuries. Following the regional abandonment of most agriculture, much of that forested area has grown back today through the process of *secondary succession* (see Chapter 6). The extensive forest clearance that occurred in North America centuries ago is now ongoing throughout much of the tropics. Loss of old-growth forest is accelerating, particularly in tropical Asia, but also elsewhere in the tropics (Table 15-1). Questions arise about the future rate of tropical forest clearance. Will tropical forest regrow, as temperate forest has in eastern North America following the abandonment of agriculture?

Deforestation in the tropics is closely linked with growth of human population. As rural populations have burgeoned, forest loss has increased (Figure 15-2). The United Nations Food and Agriculture Organization (FAO), using data generated from satellite observation, has documented that tropical deforestation usually occurs first in the more dry and open forests (which are easier for humans to access) rather than humid rain forests, and that about 2.2% of the potential closed-canopy tropical forest (for all forest types combined) is being removed each decade (FAO 2000; Wright and Muller-Landau 2006a). But tropical moist forests have also been cleared at a high rate, a rate that promises to greatly accelerate in places like the Democratic Republic of the Congo, largely due to predicted increases in human population (FAO 2000).

Rates of deforestation throughout the tropics are difficult to obtain. Anyone who travels throughout a tropical region will bear witness to deforestation, but such observations are anecdotal and difficult to quantify (Plate 15-5). The method of choice for estimating large-scale regional deforestation rates has been satellite imagery, because it provides a large-scale look at forest changes over time.

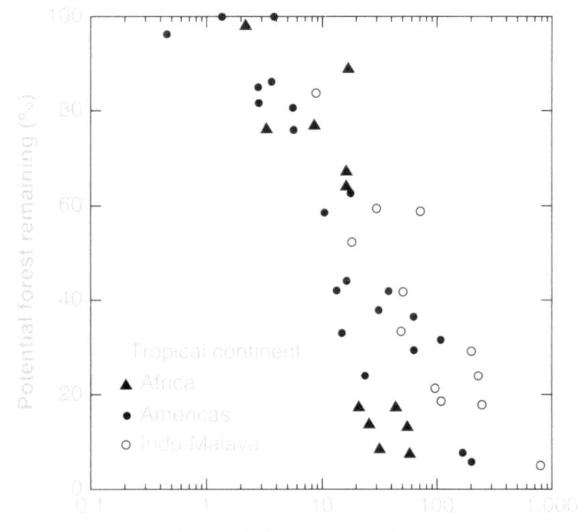

FIGURE 15-2
The relationship between the percentage of closed-canopy forest remaining and rural population density for 45 humid tropical countries. Potential forest cover (the denominator of the ordinate) includes the forest and woodland biomes of Ramankutty and Foley (1999). Extant closed-canopy forest cover (the numerator of the ordinate) was taken from FAO (2000). Humid tropical countries were defined to be those countries with closed-canopy forest being the potential vegetation over 40% or more of their national territory and whose geographic centers were at latitudes below 24°.

PLATE 15-5
This is a typical sight throughout much of the tropics—in this case, from Belize, Central America. It shows a pasture that once was closed forest. A lone kapok tree (*Ceiba pentandra*) remains uncut.

PLATE 15-6
This secondary forest is ecologically distinct from mature rain forest. Many areas in the tropics are undergoing secondary succession due to previous clearance and/or degradation by human activities.

For example, Table 15-1 includes data from a study by Achard et al. (2002). The study, which used 30-meter-resolution (99 feet) satellite imagery, concluded that from 1990 to 1997, a total of 5.8 million hectares (14,326,000 acres) of humid tropical forest were lost annually (about the size of the state of Maryland) and that an additional 2.3 million hectares (5,681,000 acres) were "visibly degraded." Those figures were 23% lower than the generally accepted rate of forest loss at the time of the study. The point remains, however, that a significant annual net loss of forest was evident. A second study by Hansen and DeFries (2004) used 8-kilometer-resolution (5 miles) satellite imagery over a much broader area (also shown in Table 15-1). The conclusions were similar to Achard et al. Both of these studies were quite different in their results from the FAO study (Table 15-1), but the FAO study was broader, measuring more kinds of forests including dry forests and other forests typically more open than humid forests. These differences among studies have to do with methodological differences and with basic definitions of what constitutes forest.

Another difficulty in estimating forest loss is that forests are dynamic. If land that has been cleared is abandoned, secondary forest will normally grow on the site (Plate 15-6). So in some areas, such as areas cleared for logging and subsequently abandoned, forests may be returning through the process of secondary succession. But secondary forests are different from old-growth forests. This dynamic between forest loss and forest regrowth is complex, and it has major implications that are essential to understand in considering the future of tropical forests. It has become a topic of significant debate (W. F. Laurance 2006; Wright and Muller-Landau 2006a and 2006b) and will be discussed further in the following sections. But first it is important to look in more detail at threats to tropical forests.

Fire, Degradation, and Deforestation

Deforestation, as the term implies, means the cutting of forest and replacing it with some other form of ecosystem. But human activities also result in forest degradation. Human-caused fire (Plate 15-7) has been shown to have a strong impact on forest degradation, even to the point of changing forest into savanna (Cochrane et al. 1999). Fires set by humans in their attempts to clear forest or simply by accident and carelessness have been studied in parts of Amazonia, with sobering results. Accidental fires are linked to road building in tropical forests because roads permit access. They also are linked to agriculture, mining,

PLATE 15-7
Distant fire in a forested area in eastern Venezuela.

and pretty much any human intrusion into forest. Initially, fires are of low intensity and travel along the forest floor, not reaching the canopy. But though the fires appear to destroy only leaf litter, they also kill tree stems, because many tropical trees have thin bark. Thicker bark and larger trees are less affected. The fire effectively thins the forest, and subsequent windthrow adds debris to the litter layer. The more open canopy permits greater solar heating, and the litter dries, making it more prone to fire. The greater openness results in growth of various vines and herbaceous species that add combustible fuel. Forests that have burned are therefore likely to burn again, far more likely than previously unburned forests. And they will burn with greater intensity. A positive feedback is established in which each subsequent fire is more severe than the previous one, and recurrent fires become more frequent. This becomes even more exacerbated in El Niño years, when there is markedly less rainfall. Indeed, heating and drought, both forecast to be more likely in the future due to climate change (see the section "Climate Change" later in this chapter), are expected to greatly increase occurrence of serious fires in areas of humid tropical forest. Forests become so degraded by fire that the land becomes, in effect, *deforested*. Bear in mind that natural fires do occur in Amazonia, but the time between fires at a given location is historically in the hundreds or thousands of years. The study by Cochrane et al. (1999) concluded that fire intervals of 90 years could eliminate some rain forest tree species, and intervals of less than 20 years would be sufficient to entirely eliminate forest from the site.

Logging

Logging and its relationship to conservation is a major and complex issue of concern throughout the global tropics. In many parts of the temperate zone, particularly in North America and Europe, sustainable logging is routinely practiced. This activity may take various forms ranging from clear-cutting to more selective removal. Sustainable logging, in its most basic form, is based on removal of trees (of economic value) that have not yet reached *senescence*, the age at which growth rate is extremely slow. The harvested trees are replaced by reseeding or replanting the area with species that grow relatively rapidly to harvestable size. In many cases, temperate zone forests may be clear-cut in 40- to 50-year cycles or less (Plate 15-8). It is, in a sense, much like basic agriculture, with trees as the crops. Arguably, sustainable logging in much of the temperate zone has not resulted in major species reductions or loss of ecosystem function, though this is not true in such places as Germany and Scandinavia.

Logging in the tropics is not as straightforward as it is at higher latitudes. Some forests are largely clear-cut. But in many cases, valuable trees must be removed individually because they are widely spread, a general characteristic of tropical forest tree distribution (see Chapter 5). Logging roads penetrate deeply into forests. Loggers make trails to access trees whose timber is deemed of value. Trees are cut and subsequently dragged over *skid* (also called *snig*) trails to the road to be loaded onto large trucks (Plate 15-9). Trucks carry logs from forest to central locations, where they can be shipped to urban areas for further processing (Plate 15-10).

PLATE 15-8
This pine forest near Savannah, Georgia, is an example of a managed forest where clear-cutting is sustainable because the growth of the pines is rapid. Note the neat rows of trees.

PLATE 15-9
Large trucks loaded with freshly cut logs are common in many places throughout the tropics. This truck was in Mato Grosso, Brazil.

PLATE 15-10
Logs at a collection area in Mato Grosso, Brazil.

Much logging activity in the tropics results in forest degradation. Logging roads, skid trails, and human activities associated with logging result in significant damage to plants other than the target species, reducing forest biomass, permitting greater light penetration, creating damage to soils, and increasing potential for serious fire (see the previous section). Largely because of logging, conservation biologists distinguish between deforested areas, as are typical in areas of agriculture, and degraded forest, where logging is practiced. Another question focuses on the impact of logging on wildlife. Logging roads and skid trails permit hunters to penetrate areas that would otherwise be difficult to access and have therefore contributed greatly to wildlife harvest (Robinson et al. 1999). Even without hunting pressure, some species may be highly sensitive to changes due to logging and others much less so. The following several case studies illustrate the complex issues associated with logging in the tropics.

CASE STUDY 1: LOGGING IN BORNEO

One question associated with tropical forest logging is its impact on species richness of trees. A study performed in Indonesian Borneo showed that tree species richness did not suffer from commercial logging (Cannon et al. 1998). (Note that politically, the island of Borneo is divided into Kalimantan, the Indonesian portion that occupies most of the island, and Brunei and East Malaysia, which occupy the northernmost part of the island.) The study focused on a logging *concession* (an area where the government has issued permits to private companies to legally perform logging) consisting of 59% dipterocarp forest and 41% swamp forest. The forest was rich with tree species, but the three most commercially valuable species (all in the dipterocarp family) made up 70% of the total precut basal area. Logging removed 62% of the dipterocarp basal area and 43% overall. Following logging, 43% of the canopy was open and, not surprisingly, pioneer plant species dominated. Logging reduced both tree density (which seems like an obvious point) and number of tree species per 0.1-hectare (0.25-acre) plot for both large and small trees. Eight years following the logging, plots contained fewer species of trees with diameters less than 20 centimeters than did similar-sized unlogged plots (again, an unsurprising result). But what is important is that in samples that contained the same numbers of trees (which required a 50% larger area for the logged areas), logged forest had a species richness that equaled that of unlogged forest (Figure 15-3). What this means is that logging did not reduce tree species richness disproportionately in comparison with unlogged forest. The researchers caution against extrapolating their results to other forests in other regions and note that other (potentially negative) logging effects were not part of the study. The study does suggest that logging does not automatically reduce tree species diversity, at least not in some tropical forests.

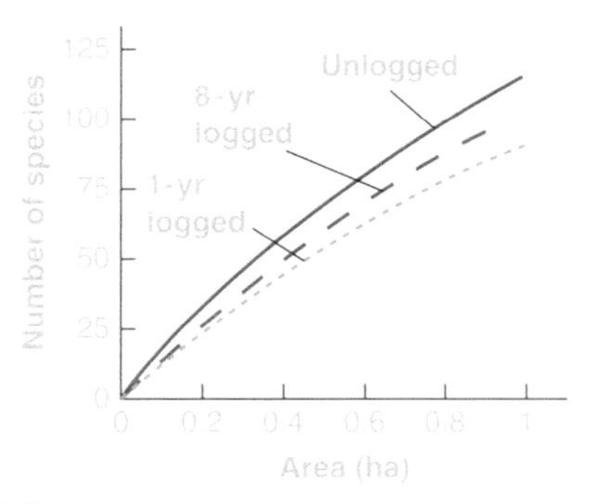

FIGURE 15-3
This graph compares the number of tree species in relation to area of two logged plots and one unlogged plot. Note that there is little difference among them.

The problem that many ecologists identify with tropical logging is its potential for continual expansion of forest degradation, even within what are deemed to be protected areas such as national parks. Logging is done for profit, international market forces are strong, and regulations are not strictly enforced. A team of researchers documented lowland forest loss and degradation from logging in Gunung Palung National Park (GPNP) in West Kalimantan, along with its 10-kilometer (6.2-mile) buffer zone, as well as Kalimantan as a whole (Curran et al. 2004). Data were collected by Geographic Information System (GIS) satellite imagery and field-based analysis. From 1985 to 2001, Kalimantan's protected lowland forests (note the word *protected*) declined by 56% (a figure that represents more than 29,000 square kilometers, or 11,194 square miles), largely to supply international lumber markets. Protected forests were obviously not being protected, nor were the buffer zones surrounding them (Figure 15-4). Increasingly, widespread degradation is of serious concern for all the reasons listed earlier that are identified as negative effects of logging.

The extent of forest reduction evident within the park boundary creates a variety of ecological concerns.

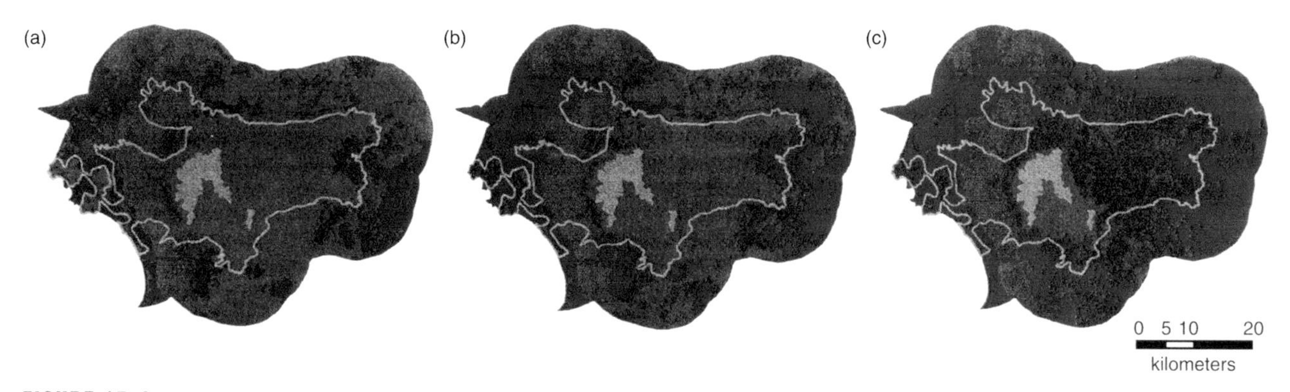

FIGURE 15-4
These three maps show the extent of forest loss in GPNP and surrounding areas outside the park. The park is outlined in yellow. (a) is 1988, (b) is 1994, and (c) is 2002. Areas in red are deforested. The gray area is montane forest, not part of the analysis. Tan areas are peat forests, and green areas are lowland forests. Note how deforestation has dramatically increased outside the park (not just from logging but for oil palm plantations), at the park buffer zone and, by 2002, was extensive within the park boundary.

Dipterocarps reproduce synchronously (see Chapter 7), and increasing degradation and fragmentation within the park will likely have an impact on dipterocarp trophic interactions. Large vertebrate species such as the Malaysian sun bear (*Helarctos malayanus*) (Plate 15-11), bearded pig (*Sus barbatus*), and orangutan (*Pongo pygmaeus*) are predicted to rapidly decline as a result of forest degradation. The authors conclude, "Effective frontier governance and sound regional land-use planning are critical to protecting even uninhabited and remote PAs [protected areas] from regional, and international, market forces" (p. 1002).

Ecologists and evolutionary biologists have attempted to predict which kinds of animal species are at most risk and which are at least risk in degraded forests. Another study from Borneo approached this problem by correlating *phylogenetic age* (the time from speciation to the present) of mammal species with sensitivity to logging (Meijaard et al. 2008). The researchers, using 24 regional field studies on mammal sensitivity to logging, placed each mammal species into one of three categories: *intolerant* (15 species), meaning severely impacted by logging; *neutral* (12 species), meaning no recorded effects from selective logging; and *tolerant* (14 species), meaning densities actually increased by more than 20% within a year following logging. The result was that the more evolutionarily ancient species of mammals were by far the most sensitive to forest degradation. Those species most recently evolved were far less sensitive. The correlation between evolutionary time of origin and logging sensitivity was strong. The evolutionary time

PLATE 15-11
Malaysian sun bear, threatened by increasing deforestation even within protected areas.

frames were that the most sensitive mammal species evolved in the Miocene or early Pliocene and the least sensitive evolved in the late Pliocene or Pleistocene. Of those species negatively affected by logging, 87% were Bornean or Sundaland endemics. None of the eight Bornean endemic mammals were tolerant of logging. Of those species that were tolerant, 50% were widespread throughout southwest Asia. This suggests that they may be strong *ecological generalists*, able to endure varied conditions. Those species that exhibited some form of ecological specialization in diet (frugivores, insectivores, carnivores) were highly sensitive to logging, but those species that were generalist omnivores were least sensitive. The authors conclude that the effect of logging on tropical mammal species, and likely on other vertebrate taxa, is selective based on the following: relatively old phylogenetic age, restricted distribution and endemism, and dietary specializaton. It is not clear as to why phylogenetic age would put a species more at risk. Perhaps such old species have become specialized to the degree that they cannot tolerate ecological change. Clearly, further research needs to focus on factors affecting species' evolutionary histories in evaluating their risk relative to anthropogenic-induced changes in their ecosystem.

CASE STUDY 2: FOREST DEGRADATION IN PAPUA NEW GUINEA, 1972–2002

Logging and forest degradation have been studied over a 30-year period in Papua New Guinea, an area considered among the most heavily forested areas in the Asian tropics, containing approximately half of the third-largest extant area of tropical rain forest in the world (Shearman et al. 2009). Like Borneo, the future of forests is not encouraging. Papua New Guinea has experienced increasing rates of forest loss from agriculture, forestry, fire, plantation development, and mining, as well as a substantial human population increase. Logging has caused forest degradation, defined in this study as loss of primary forest and conversion to secondary forest. The human population of Papua New Guinea has increased substantially as it has increased exports from commercial forestry (aka logging) and oil palm plantations. (Oil from oil palm is extracted from two palm species in the genus *Elaeis*, one from Africa and one from the Neotropics. The trees are cultivated in large plantations after extant forests are clear-cut. The palm oil is of high commercial value, is widely used in cooking, and is an important export of various tropical countries.)

The researchers considered forest to be natural woody vegetation with a contiguous tree canopy and a canopy height > 5 meters (16.5 feet). This definition is important, because definitions of what, exactly, is meant by *forest* vary from one study to another, a subject with major implications that will be further discussed later in this chapter. Data were collected with Landsat images. Efforts were also made to identify older logged areas (from the presence of snig trails and old roads), areas where mining had occurred (open pits and areas with tailings), and areas of formerly burned forest.

There was a net clearance of 15.0% of primary rain forest between 1972 and 2002, and an additional 8.8% was degraded to secondary forest through logging (Tables 15-2 and 15-3).

TABLE 15-2 The area of primary rain forest and primary rain forest accessible to commercial logging in Papua New Guina in 1972 and 2002 and change (%) due to both deforestation and degradation over this period. Swamp, mangrove, and dry evergreen forest are excluded. DF refers to the percentage of 1972 forest area deforested by 2002, DG refers to the area degraded, and Total refers to the area deforested and degraded.

	ALL RAIN FOREST					ACCESSIBLE RAIN FOREST				
			CHANGE (%)					CHANGE (%)		
REGION	1972 (HA) PRIMARY	2002 (HA) PRIMARY	DF	DG	TOTAL	1972 (HA) PRIMARY	2002 (HA) PRIMARY	DF	DG	TOTAL
Islands	4,885,727	2,699,103	21	24	45	2,877,354	1,064,717	22	41	63
Highlands	4,776,533	4,104,916	14	0	14	507,092	436,716	11	3	14
Mainland lowland	23,565,330	18,528,234	14	7	21	10,090,542	7,182,347	12	17	29
Total	33,227,590	25,332,253	15	9	24	13,474,988	8,683,780	14	22	36

TABLE 15-3 The percentages of each category of forest change in Papua New Guinea from 1972 to 2002.

ACTIVITIES	PERCENT CHANGE
Logging	48.2%
Subsistence agriculture	45.2%
Forest fires	4.4%
Plantations	1.2%
Mining	0.6%

From Shearman et al. 2009.

Human population density correlated closely with forest clearance throughout Papua New Guinea. The researchers concluded that "dramatic increases in commercial logging over the 1990s drove the rate of change in commercially accessible forest to a 30-year high of 3.4 percent/yr in 1997–1998, before lowering to 2.6 percent/yr in 2001–2002" (p. 386). The authors further note that their estimates of forest loss have exceeded those (for most of the study period) of the FAO. The conclusion of the study was that timber loss in Papua New Guinea was "considerably higher than previously thought" (p. 388), though the authors note that deforestation rate in Papua New Guinea is lower than in other tropical Asian countries. Only the Asian financial crisis of 1997–1998 was able to sharply reduce the rate of deforestation. The authors suggest that the central government in Papua New Guinea should focus sharply on addressing deforestation, particularly in light of its role as a founding member of the Coalition of Rainforest Nations, a group of 15 nations attempting to use market forces to help curtail greenhouse gas emissions that are caused by tropical deforestation and degradation (see http://www.rainforestcoalition.org/eng/).

Logging was clearly the most important factor in causing forest clearance from 1972 to 2002 in Papua New Guinea. Should this trend continue, it places at risk Papua New Guinea's reputation for being conservation-minded as part of the Coalition for Rainforest Nations.

CASE STUDY 3: FOREST CLEARANCE IN THE AMAZON BASIN (AMAZONIA)

Many North Americans are aware of extensive deforestation that has occurred in many areas of the Amazon Basin, particularly in Brazil, which occupies the majority of the region. The Brazilian Amazon comprises about 40% of all remaining tropical rain forest (Rodrigues et al. 2009), and because of its massive area, the future of the Brazilian Amazon has been a focus of concern for many years (W. F. Laurance et al. 2001). Most deforestation has been associated with the growth of cattle ranching and soybean production. Some years ago, for example, there was much discussion about the *hamburger connection*, having to do with the importation of inexpensive South American beef, raised in pastures that were once rain forest, for use in North American fast food restaurants. Some estimates suggest that cattle ranching might account for nearly 80% of Amazonian deforestation (Tollefson 2008), but that estimate seems too high, given other factors such as vast areas of soy production that also contribute. For a broad summary of deforestation patterns in Amazonia, see Fearnside (2007). Land use and legal access for activities such as logging and agriculture are complex topics throughout Amazonia (not just in Brazil). About 20% of the land has legal federal or state protection; 21% belongs to indigenous people (total population about 250,000); and 24% is in private hands, though some of the holdings many be illegal. The remaining 35% (an area larger than Germany) is categorized as *open-access*, meaning that it is not zoned or otherwise protected (data from Matthew Pearl, World Wildlife Fund, cited in Tollefson 2008).

Brazil has been the focus of attention for many years, though other Amazonian countries have also experienced major deforestation of lowland forest. In Brazil, forest clearance has been most extensive in the states of Mato Grosso, Para, and Rondônia. Much of the rest of the country remains deeply forested. While much of Brazil's clearance has been for cattle ranching and soybean production, logging has also been a major activity.

A comprehensive review of logging in the Brazilian Amazon from the years 1999 to 2002 showed that between 12,075 and 19,823 square kilometers (4,661 and 7,652 square miles) were logged per year, a figure that was somewhere between 60% to 123% greater (depending on what study it was compared with) than previously reported (Asner et al. 2005). The study employed large-scale, high-resolution, automated sensing analysis using Landsat Enhanced Thematic Mapper Plus satellite data to detect previously undetectable logged areas. The study found that, in general, such protected areas as national parks, reserves, and lands belonging to indigenous people were not subject to illegal logging, with the exception of areas in northern

TABLE 15-4 Selective logging rates from 1999 to 2002 in five major timber-producing states of the Brazilian Amazon.

STATE	1999–2000 RATES (KM^2/YR)		2000–2001 RATES (KM^2/YR)		2001–2002 RATES (KM^2/YR)	
	LOGGED	DEFORESTED	LOGGED	DEFORESTED	LOGGED	DEFORESTED
Acre	64	547	53	419	111	727
Mato Grosso*	13,015	6,176	7,878	7,504	7,207	6,880
Pará	5,939	6,671	5,343	5,237	3,791	8,697
Rondônia	773	2,465	923	2,673	946	3,605
Roraima	32	253	55	345	20	54
Total	19,823	16,112	14,252	16,178	12,075	19,963

*Only the northern 58% of Mato Grosso containing forested lands was included in the analysis.

Mato Grosso. In effect, selective logging of the Brazilian Amazon doubled previous estimates of the total amount of forest degraded by human activities (Table 15-4). The researchers estimated that the total volume harvested equated to between 10 and 15 million metric tons of carbon removed. This did not count the masses of debris such as stumps, branches, foliage, and so on that will eventually decompose, adding carbon dioxide to the atmosphere.

An argument has been made for using prudence to wisely manage selective logging in Amazonian forests (Keller et al. 2007). No one denies that logging may create what has been described as significant collateral damage to ecosystems. But in global markets, demand remains high for wood from tropical trees. For any form of sustained logging to succeed, the biological realities that accompany forest diversity and productivity must be considered as well as policies relating to economics (Keller et al. 2007). Brazil, for example, has been very active in attempting to halt illegal logging (Tollefson 2008). The limitations to sustained logging are formidable: high species richness within the world's most complex terrestrial ecosystem, slow growth and regeneration of trees, and relatively few species of trees of high economic value (such as mahogany, *Swietenia mahagoni*). Any logging will result, under the best of circumstances, in what has been called *modest biodiversity trade-offs* (Keller et al. 2007). It is reported that Brazil earned $2.3 billion and supplied 380,000 direct and indirect jobs (annually) in the legal timber industry. Keller et al. (2007) pose this challenge: "It is therefore inevitable that working forests will suffer some changes in structure and composition relative to their undisturbed state. One may either decry this change or consider the value of retaining managed forests as opposed to an alternative land use, such as cattle pasture" (p. 214). They call for the implementation of a government-run forest concession system, citing managed forest in Bolivia as an example of success.

The pragmatics of logging management aside, the fact remains that logging in any form within species-rich complex tropical forests involves a significant societal trade-off. The degradation of ecosystem services from logging and other forest clearance activities is potentially staggering (Foley et al. 2007). These include nontrivial impacts on species dynamics, carbon storage, water flow regulation, climate impacts, and changes in abundances of organisms that transmit vector-borne diseases. That said, it seems highly unlikely that collateral damage to forests from logging will cease any time soon.

Climate Change

All terrestrial ecosystems and oceanic ecosystems (including the deep-ocean benthic zone) will be affected by global climate change throughout the present century. At the root cause of climate change is the emission, largely from anthropogenic activities, of additional greenhouse gases. There is now an immense body of research ranging from climatology to ecology that documents the reality of ongoing and perhaps accelerating global climate change. The driver for climate change is global warming from additional greenhouse gases, in particular carbon dioxide. The terrestrial tropics have been a major source of added greenhouse gases essentially due to deforestation, which releases carbon dioxide when forest remnants are burned to clear the brush. It is estimated that 85% of the carbon contained in forest trees is released to the atmosphere following deforestation (Houghton et al. 2000; Soares-Filho 2006). Ecologically speaking, climate change is the proverbial "elephant in the room." Don't even think for a minute that you can ignore it. It's there and you do not know what it's going to do.

Climate change is sufficiently clear at this point to measure its effects. These include numerous range changes of organisms in relation to patterns of temperature change (Parmesan and Yohe 2003). Rates of range changes are determined by the velocity of climate change. The velocity of climate change (in kilometers per year) has been measured and varies considerably among the major terrestrial ecosystems (Loarie et al. 2009). Velocity is based on rate of temperature change, so the faster an area is warming, the greater its velocity of climate change. It is currently fastest in areas such as flooded grasslands, deserts, and mangrove forests (Figure 15-5). It is more moderate for tropical and subtropical broadleaf forests. What this means is that organisms within various ecosystems must essentially "move" to keep up with changing temperature. Distributions of organisms along various montane elevational zones will change such that organisms from lower elevations will attempt to move higher, thus remaining within their thermal comfort zone (see Chapter 9). The ecosystems of the world are being "reshuffled" as organisms are increasingly affected by climate, and most particularly temperature change, a trend that will continue indefinitely.

For species of the lowland tropics, the largest threat from climate change is biological attrition (see Chapter 9), where climate warms beyond species' thermal tolerances, and there is no region into which to migrate. There is paleoecological evidence that some species are resilient to gradual climate change (Bush and Hooghiemstra 2005). Adaptation by natural selection is, of course, possible, but such a process is measured in generations and may never occur. Strong selection pressures may just as easily result in extinction as adaptation. And without question, climate change is likely to force the reassembly of ecological communities as species react individualistically to the impact of climate change.

Evidence for how climate change results in range changes and potential extinctions comes from a broad study of lizards (Plate 15-12), with a focus on Mexico (Sinervo et al. 2010). The study looked at 48 species of Mexican lizards at 200 sites. A total of 12% of the species had become extinct in various local populations since 1975. The study further examined lizard population patterns on a global basis, estimating that 4% of local populations have gone extinct worldwide. Projections based on climate change led to the conclusion that local population extinction rates will reach 39% by 2080 and that globally 20% of all lizard species will have become extinct. These projections were tested based on local

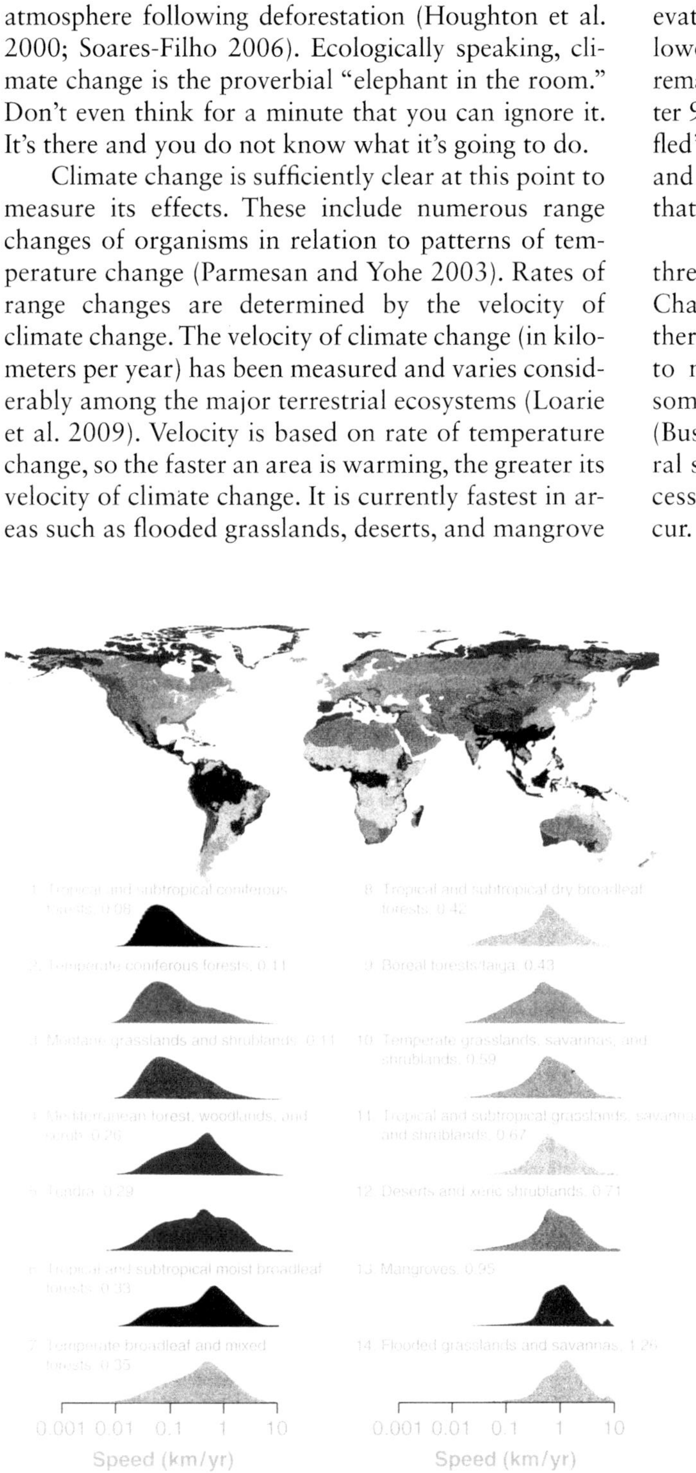

FIGURE 15-5
This map and the curves below it are estimates of the rate of temperature change in each major biome, measured in kilometers per year. Note that mangroves are experiencing a change of 0.95 kilometer/year, and tropical and subtropical moist and broadleaf forest is experiencing a rate of 0.33 kilometer/year.

PLATE 15-12
This basilisk lizard (*Basiliscus basiliscus*) in Belize is resting in the shady undergrowth. The thermal world of lizards requires constant adjustment to maintain a suitable body temperature, making them susceptible to climate change.

extinctions observed from 1975 to 2009 in Mexico and four other continents. But why are lizards in such danger of extinction?

The answer lies in the fact that they are *ectothermic*. As such, they have strict thermal tolerance ranges that permit normal activity. This range is termed *physiologically active body temperature* (T_b). The rate of change measured and projected for air temperature will have a strong impact on lizard body temperature and thus on lizard ecology. The extinctions noted for lizard populations in Mexico (in the genus *Sceloporus*) were linked to the degree of warming in spring. Lizards are reproducing then, but higher temperatures prevent the lizards from foraging as much as they should. Instead, they must seek shelter from the heat, a factor that ultimately compromises their evolutionary fitness. The authors of the study estimated the number of hours per day that temperatures exceeded what lizards could tolerate for high activity levels. Their conclusion was that lizards "cannot evolve rapidly enough to track current climate change because of constraints arising from the genetic architecture of thermal preference" (p. 895). In other words, climate may be changing too rapidly for the lizards to adapt to it (Figure 15-6).

Many tropical ecologists have considered the potential effects of climate change on tropical forests, particularly on carbon dynamics. Tropical forests are huge ecosystems with major global impact on carbon cycling. It is not clear whether tropical forests, particularly tropical moist forests, will ultimately act as carbon sinks or

$29 \leq T_b < 32°C$

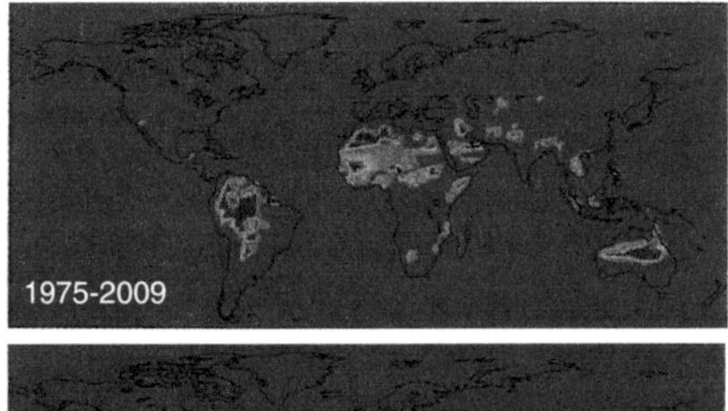

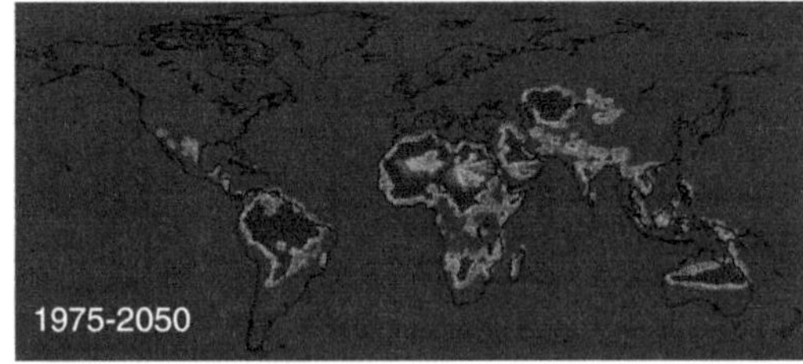

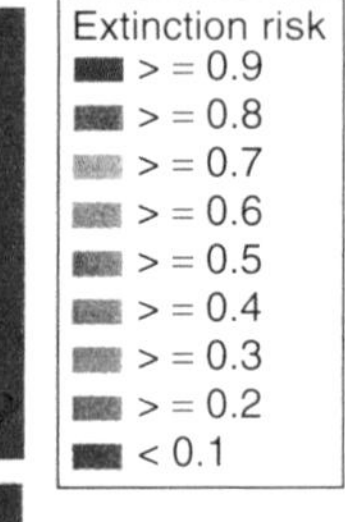

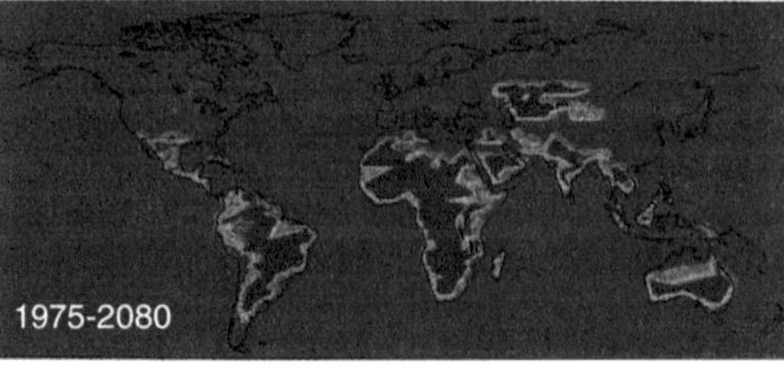

FIGURE 15-6
These three maps document extinction risk to lizards over three ranges of time, for those lizards with a T_b less than 32°C and greater than 29°C. Note how the risk expands over time.

PLATE 15-13
Carbon source or sink? This lush forest in Trinidad could be either, depending on environmental conditions.

carbon sources (see Chapter 9) (Plate 15-13). In a comprehensive review, Deborah A. Clark (2004) urged that a vigorous research program should be pursued to focus more sharply on the complexities of carbon flux in tropical forests. Clark points out that "no one has actually measured the C content of even one hectare of tropical forest" (p. 74). She identifies an unsettling correlation of sorts between carbon flux (input compared with output) and years of excessive heat and drought, both of which are forecast to increase with climate change:

> In summary, therefore, analyses of atmospheric data from the past two decades have indicated that the terrestrial tropics have strongly influenced the temporal variation in atmospheric CO_2 accumulation, and that in years of higher tropical temperatures (generally associated with rainfall minima), the C balance of the land tropics has shifted in the direction of increased emissions. It is important to remember that these tropical emissions are the sum of three fluxes: (1) C uptake through photosynthesis; (2) C emission by respiration of both plants and heterotrophic organisms (principally microbes); and (3) the net emissions from tropical land use change. (pp. 77–78)

While many studies (review Chapter 9) suggest that tropical forests are capable of acting as carbon sinks (sequestering carbon through net primary productivity), it is clear that when major events such as an El Niño occur (associated with higher than usual temperatures as well as drought), many trees perish, and subsequent decomposition liberates CO_2 to the atmosphere, making the forest a carbon source to the atmosphere, adding to greenhouse gas accumulation. Climate change is expected to exacerbate such occurrences, and models of climate change predict much lower rainfall in parts of the Amazon.

An unsettling scenario emerges that has been suggested by Nepstad et al. (2008). The Amazon Basin, you will recall, to a large degree (and because of its vastness) makes its own weather. The transpiration of water within the basin and its subsequent condensation into rainfall is a major factor in the basin's hydrodynamics and ultimately its productivity. It is estimated that at least 70% of the present forest cover would need to remain to ensure that this internal hydrologic cycling continues (Silva Dias et al. 2002). With continuing logging and land clearance throughout Amazonia, forest degradation will continue. This trend toward deforestation will be forced by increasing global demand for meat, lumber, and biofuels. (Soybeans are used in biofuels.) More greenhouse gases will be emitted by human activities associated with forest clearance, and in particular, accidental fires could greatly increase, adding that much more greenhouse gas to the atmosphere and acting to reduce rainfall even further. The combination of forest fires, drought (expected from global climate change), and logging will combine to establish a positive feedback such that Amazonian forests may reach a *tipping point* beyond which forests cannot recover. At that point, large areas such as southeastern Brazil may convert to savanna-type ecosystems, replacing what was formerly forest. It has been estimated that the tipping point will occur with an added 4°C, within the range of some models of global warming (Carlos Nobre, interviewed in *Nature* 452, p. 137, March 2008). It is not yet clear just how Amazonian forests are going to react to short-term drought (Tollefson 2010), but longer droughts and more frequent storms will likely reduce the carbon storage capacity of the forest (Plate 15-13).

Emerging Pathogens

The global decline of amphibian species, much of it traced to the fungal pathogen *Batrachochytrium dendrobatidis* (see Chapter 12), is not the only incidence of an emergent pathogen. Most people are familiar with AIDS (short for acquired immune deficiency syndrome). The disease first sporadically appeared in humans in the late 1950s but did not reach epidemic proportions until the 1980s. AIDS is caused by the HIV virus, and that virus has been linked to African simian (monkey and ape) populations because of its structural similarity to simian viruses. It is now established that HIV-1 virus is most similar to a virus from West-Central African chimpanzees (*Pan troglodytes troglodytes*). What apparently happened is that the virus "jumped" across species lines from chimpanzee to human perhaps as long ago as a century (Sharp et al. 2001; Keele et al. 2006; Takebe et al. 2008; Weiss and Heeney 2009). What had been a simian virus (SIV) apparently mutated to a human virus (HIV). The prevalent hypothesis as to how such an event may have occurred has focused on bushmeat (see the section "Bushmeat" later in this chapter). The hypothesis is that bushmeat hunters butchering chimpanzees in the forest were exposed through cuts in the skin to SIV. The virus subsequently adapted to humans, and the result became the AIDS pandemic. Other viruses such as Ebola (EBOV), a cause of severe hemorrhagic fever, have apparent origins in animals (bats are strongly suspected in the case of Ebola) from various African countries such as Gabon, the Democratic Republic of the Congo, and Sudan. Ebola (named for the Ebola River Valley in the Democratic Republic of the Congo) represents another animal virus that has mutated to infect humans, likely because of close contact with wild animals (Garrett 1995).

HIV and Ebola are clear examples of the potential for animal viruses to mutate and infect humans. They have been used to illustrate the risk imposed on humans when complex forests are degraded and ecosystems simplified, thus uncoupling the trophic dynamics that normally keep such viruses from being "released" (Garrett 1995).

The apparent tendency of viruses to adapt by jumping across species lines is both initially surprising and an obvious cause for concern. But humans who have been in continuous close contact with animals have acquired animal viruses in the past, and in fact, Jared Diamond (1997) has argued that human-adapted resistance to various viruses of animal origin, acquired during the domestication of animals, was a major reason why European colonialists succeeded so well and why other peoples were so susceptible to European-borne diseases. But adaptation to virulent viruses occurs by natural selection and requires many thousands of human deaths—a sobering thought.

Ape populations in Africa have experienced severe declines due both to hunting and to viruses such as Ebola. In western equatorial Africa, apes have experienced what has been called a *catastrophic decline*, half of it caused by hunting and half by Ebola (Walsh et al. 2003). What has been termed *massive die-offs* of gorillas (*Gorilla gorilla*) due to Ebola has been documented in the Democratic Republic of the Congo (Bermejo 2006), which has made the western gorilla population critically endangered, with about 90% of the population killed by Ebola (Vogel 2007).

Chimpanzees (Plate 15-14) were severely reduced in Tai National Park in the Ivory Coast by an outbreak of anthrax (*Bacillus anthracis*). The chimpanzees may have been exposed to anthrax from consuming an animal (such as an infected ungulate) or, more likely, ingestion of spores from contaminated water (Leendertz et al. 2004). Anthrax occurs in hoofed animals such as domestic cattle, so again, the proximity of human activity to wild animals seems to have put the wild animals at risk.

Chimpanzees also face a challenge with their own emerging virus, SIVcpz, which causes simian immuno-

PLATE 15-14
CHIMPANZEE (*PAN TROGLODYTES*)

deficiency disease. In Gombe Natonal Park in Tanzania, chimps that are positive for SIVcpz are experiencing much higher death rates than noninfected chimps (Keele et al. 2009). There are some 40 known simian immunodeficiency viruses (including two that crossed the species barrier to humans and became HIV-1 and VIV-2). Normally, chimpanzees do not become ill from any of the SIVs, indicating that they have essentially adapted to them and are resistant to them. In a nine-year study of 94 members of two chimpanzee troops, SIVcpz (the precursor virus of HIV-1) caused a 10- to 16-fold greater risk of death to infected chimps in Gombe National Park. Infected females were less likely to give birth, and infant mortality increased among infant chimps with infected mothers. This result was a surprise because up to the time of this study, there was no evidence that chimps infected with any of the SIV strains were susceptible to fatal illness caused by the virus. It is possible that SIVcpz is a relatively recently evolved virus, unlike virtually all of the other SIV strains. The Gombe chimps all belong to the same subspecies, *Pan troglodytes schweinfurthii*. The possibility exists that the Gombe chimps may have been exposed to a strain of SIV characteristic of another subspecies, *P. t. troglodytes*. It appears that ape viruses are evolving new strains and add yet another threat to great ape populations (Weiss and Heeney 2009).

Mature tropical forests with complex trophic dynamics are the best protection against the emergence of virulent pathogens. Mosquitoes in the genus *Anopheles* are vectors for malaria. Mosquito larvae from *Anopheles gambiae*, which is a vector for some of the most serious forms of malaria in Africa, survive best in deforested areas (Tuno et al. 2005). In Amazonia, there is risk from spread of malaria, dengue, leishmaniasis, and various arboviruses. Approximately 400,000 to 600,000 cases of malaria are diagnosed annually in Amazonia, and malaria is estimated to kill 1.2 million persons annually (World Health Organization 2005, cited in Foley et al. 2007). The risk of malaria has been documented to have significantly increased in Amazonia during the 1980s, an event that has been linked with changes in forest cover. A study in Peru supported a direct relationship between deforestation and the increased biting rates of the mosquito *Anopheles darlingi*, the principal vector of malaria in that region (Foley et al. 2007). The deforested area experienced an estimated 300-fold increase in the risk of malaria infection when compared with uncut forest (and controlling for human population density). There is likely to be more standing water after precipitation in deforested areas and that would stimulate mosquito population growth. Foley et al. (2007) conclude:

> In this framework, deforestation is recognized to have important benefits for society, as it increases economic opportunities and the availability of many ecosystem goods, at least in the short term. However, the loss of rainforests may also degrade many critical ecosystem services, such as carbon storage in forests and soils, regulation of water balance and river flow, modulation of atmospheric circulation and regional climate, and the amelioration of infectious diseases. (p. 30)

It would be naïve to say that tropical diseases would somehow go away if only tropical forests were left intact. They will not. But what ecologists are learning is that it is equally naïve to ignore deforestation and forest degradation as factors contributing to the potential proliferation and spread of many tropical pathogens. Deforestation exposes humans to diseases that most have not previously come in contact with. Human populations increase following deforestation and as such promote the spread of pathogens. Reduction of reservoir hosts following deforestation may result in more pathogenic species infecting humans.

Invasive Species

The tropical island of Guam, part of the Marianas Islands located just north of New Guinea, once contained a rich diversity of native bird species. That is no longer the case, and it is all due to a single invasive species, the brown tree snake (*Boiga irregularis*). This reptile, which can attain a length of over 2.5 meters (nearly 10 feet), was accidentally introduced to Guam sometime in the 1950s, and it soon became a serious ecological problem as it rapidly spread throughout the island (Plate 15-15). Experiencing what is termed *ecological release*, the snake lacked competitors or predators. No pathogens were there to attack it. The climate was ideal for it. Guam is small, only about 48 kilometers (30 miles) long and no more than 24 kilometers (15 miles) wide. The brown tree snake prospered on Guam, and soon it was consuming virtually every native bird species, resulting in the extinction of 10 native forest bird species and continuing reduction of 2 others. Once they disappear, there will be no native forest bird species on Guam. Some mammal species were also reduced as the brown tree snake population continued to grow beyond any normal population of snakes (estimated at an amazing 1,100 per square kilometer, or about 4,000 per square mile). The loss of the native vertebrate fauna of Guam is well-documented and irreversible. The snake continues to spread by accidentally getting into aircraft via wheel wells. It has been found on other Pacific Islands and in Australia.

PLATE 15-15
BROWN TREE SNAKE

Hawaii is constantly alert to accidental importation of brown tree snakes.

Invasive species are increasingly common throughout the world, including the tropics. The Nile perch (*Lates niloticus*) is a game fish over much of its range in Africa, some reaching lengths of 2 meters and weights of 200 kilograms (about 440 pounds). When Nile perch were purposefully introduced into Lake Victoria (bordering Tanzania, Kenya, and Uganda) in the 1950s, they posed immediate and continuing problems for the diverse cichlid fish community (see Chapter 2). They likely caused the extinction and, in other cases, severe reduction of many cichlid species. Today, there is sufficient fishing pressure on Nile perch that their depredations of native species are fewer, and some of the cichlid species appear to be rebounding. But in both the case of the brown tree snake, introduced accidentally, and the case of the Nile perch, introduced on purpose, an ecological meltdown in diversity has been the result.

It is difficult to say how species-rich tropical rain forests will be affected by invasive species, but the high biodiversity of these ecosystems is no guarantee against invasive species immigration. *Invasive species* are nonnative species that become highly problematic because of their unchecked proliferation. Nonnative species are referred to as *exotic species*, species that are feral and reproducing. In a survey of the plant growth forms at La Selva Biological Station in Costa Rica, Hartshorn and Hammel (1994) documented that 34% of herbaceous species were exotic species, a total of 197 exotic species. They found 23 exotic shrub species (8% of the total shrub species) to be exotic. However, among trees, lianas, and epiphytes, there were no exotic species. An exotic species can become invasive if it undergoes ecological release. Many ecologists are studying invasive species in ecosystems throughout the world, but there is still relatively little focus on lowland tropical forests and the potential for invasive species to cause reductions in biodiversity.

Bushmeat

Bushmeat is a term applied to wild animals killed for food. It is obvious that humans living in tropical forests hunt animals and always have hunted animals. Animals provide critical protein. The animals hunted tend to be large birds (Plate 15-16) and many mammal species. Fishing is also extensively practiced to obtain protein but is usually not included in the category of bushmeat. Viewed over the course of history, human populations in tropical forest have posed little long-term threat to wildlife populations. (There are exceptions, particularly on islands. For example, when the Maori colonized New Zealand, they annihilated all species of large flightless birds called *moas* [Dinornithidae] within a century [Diamond 2000].) When game is depleted from a local area, the humans relocate to another area, richer in game (see Chapter 13). There is a clear and well-documented relationship between hunting pressure and wildlife abundance (Redford and Robinson 1991).

PLATE 15-16
This crested guan (*Penelope purpurascens*) is commonly hunted as bushmeat in Central America. It is an important seed disperser of some tree species.

PLATE 15-17
Animals such as this baboon (*Papio anubis*) are considered bushmeat. Primates are routinely hunted and consumed in all tropical countries where they occur. This baboon was likely safe, residing in a carefully patrolled national park in Tanzania.

But this historical pattern has radically altered to become what is termed the *bushmeat* or *wild meat crisis* (Wright et al. 2007a). The degradation of forest and penetration by roads has made access possible to hunters who do not reside in the forest itself. Aboriginal subsistence hunting continues to occur, but in recent years the killing of native animals in many areas has been greatly exacerbated by professional hunters using technologies such as powerful guns, battery-powered lights, wire snares, and motor vehicles and boats. In many areas, primates are particularly targeted (Plate 15-17). Bushmeat has proven lucrative because of demand from many countries for exotic forms of food and for other uses of animals related to cultures around the world. Indeed, bushmeat has been recognized to be a multibillion-dollar industry (Bushmeat Task Force 2003; Brashares et al. 2004). In a story that appeared on July 18, 2010, Associated Press reported that an estimated 5 tons of bushmeat passes through the main Paris airport on a weekly basis. It further reported increasing concerns among health experts that diseases such as Ebola and monkeypox might enter with the bushmeat. France is not the only European country to offer exotic bushmeat at high-tier restaurants. The practice is increasingly common throughout Europe and parts of Asia. The result, to take one example, is that it is estimated that 1 million tons of forest animals are killed annually for bushmeat in west and central Africa alone, equivalent to a daily quarter-pound hamburger for each of the region's 30 million people (Bushmeat Task Force 2003).

A study from Ghana using 30 years' worth of data documents the impact of local bushmeat hunting (Brashares et al. 2004). Normally, the people in the region studied rely most heavily on marine and freshwater fish as a protein source. But in years when fish supply declined (due to overharvesting), they resorted to bushmeat, including hunting in six nature reserves. The study linked fish supply with hunting pressure and showed major declines in 41 wildlife species due to increased hunting when fishing was poor. The decline in wildlife was quite dramatic, a decrease of 76% over the years of the study (Figure 15-7), including some local extinctions. It also showed how deeply local people relied on bushmeat as a dietary staple and how much of an impact hunting could make on local wildlife populations. The researchers called for greater emphasis on building up regional agriculture and livestock to reduce the impact of overfishing and its resultant pressure on forest wildlife. Of course, if that were to occur, there would have to be additional deforestation to accommodate the cattle herds.

The study just discussed focused on how local populations have the capacity to significantly reduce local wildlife. It has been estimated, for example, that subsistence harvest in the Brazilian Amazon annually takes between 67,000 and 164,000 metric tons of wild meat, with most of the animals weighing more than 1 kilogram (2.2 pounds) (Robinson and Redford 1991). Local people have access to modern hunting techniques and therefore they are able to quickly deplete local populations of wildlife. But activities such as logging make such depletions even worse. Robinson et al. (1999) pointed out the dearth of regulations and enforcement of hunting laws, particularly in logging camps. They cited one case in

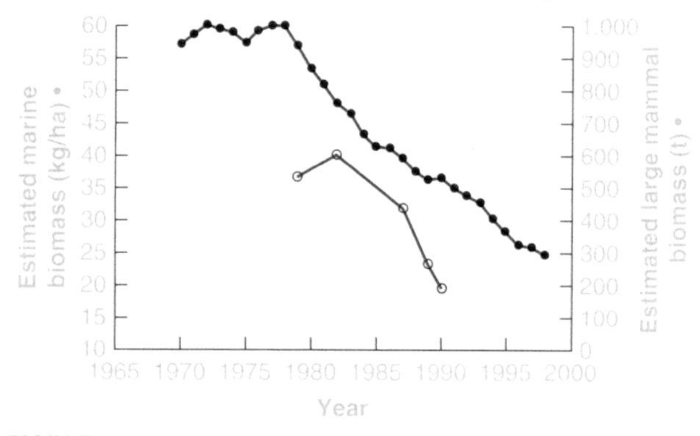

FIGURE 15-7
Estimates of marine fish biomass (open circles) and large mammal biomass (dark circles) in Ghana.

PLATE 15-18
These skins were for sale. There are two ocelots, one red howler monkey, one puma, and one jaguar. They were killed by bushmeat hunters in Venezuela.

which a single logging camp of 648 people in the Democratic Republic of the Congo took 8,251 animals in a single year, equating to 124 tons of wild meat. Some of this kill was not for food but for products such as ivory, horns, and skins (Plate 15-18), a characteristic that makes bushmeat hunting so lucrative for some.

Bushmeat hunting has also had major impacts on tropical Asian animals (Corlett 2007). Though hunting has occurred throughout the region for 40,000 years, never has the impact approximated what it has become today. Most species of large mammals have declined over the past 50 years. Subsistence hunters traditionally rely on harvesting species such as deer, pigs, monkeys, and various rodents. Market hunters have added pressure on these species plus others. There is what is described as a "colossal" regional trade in animals for such uses, not only as food but also as medicine, raw materials, and even as pets. Many of the affected species are understory browsers and/or seed dispersers. Their reduction and potential loss threatens the overall function of the forest.

There are two significant problems associated with the growth in bushmeat trade. One is the effect that the activity has on depleting animal populations. Hunting intensity is sometimes sufficient to drive local populations to extinction. The other is less obvious—that is the negative effect that the elimination of large animals has on the functioning of the forest, first eloquently described by Kent Redford in a paper with the evocative title "The Empty Forest" (1992). A *meta-analysis* (a large-scale study employing data sets from many separate sites) performed in Amazonia looked at 101 undisturbed forest sites, some of them hunted, some not (Peres and Palacios 2007). The study focused on 30 medium-sized to large mammal, bird, and reptile species. A total of 22 of the 30 species declined significantly with intense hunting. Losses of up to 74.8% were observed in hunted sites compared with sites without hunting. Animals of large body size were most negatively affected by hunting. Those animals that were primarily frugivores were most strongly affected, while those that were seed predators and browsers suffered less impact. Because frugivores were the favored targets by hunters, and because frugivores are primary seed dispersers, this study has strong implications for the overall reproductive health of the forest.

The alteration of mammal and bird abundances by hunting pressure has been shown to potentially affect complex ecological patterns such as the effect animals have on seed shadows (Stoner et al. 2007a), seed dispersal (Wang et al. 2007), and seed predation (Beckman and Muller-Landau 2007; Dirzo and Mendoza 2007). One model, developed by Helene C. Muller-Landau (2007), predicts that reduced seed dispersal will intensify competition among plant kin and ultimately increase vulnerability to natural enemies. Plant species composition is very dependent on the vertebrate community, particularly birds and mammals. In a study conducted at Manu National Park in the Peruvian Amazon, the effect of loss of large and medium-sized primates had a strong effect on various community attributes (Nuñez-Iturri and Howe 2007). Local hunting exterminated large primates (such as spider and woolly monkeys) and reduced medium-sized primates (such as capuchins) by up to 80%, many of which are critical seed dispersers. As a consequence, tree species richness was 55% lower on the hunted sites than on sites where monkeys were not hunted. The density of plant species dispersed by large and medium-sized primates was 60% lower in hunted compared with protected sites. Species richness of plant species that are dispersed abiotically as well as those dispersed by nongame species was higher on the hunted sites, probably because of reduced competition with the game-dispersed tree species whose populations had declined. In a carefully controlled series of experiments done in Veracruz, Mexico, Rudolfo Dirzo and Eduardo Mendoza (2007) documented that in areas subject to strong hunting, the mammalian fauna is altered, the large animals largely eliminated, favoring small animals such as rodents. These small mammals preferentially devoured small-seeded species of plants, providing an advantage for large-seeded species whose seed predators had been eliminated by hunting. Last, work performed in central Panama showed that hunting was sufficient to alter seed banks such that lianas were favored as well as plant species that produced large seeds or seeds dispersed by either small to medium-sized birds, bats, or wind (Wright et al. 2007a and 2007b). Hunting pressure is clearly sufficient to change the plant species composition of a tropical forest.

It is difficult to predict with high precision the long-term effects of hunting on forest ecology, particularly on how the loss of animals will affect plant species richness and community structure. Stoner et al. (2007b) summarize what is currently known:

- Hunting tends to reduce seed dispersal for animal-dispersed plant species with very large seeds.
- Hunting reduces seed predation for those plant species with large seeds.
- Hunting alters species composition for the seedling and sapling layers.

Hunting will likely continue as a problem throughout the tropics. Whatever balance there may have been between indigenous peoples and the animals they hunted is no longer the case, at least not in places where guns have replaced traditional forms of hunting.

What is to be done about the bushmeat problem? Stoner et al. (2007b) offer several suggestions. Some are obvious, and all are worth consideration. They suggest the use of alternative sources of animal protein, income supplementation for local people from sources other than selling bushmeat, community outreach and education, promoting shifting attitudes toward hunting, implementation of community-based wildlife management programs in regulated-use areas, and landscape-scale conservation planning that permits areas with animal harvest to have regular input of animals from nearby areas where hunting is not permitted (a *source–sink model*). As always with good intentions that attempt to alter societal inertia, the "devil is in the details." It is likely that the bushmeat crisis will remain unresolved for the foreseeable future (Plate 15-19).

PLATE 15-19
Bushmeat will remain a major concern for conservation biologists throughout the tropics.

Restoration and Rehabilitation of Tropical Forests: Hope for the Future?

Realistically, what is the outlook for tropical forests over the remainder of the century? No one really knows with any degree of certainty. But there is no shortage of people who are giving serious thought to the issue of conservation in the tropics and how to make goals such as biodiversity preservation and carbon sequestration (as well as other ecological services performed by forests) compatible with basic human economics.

A research team constructed a model allowing them to compare various scenarios from the present to 2050 for the Amazon Basin (Soares-Filho et al.

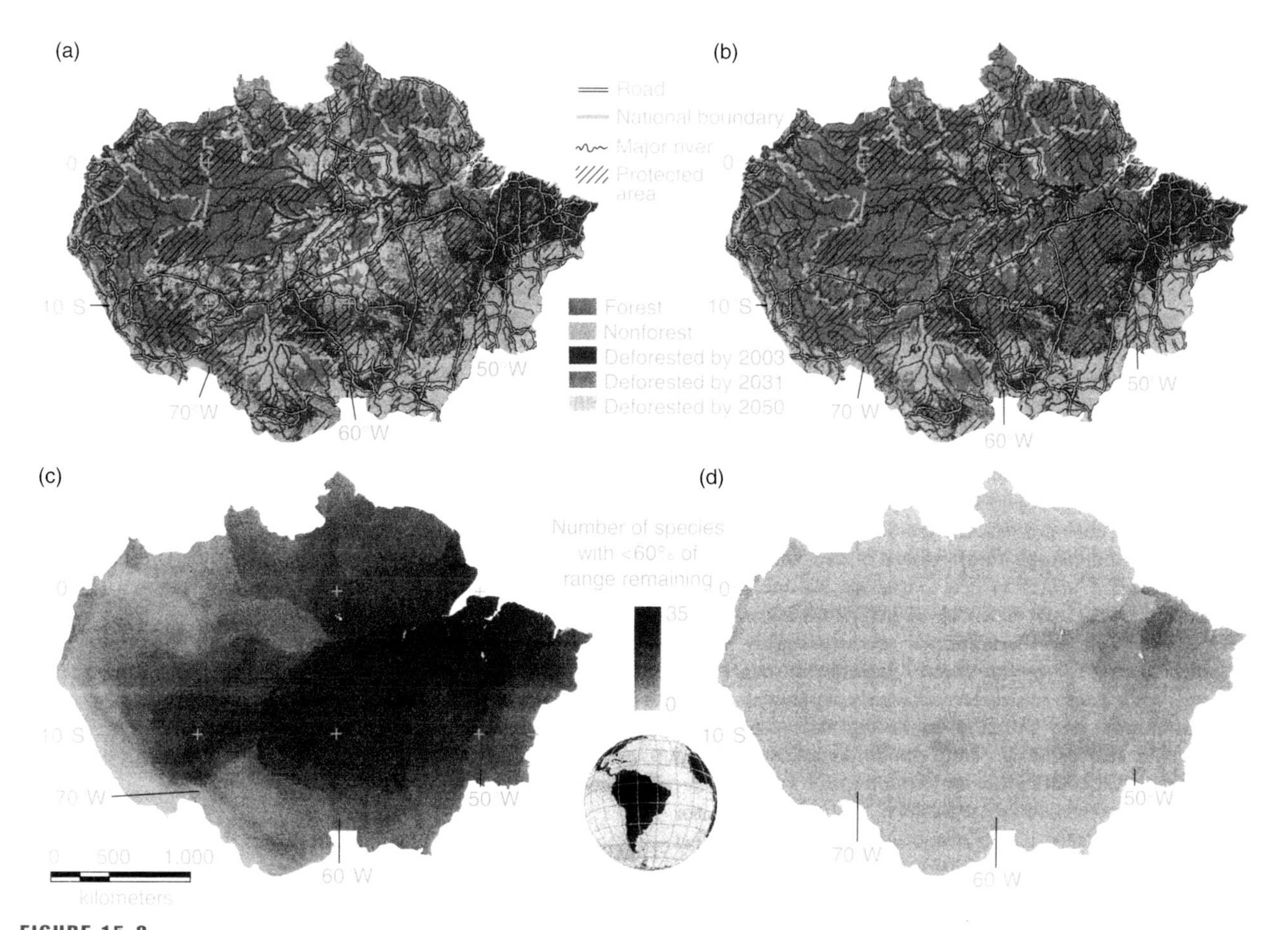

FIGURE 15-8

The two maps at the top represent the BAU scenario (map a) and the governance scenario (map b). The two maps at the bottom are projections for numbers of imperiled mammal species. Map c is the BAU scenario; map d is the governance scenario.

2006). They examined eight scenarios and looked closely at the two extremes of the continuum. One was called the *BAU model*, as in "business as usual." The other they termed the *governance scenario*, which assumes a best-case outlook involving much greater implementation of environmental legislation and more careful and sophisticated use of the land (Figure 15-8). The BAU model predicts that by 2050, there will be a loss of 40% of Amazon forests, mostly due to expansion of agriculture. This represents a current forest area of 5.3 million square kilometers (2,045,800 square miles) that, by 2050, would decline to 3.2 million square kilometers (1,235,200 square miles) (to about 53% of its original area). This would include two-thirds of the forest cover in six major watersheds. An estimated total of 32 ± 8 petagrams (Pg, or 1,000,000,000,000,000 grams) of carbon would be released into the atmosphere. To estimate the impact on biodiversity, the researchers used nonflying mammals as indicator taxa. Under the BAU projection, one-fourth of the 382 mammalian species would lose more than 40% of forest within their ranges. Thirty-five primate species would stand to lose 60% to 100% of their Amazonian range (100%, of course, means local extinctions). Ecosystems most at risk include major watersheds, savanna and closed-canopy forest, and wetland forest, which taken together compose most of Amazonia. Most deforestation would be concentrated in the eastern Amazon.

Under the governance model, the projections suggest far less severe impacts. The model predicts about a two-thirds reduction (compared with the BAU model) in the number of threatened watersheds and mammal species as well as far less carbon emission. The model assumes a planned expansion of protected areas (PAs) from 32% to 41% of the total forest area and that 100% of the forests with PAs are kept intact. Only 50% of the forests outside PAs would be subject to any form of deforestation. The researchers suggest that the governance model would be more likely if developed countries were willing to pay to make *frontier governance* more politically feasible. They urge consideration of sale of carbon credits as well as environmental certification (meaning close government oversight) of beef, soybeans, and timber. They also urge strong action to encourage conservation on private lands.

Questions arise about the degree to which conservation goals, which typically focus on biodiversity preservation and maintenance of intact forests, are compatible with such activities as human subsistence farming. There is a divide between what is perceived as good for people and what is good for nature (Terborgh 2000). If biodiversity and ecosystem services are good for all of Earth's people, why should a small percentage of Earth's people (such as subsistence farmers in the tropics) bear the economic burden of biodiversity preservation? Acronyms have become popular in this discussion: REDD stands for "reducing emissions from deforestation and degradation"; PES stands for "payments for ecosystem services." Both of these were alluded to in the previous study (see the sidebar below). Add to this EEFD, which stands for "ecological-economic farm diversification." These acronyms crop up frequently, as they apply to numerous studies, humanitarian efforts, and programs that look at the difficult problem of reconciling basic human needs with the goal of forest restoration and preservation.

REDD and PES

REDD stands for "reducing emissions from deforestation and degradation." Started in 2008, REDD is a United Nations collaborative program (under the United Nations Framework Convention on Climate Change) that is designed to aid developing countries in an ecologically sound way. Pilot programs are ongoing in Bolivia, Democratic Republic of the Congo, Indonesia, Panama, Papua New Guinea, Paraguay, United Republic of Tanzania, Vietnam, and Zambia. The REDD policy board is composed of representatives of other tropical countries as well as those named here. The main objective of REDD is to develop economic policies that add high financial value for carbon that remains stored in forests. Estimates of carbon emission reduction under REDD range from 13 to as much as 50 billion tons of carbon by 2100 (Gullison et al. 2007). Objectives include basic conservation of biodiversity, sustainable management of forests, and methods that increase carbon stocks. The developed countries of the world are expected to fund the various REDD initiatives, because it is in everyone's self-interest to reduce carbon emissions. The UN predicts that as much as $30 billion will flow annually from developed to developing countries to support REDD initiatives (see www.un-redd.org). In order for REDD to succeed, developing nations must be willing to expend resources to build the economic stability of developing countries in ways that have less ecological and climatic impact than standard practices leading to deforestation and degradation. This aspect of REDD is politically difficult and is beyond the scope of this book.

A study from Amazonia suggests that REDD policies may have unexpected ecological consequences, specifically, increased numbers of fires (Aragão and Shimabukuro 2010). During drought years, natural fires could potentially release as much carbon from forested areas protected under REDD as would be the case if deforestation had occurred. In their analysis, Aragão and Shimabukuro suggest three pathways that fire frequency may take in the Brazilian Amazon. One is that fire incidence may decrease with reduced rates of deforestation. This would follow if human activities were reduced such that the probability of human-set fires diminished. Another pathway is that fires may increase through slash-and-burn practices in secondary forests growing in already deforested areas. A third pathway is that fire may decrease by a shift from extensive unmanaged forest to managed land-use practices. Aragão and Shimabukuro examined two land-use gradients and found that fire occurrence had increased in 59% of the area where there was reduced deforestation. In areas with land management, they estimated that fire frequency was reduced by about 69%. Aragão and Shimabukuro urge that REDD policies adopt land-management practices that reduce probability of fire. These would include agricultural practices that do not result in burning for land clearance.

PES, which stands for "payments for ecosystem services," is similar to REDD in that it looks to developed nations to support sound ecological policies in developing nations, but broader in scope in that it is not focused primarily on reduction of carbon emissions but also includes a broad range of ecosystem services. The objective of PES, which is supported by conservation groups such as the World Wildlife Fund (www.worldwildlife.org), is to assess ecosystem services in selected areas, to place economic value on such services, and to formulate a policy, an economic market, or a subsidy to reward people in the area for conserving ecosystem services. The PES approach is challenging in that it must measure the value of ecosystem services, something that is often fairly difficult. It then must obtain local buy-in, sometimes in the most literal sense. In addition, it must in some cases identify and obtain sources of subsidies. Nonetheless, there are numerous PES programs in various places in the tropical world and elsewhere. For example, the World Wildlife Fund is working on the island of Lombok, in Indonesia, to conserve forests on Mount Rinjani. These forests have high economic value as watersheds as well as ecotourist value. In this example, it is suggested that 43,000 households in the area would actually agree to pay $0.60 per month (in U.S. dollars) as a "special charge" that would go to conserving the watershed forests.

The EEFD approach was investigated for local land use near Podocarpus National Park in southern Ecuador (Knoke et al. 2009). Over 500 bird species and 40 mammal species have been documented to occur within the park and surrounding area. The park is located in a relatively high montane zone where there is extensive farming. This activity has led to overused, degraded, and abandoned land parcels. The high elevation of the park results in a slow process of secondary succession, exacerbating the problem of land recovery. Farmers near the park normally use pastures for only a few years and then abandon them, promoting degradation. The EEFD model examined a small 30-hectare (74-acre) farm that included 10 hectares (24.7 acres) of previously degraded wastelands. The EEFD model focused on active reforestation of degraded land using a native species, Andean alder (*Alnus acuminata*). It is worth noting that many reforestation projects do not rely on native species but instead use various eucalyptus and pine species that, because they are exotic species, have no value with regard to biodiversity. Andean alder grows quickly and adds much nitrogen to the soil. Commercial harvesting of Andean alder may commence as early as a decade past planting, but the model predicts peaks in revenue in year 20, 30, and 40, when various crops of Andean alder are harvested.

The model requires diversification in land use, combining limited logging with agriculture and restoration. A rotational and sustainable system is established that does not require encroachment onto protected areas. The point was made that market values for agriculture and logging are uncoupled so that, for example, if the price of timber drops in a given year, the economic loss may be buffered by good crop production. Diversification hedges against market fluctuations. Reforestation acts as an investment against price fluctuations in agricultural products. The model has a major limitation in that natural forest, as such, has no tangible value to local farmers other than as potential "insurance" should they need to clear it. So if the reclamation of abandoned pasture fails, deforestation could result. This model suggests that economic advantages of good pasture management should normally buffer the possible need for deforestation.

In a comprehensive review, Lamb et al. (2005) attempt to identify ways to restore degraded tropical forest landscapes that also ensure that conservation goals are attained. Lamb et al. examined various ways to link financial and livelihood benefits to enhancement of biodiversity (Figure 15-9). Some, like traditional monoculture of exotic species, pay high benefits but result in severe biodiversity loss. Others, such as intensive restoration, enhance biodiversity but are economically weak. Responses to forest degradation typically fall into three categories: (1) to simply expand the number of PAs in the hope of protecting biodiversity; (2) to improve agricultural techniques and avoid or minimize degradation; and (3) to employ some form of active reforestation and restoration. These responses are not mutually exclusive. Degradation following agriculture, as noted in the previous example, is a major challenge. In some areas in the tropics, and in particular in Amazonia, land that once supported some form of forest has succeeded to grassland, sometimes because of invasion by exotic grass species (Plate 15-20).

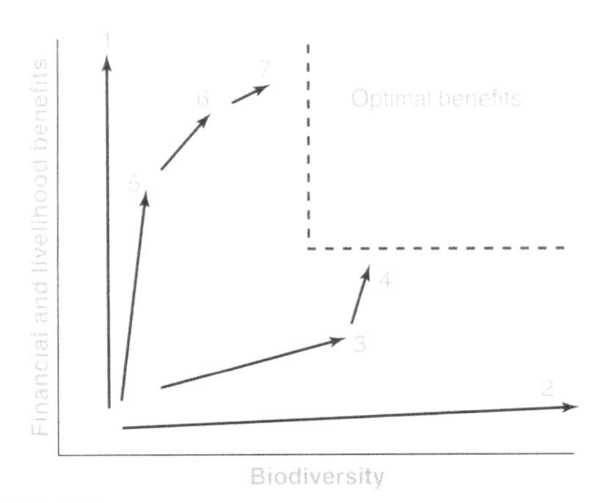

FIGURE 15-9
It is difficult to develop restoration methods at a particular site that optimize financial and livelihood benefits as well as generate improvements in biodiversity (top-right corner). Traditional monoculture plantations of exotic species (arrow 1) mostly generate just financial benefits, whereas restoration using methods that maximize diversity (arrow 2) and enhance biodiversity yields few direct financial benefits to landowners, at least in the short term. Protecting forest regrowth (arrow 3) generates improvements in both biodiversity and livelihoods, although the magnitude of the benefits depends on the population density of commercially or socially important species; these can be increased by enrichment of secondary forest with commercially attractive species (arrow 4). Restoration in landscapes where poverty is common necessitates attempting both objectives simultaneously. But, in many situations, it may be necessary to give initial priority to forms of reforestation that improve financial benefits, such as woodlots (arrow 5). In subsequent rotations, this balance might change over time (moving to arrow 6 and later to arrow 7 by using a greater variety of species). There may be greater scope for achieving multiple objectives by using several of these options at different locations within the landscape mosaic.

One avenue toward restoration is termed *restoration plantings*. These attempt to mimic ecological succession. Numbers of rapidly growing sun-adapted species of plants are introduced to hopefully shade out invasive weeds and grasses and reduce potential fire hazard. A variation of this approach is to instead introduce plant species more typical of midsuccession, essentially bypassing early succession. Both of these techniques require some proximity to established forest in order to encourage the fauna required, in particular species such as seed dispersers. One major limitation of this method is that it is usually costly. For an overview of restoration with examples from Costa Rica, see Zahawi and Holl (2010).

Another approach is *plantation establishment*, which is typically a monoculture but need not be. In a worst case, as mentioned earlier, plantations do not even consist of native plant species. Even when they do, the structural and species diversity of the habitat pales in comparison with normal forest. Plantations have strong ecological limitations. They certainly are potentially profitable, but they are no solution to the biodiversity problem—quite the contrary. The closest to a harmonious relationship between a plantation approach and promotion of species diversity is seen in the *Inga* overstory of shade coffee plantations (see Chapter 14). The use of many species rather than monocultures in plantations offers some hope toward encouraging biodiversity.

Last, just how likely is ecological restoration to accomplish the two basic objectives of conservation science: the recovery of both biodiversity and

PLATE 15-20
Exotic grass species have become established on what was once dry forest in Venezuela.

ecosystem services? A meta-analysis performed on 89 restoration assessments (including both tropical aquatic and tropical terrestrial ecosystems as well as temperate zone ecosystems) suggests that restoration is but a qualified success (Benayas et al. 2009). The overall conclusion of the meta-analysis was that in comparison with reference ecosystems that were not degraded, ecological restoration measures such as cessation of degrading activity, extirpation of damaging species, planting of trees, and reintroduction of select herbivores and carnivores did succeed in restoring biodiversity and ecosystem services, but only to a degree. Restoration increased provision of biodiversity by 44% but increased ecosystem services by only 25%. This, of course, indicates a poor correlation between biodiversity and ecosystem services, suggesting that restoring biodiversity does not automatically restore ecosystem services. The authors of the study point out that the relationship between biodiversity and ecosystem services remains poorly understood. This conclusion certainly points out that much more research is needed.

Reforestation and ecological restoration is a very complex topic, and in addition to sound ecological approaches, it requires economic, legal, and even political incentives. It is beyond the scope of this text to explore these potential approaches, but the interested reader should begin by reading Costanza et al. (1997), Pearce et al. (2003), Koellner and Schmitz (2006), and Zahawi and Holl (2010).

In their summary of how to successfully restore tropical forests, Lamb et al. (2005) state that conserving and actively managing large areas of extant secondary forest as well as dealing directly with the pervasive problem of human poverty are promising avenues toward the goal of reducing and reversing forest loss. Which brings us to a very active subject of debate.

The Wright/Laurance Debate

Conservation biologists are currently engaged in a productive debate about the future of tropical forests and biodiversity. This debate grew out of arguments presented initially by S. J. Wright and H. C. Muller-Landau (2006a and 2006b). They were answered by a number of ecologists (Brook et al. 2006; Gardner et al. 2006), but most particularly by W. F. Laurance (2006). The debate was the subject of a workshop at the Smithsonian Tropical Research Institute in Panama in 2008 and a symposium held at the Smithsonian National Museum of Natural History in 2009.

The debate has sometimes been characterized as a *cup half-full versus cup half-empty* argument. In short, Wright and Muller-Landau assert that demographic trends in human populations between now and mid-century will result in a net reduction in birth rate in most tropical regions and a large-scale movement from rural into urban areas. This projection, based on figures and forecasts from the United Nations, will alleviate the trend toward increasing deforestation and forest degradation. Secondary forests will regenerate, and there will eventually be net increase in rate of regrowth of forests (similar to what has occurred in temperate nations when agriculture is reduced and urbanization increased). Because the demographic trends cited earlier are already occurring, there will be little change in forest cover between now and 2030, and possibly after that time a net annual increase in forest cover throughout most of the tropics. The tropics will consist largely of secondary forests. While secondary forests are less species-rich than primary forests, they will nonetheless serve well for most species. Extinction rates, based on the traditional species/area curve (see Chapter 5), forecast (between now and 2030) a maximum extinction rate of 21% to 24% in Asia, 16% to 35% in Africa, and much less (no estimate is given) in the Neotropics. (Recall that the species/area curve is logarithmic. This means that a 90% reduction in area would predict a 50% reduction in species.) Most forest species would be generalists, but a few specialist species with rigid ecological requirements would be likely to survive as well. This is the *cup half-full* view (Figure 15-10).

Wright and Muller-Landau acknowledge that their view contains numerous uncertain assumptions. Figures on predicted population trends, forest cover, and rate of industrial growth are all projections with no guarantee of certainty or even high degree of accuracy. Beyond that is the major assumption that there is a close linkage between deforestation and rural populations that will be uncoupled by rural peoples' moving to urban areas. To quote Wright and Muller-Landau (2006a):

> The specifics of our projections will almost certainly prove wrong, but we believe the qualitative predictions will prove correct.

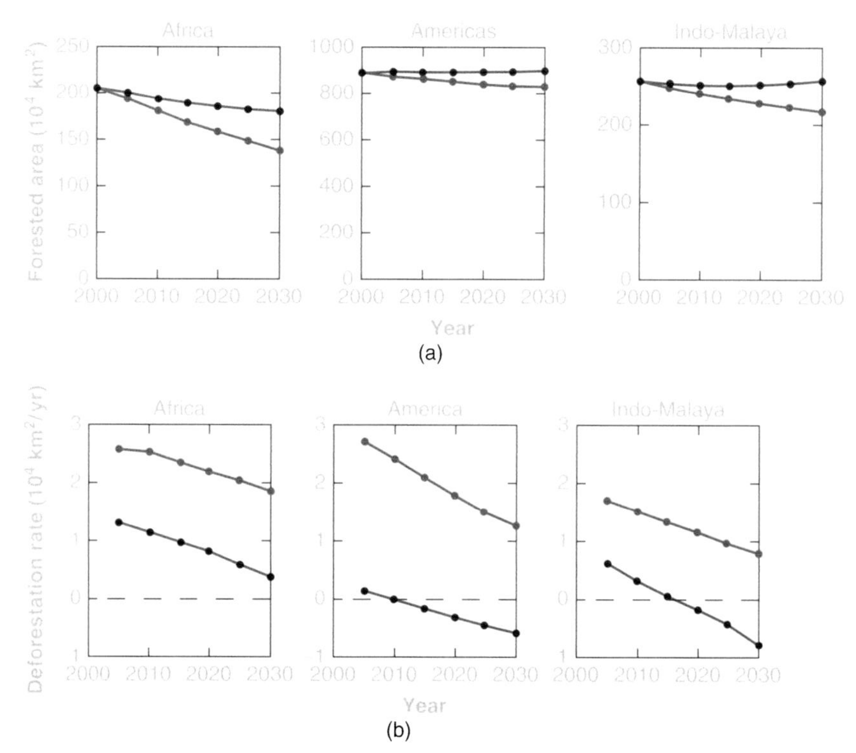

FIGURE 15-10
The relatively optimistic projections expressed by Wright and Muller-Landau.

> Specifically, we expect that in the next 25 yr the rate of net tropical deforestation will slow on all continents. Further, we predict a switch to a net increase in forest area in Latin America and Asia if not within 25 then at least within 50 yr, and in Africa within 100 yr. The fundamental causes of such changes will be stabilizing human populations and thus stabilizing demand for agricultural commodities, increased nonagricultural economic opportunities in developing countries, and increased agricultural land use efficiency due to continuing technological improvements and their more widespread use. (p. 296)

Is the cup really half-full? The response by W. F. Laurance along with others (Brook et al. 2006; Gardner et al. 2006) is to question many of the assumptions underlying the Wright and Muller-Landau position, most particularly the decoupling of rural human population with release of pressure for deforestation. W. F. Laurance argues that timber extraction, large-scale agriculture and cattle ranching, plantation growth, and other activities that have nothing directly to do with rural subsistence farmers will soon dominate much of the tropics. Even if large numbers of humans abandon the forests and move to the city, the industrialists will still invest in forest extraction industries. Therefore, the economic pressure that results in deforestation will not attenuate. This is a strong criticism of the Wright and Muller-Landau position.

Another contentious issue revolves around extinction rates and loss of biodiversity. Critics of Wright and Muller-Landau take strong issue with their extinction forecasts. W. F. Laurance as well as others have pointed out that hotspots (see Chapter 14) are not considered in any of the Wright and Muller-Landau projections. Also left out are considerations of the effects of severe fragmentation (see Chapter 14). The reliance on the species/area curve as an accurate predictor of extinction potential has also been questioned. In short, critics of Wright and Muller-Landau contend that extinction rates will be considerably more severe than are forecast by Wright and Muller-Landau. They assert that a significant data vacuum exists, arguing that there is simply not sufficient information available on the distribution, magnitude, and sensitivity of biodiverse tropical forest communities to even hazard

TABLE 15-5 Estimated percentage of endemic species committed to extinction (% extinct) in Southeast Asia by 2030 under two scenarios of forest loss. S1 assumes that the annual percentage rate of forest loss (ARFL) remains constant at levels observed between 1990 and 2000 (business as usual), while S2 assumes a linear decline in ARFL to the point of zero net deforestation by 2030 (zero net loss). Also tabulated are the estimated original forest cover, forest cover in the year 2000, and projected percentage forest cover in 2030 relative to original area for S1 and S2.

	FOREST (000 HA)			SI: BUSINESS AS USUAL		S2: ZERO NET LOSS	
LOCALITY	ORIGINAL	2000	ARFL (%)	FOREST (%)	EXTINCT (%)	FOREST (%)	EXTINCT (%)
Indonesia	181,157	91,134	−1.50	32.0	26.0	40.4	21.3
Myanmar	65,755	33,519	−1.50	32.4	26.5	41.0	21.7
Thailand	51,089	17,107	−2.90	13.8	41.1	21.9	33.4
Malaysia	32,691	13,452	−1.40	27.0	29.3	33.6	25.1
Vietnam	32,452	5,015	−0.30	14.1	40.7	14.8	40.0
Philippines	28,416	2,405	−2.10	4.5	55.7	6.2	51.7
Laos	23,057	4,495	−0.50	16.8	47.1	18.1	45.6
Cambodia	17,652	11,562	−0.60	54.7	13.7	60.0	11.7
Brunei	527	267	−0.30	46.3	19.5	48.5	18.4
Singapore	54	0.2	0.00	0.4	67.1	0.4	67.1
SEA Total	432,850	178,956	−1.40	26.5	29.7	33.3	25.3

an educated guess as to how they will fare into the present century and beyond. In particular, the role of secondary and often fragmented forests to serve adequately for species with specialized ecological requirements is not strongly supported by existing studies (review this chapter, and see Chapter 14). In their critique of Wright and Muller-Landau, Brook et al. (2006) assert that momentum has already been established that will lead to what they term *mass extinctions* in the tropics and that the scenario presented by Wright and Muller-Landau will not avoid this outcome. Their analogy for the situation in the tropics is thought-provoking:

> The situation is analogous to a motor vehicle (forest ecosystem) that has suffered from brake failure (constituent species reduced to below their minimum viable population size)—a catastrophic and potentially fatal crash cannot be avoided simply by cutting the engine (reducing the once high rate of deforestation) or shooting out the tires (active conservation interventions) because the inertia produced by its forward momentum (past forest losses) will assuredly propel it onward to collide with the brick wall of extinction. (p. 304)

The extinction rates calculated by Brook et al. (2006) both for a business-as-usual approach and zero net forest loss are shown in Table 15-5. They are considerably higher than those of Wright and Muller-Landau. This is a *cup half-empty* view.

The Wright and Muller-Landau/ W. F. Laurance debate is not a winner-take-all contest. It is difficult not to see strong elements of reality in both positions. Wright and Muller-Landau and W. F. Laurance seem to agree that primary forest will continue to be degraded indefinitely and that secondary forest will prevail in many areas. The extinction debate focuses on just how severe extinction might be in the present century. For a somewhat more detailed account with more quotes from various participants, see Butler (2007).

Birders' Exchange

This text has highlighted quite a few studies involving birds in various tropical regions. Imagine the difficulty of studying birds in the field without such things as binoculars, spotting scopes, tape recorders, cameras, backpacks, and field guides—all the basic tools of the trade for field ornithology. It is rather like trying to study astronomy without a telescope. But that is the problem facing many would-be ecologists and conservation biologists in poor nations.

In 1989, a meeting was convened at Woods Hole, Massachusetts, where ornithologists devoted to studying the ecology of long-distance migrant birds in the Neotropics met to share their research. The Latin Americans that attended made it clear that lack of basic equipment severely limited what they could do in the field. From this meeting, a concept was developed that was soon called *Birders' Exchange*, or BEX. The concept is simple. North Americans, many of them world traveler ecotourists, routinely upgrade their optics, purchasing new and improved binoculars, scopes, and so on. They then donate their used optics to BEX. Couriers, also typically ecotourists, take the optics to Latin America and quite literally give them to a person whom BEX has deemed worthy of receiving them. BEX takes applications from Latin American educators, scientists, and conservation biologists and determines what equipment they might receive. Since its origin in 1989, BEX, now under the auspices of the American Birding Association (www.aba.org/bex), has delivered over 30,000 pieces of optical and other equipment to literally thousands of deserving Latin American conservationists and scientists. BEX also oversaw the translation into Spanish of the author's book *A Neotropical Companion* and has hand-delivered several thousand copies of that book free of charge to Latin Americans interested in tropical ecology (Plate 15-21).

Birders' Exchange is an example of a grassroots nongovernmental effort designed to achieve small but additive goals throughout Latin America. Its success has become obvious. Many papers are now appearing in peer-reviewed scientific journals in which the Latin American authors gratefully acknowledge help by Birders' Exchange.

PLATE 15-21
The author (right) presents a Spanish edition of his book *A Neotropical Companion* to Olger Licuy, an expert ornithologist and local guide in Ecuador.

A Final Word

This book began with a challenge to students to think about what is unique about the tropics. Hopefully, that goal has been achieved. How does uniqueness equate to value? How much does, or should, humanity place value on the tropics? Ultimately, the tropics and all of the varied ecosystems and biota will become what we humans make of them. No amount of natural diversity can stop human intervention. That much is clear. There seems no room for doubt regarding the continuing loss of biodiversity that will occur during the current century, much if not most of it in the tropics. It is merely a question of magnitude of loss. There is no doubt that climate is changing, and the long-term effects of that reality are not yet clear. A helpful and somewhat sobering way to think about the future of biodiversity, not only in the tropics, but on the entire planet, is to compare historic biodiversity with present biodiversity. In doing so, it becomes apparent that anthropogenic factors are and will continue to force biodiversity through a series of filters ranging from local to global (Brashares 2010). Beginning with local and moving to global, these filters are human exploitation; habitat loss and/or change; invasive species; isolation and fragmentation; fire; disease; and climate change. Each extant species must pass through each of these filters in order to remain extant. For some, such as many frog species, that could be a tall order.

But that said, there are literally thousands of governmental and nongovernmental associations diligently and conscientiously working toward global conservation goals, including in the tropics. In the summer of 2010, the First Global Business of Biodiversity Symposium (www.businessofbiodiversity.co.uk) was convened (Bayon and Jenkins 2010), in which businesses looked for ways to internalize ecosystem services such that governments could regulate their use based on actual economic value of their services. Throughout the world, academicians from biology, ecology, conservation science, climatology, geology, economics, political science, anthropology, and other disciplines are focusing their research agendas on the tropics. There is urgency in that biodiversity continues to decline throughout much of the world, particularly in the tropics. The Convention on Biological Diversity that occurred in 2002 set a goal of significantly reducing the rate of biodiversity loss by 2010. A comprehensive study that looked at 31 indicators (such as extinction risk, population trends, and habitat extent) found that biodiversity is still declining at rates comparable to what they were in 2002 (Butchart et al. 2010). In fact, the study points out that pressures on biodiversity such as alien species, overexploitation, and climate change impacts are presently increasing.

The tropics contain most of the world's biological riches in the form of its myriad species. Conservation agendas directed at the tropics properly focus on biodiversity and its relationship to ecosystem services. At the same time, conservation agendas must also embrace another reality. Many millions of people are not permitted the luxury of concern over biodiversity. It is, in fact, admirable that in so many tropical areas, people of little economic means seem to be increasingly knowledgeable of the biodiversity in their nations. Most people in the tropics need to concern themselves with basic daily nutrition, their health, and simply having a place to live (Kaimowitz and Shell 2007). Recognizing this and considering it when structuring plans to implement conservation more widely has been called developing a *pro-poor approach to conserving biodiversity* (Kaimowitz and Shell 2007). A pro-poor approach emphasizes the need for grassroots local efforts, working with local people and developing a plan that is in their interests.

In her efforts to ensure the conservation of chimpanzees not only at Gombe Stream National Park in Tanzania but elsewhere in Africa, Jane Goodall strongly acknowledges the need to address the problems faced by local people, most of them quite poor and with minimal basic resources (Goodall and Pintea 2010). Lack of clean water, health care, and education usually prevent local villagers who reside near chimpanzees from viewing the chimpanzees as other than another form of bushmeat. The Jane Goodall Institute has organized TACARE (pronounced "Take Care") to integrate conservation with development of environmentally sustainable rural development. Such efforts involve establishment of schools, clinics, and greater food security. The essence of Goodall's approach is to realize that conservation of chimpanzees, and by implication other forms of biodiversity, requires serious work in human services that have been long neglected by the developed world.

PLATE 15-22
Ecotourists, such as these people who were part of a workshop taught by the author at the Canopy Tower in Panama, each strive to promote conservation goals and understanding with local people as their travels take them to various tropical countries.

PLATE 15-23
The unique Canopy Tower in Panama, converted from an old radar installation, affords ecotourists extraordinary canopy-level views of numerous rain forest animals. It is a major attraction for ecotourists and promotes strong conservation ethics.

There has never been a time when it has been so convenient to visit tropical destinations and experience the amazing ecosystems found there (Plates 15-22 and 15-23). Visitors to the tropics from wealthy nations, whether ecotourists or students, should strive to act as ambassadors and should listen to and learn from the people they meet. The more this happens, the more the cup might be half-full (Plate 15-24.)

PLATE 15-24
Sunset on the Serengeti.

Appendix

GEOLOGIC TIME CHART

This chart divides Earth's history based on the fossil record. The various eras and periods are separated by major changes in fossil assemblages. For example, at the end of the Paleozoic era, there was the most severe major extinction event in Earth's history. The dates separating the various intervals are based on radiometric dating techniques, such as potassium-argon dating. Note that most of Earth's history occurred prior to the emergence of diverse multicellular plants and animals. The history of tropical terrestrial ecosystems begins in the Carboniferous period of the Paleozoic era and continues, with many changes, to the present day.

Eon	Era	Period	Subperiod	Epoch	Age	Millions of years ago
Phanerozoic	Cenozoic	Quaternary		Holocene		0.01
				Pleistocene	Late	0.76
					Early	1.8
		Tertiary	Neogene	Pliocene	Late	3.6
					Early	5
				Miocene	Late	11
					Middle	16.5
					Early	24
			Paleogene	Oligocene	Late	28.5
					Early	34
				Eocene	Late	37
					Middle	49
					Early	55
				Paleocene	Late	61
					Early	65
	Mesozoic	Cretaceous		Late		97
				Early		144
		Jurassic		Late		160
				Middle		180
				Early		205
		Triassic		Late		228
				Middle		242
				Early		248
	Paleozoic	Permian		Late		256
				Early		295
		Pennsylvanian		Late		304
				Middle		311
				Early		324
		Mississippian		Late		340
				Early		354
		Devonian		Late		372
				Middle		391
				Early		416
		Silurian		Late		422
				Early		442
		Ordovician		Late		458
				Middle		470
				Early		495
		Cambrian		Late		505
				Middle		518
				Early		544
Precambrian: Proterozoic	Late	None defined				900
	Middle					1,600
	Early					2,400
Precambrian: Archean	Late					3,000
	Middle					3,400
	Early					3,800

Literature Cited

INTRODUCTION

Darwin, C. 1845. *Journal of Researches into the Geology and Natural History of the Various Countries Visited by the H.M.S.* Beagle *Round the World*. London: John Murray. This book was retitled *The Voyage of the Beagle* when it was republished in 1905, and that is the title it remains known by today.

Darwin, C. 1887. *Life and Letters of Charles Darwin*. London: John Murray. This book was retitled *The Autobiography of Charles Darwin* and remains available today.

Dobzhansky, T. 1973. Biology, molecular and organismic. *American Zoologist* 4: 443–452.

Hutchinson, G. E. 1965. *The Ecological Theater and the Evolutionary Play*. New Haven, CT: Yale University Press.

Kricher, J. 1989. *A Neotropical Companion*. Princeton, NJ: Princeton University Press.

Kricher, J. 2009. *The Balance of Nature: Ecology's Enduring Myth*. Princeton, NJ: Princeton University Press.

CHAPTER 1

Achard, F., et al. 2002. Determination of deforestation rates of the world's humid tropical forests. *Science* 297: 999–1002.

Brook, B. W., C. J. A. Bradshaw, L. Pin Koh, and M. N. S. Sodhi. 2006. Momentum drives the crash: mass extinction in the tropics. *Biotropica* 38: 302–305.

Canby, T. Y. 1984. El Niño's ill wind. *National Geographic* 162: 143–183.

Chazdon, R. L. 2002. Introduction. In *Foundations of Tropical Biology*, R. L. Chazdon and T. C. Whitmore (eds.). Chicago: University of Chicago Press.

Clark, D. A. 2004. Tropical forest and global warming: Slowing it down or speeding it up? *Frontiers in Ecology and Environment* 2: 73–80.

Cohen, J. E. 2006. Human population: The next half century. In *Science Magazine's State of the Planet 2006–2007*, D. Kennedy (ed.). Washington, DC: Island Press.

Foster, R. B. 1982. Famine on Barro Colorado Island. In *The Ecology of a Tropical Forest,* E. G. Leigh Jr., A. S. Rand, and D. M. Windsor (eds.). Washington, DC: Smithsonian Institution Press.

Gardner, T. A., J. Barlow, L. W. Parry, and C. A. Peres. 2006. Predicting the uncertain future of tropical forest species in a data vacuum. *Biotropica* 39: 25–30.

Graham, N. E., and W. B. White. 1988. The El Niño cycle: A natural oscillator of the Pacific Ocean–atmosphere system. *Science* 240: 1293–1301.

Head, J. J., et al. 2009. Giant boid snake from the Paleocene Neotropics reveals hotter past equatorial temperatures. *Nature* 457: 715–717.

Holdridge, L. R. 1947. Determination of world plant formations from simple climatic data. *Science* 105: 367–368.

Huber, M. 2009. Snakes tell a torrid tale. *Nature* 457: 699–700.

Jackson, S. T. 2009. Alexander von Humboldt and the general physics of the Earth. *Science* 324: 596–597.

Johnson, K. R., and B. Ellis. 2002. A tropical rain forest in Colorado 1.4 million years after the Cretaceous-Tertiary boundary. *Science* 296: 2381.

Junk, W. J., and K. Furch. 1985. The physical and chemical properties of Amazonian waters and their relationships with the biota. In *Amazonia*, G. T. Prance and T. E. Lovejoy (eds.). Oxford, UK: Pergamon Press.

Laurance, W. F. 2006. Have we overstated the tropical biodiversity crisis? *Trends in Ecology and Evolution* 22: 65–70.

Leigh, E. G., Jr. 1999. *Tropical Forest Ecology: A View from Barro Colorado Island*. New York: Oxford University Press.

Ozanne, C. M. P., et al. 2003. Biodiversity meets the atmosphere: A global view of forest canopies. *Science* 301: 183–186.

Parker, G. G., A. P. Smith, and K. P. Hogan. 1992. Access to the upper forest canopy with a large tower crane. *BioScience* 42: 664–670.

Primack, R., and R. Corlett. 2005. *Tropical Rain Forests: An Ecological and Biogeographic Comparison*. Malden, MA: Blackwell.

Richards, P. W. 1952. *The Tropical Rain Forest*. Cambridge, UK: Cambridge University Press.

Salati, E., and P. B. Vose. 1984. Amazon Basin: A system in equilibrium. *Science* 225: 129–138.

Stocks, G., L. Seales, F. Paniagua, E. Maehr, and E. M. Bruna. 2008. The geographical and institutional distribution of ecological research in the tropics. *Biotropica* 40(4): 397–404.

Suplee, C. 1999. El Niño–La Niña: Nature's vicious cycle. *National Geographic* 195: 72–95.

Walsh, P. D., et al. 2003. Catastrophic ape decline in western equatorial Africa. *Nature* 422: 611–614.

Whitmore, T. C. 2002. What shaped tropical biotas as we see them today? In *Foundations of Tropical Forest Biology*, R. L. Chazdon and T. C. Whitmore (eds.). Chicago: University of Chicago Press.

Wright, S. J., and H. C. Muller-Landau. 2006a. The future of tropical forest species. *Biotropica* 38: 287–301.

———. 2006b. The uncertain future of forest species. *Biotropica* 38: 443–445.

CHAPTER 2

Ali, J. R., and M. Huber. 2010. Mammalian biodiversity on Madagascar controlled by ocean currents. *Nature* 463: 653–656.

Antonelli, A., J. A. A. Nylander, C. Persson, and I. Sanmartin. 2009. Tracing the impact of the Andean uplift on Neotropical plant evolution. *Proceedings of the National Academy of Sciences of the United States of America* 106(24): 9749–9754. http://www.zora.uzh.ch/19409/.

Barlow, G. W. 2000. *The Cichlid Fishes: Nature's Grand Experiment in Evolution*. New York: Perseus.

Bush, M. B., and P.E. de Oliveira. 2006. The rise and fall of the Refugial Hypothesis of Amazonian speciation: A palaeoecological perspective. *Biota Neotropica* 6. Online version, http://www.scielo.br/scielo.php?script=sci_arttext&pid=S1676-06032006000100002.

Colinvaux, P. A. 1989a. Ice-age Amazon revisited. *Nature* 340: 188–189.

———. 1989b. The past and future Amazon. *Scientific American* 259: 102–108.

Cracraft, J. 1983. Cladistic analysis and vicariance biogeography. *American Scientist* 71: 273–281.

———. 1985. Historical biogeography and patterns of differentiation within the South American avifauna: Centers of endemism. In *Ornithological Monographs*, No. 36, P. A. Buckley et al. (eds.). Washington, DC: American Ornithologists' Union.

Craig, C. L. 1989. Alternative foraging modes of orb weaving spiders. *Biotropica* 21: 257–264.

Eisenberg, J. F. 1989. *Mammals of the Neotropics*, vol. 1. Chicago: University of Chicago Press.

Eisner, T., and S. Nowicki. 1983. Spider web protection through visual advertisement: Role of the stabilimentum. *Science* 219: 185–186.

Garzione, C. N., et al. 2008. Rise of the Andes. *Science* 320: 1304–1308.

Gibbons, A. 2009. A new kind of ancestor: Ardipithecus unveiled. *Science* 326: 36–40.

Haffer, J. 1969. Speciation in Amazonian forest birds. *Science* 165: 131–137.

———. 1974. *Avian Speciation in Tropical South America*, no. 14., R. E. Paynter (ed.). Cambridge, MA: Publications of the Nuttall Ornithological Club.

———. 1985. Avian zoogeography of the Neotropical lowlands. In *Neotropical Ornithology*, P. A. Buckley, M. S. Foster, E. S. Morton, R. S. Ridgely, and F. G. Buckley (eds.). Washington, DC: American Ornithologists' Union.

———. 1993. Time's cycle and time's arrow in the history of Amazonia. *Biogeographica* 69: 15–45.

Haffer, J., and J. W. Fitzpatrick. 1985. Geographic variation in some Amazonian forest birds. In *Neotropical Ornithology*, P. A. Buckley, M. S. Foster, E. S. Morton, R. S. Ridgely, and F. G. Buckley (eds.). Washington, DC: American Ornithologists' Union.

Haffer, J., and G. T. Prance. 2001. Climatic forcing of evolution in Amazonia during the Cenozoic: On the refuge theory of biotic differentiation. *Amazoniana* 16: 579–607.

Heyer, W. R., and L. R. Maxson. 1982. Distributions, relationships, and zoogeography of lowland frogs—the *Leptodactylus* complex in South America, with special reference to Amazonia. In *Biological Diversification in the Tropics*, G. T. Prance (ed.). New York: Columbia University Press.

Irion, G. 1989. Quaternary geological history of the Amazon lowlands. In *Tropical Forests,* L. B. Holm-Nielson, I. Neilson, and H. Basley (eds.). London: Academic Press.

Janzen, D. H., and P. S. Martin. 1982. Neotropical anachronisms: The fruits the gomphotheres ate. *Science* 215: 19–27.

Kinzey, W. G. 1982. Distribution of primates and forest refuges. In *Biological Diversification in the Tropics*, G. T. Prance (ed.). New York: Columbia University Press.

Krause, D. W. 2010. Washed up in Madagascar. *Nature* 463: 613–614.

Lomolino, M. V., F. S. Dove, and J. H. Brown, eds. 2004. *Foundations of Biogeography: Classic Papers with Commentaries.* Chicago: University of Chicago Press.

Marshall, L. G. 1988. Land mammals and the Great American Interchange. *American Scientist* 76: 380–388.

Marshall, L. G., S. D. Webb, J. J. Sepkowski, and D. M. Raup. 1982. Mammalian evolution and the Great American Interchange. *Science* 215: 1351–1357.

Maxson, L. R., and W. R. Heyer. 1982. Leptodactylid frogs and the Brazilian Shield: An old and continuing adaptive relationship. *Biotropica* 14: 10–14.

Mayle, F. E. 2004. Assessment of the Neotropical dry forest refugia hypothesis in the light of palaeoecological data and vegetation model simulations. *Journal of Quaternary Science* 19: 713–717.

Pearson, D. L. 1977. A pantropical comparison of bird community structure on six lowland forest sites. *Condor* 79: 232–244.

———. 1982. Historical factors and bird species richness. In *Biological Diversification in the Tropics*, G. T. Prance (ed.). New York: Columbia University Press.

Prance, G. T. 1982b. Forest refuges: Evidence from woody angiosperms. In *Biological Diversification in the Tropics*, G. T. Prance (ed.). New York: Columbia University Press.

———. 1985. The changing forests. In *Amazonia*, G. T. Prance and T. E. Lovejoy (eds.). Oxford, UK: Pergamon Press.

Prance, G. T., ed. 1982a. *Biological Diversification in the Tropics.* New York: Columbia University Press.

Price, T. 2008. *Speciation in Birds*. Greenwood Village, CO: Roberts and Company.

Primack, R., and R. Corbett. 2005. *Tropical Rain Forests: An Ecological and Biogeographical Comparison.* Malden, MA: Blackwell.

Roca, A. L., N. Georgiadis, J. Pecon-Slattery, and S. J. O'Brien. 2001. Genetic evidence for two species of elephant in Africa. *Science* 293: 1473–1476.

Seehausen, O., et al. 2008. Speciation through sensory drive in cichlid fish. *Nature* 455: 620–626.

Simpson, B. B., and J. Haffer. 1978. Speciation patterns in the Amazonian forest biota. *Annual Review of Ecology and Systematics* 9: 497–518.

Stiassny, M. L. J., and A. Meyer. 1999. Cichlids of the Rift lakes. *Scientific American* 280(2): 64–69.

Tattersal, I. 1985. *The Fossil Trail*. Oxford, UK: Oxford University Press.

Traylor, M. A., Jr., 1985. Species limits in the *Ochthoeca diadema* species-group (Tyrannidae). In *Ornithological Monographs*, no. 36, P. A. Buckley et al. (ed.). Washington, DC: American, Ornithologists' Union.

Verheyen, E., W. Salzburger, J. Snoeks, and A. Meyer. 2003. Origin of the superflock of cichlid fishes from Lake Victoria, East Africa. *Science* 300: 325–329.

Waterton, C. [1825]. 1983. *Wanderings in South America.* Reprint. London: Century Publishing.

Webb, S. D. 1978. A history of savanna vertebrates in the new world. Part II: South America and the great interchange. *Annual Review of Ecology and Systematics* 9: 393–426.

Wilf, P., et al. 2003. High plant diversity in Eocene South America: Evidence from Patagonia. *Science* 300: 122–125.

Willis, K. J., and R. J. Whittaker. 2000. Paleoecology: The refugial debate. *Science* 25: 1406–1407.

CHAPTER 3

Bawa, K. S., and L. A. McDade. 1994. Commentary. In *La Selva: Ecology and Natural History of a Neotropical Rain Forest*, L. A. McDade, K. S. Bawa, H. A. Hespenheide, and G. S. Hartshorn (eds.). Chicago: University of Chicago Press.

Clark, D. A. 1994. Plant demography. In *La Selva: Ecology and Natural History of a Neotropical Rain Forest,* L. A. McDade, K. S. Bawa, H. A. Hespenheide, and G. S. Hartshorn (eds.). Chicago: University of Chicago Press.

Darwin, C. 1862. *On the Various Contrivances bv Which British and Foreign Orchids Are Fertilised by Insects, and on the Good Effects of Intercrossing.* London: John Murray.

Dressler, R. L. 1968. Pollination by euglossine bees. *Evolution* 22: 202–210.

———. 1981. *The Orchids—Natural History and Classification.* Cambridge, MA: Harvard University Press.

Flannery, K. V., ed. 1982. *Maya Subsistence.* New York: Academic Press.
Gentry, A. H. 1991. The distribution and evolution of climbing plants. In *The Biology of Vines,* F. E. Putz and H. A. Mooney (eds.). Cambridge, UK: Cambridge University Press.
———. 1993. *A Field Guide to the Families and Genera of Woody Plants of Northwest South America.* Washington, DC: Conservation International.
Halle, F., R. A. A. Oldman, and P. B. Tomlinson. 1978. *Tropical Trees and Forests: An Architectural Analysis.* Berlin: Springer-Verlag.
Hartshorn, G. S. 1980. Neotropical forest dynamics. In *Tropical Succession*, supplement to *Biotropica* 12: 23–30.
———. 1983. Plants. In *Costa Rican Natural History*, D. H. Janzen (ed.). Chicago: University of Chicago Press.
Henderson, A., G. Galeano, and R. Bernal. 1995. *Field Guide to the Palms of the Americas.* Princeton, NJ: Princeton University Press.
Horn, H. S. 1971. *The Adaptive Geometry of Trees.* Princeton, NJ: Princeton University Press.
Jacobs, M. 1981. *The Tropical Rain Forest: A First Encounter.* Berlin: Springer-Verlag.
Janzen, D. H., and P. S. Martin. 1982. Neotropical anachronisms: The fruits the gomphotheres ate. *Science* 215: 19–27.
King, D. A. 1990. Allometry of saplings and understory trees in a Panamanian forest. *Functional Ecology* 5: 485–492.
———. 1996. The allometry and life history of tropical trees. *Journal of Tropical Ecology* 12: 25–44.
Klinge, H., W. A. Rodrigues, E. Brunig, and E. J. Fittkau. 1975. Biomass and structure in a central Amazonian rain forest. In *Tropical Ecological Systems: Trends in Terrestrial and Aquatic Research*, F. B. Golley and E. Medina (eds.) New York: Springer-Verlag.
Longman, K. A., and J. Jenik. 1974. *Tropical Forest and Its Environment.* London: Longman.
Lotschert, W., and G. Beese. 1981. *Collins Guide to Tropical Plants.* London: Collins.
Murali, K. S., and R. Sukumar. 1993. Leaf flushing phenology and herbivory in a tropical dry deciduous forest, southern India. *Oecologia* 94: 114–119.
Myers, N., for the Committee on Research Priorities in Tropical Biology of the National Research Council. 1980. *Conversion of Tropical Moist Forests.* Washington, DC: National Academy of Science.
Nadkarni, N. M. 1981. Canopy roots: Convergent evolution in rainforest nutrient cycles. *Science* 214: 1023–1024.
Ngai, J. T., and D. S. Srivatava. 2006. Predators accelerate nitrogen cycling in a bromeliad ecosystem. *Science* 314: 963.
Oster, G., and S. Oster. 1985. The great breadfruit scheme. *Natural History* 94: 35–41.
Perry, D. R. 1978. Factors influencing arboreal epiphytic phytosociology in Central America. *Biotropica* 10: 235–237.
———. 1984. The canopy of the tropical rain forest. *Scientific American* 251: 138–147.
Phillips, O. 1991. The ethnobotany and economic botany of tropical vines. In *The Biology of Vines,* F. E. Putz and H. A. Mooney (eds.). Cambridge, UK: Cambridge University Press.
Pires, J. M., and G. T. Prance. 1985. The vegetation types of the Brazilian Amazon. In *Amazonia*, G. T. Prance and T. E. Lovejoy (eds.). Oxford, UK: Pergamon Press.
Plotkin, M. 1993. *Tales of a Shaman's Apprentice: An Ethnobotanist Searches for New Medicines in the Amazon Rain Forest.* New York: Viking.
Poorter, L., F. Bongers, F. J. Sterck, and H. Woll. 2003. Architecture of 53 rain forest tree species differing in adult stature and shade tolerance. *Ecology* 84: 602–608.
Putz, F. E. 1984. The natural history of lianas on Barro Colorado Island, Panama. *Ecology* 65: 1713–1724.
Putz, F. E., and N. M. Holbrook. 1988. Tropical rain forest images. In *People of the Tropical Rain Forest*, J. S. Denslow and C. Padoch (eds.). Berkeley, CA: University of California Press.
Putz, F. E., and H. A. Mooney, eds. 1991. *The Biology of Vines.* Cambridge, UK: Cambridge University Press.
Richards, P. W. 1952. *The Tropical Rain Forest.* Cambridge, UK: Cambridge University Press.
Royer, D. L., and P. D. Wilf. 2006. Why do toothed leaves correlate with cold climates? Gas exchange at leaf margins provides new insights into a classic paleotemperature proxy. *International Journal of Plant Science* 167: 11–18.
Schnitzer, S. A., and F. Bongers. 2002. The ecology of lianas and their role in forests. *Trends in Ecology and Evolution* 17: 223–230.
Schnitzer, S. A., J. W. Dalling, and W. P. Carson. 2000. Lianas and gap-phased regeneration in a tropical forest. *Journal of Ecology* 88: 655–666.
Smith, N., S. A. Mori, A. Henderson, D. W. Stevenson, and S. V. Heald. 2004. *Flowering Plants of the Neotropics.* Princeton, NJ: Princeton University Press.
Tramer, Eliot. 1974. On latitudinal gradients in avian diversity. *Condor* 76: 123–139.
Utley, J. F., and K. Burt-Utley. 1983. Bromeliads (Piña silvestre, piñuelas, chiras, wild pineapple). In *Costa Rican Natural History*, D. H. Janzen (ed.). Chicago: University of Chicago Press.
Wallace, A. R. 1853. *Palm Trees of the Amazon and Their Uses.* London: Van Voorst.
———. 1895. *Natural Selection and Tropical Nature.* London: Macmillan.
Walterm, K. S. 1983. Orchidaceae (orquideas, orchids). In *Costa Rican Natural History*, D. H. Janzen (ed.). Chicago: University of Chicago Press.
Wiemann, M. C., S. R. Manchester, D. L. Dilcher, L. F. Hinojosa, and E. A. Wheeler. 1998. Estimation of temperature and precipitation from morphological characters of dicotyledonous leaves. *American Journal of Botany* 85: 1796–1802.
Wilf, P. 1997. When are leaves good thermometers? A new case for leaf margin analysis. *Paleobiology* 23: 373–390.
Wilson, E. O. 1991. Rain forest canopy: The high frontier. *National Geographic* 180: 78–107.
Zahl, P. A. 1975. Hidden worlds in the heart of a plant. *National Geographic* 147: 388–397.

CHAPTER 4

Askins, R. A. 1983. Foraging ecology of temperate-zone and tropical woodpeckers. *Ecology* 64: 945–956.
Bermingham, E., and C. Dick. 2001. The *Inga*—newcomer or museum antiquity? *Science* 293: 2214–2215.
Blake, J. G., F. G. Stiles, and B. A. Loiselle. 1990. Birds of La Selva Biological Station: Habitat use, trophic composition, and migrants. In *Four Neotropical Rain Forests*, A. H. Gentry (ed.). New Haven, CT: Yale University Press.

Condit, R., et al. 2002. Beta-diversity in tropical forest trees. *Science* 295: 666–669.

Connell, J. H., and E. Orias. 1964. The ecological regulation of species diversity. *American Naturalist* 98: 399–414.

DeVries, P. J. 1987. *The Butterflies of Costa Rica and Their Natural History*, vol. 1. Princeton, NJ: Princeton University Press.

———. 1994. Patterns of butterfly diversity and promising topics in natural history and ecology. In *La Selva: Ecology and Natural History of a Neotropical Rain Forest,* L. A. McDade, K. S. Bawa, H. A. Hespenheide, and G. S. Hartshorn (eds.), Chicago: University of Chicago Press.

———. 1997. *The Butterflies of Costa Rica and Their Natural History*, vol. 2. Princeton, NJ: Princeton University Press.

Dial, R., and J. Roughgarden. 1995. Experimental removal of insectivores from rain forest canopy: Direct and indirect effects. *Ecology* 76: 1821–1834.

Diamond, J. M. 1973. Distributional ecology of New Guinea birds. *Science* 179: 759–769.

Dobzhansky, T. 1950. Evolution in the tropics. *American Scientist* 38: 209–221.

Duellman, W. E. 1992. Reproductive strategies of frogs. *Scientific American* 267: 80–87.

Dyer, L.A., T. R. Walla, H. F. Greeney, J. O. Stireman III, and R. F. Hazen. 2010. Diversity of interactions: A metric for studies of biodiversity. *Biotropica* 42: 281–289.

Emmons, L. H. 1984. Geographic variation in densities and diversities of nonflying mammals in Amazonia. *Biotropica* 16: 210–222.

———. 1987. Comparative feeding ecology of felids in a Neotropical rain forest. *Behavioral Ecology and Social Biology* 20: 271–283.

———. 1990. *Neotropical Rain Forest Mammals: A Field Guide.* Chicago: University of Chicago Press.

Erwin, T. L. 1982. Tropical forests: Their richness in Coleoptera and other arthropod species. *Colleopterists' Bulletin* 36: 74–75.

———. 1983. Beetles and other insects of tropical forest canopies at Manaus, Brazil, sampled by insecticidal fogging. In *Tropical Rain Forest: Ecology and Management,* S. L. Sutton, T. C. Whitmore, and A. C. Chadwick (eds.). London: Blackwell.

———. 1988. The tropical forest canopy: The heart of biotic diversity. In *Biodiversity,* E. O. Wilson (ed.). Washington, DC: National Academy Press.

Fine, P. A. V., I. Mesones, and P. D. Coley. 2004. Herbivores promote habitat specialization by trees in Amazonian forests. *Science* 305: 663–667.

Foster, R. B., and S. P. Hubbell. 1990. The floristic composition of the Barro Colorado Island forest. In *Four Neotropical Rain Forests,* A. H. Gentry (ed.). New Haven, CT: Yale University Press.

Foster, R. B., et al. 1994. *The Tambopata-Candamo Reserved Zone of Southeastern Peru: A Biological Assessment.* Washington, DC: Conservation International.

Gentry, A. H. 1982. Neotropical floristic diversity: Phytogeographical connections between Central and South America, Pleistocene climatic fluctuations, or an accident of Andean orogeny? *Annals of the Missouri Botanical Garden* 69: 557–593.

———. 1986. Species richness and floristic composition of Choco region plant communities. *Caldasia* 15: 71–91.

———. 1988. Tree species richness of upper Amazon forests. *Proceedings of the National Academy of Science USA* 85: 156–159.

———. 1990. Floristic similarities and differences between Southern Central America and Upper and Central Amazonia. In *Four Neotropical Rain Forests,* A. H. Gentry (ed.). New Haven, CT: Yale University Press.

Hammel, B. 1990. The distribution of diversity among families, genera, and habit types in the La Selva flora. In *Four Neotropical Rain Forests,* A. H. Gentry (ed.). New Haven, CT: Yale University Press.

Hartshorn, G. S., and B. E. Hammel. 1994. Vegetation types and floristic patterns. In *La Selva: Ecology and Natural History of a Neotropical Rain Forest,* L. A. McDade, K. S. Bawa, H. A. Hespenheide, and G. S. Hartshorn (eds.). Chicago: University of Chicago Press.

Hawkins, B. A., E. R. Porter, and J. A. F. Diniz-Filho. 2003a. Productivity and history as predictors of the latitudinal diversity gradient of terrestrial birds. *Ecology* 84: 1608–1623.

Hawkins, B. A., et al. 2003b. Energy, water, and broad-scale geographic patterns of species richness. *Ecology* 84: 3105–3117.

Hawksworth, D. L. 2006. The fungal dimension of biodiversity: Magnitude, significance, and conservation. *Mycological Research* 95: 641–655.

Heithaus, E. R., T. H. Fleming, and P. A. Opler. 1975. Foraging patterns and resource utiliization in seven species of bats in a seasonal tropical forest. *Ecology* 56: 841–854.

Hilty, S. L., and W. L. Brown. 1986. *A Guide to the Birds of Colombia.* Princeton, NJ: Princeton University Press.

Hubbell, S. P., and R. B. Foster. 1986a. Canopy gaps and the dynamics of a Neotropical forest. In *Plant Ecology*, M. J. Crawley (ed.). Oxford, UK: Blackwell Scientific.

———. 1986b. Commonness and rarity in a Neotropical forest: Implications for tropical tree conservation. In *Conservation Biology: The Science of Scarcity and Diversity*, M. E. Soule (ed.). Sunderland, MA: Sinauer Associates.

Jablonski, D., K. Roy, and J. W. Valentine. 2006. Out of the tropics: Evolutionary dynamics of the latitudinal diversity gradient. *Science* 314: 102–106.

Janzen, D. H. 1976. Why are there so many species of insects? *Proceedings of XV International Congress of Entomology*, pp. 84–94.

Jaramillo, C., M. J. Rueda, and G. Mora. 2006. Cenozoic plant diversity in the Neotropics. *Science* 311: 1893–1896.

Kalko, E., and C. O. Handley Jr. 2001. Neotropical bats in the canopy: Diversity, community structure, and implications for conservation. *Plant Ecology* 153: 319–333.

Karr, J. R. 1975. Production, energy pathways, and community diversity in forest birds. In *Tropical Ecological Systems: Trends in Terrestrial and Aquatic Research*, F. B. Golley and E. Medina (eds.). New York: Springer-Verlag.

———. 1976. Within- and between-habitat avian diversity in African and Neotropical lowland habitats. *Ecological Monographs* 46: 457–481.

Knight, D. H. 1975. A phytosociological analysis of species-rich tropical forest on Barro Colorado Island, Panama. *Ecological Monographs* 45: 259–284.

Lamoreux, J. F., et al. 2006. Global tests of biodiversity concordance and the importance of endemism. *Nature* 440: 212–213.

Lovejoy, T. E. 1974. Bird diversity and abundance in Amazon forest communities. *Living Bird* 13: 127–192.

MacArthur, R. H. 1965. Patterns of species diversity. *Biology Review* 40: 510–533.

MacArthur, R. H., and E. O. Wilson. 1967. *The Theory of Island Biogeography*. Princeton, NJ: Princeton University Press.
Magurran, A. E. 2003. *Measuring Biological Diversity*. New York: Wiley-Blackwell.
May, R. M. 1988. How many species are there on Earth? *Science* 241: 1441–1449.
———. 1992. How many species inhabit the Earth? *Scientific American* 267: 42–48.
Novotny, V., P. Drozd, S. E. Miller, M. Kulfan, M. Janda, Y. Basset, and G. D. Weiblen. 2006. Why are there so many species of herbivorous insects in tropical rain forests? *Sciencexpress*, available at www.sciencexpress.org/13; accessed July 2006.
Pianka, E. R. 1966. Latitudinal gradients in species diversity: A review of concepts. *American Naturalist* 100: 33–45.
Prance, G. T. 1990. The floristic composition of the forests of Central Amazonian Brazil. In *Four Neotropical Rain Forests*, A. H. Gentry (ed.). New Haven, CT: Yale University Press.
Prance, G. T., W. A. Rodrigues, and M. F. da Silva. 1976. Inventario florestal de um hectare de mata de terra firme, km 30 da Estrada Manaus-Itacoatiara. *Acta Amazonica* 6: 9–35.
Pruvis, A., and A. Hector. 2000. Getting the measure of biodiversity. *Nature* 405: 212–219.
Remsen, J. V. 1990. Community ecology of Neotropical kingfishers. *University of California Publications* 124: 1–116.
Remsen, J. V., Jr., and T. A. Parker III. 1983. Contribution of river-created habitats to bird species richness in Amazonia. *Biotropica* 15: 223–231.
Richardson, J. E., R. T. Pennington, T. D. Pennington, and P. R. Hollingworth. 2001. Rapid diversification of a species-rich genus of Neotropical rain forest trees. *Science* 293: 2242–2245.
Robinson, S. K., and J. Terborgh. 1990. Bird communities of the Cocha Cashu Biological Station in Amazonian Peru. In *Four Neotropical Rain Forests*, A. H. Gentry (ed.). New Haven, CT: Yale University Press.
Sayer, E. J., L. M. E. Sutcliffe, R. I. C. Ross, and E. V. J. Tanner. 2010. Arthropod abundance and diversity in a lowland tropical forest floor in Panama: The role of habitat space vs. nutrient concentrations. *Biotropica* 42: 194–200.
Schoener, T. W. 1971. Large-billed insectivorous birds: A precipitous diversity gradient. *Condor* 73: 154–161.
Sherry, T. W. 1984. Comparative dietary ecology of sympatric, insectivorous Neotropical flycatchers. *Ecological Monographs* 54: 313–338.
Simpson, E. H. 1949. Measurement of diversity. *Nature* 163: 688.
———. 1992. Maintenance of diversity in tropical forests. *Biotropica* 24(2b): 283–292.
Snow, D. W. 1966. A possible selective factor in the evolution of fruiting seasons in tropical forest. *Oikos* 15: 274–281
Stork, N. E. 1988. Insect diversity: Facts, fiction, and speculation. *Biological Journal of the Linnean Society* 35: 321–337.
Terborgh, J. 1992. Maintenance of diversity in tropical forests. *Biotropica* 24: 283–292.
Terborgh, J., S. K. Robinson, T. A. Parker III, C. A. Munn, and N. Pierpont. 1990. Structure and organization of an Amazonian forest bird community. *Ecological Monographs* 60: 213–238.
Terborgh, J., and J. S. Weske. 1975. The role of competition in the distribution of Andean birds. *Ecology* 56: 562–576.
Tramer, E. J. 1974. On latitudinal gradients in avian diversity. *Condor* 76: 123–130.
Wallace, A. R. 1895. *Natural Selection and Tropical Nature*. London: Macmillan.
Weir, J. T., and D. Schluter. 2007. The latitudinal gradient in recent speciation and extinction rates of birds and mammals. *Science* 315: 1574–1576.
Wilson, E. O. 1987. The arboreal ant fauna of Peruvian Amazon forests: A first assessment. *Biotropica* 19: 245–251.
World Conservation Monitoring Centre. 1992. *Global Biodiversity: Status of the Earth's Living Resources*, B. Groombridge (ed.). London: Chapman and Hall.

CHAPTER 5

Arnold, A. E., and L. C. Lewis. 2005. Ecology and evolution of fungal endophytes and their roles against insects. In *Insect-Fungal Associations: Ecology and Evolution*, F. E. Vega and M. Blackwell (eds.). Oxford, UK: Oxford University Press.
Bell, G. 2001. Neutral macroecology. *Science* 293: 2413–2418.
Boucher, D. H. 1990. Growing back after hurricanes: Catastrophes may be critical to rain forest dynamics. *Bioscience* 40: 163–166.
Burslem, D. F. R. P, N. C. Garwood, and S. C. Thomas. 2001. Tropical forest diversity—the plot thickens. *Science* 291: 606–607.
Case, T. J., and M. L. Cody, eds. 1983. *Island Biogeography in the Sea of Cortez*. Berkeley: University of California Press.
———. 1987. Testing theories of island biogeography. *American Scientist* 75: 402–411.
Censky, E. J., K. Hodge, and J. Dudley. 1998. Over-water dispersal of lizards due to hurricanes. *Nature* 395: 556.
Chave, J. M., H. C. Muller-Landau, and S. A. Levin. 2002. Comparing classical community models: Theoretical consequences of patterns of diversity. *American Naturalist* 159: 1–22.
Clark, D. A., and D. B. Clark. 1992. Life history diversity of canopy and emergent trees in a Neotropical rain forest. *Ecological Monographs* 62: 315–344.
Comita, L. S., H. C. Muller-Landau, S. Aguilar, and S. P. Hubbell. 2010. Asymmetric density dependence shapes species abundances in a tropical tree community. *Science* 329: 330–332.
Condit, R., S. P. Hubbell, and R. B. Foster. 1992. Short term dynamics of a Neotropical forest. *Bioscience* 42: 822–828.
———. 1995. Mortality rates of 205 Neotropical tree and shrub species and the impact of a severe drought. *Ecological Monographs* 65: 419–439.
Condit, R., et al. 2000. Spatial patterns in the distribution of tropical tree species. *Science* 288: 1414–1418.
———. 2002. Beta-diversity in tropical forest trees. *Science* 295: 666–669.
Connell, J. H. 1971. On the role of natural enemies in preventing competitive exclusion in some marine animals and in rain forest trees. In *Dynamics of Numbers in Populations*, P. J. den Boer and G. Gradwell (eds.). Wageningen, The Netherlands: Center for Agricultural Publication and Documentation.
———. 1978. Diversity in tropical rain forests and coral reefs. *Science* 199: 1302–1310.
Connell, J. H., and R. O. Slatyer. 1977. Mechanisms of succession in natural communities and their role in community stability and organization. *American Naturalist* 111: 1119–1144.
Connor, E. F., and D. Simberloff. 1979. The assembly of species communities: Chance of competition. *Ecology* 60: 1132–1140.
Diamond, J. M. 1975. Assembly of species communities. In *Ecology and Evolution of Communities*, M. L. Cody and J. M. Diamond (eds.). Cambridge, MA: Harvard University Press.

Dornelas, M., S. R. Connolly, and T. P. Hughes. 2006. Coral reef diversity refutes the neutral theory of biodiversity. *Nature* 440: 80–82.

Emerson, B. C., and N. Kolm. Species diversity can drive speciation. *Nature* 434: 1015–1017.

Fargione, J., et al. 2007. From selection to complementarity: Shifts in the causes of biodiversity–productivity relationships in a long-term biodiversity experiement. *Proceedings of the Royal Society B* 274: 871–876.

Finke, D. L., and W. E. Snyder. 2008. Niche partitioning increases resource exploitation by diverse communities. *Science* 321: 1488–1490.

Gilbert, G. S. 2005. Dimensions of plant disease in tropical forests. In *Biotic Interactions in the Tropics: Their Role in the Maintenance of Species Diversity*, D. Burslem, M. Pinard, and S. Harley (eds.). Cambridge, UK: Cambridge University Press.

Graves, G. R., and C. Rahbek. 2005. Source pool geometry and the assembly of continental avifaunas. *Proceedings of the National Academy of Sciences USA* 102: 7871–7876.

Hille Ris Lambers, J., J. S. Clark, and B. Beckage. 2002. Density-dependent mortality and the latitudinal diversity gradient. *Nature* 417: 732–735.

Hubbell, S. P. 1979. Tree dispersion, abundance, and diversity in a tropical dry forest. *Science* 203: 1299–1309.

———. 2001. *The Unified Neutral Theory of Biodiversity and Biogeography*. Princeton, NJ: Princeton University Press.

———. 2006. Neutral theory and the evolution of ecological equivalence. *Ecology* 87: 1387–1398.

Hubbell, S. P., and R. B. Foster. 1986a. Commonness and rarity in a Neotropical forest: Implications for tropical tree conservation. In *Conservation Biology: The Science of Scarcity and Diversity*, M. E. Soule (ed.). Sunderland, MA: Sinauer Associates.

———. 1986b. Biology, chance, and history and the structure of tropical rain forest tree communities. In *Community Ecology*, J. Diamond and T. J. Case (eds.). New York: Harper and Row.

———. 1990. Structure, dynamics, and equilibrium status of old-growth forest on Barro Colorado Island. In *Four Neotropical Rain Forests*, A. H. Gentry (ed.). New Haven, CT: Yale University Press.

Hutchinson, G. E. 1961. The paradox of the plankton. *American Naturalist* 95: 137–145.

Janzen, D. H. 1969. Seed-eaters versus seed size, number, toxicity, and dispersal. *Evolution* 23: 1–27.

———. 1970. Herbivores and the number of tree species in a tropical forest. *American Naturalist* 104: 501–528.

Karr, J. R. 1990. Avian survival rates and the extinction process on Barro Colorado Island, Panama. *Conservation Biology* 4: 391–397.

Kelly, C. K., and M. G. Bowler. 2002. Coexistence and relative abundance in forest trees. *Science* 417: 437–440.

Kraft, N. J. B., R. Valencia, and D. D. Ackerly. 2008. Functional traits and niche-based tree community assembly in an Amazonian forest. *Science* 322: 580–582.

Kricher, J. 2009. *The Balance of Nature: Ecology's Enduring Myth*. Princeton, NJ: Princeton University Press.

Latimer, A. W., J. A. Silander Jr., and R. M. Cowling. 2005. Neutral ecological theory reveals isolation and rapid speciation in a biodiversity hot spot. *Science* 309: 1722–1725.

Leigh, E. G., Jr. 1999. *Tropical Forest Ecology: A View from Barro Colorado Island*. Oxford, UK: Oxford University Press.

Leigh, E. G., Jr., et al. 2004. Why do tropical forests have so many species of trees? *Biotropica* 36: 447–473.

Losos, E. C., and E. G Leigh, eds. 2004. *Tropical Forest Diversity and Dynamism: Findings from a Large-Scale Plot Network*. Chicago: University of Chicago Press.

MacArthur, R. H., and E. O. Wilson. 1967. *The Theory of Island Biogeography*. Princeton, NJ: Princeton University Press.

Mangan, S. A., et al. 2010. Negative plant–soil feedback predicts tree-species relative abundance in a tropical forest. *Nature* 466: 752–755.

May, R. M., and M. P. H. Stumpf. 2000. Species-area relations in tropical forests. *Science* 290: 2084–2085.

McGill, B. J. 2003. A test of the unified neutral theory of biodiversity. *Nature* 422: 881–888.

Molino, J.-F., and D. Sabatier. 2001. Tree diversity in tropical rain forests: A validation of the intermediate disturbance hypothesis. *Science* 294: 1702–1704.

Morin, P. J. 1999. *Community Ecology*. Malden, MA: Blackwell Science.

Petraitis, P. S., R. E. Latham, and R. A. Niesenbaum. 1989. The maintenance of species diversity by disturbance. *Quarterly Review of Biology* 64: 393–418.

Reice, S. R. 1994. Nonequilibrium determinants of biological community structure. *American Scientist* 82: 424–435.

Ricklefs, R. E. 2004. A comprehensive framework for global patterns in biodiversity. *Ecological Letters* 7: 1–15.

———. 2006. The unified neutral theory of biodiversity: Do the numbers add up? *Ecology* 87: 1424–1431.

Rosenzweig, M. L. 1995. *Species Diversity in Space and Time*. Cambridge, UK: Cambridge University Press.

Roxburgh, S. H., K. Shea, and J. B. Wilson. 2004. The intermediate disturbance hypothesis: Patch dynamics and mechanisms of species coexistence. *Ecology* 85: 359–371.

Salt, G. W., ed. 1984. *Ecology and Evolutionary Biology: A Round Table on Research*. Chicago: University of Chicago Press.

Sanders, H. 1968. Marine benthic diversity: A comparative study. *American Naturalist* 102: 243–283.

Simberloff, D. S., and E. O. Wilson. 1969. Experimental zoogeography of islands: The colonization of empty islands. *Ecology* 50: 278–296.

Tanner, J. E., T. P. Hughes, and J. H. Connell. 1994. Species coexistence, keystone species, and succession: A sensitivity analysis. *Ecology* 75: 2204–2219.

ter Steege, H., et al. 2006. Continental-scale patterns of canopy tree composition and function across Amazonia. *Nature* 442: 444–447.

Tilman, D. 2004. Niche tradeoffs, neutrality, and community structure: A stochastic theory of resource competition, invasion, and community assembly. *Proceedings of the National Academy of Sciences USA* 101: 10854–10861.

Vandermeer, J., I. G. de la Cerda, D. Boucher, I. Perfecto, and J. Ruiz. 2000. Hurricane disturbance and tropical tree species diversity. *Science* 290: 788–791.

Volkov, I., J. R. Banavar, F. He, S. P. Hubbell, and A. Maritan. 2005. Density dependence explains tree species richness abundance and diversity in tropical forests. *Nature* 438: 658–661.

Volkov, I., J. R. Banavar, S. P. Hubbell, and A. Maritan. 2003. Neutral theory and relative species abundance in ecology. *Nature* 424: 1035–1037.

Willis, E. O. 1974. Populations and local extinctions of birds on Barro Colorado Island, Panama. *Ecological Monographs* 44: 153–169.

Wills, C., et al. 2006. Nonrandom processes maintain diversity in tropical forests. *Science* 311: 527–531.

Wilson, E. O. 1992. *The Diversity of Life*. Cambridge, MA: Belknap Press of Harvard University.

Woodley, J. D., et al. 1981. Hurricane Allen's impact on Jamaican coral reefs. *Science* 214: 749–761.

Wootton, J. T. 2005. Field parameterization and experimental test of the neutral theory of biodiversity. *Nature* 433: 309–312.

CHAPTER 6

Aiba, S.-I., and K. Kitayama. 2002. Effects of the 1997–98 El Niño drought on rain forests of Mount Kinabalu, Borneo. *Journal of Tropical Ecology* 18: 215–230.

Alvarez-Clare, S., and K. Kitajima. 2009. Susceptibility of tree seedlings to biotic and abiotic hazards in the understory of a moist tropical forest in Panama. *Biotropica* 41: 47–56.

Andrade, J. C., and J. P. P. Carauta. 1982. The *Cecropica–Azteca* association: A case of mutualism? *Biotropica* 14: 15.

Baker, H. G. 1983. *Ceiba pentandra* (Ceyba, ceiba, Kapok tree). In *Costa Rican Natural History*, D. H. Janzen (ed.). Chicago: University of Chicago Press.

Bazzaz, F. A., and S. T. A. Pickett. 1980. Physiological ecology of tropical succession: A comparative review. *Annual Review of Ecological Systems* 11: 287–310.

Berg, C. C., and P. F. Rosselli. 2005. *Flora Neotropica*, Vol. 94: Cecropia. New York: New York Botanical Garden Press.

Brokaw, N. V. L. 1982. Treefalls: Frequency, timing, and consequences. In *The Ecology of a Tropical Rain Forest*, E. G. Leigh Jr., A. S. Rand, and D. M. Windsor (eds.). Washington, DC: Smithsonian Institute Press.

———. 1985. Gap-phase regeneration in a tropical forest. *Ecology* 66: 682–687.

———. 1987. Gap-phase regeneration of three pioneer tree species in a tropical forest. *Journal of Ecology* 75: 9–19.

Buschbacher, R., C. Uhl, and E. A. S. Serrao. 1988. Abandoned pastures in eastern Amazonia. II. Nutrient stocks in the soil and vegetation. *Journal of Ecology* 76: 682–699.

Bush, M. B., and P. A. Colinvaux. 1994. Tropical forest disturbance: Paleoecological records from Darien, Panama. *Ecology* 75: 1761–1768.

Chazdon, R. L., A. R. Brenes, and B. V. Alvarado. 2005. Effects of climate and stand age on annual tree dynamics in tropical second-growth rain forests. *Ecology* 86: 1808–1815.

Chazdon, R. L., and N. Fetcher. 1984. Photosynthetic light environments in a lowland tropical rain forest in Costa Rica. *Journal of Ecology* 72: 553–564.

Clark, D. A. 1994. Plant demography. In *La Selva: Ecology and Natural History of a Neotropical Rain Forest,* L. A. McDade, K. S. Bawa, H. A. Hespenheide, and G. S. Hartshorn (eds.). Chicago: University of Chicago Press.

Clark, D. A., and D. B. Clark. 1987. Population ecology and microhabitat distribution of *Dipteryx panamensis*, a Neotropical rain forest emergent tree. *Biotropica* 19: 236–244.

———. 1992. Life history diversity of canopy and emergent trees in a Neotropical rain forest. *Ecological Monographs* 62: 315–344.

Cochrane, M. A., et al. 1999. Positive feedbacks in the fire dynamic of close canopy tropical forests. *Science* 284: 1832–1835.

Condit, R., S. P. Hubbell, and R. B. Foster. 1992. Short-term dynamics of a Neotropical forest. *Bioscience* 42: 822–828.

———. 1993. Indentifying fast-growing native trees from the Neotropics using data from a large, permanent census plot. *Forest Ecology and Management* 62: 123–143.

———. 1995. Mortality rates of 205 Neotropical tree and shrub species and the impact of a severe drought. *Ecological Monographs* 65: 419–439.

Curran, L. M., et al. 1999. Impact of El Niño and logging on canopy tree recruitment in Borneo. *Science* 286: 2184–2188.

Dalling, J. W., and T. A. Brown. 2009. Long-term persistence of pioneer seeds in tropical rain forest soil seed banks. *American Naturalist* 173: 531–535.

Dalling, J. W., and R. C. John. 2008. Recruitment limitation and coexistence of pioneer species. In *Tropical Forest Community Ecology*, W. P. Carson and S. A. Schnitzer (eds.). New York: Blackwell Science.

Dalling, J. W., M. D. Swaine, and N. C. Garwood. 1998. Dispersal patterns and seed bank dynamics of pioneer trees in moist tropical forest. *Ecology* 79: 564–578.

D'Antonio, C. M., and P. M. Vitousek. 1992. Biological invasions by exotic grasses, the grass/fire cycle, and global change. *Annual Review of Ecology and Systematics* 23: 63–87.

De Andrade, J. C., and J. P. P. Carauta. 1982. The *Cecropia-Azteca* association: A case of mutualism? *Biotropica* 14: 15.

Denslow, J. S. 1980. Gap partitioning among tropical rain forest trees. In *Tropical Succession*, J. Ewel (ed.), supplement to *Biotropica* 12: 47–55.

Denslow, J. S., and G. S. Hartshorn. 1994. Tree-fall gap environments and forest dynamic processes. In *La Selva: Ecology and Natural History of a Neotropical Rain Forest,* L. A. McDade, K. S. Bawa, H. A. Hespenheide, and G. S. Hartshorn (eds.). Chicago: University of Chicago Press.

Diamond, J. 2004. *Collapse: How Societies Choose to Fail or Succeed.* New York: Viking.

Dyer, L. A., and A. D. N. Palmer, eds. 2004. Piper*: A Model Genus for Studies of Phytochemistry, Ecology, and Evolution.* New York: Kluwer.

Edwards, W., and A. Krockenberger. 2006. Seedling mortality due to drought and fire associated with the 2002 El Niño event in a tropical rain forest in North-East Queensland, Australia. *Biotropica* 38: 16–28.

Egler, F. E. 1954. Vegetation science concepts. I. Initial floristic composition, A factor in old-field vegetation development. *Vegetation* 4: 412–417.

Engelbrecht, B. J., et al. 2007. Drought sensitivity shapes species distribution patterns in tropical forests. *Nature* 447: 80–82.

Estrada, A., R. Coates-Estrada, and C. Vazquez-Yanes. 1984. Observations of fruiting and dispersers of *Cecropia obtusifolia* at Los Tuxtlas, Mexico. *Biotropica* 16: 315–318.

Ewel, J. 1980. Tropical succession: Manifold routes to maturity. In *Tropical Succession*, J. Ewel (ed.), supplement to *Biotropica* 12: 2–7.

———. 1983. Succession. In *Tropical Rain Forest Ecosystems: Structure and Function*, F. B. Golley (ed.). Amsterdam: Elsevier Scientific.

Ewel, J., et al. 1982. Leaf area, light transmission, roots and leaf damage in nine tropical plant communities. *Agro-Ecosystems* 7: 305–326.

Feldhaar, H., B. Fiala, R. B. Hashim, and U. Maschwitz. 2003. Patterns of the *Crematogaster-Macaranga* associaton: The ant partner makes a difference. *Insectes Sociaux* 50: 9–19.

Fetcher, N., S. F. Oberbauer, and R. L. Chazdon. 1994. Physiological ecology of plants. In *La Selva: Ecology and Natural History of a Neotropical Rain Forest,* L. A. McDade, K. S. Bawa, H. A. Hespenheide, and G. S. Hartshorn (eds.). Chicago: University of Chicago Press.

Fichtler, E., D. A. Clark, and M. Worbes. 2003. Age and long-term growth of trees in an old-growth tropical rain forest, based on analyses of tree rings and ^{14}C. *Biotropica* 35: 306–317.

Finegan, B. 1996. Pattern and process in Neotropical secondary rain forests: The first 100 years of succession. *Trends in Ecology and Evolution* 11: 119–124.

Flannery, K. V., ed. 1982. *Maya Subsistence.* New York: Academic Press.

Fleming, T. H. 1983. *Piper* (Candela, candelillos, piper). In *Costa Rican Natural History*, D. H. Janzen (ed.). Chicago: University of Chicago Press.

———. 1985a. A day in the life of a *Piper*-eating bat. *Natural History* 94: 52–59.

———. 1985b. Coexistence of five sympatric *Piper* (Piperaceae) species in a tropical dry forest. *Ecology* 66: 688–700.

Food and Agriculture Organization of the United Nations (FAO). 2001. *FAO Forest Paper 140*. Rome: FAO.

———. 2006. Net rate of forest loss in Africa second highest in the world. Available at http://www.fao.org/newsroom/en/news/2006/1000261/index.html.

Foster, D. R., and J. D. Aber, eds. 2006. *Forests in Time: The Environmental Consequences of 1,000 Years of Change in New England*. New Haven, CT: Yale University Press.

Foster, R. B. 1990. Long-term change in the successional forest community of the Rio Manu floodplain. In *Four Neotropical Rain Forests*, A. H. Gentry (ed.). New Haven, CT: Yale University Press.

Gentry, A. H., and J. Terborgh. 1990. Composition and dynamics of the Cocha Cashu "mature" floodplain forest. In *Four Neotropical Rain Forests*, A. H. Gentry (ed.). New Haven, CT: Yale University Press.

Gilbert, G. S., K. E. Harms, D. N. Hamill, and S. P. Hubbell. 2001. Effects of seedling size, El Niño drought, seedling density, and distance to nearest conspecific adult on 6-year survival of *Ocotea whitei* seedlings in Panamá. *Oecologia* 127: 509–516.

Greenberg, R. 1987a. Development of dead leaf foraging in a tropical migrant warbler. *Ecology* 68:130–141.

———. 1987b. Seasonal foraging specialization in the worm-eating warbler. *Condor* 89: 158–168.

Hammond, N. 1982. *Ancient Maya Civilization.* New Brunswick, NJ: Rutgers University Press.

Hart, R. D. 1980. A natural ecosystem analog approach to the design of a successional crop system for tropical forest environments. In *Tropical Succession*, J. Ewel (ed.), supplement to *Biotropica* 12: 73–82.

Hartshorn, G. S. 1978. Tree falls and tropical forest dynamics. In *Tropical Trees as Living Systems*, P. B. Tomlinson and M. H. Zimmerman (eds.). London: Cambridge University Press.

Holthuijzen, A. M. A., and J. H. A. Boerboom. 1982. The *Cecropia* seedbank in the Surinam lowland rain forest. *Biotropica* 14: 62–67.

Hooper, E. R., P. Legendre, and R. Condit. 2004. Factors affecting community composition of forest regeneration in deforested, abandoned land in Panama. *Ecology* 85: 3313–3326.

Hubbell, S. P., and R. B. Foster. 1986a. Biology, chance, and history and the structure of tropical rain forest tree communities. In *Community Ecology*, J. Diamond and T. J. Case (eds.). New York: Harper and Row.

———. 1986b. Canopy gaps and the dynamics of a Neotropical forest. In *Plant Ecology*, M. J. Crawley (ed.). Oxford, UK: Blackwell Scientific.

———. 1986c. Commonness and rarity in a Neotropical forest: Implications for tropical tree conservation. In *Conservation Biology: The Science of Scarcity and Diversity*, M. E. Soule (ed.). Sunderland, MA: Sinauer Associates.

———. 1990. Structure, dynamics, and equilibrium status of old-growth forest on Barro Colorado Island. In *Four Neotropical Rain Forests*, A. H. Gentry (ed.). New Haven, CT: Yale University Press.

———. 1992. Short-term dynamics of a Neotropical forest: Why ecological research matters to tropical conservation and management. *Oikos* 63: 48–61.

Hubbell, S. P., et al. 1999. Light-gap disturbances, recruitment limitation, and tree diversity in a Neotropical forest. *Science* 283: 554–557.

Janzen, D. H. 1969. Allelopathy by myrmecophytes: The ant *Azteca* as an allelopathic agent of *Cecropia*. *Ecology* 50: 147–153.

———. 1983. *Mimosa pigra* (Zarza, dormilona). In *Costa Rican Natural History*, D. H. Janzen (ed.). Chicago: University of Chicago Press.

Jordan, C. F. 1985. *Nutrient Cycling in Tropical Forest Ecosystems*. New York: J. Wiley.

Klinge, H., W. A. Rodrigues, E. Brunig, and E. J. Fittkau. 1975. Biomass and structure in a central Amazonian rain forest. In *Tropical Ecological Systems: Trends in Terrestrial and Aquatic Research*, F. B. Golley and E. Medina (eds.). New York: Springer-Verlag.

Knight, D. H. 1975. A phytosociological analysis of species-rich tropical forest on Barro Colorado Island, Panama. *Ecological Monographs* 45: 259–284.

LaFay, H. 1975. The Maya: Children of time. *National Geographic* 148: 728–767.

Lee, D. W., J. B. Lowry, and B. C. Stone. 1979. Abaxial anthocyanin layer in leaves of tropical rain forest plants: Enhancer of light capture in deep shade. *Biotropica* 11: 70–77.

Leigh, E. G., Jr. 1975. Structure and climate in tropical rain forest. *Annual Review of Ecological Systems* 6: 67–86.

———. 1999. *Tropical Forest Ecology: A View from Barro Colorado Island.* Oxford, UK: Oxford University Press.

Lieberman, D., M. Lieberman, R. Peralta, and G. S. Hartshorn. 1985. Mortality patterns and stand turnover rates in a wet tropical forest in Costa Rica. *Journal of Ecology* 73: 915–924.

Lotschert, W., and G. Beese. 1981. *Collins Guide to Tropical Plants.* London: Collins.

McDade, L. A., K. S. Bawa, H. A. Hespenheide, and G. S. Hartshorn, eds. 1994. *La Selva: Ecology and Natural History of a Neotropical Rain Forest.* Chicago: University of Chicago Press.

Metcalf, C. J., C. C. Horvitz, S. Tuljapurkar, and D. A. Clark. 2009. A time to grow and a time to die: A new way to analyze the dynamics of size, light, age, and death of tropical trees. *Ecology* 90: 2766–2778.

Moses, K. 2009. Borneo ablaze: Forest fires threaten world's largest remaining population of orangutans. Available at http://news.mongabay.com/2009/0816-moses_borneo.html.

Murray, K. G. 1988. Avian seed dispersal of three Neotropical gap-dependent plants. *Ecological Monographs* 58: 271–298.

Nations, J. D. 1988. The Lacandon Maya. In *People of the Tropical Rain Forest*, J. S. Denslow and C. Padoch (eds.). Berkeley: University of California Press.

Opler, P. A., H. G. Baker, and G. W. Frankie. 1980. Plant reproductive characteristics during secondary succession in Neotropical lowland forest ecosystems. In *Tropical Succession*, J. Ewel (ed.), supplement to *Biotropica* 12: 40–46.

Phillips, O. L., R. V. Martinez, A. M. Mendoza, T. R. Baker, and P. N. Vargas. 2005. Large lianas as hyperdynamic elements of the tropical forest canopy. *Ecology* 86: 1250–1258.

Pickett, S. T. A., and M. L. Cademasso. 2005. Vegetation dynamics. In *Vegetation Ecology*, E. V. D. Marrel (ed.). Malden, MA: Blackwell.

Poole, R. W., and B. J. Rathcke. 1979. Regularity, randomness, and aggregation in flowering phenologies. *Science* 203: 470–471.

Poorter, L., and L. Markesteijn. 2008. Seedling traits determine drought tolerance of tropical tree species. *Biotropica* 40: 321–331.

Putz, F. E. 1984. The natural history of lianas on Barro Colorado Island, Panama. *Ecology* 65: 1713–1724.

Rankin-de Merona, J. M., R. W. Hutchings, and T. E. Lovejoy. 1990. Tree mortality and recruitment over a five-year period in undisturbed upland rain forest in the Central Amazon. In *Four Neotropical Rain Forests*, A. H. Gentry (ed.). New Haven, CT: Yale University Press.

Reagan, D. P., and R. B. Waide, eds. 1996. *The Food Web of a Tropical Rain Forest*. Chicago: University of Chicago Press.

Sanford, R. L., Jr., J. Saldarriaga, K. Clark, C. Uhl, and R. Herrera. 1985. Amazonian rain-forest fires. *Science* 227: 53–55.

Sezen, U. U., R. L. Chazdon, and K. E. Holsinger. 2005. Genetic consequences of tropical second-growth forest regeneration. *Science* 307: 891.

———. 2007. Multigenerational genetic analysis of tropical secondary regeneration in a canopy palm. *Ecology* 88: 3065–3075.

Shure, D. J., D. L. Phillips, and P. E. Bostick. 2006. Gap size and succession in cutover southern Appalachian forests: An 18 year study of vegetation dynamics. *Plant Ecology* 185: 299–318.

Silman, M. R., J. W. Terborgh, and R. A. Kiltie. 2003. Population regulation of a dominant rain forest tree by a major seed predator. *Ecology* 84: 431–438.

Stiles, F. G. 1975. Ecology, flowering phenology, and hummingbird pollination of some Costa Rican *Heliconia* species. *Ecology:* 56: 285–301.

———. 1977. Coadapted competitors: The flowering seasons of hummingbird-pollinated plants in a tropical forest. *Science* 198: 1177–1178.

———. 1983. *Heliconia latispatha* (Platanillo, wild plantain). In *Costa Rican Natural History*, D. H. Janzen (ed.). Chicago: University of Chicago Press.

Titiz, B., and R. L. Sanford Jr. 2007. Soil charcoal in old-growth rain forests from sea level to the continental divide. *Biotropica* 39: 673–682.

Toledo, V. M. 1977. Pollination of some rain forest plants by non-hovering birds in Veracruz, Mexico. *Biotropica* 9: 262–267.

Uhl, C. 1988. Restoration of degraded lands in the Amazon Basin. In *Biodiversity*, E. O. Wilson (ed.). Washington, DC: National Academy Press.

Uhl, C., R. Buschbacher, and E. A. S. Serrao. 1988a. Abandoned pastures in eastern Amazonia. I. Patterns of plant succession. *Journal of Ecology* 76: 663–681.

Uhl, C., K. Clark, N. Dezzeo, and P. Maquirino. 1988b. Vegetation dynamics in Amazonian treefall gaps. *Ecology* 69: 751–763.

Uhl, C., and C. F. Jordan. 1984. Succession and nutrient dynamics following forest cutting and burning in Amazonia. *Ecology* 65: 1476–1490.

Uhl, C., J. B. Kauffman, and D. L. Cummings. 1988c. Fire in the Venezuelan Amazon. 2: Environmental conditions necessary for forest fires in the evergreen rain forest of Venezuela. *Oikos* 53: 176–184.

Uhl, C., D. Nepstad, R. Buschbacher, K. Clark, B. Kauffman, and S. Subler. 1990. Studies of ecosystem response to natural and anthropogenic disturbances provide guidelines for designing sustainable land-use systems in Amazonia. In *Alternatives to Deforestation: Steps toward Sustainable Use of the Amazon Rain Forest*, A. B. Anderson (ed.). New York: Columbia University Press.

Van Breugel, M., F. Bongers, and M. Martinez-Ramos. 2007. Species dynamics during early secondary forest succession: Mortality and species turnover. *Biotropica* 35: 610–619.

Williams-Linera, G. 1983. Biomass and nutrient content in two successional stages of tropical wet forest in Uxpanapa, Mexico. *Biotropica* 15: 275–284.

Wright, S. J., C. Carrasco, O. Calderon, and S. Paton. 1999. The El Niño Southern Oscillation, variable fruit production, and famine in a tropical forest. *Ecology* 80: 1632–1647.

Wright, S. J., H. C. Muller-Landau, R. Condit, and S. P. Hubbell. 2003. Gap-dependent recruitment, realized vital rates, and size distributions of tropical trees. *Ecology* 84: 3174–3185.

Yarranton, G. A., and R. G. Morrison. 1974. Spatial dynamics of a primary succession: nucleation. *Journal of Ecology* 62: 417–428.

CHAPTER 7

Andersson, M. 1994. *Sexual Selection*. Princeton, NJ: Princeton University Press.

Anstett, M. C., M. Hossaert-McKey, and F. Kjellberg. 1999. Figs and fig pollinators: Evolutionary conflicts in a coevolved mutualism. *Trends in Ecology and Evolution* 12: 94–99.

Bascompte, J. 2009. Mutualistic networks. *Frontiers in Ecology and Environment* 7: 429–436.

Bascompte, J., P. Jordano, and J. M. Olesen. 2006. Asymmetric coevolutionary networks facilitiate biodiversity maintenance. *Science* 312: 431–433.

Berens, D. G., N. Farwig, G. Schaab, and K. Bohning-Gaese. 2008. Exotic guavas are foci of forest regeneration in Kenyan farmland. *Biotropica* 40: 104–112.

Bernays, E., and M. Graham. 1988. On the evolution of host specificity in phytophagous arthropods. *Ecology* 69: 886–899.

Blundell, A. C., and D. R. Peart. 2004. Density-dependent population dynamics of a dominant rain forest canopy tree. *Ecology* 85: 704–715.

Bradford, M. G., A. J. Dennis, and D. A. Westcott. 2008. Diet and dietary preferences of the southern cassowary (*Casuarius casuarius*) in North Queensland, Australia. *Biotropica* 40: 338–343.

Bravo, S. P. 2008. Seed dispersal and ingestion of insect-infested seeds by black howler monkeys in flooded forests in the Parana River, Argentina. *Biotropica* 40: 471–476.

Chapela, I. H., S. A. Rehner, T. R. Schultz, and U. G. Mueller. 1994. Evolutionary history of the symbiosis between fungus-growing ants and their fungi. *Science* 266: 1691–1694.

Clark, C. J., J. R. Poulsen, B. M. Bolker, E. F. Connor, and V. T. Parker. 2005. Comparative seed shadows of bird-, monkey-, and wind-dispersed trees. *Ecology* 86: 2684–2694.

Clark, J. S., M. Silman, R. Kern, E. Macklin, and J. HilleRisLambers. 1999. Seed dispersal near and far: Patterns across temperate and tropical forests. *Ecology* 80: 1475–1494.

Dalling, J. W., and R. Wirth. 1998. Dispersal of *Miconia argentea* seeds by the leaf cutting ant *Atta colombica*. *Journal of Tropical Biology* 14: 705–710.

Darwin, C. 1859. *On the Origin of Species by Means of Natural Selection of Favored Races in the Struggle for Life*. London: John Murray.

———. 1871. *The Descent of Man and Selection in Relation to Sex*. London: John Murray.

Desmond, A., and J. Moore. 2009. *Darwin's Sacred Cause*. Boston: Houghton Mifflin.

Ehrlich, P. R., and P. H. Raven. 1964. Butterflies and plants: A study in coevolution. *Evolution* 18: 586–608.

Emmons, L. H. 1980. Ecology and resource partitioning among nine species of African rain forest squirrels. *Ecological Monographs* 50: 31–54.

Eshiamwata, G. W., D. G. Berens, B. Bleher, W. R. J. Dean, and K. Bohning-Gaese. 2006. Bird assemblages in isolated *Ficus* trees in Kenyan farmland. *Journal of Tropical Ecology* 22: 723–726.

Fittkau, E. J., and H. Klinge. 1973. On biomass and trophic structure of the central Amazonian rain forest ecosystem. *Biotropica* 5: 1–14.

Fleming, T. H., R. Breitwisch, and G. H. Whitesides. 1987. Patterns of tropical vertebrate frugivore diversity. *Annual Review of Ecological Systems* 18: 91–109.

Futuyma, D. J., and M. Slatkin, eds. 1983. *Coevolution*. Sunderland, MA: Sinauer.

Galetti, M., C. I. Donatti, M. A. Pizo, and H. C. Giacomini. 2008. Big fish are the best: Seed dispersal of *Bactris glaucescens* by the pacu fish (*Piaractus mesopotamicus*) in the Pantanal, Brazil. *Biotropica* 40: 386–389.

Gentry, A. H. 1990. Floristic similarities and differences between Southern Central America and Upper and Central Amazonia. In *Four Neotropical Rain Forests*, A. H. Gentry (ed.). New Haven, CT: Yale University Press.

Goulding, M. 1980. *The Fishes and the Forest: Explorations in Amazonian Natural History*. Berkeley: University of California Press.

———. 1985. Forest fishes of the Amazon. In *Amazonia*, G. T. Prance and T. E. Lovejoy (eds.). Oxford, UK: Pergamon Press.

———. 1990. *The Flooded Forest*. London: Guild Publishing.

———. 1993. Flooded forests of the Amazon. *Scientific American* 266: 114–120.

Goulding, M., R. Barthem, and E. Ferreira. 2003. *The Smithsonian Atlas of the Amazon*. Washington, DC: Smithsonian Books.

Gutierrez, P. C. 1994. Mitochondrial-DNA polymorphism in the oilbird (*Steatornis caripensis*, Steatornithidae) in Venezuela. *Auk* 111: 573–578.

Hamilton, W. D. 1979. Wingless and fighting males in fig wasps and other insects. In *Reproduction, Competition, and Selection in Insects*, M. S. Blum and N. A. Blum (eds.). New York: Academic Press.

Harrison, R. D. 2003. Fig wasp dispersal and the stability of a keystone plant resource in Borneo. *Proceedings of the Royal Society of London B* 270: S76–S79.

Heithaus, E. R., T. H. Fleming, and P. A. Opler. 1975. Foraging patterns and resource utilization in seven species of bats in a seasonal tropical forest. *Ecology* 56: 841–854.

Heithaus, E. R., P. A. Opler, and H. B. Baker. 1974. Bat activity and pollination of *Bauhinia pauletia:* Plant–pollinator coevolution. *Ecology* 55: 412–419.

Herre, E. A, C. A. Machado, E. Bermingham, N. D. Nason, D. M. Windsor, S. S. McCafferty, W. Van Houten, and K. Bachmann. 1966. Molecular phylogenies of figs and their pollinator wasps. *Journal of Biogeography* 23: 521–530.

Herz, H., W. Byeschlag, and B. Holldobler. 2007. Herbivory rate of leaf-cutting ants in a tropical moist forest in Panama. *Biotropica* 39: 482–488.

Holbrook, K., and T. B. Smith. 2000. Seed dispersal and movement patterns in two species of *Ceratogymna* hornbills in a West African tropical lowland forest. *Oecologia* 125: 249–257.

Hölldobler, B., and E. O. Wilson. 1990. *The Ants*. Cambridge, MA: Belknap Press of Harvard University.

Howe, H. F. 1977. Bird activity and seed dispersal of a tropical wet forest tree. *Ecology* 58: 539–550.

———. 1982. Fruit production and animal activity in two tropical trees. In *The Ecology of a Tropical Forest*, E. G. Leigh Jr., A. S. Rand, and D. M. Windsor (eds.). Washington, DC: Smithsonian Institution Press.

Howe, H. F., and G. F. Estabrook. 1977. On intraspecific competition for avian dispersers in tropical trees. *American Naturalist* 111: 817–832.

Howell, D. J. 1974. Bats and pollen: Physiological aspects of the syndrome of chiropterophily. *Comparative Biochemistry and Physiology A: Physiology* 48: 263–276.

———. 1976. Plant-loving bats, bat-loving plants. *Natural History* 85: 52–59.

Hubbell, S. P., J. J. Howard, and D. F. Wiemer. 1984. Chemical leaf repellency to an attine ant: Seasonal distribution among potential host plant species. *Ecology* 65: 1067–1076.

Hubbell, S. P., D. F. Wiemer, and A. Adejare. 1983. An antifungal terpenoid defends a Neotropical tree *(Hymenaea)* against attack by fungus-growing ants (*Atta*). *Oecologia* 60: 321–327.

Irwin, R. E., L. S. Adler, and A. K. Brody. 2004. The dual role of floral traits: Pollinator attraction and plant defense. *Ecology* 85: 1503–1511.

Irwin, R. E., and A. K. Brody. 2000. Consequences of nectar robbing for realized male function in a hummingbird-pollinated plant. *Ecology* 81: 2637–2643.

Irwin, R. E., A. K. Brody, and N. M. Waser. 2001. The impact of floral larceny on individuals, populations, and communities. *Oecologia* 129: 525–533.

Janzen, D. H. 1970. Herbivores and the number of tree species in tropical forests. *American Naturalist* 104: 501–528.

———. 1971. Euglossine bees as long-distance pollinators of tropical plants. *Science* 171: 203–206.

———. 1975. *Ecology of Plants in the Tropics*. London: Edward Arnold.

———. 1979. How to be a fig. *Annual Review of Ecological Systems* 10: 13–52.

———. 1980. When is it coevolution? *Evolution* 34: 611–612.

Janzen, D. H., ed. 1983. *Costa Rican Natural History*. Chicago: University of Chicago Press.

Johnson, S. D., A. L. Hargreaves, and M. Brown. 2006. Dark, bitter-tasting nectar functions as a filter of flower visitors in a bird-pollinated plant. *Ecology* 87: 2709–2716.

Jordano, P. 1983. Fig-seed predation and dispersal by birds. *Biotropica* 15: 38–41.

———. 2000. Fruits and frugivory. In *Seeds: The Ecology of Regeneration in Natural Plant Communities*, M. Fenner (ed.). Wallingford, UK: Commonwealth Agricultural Bureau International.

Jousselin, E., J. Y. Raspus, and F. Kjellberg. 2003. Convergence and coevolution in a mutualism: Evidence from molecular phylogeny of *Ficus. Evolution* 57: 1255–1269.

Kubitzki, K. 1985. The dispersal of forest plants. In *Amazonia*, G. T. Prance and T. E. Lovejoy (eds.). Oxford, UK: Pergamon Press.

Leck, C. F. 1969. Observations of birds exploiting a Central American fruit tree. *Wilson Bulletin* 81: 264–269.

Leigh, E. G., Jr., and D. M. Windsor. 1982. Forest production and regulation of primary consumers on Barro Colorado Island. In *The Ecology of a Tropical Forest*, E. G. Leigh Jr., A. S. Rand, and D. M. Windsor (eds.). Washington, DC: Smithsonian Institution Press.

Levey, D. J. 1987. Seed size and fruit-handling techniques of avian frugivores. *American Naturalist* 129: 471–485.

Levey, D. J., T. C. Moermond, and J. S. Denslow. 1984. Fruit choice in Neotropical birds: The effect of distance between fruits on preference patterns. *Ecology* 65: 844–850.

———. 1994. Frugivory: An overview. In *La Selva: Ecology and Natural History of a Neotropical Rain Forest*, L. A. McDade, K. S. Bawa, H. A. Hespenheide, and G. S. Hartshorn (eds.). Chicago: University of Chicago Press.

Levey, D. J., and F. G. Stiles. 1994. Birds: ecology, behavior, and taxonomic affinities. In *La Selva: Ecology and Natural History of a Neotropical Rain Forest*, L. A. McDade, K. S. Bawa, H. A. Hespenheide, and G. S. Hartshorn (eds.). Chicago: University of Chicago Press.

Lill, A. 1974. The evolution of clutch size and male "chauvinism" in the white-bearded manakin. *Living Bird* 13: 211–231.

Little, A. E. F., and C. R. Currie. 2008. Black yeast symbionts compromise the efficiency of antibiotic defenses in fungus-growing ants. *Ecology* 89: 1216–1222.

Loiselle, B. A., and J. G. Blake. 1992. Population variation in a tropical bird community. *Bioscience* 42: 829–837.

Lowe-McConnell, R. H. 1987. *Ecological Studies in Tropical Fish Communities*. Cambridge, UK: Cambridge University Press.

Lucas, C. M. 2008. Within flood season variation in fruit consumption and seed dispersal by two characin fishes of the Amazon. *Biotropica* 40: 581–589.

Martin, M. M. 1970. The biochemical basis of the fungus–attine ant symbiosis. *Science* 169: 16–20.

Moermond, T. C., and J. S. Denslow. 1985. Neotropical avian frugivores: Patterns of behavior, morphology, and nutrition, with consequences for fruit selection. In *Neotropical Ornithology*, P. A. Buckley, M. S. Foster, E. S. Morton, R. S. Ridgely, and F. G. Buckley (eds.). Washington, DC: American Ornithologists' Union.

Morton, E. S. 1973. On the evolutionary advantages and disadvantages of fruit eating in tropical birds. *American Naturalist* 107: 8–22.

Mueller, U. G., S. A. Rehner, and T. R. Schultz. 1998. The evolution of agriculture in ants. *Science* 281: 2034–2038.

Murray, K. G. 1988. Avian seed dispersal of three Neotropical gap-dependent plants. *Ecological Monographs* 58: 271–298.

Nason, J. D., E. A. Herre, and J. L. Hamrick. 1998 The breeding structure of a tropical keystone plant resource. *Nature* 391, 685–687.

Paine, C.E.T, K. E. Harms, S. A. Schnitzer, and W. P. Carson. 2008. Weak competition among tropical tree seedlings: Implications for species coexistence. *Biotropica* 40: 432–440.

Poulsen, J. R., C. J. Clark, and T. B. Smith. 2001. Seed dispersal by a diurnal primate community in the Dja Reserve, Cameroon. *Journal of Tropical Ecology* 17: 787–808.

Poulsen, M., and J. J. Boomsma. 2005. Mutualistic fungi control crop diversity in fungus-growing ants. *Science* 307: 741–744.

Prance, G. T. 1985. The pollination of Amazonian plants. In *Amazonia*, G. T. Prance and T. E. Lovejoy (eds.). Oxford, UK: Pergamon Press.

Roca, R. L. 1994. *Oilbirds of Venezuela: Ecology and Conservation*, no. 24, R. A. Paynter Jr. (ed.). Cambridge, MA: Publications of the Nuttall Ornithological Club.

Rockwood, L. L. 1976. Plant selection and foraging patterns in two species of leaf-cutting ants (*Atta*). *Ecology* 57: 48–61.

Rønsted, N., G. D. Weiblen, J. M. Cook, N. Salamin, C. A. Machado, and V. Savolainen. 2005. 60 million years of co-divergence in the fig–wasp symbiosis. *Proceedings of the Royal Society of London B* 272: 2593–2599.

Schreier, B. M., A. H. Harcourt, S. A. Coppeto, and M. F. Somi. 2009. Interspecific competition and niche separation in primates: A global analysis. *Biotropica* 41: 283–291.

Snow, B. K., and D. W. Snow. 1979. The ochre-bellied flycatcher and the evolution of lek behavior. *Condor* 81: 286–292.

Snow, D. W. 1961. The natural history of the oilbird, *Steatornis caripensis*, in Trinidad, West Indies. Part 1. General behavior and breeding habits. *Zoologica* 46: 27–48.

———. 1962. The natural history of the oilbird, *Steatornis caripensis*, in Trinidad, West Indies. Part 2. Population, breeding ecology, food. *Zoologica* 47: 199–221.

———. 1971. Observations on the purple-throated fruit-crow in Guyana. *Living Bird* 10: 5–18.

———. 1976. *The Web of Adaptation*. Ithaca, NY: Cornell University Press.

———. 1982. *The Cotingas: Bellbirds, Umbrellabirds, and Other Species*. Ithaca, NY: Cornell University Press.

Stiles, E. W. 1980. Patterns of fruit presentation and seed dispersal in bird-disseminated woody plants in the eastern deciduous forest. *American Naturalist* 116: 670–688.

———. 1984. Fruit for all seasons. *Natural History* 93: 42–53.

Temeles, E. J., C. R. Koulouris, S. E. Sander, and W. J. Kress. 2009. Effect of flower shape and size on foraging performance and trade-offs in a tropical hummingbird. *Ecology* 90: 1147–1161.

Temeles, E. J., and W. J. Kress. 2003. Adaptation in a plant–hummingbird association. *Science* 300: 630–633.

Temeles, E. J., I. L. Pan, J. L. Brennan, and J. N. Horwitt. 2000. Evidence for ecological causation of sexual dimorphism in a hummingbird. *Science* 289: 441–443.

Thompson, J. N. 2005. *The Geographical Mosaic of Coevolution*. Chicago: University of Chicago Press.

———. 2006. Mutualistic webs of species. *Science* 312: 372–373.

Trail, P. W. 1985a. Courtship disruption modifies mate choice in a lek-breeding bird. *Science* 227: 778–779.

———. 1985b. A lek's icon: The courtship display of a Guianan cock-of-the-rock. *American Birds* 39: 235–240.

Weber, N. A. 1972. The attines: The fungus-culturing ants. *American Scientist* 60: 448–456.

Wetterer, J. K. 1994. Nourishment and evolution in fungus-growing ants and their fungi. In *Nourishment and Evolution in Insect Societies,* J. H. Hunt and C. A. Nalepa (eds.). Boulder, CO: Westview Press of Oxford and IBH Publishing.

Wheelwright, N. T. 1983. Fruits and the ecology of resplendent quetzals. *Auk* 100: 286–301.

———. 1985. Fruit size, gape width, and the diets of fruit-eating birds. *Ecology* 66: 808–818.

———. 1988. Fruit-eating birds and bird-dispersed plants in the tropics and temperate zone. *Trends in Ecology and Evolution* 3: 270–274.

Wheelwright, N. T., W. A. Haber, K. G. Murray, and C. Guindon. 1984. Tropical fruit-eating birds and their food plants: A survey of a Costa Rican lower montane forest. *Biotropica* 16: 173–192.

Wheelwright, N. T., and G. H. Orians. 1982. Seed dispersal by animals: Constraints with pollen dispersal, problems of terminology, and constraints on coevolution. *American Naturalist* 119: 402–413.

Wiebes, J. T. 1979. Co-evolution of figs and their insect pollinators. *Annual Review of Ecological Systems* 10: 1–12.

Wilson, E. O. 1971. *The Insect Societies.* Cambridge, MA: Belknap Press of Harvard University.

Youngsteadt, E. 2008. All that makes fungus gardens grow. *Science* 320: 1006–1007.

Zahawi, R. A. 2008. Instant trees: Using giant vegetative stakes in tropical forest restoration. *Forest Ecology and Management* 255: 3013–3016.

CHAPTER 8

Bates, H. W. 1862. Contributions of an insect fauna of the Amazon Valley. *Transactions of the Linnean Society of London* 23: 495–566.

Beatty, C. D., K. Beirinckx, and T. N. Sherratt. 2004. The evolution of Müllerian mimicry in multispecies communities. *Nature* 431: 63–67.

Beehler, B. M. 2008. *Lost Worlds: Adventures in the Tropical Rain Forest.* New Haven, CT: Yale University Press.

Benson, W. W. 1972. Natural selection for Müllerian mimicry in *Heliconius erato* in Costa Rica. *Science* 176: 936–939.

———. 1985. Amazon ant plants. In *Amazonia,* G. T. Prance and T. E. Lovejoy (eds). Oxford, UK: Pergamon Press.

Benson, W. W., K. S. Brown, and L. E. Gilbert. 1976. Coevolution of plants and herbivores: Passion flower butterflies. *Evolution* 29: 659–680.

Bentley, B. L. 1976. Plants bearing extrafloral nectaries and the associated ant community: Interhabitat differences in the reduction of herbivore damage. *Ecology* 54: 815–820.

———. 1977. Extrafloral nectaries and protection by pugnacious bodyguards. *Annual Review of Ecological Systems* 8: 407–427.

Brady, S. G. 2003. Evolution of the army ant syndrome: The origin and long-term evolutionary stasis of a complex of behavioral and reproductive adaptations. *Proceedings of the National Academy of Sciences USA* 100: 6575–6579.

Brower, A.V.Z. 1996. Parallel race formation and the evolution of mimicry in *Heliconius* butterflies: A phylogenetic hypothesis from mitochondrial DNA sequences. *Evolution* 50: 195–221.

Brower, L. P. 1969. Ecological chemistry. *Scientific American* 220: 22–29.

Brower, L. P., and J.V.Z. Brower. 1964. Birds, butterflies, and plant poisons: A study in ecological chemistry. *Zoologica* 49: 137–159.

Brower, L. P., J. V. Z. Brower, and C. T. Collins. 1963. Experimental studies of mimicry: 7. Relative palatability of Müllerian mimicry among Neotropical butterflies of the subfamily Heliconiinae. *Zoologica* 48: 65–84.

Brown, B. J., and J. L. Ewel. 1987. Herbivory in complex and simple tropical successional ecosystems. *Ecology* 68: 108–116.

Chamberlain, N. L., R. I. Hill, D. D. Kaplan, L. E. Gilbert, and M. R. Kronforst. 2009. Polymorphic butterfly reveals the missing link in ecological speciation. *Science* 326: 847–850.

Chamberlain, S. A., and J. N. Holland. 2009. Quantitative synthesis of context dependency in ant–plant protectionist mutualisms. *Ecology* 90: 2384–2392.

Coley, P. D. 1982. Rates of herbivory on different tropical trees. In *The Ecology of a Tropical Forest,* E. G. Leigh Jr., A. S. Rand, and D. M. Windsor (eds.). Washington, DC: Smithsonian Institution Press.

———. 1983. Herbivory and defensive characteristics of tree species in a lowland tropical forest. *Ecological Monographs* 53: 209–233.

———. 1984. Plasticity, costs, and anti-herbivore effects of tannins in a Neotropical tree, *Cecropia peltata* (Moraceae). *Bulletin of the Ecological Society of America* 65: 229.

Coley, P. D., J. P. Bryant, and F. S. Chapin III. 1985. Resource availability and plant antiherbivore defense. *Science* 230: 895–899.

Coley, P. D., et al. 2003. Using ecological criteria to design plant collection strategies for drug discovery. *Frontiers in Ecology and the Environment* 8: 421–428.

———. 2005. Divergent defensive strategies of young leaves in two species on *Inga. Ecology* 86: 2633–2643.

Crump, M. L. 1983. *Dendrobates granuliferus* and *Dendrobates pumilio,* In *Costa Rican Natural History,* D. H. Janzen (ed.). Chicago: University of Chicago Press.

Daly, J. W., H. M. Garraffo, and T. F. Spande. 1993. Amphibian alkaloids. In *The Alkaloids,* vol. 43, G. A. Cordell (ed.). San Diego, CA: Academic Press.

Davidson, D. W., S. C. Cook, R. R. Snelling, and T. H. Chua. 2003. Explaining the abundance of ants in lowland tropical rain forest canopies. *Science* 300: 969–972.

de Ruiter, P. C., V. Wolters, and J. C. Moore, eds. 2007. *Dynamic Food Webs: Multispecies Assemblages, Ecosystem Development, and Environmental Change.* Burlington, MA: Elsevier.

DeVries, P. J. 1987. *The Butterflies of Costa Rica and Their Natural History.* Princeton, NJ: Princeton University Press.

———. 1990. Enhancement of symbiosis between butterfly caterpillars and ants by vibrational communication. *Science* 248: 1104–1106.

———. 1992. Singing caterpillars, ants, and symbiosis. *Scientific American* 267: 76–82.

DeVries, P. J., and I. Baker. 1989. Butterfly exploitation of an ant–plant mutualism: Adding insult to herbivory. *Journal of the New York Entomological Society* 97: 332–340.

Dirzo, R., and E. Mendoza. 2007. Size-related seed predation in a heavily defaunated Neotropical rain forest. *Biotropica* 39: 355–362.

Dirzo, R., and A. Miranda. 1991. Altered patterns of herbivory and diversity in a forest understory: A case study of the possible consequences of contemporary defaunation. In *Plant–Animal*

Interactions: Evolutionary Ecology in Tropical and Temperate Regions, P. W. Price, P. W. Lewinsohn, G. W. Fernandes, and W. W. Benson (eds.). New York: John Wiley.

Dumbacher, J. P., B. M. Beehler, T. F. Spande, H. M. Garraffo, and J. W. Daly. 1992. Homobatrachotoxin in the genus *Pitahui*: Chemical defense in birds? *Science* 258: 799–801.

DuVal, E. H., H. W. Greene, and K. L. Manno. 2006. Laughing falcon (*Herpetotheres cachinnans*) predation on coral snakes (*Micrurus nigrocinctus*). *Biotropica* 38: 566–568.

Dyer, L. A., et al. 2007. Host specificity of Lepidoptera in tropical and temperate forests. *Nature* 448: 696–699.

Ehrlich, P. R., and P. H. Raven. 1964. Butterflies and plants: A study in coevolution. *Evolution* 18: 586–608.

———. 1967. Butterflies and plants. *Scientific American* 216: 104–113.

Futuyma, D. J. 1983. Evolutionary interaction among herbivorous insects and plants. In *Coevolution*, D. J. Futuyma and M. Slatkin (eds.). Sunderland, MA: Sinauer.

Gilbert, G. S., and C. O. Webb. 2007. Phylogenetic signal in plant pathogen–host range. *Proceedings of the National Academy of Science* 104: 4979–4983.

Gilbert, L. E. 1971. Butterfly–plant coevolution: Has *Passiflora adenopoda* won the selectional race with Heliconiinae butterflies? *Science* 172: 585–586.

———. 1975. Ecological consequences of a coevolved mutualism between butterflies and plants. In *Coevolution of Animals and Plants*, L. E. Gilbert and P. H. Raven (eds.). Austin, TX: University of Texas Press.

———. 1982. The coevolution of a butterfly and a vine. *Scientific American* 247: 110–121.

———. 1983. Coevolution and mimicry. In *Coevolution*, D. J. Futuyma and M. Slatkin (eds.). Sunderland, MA: Sinauer.

Glander, K. E. 1977. Poison in a monkey's Garden of Eden. *Natural History* 86: 35–41.

———. 1982. The impact of plant secondary compounds on primate foraging behavior. *American Journal of Physical Anthropology* 25: 1–18.

Gotwald, W. H., Jr. 1982. Army ants. In *Social Insects*, vol. 4, H. R. Hermann (ed.). New York: Academic Press.

———. 1995. *Army Ants: The Biology of Social Predation*. Ithaca, NY: Cornell University Press.

Gould, S. J., and N. Eldredge. 1977. Punctuated equilibria: The tempo and mode of evolution reconsidered. *Paleobiology* 3: 115–151.

Greene, H. W., and R. W. McDiarmid. 1981. Coral snake mimicry: Does it occur? *Science* 213: 1207–1211.

Gross, T., L. Rudolf, S. A. Levin, and U. Dieckmann. 2009. Generalized models reveal stabilizing factors in food webs. *Science* 325: 747–750.

Hansen, M. 1983. Yuca (Yuca, cassava). In *Costa Rican Natural History*, D. H. Janzen (ed.). Chicago: University of Chicago Press.

Harborne, J. B. 1982. *Introduction to Ecological Biochemistry*. New York: Academic Press.

Heil, M, J. Rattke, and W. Boland. 2005. Postsecretory hydrolysis of nectar sucrose and specialization in ant/plant mutualism. *Science* 308: 560–563.

Heisler, I. L. 1983. *Nyssodesmus python* (Milpes, large forest-floor millipede). In *Costa Rican Natural History*, D. H. Janzen (ed.). Chicago: University of Chicago Press.

Helson, J. E., T. L. Capson, T. Johns, A. Aiello, and D. M. Windsor. 2009. Ecological and evolutionary bioprospecting: Using aposematic insects as guides to rain forest plants active against disease. *Frontiers in Ecology and Environment* 7: 130–134.

Henry, M., and S. Jouard. 2007. Effect of bat exclusion on patterns of seed rain in tropical rain forest in French Guiana. *Biotropica* 39: 510–518.

Herz, H., W. Beyschlag, and B. Holldobler. 2007a. Assessing herbivory rates of leaf-cutting ants (*Atta colombica*) colonies through short-term refuse deposition counts. *Biotropica* 39: 476–481.

———. 2007b. Herbivory rate of leaf-cutting ants in a tropical moist forest in Panama at the population and ecosystem scales. *Biotropica* 39: 482–488.

Hölldobler, B., and E. O. Wilson. 1990. *The Ants*. Cambridge, MA: Belknap Press of Harvard University.

Hubbell, S. P., J. J. Howard, and D. F. Wiemer. 1984. Chemical leaf repellency to an attine ant: Seasonal distribution among potential host plant species. *Ecology* 65: 1067–1076.

Hubbell, S. P., D. F. Wiemer, and A. Adejare. 1983. An antifungal terpenoid defends a Neotropical tree *(Hymenaea)* against attack by fungus-growing ants *(Atta)*. *Oecologia* 60: 321–327.

Janzen, D. H. 1966. Coevolution and mutualism between ants and acacias in Central America. *Evolution* 20: 249–275.

———. 1975. *Ecology of Plants in the Tropics*. London: Edward Arnold.

———. 1980. Two potential coral snake mimics in a tropical deciduous forest. *Biotropica* 12: 77–78.

———. 1985. Plant defences against animals in the Amazonian rain forest. In *Amazonia*, G. T. Prance and T. E. Lovejoy (eds.). Oxford, UK: Pergamon Press.

Kalko, M. B., A. R. Smith, and E. K. V. Kalko. 2008. Bats limit arthropods and herbivory in a tropical forest. *Science* 320: 71.

Kapan, D. D. 2001. Three-butterfly system provides a field test of Müllerian mimicry. *Nature* 409: 338–340.

Keeler, K. H. 1980. Distribution of plants with extrafloral nectaries in temperate communities. *American Midland Naturalist* 104: 274–280.

Krieger, R. I., L. P. Feeny, and C. F. Wilkinson. 1971. Detoxication enzymes in the guts of caterpillars: An evolutionary answer to plant defenses? *Science* 172: 579–581.

Kursar, T. A., and P. D. Coley. 2003. Convergence in defense syndromes of young leaves in tropical rainforests. *Biochemical Systematics and Ecology* 31: 929–949.

Kursar, T. A., et al. 2008. Linking insights from basic research with bioprospecting in order to promote conservation, enhance research capacity, and provide economic uses of biodiversity. In *Tropical Forest Community Ecology*, W. Carson and S. Schnitzer (eds.). New York: Wiley-Blackwell.

Levin, D. A. 1971. Plant phenolics: An ecological perspective. *American Naturalist* 105: 157–181.

———. 1976. Alkaloid-bearing plants: An ecogeographic perspective. *American Naturalist* 110: 261–284.

Malvárez, J., C. A. Salazar, E. Bermingham, C. Salcedo, C. D. Jiggins, and M. Linares. 2006. Speciation by hybridization in *Heliconius* butterflies. *Nature* 441: 868–871.

Marquis, R. J. 1984. Leaf herbivores decrease fitness of a tropical plant. *Science* 226: 537–539.

Martin, M. M., and J. S. Martin. 1984. Surfactants: Their role in preventing the precipitation of proteins by tannins in insect guts. *Oecologia* 61: 342–345.

Maxson, L. R., and C. W. Myers. 1985. Albumin evolution in tropical poison frogs (Dendrobatidae): A preliminary report. *Biotropica* 17: 50–56.

Milton, K. 1979. Factors influencing leaf choice by howler monkeys: A test of some hypotheses of food selection by generalist herbivores. *American Naturalist* 114: 362–378.

———. 1981 Food choice and digestive strategies of two sympatric primate species. *American Naturalist* 117: 496–505.

———. 1982. Dietary quality and demographic regulation in a howler monkey population. In *The Ecology of a Tropical Forest*, E. G. Leigh Jr., A. S. Rand, and D. M. Windsor (eds.). Washington, DC: Smithsonian Institution Press.

Moffett, M. W. 1995. Poison-dart frogs. *National Geographic* 187: 98–111.

Moynihan, M. 1962. The organization and probable evolution of some mixed species flocks of Neotropical birds. *Smithsonian Miscellaneous Collection* 143: 1–140.

Munn, C. A. 1984. Birds of different feather also flock together. *Natural History* 11: 34–42.

———. 1985. Permanent canopy and understory flocks in Amazonia: Species composition ancd population density. In *Neotropical Ornithology*, P. A. Buckley, M. S. Foster, E. S. Morton, R. S. Ridgely, and F. G. Buckley (eds.). Washington, DC: American Ornithologists' Union.

Munn, C. A., and J. W. Terborgh. 1979. Multi-species territoriality in Neotropical foraging flocks. *Condor* 81: 338–347.

Myers, C. W., and J. W. Daly. 1983. Dart-poison frogs. *Scientific American* 248: 120–133.

Nathanson, J. A. 1984. Caffeine and related methylxanthines: Possible naturally occurring pesticides. *Science* 226: 184–186.

Ness, J. H., W. F. Morris, and J. L. Bronstein. 2009. For ant-protected plants, the best defense is a hungry offense. *Ecology* 90: 2823–2831.

Nijhout, H. 1994. Developmental perspectives on evolution of butterfly mimicry. *Bioscience* 44: 148–157.

Novotny, V., et al. 2007. Low beta diversity of herbivorous insects in tropical forests. *Nature* 448: 692–695.

Oliveira, P. S., and H. F. Leitao-Filho. 1987. Extrafloral nectaries: Their taxonomic distribution and abundance in the woody flora of cerrado vegetation in southeast Brazil. *Biotropica* 19: 140–148.

Oliveira, F., F. Rohe, and M. Gordo. 2010. Hunting strategy of the margay (*Leopardus wiedii*) to attract the wild pied tamarin (*Saguinus bicolor*). *Neotropical Primates* 16: 32–34.

Opler, P. A. 1981. Polymorphic mimicry by a neuropteran. *Biotropica* 13: 165–176.

Pfennig, D. W., W. R. Harcombe, and K. S. Pfennig. 2001. Frequency-dependent Batesian mimicry. *Nature* 410: 323.

Pimm, S. 2002. *Food Webs*. Chicago: University of Chicago Press.

Rathcke, B. J., and R. W. Poole. 1975. Coevolutionary race continues: Butterfly larval adaptation to plant trichomes. *Science* 187: 175–176.

Reagan, D. P., G. R. Camilo, and R. B. Waide. 1996. The community food web: Major properties and patterns of organization. In *The Food Web of a Tropical Rain Forest,* D. P. Reagan and R. B. Waide (eds.). Chicago: University of Chicago Press.

Reagan, D. P., and R. B. Waide. 1996. *The Food Web of a Tropical Rain Forest.* Chicago: University of Chicago Press.

Redford, K. 1992. The empty forest. *Bioscience* 42: 412–422.

Rowland, H. M., E. Ihalainen, L. Lindstrom, J. Mappes, and M. P. Speed. 2007. Co-mimics have a mutualistic relationship despite unequal differences. *Nature* 448: 64–67.

Rudgers, J. A., and M. C. Gardener. 2004. Extrafloral nectar as a resource mediating multispecies interactions. *Ecology* 85: 1495–1502.

Ryan, C. A. 1979. Proteinase inhibitors. In *Herbivores: Their Interaction with Secondary Plant Metabolites*, G. A. Rosenthal and D. H. Janzen (eds.). New York: Academic Press.

Sakai, S. 2002. General flowering in lowland mixed dipterocarp forests in South-East Asia. *Biological Journal of the Linnean Society* 75: 233–247.

Smiley, J. T. 1985. *Heliconius* caterpillar mortality during establishment on plants with and without attending ants. *Ecology* 66: 845–849.

Smiley, J. T., and C. S. Wisdom. 1985. Determinants of growth rate on chemically heterogeneous host plants by specialist insects. *Biochemical Systematics and Ecology* 13: 305–312.

Smith, S. M. 1975. Innate recognition of coral snake pattern by a possible avian predator. *Science* 187: 759–760.

———. 1977. Coral snake pattern rejection and stimulus generalisation by naive great kiskadees (Aves: Tyrannidae) *Nature* 265: 535–536.

Steppuhn, A., K. Gase, B. Krock, R. Halitschke, and I. T. Baldwin. 2004. Nicotine's defensive function in nature. *PLoS Biology* 2(8): e217. doi:10.1371/journal.pbio.0020217.

Terborgh, J. 1986. Keystone plant resources in the tropical forest. In *Conservation Biology: The Science of Scarcity and Diversity*, M. E. Soule (ed.). Sunderland, MA: Sinauer.

Terborgh, J., et al. 2008. Tree recruitment in an empty forest. *Ecology* 89: 1757–1768.

Turner, J. R. G. 1971. Studies of Müllerian mimicry and its evolution in Burnet moths and heliconid butterflies. In *Ecological Genetics and Evolution*, R. Creed (ed.). Oxford, UK: Blackwell Scientific.

———. 1975. A tale of two butterflies. *Natural History* 84: 29–37.

———. 1981. Adaptation and evolution in *Heliconius:* A defense of neo-Darwinism. *Annual Review of Ecological Systems* 12: 99–121.

———. 1988. The evolution of mimicry: A solution to the problem of punctuated equilibrium. *American Naturalist* 131, Supplement: S42–S66.

Van Valen, L. 1973. A new evolutionary law. *Evolutionary Theory* 1: 1–30.

Wasko, D. K., and M. Sasa. 2009. Activity patterns of a Neotropical ambush predator: Spatial ecology of the fer-de-lance (*Bothrops asper*, Serpentes: Viperidae) in Costa Rica. *Biotropica* 41: 241–249.

Webb, C. O., J. A. Losos, and A. A. Agrawal, eds. 2006. Integrating phylogenics into community ecology. Special issue. *Ecology* 87, Supplement: S1–S165.

Weiblen, G. D., C. O. Webb, V. Novotny, Y. Basset, and S. E. Miller. 2006. Phylogenetic dispersion of host use in a tropical insect herbivore community. *Ecology* 87(7), Supplement: S62–S75.

Whittaker, R. H., and P. P. Feeny. 1971. Allelochemics: Chemical interactions between species. *Science* 171: 757–770.

Williams-Guillen, K., I. Perfecto, and J. Vanermeer. 2008. Bats limit insects in a Neotropical agroforestry system. *Science* 320: 70.

Willis, E. O., and Y. Oniki. 1978. Birds and army ants. *Annual Review of Ecological Systems* 9: 243–263.

Willson, S. K. 2004. Obligate army-ant-following birds: A study of ecology, spatial movement patterns, and behavior in Amazonian Peru. *Ornithological Monographs* 55: 1–67.

Wrege, P. H., M. Wikelski, J. T. Mandel, T. Rassweiler, and I. D. Couzin. 2005. Antbirds parasitize foraging army ants. *Ecology* 86: 555–559.

Yasuda, M., et al. 1999. The mechanism of general flowering in Dipterocarpaceae in the Malay Peninsula. *Journal of Tropical Ecology* 15: 437–449.

Zucker, W. V. 1983. Tannins. Does structure determine function? An ecological perspective. *American Naturalist* 121: 355–365.

CHAPTER 9

Baker, D. F. 2007. Reassessing carbon sinks. *Science* 316: 1708–1709.

Beerling, D. 2008. *The Emerald Planet: How Plants Changed Earth's History.* Oxford, UK: Oxford University Press.

Brown, S., and A. E. Lugo. 1982. The storage and production of organic matter in tropical forests and their role in the global carbon cycle. *Biotropica* 14: 161–187.

Bunker, D. E., F. DeClerck, J. C. Bradford, R. K. Colwell, I. Perfecto, O. L. Phillips, M. Sankaran, and S. Naeem. 2005. Species loss and aboveground carbon storage in a tropical forest. *Science* 310: 1029–1031.

Chambers, J. Q., N. Higuchi, E. S. Tribuzy, and S. E. Trumbore. 2001. Carbon sink for a century. *Nature* 410: 429.

Chave, J., et al. 2008. Assessing evidence for a pervasive alteration of tropical tree communities. *PLoS Biology* 6: e45. doi: 10.1371/journal.pbio.0060045.

Clark, D. A. 2002. Are tropical forests an important carbon sink? Reanalysis of the long-term plot data. *Ecological Applications* 12: 3–7.

———. 2004. Tropical forests and global warming: Slowing it down or speeding it up? *Frontiers in Ecology and the Environment* 2: 73–80.

Clark, D. A., S. Brown, D. W. Kicklighter, J. Q. Chambers, J. R. Thomplinson, J. Ni, and E. A. Holland. 2001. Net primary productivity in tropical forests: An evaluation and synthesis of existing field data. *Ecological Applications* 11: 371–384.

Clark, D. A., and D. B. Clark. 1994. Climate induced annual variation in canopy tree growth in a Costa Rican tropical rain forest. *Journal of Ecology* 82: 865–872.

Cochrane, M. A. 2003. Fire science for rain forests. *Nature* 421: 913–919.

Colwell, R. K., G. Brehm, C. L. Cardelus, A. C. Gilman, and J. T. Longino. 2008. Global warming, elevational range shifts, and lowland biotic attrition in the wet tropics. *Science* 322: 258–261.

Cubash, U., et al. 2001. Projections for future climate change. In *Climate Change 2001: The Scientific Basis*, J. T. Houghton et al. (eds.). Cambridge UK: Cambridge University Press.

Engelbrecht, B. M. J., et al. 2007. Drought sensitivity shapes species distribution patterns in tropical forests. *Nature* 447: 80–82.

Feeley, K. J., et al. 2007. The role of gap phase processes in the biomass dynamics of tropical forests. *Proceedings of the Royal Society B* 274: 2857–2864.

Fisher, J. I., G. C. Hurtt, R. Q. Thomas, and J. Q. Chambers. 2008. Clustered distributions lead to bias in large-scale estimates based on forest sample plots. *Ecology Letters* 11: 554–563.

Grace, J., and Y. Malhi. 2002. Carbon dioxide goes with the flow. *Nature* 416: 594–595.

Graham, E. A., S. S. Mulkey, S. J. Wright, K. Kitajima, and N. G. Phillips. 2003. Cloud cover limits productivity in a rain forest tree during tropical rainy seasons. *Proceedings of the National Academy of Sciences USA* 100: 572–576.

Holloway, M. 1993. Sustaining the Amazon. *Scientific American* 269: 90–99.

Houghton, J. T., et al., eds. 2001. *Climate Change 2001: The Scientific Basis.* Cambridge UK: Cambridge University Press.

Hughen, K. A., T. I. Eglinton, L. Xu, and M. Makou. 2004. Abrupt tropical vegetation response to rapid climate changes. *Science* 304: 1955–1959.

Jordan, C. F. 1985. *Nutrient Cycling in Tropical Forest Ecosystems.* New York: John Wiley.

Kitayama, K., and S.-J. Aiba. 2002. Ecosystem structure and productivity of tropical rain forests along altitudinal gradients with contrasting soil phosphorus pools on Mt. Kinabalu, Borneo. *Journal of Ecology* 90: 37–51.

Kormondy, E. J. 1996. *Concepts of Ecology*, 4th ed. Upper Saddle River, NJ: Prentice Hall.

Laurance, W. F., et al. 2004. Pervasive alteration of tree communities in undisturbed Amazonian forests. *Nature* 428: 171–175.

———. 2005. Altered tree communities in undisturbed Amazonian forests: A consequence of global change? *Biotropica* 37: 160–162.

Lawrence, W. T., Jr. 1996. Plants: The food base. In *The Food Web of a Tropical Rain Forest*, D. P. Reagan and R. B. Waide (eds.). Chicago: University of Chicago Press.

Le Dantec, V., E. Dufrene, and B. Saugler. 2000. Interannual and spatial variation in maximum leaf area index of temperate deciduous stands. *Forest Ecology and Management* 134: 71–81.

Leigh, E. G., Jr. 1975. Structure and climate in tropical rain forest. *Annual Review of Ecology and Systematics* 6: 67–86.

Lewis, S. L. 2006. Tropical forests and the changing Earth system. *Philosophical Transactions of the Royal Society of London B* 261: 195–210.

Lewis, S. L., et al. 2009. Increasing carbon storage in intact African tropical forests. *Nature* 457: 1003–1006.

Lovejoy, T. E., and L. Hannah, eds. 2005. *Climate Change and Biodiversity*. New Haven, CT: Yale University Press.

Malhi, Y., and J. Grace. 2000. Tropical forests and atmospheric carbon dioxide. *Trends in Ecology and Evolution* 15: 332–337.

Malhi, Y., and J. Wright. 2004. Spatial patterns and recent trends in the climate of tropical rain forest regions. *Philosophical Transactions of the Royal Society of London B* 359: 311–329.

Mayle, F. E., R. Burbridge, and T. J. Killeen. 2000. Millennial-scale dynamics of southern Amazonian rain forests. *Science* 290: 2291–2294.

Mayorga, E., et al. 2005. Young organic matter as a source of carbon dioxide outgassing from Amazonian rivers. *Nature* 436: 538–541.

Mellilo, J. M., A. D. McGuire, D. W. Kicklighter, B. I. Moore, C. J. Vorosmarty, and A. L. Schloss. 1993. Global climate change and terrestrial net primary production. *Nature* 363: 234–240.

Moran, E. F., E. Brondizio, P. Mausel, and Y. Wu. 1994. Integrating Amazonian vegetation, land-use, and satellite data. *Bioscience* 44: 329–338.

Nelson, B. W. 2005. Pervasive alteration of tree communities in undisturbed Amazonian forests. *Biotropica* 37: 158–159.

Page, S. E., F. Siegert, J. O. Rieley, H.-D. V. Boehm, A. Jaya, and S. Limin. 2002. The amount of carbon released from peat and forest fires in Indonesia during 1997. *Nature* 420: 61–65.

Phillips, O. L., et al. 1998. Changes in the carbon balance of tropical forests: Evidence from long-term plots. *Science* 282: 439–442.

———. 2002. Increasing dominance of large lianas in Amazonian forests. *Nature* 418: 770–774.

———. 2009. Drought sensitivity of the Amazon rain forest. *Science* 323: 1344–1347.

Poorter, L., and K. Kitajima. 2007. Carbohydrate storage and light requirements of tropical moist and dry forest tree species. *Ecology* 88: 1000–1011.

Raich, J. W., A. E. Russell, K. Kitayama, W. J. Parton, and P. M. Vitousek. 2006. Temperature influences carbon accumulation in moist tropical forests. *Ecology* 87: 76–87.

Raich, J. W., et al. 1991. Potential net primary productivity in South America: Application of a global model. *Ecology Applications* 1: 399–429.

Raymond, P. A. 2005. The age of the Amazon's breath. *Nature* 436: 469–470.

Richey, J. E., J. M. Melack, A. K. Aufdenkampe, V. M. Ballester, and L. L. Hess. 2002. Outgassing from Amazonian rivers and wetlands as a large tropical source of atmospheric CO_2. *Nature* 416: 617–620.

Rosenzweig, C., et al. 2008. Attributing physical and biological impacts to anthropogenic climate change. *Nature* 453: 353–357.

Round, P. D., and G. A. Gale. 2008. Changes in the status of *Lophura* pheasants in Khao Yai National Park, Thailand: a response to global warming? *Biotropica* 40: 225–230.

Saleska, S. R., K. Didan, A. R. Huete, and H. R. da Rocha. 2007. Amazon forests green-up during 2005 drought. *Science* 318: 612.

Saleska, S. R., et al. 2003. Carbon in Amazon forests: Unexpected seasonal fluxes and disturbance-induced losses. *Science* 302: 1554–1557.

Schiermeier, Q. 2006. Methane finding baffles scientists. *Nature* 439: 128.

Schuur, E. A. G. 2003. Productivity and global climate revisited: The sensitivity of tropical forest growth to precipitation. *Ecology* 84: 1165–1170.

Selaya, N. G., and N. P. R. Anten. 2010. Leaves of pioneer and later-successional trees have similar lifetime carbon gain in tropical secondary forest. *Ecology* 91: 1102–1113.

Sierra, C. A., M. E. Harmon, F. H. Moreno, S. A. Orrego, and J. I. Del Valle. 2007. Spatial and temporal variability of net ecosystem production in a tropical forest: Testing the hypothesis of a significant carbon sink. *Global Change Biology* 13: 838–853.

Smil, V. 1979. Energy flows in the developing world. *American Scientist* 67: 522–531.

Snitzer, S. A., and F. Bongers. 2002. The ecology of lianas and their role in tropical forests. *Trends in Ecology and Evolution* 17: 223–230.

Stephens, B. B., et al. 2007. Weak northern and strong tropical land carbon uptake from vertical profiles of atmospheric CO_2. *Science* 316: 1732–1735.

Svenning, J.-C. and R. Condit. 2008. Biodiversity in a warmer world. *Science* 322: 206–207.

Taylor, B. W., A. S. Flecker, and R. O. Hall Jr. 2006. Loss of a harvested fish species disrupts carbon flow in a diverse tropical river. *Science* 313: 833–836.

Tewksbury, J. L., R. B. Huey, and C. A. Deutsch. 2008. Putting the heat on tropical animals. *Science* 320: 1296–1297.

Vitousek, P. M., P. R. Ehrlich, A. E. Ehrlich, and P. A. Matson. 1986. Human appropriation of the products of photosynthesis. *Bioscience* 36: 368–373.

Vogt, K. A., C. C. Grier, and D. J. Vogt. 1986. Production, turnover, and nutrient dynamics of above- and below-ground detritus of world forests. In *Advances in Ecological Research*, vol. 15, A. Macfadyen and E. D. Ford (eds.). London: Academic Press.

Whittaker, R. H. 1975. *Communities and Ecosystems*, 2nd ed. New York: Macmillan.

Whittaker, R. H., and G. E. Likens. 1975. *The Biosphere and Man. Ecological Studies 14*. Heidelberg: Springer.

Wright, S. J., O. Calderon, A. Hernandez, and S. Paton. 2004. Are lianas increasing in importance in tropical forests? A 17-year record from Panama. *Ecology* 85: 484–489.

CHAPTER 10

Abe, T., and M. Higashi. 2001. Isoptera. In *Encyclopedia of Biodiversity*, vol. 3, S. A. Levin (ed.). San Diego, CA: Academic Press.

Ackerman, I. L., R. Constantino, H. G. Gauch Jr., J. Lehmann, S. J. Riha, and E. C. M. Fernandes. 2009. Termite (Insecta: Isoptera) species composition in a primary rain forest and agroforests in Central America. *Biotropica* 41: 226–233.

Arnold, A. E., and F. Lutzoni. 2007. Diversity and host range of foliar fungal endophytes: Are tropical leaves biodiversity hotspots? *Ecology* 88: 541–549.

Bates, H. W. 1862. Contributions of an insect fauna of the Amazon Valley. *Transactions of the Linnean Society of London* 23: 495–566.

Bentley, B. L., and E. J. Carpenter. 1980. The effects of desiccation and rehydration on nitrogen fixation by epiphylls in a tropical rain forest. *Microbiology Ecology* 6: 109–113.

———. 1984. Direct transfer of newly fixed nitrogen from free-living epiphyllous microorganisms to their host plant. *Oecologia* 63: 52–56.

Bravo, A., K. E. Harms, R. D. Stevens, and L. H. Emmons. 2008. *Collpas*: Activity hotspots for frugivorous bats (Phyllostomidae) in the Peruvian Amazon. *Biotropica* 40: 203–210.

Brightsmith, D. J., J. Taylor, and T. D. Phillips. 2008. The roles of soil characteristics and toxin adsorption in avian geophagy. *Biotropica* 40: 766–774.

Campos-Arceiz, A. 2009. Shit happens (to be useful)! Use of elephant dung as habitat by amphibians. *Biotropica* 41: 406–407.

Cleveland, C. C., S. C. Reed, and A. R. Townsend. 2006. Nutrient regulation of organic matter decomposition in a tropical rain forest. *Ecology* 87: 492–503.

Cleveland, C. C., et al. 1999. Global patterns of terrestrial biological nitrogen (N_2) fixation in natural ecosystems. *Global Biogeochemical Cycles* 13: 623–645.

Cuevas, E. 2001. Soil versus biological controls on nutrient cycling in terre firme forests. In *Biogeochemistry of the Amazon Basin*, M. E. McClain, R. L. Victoria, and J. E. Richey (eds.). Oxford, UK: Oxford University Press.

Dale, V. H., S. M. Pearson, H. L. Offerman, and R. V. O'Neill. 1994. Relating patterns of land-use change to faunal biodiversity in the Central Amazon. *Conservation Biology* 8: 1027–1036.

Dalling, J. W., M. D. Swaine, and N. G. Garwood. 1998. Dispersal patterns and seed bank dynamics of pioneer trees in moist tropical soil. *Ecology* 79: 564–578.

Dance, A. 2008. What lies beneath. *Nature* 455: 724–725.

Davidson E. A., et al. 2007. Recuperation of nitrogen cycling in Amazonian forests following agricultural abandonment. *Nature* 447: 995–998.

Eichhorn, M. P., S. G. Compton, and S. E. Hartley. 2008. The influence of soil type on rain forest insect herbivore communities. *Biotropica* 40: 707–713.

Falkowski, P. G., T. Fenchel, and E. F. Delong. 2008. The microbial engines that drive Earth's biogeochemical cycles. *Science* 320: 1034–1039.

Feeley, K. J., and J. W. Terborgh. 2005. The effects of herbivore density on soil nutrients and tree growth in tropical forest fragments. *Ecology* 86: 116–124.

Fine, P. V. A., et al. 2006. The growth-defense trade-off and habitat specialization by plants in Amazonian forests. *Ecology* 87(7), Supplement: S150–S162.

Forman, R. T. T. 1975. Canopy lichens with blue-green algae: A nitrogen source in a Colombian rain forest. *Ecology* 56: 1176–1184.

Gallery, R. E., J. W. Dalling, and A. E. Arnold. 2007. Diversity, host affinity, and distribution of seed-infecting fungi: A case study with *Cecropia*. *Ecology* 88: 582–588.

Golley, F. B. 1975. *Mineral Cycling in a Tropical Moist Forest Ecosystem*. Athens, GA: University of Georgia Press.

———. 1983. Nutrient cycling and nutrient conservation. In *Tropical Rain Forest Ecosystems: Structure and Function*, F. B. Golley (ed.). Amsterdam: Elsevier Scientific.

Golley, F. B., J. T. McGinnis, R. G. Clements, G. I. Child, and H. J. Duever. 1969. The structure of tropical forests in Panama and Colombia. *BioScience* 19: 693–696.

Golley, F. B., H. T. Odum, and R. F. Wilson. 1962. The structure and metabolism of a Puerto Rican red mangrove forest in May. *Ecology* 43: 9–19.

Herre, E. A., et al. 2007. Ecological implications of anti-pathogen effects of tropical fungal endophytes and mycorrhizae. *Ecology* 88: 550–558.

Herrera, R. 1985. Nutrient cycling in Amazonian forests. In *Amazonia*, G. T. Prance and T. E. Lovejoy, (eds.). Oxford, UK: Pergamon Press.

Hubbell, S. P., J. J. Howard, and D. F. Wiemer. 1984. Chemical leaf repellency to an attine ant: Seasonal distribution among potential host plant species. *Ecology* 65: 1067–1076.

Hubbell, S. P., D. F. Wiemer, and A. Adejare. 1983. An antifungal terpenoid defends a Neotropical tree *(Hymenaea)* against attack by fungus-growing ants (*Atta*). *Oecologia* 60: 321–327.

Hyodo, F., I. Tayasu, T. Inoue, J. I. Azuma, T. Kudo, and T. Abe. 2003. Differential role of symbiotic fungi in lignin degradation and food provision for fungus-growing termites (Macrotermitinae: Isoptera). *Functional Ecology* 17: 186–193.

Irion, G. 1978. Soil infertility in the Amazonian rain forest. *Naturwissenschaften* 65: 515–519.

Janos, D. P. 1980. Mycorrhizae influence tropical succession. *Biotropica* 12, Supplement: 56–64.

———. 1983. Tropical mycorrhizae, nutrient cycles, and plant growth. In *Tropical Rain Forest: Ecology and Management*, S. L. Sutton, T. C. Whitmore, and A. C. Chadwick (eds.). Oxford, UK: Blackwell Scientific.

Janos, D. P., C. T. Sahley, and L. H. Emmons. 1995. Rodent dispersal of vesicular-arbuscular mycorrhizal fungi in Amazonian Peru. *Ecology* 76: 1852–1858.

Janzen, D. H. 1974. Tropical blackwater rivers, animals, and mast fruiting in the Dipterocarpaceae. *Biotropica* 6: 69–103.

Janzen, D. H., and C. Vasquez-Yanes. 1991. Aspects of tropical seed ecology of relevance to management of tropical forested wildlands. In *Rain Forest Regeneration and Management*, A. Gomez-Pompa, T. C. Whitmore, and M. Hadley (eds.). Paris: UNESCO.

John, R., et al. 2007. Soil nutrients influence spatial distributions of tropical tree species. *Proceedings of the National Academy of Sciences* 104: 864–869.

Jordan, C. F. 1982. Amazon rain forests. *American Scientist* 70: 394–401.

———. 1985a. *Nutrient Cycling in Tropical Forest Ecosystems*. New York: J. Wiley.

———. 1985b. Soils of the Amazon rain forest. In *Amazonia*, G. T. Prance and T. E. Lovejoy (eds.). Oxford, UK: Pergamon Press.

Jordan, C. F., F. Golley, J. D. Hall, and J. Hall. 1979. Nutrient scavenging of rainfall by the canopy of an Amazonian rain forest. *Biotropica* 12: 61–66.

Jordan, C. F., and R. Herrera. 1981. Tropical rain forests: Are nutrients really critical? *American Naturalist* 117: 167–180.

Jordan, C. F., and J. R. Kline. 1972. Mineral cycling: Some basic concepts and their application in a tropical rain forest. *Annual Review of Ecological Systems* 3: 33–50.

Klinge, H., W. A. Rodrigues, E. Brunig, and E. J. Fittkau. 1975. Biomass and structure in a central Amazonian rain forest. In *Tropical Ecological Systems: Trends in Terrestrial and Aquatic Research*, F. B. Golley and E. Medina (eds.). New York: Springer-Verlag.

Kurokawa, H., and T. Nakashizuka. 2008. Leaf herbivory and decomposability in a Malaysian rain forest. *Ecology* 89: 2645–2656.

Lal, R. 1990. Tropical soils: Distribution, properties, and management. In *Tropical Resources: Ecology and Developemnt*, J. I. Furtado, W. B. Morgan, J. R. Pfafflin, and K. Ruddle (eds.). London: Harwood Associates.

Langley, J. A., and B. A. Hungate. 2003. Mycorrhizal controls on belowground litter quality. *Ecology* 84: 2302–2312.

Lavelle, P., et al. 1993. A hierarchical model for decomposition in terrestrial ecosystems: Application to soils of the humid tropics. *Biotropica* 25: 130–150.

Lloyd, J., M. I. Bird, E. M. Veenendaal, and B. Kruijt. 2001. Should phosphorus availability be constraining moist tropical forest response to increasing CO_2 concentrations? In *Global Biogeochemical Cycles in the Climate System*, E.-D. Schulze et al. (eds.). New York: Academic Press.

Lodge, D. J. 1996. Microorganisms. In *The Food Web of a Tropical Rain Forest*, D. P. Reagan and R. B. Waide (eds.). Chicago: University of Chicago Press.

Lubin, Y. D. 1983. *Nasutitermes* (Comejan, hormiga blanca, nausute termite, arboreal termite). In *Costa Rican Natural History*, D. H. Janzen (ed.). Chicago: University of Chicago Press.

Lubin, Y. D., and G. G. Montgomery. 1981. Defenses of *Nasutitermes* termites (Isoptera, Termitidae) against *Tamandua* anteaters (Edentata, Myrmecophagidae). *Biotropica* 13: 66–76.

Lucas, Y., F. J. Luizao, A. Chauvel, J. Rouiller, and D. Nahon. 1993. The relation between biological activity of the rain forest and mineral composition of soils. *Science* 260: 521–523.

Martinelli, L. A., et al. 1999. Nitrogen stable isotopic composition of leaves and soil: Tropical versus temperate forests. *Biogeochemistry* 46: 45–65.

Martius, C. 1994. Diversity and ecology of termites in Amazonian forests. *Pedobiologia* 38: 407–428.

McGuire, K. L. 2007. Common ectomycorrhizal networks may maintain monodominance in a tropical rain forest. *Ecology* 88: 567–574.

Meyer, J. L. 1990. A blackwater perspective on riverine ecosystems. *Bioscience* 40: 643–651.

Montagnini, F., and C. F. Jordan. 2005. *Tropical Forest Ecology: The Basis for Conservation and Management*. Berlin: Springer.

Motavalli, P. P., C. A. Palm, W. J. Parton, E. T. Elliott, and S. D. Frey. 1995. Soil pH and organic C dynamics in tropical forest soils: Evidence from laboratory and simulation studies. *Soil Biology and Biochemistry* 27: 1589–1599.

Moutinho, P., D. C. Nepstad, and E. A. Davidson. 2003. Influence of leaf-cutting ant nests on secondary forest growth and soil properties in Amazonia. *Ecology* 84: 1265–1276.

Moyersoen, B., I. J. Alexander, and A. H. Fitter. 1998. Phosphorus nutrition and ectomycorrhizal and arbuscular mycorrhizal tree seedlings from a lowland tropical rain forest in Korup National Park, Cameroon. *Journal of Tropical Ecology* 14: 47–61.

Nadkarni, N. 1981. Canopy roots: Convergent evolution in rainforest nutrient cycles. *Science* 214: 1023–1024.

Nicholaides, J. J., III, et al. 1985. Agricultural alternatives for the Amazon Basin. *Bioscience* 35: 279–285.

Paoli, G. D., L. M. Curran, and D. R. Zak. 2006. Phosphorus efficiency of Bornean rain forest productivity: Evidence against the unimodal efficiency hypothesis. *Ecology* 86: 1548–1561.

Parker, G. G. 1994. Soil fertility, nutrient acquisition, and nutrient cycling. In *La Selva: Ecology and Natural History of a Neotropical Rain Forest,* L. A. McDade, K. S. Bawa, H. A. Hespenheide, and G. S. Hartshorn (eds.). Chicago: University of Chicago Press.

Poulsen, A. D., H. Tuomisto, and H. Balslev. 2006. Edaphic and floristic variation within a 1-ha plot of lowland Amazonian rain forest. *Biotropica* 38: 468–478.

Powell, L. L., T. U. Powell, G. V. N. Powell, and D. J. Brightsmith. 2009. Parrots take it with a grain of salt: Available sodium content may drive *collpa* (clay lick) selection in southeastern Peru. *Biotropica* 41: 279–282.

Prestwich, G. D., and B. L. Bentley. 1981. Nitrogen fixation by intact colonies of the termite *Nasutitermes corniger. Oecologia* 49: 249–251.

Prestwich, G. D., Bentley, B. L., and E. J. Carpenter. 1980. Nitrogen sources for Neotropical nasute termites: Fixation and selective foraging. *Oecologia* 46: 397–401.

Pringle, R. M., D. F. Doak, A. K. Brody, R. Jocqué, and T. M. Palmer. 2010. Spatial pattern enhances ecosystem functioning in an African savanna. *PLoS Biology* 8(5): e1000377; doi: 10–1371/journal. pbio. 1000377.

Ramos-Zapata, J., R. Orellana, P. Guaderrama, S. Medina-Peralta. 2009. Contribution of mycorrhizae to early growth and phosphorus uptake by a neotropical palm. *Journal of Plant Nutrition* 32: 855–866.

Reed, S. C., C. C. Cleveland, and A. R. Townsend. 2007. Controls over leaf litter and soil nitrogen fixation in two lowland tropical rain forests. *Biotropica* 39: 585–592.

———. 2008. Tree species control rates of free-living nitrogen fixation in a tropical rain forest. *Ecology* 89: 2924–2934.

Richards, P. W. 1952. *The Tropical Rain Forest.* Cambridge, UK: Cambridge University Press.

St. John, T. V. 1985. Mycorrhizae. In *Amazonia*, G. T. Prance and T. E. Lovejoy (eds.). Oxford, UK: Pergamon Press.

Salati, E., and P. B. Vose. 1984. Amazon Basin: a system in equilibrium. *Science* 225: 129–138.

Salick, J., R. Herrera, and C. F. Jordan. 1983. Termitaria: Nutrient patchiness in nutrient-deficient rain forests. *Biotropica* 15: 1–7.

Sanford, R. L., Jr. 1987. Apogeotropic roots in an Amazon rain forest. *Science* 235: 1062–1064.

Santiago, L. S. 2007. Extending the leaf economics spectrum to decomposition: Evidence from a tropical forest. *Ecology* 88: 1126–1131.

Scott, D. A., J. Proctor, and J. Thompson. 1992. Ecological studies on a lowland evergreen rain forest on Maraca Island, Roraima, Brazil, II: Litter and nutrient cycling. *Journal of Ecology* 80: 705–717.

Sollins, P., F. Sancho M., R. Mata Ch., and R. L. Sanford Jr. 1994. Soils and soil processes. In *La Selva: Ecology and Natural History of a Neotropical Rain Forest,* L. A. McDade, K. S. Bawa, H. A. Hespenheide, and G. S. Hartshorn (eds.). Chicago: University of Chicago Press.

Sternberg, L. da S. L., M. C. Pinzon, M. Z. Moreira, P. Moutinho, E. I. Rojas, and E. A. Herre. 2007. Plants use macronutrients accumulated in leaf-cutting ant nests. *Proceedings of the Royal Society B* 274: 315–321.

Thompson, J., J. Proctor, V. Viana, W. Milliken, J. A. Ratter, and D. A. Scott. 1992. Ecological studies on a lowland evergreen rain forest on Maraca Island, Roraima, Brazil. I: Physical environment, forest structure, and leaf chemistry. *Journal of Ecology* 80: 689–703.

Townsend, A. R., C. C. Cleveland, G. P. Asner, and M. M. C. Bustamante. 2007. Controls over foliar N:P ratios in tropical rain forests. *Ecology* 88: 107–118.

Uhl, C., D. Nepstad, R. Buschbacher, K. Clark, B. Kauffman, and S. Subler. 1990. Studies of ecosystem response to natural and anthropogenic disturbances provide guidelines for designing sustainable land-use systems in Amazonia. In *Alternatives to Deforestation: Steps toward Sustainable Use of the Amazon Rain Forest*, A. B. Anderson (ed.). New York: Columbia University Press.

Vitousek, P. M. 1984. Litterfall, nutrient cycling, and nutrient limitation in tropical forests. *Ecology* 65: 285–298.

Vitousek, P. M., and R. L. Sanford. 1986. Nutrient cycling in moist tropical forest. *Annual Review of Ecological Systems* 17: 137–167.

Whitfield, J. 2007. Underground networking. *Nature* 449: 136–138.

Wieder, W. R., C. C. Cleveland, and A. R. Townsend. 2009. Controls over leaf litter decomposition in wet tropical forests. *Ecology* 90: 3333–3341.

Wilson, E. O. 1971. *The Insect Societies.* Cambridge, MA: Belknap Press of Harvard University.

Wright, I. J., et al. 2004. The worldwide leaf economics spectrum. *Nature* 428: 821–827.

Zimmerman, P. R., J. P. Greenberg, S. O. Wandiga, and P. J. Crutzen. 1982. Termites: A potentially large source of atmospheric methane, carbon dioxide, and molecular hydrogen. *Science* 218: 563–565.

CHAPTER 11

Aber, J. D., and J. M. Melillo. 2001. *Terrestrial Ecosystems*, 2nd ed. San Diego, CA: Harcourt Science and Technology Company.

Anderson, T. M, M. E. Ritchie, and S. J. McNaughton. 2007. Rainfall and soils modify plant community response to grazing in Serengeti National Park. *Ecology* 88: 1191–1201.

Beard, J. S. 1953. The savanna vegetation of northern tropical America. *Ecological Monographs* 23: 149–215.

Blake, S., S. L. Deem, E. Mossimbo, F. Maiseis, and P. Walsh. 2009. Forest elephants: Tree planters of the Congo. *Biotropica* 41: 459–468.

Blydenstein, J. 1967. Tropical savanna vegetation of the *Llanos* of Colombia. *Ecology* 48: 1–15.

Boone, R. B., S. J. Thirgood, and J. G. C. Hopcraft. 2006. Serengeti wildebeest migratory patterns molded from rainfall and new vegetation growth. *Ecology* 87: 1987–1994.

Boulière, F., and H. Hadley. 1970. The ecology of tropical savannas. *Annual Review of Ecological Systems* 1: 125–152.

Brown, J. S., and B. P. Kotler. 2004. Hazardous duty pay and the foraging cost of predation. *Ecology Letters* 7: 999–1014.

Crooks, J. A. 2002. Characterizing ecosystem-level consequences of biological invasions: The role of ecosystem engineers. *Oikos* 97: 153–166.

Diamond, J. M. 2005. *Guns, Germs, and Steel.* New York: W.W. Norton.

Dublin, H. T., A. R. E. Sinclair, and J. McGlade. 1990. Elephants and fire as causes of multiple stable states in the Serengeti–Mara woodlands. *Journal of Animal Ecology* 59: 1147–1164.

Edwards, E. J., C. P. Osborne, C. A. E. Stromberg, S. A. Smith, C_4 Consortium. 2010. The origins of C_4 grasslands: Integrating evolutionary and ecosystem science. *Science* 328: 587–591.

Estes, R. D. 1976. The significance of breeding synchrony in the wildebeest. *East African Wildlife Journal* 14: 135–152.

Fisher, M. J., et al. 1994. Carbon storage by introduced deep-rooted grasses in South American savannas. *Nature* 371: 236–238.

Fornara, D. A., and J. T. Du Toit. 2007. Browsing lawns? Responses of *Acacia nigrescens* to ungulate browsing in an African savanna. *Ecology* 88: 200–209.

Fryxell, J. M., J. F. Wilmshurst, and A.R.E. Sinclair. 2004. Predictive models of movement by Serengeti grazers. *Ecology* 85: 2429–2435.

Goheen, J. R., F. Keesing, B. F. Allan, D. Ogada, and R. S. Ostfeld. 2004. Net effects of large mammals on *Acadia* seedling survival in an African savanna. *Ecology* 85: 1555–1561.

Hammond, B. 1999. *Saccharum spontaneum* (Graminae) in Panama: The physiology and ecology of invasion. *Journal of Sustainable Forestry* 8: 23–38.

Hoffman, W. A., et al. 2009. Tree topkill, not mortality, governs the dynamics of savanna–forest boundaries under frequent fire in central Brazil. *Ecology* 90: 1326–1337.

Huber, O. 1982. Significance of savanna vegetation in the Amazon territory of Venezuela. In *Biological Diversification in the Tropics*, G. T. Prance (ed.). New York: Columbia University Press.

———. 1987. Neotropical savannas: Their flora and vegetation. *Trends in Ecology and Evolution* 2: 67–71.

Inchausti, P. 1995. Competition between perennial grasses in a Neotropical savanna: The effects of fire and of hydric-nutritional stress. *Journal of Ecology* 83: 231–243.

Jones, C. G., J. H. Lawton, and M. Shachak. 1994. Organisms as ecosystem engineers. *Oikos* 69: 373–386.

Kauffman, J. B., D. L. Cummings, and D. E. Ward. 1994. Relationships of fire, biomass, and nutrient dynamics along a vegetation gradient in the Brazilian cerrado. *Journal of Ecology* 82: 519–531.

Kushlan, J. A., G. Morales, and P. C. Frohring. 1985. Foraging niche relations of wading birds in tropical wet savannas. In *Neotropical Ornithology*, P. A. Buckley, M. S. Foster, E. S. Morton, R. S. Ridgely, and F. G. Buckley (eds.). Washington, DC: American Ornithologists' Union.

Lack, D. 1971. *Ecological Isolation in Birds.* Cambridge, MA: Harvard University Press.

Marris, E. 2005. The forgotten ecosystem. *Nature* 437: 944–945.

———. 2007. Making room. *Nature* 448: 860–863.

Martin, P. S., and R. G. Klein. 1984. *Quaternary Extinctions.* Tucson: University of Arizona Press.

McCauley, D. J., F. Keesing, T. P. Young, B. F. Allan, and R. M. Pringle. 2006. Indirect effects of large herbivores on snakes in an African savanna. *Ecology* 87: 2657–2663.

McNaughton, S. J. 1976. Serengeti migratory wildebeest: Facilitation of energy flow by grazing. *Science* 191: 92–94.

———. 1983. Serengeti grassland ecology: The role of composite environmental factors and contingency in community organization. *Ecological Monographs* 53: 291–320.

———. 1984. Grazing lawns: Animals in herds, plant form, and coevolution. *American Naturalist* 124: 863–886.

———. 1985. Ecology of a grazing ecosystem: The Serengeti. *Ecological Monographs* 55: 259–294.

McNeil, D. G. 2010. Virus deadly in livestock is no more, U.N. declares. *New York Times*, October 15, 2010.

Murray, M. G. 1995. Specific nutrient requirements and migration of wildebeest. In *Serengeti II: Dynamics, Management, and Conservation of an Ecosystem*, A. R. E. Sinclair and P. Arcese (eds.). Chicago: University of Chicago Press.

Pires, J. M., and G. T. Prance. 1985. The vegetation types of the Brazilian Amazon. In *Amazonia*, G. T. Prance and T. E. Lovejoy (eds.). Oxford, UK: Pergamon Press.

Pringle, R. M. 2008. Elephants as agents of habitat creation for small vertebrates at the patch scale. *Ecology* 89: 26–33.

Pringle, R. M., T. P. Young, D. I. Rubenstein, and D. J. McCauley. 2007. Herbivore-initiated interaction cascades and their modulation by productivity in an African savanna. *Proceedings of the National Academy of Sciences USA* 104: 193–197.

Redfern, J. V., R. Grant, H. Biggs, and W. M. Getz. 2003. Surface-water constraints on herbivore foraging in the Kruger National Park, South Africa. *Ecology* 84: 2092–2107.

Rich, P. V., G. F. van Tets, and F. Knight. 1985. *Kadimakara: Extinct Vertebrates of Australia.* Princeton, NJ: Princeton University Press.

Riginos, C. 2009. Grass competition suppresses savanna tree growth across multiple demographic stages. *Ecology* 90: 335–340.

Riginos, C., and J. B. Grace. 2008. Savanna tree density, herbivores, and the herbaceous community: Bottom-up vs. top-down effects. *Ecology* 89: 2228–2238.

Salgado-Labouriau, M. L. 1998. Late quaternary palaeoclimate in the savannas of South America. *Journal of Quaternary Science* 12: 371–379.

Sankaran, M., et al. 2005. Determinants of woody cover in African savannas. *Nature* 438: 846–849.

Sarmiento, G. 1983. The savannas of tropical America. In *Tropical Savannas*, F. Boulière (ed.). New York: Elsevier.

Sarmiento, G., and M. Monasterio. 1975. A critical consideration of the environmental conditions associated with the occurrence of savanna ecosystems in tropical America. In *Tropical Ecological Systems: Trends in Terrestrial and Aquatic Research*, F. B. Golley and E. Medina (eds.). New York: Springer-Verlag.

Shell, D., and A. Salim. 2004. Forest tree persistence, elephants, and stem scars. *Biotropica* 36: 505–521.

Silva, J. F., J. Raventos, H. Caswell, and M. C. Trevisan. 1991. Population responses to fire in a tropical savanna grass, *Andropogon semiberbis*: A matrix model approach. *Journal of Ecology* 79: 345–356.

Sinclair, A. R. E. 1995. Serengeti past and present. In *Serengeti II: Dynamics, Management, and Conservation of an Ecosystem*, A. R. E. Sinclair and P. Arcese (eds.). Chicago: University of Chicago Press.

Sinclair, A. R. E., and P. Arcese, eds. 1995. *Serengeti II: Dynamics, Management, and Conservation of an Ecosystem*. Chicago: University of Chicago Press.

Sinclair, A. R. E., S.A.R. Mduma, and P. Arcese. 2000. What determines phenology and synchrony of ungulate breeding in Serengeti? *Ecology* 81: 2100–2111.

Sinclair, A. R. E., and M. Norton-Griffiths, eds. 1979. *Serengeti: Dynamics of an Ecosystem*. Chicago: University of Chicago Press.

Solbrig, O. T., and M. D. Young. 1992. Toward a sustainable and equitable future for savannas. *Environment* 34: 6–15; 32–35.

Van Aarde, R. J., and T. P. Jackson. 2007. Megaparks for metapopulations: Addressing the causes of locally high elephant numbers in southern Africa. *Biological Conservation* 134: 289–297.

van Langevelde, F., et al. 2003. Effects of fire and herbivory on the stability of savanna ecosystems. *Ecology* 84: 337–350.

Vickers-Rich, P., and T. H. Rich. 1999. *Wildlife of Gondwana*. Bloomington: Indiana University Press.

Walter, H. 1971. *Ecology of Tropical and Subtropical Vegetation*. New York: Van Nostrand Reinhold. OK

———. 1973. *Vegetation of the Earth in Relation to Climate and the Ecophysiological Conditions*. London: English Universities Press.

Western, D. 1975. Water availability and its influence on the structure and dynamics of a savannah large mammal community. *East African Wildlife Journal* 13: 265–286.

Wilmshurst, J. F., J. M. Fryxell, and P. E. Colucci. 1999. What constrains daily intake in Thomson's gazelles? *Ecology* 80: 2338–2347.

Wishnie, M., J. Deago, E. Mariscal, and A. Sautu. 2002. The efficient control of *Saccharum spontaneum* (L.) (Graminae) in mixed plantations of six native species of tree and teak (*Tectona grandis*) in the Panama Canal Watershed, Republic of Panama: 2nd annual report. PRORENA Working Paper ECO-03-03-En; available at http://prorena.research.yale.edu/publicaciones_files/ECO-03-03-En.pdf.

CHAPTER 12

Alongi, D. M. 2009. *The Energetics of Mangrove Forests*. New York: Springer.

Anderson, A. B., P. Magee, A. Gely, and M.A.G. Jardim. 1995. Forest management patterns in the floodplain of the Amazon estuary. *Conservation Biology* 9: 47–61.

Andrews, M. 1982. *Flight of the Condor*. Boston: Little Brown.

Balick, M. J. 1985. Useful plants of Amazonia: A resource of global importance. In *Amazonia*, G. T. Prance and T. E. Lovejoy (eds.). Oxford, UK: Pergamon Press.

Becker, C. G., C. R. Fonseca, C. F. B. Haddad, R. F. Batista, and P. I. Prado. 2007. Habitat split and the global decline of amphibians. *Science* 318: 1775–1777.

Berger, L., et al. 1998. Chytridiomycosis causes amphibian mortality associated with population declines in the rain forests of Australia and Central America. *Proceedings of the National Academy of Sciences USA* 95: 9031–9036.

Bridges, E. L. 1949. *Uttermost Part of the Earth: Indians of Tierra del Fuego*. New York: E. P. Dutton. Reprint New York: Dover Publications).

Bustamante, M. R., S. R. Ron, and L. A. Coloma. 2005. Cambios en la diversidad en siete comunidades de anuros en los Andes de Ecuador. *Biotropica* 37: 180–189.

Chazdon, R. L., and D. F. R. P. Burslem, the Earl of Cranbook. 2002. Tropical naturalists of the sixteenth through nineteenth centuries. In *Foundations of Tropical Forest Biology*, R. L. Chazdon and T. C. Whitmore (eds.). Chicago: University of Chicago Press.

Clerjacks, A., K. Wesche, and I. Hensen. 2007. Lateral expansion of *Polylepis* forest fragments in central Ecuador. *Forest Ecology and Management* 242: 477–486.

Colwell, R. K., G. Brehm, C. L. Cardelús, A. C. Gilman, and J. T. Longino. 2008. Global warming, elevational range shifts, and lowland biotic attrition in the wet tropics. *Science* 322: 258–261.

Crump, M. L., F. R. Hensley, and K. L. Clark. 1992. Apparent decline of the golden toad: Underground or extinct? *Copeia* 1992: 413–420.

De Barcellos Falkenberg, D., and J. C. Voltilini. 1995. The montane cloud forest in southern Brazil. In *Tropical Montane Cloud Forests: Ecological Studies 110*, L. S. Hamilton, J. O. Juvik, and F. N. Scatena (eds.). New York: Springer-Verlag.

Doumenge, C, D. Gilmour, M. R. Perez, and J. Blockhus. 1995. Tropical montane cloud forests: Conservation status and management issues. In *Tropical Montane Cloud Forests: Ecological Studies 110*, L. S. Hamilton, J. O. Juvik, and F. N. Scatena (eds.). New York: Springer-Verlag.

Dyk, J. V. 1995. The Amazon. *National Geographic* 187: 2–39.

Eisenberg, J. F. 1989. *Mammals of the Neotropics: Volume 1, The Northern Neotropics, Panama, Colombia, Venezuela, Guyana, Suriname, French Guiana*. Chicago: University of Chicago Press.

Ellison, A. M., and E. L. Farnsworth. 1993. Seedling survivorship, growth, and response to disturbance in Belizean mangal. *American Journal of Botany* 80: 1137–1145.

Eterovick, P. C., A. C. O. de Queiroz Carnaval, D. M. Borges-Nojosa, D. L. Silvano, M. V. Sagalla, and I. Sazima. 2005. Amphibian declines in Brazil: an overview. *Biotropica* 37: 166–179.

Gentry, A. H. 1986. Endemism in tropical versus temperate plant communities. In *Conservation Biology: The Science of Scarcity and Diversity*, M. E. Soule (ed.). Sunderland, MA: Sinauer.

George, U. 1989. Tepuis—Venezuela's islands in time. *National Geographic* 175: 526–562.

Goulding, M. 1980. *The Fishes and the Forest: Explorations in Amazonian Natural History*. Berkeley: University of California Press.

———. 1985. Forest fishes of the Amazon. In *Amazonia*, G. T. Prance and T. E. Lovejoy (eds.). Oxford, UK: Pergamon Press.

———. 1993. Flooded forests of the Amazon. *Scientific American* 266: 114–120.

Goulding, M., N. J. H. Smith, and D. J. Mahar. 1996. *Floods of Fortune: Ecology and Economy along the Amazon*. New York: Columbia University Press.

Grajal, A., S. D. Strahl, R. Parra, M. G. Dominguez, and A. Neher. 1989. Foregut fermentation in the hoatzin, a Neotropical leaf-eating bird. *Science* 245: 1236–1238.

Graves, G. R., J. P. O'Neill, and T. A. Parker III. 1983. *Grallaricula ochraceifrons*, a new species of antpitta from northern Peru. *Wilson Bulletin* 95: 1–6.

Grubb, P. J. 1971. Interpretation of the "Massenerhebung" effect on tropical mountains. *Nature* 229: 44–45.

———. 1977. Control of forest growth and distribution on wet tropical mountains. *Annual Review of Ecological Systems* 8: 83–107.

Haber, W. A. 2000. Plants and vegetation. In *Monteverde: Ecology and Conservation of a Tropical Cloud Forest,* N. M. Nadkarni and N. T. Wheelright (eds.). New York: Oxford University Press.

Hamilton, L. S., J. O. Juvik, and F. N. Scatena, eds. 1995. *Tropical Montane Cloud Forests: Ecological Studies 110.* New York: Springer-Verlag.

Hammel, B. 1990. The distribution and diversity among families, genera, and habit types in the La Selva flora. In *Four Neotropical Rain Forests*, A. H. Gentry (ed.). New Haven, CT: Yale University Press.

Hartshorn, G., and Peralta, R. 1988. Preliminary description of primary forests along the La Selva–Volcán Barva altitudinal transect, Costa Rica. In *Tropical Rain Forests*, F. Almeda and C. M. Pringle (eds.). San Francisco: California Academy of Sciences and AAAS Pacific Division.

Henderson, A., S. P. Churchill, and J. Luteyn. 1991. Neotropical plant diversity. *Nature* 229: 44–45.

Hilty, S. L. 2003. *Birds of Venezuela*, 2nd ed. Princeton, NJ: Princeton University Press.

Isler, M. L., and P. R. Isler. 1987. *The Tanagers: Natural History, Distribution, and Identification.* Washington, DC: Smithsonian Institution Press.

Junk, W. J. 1970. Investigations on the ecology and production-biology of the "floating meadows" (*Paspalo-Echinochloetum*) on the middle Amazon. *Amazonia* 2: 449–495.

Junk, W. J., and K. Furch. 1985. The physical and chemical properties of Amazonian waters and their relationships with the biota. In *Amazonia*, G. T. Prance and T. E. Lovejoy (eds.). Oxford, UK: Pergamon Press.

Kalliola, R., J. Salo, M. Puhakka, and M. Rajasilta. 1991. New site formation and colonizing vegetation in primary succession on the western Amazon floodplains. *Journal of Ecology* 79: 877–901.

Klesecker, J. M., A. R. Blaustein, and L. K. Belden. 2001. Complex causes of amphibian population declines. *Nature* 410: 681–684.

La Marca, H., et al. 2005. Catastrophic population declines and extinctions in Neotropical harlequin frogs (Bufonidae: *Atelopus*). *Biotropica* 37: 190–201.

Leo, M. 1995. The importance of tropical montane cloud forest for preserving vertebrate endemism in Peru: The Rio Abiseo National Park as a case study. In *Tropical Montane Cloud Forests: Ecological Studies 110*, L. S. Hamilton, J. O. Juvik, and F. N. Scatena (eds.). New York: Springer-Verlag.

Lips, K. R., P. A. Burrowes, J. R. Mendelson III, and G. Parra-Olea. 2005. Amphibian population declines in Latin America: a synthesis. *Biotropica* 37: 222–226.

Lips, K. R., et al. 2006. Emerging infectious disease and the loss of biodiversity in a Neotropical amphibian community. *Proceedings of the National Academy of Sciences USA* 103: 3165–3170.

Loiselle, B. A., and J. G. Blake. 1992. Population variation in a tropical bird community. *Bioscience* 42: 829–837.

Long, A. J. 1995. The importance of tropical montane cloud forests for endemic and threatened birds. In *Tropical Montane Cloud Forests: Ecological Studies 110*, L. S. Hamilton, J. O. Juvik, and F. N. Scatena (eds.). New York: Springer-Verlag.

Lopez, O. R., and T. A. Kursar. 1999. Flood tolerance of four tropical tree species. *Tree Physiology* 19: 925–932.

———. 2003. Does flood tolerance explain tree species distribution in tropical seasonally flooded habitats? *Oecologia* 136: 193–204.

Lowe-McConnell, R. H. 1987. *Ecological Studies in Tropical Fish Communities.* Cambridge, UK: Cambridge University Press.

Lugo, A. E., and S. C. Snedaker. 1974. The ecology of mangroves. *Annual Review of Ecological Systems* 5: 39–64.

Meade, R. H., and L. Koehnken. 1991. Distribution of the river dolphin, tonina *Inia geoffrensis*, in the Orinoco River Basin of Venezuela and Colombia. *Interciencia* 16: 300–312.

Meade, R. H., J. M. Rayol, S. C. Da Conceicao, and J. R. G. Natividade. 1991. Backwater effects in the Amazon River Basin of Brazil. *Environmental Geology and Water Sciences* 18: 105–114.

Morrison, T. 1974. *Land above the Clouds.* London: Andre Deutsch.

———. 1976. *The Andes.* Amsterdam: Time-Life International.

Muller-Karger, F. E., C. R. McClain, and P. L. Richardson. 1988. The dispersal of the Amazon's water. *Nature* 333: 56–59.

Murray, K. G., S. Kinsman, and J. L. Bronstein. 2000. Plant–animal interactions. In *Monteverde: Ecology and Conservation of a Tropical Cloud Forest*, N. M. Nadkarni and N. T. Wheelright (eds.). New York: Oxford University Press.

Nadkarni, N. M., and N.T. Wheelright, eds. 2000. *Monteverde: Ecology and Conservation of a Tropical Cloud Forest.* New York: Oxford University Press.

Nordin, C. F., Jr., and R. H. Meade. 1982. Deforestation and increased flooding in the upper Amazon. *Science* 215: 426–427.

———. 1985. The Amazon and the Orinoco. In *McGraw-Hill Yearbook of Science and Technology 1986.* New York: McGraw-Hill.

Ojasti, J. 1991. Human exploitation of capybara. In *Neotropical Wildlife Use and Conservation*, J. G. Robinson and K. H. Redford (eds.). Chicago: University of Chicago Press.

O'Neill, J. P., and G. R. Graves. 1977. A new genus and species of owl (Aves: Strigidae) from Peru. *Auk* 94: 409–416.

Parker, T. A., III, and J. P. O'Neill. 1985. A new species and new subspecies of *Thryothorus* wren from Peru. In *Neotropical Ornithology*, P. A. Buckley, M. S. Foster, E. S. Morton, R. S. Ridgely, and F. G. Buckley (eds.). Washington, DC: American Ornithologists' Union.

Parra-Olea, G., E. Martinez-Meyer, and G. Perez-Ponce de Leon. 2005. Forecasting climate change effects on salamander distribution in the highlands of Central Mexico. *Biotropica* 37: 202–208.

Phillips, O., A. H. Gentry, C. Reynel, P. Wilkin, and C. Galvez-Durand B. 1994. Quantitative ethnobotany and Amazonian conservation. *Conservation Biology* 8: 225–248.

Piou, C., I. C. Feller, U. Berger, and F. Chi. 2006. Zonation patterns of Belizean offshore mangrove forests 41 years after a catastrophic hurricane. *Biotropica* 38: 365–374.

Pounds, J. A. 1990. Disappearing gold. *BBC Wildlife* 8: 812–817.

———. 2000a. Amphibians and reptiles. In *Monteverde: Ecology and Conservation of a Tropical Cloud Forest*, N. M. Nadkarni and N. T. Wheelright (eds.). New York: Oxford University Press.

———. 2000b. Monteverde salamanders, golden toads, and the emergence of the global amphibian crisis. In *Monteverde: Ecology and Conservation of a Tropical Cloud Forest*, N. M.

Nadkarni and N. T. Wheelright (eds.). New York: Oxford University Press.

Powell, G. V. N., R. D. Bjork, S. Barrios, and V. Espinoza. 2000. Elevational migrations and habitat linkages: Using the resplendent quetzal as an indicator for evaluating the design of the Monteverde Reserve Complex. In *Monteverde: Ecology and Conservation of a Tropical Cloud Forest*, N. M. Nadkarni and N. T. Wheelright (eds.). New York: Oxford University Press.

Pringle, C. M. 1988. History of conservation efforts and initial exploration of the lower extension of Parque Nacional Braulio Carrillo, Costa Rica. In *Tropical Rain Forests*, F. Almeda and C. M. Pringle (eds.). San Francisco: California Academy of Sciences and AAAS Pacific Division.

Pringle, C. M., et al. 1984. Natural history observations and ecological evaluation of the La Selva Protection Zone, Costa Rica. *Brenesia* 22: 189–206.

Ridgely, R. S., and P. J. Greenfield. 2001. *The Birds of Ecuador Field Guide*. New York: Comstock Associates.

Ridgely, R. S., and G. Tudor. 1994. *The Birds of South America*, vol. 2. Austin: University of Texas Press.

Robbins, M. B., G. H. Rosenberg, and F. S. Molina. 1994. A new species of cotinga (Cotingidae: *Doliornis*) from the Ecuadorian Andes, with comments on plumage sequences in *Doliornis* and *Ampelion*. *Auk* 111: 1–7.

Rodriguez, G. 1987. Structure and production in Neotropical mangroves. *Trends in Ecology and Evolution*: 2: 264–267.

Ron, S. R. 2005. Predicting the distribution of the amphibian pathogen *Batrachochytrium dendrobatidis* in the New World. *Biotropica* 37: 209–221.

Rutzler, K., and I. C. Feller. 1987. Mangrove swamp communities. *Oceanus* 30: 16–24.

———. 1996. Caribbean mangrove swamps. *Scientific American* 274: 94–99.

Saenger, P. 2009. *Mangrove Ecology, Silviculture, and Conservation*. Heidelberg: Springer Netherlands.

Salati, E., and P. B. Vose. 1984. Amazon Basin: A system in equilibrium. *Science* 225: 129–138.

Savage, J. M. 1966. An extraordinary new toad (*Bufo*) from Costa Rica. *Revista de Biologia Tropical* 14: 153–167.

———. 2000. The discovery of the golden toad. In *Monteverde: Ecology and Conservation of a Tropical Cloud Forest*, N. M. Nadkarni and N. T. Wheelright (eds.). New York: Oxford University Press.

Schulenberg, T. S., and M. D. Williams. 1982. A new species of antpitta *(Grallaria)* from northern Peru. *Wilson Bulletin* 94: 105–113.

Sibley, C. G., and J. E. Ahlquist. 1983. Phylogeny and classification of birds based on the data of DNA–DNA hybridization. In *Current Ornithology*, R. F. Johnston (ed.). New York: Plenum Press.

———. 1990. *Phylogeny and Classification of Birds: A Study in Molecular Evolution*. New Haven, CT: Yale University Press.

Smith, T. J. 1987. Seed predation in relation to tree dominance and distribution in mangrove forests. *Ecology* 68: 266–273.

Spalding, M, M. Kainuma, and L. Collins. 2010. *World Atlas of Mangroves*. London: Earthscan Publications, Ltd.

Stevenson, R., and W. A. Haber. 2000. Migration of butterflies through Monteverde In *Monteverde: Ecology and Conservation of a Tropical Cloud Forest*, N. M. Nadkarni and N. T. Wheelright (eds.). New York: Oxford University Press.

Stiles, F. G. 1988. Altitudinal movements of birds on the Caribbean slope of Costa Rica: Implications for conservation. In *Tropical Rain Forests*, F. Almeda and C. M. Pringle (eds.). San Francisco: California Academy of Sciences and AAAS Pacific Division.

Stokstad, E. 2004. Global survey documents puzzling decline of amphibians. *Science* 306: 391.

Stotz, D. F., J. W. Fitzpatrick, T. A. Parker III, and D. K. Moskovits. 1996. *Neotropical Birds: Ecology and Conservation*. Chicago: University of Chicago Press.

Strahl, S. D. 1985. Correlates of reproductive success in communal hoatzins (*Opisthocomus hoazin*). In *Abstracts of the 103rd American Ornithologists' Union Meeting*. Washington, DC: American Ornithologists' Union.

Strahl, S. D., and A. Schmitz. 1990. Hoatzins: Cooperative breeding in a folivorous Neotropical bird. In *Cooperative Breeding in Birds*, P. B. Stacey and W. D. Koenig (eds.). Cambridge, UK: Cambridge University Press.

Stuart, S. N., et al. 2004. Status and trends of amphibian declines and extinctions worldwide. *Science* 306: 1783–1786.

Timm, R. M., and R. K. LaVal. 2000. Mammals. In *Monteverde: Ecology and Conservation of a Tropical Cloud Forest*, N. M. Nadkarni and N. T. Wheelright (eds.). New York: Oxford University Press.

Tomlinson, P. B. 1995. *The Botany of Mangroves*. Cambridge, UK: Cambridge University Press.

Voyles, J., et al. 2009. Pathogenesis of chytridiomycosis, a cause of catastrophic amphibian declines. *Science* 326: 582–585.

Walsh, G. E. 1974. Mangroves: A review. In *Ecology of Halophytes*, R. J. Reimold and W. H. Queens (eds.). New York: Academic Press.

Weldon, C., L. H. du Preez, A. D. Hyatt, R. Muller, and R. Spears. 2004. Origin of the amphibian chytrid fungus. *Emerging Infectious Diseases* 10: 2100–2105.

Wheelwright, N. T. 1983. Fruits and the ecology of resplendent quetzals. *Auk* 100: 286–301.

Young, B. E., and D. B. McDonald. 2000. Birds. Chapter 6 in *Monteverde: Ecology and Conservation of a Tropical Cloud Forest*, N. M. Nadkarni and N. T. Wheelright (eds.). New York: Oxford University Press.

CHAPTER 13

Ayensu, E. S., ed. 1980. *The Life and Mysteries of the Jungle*. New York: Crescent Books.

Bahn, P. G. 1992. Amazon rocks the cradle. *Nature* 355: 588–589.

Balick, M. J. 1985. Useful plants of Amazonia: A resource of global importance. In *Amazonia*, G. T. Prance and T. E. Lovejoy (eds.). Oxford, UK: Pergamon Press.

Balick, M. J., R. Arvigo, and L. Romero. 1994. The development of an ethnobiomedical forest reserve in Belize: Its role in the preservation of biological and cultural diversity. *Conservation Biology* 8: 316–317.

Balick, M. J., and P. A. Cox. 1997. *Plants, People, and Culture: The Science of Ethnobotany*. San Francisco: W.H. Freeman.

Balter, M. 2007. Seeking agriculture's ancient roots. *Science* 316: 1830–1835.

Beckerman, S. 1987. Swidden in Amazonia and the Amazon rim. In *Comparative Farming Systems*, B. L. Turner and S. B. Brush (eds.). New York: Guilford Press.

Bellwood, P. 2004. *First Farmers: The Origins of Agricultural Societies*. Malden, MA: Wiley-Blackwell.

Bettinger, R. L. 2009. *Hunter Gatherer Foraging: Five Simple Models*. Clinton Corners, NY: Eliot Werner Publications.

Borofsky, R. 2005. *Yanomami: The Fierce Controversy and What We Can Learn From It*. Berkeley: University of California Press.

Boucher, D. H. 1991. Cocaine and the coca plant: Traditional and illegal uses. *Bioscience* 41: 72–76.

Brooke, J. 1993. Gold miners and Indians: Brazil's frontier war. *New York Times*, September 7, 1993.

Bush, M. A., and M. R. Silman. 2007. Amazonian exploitation revisited: Ecological asymmetry and the policy pendulum. *Frontiers in Ecology and Environment* 5: 457–465.

Calloway, J. C., G. S. Brito, and E. S. Neves. 2005. Phytochemical analyses of *Banisteriopsis caapi* and *Psychotria viridis*. *Journal of Psychoactive Drugs* 37: 145–150.

Carneiro, R. L. 1988. Indians of the Amazonian forest. In *People of the Tropical Rain Forest*, J. S. Denslow and C. Padoch (eds.). Berkeley: University California Press.

Chagnon, N. 1992. *Yanomamo: The Last Days of Eden*. New York: Harcourt, Brace, Jovanovich.

Collins, M., ed. 1990. *The Last Rain Forests*. New York: Oxford University Press.

Cordain, L., J. Miller, S. B. Eaton, N. Mann, S. H. A. Holt, and J. D. Speth. 2000. Plant–animal subsistence ratios and macronutrient energy estimations in worldwide hunter–gatherer diets. *American Journal of Clinical Nutrition* 7: 682–692.

Cowan, C. W., and P. J. Watson, eds. 2006. *The Origins of Agriculture: An International Perspective*. Tuscaloosa: University of Alabama Press.

Cox, P. A., and M. J. Balick. 1994. The ethnobotanical approach to drug discovery. *Scientific American* 270: 82–87.

Cruz-Angon, A., T. S. Sillett, and R. Greenberg. 2008. An experimental study of habitat selection by birds in a coffee plantation. *Ecology* 89: 921–927.

Denevan, W. M. 1976. The aboriginal population of Amazonia. In *The Native Population of the Americas in 1492*, W. M. Denevan (ed.). Madison: University Wisconsin Press.

———. 2003. The native population of Amazonia in 1492 reconsidered. *Revista Indias* 62: 175–188.

Diamond, J. M. 1997. *Guns, Germs, and Steel*. New York: W.W. Norton.

Dufour, D. L. 1990. Use of tropical rain forests by native Amazonians. *Bioscience* 40: 652–659.

Erickson, C. L. 2000. An artificial landscape-scale fishery in the Bolivian Amazon. *Nature* 408: 190–193.

Ewel, J., C. Berish, B. Brown, N. Price, and J. Raich. 1981. Slash and burn impacts on a Costa Rican wet forest site. *Ecology* 62: 816–829.

Fedick, S. L., and A. Ford. 1990. The prehistoric agricultural landscape of the Central Maya Lowlands: An examination of local variability in a regional context. *World Archaelogy* 22: 18–33.

Flannery, T. 1994. *The Future Eaters: An Ecological History of Australian Lands and People*. Port Melbourne, Australia: Reed Books.

———. 2001. *The Eternal Frontier: An Ecological History of North America and Its Peoples*. New York: Atlantic Monthly Press.

Furst, P. T., and M. D. Coe 1977. Ritual enemas. *Natural History* 86: 88–91.

Gibbons, A. 1990. New view of early Amazonia. *Science* 248: 1488–1490.

———. 2009. A new kind of ancestor: Ardipithecus unveiled. *Science* 326: 36–40.

Glaser, B. 2007. Prehistorically modified soils in central Amazonia: A model for sustainable agriculture in the twenty-first century. *Philosophic Transactions of the Royal Society B* 362: 187–196.

Goebel, T., M. R. Waters, and D. H. O'Rourke. 2008. The late Pleistocene dispersal of modern humans in the Americas. *Science* 319: 1497–1502.

Gomez-Pompa, A., and A. Kaus. 1990. Traditional management of tropical forests in Mexico. In *Alternatives to Deforestation: Steps toward Sustainable Use of the Amazonian Rain Forest*, A. B. Anderson (ed.). New York: Columbia University Press.

Gomez-Pompa, A., H. L. Morales, E. J. Avilla, and J. J. Avilla. 1982. Experiences in traditional hydraulic agriculture. In *Maya Subsistence*, K. V. Flannery (ed.). New York: Academic Press.

Gottlieb, O. R. 1985. The chemical uses and chemical geography of Amazon plants. In *Amazonia*, G. T. Prance and T. E. Lovejoy (eds.). Oxford, UK: Pergamon Press.

Goulding, M., N. J. H. Smith, and D. J. Mahar. 1996. *Floods of Fortune: Ecology and Economy along the Amazon*. New York: Columbia University Press.

Grayson, D. K., and D. J. Meltzer. 2003. A requiem for North American overkill. *Journal of Archaeological Science* 5: 585–593.

Greenberg, R., P. Bichier, A. Cruz-Angon, and R. Reitsma. 1997a. Bird populations in shade and sun coffee plantations in Central Guatemala. *Conservation Biology* 11: 448–459.

Greenberg, R., P. Bichier, and J. Sterling. 1997b. Bird populations in rustic and planted shade coffee plantations of Eastern Chiapas, Mexico. *Biotropica* 29: 501–514.

Greenberg, R., I. Perfecto, and S. M. Philpott. 2008. Agroforests as model systems in tropical ecology. *Ecology* 89: 913–914.

Griscom, L. 1932. The distribution of bird-life in Guatemala. *Bulletin of the American Museum of Natural History* 64: 1–439.

Hammond, N. 1982. *Ancient Maya Civilization*. New Brunswick, NJ: Rutgers University Press.

Heckenberger, M. J., et al. 2003. Amazonia 1492: Pristine forest or cultural parkland? *Science* 301: 1710–1714.

Heckenberger, M. J., et al. 2008. Pre-Columbian urbanism, anthropogenic landscapes, and the future of the Amazon. *Science* 321: 1214–1217.

Hill, K., and A. M. Hurtado. 1989. Hunter-gatherers of the New World. *American Scientist* 77: 436–443.

Hillocks, R. J., J. M. Thresh, and A. Bellotti. 2003. *Cassava: Biology, Production, and Utilization*. Wallingford, UK: CABI.

Horn, S. P. 1993. Postglacial vegetation and fire history in the Chirripo Paramo of Costa Rica. *Quaternary Research* 40: 107–116.

Horn, S. P., and R. L. Sanford Jr. 1992. Holocene fires in Costa Rica. *Biotropica* 24: 354–361.

Iltis, H. H., J. F. Doebley, R. Guzman, and B. Pazy. 1979. *Zea diploperennis* (Gramineae): A new teosinte from Mexico. *Science* 203: 186–188.

Joyce, C. 1992. Western medicine men return to the field. *Bioscience* 42: 399–403.

Kellman, M., and R. Tackberry. 1997. *Tropical Environments: The Functioning and Management of Tropical Ecosystems*. London: Routledge.

Kelly, R. L. 2007. *The Foraging Spectrum: Diversity in Hunter-Gatherer Lifeways*. Clinton Corners, NY: Eliot Werner Publications.

Kricher, J. 2000. Evaluating shade-grown coffee and its importance to birds. *Birding* 32: 57–60.

———. 2009. *The Balance of Nature: Ecology's Enduring Myth.* Princeton, NJ: Princeton University Press.

Lee, R. B., and R. Daly, eds. 2004. *The Cambridge Encyclopedia of Hunters and Gatherers.* Cambrige, UK: Cambridge University Press.

Mann, C. C. 2008. Ancient earthmovers of the Amazon. *Science* 321: 1148–1152.

Martin, P. S., and R. G. Klein. 1984. *Quaternary Extinctions.* Tucson: University of Arizona Press.

Mas, A. H., and T. V. Dietsch. 2004. Linking shade coffee certification and biodiversity conservation: Butterflies and birds in Chiapas, Mexico. *Ecological Applications* 14: 642–654.

McLean, J. 2000. Status of shade grown coffee. *Birding* 32: 61–65.

Meggars, B. J. 1985. Aboriginal adaptations to Amazonia. In *Amazonia,* G. T. Prance and T. E. Lovejoy (eds.). Oxford, UK: Pergamon Press.

———. 1988. The prehistory of Amazonia. In *People of the Tropical Rain Forest,* J. S. Denslow and C. Padoch (eds.). Berkeley: University California Press.

———. 2003. Natural versus anthropogenic sources of Amazonian biodiversity: The continuing quest for El Dorado. In *How Landscapes Change,* G. A. Bradshaw and P. A. Marquet (eds.). Berlin: Springer.

Odum, H. T. 1971. *Environment, Power, and Society.* New York: Wiley-Interscience.

Padmaja, G. 1995. Cyanide detoxification in cassava for food and feed uses. *Critical Review of Food Science Nutrition* 35: 299–339.

Padoch, C. 1988. People of the floodplain and forest. In *People of the Tropical Rain Forest,* J. S. Denslow and C. Padoch (eds.). Berkeley: University of California Press.

Panter-Brick, C., R. H. Layton, and P. Rowley-Conway, eds.. 2001. *Hunter-Gatherers: An Interdisciplinary Perspective.* Cambridge, UK: Cambridge University Press.

Peres, C. A. 1994. Indigenous reserves and nature conservation in Amazonian forests. *Conservation Biology* 8: 586–588.

Perfecto, I., R. A. Rice, R. Greenberg, and M. E. Van der Voort. 1996. Shade coffee: A disappearing refuge for biodiversity. *Bioscience* 46: 598–608.

Perfecto, I., J. Vandermeer, A., A. Mas, and L. S. Pinto. 2005. Biodiversity, yield, and shade coffee certification. *Ecological Economics* 54: 435–446.

Perfecto, I., et al. 2004. Greater predation in shaded coffee farms: The role of resident Neotropical birds. *Ecology* 85: 2677–2681.

Perry, L., et al. 2006. Early maize agriculture and interzonal interaction in southern Peru. *Nature* 440: 76–79.

Peters, C. M. 2000. Precolumbian silviculture and indigenous management of Neotropical forests. In *Imperfect Balance,* D. L. Lentz (ed.). New York: Columbia University Press.

Phillips, O., A. H. Gentry, C. Reynel, P. Wilkin, and C. B. Galvez-Durand. 1994. Quantitative ethnobotany and Amazonian conservation. *Conservation Biology* 8: 225–248.

Piperno, D. R., and D. M. Pearsall. 1998. *The Origin of Agriculture in the Lowland Neotropics.* New York: Academic Press.

Plotkin, M. J. 1993. *Tales of a Shaman's Apprentice: An Ethnobotanist Searches for New Medicines in the Amazon Rain Forest.* New York: Viking.

Plotkin, M. J., and L. Famolare, eds. 1992. *Sustainable Harvest and Marketing of Rain Forest Products.* Washington, DC: Island Press.

Posey, D. A. 1982. Keepers of the forest. *Garden* 6: 18–24.

Pringle, H. 1998. The slow birth of agriculture. *Science* 282: 1446.

Redford, K. H. 1992. The empty forest. *Bioscience* 42: 412–422.

Redford, K. H., and J. G. Robinson. 1987. The game of choice: Patterns of Indian and colonial hunting in the Neotropics. *American Anthropologist* 89: 650–667.

Robinson, J. G., and K. H. Redford, eds. 1991. *Neotropical Wildlife Use and Conservation.* Chicago: University of Chicago Press.

Roosevelt, A. C. 1989. Lost civilizations of the lower Amazon. *Natural History* 98: 75–83.

Roosevelt, A. C., R. A. Housley, M. Imazio da Silveira, S. Maranca, and R. Johnson. 1991. Eighth millennium pottery from a prehistoric shell midden in the Brazilian Amazon. *Science* 254: 1621–1624.

Schultes, R. E. 1992. Ethnobotany and technology in the northwest Amazon: A partnership. In *Sustainable Harvest and Marketing of Rain Forest Products,* M. Plotkin and L. Famolare (eds.). Washington, DC: Island Press.

Schultes, R. E., and A. Hoffmann. 1992. *Plants of the Gods: Their Sacred, Healing, and Hallucinogenic Powers.* Rochester, VT: Healing Arts Press.

Schultes, R. E., and R. F. Raffauf. 1990. *The Healing Forest: Medicinal and Toxic Plants of the Northwest Amazonia.* Portland, OR: Dioscorides Press.

Schultes, R. E., and S. von Reis. 2008. *Ethnobotany: Evolution of a Discipline.* Portland, OR: Timber Press.

Siemens, A. H. 1982. Pre-Hispanic agricultural use of the wetlands of northern Belize. In *Maya Subsistence,* F. V. Fleming (ed.). New York: Academic Press.

Stone, R. 2009. Answers from Angkor. *National Geographic* 216: 26–55.

Tejada-Cruz, C., E. Silva-Rivera, J. R. Barton, and W. J. Sutherland. 2010. Why shade coffee does not guarantee biodiversity conservation. *Ecology and Society* 15; available at www.ecologyandsociety.org/vol15/iss1/art13/.

Thompson, L. G. 2000. Ice core evidence for climate change in the tropics: Implications for our future. *Quaternary Science Review* 19: 19–35.

Turner, B. L., II, and P. D. Harrison. 1981. Prehistoric raised-field agriculture in the Maya lowlands. *Science* 213: 339–405.

Turner, B. L., II, and P. D. Harrison, eds. 1983. *Pulltrouser Swamp: Ancient Maya Habitat, Agriculture, and Settlement in Northern Belize.* Austin: University of Texas Press.

Uhl, C. 1987. Factors controlling succession following slash and burn agriculture in Amazonia. *Ecology* 75: 377–407.

Van Bael, S. A., et al. 2008. Birds as predators in tropical agroforestry systems. *Ecology* 89: 928–934.

Vega, F. E. 2008. The rise of coffee. *American Scientist* 96: 138–145.

Vickers, W. T. 1988. Game depletion hypothesis of Amazonian adaptation: Data from a native community. *Science* 239: 1521–1522.

———. 1991. Hunting yields and game composition over ten years in an Amazon Indian territory. In *Wildlife Use and Conservation,* J. G. Robinson and K. H. Redford (eds.). Chicago: University of Chicago Press.

Vitousek, P. M., P. R. Ehrlich, A. H. Ehrlich, and P. A. Matson. 1986. Human appropriation of the products of photosynthesis. *BioScience* 36(6): 363–373.

Vitousek, P. M., H. A. Mooney, J. Lubchenco, and J. M. Melillo. 1997. Human domination of earth's ecosystems. *Science* 277: 494–499.

Waterton, C. [1825]. 1983. *Wanderings in South America.* Reprint, London: Century Publishing.
White, P., and T. Denham, eds. 2006. *The Emergence of Agriculture: A Global View.* London: Routledge.
Willis, K. J., L. Gillson, and T. M. Brncic. 2004. How "virgin" is virgin rain forest? *Science* 304: 402–403.
Wilson, E. O. 1978. *On Human Nature.* Cambridge, MA: Harvard University Press.
Wunderle, J. M., Jr. 1999. Avian distribution in Dominican shade coffee plantations: Area and habitat relationships. *Journal of Field Ornithology* 70: 58–70.
Wunderle, J. M., Jr., and S. C. Latta. 1998. Avian resources in Dominican shade coffee plantations. *Wilson Bulletin* 110: 271–281.
———. 2000. Winter site fidelity of Nearctic migrants in shade coffee plantations of different sizes in the Dominican Republic. *Auk* 117: 596–614.

CHAPTER 14

Achard, F., et al. 2002. Determination of deforestation rates in the world's humid tropical forests. *Science* 297: 999–1002.
Arroyo-Rodriguez, V., S. Mandujano, J. Benitez-Malvido, and C. Cuende-Fanton. 2007. The influence of large tree density on howler monkey (*Alouatta palliata mexicana*) presence in very small rain forest fragments. *Biotropica* 39: 760–766.
Balmford, A., et al. 2001. Conservation conflicts across Africa. *Science* 291: 2616–2619.
Bierregaard, R. O., Jr., T. E. Lovejoy, V. Kapos, A. Augusto de Santos, and R. W. Hutchings. 1992. The biological dynamics of tropical rain forest fragments. *Bioscience* 42: 859–866.
Bierregaard, R. O., Jr., et al. 2001b. Principles of forest fragmentation and conservation in the Amazon. In *Lessons from Amazonia: The Ecology and Conservation of a Fragmented Forest*, R. O. Bierregaard Jr. et al. (eds.). New Haven, CT: Yale University Press.
Bierregaard, R. O., Jr., C. Gascon, T. E. Lovejoy, and R. Mesquita, eds. 2001a. *Lessons from Amazonia: The Ecology and Conservation of a Fragmented Forest.* New Haven, CT: Yale University Press.
Bradshaw, C. J. A., N. S. Sodhi, and B. W. Brook. 2009. Tropical turmoil: A biodiversity tragedy in progress. *Frontiers of Ecology and the Environment* 7: 79–87.
Brook, B. W., N. S. Sodhi, and P. K. L. Ng. 2003. Catastrophic extinctions follow deforestation in Singapore. *Nature* 424: 420–423.
Brown, K. S., Jr., and G. G. Brown. 1992. Habitat alteration and species loss in Brazilian forests. In *Tropical Deforestation and Species Extinction*, T. C. Whitmore and J. A. Sayer (eds.). London: Chapman & Hall.
Bruna, E. M. 1999. Seed germination in rain forest fragments. *Nature* 402: 139.
———. 2003. Are plant populations in fragmented habitats recruitment limited? Tests with an Amazonian herb. *Ecology* 84: 932–947.
Bruna, E. M., and M. K. Oli. 2005. Demographic effects of habitat fragmentation on a tropical herb: Life-table response experiments. *Ecology* 86: 1816–1824.
Cincotta, R. P., J. Wisnewski, and R. Engelman. 2000. Human population in the biodiversity hotspots. *Nature* 404: 990–991.
Costanza, R., et al. 1997. The economic value of the world's ecosystems. *Nature* 387: 253–260.
Cramer, J. M., R. C. G. Mesquita, T. V. Bentos, B. Moser, and G. B. Williamson. 2007. Forest fragmentation reduces seed dispersal of *Duckeodendron cestroides*, a Central Amazon endemic. *Biotropica* 39: 709–718.
Dale, V. H., S. M. Pearson, H. L. Offerman, and R. V. O'Neill. 1994. Relating patterns of land-use change to faunal biodiversity in the Central Amazon. *Conservation Biology* 8: 1027–1036.
Desouza, O., J. H. Schoereder, V. Brown, and R. O. Bierregaard Jr. 2001. The theoretical overview of the processes determining species richness in forest fragments. In *Lessons from Amazonia: The Ecology and Conservation of a Fragmented Forest*, R. O. Bierregaard Jr. et al. (eds.). New Haven, CT: Yale University Press.
Develey, P. F., and P. C. Stouffer. 2001. Effects of roads on movements of understory birds in mixed-species flocks in central Amazonian Brazil. *Conservation Biology* 15: 1416–1422.
Diamond, J. M. 1976. Island biogeography and conservation: Strategy and limitations. *Science* 193: 1027–1029.
Dirzo, R., and P. H. Raven. 2003. Global state of biodiversity and loss. *Annual Review of Environmental Resources* 28: 137–167.
Dobson, A., et al. 2006. Habitat loss, trophic collapse, and the decline of ecosystem services. *Ecology* 87: 1915–1924.
Ehrlich, P. R., and E. O. Wilson. 1991. Biodiversity studies: Science and policy. *Science* 253: 758–762.
Feeley, K. J., and J. W. Terborgh. 2006. Habitat fragmentation and effects of herbivore (howler monkey) abundances on bird species richness. *Ecology* 87: 144–150.
Ferraz, G., J. D. Nichols, J. E. Hines, P. C. Stouffer, R. O. Bierregaard Jr., and T. E. Lovejoy. 2007. A large-scale deforestation experiment: Effects of patch area and isolation on Amazon birds. *Science* 315: 238–241.
Foley, J. A., et al. 2007. Amazonia revealed: Forest degradation and loss of ecosystem goods and services in the Amazon Basin. *Frontiers in Ecology and the Environment* 5: 25–32.
Galindo-Leal, C., and I. de Gusmao Camara, eds. 2003. *The Atlantic Forest of South America: Biodiversity, Threats, and Outlook.* Washington, DC: Island Press.
Gascon, C., G. B. Williamson, and G. A. B. da Fonseca. 2000. Receding forest edges and vanishing reserves. *Science* 288: 1356–1358.
Gentry, A. H. 1986a. Endemism in tropical versus temperate plant communities. In *Conservation Biology: The Science of Scarcity and Diversity*, M. E. Soule (ed.). Sunderland, MA: Sinauer.
———. 1986b. Species richness and floristic composition of Choco region plant communities. *Caldasia* 15: 71–91.
Holloway, M. 1993. Sustaining the Amazon. *Scientific American* 269: 90–99.
Kareiva, P., and M. Marvier. 2003. Conserving biodiversity coldspots. *American Scientist* 91: 344–351.
Karr, J. R., and K. E. Freemark. 1983. Habitat selection and environmental gradients: Dynamics in the "stable" tropics. *Ecology* 64: 1481–1494.
Klein, B. C. 1989. Effects of forest fragmentation on dung and carrion beetle communities in Central Amazonia. *Ecology* 70: 1715–1725.
Laurance, W. F. 1999. Reflections on the tropical deforestation crisis. *Biological Conservation* 91: 109–117.
———. 2001. Fragmentation and plant communities. In *Lessons from Amazonia: The Ecology and Conservation of a Fragmented Forest*, R. O. Bierregaard Jr. et al. (eds.). New Haven, CT: Yale University Press.

Laurance, W. F., et al. 2006. Rain forest fragmentation and the proliferation of successional trees. *Ecology* 87: 469–482.

Lens, L., S. Van Dongen, K. Norris, M. Githiru, and E. Matthysen. 2002. Avian persistence in fragmented rain forest. *Science* 298: 1236–1238.

Loreau, M., et al. 2001. Biodiversity and ecosystem functioning: Current knowledge and future challenges. *Science* 294: 804–808.

Lovejoy, T. E., and R. O. Bierregaard Jr. 1990. Central Amazonian forests and the minimum critical size of ecosystems project. In *Four Neotropical Rain Forests,* A. H. Gentry (ed.). New Haven, CT: Yale University Press.

Lovejoy, T. E., et al. 1986. Edge and other effects of isolation on Amazon forest fragments. In *Conservation Biology: The Science of Scarcity and Diversity*, M. E. Soule (ed.). Sunderland, MA: Sinauer.

Martin, P. H., C. D. Canham, and P. L. Marks. 2009. Why forests appear resistant to exotic plant invasions: Intentional introductions, stand dynamics, and the role of shade tolerance. *Frontiers in Ecology and the Environment* 7: 142–149.

May, R. M. 1988. How many species are there on Earth? *Science* 241: 1441–1449.

———. 1992. How many species inhabit the Earth? *Scientific American* 267: 42–48.

McConkey, K. R., and D. R. Drake. 2006. Flying foxes cease to function as seed dispersers long before they become rare. *Ecology* 87: 271–276.

Michalski, F., I. Nishi, and C.A. Peres. 2007. Disturbance-mediated drift in tree functional groups in Amazonian forest fragments. *Biotropica* 39: 691–701.

Myers, N. 1988. Threatened biotas: "Hotspots" in tropical forests. *The Environmentalist* 8: 187–208.

Myers, N., R. A. Mittermeier, C. G. Mittermeier, G. A. B. da Fonseca, and J. Kent. 2000. Biodiversity hotspots for conservation priorities. *Nature* 403: 853–858.

Pimm, S. L., and P. R. Raven. 2000. Extinction by the numbers. *Nature* 403: 843–845.

Prance, G. T., ed. 1982. *Biological Diversification in the Tropics.* New York: Columbia University Press.

Quinn, A. F., and A. Hastings. 1987. Extinction in subdivided habitats. *Conservation Biology* 1: 198–208.

Quintero, I., and T. Roslin. 2005. Rapid recovery of dung beetle communities following habitat fragmentation in Central Amazonia. *Ecology* 86: 3303–3311.

Rankin-de Merona, J. M., and R. W. Hutchings. 2001. Deforestation effects at the edge of an Amazonian forest fragment. In *Lessons from Amazonia: The Ecology and Conservation of a Fragmented Forest*, R. O. Bierregaard Jr. et al. (eds.). New Haven, CT: Yale University Press.

Robinson, W. D. 1999. Long-term changes in the avifauna of Barro Colorado Island, Panama, a tropical forest isolate. *Conservation Biology* 13: 85–97.

Rudel, T. K. 2005. *Tropical Forests: Regional Paths of Destruction and Regeneration in the Late Twentieth Century.* New York: Columbia University Press.

Schelhas, J., and R. Greenberg, eds. 1996. *Forest Patches in Tropical Landscapes.* Washington, DC: Island Press.

Simberloff, D. S. 1984. Mass extinction and the destruction of moist tropical forests. *Zhurnal Obshchei Biologii* 45: 767–778.

Simberloff, D. S., and L. G. Abele. 1976. Island biogeography theory and conservation practice. *Science* 191: 285–286.

Spray, S. L., and M. D. Moran, eds. 2006. *Tropical Deforestation.* New York: Rowman and Littlefield.

Stattersfield, A. J., and D. R. Capper. 2000. *Threatened Birds of the World.* Cambridge, UK: BirdLife International.

Stotz, D. F., J. W. Fitzpatrick, T. A. Parker III, and D. K. Moskovits. 1996. *Neotropical Birds: Ecology and Conservation.* Chicago: University of Chicago Press.

Stouffer, P. C., and R. O. Bierregaard Jr. 1995. Use of Amazonian forest fragments by understory insectivorous birds. *Ecology* 76: 2429–2445.

Stratford, J. A., and W. D. Robinson. 2005. Gulliver travels to the fragmented tropics: Geographic variation in mechanisms of avian extinction. *Frontiers of Ecology and Environment* 3: 91–98.

Terborgh, J. 1974. Preservation of natural diversity: The problem of extinction prone species. *Bioscience* 24: 715–722.

———. 1986. Keystone plant resources in the tropical forest. In *Conservation Biology: The Science of Scarcity and Diversity*, M. E. Soule (ed.). Sunderland, MA: Sinauer.

Terborgh, J., S. K. Robinson, T. A. Parker III, C. A. Munn, and N. Pierpont. 1990. Structure and organization of an Amazonian forest bird community. *Ecological Monographs* 60: 213–238.

Terborgh, J., et al. 2001. Ecological meltdown in predator-free forest fragments. *Science* 294: 1923–1926.

Tuno, N., W. Okeka, N. Minakawa, M. Takagi, and G. Yan. 2005. Survivorship of *Anopheles gambiae* sensu stricto (Diptera: Culicidae) larvae in western Kenyan highland forest. *Journal of Medical Entomology* 33: 200–209.

Willis, E. O. 1974. Populations and local extinctions of birds on Barro Colorado Island, Panama. *Ecological Monographs* 44: 153–169.

Wilson, E. O. 2002. *The Future of Life.* New York: A.A. Knopf.

Wilson, E. O., ed. 1988. *Biodiversity.* Washington, DC: National Academy Press.

Wood, C. H., and R. Porro, eds. 2002. *Deforestation and Land Use in the Amazon.* Gainesville: University Press of Florida.

CHAPTER 15

Achard, F., et al. 2002. Determination of deforestation rates of the world's humid tropical forests. *Science* 297: 999–1002.

Aragão, L. E. O., and Y. E. Shimabukuro. 2010. The incidence of fire in Amazonian forests with implications for REDD. *Science* 328: 1275–1278.

Asner, G. P., D. E. Knapp, E. N. Broadbent, P. J. C. Oliveira, M. Keller, and J. N. Silva. 2005. Selective logging in the Brazilian Amazon. *Science* 310: 480–482.

Bayon, R., and M. Jenkins. 2010. The business of biodiversity. *Nature* 466: 184–185.

Beckman, N. G., and H. C. Muller-Landau. 2007. Differential effects of hunting on pre-dispersal seed predation and primary and secondary seed removal of two Neotropical tree species. *Biotropica* 39: 328–339.

Benayas, J. M., A. C. Newton, A. Diaz, and J. M. Bullock. 2009. Enhancement of biodiversity and ecosystem services by ecological restoration: A meta-analysis. *Science* 325: 1121–1124.

Bermijo, M., J. D. Rodríguez-Teijeiro, G. Illera, A. Borroso, C. Vila, and P. D. Walsh. 2006. Ebola outbreak killed 5000 gorillas. *Science* 314: 1564.

Brashares, J. S. 2010. Filtering wildlife. *Science* 329: 402–403.

Brashares, J. S., P. Arcese, M. K. Sam, P. B. Coppolillo, A. R. E. Sinclair, and A. Balmford. 2004. Bushmeat hunting, wildlife declines, and fish supply in West Africa. *Science* 306: 1180–1183.

Brook, B. W., C. J. A. Bradshaw, L. Pin Koh, and M. N. S. Sodhi. 2006. Momentum drives the crash: Mass extinction in the tropics. *Biotropica* 38: 302–305.
Bush, M. B., and H. Hooghiemstra. 2005. Tropical biotic responses to climate change. In *Climate Change and Biodiversity*, T. E. Lovejoy and L. Hannah (eds.). New Haven, CT: Yale University Press.
Bushmeat Task Force. 2003. The law of the jungle. *Nature* 421: 8–9. Note: The Bushmeat Task Force can be found at www.bushmeat.org.
Butchart, S. H. M., et al. 2010. Global biodiversity: Indicators of recent decline. *Science* 328: 1164–1168.
Butler, R. A. 2007. Just how bad is the biodiversity extinction crisis: A debate erupts in the halls of conservation science. Available at http://news.mongabay.com/2007/0206-biodiversity.html.
Cannon, C. H., D. R. Peart, and M. Leighton. 1998. Tree species diversity in commercially logged Bornean rainforest. *Science* 281: 1366–1368.
Chazdon, R. L. 2003. Tropical forest recovery: Legacies of human impact and natural disturbances. *Perspectives in Plant Ecology and Evolutionary Systematics* 6: 51–71.
Clark, D. A. 2004. Tropical forests and global warming: Slowing it down or speeding it up? *Frontiers in Ecology and Environment* 2: 73–80.
Clark, D. B. 1996. Abolishing virginity. *Journal of Tropical Ecology* 12: 735–739.
Cochrane, M. A., et al. 1999. Positive feedback in the fire dynamics of close canopy tropical forests. *Science* 284: 1832–1835.
Corlett, R. T. 2007. The impact of hunting on the mammalian fauna of tropical Asian forests. *Biotropica* 39: 292–303.
Costanza, R., et al. 1997. The value of the world's ecosystem services and natural capital. *Nature* 387: 253–260.
Curran, L. M., et al. 2004. Lowland forest loss in protected areas of Indonesian Borneo. *Science* 303: 1000–1003.
Danielsen, F., et al. 2005. The Asian tsunami: A protective role for coastal vegetation. *Science* 310: 643.
Dean, W. 1995. *With Broadax and Firebrand: The Destruction of the Brazilian Atlantic Forest*. Los Angeles: University of California Press.
Diamond. J. 1997. *Guns, Germs, and Steel*. New York: W.W. Norton.
———. 2000. Blitzkrieg against the moas. *Science* 287: 2170–2171.
Dirzo, R., and E. Mendoza. 2007. Size-related differential seed predation in a heavily defaunated Neotropical rain forest. *Biotropica* 39: 355–362.
Fearnside, P. M. 2007. Deforestation in Amazonia. In *Encyclopedia of Earth*; available at http://www.eoearth.org/article/Deforestation_in_Amazonia.
Foley, J. A., et al. 2007. Amazonia revealed: Forest degradation and loss of ecosystem goods and services in the Amazon Basin. *Frontiers of Ecology and Environment* 5: 25–32.
Food and Agriculture Organization (FAO). 2000. Global forest resource assessment 2000—main report. FAO Forestry Paper 140. New York: United Nations Food and Agriculture Organization.
Galindo-Leal, C., and I. de Gusmao Camara. 2003. *The Atlantic Forest of South America: Biodiversity Status, Threats, and Outlook (State of the Hotspots)*. New York: Island Press.
Gardner, T. A., J. Barlow, L. W. Parry, and C. A. Peres. 2006. Predicting the uncertain future of tropical forest species in a data vacuum. *Biotropica* 39: 25–30.
Garrett, L. 1995. The coming plague: Newly emerging diseases in a world out of balance. New York: Penguin.
Goodall, J., and L. Pintea. 2010. Securing a future for chimpanzees. *Nature* 466: 180–181.
Gullison, R. E., et al. 2007. Tropical forests and climate policy. *Science* 316: 985.
Hansen, M. C., and R. DeFries. 2004. Detecting long-term global forest change using continuous fields of tree-cover maps from 8-km advanced very high resolution radiometer (AVHRR) data for years 1982–99. *Ecosystems* 7: 695–716.
Hartshorn, G. S., and B. E. Hammel. 1994. Vegetation types and floristic patterns. In *La Selva: Ecology and Natural History of a Neotropical Rain Forest*, L. A. McDade, K. S. Bawa, H. A. Hespenheide, and G. S. Hartshorn (eds.). Chicago: University of Chicago Press.
Houghton, R. A., D. L. Skole, C. A. Nobre, J. L. Hackler, K. T. Lawrence, and W. H. Chomentowski. 2000. Annual fluxes of carbon from deforestation and regrowth in the Brazilian Amazon. *Nature* 403: 301–304.
Kaimowitz, D., and D. Shell. 2007. Conserving what and for whom? Why conservation should help meet basic human needs in the tropics. *Biotropica* 39: 567–574.
Keele, B. F., et al. 2006. Chimpanzee reservoirs of pandemic and nonpandemic HIV-1. *Science Online* 2006-05-25 (5786): 523; doi: 10.1126/science.1126531.PMID16728595.PMC2442710.
———. 2009. Increased mortality and AIDS-like immunopathology in wild chimpanzees infected with SIVcpz. *Nature* 460: 515–519.
Keller, M., G. A. Asner, G. Blate, J. McGlocklin, F. Merry, M. Pena-Claros, and J. Zweede. 2007. Timber production in selectively logged tropical forests in South America. *Frontiers in Ecology and Environment* 5: 213–216.
Knoke, T., et al. 2009. Can tropical farmers reconcile subsistence needs with forest conservation? *Frontiers in Ecology and Environment* 7: 548–554.
Koellner, T., and O. J. Schmitz. 2006. Biodiversity, ecosystem function, and investment risk. *BioScience* 56: 977–985.
Lamb, D., P. D. Erskine, and J. A. Parotta. 2005. Restoration of degraded tropical forest landscapes. *Science* 310: 1628–1632.
Laurance, S. G. W. 2006. Rainforest roads and the future of forest dependent wildlife: A case study of understory birds. In *Emerging Threats to Tropical Forests*, W. F. Laurance and C. A. Peres (eds.). Chicago: University of Chicago Press.
Laurance, W. F. 2006. Have we overstated the tropical biodiversity crisis? *Trends in Ecology and Evolution* 22: 65–70.
Laurance, W. F., et al. 2001. The future of the Brazilian Amazon. *Science* 291: 438–439.
Laurance, W. F., and C. A. Peres, eds. 2006. *Emerging Threats to Tropical Forests*. Chicago: University of Chicago Press.
Leendertz, F. H., et al. 2004. Anthrax kills wild chimpanzees in a tropical rainforest. *Nature* 430: 451–452.
Loarie, S. R., P. B. Duffy, H. Hamilton, G. P. Asner, C. B. Field, and D. D. Ackerly. 2009. The velocity of climate change. *Nature* 462: 1052–1055.
Meijaard, E., D. Sheil, A. J. Marshall, and R. Nasi. 2008. Phylogenetic age is positively correlated with sensitivity to timber harvest in Bornean mammals. *Biotropica* 40: 76–85.
Montagnini, F., and C. F. Jordan. 2005. *Tropical Forest Ecology: The Basis for Conservation and Management*. Berlin: Springer.
Muller-Landau, H. C. 2007. Predicting the long-term effects of hunting on plant species composition and diversity in tropical forests. *Biotropica* 39: 372–384.

Nepstad, D. C., C. M. Stickler, B. Soares-Filho, and F. Merry. 2008. Interactions among Amazon land use, forests and climate: Prospects for a near-term forest tipping point. *Philosophical Transactions of the Royal Society B* 27: 1737–1746.

Ninan, K. N., ed. 2009. *Conserving and Valuing Ecosystem Services and Biodiversity: Economic, Institutional, and Social Challenges*. London: Earthscan Publications.

Nuñez-Iturri, G., and H. F. Howe. 2007. Bushmeat and the fate of trees with seeds dispersed by large primates in a lowland rain forest in western Amazonia. *Biotropica* 39: 348–354.

Parmesan, C., and G. Yohe. 2003. A globally coherent fingerprint of climate change impacts across natural systems. *Nature* 421: 37–42.

Pearce, D., F. E. Putz, and J. K. Vanclay. 2003. Sustainable forestry in the tropics: Panacea or folly? *Forest Ecological Management* 172: 229–247.

Peres, C. A., and E. Palacios. 2007. Basin-wide effects of game harvest on vertebrate population densities in Amazonian forests: Implications for animal-mediated seed dispersal. *Biotropica* 39: 304–315.

Putz, F. E., and K. H. Redford. 2010. The importance of defining "'forest": Tropical forest degradation, deforestation, long-term phase shifts, and further transitions. *Biotropica* 42: 10–20.

Ramankutty, N., and J. A. Foley. 1999. Estimating historical changes in global land cover: Croplands from 1700 to 1992. *Global Biogeochemical Cycles* 13: 997–1027.

Redford, K. H. 1992. The empty forest. *Bioscience* 42: 412–422.

Robinson, J. G., and K. H. Redford. 1991. *Neotropical Wildlife Use and Conservation*. Chicago: University of Chicago Press.

Robinson, J. G., K. H. Redford, and E. L. Bennett. 1999. Wildlife harvest in logged tropical forests. *Science* 284: 595–596.

Rodrigues, A. S. L., R. M. Ewers, L. Parry, C. Souza Jr., A. Verissimo, and A. Balmford. 2009. Boom-and-bust development patterns across the Amazon deforestation frontier. *Science* 324: 1435–1437.

Rudel, T. K. 2005. *Tropical Forests*. New York: Columbia University Press.

Sharp, P. M., E. Bailes, R. R. Chaudhuri, C. M. Rodenburg, M. O. Santiago, and B. H. Hahn. 2001. The origins of acquired immune deficiency syndrome viruses: Where and when? *Philosophical Transactions of the Royal Society B* 356: 867–876.

Shearman, P. L., J. Ash, B. Mackey, J. E. Bryan, and B. Lokes. 2009. Forest conversion and degradation in Papua New Guinea 1972–2002. *Biotropica* 41: 379–390.

Silva Dias, M. A .F., et al. 2002. Cloud and rain processes in a biosphere–atmosphere interaction context in the Amazon region. *Journal of Geophysical Research* 107(8072); doi: 10.1029/2001JD000335.

Sinervo, B., et al. 2010. Erosion of lizard diversity by climate change and altered thermal niches. *Science* 328: 894–899.

Soares-Filho, B. S., et al. 2006. Modelling conservation in the Amazon Basin. *Nature* 440: 520–523.

Spray, S. L., and M. D. Moran, eds. 2006. *Tropical Deforestation*. New York: Rowman and Littlefield.

Stoner, K. E., P. Riba-Hernández, K. Vulinec, and J. E. Lambert. 2007a. The role of mammals in creating and modifying seed shadows in tropical forests and some possible consequences of their elimination. *Biotropica* 39: 316–327.

Stoner, K. E., K. Vulinec, S. J. Wright, and C. A. Peres. 2007b. Hunting and plant community dynamics in tropical forests: A synthesis and future directions. *Biotropica* 39: 385–392.

Takebe, Y., R. Uenishi, and X. Li. 2008. Global molecular epidemiology of HIV: Understanding the genesis of AIDS pandemic. *Advances in Pharmacology* 56: 1–25.

Terborgh, J. 2000. The fate of tropical forests: A matter of stewardship. *Conservation Biology* 14: 1358–1361.

Tollefson, J. 2008. Brazil goes to war against logging. *Nature* 452: 134–135.

———. 2010. Amazon drought raises research doubts. *Nature* 466: 423.

Tuno, N., O. Wilberforce, N. Minakawa, M. Takagi, and G. Yan. 2005. Survivorship of *Anopheles gambiae* sensu stricto (Diptera: Culicidae) larvae in Western Kenya highland forest. *Journal of Medical Entomology* 42: 270–277.

Uriarte, M., L. Schneider, and T. K. Rudel. 2009. Land transformations in the tropics: Going beyond the case studies. *Biotropica* 42: 1–2.

Vogel, G. 2007. Scientists say Ebola has pushed western gorillas to the brink. *Science* 317: 1484.

Walsh, P. D., et al. 2003. Catastrophic ape decline in western equatorial Africa. *Nature* 422: 611–614.

Wang, B. C., W. L. Sork, M. T. Leonig, and T. B. Smith. 2007. Hunting of mammals reduces seed removal and dispersal of the Afrotropical tree *Antrocaryon klaineanum* (Anacardiaceae). *Biotropica* 39: 340–347.

Weiss, R. A., and J. H. Heeney. 2009. An ill wind for wild chimps? *Nature* 460: 470–471.

Wood, C. H., and R. Porro, eds. 2002. *Deforestation and Land Use in the Amazon*. Gainesville: University Press of Florida.

Wright, S. J., A. Hernandez, and R. Condit. 2007a. The bushmeat harvest alters seedling banks by favoring lianas, large seeds, and seeds dispersed by bats, birds, and wind. *Biotropica* 39: 363–371.

Wright, S. J., and H. C. Muller-Landau. 2006a. The future of tropical forest species. *Biotropica* 38: 287–301.

———. 2006b. The uncertain future of forest species. *Biotropica* 38: 443–445.

Wright, S. J., et al. 2007b. The plight of large animals in tropical forests and the consequences for plant regeneration. *Biotropica* 39: 289–291.

Zahawi, R. A., and K. D. Holl. 2010. Bridging the gap between scientific research and tropical forest restoration: A multifaceted research, conservation, education, and outreach program in southern Costa Rica. *Ecological Restoration* 28: 143–146.

Illustration Credits

All photos by the author (John Kricher) unless otherwise noted.

CHAPTER 1

PLATE 1-1 (a) http://en.wikipedia.org/wiki/File:Darwin_1881.jpg
PLATE 1-1 (b) Photo © Charles H. Smith, 2009.
PLATE 1-2 http://commons.wikimedia.org/wiki/File:Humboldt.jpg
FIGURE 1-1 From Bates, H. W. 1892. *The Naturalist on the River Amazons*. London: John Murray, pp. 335–337.
PLATE 1-5 From the collection of Carol Gould.
PLATE 1-6 Photo by Lena Strewe.
PLATE 1-7 Photo © rhpayne.
FIGURE 1-2 Adapted from Stocks, G., L. Seales, F. Paniagua, E. Maehr, and E. M. Bruna. 2008. The geographical and institutional distribution of ecological research in the tropics. *Biotropica* 40(4): 397–404, Figure 1. Reproduced with permission of Blackwell Publishing Ltd.
FIGURE 1-3 Adapted from Parker, G. G., A. P. Smith, and K. P. Hogan. 1992. Access to the upper forest canopy with a large tower crane. *BioScience* 42: 664–670, Figure 1.
FIGURE 1-4 Adapted from Davis, T. A., and Richards, P. W. 1933–1934. The vegetation of Moraballi Creek, British Guiana: An ecological study of a limited area of tropical rain forest. Parts I and II. *Journal of Ecology* 21: 350–384; 22: 106–155.
FIGURE 1-5 Adapted from Richards, P. W. 1939. Ecological studies on the rain forest of Southern Nigeria. I. The structure and floristic composition of the primary forest. *Journal of Ecology* 27: 1–61.
FIGURE 1-6 From Johnson, K. R., and B. Ellis. 2002. A tropical rainforest in Colorado 1.4 million years after the Cretaceous–Tertiary boundary. *Science* 296: 2381, Figure 2. Reprinted with permission from AAAS.
PLATE 1-13 Courtesy of the National Park Service, U.S. Department of the Interior. *Jurassic Landscape and Dinosaur Mural for Dinosaur National Monument*, 2009, by Liz Bradford.
FIGURE 1-7 Photo: ESA/Earth Observation/Envisat.
FIGURE 1-9 Adapted from MacArthur, R. H. 1972. *Geographical Ecology: Patterns in the Distribution of Species*. New York: Harper and Row, Figure 8-1.
FIGURE 1-10 From Holdridge, L. R. 1947. Determination of world plant formations from simple climatic data. *Science* 105: 367–368 (1947). Reprinted with permission from AAAS. Image adapted from Peter Halasz.
FIGURE 1-11 Adapted from Whittaker, R. H. 1975. *Communities and Ecosystems*, 2nd ed. New York: Macmillan Publishing, Figure 4-10.
FIGURE 1-12 Adapted from Walter, H. 1975. *Vegetation of the Earth in Relation to Climate and the Eco-physiological Conditions*. London: The English Universities Press Ltd., Figure 15.
FIGURE 1-13 (a) Adapted from Bailey, R. G. 1998. *Ecoregions*. New York: Springer, Figure 8-1. (b) Adapted from Walter, H. 1975. *Vegetation of the Earth in Relation to Climate and the Eco-physiological Conditions*. London: The English Universities Press Ltd., Figure 15.
FIGURE 1-14 Adapted from Emmel, T. C. 1976. *Population Biology*. New York: Harper and Row, Figure 10-2.
FIGURE 1-15 Adapted from Emmel, T. C. 1976. *Population Biology*. New York: Harper and Row, Figure 10-3.
FIGURE 1-16 Adapted from Emmel, T. C. 1976. *Population Biology*. New York: Harper and Row, Figure 10-4.
FIGURE 1-17 Adapted from Emmel, T. C. 1976. *Population Biology*. New York: Harper and Row, Figure 10-5.
FIGURE 1-18 Adapted from Lutgens, F. K., E. J. Tarbuck, and D. Tasa. 2010. *The Atmosphere: An Introduction to Meteorology*, 11th ed. Upper Saddle River, NJ: Pearson Education, Figure 7-10.
PLATE 1-27 (a) Photo © Margaret E. Poggio. (b) Photo by Anne Thoul.
FIGURE 1-19 Adapted from Cohen, J. E. 2006. Human population: The next half-century, *Science Magazine's State of the Planet: 2006–2007*, D. Kennedy (ed.). Washington, DC: Island Press, Figure 1.
TABLE 1-1 Used with permission from Achard, F., et al. 2002. Determination of deforestation rates of the world's humid tropical forests. *Science* 297: 999–1002, Table 1. Reprinted with permission from AAAS.
PLATE 1-28 Photo by Bruce Hallett.

CHAPTER 2

PLATE 2-1 Photos by William E. Davis Jr.
FIGURE 2-1 Adapted from http://taxonomy.zoology.gla.ac.uk/~rdmp1c/teaching/L1/Evolution/l1/56.gif.
FIGURE 2-4 Adapted by permission from Macmillan Publishers Ltd: NATURE. (a) Krause, D. W. Washed up in Madagascar. *Nature* 463: 613–614, Figure 1, copyright 2010; (b) Ali, J. R., and M. Huber. Mammalian biodiversity on Madagascar controlled by ocean currents. *Nature* 463: 653–656, Figure 1, copyright 2010.
FIGURE 2-7 Adapted from http://pubs.usgs.gov/gip/dynamic/graphics/midatlantic_ridge.gif.
FIGURE 2-8 Adapted from a painting by Zdeněk Burian.
PLATE 2-5 (a) Photo by Frederick J. Dodd, International Zoological Expeditions. (d) Courtesy of John and Jenny Wright, Crater Lakes Rainforest Cottages, Cairns, Australia. (e) Photo © R. Brown/VIREO. (f) Barbara Hardy Centre, http://www.unisa.edu.au/barbarahardy/Photographer: John Hodgson.
PLATE 2-9 Photo by Frederick J. Dodd, International Zoological Expeditions.
PLATE 2-10 Photo © P. Robles Gil/VIREO.
PLATE 2-11 Photo © Bernard d'Abrera, 2010. Digital file provided by Professor James Mallet, http://www.ucl.ac.uk/taxome/jim/.
PLATE 2-12 (b) Photo © Andrea K. Turkalo, Wildlife Conservation Society.
FIGURE 2-9 Adapted from Roca, A. L., N. Georgiadis, J. Pecon-Slattery, and S. J. O'Brien. 2001. Genetic evidence for two species of elephant in Africa. *Science* 293: 1473–1476, Figure 3. Reprinted with permission from AAAS.
FIGURE 2-10 Adapted from http://whozoo.org/mammals/Primates/primatephylogeny.htm.
FIGURE 2-12 Adapted from http://upload.wikimedia.org/wikipedia/commons/thumb/c/c8/Great_Rift_Valley_map-fr.svg/465px-Great_Rift_Valley_map-fr.svg.png.
FIGURE 2-13 From Garzione, C. N., et al. 2008. Rise of the Andes. *Science* 320: 1304–1307, Figure 1, p. 620. Reprinted with permission from AAAS.

FIGURE 2-14 Adapted from http://tapirs.org/tapirs/bairds.html.

FIGURE 2-15 Used with permission from Traylor, M. A., Jr. 1985. Species limits in the *Ochthoeca diadema* species-group (Tyrannidae). In P. A. Buckley et al. (eds), *Ornithological Monographs* No. 36. Washington, DC: American Ornithologists' Union, Plate VI.

FIGURE 2-16 Used with permission from Price, T. 2008. *Speciation in Birds.* Greenwood Village, CO: Roberts and Company. Images in *Speciation in Birds* are acknowledged as follows: The Tyrian metaltail and the Neblina metaltail were redrawn from original illustrations by David Alker; distributions are from del Hoya, J., A. Elliott, and J. Sargatal. 1999. *Handbook of Birds of the World*, vol. 5. Barcelona: Lynx Editions.

FIGURE 2-17 Adapted from Cracraft, J. 1985. Historical biogeography and patterns of differentiation within South American avifauna: centers of endemism. In *Ornithological Monographs* No. 36, P. A. Buckley et al. (eds.). Washington, DC: American Ornithologists' Union, Figure 1.

FIGURE 2-18 Adapted from Cracraft, J. 1985. Historical biogeography and patterns of differentiation within South American avifauna: Centers of endemism. In *Ornithological Monographs* No. 36, P. A. Buckley et al. (eds.). Washington, DC: American Ornithologists' Union, Figures 4 and 5.

FIGURE 2-19 Adapted from Haffer, J. 1985. Avian zoogeography of the Neotropical lowlands. In *Neotropical Ornithology*, P. A. Buckley, M. S. Foster, E. S. Morton, R. S. Ridgely, and F. G. Buckley (eds.). Washington, DC: American Ornithologists' Union, Figures 10 and 11.

FIGURE 2-20 Adapted from Haffer, J. 1974. *Avian Speciation in Tropical South America*, no. 14., R. E. Paynter (ed.). Cambridge, MA: Publications of the Nuttall Ornithological Club, Figure 13-2.

PLATE 2-15 From Wilf, P., et al. 2003. High plant diversity in Eocene South America: Evidence from Patagonia. *Science* 300: 122–125, Figure 1. Reprinted with permission from AAAS. Specimen (a) from Wilf, P., et al. 2009. *Papuacedrus* (Cupressaceae) in Eocene Patagonia: A new fossil link to Australasian rainforests. *American Journal of Botany* 96: 2031–2047. Specimen (g) from Zamaloa, M. C., M. A. Gandolfo, C. C. González, E. J. Romero, N. R. Cúneo, and P. Wilf, 2006. Casuarinaceae from the Eocene of Patagonia, Argentina. *International Journal of Plant Sciences* 167: 1279–1289.

FIGURE 2-21 Adapted from Benton, M. J. 2005. *Vertebrate Paleontology*, 3rd ed. Malden, MA: Blackwell, Figure 10-21, page 322. Reproduced with permission from Blackwell Publishing Ltd.

FIGURE 2-22 Art by Mauricio Anton.

FIGURE 2-23 Art by Mauricio Anton.

CHAPTER 3

FIGURE 3-1 Adapted from Tramer, E. J. 1974. On latitudinal diversity in avian diversity. *Condor* 76: 123–130, Figure 1.

FIGURE 3-2 Adapted from Tramer, E. J. 1974. On latitudinal diversity in avian diversity. *Condor* 76: 123–130, Figure 3.

FIGURE 3-3 Page 6, Figure 1-2, Forest on Barro Colorado, from *Tropical Forest Ecology: A View from Barro Colorado Island*, by Leigh, E. G., Jr. (1999). By permission of Oxford University Press, Inc.

FIGURE 3-4 Adapted from Whitmore, J. T. 1990. *An Introduction to Tropical Rain Forests.* Oxford, UK: Clarendon Press, Figure 2-24.

PLATE 3-7 Photo by Gregory Gilbert.

FIGURE 3-5 Adapted from Whitmore, J. T. 1990. *An Introduction to Tropical Rain Forests.* Oxford, UK: Clarendon Press, Figure 3-24.

FIGURE 3-6 Adapted from Poorter, L., F. Bongers, F. J. Sterck, and H. Woll. 2003. Architecture of 53 rain forest tree species differing in adult stature and shade tolerance. *Ecology* 84: 602–608, Figure 2. Copyright 2003 by Ecological Society of America. Reproduced with permission of Ecological Society of America via Copyright Clearance Center.

PLATE 3-18 (a) Photo by Scott W. Shumway.

PLATE 3-19 Photo by Scott W. Shumway.

PLATE 3-20 Photo by Scott W. Shumway.

PLATE 3-22 Photo courtesy of kkaplan/Shutterstock.

PLATE 3-24 Photo © T. S. Yeoh.

PLATE 3-26 Photo by Eduardo López.

FIGURE 3-7 Art by Mauricio Anton.

PLATE 3-27 Photo by Scott W. Shumway.

PLATE 3-31 Photo © avlxyz (http://flic.kr/p/8scUJ).

FIGURE 3-8 Adapted from Whitmore, T. C. 1984. *Tropical Rain Forests of the Far East*, 2nd ed. Oxford, UK: Clarendon Press, Figure 3-24.

PLATE 3-39 Photo by Scott W. Shumway.

PLATE 3-40 Photo by Scott W. Shumway.

PLATE 3-41 Photo by Scott W. Shumway.

CHAPTER 4

FIGURE 4-3 Adapted from T. C. Whitmore 1984. *Tropical Rain Forests of the Far East.* London: Clarendon, and A. L. Gentry. 1988. Tree species richness of upper Amazonian forests. *Proceedings of the National Academy of Sciences USA* 85:156–159. Copyright 1988 National Academy of Sciences, USA.

TABLE 4-2 Adapted from Karr, J. R. 1976. Within- and between-habitat avian diversity in African and Neotropical lowland habitats. *Ecological Monographs* 46: 457–481, Table 8.

FIGURE 4-4 Adapted from Karr, J. R. 1976. Within- and between-habitat avian diversity in African and Neotropical lowland habitats. *Ecological Monographs* 46: 457–481, Figure 6.

FIGURE 4-5 Adapted from Condit et al. 2002. Beta-diversity in tropical forest trees. *Science* 295: 666–669, Figure 1. Reprinted with permission from AAAS.

PLATE 4-2 Photo courtesy of the Smithsonian Tropical Research Institute, Marcos Guerra.

FIGURE 4-6 Adapted from Foster, R. B., and S. P. Hubbell. 1990. The floristic composition of the Barro Colorado Island forest. In *Four Neotropical Rain Forests*, A. H. Gentry (ed.). New Haven: Yale University Press, Figure 6.3.

FIGURE 4-7 Adapted from Foster, R. B., and S. P. Hubbell. 1990. The floristic composition of the Barro Colorado Island forest. In *Four Neotropical Rain Forests*, A. H. Gentry (ed.). New Haven: Yale University Press, Figure 7.2.

FIGURE 4-8 Adapted from Foster, R. B., and S. P. Hubbell. 1990. The floristic composition of the Barro Colorado Island forest. In *Four Neotropical Rain Forests*, A. H. Gentry (ed.). New Haven: Yale University Press, Figure 7.1.

PLATE 4-6 Photo © Dr. James L. Castner.

PLATE 4-7 Photo © Dr. James L. Castner.

FIGURE 4-9 Adapted from Hawkins, B. A., E. R. Porter, and J. A. F. Diniz-Filho. 2003a. Productivity and history as predictors of the latitudinal diversity gradient of terrestrial birds. *Ecology* 84: 1608–1623, Figure 3. Copyright 2003 by Ecological Society of America. Reproduced with permission of Ecological Society of America via Copyright Clearance Center.

FIGURE 4-10 Adapted from Hawkins, B. A., E. R. Porter, and J. A. F. Diniz-Filho. 2003a. Productivity and history as predictors of the latitudinal diversity gradient of terrestrial birds. *Ecology* 84: 1608–1623, Figure 4. Copyright 2003 by Ecological Society of America. Reproduced with permission of Ecological Society of America via Copyright Clearance Center.

PLATE 4-8 Photo by Bruce M. Beehler.

PLATE 4-9 Photo by Bruce M. Beehler.

PLATE 4-10 Photo by Bruce M. Beehler.

FIGURE 4-12 Adapted from Jablonski, D., K. Roy, and J. W. Valentine. 2006. Out of the tropics: Evolutionary dynamics of the latitudinal diversity gradient. *Science* 314: 102–106, Figure 2. Reprinted with permission from AAAS.

FIGURE 4-13 Adapted from Jablonski, D., K. Roy, and J. W. Valentine. 2006. Out of the tropics: Evolutionary dynamics of the latitudinal diversity gradient. *Science* 314: 102–106, Figure 3. Reprinted with permission from AAAS.

FIGURE 4-14 Adapted from Jablonski, D., K. Roy, and J. W. Valentine. 2006. Out of the tropics: Evolutionary dynamics of the latitudinal diversity gradient. *Science* 314: 102–106, Figure 1. Reprinted with permission from AAAS.

FIGURE 4-15 Adapted by permission from Macmillan Publishers Ltd: NATURE. Lamoreux, J. F., et al. Global tests of biodiversity concordance and the importance of endemism. *Nature* 440: 212–213, Figure 1, copyright 2006.

TABLE 4-3 Adapted by permission from Macmillan Publishers Ltd: NATURE. Lamoreux, J. F., et al. Global tests of biodiversity concordance and the importance of endemism. *Nature* 440: 212–213, Table 1, copyright 2006.

FIGURE 4-16 Adapted by permission from Macmillan Publishers Ltd: NATURE. Lamoreux, J. F., et al. Global tests of biodiversity concordance and the importance of endemism. *Nature* 440: 212–213, Figure 2b, copyright 2006.

FIGURE 4-17 Adapted from Weir, J. T., and D. Schluter. 2007. The latitudinal gradient in recent speciation and extinction rates of birds and mammals. *Science* 315: 15741576, Figure 1a. Reprinted with permission from AAAS.

FIGURE 4-18 Adapted from Weir, J. T., and D. Schluter. 2007. The latitudinal gradient in recent speciation and extinction rates of birds and mammals. *Science* 315: 1574–1576, Figure 2. Reprinted with permission from AAAS.

FIGURE 4-19 Adapted from Richardson, J. E., R. T. Pennington, T. D. Pennington, and P. R. Hollingworth. 2001. Rapid diversification of a species-rich genus of Neotropical rain forest trees. *Science* 293: 2242–2245, Figure 1. Reprinted with permission from AAAS.

FIGURE 4-20 Adapted from Richardson, J. E., R. T. Pennington, T. D. Pennington, and P. R. Hollingworth. 2001. Rapid diversification of a species-rich genus of Neotropical rain forest trees. *Science* 293: 2242–2245, Figure 3. Reprinted with permission from AAAS.

FIGURE 4-21 Adapted from Hawkins, B. A., E. R. Porter, and J. A. F. Diniz-Filho. 2003a. Productivity and history as predictors of the latitudinal diversity gradient of terrestrial birds. *Ecology* 84: 1608–1623, Figure 1. Copyright 2003 by Ecological Society of America. Reproduced with permission of Ecological Society of America via Copyright Clearance Center.

FIGURE 4-22 Adapted from Hawkins, B. A., E. R. Porter, and J. A. F. Diniz-Filho. 2003a. Productivity and history as predictors of the latitudinal diversity gradient of terrestrial birds. *Ecology* 84: 1608–1623, Figure 9. Copyright 2003 by Ecological Society of America. Reproduced with permission of Ecological Society of America via Copyright Clearance Center.

FIGURE 4-23 Adapted from Hawkins, B. A., et al. 2003b. Energy, water, and broad-scale geographic patterns of species richness. *Ecology* 84: 3105-3117, Figure 1. Copyright 2003 by Ecological Society of America. Reproduced with permission of Ecological Society of America via Copyright Clearance Center.

FIGURE 4-24 Adapted from Hawkins, B.A., et al. 2003b. Energy, water, and broad-scale geographic patterns of species richness. *Ecology* 84: 3105-3117, Figure 3. Copyright 2003 by Ecological Society of America. Reproduced with permission of Ecological Society of America via Copyright Clearance Center.

FIGURE 4-25 Adapted from Jaramillo, C., M. J. Rueda, and G. Mora. 2006. Cenozoic plant diversity in the Neotropics. *Science* 311: 1893–1896, Figure 2. Reprinted with permission from AAAS.

FIGURE 4-26 Adapted from Jaramillo, C., M. J. Rueda, and G. Mora. 2006. Cenozoic plant diversity in the Neotropics. *Science* 311: 1893–1896, Figure 3. Reprinted with permission from AAAS.

FIGURE 4-27 Schulenberg, Thomas S.; *Birds of Peru*. © 2007 by Princeton University Press. Reprinted by permission of Princeton University Press.

FIGURE 4-28 © Zimmerman et al., *Birds of Kenya*, Plate 59, p. 137, and Christopher Helm, an imprint of A&C Black Publishers Ltd.

FIGURE 4-29 Adapted from Schoener, T. W. 1971. Large-billed insectivorous birds: A precipitous diversity gradient. *Condor* 73: 154–161, Figure 1.

PLATE 4-12 Photo © Dr. James L. Castner

PLATE 4-13 Photo by Vlastik Bracha.

PLATE 4-14 Photo by Ryan Shaw.

PLATE 4-15 Photo by Bob Iannucci.

FIGURE 4-30 Adapted from Novotney, V. P., et al. 2006. Why are there so many species of herbivorous insects in tropical rain forests? *Sciencexpress*, www.sciencexpress.org/13 July 2006, Figure 1A.

FIGURE 4-31 Adapted from Novotney, V. P., et al. 2006. Why are there so many species of herbivorous insects in tropical rain forests? *Sciencexpress*, www.sciencexpress.org/13 July 2006, Figure 2.

PLATE 4-20 Photo by Frederick J. Dodd, International Zoological Expeditions.

FIGURE 4-32 de Schauensee, Rodolphe Meyer; *A Guide to the Birds of Venezuela*. © 1978 Princeton University Press. 2nd ed. © 2003 by Steven L. Hilty. Reprinted by permission of Princeton University Press.

FIGURE 4-33 Adapted from Diamond, J. M. 1973. Distributional ecology of New Guinea birds. *Science* 179: 759–769, Figure 4. Reprinted with permission from AAAS.

FIGURE 4-34 Adapted from Diamond, J. M. 1973. Distributional ecology of New Guinea birds. *Science* 179: 759–769, Figure 6. Reprinted with permission from AAAS.
PLATE 4-21 Photo by Scott W. Shumway.
PLATE 4-23 Photo by Scott W. Shumway.
PLATE 4-25 Photo by Tom Boyden/Lonely Planet Images.
FIGURE 4-35 Adapted from from Marquis, R. J. 2004. Herbivores rule. *Science* 305: 619–620; illustration by Taina Litwak.
FIGURE 4-36 Adapted from Fine, P. A. V., I. Mesones, and P. D. Coley. 2004. Herbivores promote habitat specialization by trees in Amazonian forests. *Science* 305: 663–667, Figure 2. Reprinted with permission from AAAS.
FIGURE 4-37 Adapted from Dyer, L. A., T. R. Walla, H. F. Greeney, J. O. Stireman III, and R. F. Hazen. 2010. Diversity of interactions: A metric for studies of biodiversity. *Biotropica* 42: 281–289, Figure 1. Reproduced with permission of Blackwell Publishing Ltd.

CHAPTER 5

PLATE 5-1 Photo courtesy Chris Wills.
FIGURE 5-7 Adapted from Hubbell, S. P. 2001. *The Unified Neutral Theory of Biodiversity and Biogeography*. Princeton, NJ: Princeton University Press, Figure 5-9.
FIGURE 5-8 Adapted by permission from Macmillan Publishers Ltd: NATURE. ter Steege, H., et al. Continental-scale patterns of canopy tree composition and function across Amazonia. *Nature* 442: 444–447, Figure 1, copyright 2006.
FIGURE 5-9 Used with permission from Condit, R., et al. 2000. Spatial patterns in the distribution of tropical tree species. *Science* 288: 1414–1418, Figure 1. Reprinted with permission from AAAS.
FIGURE 5-10 Used with permission from Condit, R., et al. 2000. Spatial patterns in the distribution of tropical tree species. *Science* 288: 1414–1418, Figures 2 and 3. Reprinted with permission from AAAS.
FIGURE 5-11 Used with permission from the Center for Tropical Forest Science (CTFS)—Smithsonian Institution Global Earth Observatory Program.
FIGURE 5-12 Adapted by permission from Macmillan Publishers Ltd: NATURE. Lambers, J. H. R., J. S. Clark, and B. Beckage. Density dependent mortality and the latitudinal gradient in species diversity. *Nature* 417: 732–734, Figure 2, copyright 2002.
FIGURE 5-13 Adapted by permission from Macmillan Publishers Ltd: NATURE. Kelly, C. K., and M. G. Bowler. Coexistence and relative abundance in forest trees. *Nature* 417: 437–440, Figure 3, copyright 2002.
FIGURE 5-14 Adapted from Connell, J. H. 1978. Diversity in tropical rain forests and coral reefs. *Science* 199: 1302–1310, Figure 1. Reprinted with permission from AAAS.
FIGURE 5-15 Adapted from Vandermeer, J., I. G. de la Cerda, D. Boucher, I. Perfecto, and J. Ruiz. 2000. Hurricane disturbance and tropical tree species diversity. *Science* 290: 788–791, Figure 2. Reprinted with permission from AAAS.
FIGURE 5-16 Adapted from Molino, J.-F., and D. Sabatier. 2001. Tree diversity in tropical rain forests: A validation of the intermediate disturbance hypothesis. *Science* 294: 1702–1704, Figure 2. Reprinted with permission from AAAS.
FIGURE 5-17 Adapted from Wilson, E. O. 1992. *The Diversity of Life*. New York: W.W. Norton & Company.
PLATE 5-10 Photo by Frederick J. Dodd, International Zoological Expeditions.
FIGURE 5-18 Adapted from MacArthur, R. H., and E. O. Wilson. 1967. *The Theory of Island Geography*. Princeton, NJ: Princeton University Press, Figures 7 and 8.
PLATE 5-12 Photo by Frederick J. Dodd, International Zoological Expeditions.
PLATE 5-13 Photo © J. Dunning/VIREO.
TABLE 5-1 Adapted from Hubbell, S. P. 2001. *The Unified Neutral Theory of Biodiversity and Biogeography*. Princeton, NJ: Princeton University Press, Table 5-1.
FIGURE 5-20 Adapted from Hubbell, S. P. 2001. *The Unified Neutral Theory of Biodiversity and Biogeography*. Princeton, NJ: Princeton University Press, Figure 5-10.
FIGURE 5-21 Adapted from Hubbell, S. P. 2001. *The Unified Neutral Theory of Biodiversity and Biogeography*. Princeton, NJ: Princeton University Press, Figure 10-1.
FIGURE 5-22 Adapted from (a) Blackburn, T. M., and J. Gaston. 1996. *Philosophical Transactions Royal Society London Biological Sciences* 351: 897; (b) Schoener, T. W. 1987. *Oecologia* 74: 161; and (c) Brown, J. H. 1995. *Macroecology*. Chicago: University of Chicago Press. (d) Adapted from Bell, G. 2001. Neutral macroecology. *Science* 293: 2413–2418, Figure 1. Reprinted with permission from AAAS.
FIGURE 5-23 Adapted from Hubbell, S. P. 2006. Neutral theory and the evolution of ecological equivalence. *Ecology* 87: 1387–1398, Figure 2.
FIGURE 5-24 Adapted from Kraft, N. J. B., R. Valencia, and D. D. Ackerly. 2008. Functional traits and niche-based tree community assembly in an Amazonian forest. *Science* 322: 580–582, Figure 1. Reprinted with permission from AAAS.
FIGURE 5-25 Adapted by permission from Macmillan Publishers Ltd: NATURE. Emerson, B. C., and N. Kolm. Species diversity can drive speciation. *Nature* 434: 1015–1017, Figure 3, copyright 2005.

CHAPTER 6

FIGURE 6-1 Adapted from Finegan, B. 1996. Pattern and process in neotropical secondary rain forests: The first 100 years of succession. *Trends in Ecology and Evolution* 11: 119–124.
FIGURE 6-2 Adapted from van Breugel, M., F. Bongers, and M. Martinez-Ramos. 2007. Species dynamics during early secondary forest succession: Recruitment, mortality and species turnover. *Biotropica* 35(5): 610–619, Figure 1. Reproduced with permission of Blackwell Publishing Ltd.
FIGURE 6-3 Used with permission from Sezen, U. U., R. L. Chazdon, and K. E. Holsinger. 2007. Multigenerational genetic analysis of tropical secondary regeneration in a canopy palm. *Ecology* 88: 3065–3075, Figure 1. Copyright 2007 by Ecological Society of America. Reproduced with permission of Ecological Society of America via Copyright Clearance Center.
FIGURE 6-4 Adapted from Sezen, U. U., R. L. Chazdon, and K. E. Holsinger. 2007. Multigenerational genetic analysis of tropical secondary regeneration in a canopy palm. *Ecology* 88: 3065–3075, Figure 2. Copyright 2007 by Ecological Society of America. Reproduced with permission of Ecological Society of America via Copyright Clearance Center.

FIGURE 6-5 Adapted from Chazdon, R. L., A. R. Brenes, and B. V. Alvarado. 2005. Effects of climate and stand age on annual tree dynamics in tropical second-growth rain forests. *Ecology* 86: 1808–1815, Figure 1. Copyright 2005 by Ecological Society of America. Reproduced with permission of Ecological Society of America via Copyright Clearance Center.

FIGURE 6-6 Adapted from Chazdon, R. L., A. R. Brenes, and B. V. Alvarado. 2005. Effects of climate and stand age on annual tree dynamics in tropical second-growth rain forests. *Ecology* 86: 1808–1815, Figure 2. Copyright 2005 by Ecological Society of America. Reproduced with permission of Ecological Society of America via Copyright Clearance Center.

TABLE 6-1 Adapted from Chazdon, R. L., A. R. Brenes, and B. V. Alvarado. 2005. Effects of climate and stand age on annual tree dynamics in tropical second-growth rain forests. *Ecology* 86: 1808–1815, Table 2. Copyright 2005 by Ecological Society of America. Reproduced with permission of Ecological Society of America via Copyright Clearance Center.

FIGURE 6-7 Adapted from Chazdon, R. L., A. R. Brenes, and B. V. Alvarado. 2005. Effects of climate and stand age on annual tree dynamics in tropical second-growth rain forests. *Ecology* 86: 1808–1815, Figure 4. Copyright 2005 by Ecological Society of America. Reproduced with permission of Ecological Society of America via Copyright Clearance Center.

PLATE 6-11 USDA APHIS PPQ Archive, USDA APHIS PPQ, Bugwood.org.

PLATE 6-14 Photo by Scott W. Shumway.

PLATE 6-15 Photo by Frederick J. Dodd, International Zoological Expeditions.

FIGURE 6-8 Adapted from Titiz, B., and R. L. Sanford Jr. 2007. Soil charcoal in old-growth rain forests from sea level to the continental divide. *Biotropica* 39: 673–682, Figure 3. Reproduced with permission of Blackwell Publishing Ltd.

PLATE 6-18 Photo by Yuda Prawira, TSA-KALTENG.

PLATE 6-19 James H. Miller, USDA Forest Service, Bugwood.org.

TABLE 6-2 Adapted from Hooper, E. R., et al. 2004. Factors affecting community composition of forest regeneration in deforested, abandoned land in Panama. *Ecology* 84(12): 3313–3326, Table 1. Copyright 2004 by Ecological Society of America. Reproduced with permission of Ecological Society of America via Copyright Clearance Center.

FIGURE 6-9 Adapted from Hart, R. D. 1980. A natural ecosystem analog approach to the design of a successional crop system for tropical forest environments. In *Tropical Succession*, supplement to *Biotropica* 12: 73–82. Reproduced with permission of Blackwell Publishing Ltd.

FIGURE 6-10 Adapted from Reagan, D. P., and R. B. Waide, eds. 1996. *The Food Web of a Tropical Rain Forest.* Chicago: University of Chicago Press, Figure 2.2.

PLATE 6-23 Photo by Camila Pizano.

FIGURE 6-11 From Hubbell, S. P., et al. 1999. Light-gap disturbances, recruitment limitation, and tree diversity in a Neotropical forest. *Science* 283: 554–557, Figure 1. Reprinted with permission from AAAS.

FIGURE 6-12 Adapted from Hubbell, S. P., et al. 1999. Light-gap disturbances, recruitment limitation, and tree diversity in a Neotropical forest. *Science* 283: 554–557, Figure 2a. Reprinted with permission from AAAS.

FIGURE 6-13 Adapted from Hubbell, S. P., et al. 1999. Light-gap disturbances, recruitment limitation, and tree diversity in a Neotropical forest. *Science* 283: 554–557, Figure 4. Reprinted with permission from AAAS.

FIGURE 6-14 Adapted from Wright, S. J., et al. 2003. Gap-dependent recruitment, realized vital rates, and size distributions of tropical trees. *Ecology* 84(12): 3174–3185, Figure 2. Copyright 2003 by Ecological Society of America. Reproduced with permission of Ecological Society of America via Copyright Clearance Center.

PLATE 6-25 Photo by Rolando Peréz/Smithsonian Tropical Research Institute.

CHAPTER 7

PLATE 7-3 Photo by Frederick J. Dodd, International Zoological Expeditions.

PLATE 7-5 Kingdon, Jonathan: *The Kingdon Pocket Guide to African Mammals.* 2004. Princeton University Press. Reprinted by permission of Princeton University Press.

FIGURE 7-1 Adapted from Emmons, L. H. 1980. Ecology and resource partitioning among nine species of African rain forest squirrels. *Ecological Monographs* 50: 31–54.

TABLE 7-1 Adapted from Emmons, L. H. 1980. Ecology and resource partitioning among nine species of African rain forest squirrels. *Ecological Monographs* 50: 31–54, Table 1.

FIGURE 7-2 Adapted from Emmons, L. H. 1980. Ecology and resource partitioning among nine species of African rain forest squirrels. *Ecological Monographs* 50: 31–54, Figure 11.

FIGURE 7-3 Adapted from Schreier, B. M., A. H. Harcourt, S. A. Coppeto, and M. F. Somi. 2009. Interspecific competition and niche separation in primates: A global analysis. *Biotropica* 41: 283–291, Figure 2. Reproduced with permission of Blackwell Publishing Ltd.

PLATE 7-7 Photo © A. J. Haverkamp.

TABLE 7-2 Adapted from Blundell, A. C., and D. R. Peart. 2004. Density-dependent population dynamics of a dominant rain forest canopy tree. *Ecology* 85: 704–715, Table 2. Copyright 2004 by Ecological Society of America. Reproduced with permission of Ecological Society of America via Copyright Clearance Center.

FIGURE 7-4 Adapted from Blundell, A. C., and D. R. Peart. 2004. Density-dependent population dynamics of a dominant rain forest canopy tree. *Ecology* 85: 704–715, Figure 4. Copyright 2004 by Ecological Society of America. Reproduced with permission of Ecological Society of America via Copyright Clearance Center.

FIGURE 7-6 Adapted from Clark, C. J., J. R. Poulsen, B. M. Bolker, E. F. Connor, and V. T. Parker. 2005. Comparative seed shadows of bird-, monkey-, and wind-dispersed trees. *Ecology* 86: 2684–2694, Figure 1. Copyright 2005 by Ecological Society of America. Reproduced with permission of Ecological Society of America via Copyright Clearance Center.

PLATE 7-10 (a) Photo by Edward Harper. (b) Photo by Bruce Hallett.

PLATE 7-13 Photo © D. Huntington/VIREO.

PLATE 7-14 (a) Photo © www.glennbartley.com.

FIGURE 7-7 Adapted from Kricher, J. C. 1989. *A Neotropical Companion: An Introduction to the Animals, Plants, and Ecosystems of the New World Tropics.* Princeton, NJ: Princeton University Press.

PLATE 7-20 Photo © rhpayne.

FIGURE 7-8 Adapted from Murray, K. G. 1988. Avian seed dispersal of three Neotropical gap-dependent plants. *Ecological Monographs* 58: 271–298, Figure 8.

PLATE 7-21 Photo © R. & N. Bowers/VIREO.

PLATE 7-23 Photos by Pepper Trail.

FIGURE 7-9 Adapted from Berens, D. G., N. Farwig, G. Schaab, and K. Bohning-Gaese. 2008. Exotic guavas are foci of forest regeneration in Kenyan farmland. *Biotropica* 40: 104–112, Figure 1. Reproduced with permission of Blackwell Publishing Ltd.

PLATE 7-25 Photo © Andrea Florence/ardea.com.

FIGURE 7-10 Used with permission from Bascompte, J. 2009. Mutualistic networks. *Frontiers in Ecology and Environment* 7: 429–436, Figure 2. Copyright 2009 by Ecological Society of America. Reproduced with permission of Ecological Society of America via Copyright Clearance Center.

PLATE 7-29 Photo © Doug Wechsler/VIREO.

PLATE 7-30 From Temeles, E. J., and W. J. Kress. 2003. Adaptation in a plant–hummingbird association. *Science* 300: 630–633, Figure 1. Reprinted with permission from AAAS.

PLATE 7-31 Photo © D. Huntington/VIREO.

PLATE 7-32 Photo by Ryan Somma.

PLATE 7-33 (a) Photo by Tian Yake. (b) Photo by G. Dimijian, MD/Photo Researchers.

PLATE 7-37 Photo by James K. Wetterer.

PLATE 7-38 Photo by James K. Wetterer.

FIGURE 7-11 Adapted from Youngsteadt, E. 2008. All that makes fungus gardens grow. *Science* 320: 1006–1007.

CHAPTER 8

FIGURE 8-2 Adapted from Reagan, D. P., and R. B. Waide, eds. 1996. *The Food Web of a Tropical Rain Forest*. Chicago: University of Chicago Press.

PLATE 8-2 Photo © J. Dunning/VIREO.

PLATE 8-3 Photo by Frederick J. Dodd, International Zoological Expeditions.

PLATE 8-4 Photo © Dr. James L. Castner.

PLATE 8-5 Photos by James K. Wetterer.

FIGURE 8-3 Adapted from Maxson, L. R., and C. W. Myers. 1985. Albumin evolution in tropical poison frogs (Dendrobatidae): A preliminary report. *Biotropica* 17: 50–56. Reproduced with permission of Blackwell Publishing Ltd.

PLATE 8-9 (a) Photo by Scott Shumway. (b) Photo © rhpayne.

PLATE 8-10 Photo by Bruce M. Beehler.

PLATE 8-11 Photo by James K. Wetterer.

PLATE 8-14 Used with permission from DuVal, E. H., H. W. Greene, and K. L. Manno. 2006. Laughing falcon (*Herpetotheres cachinnans*) predation on coral snakes (*Micrurus nigrocinctus*). *Biotropica* 38: 566–568, Figure 1. Reproduced with permission of Blackwell Publishing Ltd.

PLATE 8-16 Photo by Frederick J. Dodd, International Zoological Expeditions.

TABLE 8-1 Adapted from Coley, P. D. 1983. Herbivory and defensive characteristics of tree species in a lowland tropical forest. *Ecological Monographs* 53: 209–233, Table 2.

FIGURE 8-6 Adapted from Heil, M., J. Rattke, and W. Boland. 2005. Postsecretory hydrolysis of nectar sucrose and specialization in ant/plant mutualism. *Science* 308: 560–563, Figure 3. Reprinted with permission from AAAS.

FIGURE 8-7 Adapted from Davidson, D. W., S. C. Cook, R. R. Snelling, and T. H. Chua. 2003. Explaining the abundance of ants in lowland tropical rain forest canopies. *Science* 300: 969–972, Figure 1. Reprinted with permission from AAAS.

FIGURE 8-8 Adapted from Davidson, D. W., S. C. Cook, R. R. Snelling, and T. H. Chua. 2003. Explaining the abundance of ants in lowland tropical rain forest canopies. *Science* 300: 969–972, Figure 2. Reprinted with permission from AAAS.

TABLE 8-2 Adapted from Herz, H., W. Beyschlag, and B. Holldobler. 2007. Herbivory rate of leaf-cutting ants in a tropical moist forest in Panama at the population and ecosystem scales. *Biotropica* 39: 482–488, Table 2. Reproduced with permission of Blackwell Publishing Ltd.

TABLE 8-3 Adapted from Coley, P. D., et al. 2005. Divergent defensive strategies of young leaves in two species on *Inga*. *Ecology* 86: 2633–2643, Table 1. Copyright 2005 by Ecological Society of America. Reproduced with permission of Ecological Society of America via Copyright Clearance Center.

FIGURE 8-9 Adapted from Coley, P. D., et al. 2005. Divergent defensive strategies of young leaves in two species on *Inga*. *Ecology* 86: 2633–2643, Figure 1. Copyright 2005 by Ecological Society of America. Reproduced with permission of Ecological Society of America via Copyright Clearance Center.

FIGURE 8-10 Adapted from Weiblen, G. D., C. O. Webb, V. Novotny, Y. Basset, and S. E. Miller. 2006. Phylogenetic dispersion of host use in a tropical insect herbivore community. *Ecology* 87(7) Supplement: S62–S75, Figure 3. Copyright 2005 by Ecological Society of America. Reproduced with permission of Ecological Society of America via Copyright Clearance Center.

PLATE 8-24 (a) Photo by Tian Yake. (b) Photo by Ruth Whitten.

TABLE 8-4 Adapted from Terborgh, J. 1986. Keystone plant resources in the tropical forest. In *Conservation Biology: The Science of Scarcity and Diversity*, M. E. Soule (ed.). Sunderland, MA: Sinauer, Table 3.

TABLE 8-5 Adapted from Terborgh, J., et al. 2008. Tree recruitment in an empty forest. *Ecology* 89: 1757–1768, Table 1. Copyright 2008 by Ecological Society of America. Reproduced with permission of Ecological Society of America via Copyright Clearance Center.

TABLE 8-6 Adapted from Terborgh, J., et al. 2008. Tree recruitment in an empty forest. *Ecology* 89: 1757–1768, Table 4. Copyright 2008 by Ecological Society of America. Reproduced with permission of Ecological Society of America via Copyright Clearance Center.

FIGURE 8-11 Adapted by permission from Macmillan Publishers Ltd: NATURE. Dyer, L. A., et al. Host specificity of Lepidoptera in tropical and temperate forests. *Nature* 448: 696–699, Figure 3, copyright 2007.

FIGURE 8-12 Adapted from Gilbert, G. S., and C. O. Webb. 2007. Phylogenetic signal in plant pathogen–host range. *Proceedings of the National Academy of Science* 104: 4979–4983, Figure 1. Copyright 2007 National Academy of Sciences, USA.

PLATE 8-29 Photo © Dr. James L. Castner.

PLATE 8-32 Photo © B. Miller/VIREO.

FIGURE 8-13 Adapted by permission from Macmillan Publishers Ltd: NATURE. Malvárez, J., C. A. Salazar, E. Bermingham, C. Salcedo, C. D. Jiggins, and M. Linares. Speciation by hybridization in *Heliconius* butterflies. *Nature* 441: 868–871, Figure 1, copyright 2006.

FIGURE 8-14 Adapted by permission from Macmillan Publishers Ltd: NATURE. Malvárez, J., C. A. Salazar, E. Bermingham,

C. Salcedo, C. D. Jiggins, and M. Linares. Speciation by hybridization in *Heliconius* butterflies. *Nature* 441: 868–871, Figure 2, copyright 2006.

PLATE 8-34 Photo © Dr. James L. Castner.

FIGURE 8-15 Used with permission from Turner, J.R.G. 1975. A tale of two butterflies. *Natural History* 84: 29–37.

FIGURE 8-16 Adapted from Helson, J. E., T. L. Capson, T. Johns, A. Aiello, and D. M. Windsor. 2009. Ecological and evolutionary bioprospecting: Using aposematic insects as guides to rain forest plants active against disease. *Frontiers in Ecology and Environment* 7: 130–134, Figure 2.

FIGURE 8-17 From Kalka, M. B., A. R. Smith, and E. K. V. Kalko. 2008. Bats limit arthropods and herbivory in a tropical forest. *Science* 320: 71, Figure 1. Reprinted with permission from AAAS. (c) Photo by Christian Ziegler.

PLATE 8-35 Photo by Dr. Arthur Anker.

PLATE 8-36 (a) Photo © Doug Wechsler/VIREO. (b) Photo © J. Dunning/VIREO. (c) Photo © C. H. Greenewalt/VIREO.

FIGURE 8-18 Adapted from Wrege, P. H., M. Wikelski, J. T. Mandel, T. Rassweiler, and I. D. Couzin. 2005. Antbirds parasitize foraging army ants. *Ecology* 86: 555–559, Figure 2. Copyright 2005 by Ecological Society of America. Reproduced with permission of Ecological Society of America via Copyright Clearance Center.

PLATE 8-37 Photo by Scott Shumway.

CHAPTER 9

FIGURE 9-2 Adapted from http://www.esrl.noaa.gov/

FIGURE 9-3 Adapted from http://earthobservatory.nasa.gov/

FIGURE 9-4 Adapted from http://www.globalchange.umich.edu/

PLATE 9-4 Photo by Martha Vaughan.

FIGURE 9-5 Adapted from Raich, J. W., et al. 1991. Potential net primary productivity in South America: Application of a global model. *Ecological Applications* 1: 399–429, Figure 1. Copyright 1991 by Ecological Society of America. Reproduced with permission of Ecological Society of America via Copyright Clearance Center.

FIGURE 9-6 From Raich, J. W., et al. 1991. Potential net primary productivity in South America: Application of a global model. *Ecological Applications* 1: 399–429, Figures 2 and 3. Copyright 1991 by Ecological Society of America. Reproduced with permission of Ecological Society of America via Copyright Clearance Center.

TABLE 9-1 Adapted from Raich, J. W., et al. 1991. Potential net primary productivity in South America: Application of a global model. *Ecological Applications* 1:399–429, Table 4. Copyright 1991 by Ecological Society of America. Reproduced with permission of Ecological Society of America via Copyright Clearance Center.

TABLE 9-2 Adapted from Raich, J. W., et al. 1991. Potential net primary productivity in South America: Application of a global model. *Ecological Applications* 1: 399–429, Table 6. Copyright 1991 by Ecological Society of America. Reproduced with permission of Ecological Society of America via Copyright Clearance Center.

FIGURE 9-7 From Raich, J. W., et al. 1991. Potential net primary productivity in South America: Application of a global model. *Ecological Applications* 1: 399–429, Figure 6. Copyright 1991 by Ecological Society of America. Reproduced with permission of Ecological Society of America via Copyright Clearance Center.

FIGURE 9-8 Adapted from Raich, J. W., et al. 1991. Potential net primary productivity in South America: Application of a global model. *Ecological Applications* 1: 399–429, Figure 7. Copyright 1991 by Ecological Society of America. Reproduced with permission of Ecological Society of America via Copyright Clearance Center.

FIGURE 9-9 Adapted from Poorter, L., and K. Kitajima. 2007. Carbohydrate storage and light requirements of tropical moist and dry forest tree species. *Ecology* 88: 1000–1011, Figure 3. Copyright 2007 by Ecological Society of America. Reproduced with permission of Ecological Society of America via Copyright Clearance Center.

FIGURE 9-10 Adapted from Phillips, O. L., et al. 1998. Changes in the carbon balance of tropical forests: Evidence from long-term plots. *Science* 282: 439–442, Figure 2. Reprinted with permission from AAAS.

FIGURE 9-11 Adapted by permission from Macmillan Publishers Ltd: NATURE. Lewis, S. L., et al. Increasing carbon storage in intact African tropical forests. *Nature* 457: 1003–1006, Figure 1, copyright 2009.

TABLE 9-3 Adapted by permission from Macmillan Publishers Ltd: NATURE. Lewis, S. L., et al. Increasing carbon storage in intact African tropical forests. *Nature* 457: 1003–1006, Table 1, copyright 2009.

FIGURE 9-12 Adapted by permission from Macmillan Publishers Ltd: NATURE. Lewis, S. L., et al. Increasing carbon storage in intact African tropical forests. *Nature* 457: 1003–1006, Figure 2, copyright 2009.

FIGURE 9-13 Adapted by permission from Macmillan Publishers Ltd: NATURE. Cochrane, M. A. Fire science for rain forests. *Nature* 421: 913–919, Figure 1, copyright 2003.

FIGURE 9-14 Adapted by permission from Macmillan Publishers Ltd: NATURE. Cochrane, M. A. Fire science for rain forests. *Nature* 421: 913–919, Figure 3, copyright 2003.

FIGURE 9-15 Adapted by permission from Macmillan Publishers Ltd: NATURE. Raymond, P. A. The age of the Amazon's breath. *Nature* 436: 469–470, Figure 1, copyright 2005.

FIGURE 9-16 Adapted by permission from Macmillan Publishers Ltd: NATURE. Richey, J. E., J. M. Melack, A. K. Aufdenkampe, V. M. Ballester, and L. L. Hess. Outgassing from Amazonian rivers and wetlands as a large tropical source of atmospheric CO_2. *Nature* 416: 617–620, Figure 1, copyright 2002.

FIGURE 9-17 Adapted by permission from Macmillan Publishers Ltd: NATURE. Richey, J. E., J. M. Melack, A. K. Aufdenkampe, V. M. Ballester, and L. L. Hess. Outgassing from Amazonian rivers and wetlands as a large tropical source of atmospheric CO_2. *Nature* 416: 617–620, Figure 4, copyright 2002.

PLATE 9-12 Photo by Brad Taylor.

FIGURE 9-18 Adapted from Taylor, B. W., A. S. Flecker, and R. O. Hall Jr. 2006. Loss of a harvested fish species disrupts carbon flow in a diverse tropical river. *Science* 313: 833–836, Figure 1. Reprinted with permission from AAAS.

FIGURE 9-19 Used with permission from Taylor, B.W., A. S. Flecker, and R. O. Hall Jr. 2006. Loss of a harvested fish species disrupts carbon flow in a diverse tropical river. *Science* 313: 833–836, Figure 2. Reprinted with permission from AAAS.

FIGURE 9-20 Adapted by permission from Macmillan Publishers Ltd: NATURE. Engelbrecht, B. M. J., et al. Drought sensitivity shapes species distribution patterns in tropical forests. *Nature* 447: 80–82, Figure 1a, copyright 2007.

FIGURE 9-21 Adapted by permission from Macmillan Publishers Ltd: NATURE. Engelbrecht, B.M.J., et al. Drought sensitivity shapes species distribution patterns in tropical forests. *Nature* 447: 80–82, Figure 2, copyright 2007.

FIGURE 9-22 From Saleska, S. R., K. Didan, A. R. Huete, and H. R. da Rocha. 2007. Amazon forests green-up during 2005 drought. *Science* 318: 612, Figure 1. Reprinted with permission from AAAS.

FIGURE 9-23 From Phillips, O. L., et al. 2009. Drought sensitivity of the Amazon rain forest. *Science* 323: 1344–1347, Figure 3. Reprinted with permission from AAAS.

FIGURE 9-24 Adapted from Schuur, E. A. G. 2003. Productivity and global climate revisited: The sensitivity of tropical forest growth to precipitation. *Ecology* 84: 1165–1170, Figure 1. Copyright 2003 by Ecological Society of America. Reproduced with permission of Ecological Society of America via Copyright Clearance Center.

FIGURE 9-25 Reprinted by permission from Macmillan Publishers Ltd: NATURE. Rosenzweig, C., et al. Attributing physical and biological impacts to anthropogenic climate change. *Nature* 453: 353–357, Figure 2, copyright 2008.

FIGURE 9-26 Adapted from Raich, J. W., A. E. Russell, K. Kitayama, W. J. Parton, and P. M. Vitousek. 2006. Temperature influences carbon accumulation in moist tropical forests. *Ecology* 87: 76–87, Figure 2. Copyright 2006 by Ecological Society of America. Reproduced with permission of Ecological Society of America via Copyright Clearance Center.

TABLE 9-4 Adapted by permission from Macmillan Publishers Ltd: NATURE. Laurance, W. F., et al. Pervasive alteration of tree communities in undisturbed Amazonian forests. *Nature* 428: 171–175, Table 1, copyright 2004.

FIGURE 9-27 Adapted by permission from Macmillan Publishers Ltd: NATURE. Laurance, W. F., et al. Pervasive alteration of tree communities in undisturbed Amazonian forests. *Nature* 428: 171–175, Figure 3, copyright 2004.

FIGURE 9-28 Adapted by permission from Macmillan Publishers Ltd: NATURE. Phillips, O. L., et al. Increasing dominance of large lianas in Amazonian forests. *Nature* 418: 770–774, copyright 2002.

FIGURE 9-29 Adapted from Colwell, R. K., G. Brehm, C. L. Cardelus, A. C. Gilman, and J. T. Longino. 2008. Global warming, elevational range shifts, and lowland biotic attrition in the wet tropics. *Science* 322: 258–261, p. 260, Figure 3A-B. Reprinted with permission from AAAS.

PLATE 9-15 Photo © J. G. Holmes/VIREO.

CHAPTER 10

PLATE 10-6 (b) Photo by Martha Vaughan.

FIGURE 10-3 Adapted from Moutinho, P., D. C. Nepstad, and E. A. Davidson. 2003. Influence of leaf-cutting ant nests on secondary forest growth and soil properties in Amazonia. *Ecology* 84: 1265–1276, Figure 1. Copyright 2003 by Ecological Society of America. Reproduced with permission of Ecological Society of America via Copyright Clearance Center.

TABLE 10-1 Adapted from Moutinho, P., D. C. Nepstad, and E. A. Davidson. 2003. Influence of leaf-cutting ant nests on secondary forest growth and soil properties in Amazonia. *Ecology* 84: 1265–1276, Table 2. Copyright 2003 by Ecological Society of America. Reproduced with permission of Ecological Society of America via Copyright Clearance Center.

PLATE 10-9 Photos by Wittaya Kaonongbua.

FIGURE 10-4 Adapted from Gallery, R. E., J. W. Dalling, and A. E. Arnold. 2007. Diversity, host affinity, and distribution of seed-infecting fungi: A case study with *Cecropia*. *Ecology* 88: 582–588, Figure 1. Copyright 2007 by Ecological Society of America. Reproduced with permission of Ecological Society of America via Copyright Clearance Center.

FIGURE 10-5 Adapted from McGuire, K. L. 2007. Common ectomycorrhizal networks may maintain monodominance in a tropical rain forest. *Ecology* 88: 567–574, Figure 1. Copyright 2007 by Ecological Society of America. Reproduced with permission of Ecological Society of America via Copyright Clearance Center.

FIGURE 10-6 Adapted from Herre, E. A., et al. 2007. Ecological implications of anti-pathogen effects of tropical fungal endophytes and mycorrhizae. *Ecology* 88: 550–558, Figure 1. Copyright 2007 by Ecological Society of America. Reproduced with permission of Ecological Society of America via Copyright Clearance Center.

FIGURE 10-7 Adapted from Herre, E. A., et al. 2007. Ecological implications of anti-pathogen effects of tropical fungal endophytes and mycorrhizae. *Ecology* 88: 550–558, Figure 3. Copyright 2007 by Ecological Society of America. Reproduced with permission of Ecological Society of America via Copyright Clearance Center.

PLATE 10-10 (c) Photo © rhpayne.

FIGURE 10-8 Adapted from Wieder, W. R., C. C. Cleveland, and A. R. Townsend. 2009. Controls over leaf litter decomposition in wet tropical forests. *Ecology* 90: 3333–3341, Figure 1. Copyright 2009 by Ecological Society of America. Reproduced with permission of Ecological Society of America via Copyright Clearance Center.

FIGURE 10-9 Adapted from Cleveland, C. C., S. C. Reed, and A. R. Townsend. 2006. Nutrient regulation of organic matter decomposition in a tropical rain forest. *Ecology* 87: 492–503, Figure 1. Copyright 2006 by Ecological Society of America. Reproduced with permission of Ecological Society of America via Copyright Clearance Center.

FIGURE 10-10 Adapted from Cleveland, C. C., S. C. Reed, and A. R. Townsend. 2006. Nutrient regulation of organic matter decomposition in a tropical rain forest. *Ecology* 87: 492–503, Figure 5. Copyright 2006 by Ecological Society of America. Reproduced with permission of Ecological Society of America via Copyright Clearance Center.

PLATE 10-15 Photo by Gregory S. Gilbert.

FIGURE 10-11 Adapted from Reed, S. C., C. C. Cleveland, and A. R. Townsend. 2008. Tree species control rates of free-living nitrogen fixation in a tropical rain forest. *Ecology* 89: 2924–2934, Figure 1. Copyright 2008 by Ecological Society of America. Reproduced with permission of Ecological Society of America via Copyright Clearance Center.

FIGURE 10-12 Adapted from Reed, S. C., C. C. Cleveland, and A. R. Townsend. 2008. Tree species control rates of free-living nitrogen fixation in a tropical rain forest. *Ecology* 89: 2924–2934, Figure 2. Copyright 2008 by Ecological Society of America. Reproduced with permission of Ecological Society of America via Copyright Clearance Center.

FIGURE 10-13 Adapted from Townsend, A. R., C. C. Cleveland, G. P. Asner, and M. M. C. Bustamante. 2007. Controls over foliar N:P ratios in tropical rain forests. *Ecology* 88: 107–118, Figure 5. Copyright 2007 by Ecological Society of America.

Reproduced with permission of Ecological Society of America via Copyright Clearance Center.

FIGURE 10-14 Adapted from Reed, S. C., C. C. Cleveland, and A. R. Townsend. 2008. Tree species control rates of free-living nitrogen fixation in a tropical rain forest. *Ecology* 89: 2924–2934, Figure 3. Copyright 2008 by Ecological Society of America. Reproduced with permission of Ecological Society of America via Copyright Clearance Center.

FIGURE 10-15 Adapted from Paoli, G. D., L. M. Curran, and D. R. Zak. 2006. Phosphorus efficiency of Bornean rain forest productivity: Evidence against the unimodal efficiency hypothesis. *Ecology* 86: 1548–1561, Figure 4. Copyright 2006 by Ecological Society of America. Reproduced with permission of Ecological Society of America via Copyright Clearance Center.

FIGURE 10-16 Adapted from Powell, L. L., T. U. Powell, G. V. N. Powell, and D. J. Brightsmith. 2009. Parrots take it with a grain of salt: Available sodium content may drive *collpa* (clay lick) selection in Southeastern Peru. *Biotropica* 41: 279–282, Figure 1. Reproduced with permission of Blackwell Publishing Ltd.

FIGURE 10-17 Adapted from Bravo, A., K. E. Harms, R. D. Stevens, and L. H. Emmons. 2008. *Collpas*: Activity hotspots for frugivorous bats (Phyllostomidae) in the Peruvian Amazon. *Biotropica* 40: 203–210, Figure 1. Reproduced with permission of Blackwell Publishing Ltd.

FIGURE 10-18 Adapted from Kurokawa, H., and T. Nakashizuka. 2008. Leaf herbivory and decomposability in a Malaysian rain forest. *Ecology* 89: 2645–2656, Figure 2. Copyright 2008 by Ecological Society of America. Reproduced with permission of Ecological Society of America via Copyright Clearance Center.

FIGURE 10-19 Adapted from Kurokawa, H., and T. Nakashizuka. 2008. Leaf herbivory and decomposability in a Malaysian rain forest. *Ecology* 89: 2645–2656, Figure 3. Copyright 2008 by Ecological Society of America. Reproduced with permission of Ecological Society of America via Copyright Clearance Center.

FIGURE 10-20 Adapted from Feeley, K. J., and J. W. Terborgh. 2005. The effects of herbivore density on soil nutrients and tree growth in tropical forest fragments. *Ecology* 86: 116–124, Figure 2. Copyright 2005 by Ecological Society of America. Reproduced with permission of Ecological Society of America via Copyright Clearance Center.

FIGURE 10-21 Adapted from Feeley, K. J., and J. W. Terborgh. 2005. The effects of herbivore density on soil nutrients and tree growth in tropical forest fragments. *Ecology* 86: 116–124, Figure 3. Copyright 2005 by Ecological Society of America. Reproduced with permission of Ecological Society of America via Copyright Clearance Center.

CHAPTER 11

FIGURE 11-1 Adapted from http://oregonstate.edu/instruct/css/330/two/Unit3Notes.htm.

PLATE 11-1 (b) Photo by Wally Gobetz.

PLATE 11-3 Photo by Martin Heigan.

FIGURE 11-2 Adapted from Redfern, J. V., R. Grant, H. Biggs, and W. M. Getz. 2003. Surface-water constraints on herbivore foraging in the Kruger National Park, South Africa. *Ecology* 84: 2092–2107, Figure 4. Copyright 2003 by Ecological Society of America. Reproduced with permission of Ecological Society of America via Copyright Clearance Center.

FIGURE 11-3 Adapted from Wilmshurst, J. F., J. M. Fryxell, and P. E. Colucci. 1999. What constrains daily intake in Thomson's gazelles? *Ecology* 80: 2338–2347, Figure 6. Copyright 1999 by Ecological Society of America. Reproduced with permission of Ecological Society of America via Copyright Clearance Center.

FIGURE 11-4 Adapted from Wilmshurst, J. F., J. M. Fryxell, and P. E. Colucci. 1999. What constrains daily intake in Thomson's gazelles? *Ecology* 80: 2338–2347, Figure 7. Copyright 1999 by Ecological Society of America. Reproduced with permission of Ecological Society of America via Copyright Clearance Center.

FIGURE 11-5 Used with permission from Fryxell, J. M., J. F. Wilmshurst, and A. R. E. Sinclair. 2004. Predictive models of movement by Serengeti grazers. *Ecology* 85: 2429–2435. Copyright 2004 by Ecological Society of America. Reproduced with permission of Ecological Society of America via Copyright Clearance Center.

FIGURE 11-6 Adapted from Fryxell, J. M., J. F. Wilmshurst, and A. R. E. Sinclair. 2004. Predictive models of movement by Serengeti grazers. *Ecology* 85: 2429–2435, Figure 2. Copyright 2004 by Ecological Society of America. Reproduced with permission of Ecological Society of America via Copyright Clearance Center.

FIGURE 11-7 Used with permission from Fornara, D. A., and J. T. Du Toit. 2007. Browsing lawns? Responses of *Acacia nigrescens* to ungulate browsing in an African savanna. *Ecology* 88: 200–209. Copyright 2007 by Ecological Society of America. Reproduced with permission of Ecological Society of America via Copyright Clearance Center.

FIGURE 11-8 Used with permission from Boone, R. B., S. J. Thirgood, and J. G. C. Hopcraft. 2006. Serengeti wildebeest migratory patterns molded from rainfall and new vegetation growth. *Ecology* 87: 1987–1994; adapted from Murray (1995) and Sinclair (1995). Copyright 2006 by Ecological Society of America. Reproduced with permission of Ecological Society of America via Copyright Clearance Center.

FIGURE 11-9 Used with permission from Boone, R. B., S. J. Thirgood, and J. G. C. Hopcraft. 2006. Serengeti wildebeest migratory patterns molded from rainfall and new vegetation growth. *Ecology* 87: 1987–1994. Copyright 2006 by Ecological Society of America. Reproduced with permission of Ecological Society of America via Copyright Clearance Center.

FIGURE 11-10 Adapted from Sinclair, A. R. E., S. A. R. Mduma, and P. Arcese. 2000. What determines phenology and synchrony of ungulate breeding in Serengeti? *Ecology* 81: 2100–2111. Copyright 2000 by Ecological Society of America. Reproduced with permission of Ecological Society of America via Copyright Clearance Center.

FIGURE 11-11 Adapted from Anderson, T. M, M. E. Ritchie, and S. J. McNaughton. 2007. Rainfall and soils modify plant community response to grazing in Serengeti National Park. *Ecology* 88: 1191–1201, Figure 4. Copyright 2007 by Ecological Society of America. Reproduced with permission of Ecological Society of America via Copyright Clearance Center.

FIGURE 11-12 Adapted from McCauley, D. J., F. Keesing, T. P. Young, B. F. Allan, and R. M. Pringle. 2006. Indirect effects of large herbivores on snakes in an African savanna. *Ecology* 87: 2657–2663, Figure 1. Copyright 2006 by Ecological

Society of America. Reproduced with permission of Ecological Society of America via Copyright Clearance Center.

FIGURE 11-13 Adapted from Riginos, C., and J. B. Grace. 2008. Savanna tree density, herbivores, and the herbaceous community: Bottom-up vs. top-down effects. *Ecology* 89: 2228–2238, Figure 1. Copyright 2008 by Ecological Society of America. Reproduced with permission of Ecological Society of America via Copyright Clearance Center.

FIGURE 11-14 Adapted from Riginos, C., and J. B. Grace. 2008. Savanna tree density, herbivores, and the herbaceous community: Bottom-up vs. top-down effects. *Ecology* 89: 2228–2238, Figures 4b, 4f, and 4i. Copyright 2008 by Ecological Society of America. Reproduced with permission of Ecological Society of America via Copyright Clearance Center.

FIGURE 11-15 Adapted from Goheen, J. R., F. Keesing, B. F. Allan, D. Ogada, and R. S. Ostfeld. 2004. Net effects of large mammals on *Acacia* seedling survival in an African savanna. *Ecology* 85: 1555–1561, Figure 2. Copyright 2004 by Ecological Society of America. Reproduced with permission of Ecological Society of America via Copyright Clearance Center.

PLATE 11-24 Photo by Robert M. Pringle (top left). Used with permission from Pringle, R. M. 2008. Elephants as agents of habitat creation for small vertebrates at the patch scale. *Ecology* 89: 26–33 (images top right, bottom left, bottom right). Copyright 2008 by Ecological Society of America. Reproduced with permission of Ecological Society of America via Copyright Clearance Center.

FIGURE 11-16 Adapted from van Langevelde, F., et al. 2003. Effects of fire and herbivory on the stability of savanna ecosystems. *Ecology* 84: 337–350, Figure 1. Copyright 2003 by Ecological Society of America. Reproduced with permission of Ecological Society of America via Copyright Clearance Center.

PLATE 11-25 Used with permission from Riginos, C. 2009. Grass competition suppresses savanna tree growth across multiple demographic stages. *Ecology* 90: 335–340. Copyright 2009 by Ecological Society of America. Reproduced with permission of Ecological Society of America via Copyright Clearance Center.

FIGURE 11-17 Adapted from Riginos, C. 2009. Grass competition suppresses savanna tree growth across multiple demographic stages. *Ecology* 90: 335–340, Figure 2. Copyright 2009 by Ecological Society of America. Reproduced with permission of Ecological Society of America via Copyright Clearance Center.

FIGURE 11-18 Adapted from Redfern, J. V., R. Grant, H. Biggs, and W. M. Getz. 2003. Surface-water constraints on herbivore foraging in the Kruger National Park, South Africa. *Ecology* 84: 2092–2107, Figure 6. Copyright 2003 by Ecological Society of America. Reproduced with permission of Ecological Society of America via Copyright Clearance Center.

PLATE 11-32 Photo © Doug Weschsler/VIREO.

PLATE 11-35 Photo © T. J. Ulrich/VIREO

CHAPTER 12

FIGURE 12-1 Adapted from van der Hammen, T. The Pleistocene changes of vegetation and climate in tropical South America. In T. C. Whitmore and R. L. Chazdon (eds.). *Foundations of Tropical Forest Biology,* Part Two, Figure 2. Chicago: University of Chicago Press.

PLATE 12-5 Photo © G. Beaton/VIREO.

PLATE 12-6 Photo courtesy of Matt Hart/Shutterstock.

PLATE 12-7 Photo by Edward Harper.

PLATE 12-8 Photo © G. Bartley/VIREO.

PLATE 12-9 Photo © B. Miller/VIREO.

PLATE 12-10 Photo courtesy of Nicola Bilic/Shutterstock.

FIGURE 12-2 Adapted from Stuart, S. N., J. S. Chanson, N. A. Cox, B. E. Young, A. S. L. Rodrigues, D. L. Fischman, and R. W. Waller. 2004. Status and trends of amphibian declines and extinctions worldwide. *Science* 306: 1783-1786, Figure 3, p. 1785. Reprinted with permission from AAAS.

FIGURE 12-3 Used with permission from Ron, S. R. 2005. Predicting the distribution of the amphibian pathogen *Batrachochytrium dendrobatidis* in the New World. *Biotropica* 37: 209–221, Figure 1. Reproduced with permission of Blackwell Publishing Ltd.

PLATE 12-11 (a) Photo by Laurent Heroux. (b) Photo by Ana Cecilia Morandini.

FIGURE 12-4 Adapted from GraphicMaps.com

PLATE 12-13 Photo © Wolfgang Kaehler/CORBIS.

PLATE 12-15 Photo © G. Bartley/VIREO.

PLATE 12-17 (b) Photo by Bruce Hallett.

PLATE 12-18 Photo by Edward Harper.

PLATE 12-19 Photo © rhpayne.

PLATE 12-20 Photo © G. Bartley/VIREO.

PLATE 12-21 Photo © J. Dunning/VIREO

PLATE 12-23 From http://commons.wikimedia.org/wiki/File:Puya_Raimondii.JPG#file

PLATE 12-24 Photo © S. Imberti/VIREO

PLATE 12-25 Photo © K. Schafer/VIREO.

PLATE 12-26 Photo © M. P. Kahl/VIREO.

PLATE 12-27 Photo by Edward Harper.

PLATE 12-28 Photo © J. McKean/VIREO.

FIGURE 12-5 Adapted from http://gosouthamerica.com

FIGURE 12-7 Used with permission from Lowe-McConnell, R. H. 1987. *Ecological studies in tropical fish communities*. New York: Cambridge University Press. Copyright © 1987 Cambridge University Press. Reprinted with permission of Cambridge University Press.

PLATE 12-32 Photo by Thiago T. Santos (http://www.flickr.com/x/t/0098009/photos/thsant/530736563/).

PLATE 12-41 Photo by Brian Ritchie.

PLATE 12-45 Photo © G. Bartley/VIREO.

CHAPTER 13

FIGURE 13-1 Adapted from Goebel, T., M. R. Waters, and D. H. O'Rourke. 2008. The late Pleistocene dispersal of modern humans in the Americas. *Science* 319: 1497–1502, Figure 3, p. 1500. Reprinted with permission from AAAS.

FIGURE 13-2 Adapted from Willis, K. J., L. Gillson, and T. M. Brncic. 2004. How "virgin" is virgin rain forest? *Science* 304: 402–403, p. 403. Reprinted with permission from AAAS.

PLATE 13-2 Photo courtesy of Hu Xiao Fang/Shutterstock.

FIGURE 13-3 From Heckenberger, M. J., A. Kuikuro, U. T. Kuikuro, J. C. Russell, M. Schmidt, C. Fausto, and B. Franchetto. 2003. Amazonia 1492: Pristine forest or cultural parkland? *Science* 301: 1710–1714, Figure 3, p. 1712. Reprinted with permission from AAAS.

FIGURE 13-4 From Heckenberger, M. J., J. C. Russell, C. Fausto, J. R. Toney, M. J. Schmidt, E. Pereira, B. Franchetto, and

A. Kuikuro. 2008. Pre-Columbian urbanism, anthropogenic landscapes, and the future of the Amazon. *Science* 321: 1214–1217, Figures 1A and 1B, p. 1215. Reprinted with permission from AAAS.

PLATE 13-3 Photo by Gerhard Bechtold.

PLATE 13-5 Photo by Scott W. Shumway.

PLATE 13-6 Department of Anthropology, Smithsonian Institution; Neg. # 83-10122.

PLATE 13-7 Images courtesy of The Amazon Conservation Team.

FIGURE 13-5 Adapted from Balter, M. 2007. Seeking agriculture's ancient roots. *Science* 316: 1830–1835, p. 1835. Reprinted with permission from AAAS.

FIGURE 13-6 Adapted from Odum, H. T. 1971. *Environment, Power, and Society*. New York: Wiley-Interscience.

FIGURE 13-7 Adapted from Van Bael, S. A., et al. 2008. Birds as predators in tropical agroforestry systems. *Ecology* 89: 928–934, Figure 1. Copyright 2008 by Ecological Society of America. Reproduced with permission of Ecological Society of America via Copyright Clearance Center.

TABLE 13-1 From Perfecto, I., et al. 2004. Greater predation in shaded coffee farms: The role of the resident neotropical birds. *Ecology* 85:2677-2681, Table 1. Copyright 2004 by Ecological Society of America. Reproduced with permission of Ecological Society of America via Copyright Clearance Center.

PLATE 13-20 Photo courtesy of the Missouri Botanical Garden.

PLATE 13-21 Photo by James Savage-Hanford.

CHAPTER 14

PLATE 14-3 Photo by Rob Bierregaard.

PLATE 14-4 Photo by Mark W. Moffett/Minden Pictures.

FIGURE 14-1 Adapted from Bierregaard, R. O., Jr., C. Gascon, T. E. Lovejoy, and R. Mesquita, eds. 2001. *Lessons from Amazonia: The Ecology and Conservation of a Fragmented Forest.* New Haven, CT: Yale University Press.

TABLE 14-1 Adapted from Bierregaard, R. O., Jr., C. Gascon, T. E. Lovejoy, and R. Mesquita, eds. 2001. *Lessons from Amazonia: The Ecology and Conservation of a Fragmented Forest.* New Haven, CT: Yale University Press, Table 4.3.

TABLE 14-2 Adapted from Quintero, I., and T. Roslin. 2005. Rapid recovery of dung beetle communities following habitat fragmentation in Central Amazonia. *Ecology* 86: 3303–3311, Table 1. Copyright 2005 by Ecological Society of America. Reproduced with permission of Ecological Society of America via Copyright Clearance Center.

FIGURE 14-2 Adapted by permission from Macmillan Publishers Ltd: NATURE. Bruna, E. M. Seed germination in rain forest fragments. *Nature* 402: 139, Figure 1, copyright 1999.

PLATE 14-7 (a) Photo © A. Whittaker/VIREO.

FIGURE 14-3 Adapted from Feeley, K. J., and J. W. Terborgh. 2006. Habitat fragmentation and effects of herbivore (howler monkey) abundances on bird species richness. *Ecology* 87: 144–150, Figure 2. Copyright 2006 by Ecological Society of America. Reproduced with permission of Ecological Society of America via Copyright Clearance Center.

PLATE 14-8 Photo by Piers Calvert.

FIGURE 14-4 Adapted from Laurance, W. F., et al. 2006. Rain forest fragmentation and the proliferation of successional trees. *Ecology* 87: 469–482, Figure 2. Copyright 2006 by Ecological Society of America. Reproduced with permission of Ecological Society of America via Copyright Clearance Center.

FIGURE 14-5 Adapted from Michalski, F., I. Nishi, and C. A. Peres. 2007. Disturbance-mediated drift in tree functional groups in Amazonian forest fragments. *Biotropica* 39: 691–701, Figure 3. Reproduced with permission of Blackwell Publishing Ltd.

FIGURE 14-6 Adapted from Gascon, C., G. B. Williamson, and G. A. B. da Fonseca. 2000. Receding forest edges and vanishing reserves. *Science* 288: 1356–1358, p. 1356. Reprinted with permission from AAAS.

FIGURE 14-8 Adapted from Prance, G. T., ed. 1982. *Biological Diversification in the Tropics.* New York: Columbia University Press.

PLATE 14-11 Photo © P. Robles Gil/ VIREO.

FIGURE 14-9 Adapted from coolmelbourne.org.

PLATE 14-12 Photo by David Lewis.

PLATE 14-13 Photo © Flávio Cruvinel Brandão.

FIGURE 14-10 Adapted from Loreau, M., et al. 2001. Biodiversity and ecosystem functioning: Current knowledge and future challenges. *Science* 294: 804–808, Figure 4. Reprinted with permission from AAAS.

FIGURE 14-11 Adapted from McConkey, K. R., and D. R. Drake. 2006. Flying foxes cease to function as seed dispersers long before they become rare. *Ecology* 87: 271–276, Figure 1. Copyright 2006 by Ecological Society of America. Reproduced with permission of Ecological Society of America via Copyright Clearance Center.

PLATE 14-14 Photos by Carolyn M. Miller.

FIGURE 14-12 Adapted by permission from Macmillan Publishers Ltd: NATURE. Myers, N., R. A. Mittermeier, C. G. Mittermeier, G. A. B. da Fonseca, and J. Kent. Biodiversity hotspots for conservation priorities. *Nature* 403: 853–858, Figure 1, copyright 2000.

FIGURE 14-13 Adapted by permission from Macmillan Publishers Ltd: NATURE. Cincotta, R. P., J. Wisnewski, and R. Engelman. Human population in the biodiversity hotspots. *Nature* 404: 990–991, Figure 1, copyright 2000.

PLATE 14-15 Photo by William E. Davis, Jr.

PLATE 14-16 Photo by Sheila Ellen Thomson.

FIGURE 14-14 Adapted from Kareiva, P., and M. Marvier. 2003. Conserving biodiversity coldspots. *American Scientist* 91: 344–351, Figure 6.

CHAPTER 15

FIGURE 15-1 Adapted from Putz, F. E., and K. H. Redford. 2010. The importance of defining "forest": Tropical forest degradation, deforestation, long-term phase shifts, and further transitions. *Biotropica* 42: 10–20, Figure 1. Reproduced with permission of Blackwell Publishing Ltd.

TABLE 15-1 Adapted from Wright, S. J., and H. C. Muller-Landau. 2006a. The future of tropical forest species. *Biotropica* 38: 287–301, Table 1. Reproduced with permission of Blackwell Publishing Ltd.

FIGURE 15-2 Adapted from Wright, S. J., and H. C. Muller-Landau. 2006a. The future of tropical forest species. *Biotropica* 38: 287–301, Figure 2. Reproduced with permission of Blackwell Publishing Ltd.

FIGURE 15-3 Adapted from Cannon, C. H., D. R. Peart, and M. Leighton. 1998. Tree species diversity in commercially logged Bornean rainforest. *Science* 281: 1366–1368, Figure 1A. Reprinted with permission from AAAS.

FIGURE 15-4 Adapted from Curran, L. M., et al. 2004. Lowland forest loss in protected areas of Indonesian Borneo. *Science*

303: 1000–1003, Figures 2A, 2B, and 2C. Reprinted with permission from AAAS.

PLATE 15-11 Photo © A. J. Haverkamp.

TABLE 15-2 Adapted from Shearman, P. L., J. Ash, B. Mackey, J. E. Bryan, and B. Lokes. 2009. Forest conversion and degradation in Papua New Guinea 1972–2002. *Biotropica* 41: 379–390, Table 1. Reproduced with permission of Blackwell Publishing Ltd.

TABLE 15-4 Adapted from Asner, G. P., D. E. Knapp, E. N. Broadbent, P. J. C. Oliveira, M. Keller, and J. N. Silva. 2005. Selective logging in the Brazilian Amazon. *Science* 310: 480–482, Table 1. Reprinted with permission from AAAS.

FIGURE 15-5 Adapted by permission from Macmillan Publishers Ltd: NATURE. Loarie, S. R., P. B. Duffy, H. Hamilton, G.P. Asner, C. B. Field, and D. D. Ackerly. The velocity of climate change. *Nature* 462: 1052–1055, Figure 3, copyright 2009.

FIGURE 15-6 Adapted from Sinervo, B., et al. 2010. Erosion of lizard diversity by climate change and altered thermal niches. *Science* 328: 894–899, Figure 3. Reprinted with permission from AAAS.

PLATE 15-14 Photo by Bruce Hallett.

PLATE 15-15 Michael & Patricia Fogden/www.fogdenphotos.com.

FIGURE 15-7 Adapted from Brashares, J. S., P. Arcese, M. K. Sam, P. B. Coppolillo, A. R. E. Sinclair, and A. Balmford. 2004. Bushmeat hunting, wildlife declines, and fish supply in West Africa. *Science* 306: 1180–1183, Figure 3. Reprinted with permission from AAAS.

PLATE 15-19 Photo by Karl Ammann, http://karlammann.com.

FIGURE 15-8 Adapted by permission from Macmillan Publishers Ltd: NATURE. Soares-Filho, B. S., et al. Modelling conservation in the Amazon Basin. *Nature* 440: 520–523, Figure 1, copyright 2006.

FIGURE 15-9 Adapted from Lamb, D., P. D. Erskine, and J. A. Parotta. 2005. Restoration of degraded tropical forest landscapes. *Science* 310: 1628–1632, Figure 2. Reprinted with permission from AAAS.

FIGURE 15-10 Adapted from http://news.mongabay.com/2007/0206-biodiversity.html.

TABLE 15-5 Adapted from Brook, B. W., C. J. A. Bradshaw, L. Pin Koh, and M.N.S. Sodhi. 2006. Momentum drives the crash: Mass extinction in the tropics. *Biotropica* 38: 302–305, Table 1. Reproduced with permission of Blackwell Publishing Ltd.

PLATE 15-21 Photo by Martha Vaughan.

Index